XIII International Scientific-Technical Conference on Dynamic of Technical Systems (DTS 2017)

MATEC Web of Conferences Volume 132 (2017)

Rotov-on-Don, Russia
13 - 15 September 2017

Editors:

Tooraj Abbasian Najafabadi
Igor Sevostianov
Yeghiazaryan Karen Onikovich
Nguyen Dong Ahn
Vladimir Mladenovic

ISBN: 978-1-5108-5102-3

Printed from e-media with permission by:

Curran Associates, Inc.
57 Morehouse Lane
Red Hook, NY 12571

Some format issues inherent in the e-media version may also appear in this print version.

Printed by Curran Associates, Inc. (2017)

For additional information, please contact EDP Sciences – Web of Conferences at the address below.

EDP Sciences – Web of Conferences
17, Avenue du Hoggar
Parc d'Activité de Courtabœuf
BP 112
F-91944 Les Ulis Cedex A
France

Phone: +33 (0) 1 69 18 75 75
Fax: +33 (0) 1 69 28 84 91

contact-edps@webofconferences.org

Additional copies of this publication are available from:

Curran Associates, Inc.
57 Morehouse Lane
Red Hook, NY 12571 USA
Phone: 845-758-0400
Fax: 845-758-2633
Email: curran@proceedings.com
Web: www.proceedings.com

TABLE OF CONTENTS

MODELLING THE REDUCED COEFFICIENT OF VOLUMETRIC RIGIDITY IN HIGH-PRESSURE RUBBER-CORD SHELL HOSES 1
Rybak Aleksandr, Lyakhnitskaya Olga

THE TECHNOLOGY OF RECEIVING FINELY DISPERSED POWDERS OF MAGNETIC MATERIALS IN THE SCREW DRUM 5
Vernigorov Yuriy, Frolova Natalia

ANALYSIS OF THE ZINC COVERING QUALITY FORMED BY THE VIBRATION CHEMICAL-MECHANICAL SYNTHESIS 8
Ivanov Vladimir, Lebedev Valeriy, Davydova Irina, Atoyan Tatiana

THE RESEARCH OF TECHNOLOGICAL CHARACTERISTIC OF THE VIBROWAVE MECHANICAL AND CHEMICAL OXIDE COATING FORMATION 11
Ivanov Vladimir, Babichev Anatoly, Pogorelov Nick

EVALUATION OF CUTTING TEMPERATURE OF A TOOL WITH A CURVED CUTTING EDGE 15
Moiseev Denis, Ryzhkin Anatoliy, Volokitin Gennadiy

IMPROVING THE EFFICIENCY OF THE MILLING PROCESS BY REDUCING THE OSCILLATION OF THE TECHNOLOGICAL SYSTEM 20
Kozlov Alexander M., Kiryuschenko Evgeny V., Kozlov Andrey A.

FUZZY CONTROLLERS IN THE CONTROL SYSTEM OF A BRUSHLESS ELECTRIC MOTOR USING HIL TECHNOLOGY 23
Kalach Gennady, Kazachek Nina, Volosnyh Georgi

A FEASIBILITY STUDY FOR IMPLEMENTING ADVANCED CONTROL OF A HEATING SYSTEM 27
Fomin Valerii

DEVELOPMENT OF A MATHEMATICAL MODEL OF A BIOMECHATRONIC FINGER PROSTHESIS AS PART OF A MULTIFUNCTIONAL PROSTHETIC HAND 32
Lysenko Aleksey, Khashev Denis, Uryanskaya Ekaterina

DYNAMIC MONITORING OF THE DRIVING WHEELS' PRE-SLIDING OF THE RAILWAY TRANSPORT 36
Chuveyko Mikhail, Nosachev Sergey

BUILDING EMBEDDED QUASI-TIME-OPTIMAL CONTROLLER FOR TWO-WHEELED SELF-BALANCING ROBOT 41
Phan Hai N., Nguyen Chiem X.

THE INFLUENCE OF EXTERNAL MECHANICAL IMPACT ON THE DYNAMICS OF AN ELECTROMAGNETIC ACTUATOR 45
Bolshenko Irina, Batishchev Denis, Klimenko Yuri

ALGORITHM FOR INCREASING THE ACCURACY OF THREE-AXIS MEMS ORIENTATION SENSORS 49
Goryanina Ksenia, Katin Oleg, Vernezi Miroslav

NONLINEAR ORIENTATION EFFECT IN LIQUID CRYSTALS TO CREATE A LINEAR DISPLACEMENT SENSOR 53
Denisova Olga, Abramishvili Roman

INDUSTRIAL CONTROL SYSTEM OF ROLL STAMPING 58
Pilipenko Aleksandr, Pilipenko Anastasia, Pilipenko Olga

AUTOMATION OF A DYNAMIC SYSTEM WITH COMPUTER SIMULATION OF HEAT LOSSES 63
Pilipenko Aleksandr V., Petrov Sergey P., Pilipenko Olga V.

ANALYSIS OF THE INFLUENCE OF POSITIVE FEEDBACK ON THE QUALITY OF THE CONTROL SYSTEM 68
Tsybriy Irina, Ignatenko Vitaliy

THE ANALITICAL MODELING OF THE MOTION PROCESS OF THE INVERTED PENDULUM SYSTEM 71
Obukhov Pavel, Mogilin Yaroslav

SIMULATION OF PHYSICAL PROCESSES IN DEVICES WITH NON-MAGNETIC, MILK-DISPERSED SECONDARY PART 73
Minkin Maksim, Kuimov Denis, Mozgovoj Andrej

TECHNOLOGICAL PROCESSES INTENSIFICATION IN DEVICES WITH MAGNETO-FLUIDIZED BED ... 79
Egorov Ivan, Egorov Nikolay
INVESTIGATION OF THE INFLUENCE OF CONTACT FRICTION FORCES IN TESTS ON THE STRENGTH OF COMPRESSING WALL MATERIALS ... 83
Lukinova Nataliya, Matrosov Audrey, Serebryanaya Irina, Terekhina Yuliya, Nizhnik Dariya
METALLIC MATERIALS FOR MEDICAL USE ... 87
Illarionov Anatoly, Belikov Sergey, Grib Stella, Yurovskikh Artem
MORPHOLOGICAL FEATURES AND MECHANICS OF DESTRUCTION OF MATERIALS WITH DIFFERENT STRUCTURES UNDER IMPACT DROP CYCLIC LOADING ... 91
Varavka Valery N., Kudryakov Oleg V., Zabiyaka Igor Yu., Shapovalova Antonina A.
ON ACCOUNTING FOR TEMPERING EFFECT IN UNSTEADY CREEP MODEL FOR METALS ... 95
Volkov Ivan, Igumnov Leonid, Ryabov Aleksandr, Shishulin Denis
DEFINITION OF THE STRAIN-STRESS DISTRIBUTION OF POROUS GLASS IN THE RETARDED COOLING TEMPERATURE RANGE ... 100
Grushko Irina
DEVELOPING OF COMPLEX FOR HOT PLASTIC DEFORMATION MODELING OF STEEL TYPE 20-30CRNIMOV FOR HEAVY FORGING ... 104
Dub Vladimir, Churyomov Alexandr, Soloviev Igor, Popkov Anton
QUENCHING TEMPERATURE INFLUENCE ON ELASTIC AND HARDNESS BEHAVIOR IN A BIOCOMPATIBLE TI-BASED ALLOY ... 108
Illarionov Anatoliy, Karabanalov Maksim, Korelin Andrey, Novokreschenov Stepan
INHOMOGENEITY OF DEFORMED STATE DURING COMPRESSION TESTING OF TITANIUM IMPLANT ... 112
Loginov Yury, Stepanov Stepan, Khanykova Catherine
NUMERICAL-ANALYTICAL STUDY ON EFFECT OF THREE-PHASE POROELASTIC MEDIUM MODEL PARAMETERS ON DYNAMIC DISPLACEMENTS AND POROUS PRESSURES RESPONSE ... 117
Igumnov Leonid, Belov Aleksandr, Vorobtsov Igor, Petrov Andrey
SOLUTION OF A DUAL INTEGRAL EQUATION ARISING IN THE CONTACT PROBLEMS OF ELASTICITY THEORY WITH THE FULL FOURIER SERIES AS THE RIGHT-HAND SIDE ... 121
Vasiliev Andrey, Volkov Sergei
ON THE PROBLEM OF SHEAR OF A FUNCTIONALLY GRADED HALF-SPACE BY A PUNCH ... 125
Zelentsov Vladimir
MECHANISM OF PITTING CORROSION PROTECTION OF METALS AND ALLOYS IN NEW-GENERATION WATER TREATMENT PLANTS ... 130
Grachev Vladimir, Nechaev Igor, Rozen Andrei E., Rozen Andrei A.
COMPARATIVE OBSERVATION OF SILVER NANO AND MICROSTRUCTURES DEPOSITED FROM AEROSOL AND FOG ... 135
Zheltova Anna, Biryukov Yurii, Ryndya Sergey, Smolyanskiy Alexander
FRACTAL MICROGEOMETRY OF ORIGINAL AND HYDGROGEN-CHARGED GRAPHENE MULTILAYER FILMS SURFACE: RESEARCH BY THE LOW-TEMPERATURE SORPTION OF NITROGEN METHOD ... 141
Vasilyeva Svetlana, Zheltova Anna, Merkushkin Alexey, Smolyanskiy Alexander
THE ELECTROMECHANICAL CONVERTER IN THE SYSTEMS OF DESULFURISATION OF CRUDE OIL ... 145
Kuimov Denis, Minkin Maxim
MAXWELL HOMOGENIZATION SCHEME IN MICROMECHANICS: AN OVERVIEW ... 149
Sevostianov Igor
INCREASE ROBUSTNESS OF THE METHOD OF DIAGNOSTICS AND IDENTIFICATION OF HIGH-PRECISION POSITIONING SYSTEMS ... 153
Lankin Anton, Grechikhin Valery, Michail Lankin
MATHEMATICAL MODELING OF WAVE PROCESSES AND TRANSPORT OF BOTTOM MATERIALS IN COASTAL WATER AREAS TAKING INTO ACCOUNT COASTAL STRUCTURES ... 158
Protsenko Sofiya, Sukhinova Tatyana
INVESTIGATION OF NONLINEAR 2D BOTTOM TRANSPORTATION DYNAMICS IN COASTAL ZONE ON OPTIMAL CURVILINEAR BOUNDARY ADAPTIVE GRIDS ... 164
Sukhinov Alexander, Chistyakov Alexander, Sidoryakina Valentina
ON THE DEVELOPING OF THE MATHEMATICAL MODEL OF ALL-WHEEL DRIVE VEHICLE ... 171
Efimov Andrey, Kireev Sergey, Korchagina Marina

DEVELOPMENT OF STRUCTURE OF A SUBSYSTEM OF INFORMATION SECURITY OF THE AUTOMATED SYSTEM 176
Bakhracheva Yulia
THE ANALYSIS OF PASSENGERS TRAVEL TIME TO THE DISTANT FROM A CENTRAL AREA TRANSPORTATION DISTRICTS IN THE CITY OF ROSTOV-ON-DON 180
Chernova Anna, Semchugova Elena, Negrov Nikolai
HARDWARE IN THE LOOP SIMULATION OF OBJECTS INTERNAL STRUCTURES INHOMOGENEITIES IN MULTI-ANGLE ELECTRICAL IMPEDANCE TOMOGRAPHY 183
Aleksanyan Grayr, Gorbatenko Nikolay, Kucher Artem, Shcherbakov Ivan
RESEARCH OF THE CONDUCTIVITY OF ORGANIC AND INORGANIC MEDIA IN MULTI-ANGLE MULTI-FREQUENCY ELECTRICAL IMPEDANCE TOMOGRAPHY 187
Aleksanyan Grayr, Shcherbakov Ivan, Kucher Artem, Priyma Maria
MODELLING OF FUZZY EXPERT INFORMATION IN THE PROBLEM OF A MACHINE TECHNOLOGICAL ADJUSTMENT 191
Dimitrov Valery, Borisova Lyudmila, Nurutdinova Inna
OPTIMIZATION OF THE AUTOMATED COLORIMETRIC MEASUREMENT SYSTEM FOR PH OF LIQUID 195
Katin Oleg, Lukyanov Aleksander, Goryanina Ksenia
MODEL IMPLEMENTATION OF THE SIMULATION ENVIRONMENT OF VOTING ALGORITHMS, AS A DYNAMIC SYSTEM FOR INCREASING THE RELIABILITY OF THE CONTROL COMPLEX OF AUTONOMOUS UNMANNED OBJECTS 199
Kovalev Igor, Losev Vasiliy, Saramud Mikhail, Petrosyan Mariam
ARTIFICIAL INTELLIGENCE METHOD FOR ELECTRIC DRIVES MODE OPERATING AND TECHNICAL CONDITION DETERMINATION 204
Kruglova Tatyana, Bulgakov Alexey, Vlasov Alexander, Shmelev Ivan
INVESTIGATION OF THE INFLUENCE OF THE MICROCAPILLARY STRUCTURE OF NATURAL SKINS ON RELATIVE HUMIDITY IN VACUUM-SORPTION HUMIDIFICATION 209
Larina Ludmila, Ruslyakov Dmitryi, Tikhonova Olga
DEVELOPMENT OF THE CONTROL ALGORITHM OF PROCESSES OF INTENSIVE HYGROTHERMAL IMPACT ON CAPILLARY AND POROUS MATERIALS IN THE CONDITIONS OF THE VACUUM 214
Larina Ludmila, Ruslyakov Dmitriy, Tikhonova Olga, Morozov Sergey
MODEL OF MOTION OF THE COMBINED MANIPULATOR 218
Poletaykin Vladimir, Kolesnikov Pavel
COMPLEX OF PARALLEL PROGRAMS FOR MODELING OIL PRODUCTS TRANSPORT IN COASTAL SYSTEMS 223
Semenyakina Alena, Protsenko Sophia
AUTOMATED CALCULATION OF THE ABSORBER TEMPERATURE FIELD OF AN INFLATABLE COLLECTOR 229
Smirnova Olga, Avetisyan Yuri, Dmitrienko Nadezhda, Balabaeva Oxana
ECOLOGICAL AND HYDROPHYSICAL RESEARCH OF IMPACT THE VERTICAL TURBULENT EXCHANGE COEFFICIENT ON THE CONCENTRATION OF DISSOLVED OXYGEN IN THE BOTTOM LAYER OF SHALLOW WATER 233
Sukhinov Alexander, Nikitina Alla, Belova Yulia, Bednaya Tatyana
SUBJECT-INDEPENDENT MODELING AND REPRESENTATION DATA ON THE FORMATION AND DISTRIBUTION OF INNOVATIVE VALUE 238
Frolov Aleksey, Maslova Irina
THE CALCULATION PROBLEM OF THERMODYNAMIC PROCESSES IN A STEAM TURBINE 243
Chistyakov Alexander, Nikitina Alla, Protsenko Elena
REGULATOR SYNTHESIS FOR THE SELF-SENSING CONTROL SYSTEM OF THE PROPORTIONAL ELECTROMAGNET DC BASED ON REDUCED-ORDER MODELS 249
Shaykhutdinov Danil, Gorbatenko Nikolay, Manackov Yuri, Shirokov Konstantin
INVESTIGATION OF ELECTROSTATIC FIELDS BY MODELING METHODS 254
Merkulova Alla, Cherunova Irina, Danilenko Irina
THE HYDRODYNAMIC DUST RETENTION MODELING PROCESS WITH A FOAM LAYER FOR OPEN LINEAR DYNAMIC TECHNOLOGICAL SYSTEMS AT CONSTRUCTION INDUSTRY ENTERPRISES 258
Bespalov Vadim, Gurova Oksana, Tsarevskaya Irina, Alekseenko Ludmila

MODELING OF DYNAMIC CONDITIONS OF OPERATION OF WIND TURBINES (WT) AND CHOICE OF ENVIRONMENTALLY EFFICIENT AND ENERGY ECONOMIC TECHNOLOGY OF WIND POWER ENERGY TRANSFORMATION 262
Bespalov Vadim, Gurova Oksana, Akay Oksana, Izotov Mikhail

BLUETOOTH API IMPLEMENTATION INTO ANDROID 266
Konev Sergey, Stus Anastasia, Kasyanenko Elena, Dolgov Vasiliy

THE DEVELOPMENT OF A MOBILE MONITORING SYSTEM FOR AGRICULTURAL OBJECT 269
Konev Sergey, Vernezi Miroslav, Donskoy Daniil, Lukyanov Alexander

STRUCTURAL-PARAMETRIC OPTIMIZATION OF THE EXPERIMENTAL DATA DECOMPOSITION INTO APPROXIMATED FRAGMENTS 273
Neydorf Rudolf, Iarakhmedov Orkhan

DYNAMIC DOCUMENT OBJECT MODEL FORMATION TECHNIQUE FOR CORPORATE WEBSITE PROTECTION AGAINST AUTOMATIC COPING OF INFORMATION 279
Galushka Vasily, Marshakov Daniil, Fathi Vladimir

FUSION OF DEEP FEATURES AND WEIGHTED VLAD VECTORS BASED ON MULTIPLE FEATURES FOR IMAGE RETRIEVAL 283
Wang Yanhong., Cen Yigang., Liang Liequan., Zhang Linna., Voronin Viacheslav., Mladenovic Vladimir.

PLANNING THE TRAJECTORY OF MANIPULATOR GRIPPER WITH A PARALLEL STRUCTURE AND FLEXIBLE LINKS 288
Dubovskov Vadim, Valyukevich Yuriy

ALGORITHM FOR CONTROLLING PARAMETERS OF LINEAR FILTER BASED ON FUZZY EXPERT SYSTEM 292
Eliseev Alexandr, Prygunov Alexandr

ALGORITHM FOR DETECTING VIOLATIONS OF TRAFFIC RULES BASED ON COMPUTER VISION APPROACHES 296
Ibadov Samir, Ibadov Ragim, Kalmukov Boris, Krutov Vladimir

MONITORING OF THE TECHNICAL REQUIREMENTS COMPLIANCE IN THE SOLID HOUSEHOLD WASTE LANDFILLS FUNCTIONING THROUGH THE METHOD OF VISUAL DETECTION 300
Richter Andrey, Kazaryan Maretta, Shahramanyan Michael

THE SYNTHESIS OF ADAPTIVE MULTI-MODE REGULATORS BASED ON COMBINED CONTROL OF THE COMBINED MAXIMUM PRINCIPLE 305
Kostoglotov Andrey, Lazarenko Sergey, Losev Vitaliy, Pugachev Igor

MODEL OF MULTICOMPONENT MICRO-DOPPLER SIGNAL IN ENVIRONMENT MATLAB 310
Kucheryavenko Alexander, Fedosov Valentine

THE COMBINED MAXIMUM PRINCIPLE IN THE PROBLEM OF SYNTHESIS OF AN ADAPTIVE DYNAMIC FILTER UNDER CONDITIONS OF DISTURBANCES IN THE MEASUREMENT PROCESS 314
Kostoglotov Andrey, Kuzin Aleksandr, Lazarenko Sergey, Pugachev Igor

RESEARCH AND COMPARATIVE ANALYSIS OF METHODS LINEAR INTERPOLATION TRAJECTORY OF THE WORKING TOOL OF A TWO-TIER MECHANISM 319
Naumov Ivan, Paschenko Anton, Koltcov Aleksandr

SYNTHESIS OF AN INTERPOLATION ALGORITHM STRAIGHT-LINE SEGMENT AT PLANNING THE TRAJECTORY OF A DEGENERATE TWO-TIER MECHANISM 322
Naumov Ivan, Tolstunov Oleg, Perekhrestenko Larisa

MULTI-CHANNEL MONITORING APPROACH OF E1 STREAMS 325
Nikitin Anton, Makov Sergey, Klimenko Sergey

A MATHEMATICAL MODEL FOR ESTIMATING AND FORECASTING THE STATE OF A DIGITAL SUBSTATION BASED ON THE WAVELET TRANSFORM METHOD 329
Oganyan Roman, Shaykhutdinov Danil, Narakidze Nuri, Shurygin Dmitriy

DEFECT DETECTION ON VIDEOS USING NEURAL NETWORK 334
Sizyakin Roman, Gapon Nikolay, Shraifel Igor, Tokareva Svetlana, Bezuglov Dmitriy

IMAGE INPAINTING USING A NEURAL NETWORK 339
Gapon Nikolay, Ponamorenko Mikola, Pismenskova Marina, Tokareva Olga

CLASSIFICATION OF A TWO-DIMENSIONAL POSE USING A HUMAN SKELETON 343
Pismenskova Marina, Balabaeva Oxana, Voronin Viacheslav, Fedosov Valentin

METHOD TO REDUCE THE EFFECT OF MIAGRAFIC AND SENSORY NOISE WITH ISOLATING THE ISOLINE ON ECG SIGNAL 347
Semenishchev Evgeny, Chernyshov Dmitry, Svirin Ilya
ALGORITHM COMBINATION OF DEBLURRING AND DENOISING ON VIDEO FRAMES USING THE METHOD SEARCH OF LOCAL FEATURES ON IMAGE 352
Semenishchev Evgeny
Author Index

edp sciences

MATEC Web of Conferences

Statement of Peer review

In submitting conference proceedings to MATEC Web of Conferences, I certify to the Publisher that I adhere to the **Policy on Publishing Integrity** of the journal in order to safeguard good scientific practice in publishing.

1. All articles have been subjected to peer review administered by the proceedings editors.
2. Reviews have been conducted by expert referees, who have been requested to provide unbiased and constructive comments aimed, whenever possible, at improving the work.
3. Proceedings editors have taken all reasonable steps to ensure the quality of the materials they publish and their decision to accept or reject a paper for publication has been based only on the merits of the work and the relevance to the journal.

Title, date and place of the conference

XIII International Scientific-Technical Conference "Dynamic of Technical Systems" - DTS-2017

Sept. 13 - 15, 2017

Rostov-on-Don, Russia

Proceedings editor(s):

Tooraj Abbasian Najafabadi
Igor Sevostianov
Yeghiazaryan Karen Onikovich
Nguyen Dong Ahn
Vladimir Mladenovic

Date and editor's signature

12.10.2017

Modelling the Reduced Coefficient of Volumetric Rigidity in High-Pressure Rubber-Cord Shell Hoses

Aleksandr Rybak[1] and *Olga* Lyakhnitskaya[1,*]

[1]Don State Technical University, Rostov-on-Don, Russian Federation

Abstract. In recent years, methods for calculating the hydraulic power drive have become increasingly widespread, taking into account the violation of the continuity equation during its operation at unsteady modes due to the use of the theory of volumetric rigidity in modeling. In this article approach to calculation of the reduced coefficient of volume rigidity of high-pressure hoses is considered.

1 Introduction

Hydraulic systems are increasingly used in modern machines and mechanisms. Typically, the working fluid in such systems operates under high pressure. To transport this liquid, metal pipelines and high-pressure hoses (HPH) are used. The working fluid can be various kinds of mineral and synthetic oils, liquid fuels, water-oil emulsions or water.

One of the most important characteristics of a hydraulic drive is the quality of its operation at non-steady-state conditions, when the equation of continuity of flow is violated because of the compressibility of the working fluid and the deformation of the hydraulic system [1-3]. In this case, in modelling, a certain dummy flow of fluid is introduced into the equation of continuity, which compensates for these effects.

However, in recent years, another method of modelling hydraulic power drives has been increasingly used, to take account of the violation of the continuity equation during its operation at unsteady modes in modelling - the theory of volumetric rigidity [4,5].

Thus, instead of the reduced modulus of elasticity of the system, the reduced coefficient of its volume rigidity is used [6,7]. Several formalized dependencies have been obtained to determine the reduced coefficient of volume hardness of various hydraulic devices [8-15]. However, for rubber-cord shell high-pressure hoses, such dependencies have not yet been obtained, since their elastic properties still do not lend themselves to a rigorous analytical description. Here, we address this omission.

2 Theory

The volumetric rigidity of a hydraulic drive is the property of the hydraulic system to resist changing its internal volume when the pressure changes. It is characterized by "reduced coefficient of volumetric rigidity" of a system. The given volume rigidity will allow us to judge pliability of system in general and each her element separately similar to a concept of rigidity of solid bodies which indicates a change of the linear sizes of elastic elements at impact on them of external forces [1].

Liquid compresses when forcing in system with increased pressure that imposes the influence on property of the system in general. Therefore, the "volume rigidity" characterized by a coefficient of volumetric rigidity may serve as an integral criterion of dynamic properties of hydraulic system considering that change of volume of hydraulic system results from the increase liquid pressure boost that at the same time compresses [6].

$$C_{\text{пр}} = -\frac{dp}{dV}, \qquad (1)$$

where $C_{\text{пр}}$ - the reduced coefficient of volumetric rigidity of the considered site of the hydraulic actuator; dp – pressure increment of the considered site; dV – interior content increment of the considered site corresponding to an increment of pressure upon dp.

We will define coefficient of volumetric rigidity of plant and equipment, hydraulic actuator as pressure increment caused by additional injection of a unit of volume, power liquid given to power liquid condition in hydraulic reservoir and environmental condition.

In addition to metal pipelines, high-pressure hoses (HPH) are also in current use. They are characterised by being composed of a rubber-cord shell and are used in the machine-building industry, in agriculture, in various types of mining and mining machines, in road-building machinery and in the metallurgical industry and other industries. They are also used in vehicles as power steering hoses and brake hoses. HPHs are thus essential to a wide range of different applications The technique and experimental investigations made for a narrow class of high pressure hoses [16-18], but a general mathematical formula, suitable for calculating the reduced volumetric coefficient of stiffness of any type and size do not exist. The aim of this work is to find the single formula for

* Corresponding author: lolyv@yandex.ru

 DOI: 10.1051/matecconf/201713201001

calculating the reduced volumetric stiffness of any high pressure hoses.

3 Analysis

According to GOST 6286-73 [19] the high-pressure hoses depending on the strength of the wire used, must be made in three groups: A, B and B.

Group A is manufactured with the use of a wire with a breaking force of not less than 15 kg.

Group B is manufactured using a wire with a breaking force of not less than 17.5 kg.

Group B is manufactured using a wire with a breaking force of at least 20 kg.

Group Z should are made of a lattin wire and a breaking force of at least 20 kg.

The high-pressure hoses of each group, depending on the design, can be of two types:

Type I- with one metal braiding (fig. 1)

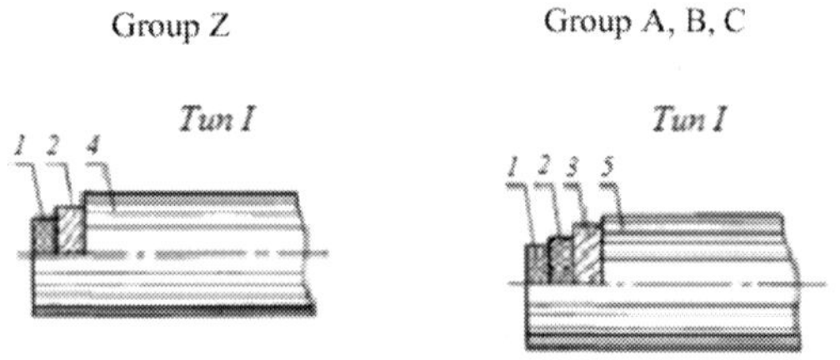

Fig. 1. The high-pressure hoses type I construction with metallic braidings. Z group: 1 — hose tube; 2 — metal braiding; 3 — squeegee; 4 — outer rubber layer. A, B, C group: 1 — hose tube; 2 — textile braiding; 3 — metal braiding; 4 — provisional rubber layer; 5 — outer rubber layer.

Type II- with two metal braiding (fig. 2)

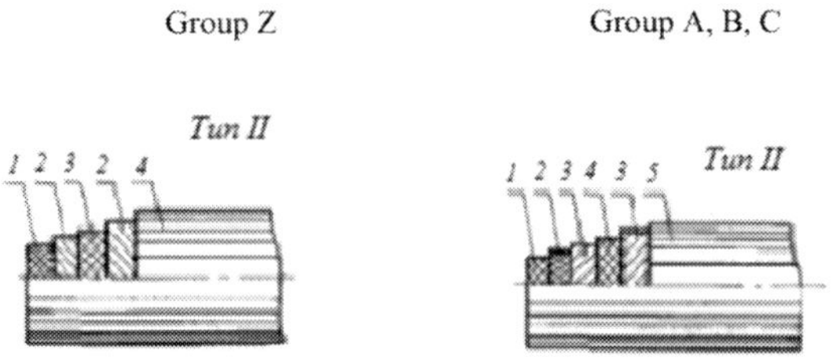

Fig. 2. The high-pressure hoses type II construction with metallic braidings. Z group: 1 — hose tube; 2 — metal braiding; 3 — squeegee; 4 — outer rubber layer. A, B, C group: 1 — hose tube; 2 — textile braiding; 3 — metal braiding; 4 — provisional rubber layer; 5 — outer rubber layer.

In the HPH, the first layer (the inner rubber hose chamber), which is firmly connected to the frame, receives a radially directed hydrostatic pressure *(p_1)* applied to the surface of radius *(r_1)*. On the outer surface of the chamber of radius *(r_2)*, there is a pressure *(p_2)* arising from the action of the frame. The rubber chamber is formed from a material with a Poisson's ratio close to 0.5 [4]. The second frame ply, along its inner surface of radius *(r_2)*, experiences a radially directed hydrostatic pressure *(p_2)*. On the outer surface of the chamber, of radius r_3, there is a pressure p_3. The last pressure is assumed equal to 0, thus excluding the influence of the third layer-the external rubber lining. From the assumptions made, a calculation formula can be obtained for a single-loop HPH:

$$p_1 = 0.184 \cdot K_B \cdot N \cdot n \cdot \frac{1}{(r+r_1)^2} \qquad (2),$$

where K_B — braid durability to a break-up of N/thread, N — quantity of the streams of a wire put by the spool of the braiding machine, n — number of wires in the spool.

The second layer is the frame consisting of a row concentric or spirulate located laying which elements have some possibility of a shift, unlike the first layer. The special layer and structural design represent in homogeneous and anisotropic material.

We will accept the following assumptions for simplification of calculations:

1. Forces arising in a hose wall are perceived generally by a frame as the most rigid part of a hose;
2. The frame is considered as thin-walled cylinder closed from edges the outer rubber cover isn't taken into consideration;
3. The outer rubber cover is not taken into consideration.

According to Sukharev and Lepetov [20] the significant difference in the loadings perceived by the first and last braidings is revealed. They represented the distribution of loads in the frame of a hose having ***i*** homogeneous braidings in the form:

$$p = p_1 + p_2 + p_3 + \cdots + p_i, \qquad (3)$$

where $p_1, p_2, p_3, \ldots, p_i$ are the fractions of the hydraulic pressure p perceived by the corresponding metal braidings.

The equation of joint deformations takes the form:

$$p \cdot r_1^2 = p_{1\_2} \cdot (r_2^2 + 2r_1^2),$$

where r_1 and r_2 are the average calculated diameters of the overlap of the first and second braidings (in centimeters), and p_{1_2} is the intermediate pressure between the first and second braidings (10^5 Pa).

It follows from (2) that

$$p_1 = p - p_{1\_2}$$

or

$$p_2 = p_{1\_2}$$

From here:

$$p_1 = p \cdot \left(1 - \frac{r_1^2}{r_2^2 + 2r_1^2}\right)$$

$$p_2 = p \cdot \left(\frac{r_1^2}{r_2^2 + 2r_1^2}\right)$$

Expressing p_2 in terms of p_1, we obtain the calculated formula for finding p_2:

$$p_2 = p_1 \cdot \left(\frac{r_1^2}{r_2^2 + 2r_1^2}\right)$$

Using formula (2), we thus obtain a formula for calculating the pressure p_2:

$$p_2 = 0.184 \cdot k_B \cdot N \cdot n \cdot \left(\frac{1}{(r+r_1)^2} + \frac{1}{r_1^2 + r_2^2}\right) \cdot \frac{C_1}{C_2^2}$$

A variety of correction coefficient is introduced for correction of deviations hose multilayered frame from conditions in thin-walled isotropic covers and for accounting of strength and deformation properties of the material of a frame. We will enter C1 and C2 coefficients for braided HPH [20]. The multiplier C_1 considers non-simultaneity of a gap in parallel going braidings in a stream:

$$C_1 = 1 - 0{,}015(n - 1),$$

and multiplier C2- elongation ε wire material in braid:

$$C_2 = 1 + \frac{\varepsilon}{i}.$$

Hose process pressures containing i of braidings are calculated in the same way.

4 Results

Calculation data may be significantly simplified if conditionally to assume that in thin layer braidings of value of superimposing of average radii are close among themselves, in other word $r_1 = r_2 = \cdots = r_i$ equation (2) will become:

$$p_i = 0.184 \cdot K_B \cdot N \cdot n \cdot \sum_1^i \frac{1}{2^{i+1} \cdot (r + r_1)^2}$$

We will receive the generalized formula for HPH containing i of braidings, having entered coefficient

$$\theta_i = \frac{0.735}{2^{i+1}}$$

$$p_i = \theta_i \cdot K_B \cdot N \cdot n \cdot \frac{1}{(r+r_1)^2} \cdot \frac{C_1}{C_2^2} \quad (4)$$

We will find pressure increment, having differentiated (4)

$$dp_i = -\theta_i \cdot K_B \cdot N \cdot n \cdot \frac{dr}{(r+r_1)^3} \cdot \frac{C_1}{C_2^2} \quad (5)$$

The internal volume of pressure will change due to the deformation which has resulted because of change of pressure in the pipeline [3]:

$$V\text{тр} = \pi \cdot r^2 \cdot l,$$

where d and l are the current values of the diameter and length of the pipe section in question, respectively.

The increment in the volume of the pipe is determined by the formula

$$dV_{\text{тр}} = d(\pi \cdot r^2 \cdot l) = 2 \cdot \pi \cdot r \cdot l dr + \pi \cdot r^2 \cdot dl, \quad (6)$$

where dr and dl — increment of the radius of a pipe and its length corresponding to pressure increment dp.

We will find a formula for calculation of the reduced factor of volume rigidity of i-braided HPH having substituted the found increments (5) and (6) in a formula (1)

$$C_{\text{пр}} = \frac{\theta_i \cdot K_B \cdot N \cdot n}{\pi \cdot l \cdot r \cdot (r + r_1)^3} \cdot \frac{C_1}{C_2^2}$$

5 Conclusions

The analytical formula of calculation of the reduced factor of volume rigidity of HPH is received. It is required to carry out a series of tests for her confirmation and specification.

References

1. A.T. Rybak, V.P. Zharov, V.I. Mirnyj, S.A. Zatolokin *DTS-2007: Proc. 8th Int. Conf.*, **1**, 66 – 70 (2007)
2. A.T. Rybak, *Proceedings of the International Scientific-Technical and Scientific-Methodical Conference: Hydrogasodynamics, hydraulic machines and hydropneumatic systems*, 186-189, (2006)
3. A.T. Rybak, I. V. Boguslavskiy, Vestnik of the DSTU, **2**, 249-257 (2010)
4. A.T. Rybak, Vestnik of the DSTU, **6**, 200-207 (2006)
5. A.T. Rybak, Theory and methodology of calculation and design of drive systems of technological machines and aggregates of agroindustrial complex, thesis abstract of Doctor of Technical Sciences, 39 (2011)
6. A.T. Rybak, I. V. Boguslavskiy, Bulletin of Machine Building., **10**,39-46 (2010)
7. A.T. Rybak, *Modeling and calculation of hydromechanical systems at the design stage*, 167 (DSTU, Rostov n/D, 2006) [in Russian]
8. A.R. Temirkanov, A.T. Rybak, Vestnik of the DSTU, **4**, 176-185 (2014)
9. S.A. Zatolokin, A.T. Rybak, Vestnik of the DSTU, **10**, 84-90 (2010)
10. S.A. Zatolokin, Perfection of the theory and methods of designing hydromechanical systems with a pump-accumulator constant-pressure flow source: the author's abstract of the dissertation of the candidate of technical sciences, 18, (2009)
11. A.V. Anisimov, A.I. Butov, V.L. Kondrashev, K.V. Dolgov, K.E. *7th International Scientific and Technical Conference New technologies for controlling the movement of technical objects*, 5, 129-134 (2004)
12. G.M. Vodianik, A.V. Anisimov, N.V. Divulin, *6th International Scientific and Technical Conference New technologies for controlling the movement of technical objects*,**4**, 144-148 (2003)
13. A.T. Rybak, V.P. Zharov, A.V. Korchagin, *International Scientific and Technical Conference*

Progressive technological processes in metallurgy and machine building. Ecology and life support. Information technologies in industry and education, 46-49, (2005)

14. A.T. Rybak, V.I. Mirnyj, Izvestiya Institute of Management and Innovation of the Aviation Industry, **3-4**, 75-79, (2005)
15. A.T. Rybak, Vestnik of the DSTU,**7**, 218-223 (2007)
16. A.T. Rybak, A.A. Serdyukov, *International Scientific and Practical Conference Innovative technologies in mechanical engineering*, 216-218 (2008)
17. A.T. Rybak, O. V. Lyakhnitskaya, E. N. Shein, *X International Scientific and Methodological Symposium Contemporary problems of multilevel education*, 374-381 (2015)
18. A.T. Rybak, O. V. Lyakhnitskaya, *Proceedings of the 7th International Seminar System analysis, management and information processing,* 62–67, (2016)
19. GOST 6286-73
20. V. A. Lepetov, Chemistry, 350-367 (1976)

The technology of receiving finely dispersed powders of magnetic materials in the screw drum

Yuriy Vernigorov[1,*], *Natalia* Frolova[1]

[1]Don State Technical University, Rostov-on-Don, Russia

Abstract. It is experimentally shown the possibility of obtaining a finely dispersed powder using mills of non-impact type on the basis of the screw drum. The dependence of grinding the powder to the weight of the load is proved.

1 Introduction

Magnetically hard materials are widely used for the manufacture of sources of permanent magnetic fields, which in some cases are technically and economically more advantageous than DC electromagnets. By the method of production, the permanent magnets are divided into cast and powder ones. The requirements of miniaturization of modern magnetic elements, as well as the possibility of variation of their physical properties in a large range, have determined the wide spread of powder technology for manufacturing permanent magnets. While coarse grinding, the proportion of energy consumption to the cost of the final product is negligible. For many technical details it is necessary to have powders with a very small particle size, of about 10^{-8} m. For this purpose, fine grinding is used, which is produced in spherical vibrating, rotating or planetary centrifugal, vortex and hammer mills [1]. The production of highly dispersed powders, which is always associated with higher energy costs, simply cannot afford to use inadequate methods of dispersion. However, the moving elements of the mills used in industry today bear their negative significance: due to vibration, large shock loads, and also a complex control device for such mills, the quality of the powders obtained and the reliability of the grinding units themselves are significantly reduced. In impact mills, the destruction of material particles occurs as a result of impact loads. These loads can occur in a wide variety of conditions and circumstances. For example, when the grinding bodies fall, when a particle collides with a fixed obstacle, mutual collisions of particles in flight are also possible. But in any case, a piece of material or the self-grinding body must have such kinetic energy that it would be enough to overcome the internal bonds between the particles. Obviously, in such a field of technology as shock disintegration, there can be no final solutions. Modern equipment that implements the centrifugal-impact crushing-grinding method is also just another stage in the development of the technique of shock disintegration of solid materials.

2 Objective

Experimental study of the effect of the weight of the load of the powder charge on the granulometric composition of ferromagnetic materials by grinding by magnet-wave technology in a screw drum.

3 Main part

A block diagram of the experimental setup is shown in Fig. 1

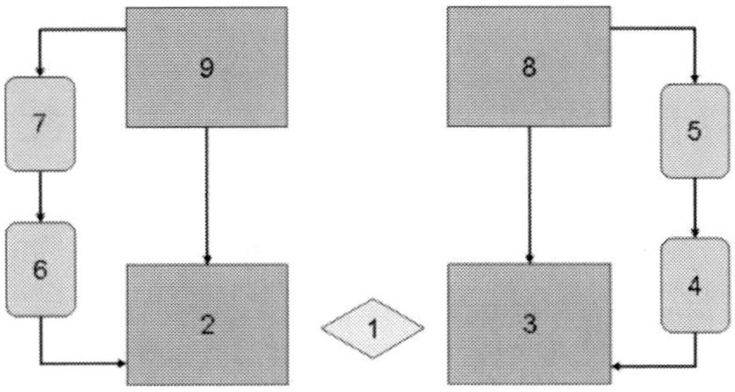

Fig.1. Block diagram of a mill based on a screw drum. 1 - screw drum, 2 - permeameter, 3 - permanent magnet, 4 - direct current ammeter, 5 - DC voltmeter, 6 - AC ammeter, 7 - AC voltmeter, 8 - magnet power supply 3, 9 - power supply of permeameter.

Appearance and type of the inter-pole space of the experimental setup are shown in Fig.2

A screw drum [5] is placed in the common inter-pole space of permanent field of electromagnets and permeameter [5], the circuit of which is shown in Fig.3. The drum is made of steel grade 12X18HT GOST 5582-75 and had a working chamber volume of 290 cm^3.

* Corresponding author: jvernigorov@donstu.ru

(a)

(b)

Fig. 2. Appearance (a) and type of the inter-polar space (b) of the experimental setup.

The shape of the drum makes it possible to keep the charge in a magnetic field until a predetermined granulometric composition is obtained. The charge is placed in the drum, the magnetic fields of the specified induction values and induction gradient are switched on.

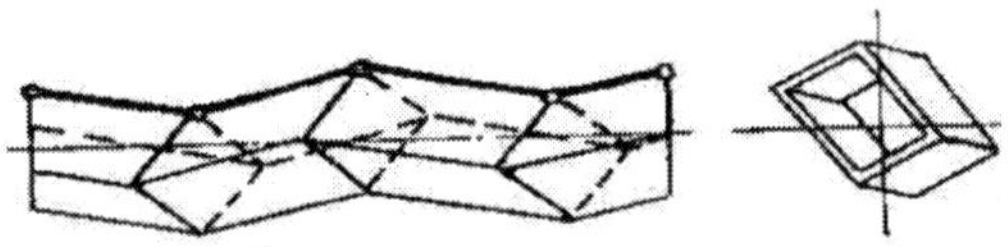

Fig.3. Screw Drum Scheme

In a stationary drum under the action of ponderomotive forces, the powder is transferred to a magneto-vibrating state, in which the particles are destroyed due to their collision with each other and with the walls of the drum. The efficiency of the destruction of particles in the magneto-vibrating layer depends on the material and the granulometric composition of the original powder, as well as on the induction and the gradient of magnetic field induction. When the bodies are destroyed by impact, the process can be represented in two stages. The first begins when the surfaces of the bodies come into contact. By the end of the first stage, the bodies stop coming closer and part of their mechanical energy passes into the energy of deformation. At the second stage, the energy of deformation is transferred to the kinetic energy of the bodies [2]. When the real bodies collide, the mechanical energy is not completely restored. Part of the energy is expended on residual deformation, heating, destruction of bodies, etc. To determine the impact forces, it is necessary to take into account the mechanical properties of the materials of the bodies. The solution of the problem is substantially complicated not only because of the difficulties of a purely mathematical nature, but also because of the lack of sufficient data on the parameters determining the behavior of the materials of bodies under impact loads [3].

All of the above mentioned led to the need to experimentally test the influence of the charge load mass on the granulometric composition of the material obtained after milling. The results of the experiment are shown in Fig. 4.

The analysis of the presented dependences shows that the process of powder grinding with loading masses from 100 grams has little effect on the size of the particles obtained, the optimal loading weight for the chosen parameters of the magnetic field and the geometry of the screw drum is a mass of 30-60 grams, which amounts to 30% of the drum loading.

Fine grinding is one of the most expensive operations in large-tonnage production. Considering the production volumes, even a relatively small increase in the efficiency of the grinding unit can give a significant economic effect. The described mill on the basis of a screw drum, the principle of operation of which is based on the use of magnetic field energy, does not have mechanically moving elements, which reduces energy costs [4].

4 Conclusions

1. It has been experimentally proved that due to a collision between particles of a ferromagnetic material in a magneto-vibrating layer in a screw drum, grinding of $SmCo_5$ powder occurs.
2. When designing the grinding devices which use the impact mechanism of the destruction of particles in the magneto-vibrating layer, the mass of charge to be loaded should be chosen in accordance with the results obtained in the work. In addition, the initial granulometric composition of the material, as well as its mechanical and magnetic properties, should be taken into account.
3. When grinding in the magnetically-vibrating layer of disperse ferromagnets due to impact interaction between particles, the grinding process is accompanied by an aggregation process, which requires additional energy expenditure.
4. Shock grinding in a magneto-vibrating layer in a screw drum allows to obtain a powder of a fine fraction, while an increase in the weight of the charge leads to more intensive aggregation.

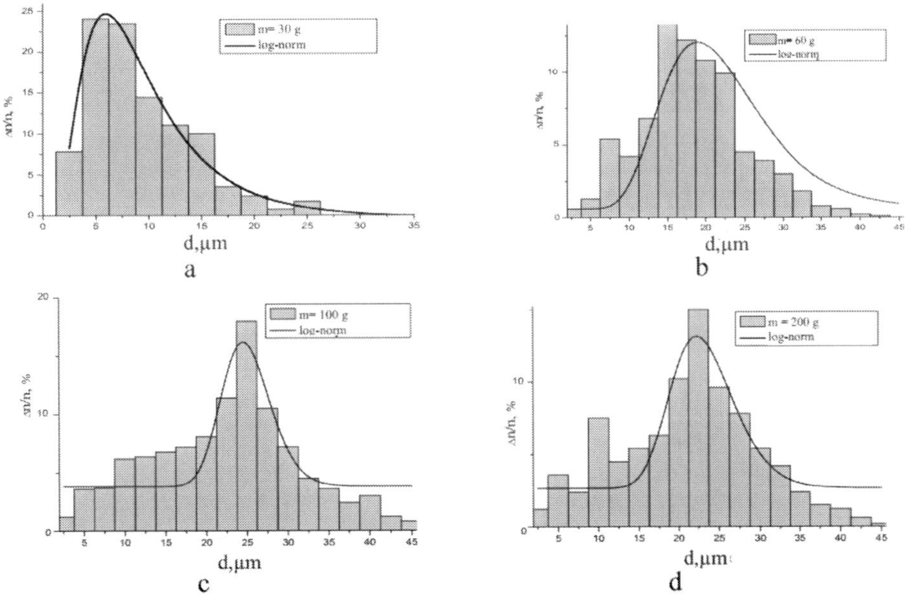

Fig. 4. Histograms and log-normal particle size distribution curves for different load masses

References

1. A. Libenson, V. Yu. Lopatin, and G. V. Komarnitskii, *Processes in Powder Metallurgy* (Moscow Institute of Steel and Alloys, Moscow, 2001), Vol. **1** [in Russian]

2. Yu. M. Vernigorov, N.N. Frolova, Vestnik DSTU, **11**, №7 (58) (2011)

3. G.S. Khodakov, *The physics of grinding* (Publishing House of Literature on Construction, Moscow, 1985)

4. Yu. M. Vernigorov, N.N. Frolova, G.V. Serga, A.B. Gordeeva, *Scientific research and their practical application. The current state and development*, **5**, 52 (2011) [in Russian]

5. G.V Serga, V.G. Serga, *Screw rotors*: Monograph (KubSAU, Krasnodar, 2003)

Analysis of the zinc covering quality formed by the vibration chemical-mechanical synthesis

Vladimir Ivanov[1], *Valeriy* Lebedev[1,*], *Irina* Davydova[1], *Tatiana* Atoyan[1]

[1]Don State Technical University, Rostov-on-Don, Russia

Abstract. The analysis of the results of experimental studies of production of zinc coverings by the way of vibration chemical and mechanical synthesis words.

The specific of formation of a zinc covering in vibration technological systems is that in the conditions of dynamic influence of the free moving metal or glassy indenters, activation of all components of the elements participating in formation of a covering is sharply increased. Molecules and atoms gain the padding energy which promoet overcome the potential hill influencing on the speed of chemical reactions. Catalytic properties are given to the surfaces of material relation to the reaction of restitution of ions of a double electric layer with the established equilibrium potential. In the result in the boundary layer, electrolytic potential in relation to equilibrium is displaced up to the sizes at which the reaction of restitution of ions of zinc` ions is possible.

The solid bodies which are immediately participating in chemical and mechanical process of formation of a zinc covering always have the ability to absorb in a varying degree (to adsorb) from environment on the surface molecules, atoms or ions. [1]. The numerous researches in this area led to a number of interesting conclusions for many almost important processes.

For the right understanding of adsorption process it is necessary to consider that various atoms of the surface layer of metal` adsorbent are not in identical conditions. It occurs because the surface of a solid body is not smooth, but has numerous micro roughness. At the atoms located on various sites of a surface the degree of saturation of valence forces is various, therefore, also their ability to the interaction with atoms and molecules from solution or environment is not identical. The most fissile section of a surface do not adsorb the molecule from solution any more, and its chemical properties have the initial value, i.e. adsorption in this case is specific. [2,3].

In the process of the chemical and mechanical synthesis operated by power influence of vibration system, the surface of metal is exposed to numerous multidirectional blows, elasto-plastic deformation and, as a result, activation of the surface; at the same time such phenomena as adsorption and diffusion are talien place [4].

At the first moment of formation of a covering Zincum ion, from solution passed a double layer, exempted from a hydrated envelope, is adsorbed on the fissile surfaces of ledges. Here Zincum ion (Zn ++) is discharged (accepts electrons) and (Zn) is besieged on the surface of metal at the expense of an oxidizing the restored processes in the form of a zinc covering. The formation of the first crystals of the besieged metal leads to the emergence of the micro/nano covering metal elements. As a result on a potential difference of Fe-Zn the field of minerals is imposed. The increase of an internal energy of the surface layers of metal, as a result of elasto-plastic deformation, leads to increase in the adsorption activity of a metal surface, ion activity and molecules, increase in crystallization centers and, density of micro currents. At the same time in the surface layer the micro/nano structures of a nucleus of a zinc covering are formed, at the same time appearing EMF has significant effect on further body height of a covering. The formation of crystalline structures of a covering happens not only on ledges, but also on sides, and in deepenings of a microprofile of a surface. As a result metal is covered more evenly, crystals are closed, forming an intact coating [5].

The research of micro and nano structure of a surface of a zinc covering on sampels from carbon steel shown that the covering received as a result of vibration chemical and mechanical synthesis possesses chaotically focused surface morphology where the open sections of a zinc covering are ≈ 100 microns (Fig. 1a). At the scale of 10 microns it is visible that on border of large sections of a covering structures of 2-5 microns in size are created (the allocated sections in Fig. 3b). In Fig. 1d the arrow shows the structure of a covering of the transition zone consisting of crystals from 1 micron in size. The increasing of this section up to the scale of 1 micron, shows the change of morphology and appearing on the top of a crystal of communications in the form of threads of 0,2-0,5

* Corresponding author: va.lebidev@yandex.ru

DOI: 10.1051/matecconf/201713201003

microns in size. A rescaling up to 200 nanometers (Fig. 3e) allowed to establish the amount of intercrystal communications and the sizes of separate elements ranging from 50 up to 100 nanometers. In Fig. 3g the scale of 100 nanometers intercrystal communications and separate crystals of a covering at a background in the form of light oval sites clearly are visible. Perhaps to receive such character of an arrangement of crystals of zincum in a covering only is the result of vibration influence.

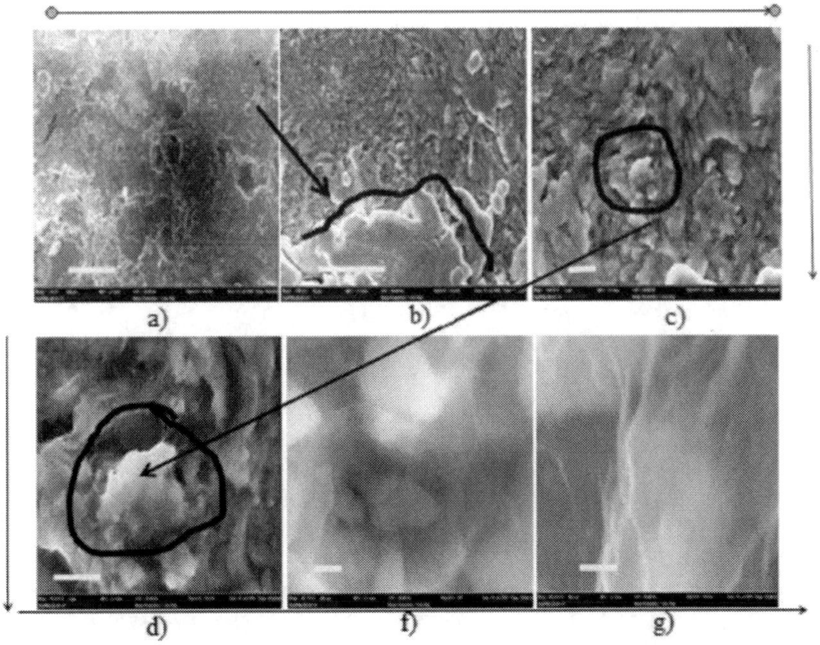

Fig. 1. The morphology of a surface of a zinc covering received by vibration of mechanical - chemical synthesis at various increase (a - the scale of 100 microns, b- scale is 10 microns, c - scale is 2 microns, d- scale is1 micron, e -scale is 200 nanometers, g-scale is 100 nanometers) and a process conditions: amplitude of fluctuation of the building bag of A=3,5 mm, frequency f = 33,3 Hz, time t = 60 min., actuation medium porcelain spheres D= 3-10 mm., working solution of chloride zincum - 250 g/l, zincum powder – 100 g/l/.

The layer of a zinc covering formed in an incipient state is a basis for the subsequent its body height, it is very important to provide the maximal activity of a surface. The formation of a large number of shallow crystals provides more complete overlapping of the main metal. The smaller porosity, the increased rust resistance and larger anchoring strength with the main metal is a reached as the result.

In Fig. 2 the electronic image of texture of the covering received in the course of vibration chemical and mechanical synthesis and a galvanic way is submitted. The structure of galvanic zinc coverings is porous, the sharp border dividing metal and a covering is observed, the micro emptiness on all surfaces of a micro section which existence does not promote giving to a covering of high corrosion behavior [6] are legibly expressed. Unlike them forthe tightly paked structure with lack of sharp border between metal and a covering is characteristic of the zinc coverings received in the course of vibration chemical and mechanical synthesis, thickness of a covering reaches 30-40 microns, the covering and the substrate have the identical metal nature the navodorozhenny layer inherent to isabsent.

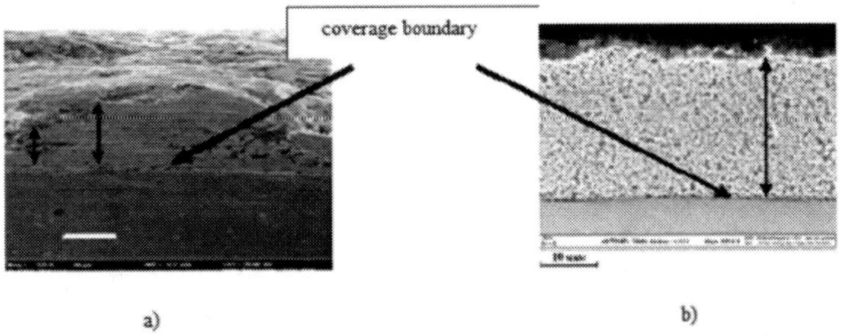

Fig.2. A micro section of the zinc covering received:
a) vibration chemical and mechanical synthesis, micron scale-20; b) - in the galvanic way, micron scale-10.

One of key indicators of quality of zinc coverings is rust resistance. In Fig. 3 the results of corrosion tests in the camera of salty fog within 20 days are presented, it is model from steel of St 3 with a zinc covering, put in the course of vibration chemical and mechanical synthesis-1 and a galvanic way-2. In the a result of corrosion tests are established: the zinc covering created by vibration chemical and mechanical synthesis surpasses the zinc covering received by a traditional method by 1,7-2,0 times in rust resistance; destruction of zinc coverings is influenced by porosity: than less thickness of a covering, subjects it is higher that promotes more mild penetration a severe atmosphere to the main metal.

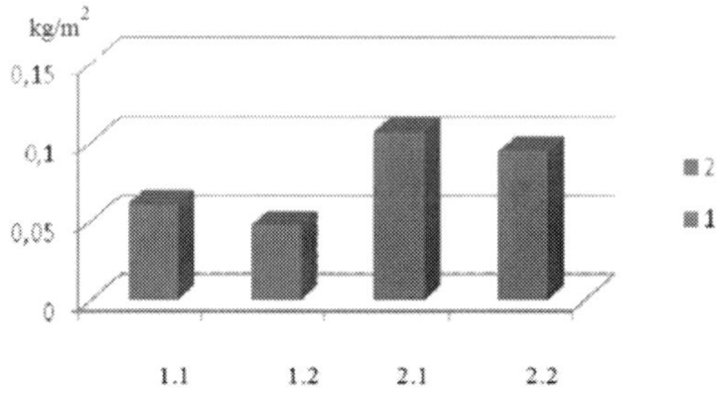

Fig. 3. Change of corrosion rate of a zinc covering depending on a method of drawing and its thickness:1.1-6 microns.; 1.2 - 14 microns.; 2.1-6 microns., 2.2 - 14 microns.

Conclusions

The carried-out analysis confirms the effectiveness of receiving zinc coverings by the way of vibration chemical and mechanical synthesis. As the main advantages in comparing with traditional methods of their drawing, it is possible to allocate: combination of processes of deposition and machining of a surface; simplicity of a design of an inventory owing to lack of current lead; lack of a hydrogen charging of the surface layer; a possibility of the covering "difficult covered" metals; use of a trace amount of electrolyte; less rigid conditions to a surface pre-treatment.

References

1. A.P. Babichev, V.V. Ivanov, S.N. Hudalya, P.D. Motrenko, Yu. P. Ankudimov, I.A. Babichev, *A vibration mechanochemistry in processes of the finishing strengthening processing and coverings of details of cars* (monograph), 204 (DGTU publishing center, Rostov n / D, 2012)
2. A.P. Babichev, V.V. Ivanov, V.E. Burlakova, O.S. Filipchuk. Uprochnyayushchiye tekhnologii i pokrytiya, No. 7, 46-49 (2014) [in Russian]
3. V.V. Ivanov, *Vibration mechanochemical coating* (monograph), ISBN 978-3-659-49674-5, (LAMBERT Academic Publishing, Saarbrucken, Germany, 2014)
4. V. A. Lebedev, V. V. Ivanov and V.P. Fedorov, *IOP Conf. Series: Materials Science and Engineering*, 124 (2016)
5. V.V. Ivanov, G.A. Galchenko, Avtomobilnaya Promyshlennost, № 2, 30-32 (2016)
6. V.V. Ivanov, N.S. Dontsov, A.V. Kirichek, Key Engineering Materials, **736**, 105-109 (2017)
7. V.V. Ivanov, S.N. Popov, V. Yu. Marchenko and E. V. Marchenko, Mir gal'vaniki, **1**(29), 34-38 (2015) [in Russian]
8. V.A. Lebedev, V.V. Ivanov, S.Y. Stine and I.V. Davydova, Uprochnyayushchiye tekhnologii i pokrytiya, **8**, 19-23 (2015) [in Russian]
9. V.V. Ivanov and G.A. Galchenko, Automobile Industry (2016)
10. E.A. Afanasiev and V.I. Serebryakovsky, Bulletin of Kursk State Agricultural "Agroengineering" Academy" (2014)
11. V.V. Ivanov, V.Ju. Valyavin, S.I. Popov, Ju. V. Marchenko and E.V. Marchenko, Mir gal'vaniki (2015)
12. David A. Davidson, Metal Finishing, **106**, Iss. 5, 30-34 (2008)
13. LRK Gillespie, *Mass Finishing Handbook*, (Industrial Press, 2006)
14. W. Nebiolo, Loading vibratory finishing bowls for optimal quality, Gardner Business Media, Inc., **78**, iss. 06, ISSN 0032-9940/82, 26-29 (2014)
15. A.V. Kirichek, D.L, Solovyev, A.G. Lazutkin, *Technology and equipmentstatistical-width surface treatment plastic deformation* (monograph), 282-287 (92 n.). (Engineering, Moscow, 2004)

The research of technological characteristic of the vibrowave mechanical and chemical oxide coating formation

Vladimir Ivanov[1], *Anatoly* Babichev[1,*], and *Nick* Pogorelov[1]

[1]Don State Technical University, Rostov-on-Don, Russia

Abstract. The article presents experimental research devoted to the possibilities of usage for the vibrowave technology for the processing oxide coating on the details made of aluminum alloy. The technology of oxide coating formation is worked out in the process of vibrowave technology influence, that reflects the essence of complex impact of mechanical and chemical components on the formation go the top surface layer.

1 Introduction

Vibrowave processing allows to provide the effect of joint anergy influence, resulting from mechanical and chemical synthesis of the surface formation. Vibrowave processing provides alteration of physicochemical characteristic of the metal top layer, besides its activation. Taking into account its universal character, vibrowave equipment can be used not only for top layer formation but also finishing, scraping and hardening actions. that's of high importance due to decrease production net cost [1, 3].

Vibrowave processing can have the following parameters: regimes of main camera swaying; the time of details processing, density of the processed material; rough surface of the processed detail; capacity of the effective environment in the working camera; effective environment characteristic; the amount of oxid liquid in the working camera, liquor temperature [2].

The main aim of the given experimental research was the mastering of the most optimal highly-technological process of decorate surface inscription under vibrowave process system conditions.

The given research presupposes solution of the following tasks.

1. The selection of processing regimes and parameters, equipment, effective environment of the liquids in vibrowave abrasive finishing conditions .
2. The mastering of process coating inscription, for surface hardening of aluminum alloy and firmness definition.

2 Research results

The research to identify the degree of amplitude of swaying influence on the working camera on the thickness of oxide layer, they were carried out on the adenol aluminum samples (HB 28) and aluminium alloy (HB 50), swaying frequency of the working camera f = 33,3 Гц; processing duration, τ = 30 min, working material – polyethylene ball D 3-4 mm. Swaying amplitude was changing within the range, the most frequently used during the practice from 1 to 5 mm. The amount of oxide liquid was calculated 30 % from the bulk volume of the working camera, taking into account that it's filled with working material and details not more than for 0,8 of the total volume. The definition of the oxide layer thickness was carried out by the method of dropping [3].

The received data is presented in the table 1, fig. 1.

Table 1. The influence of that swaying amplitude on the thickness layer (mcm)

Material	Swaying amplitude, mm				
	1	2	3	4	5
АДО	5	5	4,9	4,7	4,4
АЛ9	4,2	4,2	4,1	3,9	3,5

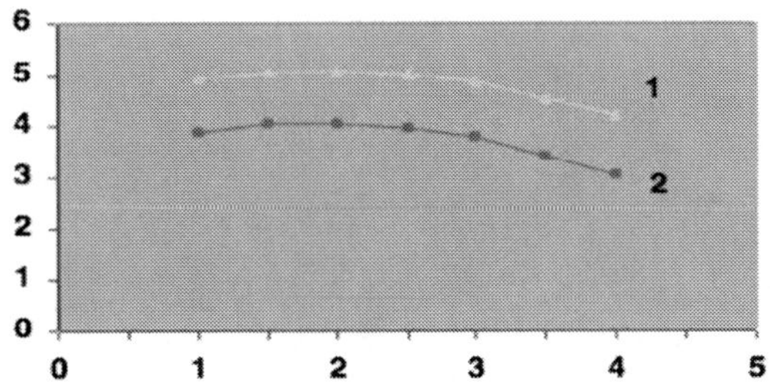

Fig. 1. The dependence of oxide layer thickness from the swaying amplitude of the working camera: 1 – aluminum; 2 – aluminum alloy

* Corresponding author: vibrotech@mail.ru

DOI: 10.1051/matecconf/201713201004

As the researches have shown, the increase of swaying amplitude up to A = 2,5 mm, it does not effect growth of oxide layer. With the swaying amplitude 3 mm, however there is a certain decrease of oxide layer thickness, which does not cause the change of layer density [4,5].

The research of the density for the vibrowave mechanical and chemical oxide layer (VMCOL) had its own specifics. The adenol alloy shields size 10 by 10, thickness 2 mm were used as the research samples. Before the placing VMCOL, the sample surface were polished to get mirror effect [6, 7]. Tridimensional SPM image of the micro print of the initial surface fig. 2. The indenter prints are quite visible on the surface and the absence of the coating on the sample surface.

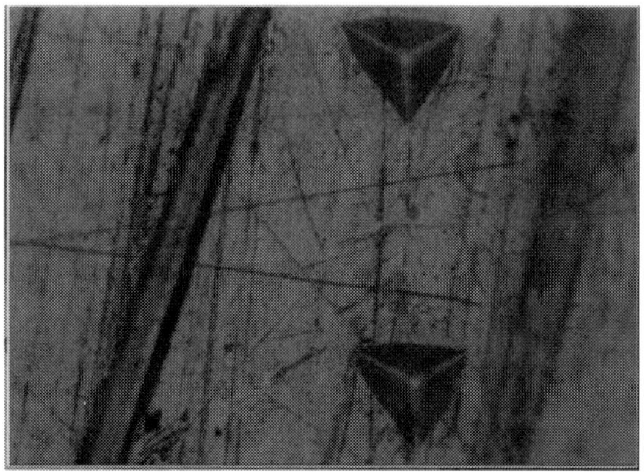

Fig. 2. Tridimensional SPM micro print image

On the fig. 3 there are shown curves of the loading and unloading and numerical indicators of the density of the initial surface.

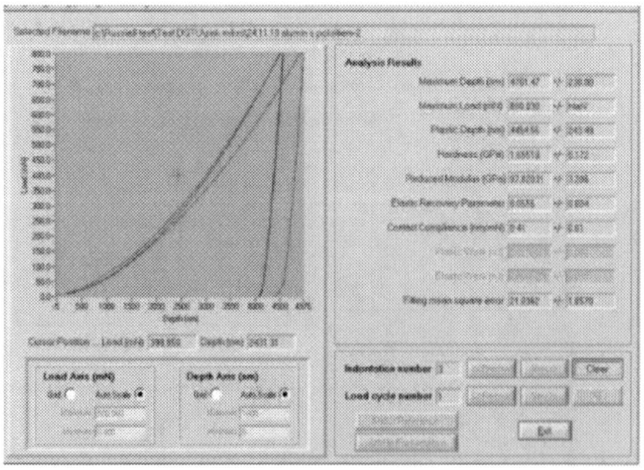

Fig. 3. The initial sample adenol aluminum. The square 500-1000 nm.

Density 1.65 GPa. Curve of the «loading – unloading» The time interval for VMCOL formation was 5, 10, 15 min. All these factors created conditions for formation morphological oxide coating with different thickness.

After the coating within 5 min micro density increased up to the point 1,78 GPa on the tridimensional SPM image six micro prints, VMCOL in the initial stage of its formation is visible as light yellow parts of irregular shape (fig. 4).

a)

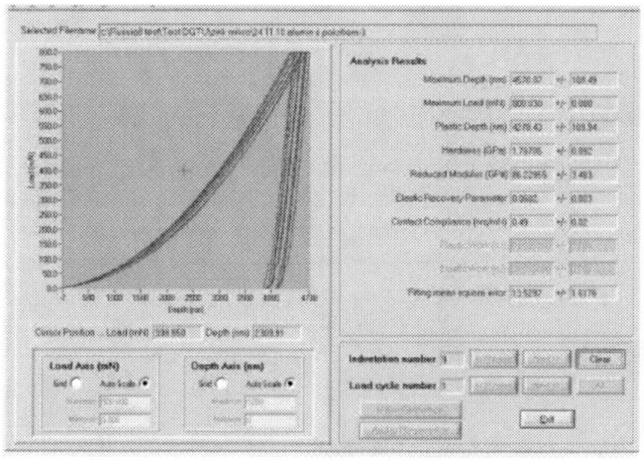

b)

Fig. 4. VMCOL top layer on the adenol aluminum alloy square 500-1000 nm; a – T=5 min, density 1.78 GPa.;
b – tridimensional SPM image.

The increase time for oxide layer formation up to 10 min has lead to the increase of oxide layer thickness and accordingly micro density up to 2.03GPa. (fig. 4 a, b)

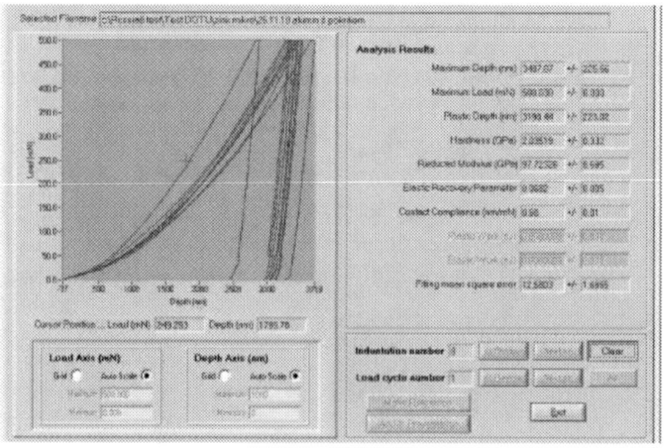

a)

b)

Fig. 5. Top layer VMCOL on the adenol aluminum alloy square 500-1000 nm; a – T=10 min. density 2.03GPa; b – tridimensional SPM image

As it's seen on the fig. 6 a,b the time increase up to 15 min. has led to the formation of the layer with max thickness 5 mcm, density 2.18 GPa. on the aluminum surface.

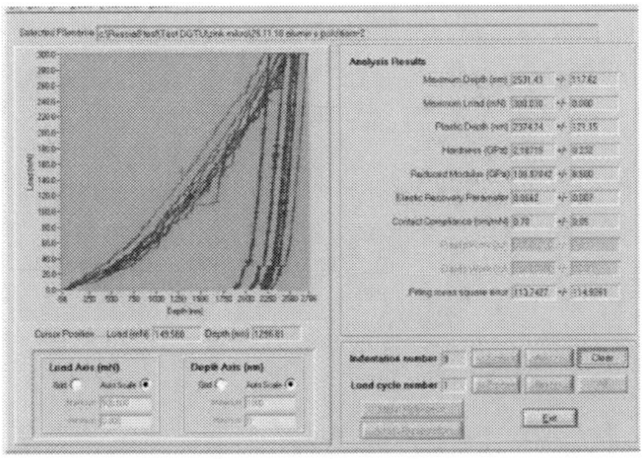

a)

b)

Fig. 6. VMCOL top layer on the aluminum alloy square 500-1000 nm; a – T=15 min. density 2.18 GPa.; b – tridimensional SPM image.

The analyses of the received results has shown that micro density VMCOL is 0,54 GPa higher than initial sample (fig. 7), alongside with this maximal density of the top layer is 0,25 GPa within the time period 5 and 15 min. of the processing.

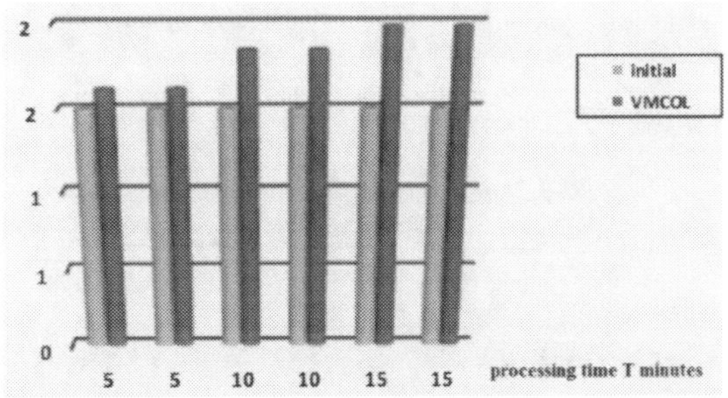

Fig. 7. Diagram of VMCOL density change

As the result of the carried out research the optimal technological process of placing of the vibrowave mechanical and chemical oxide coating has been defined, it consists of the following:

1. Loading of the details in the vibrowave camera.
2. Vibrowave mechanical and chemical oxide coating.
3. Washing down in the cold running water.
4. Washing down in the hot running water.
5. Unloading of the details.
6. Drying.
7. Final control.

Conclusion

Optimal regimes of the VMCOL have been specified. Swaying amplitude – 2,5 mm. Swaying frequency – 16 Hz. Processing time – 15 min. Working material is the polyethylene ball ø 2-2.5 mm. Layer thickness 5 mcm. density 2.18 GPa.

The research of the possibilities and usage of vibrowave mechanical and chemical oxide coating has shown, that protective and decorate coating placed on the details can be of different colors. Due to the increase layer density, protective capability is 1.5 times higher than oxide layer, received by the standard way.

References

1. A.P. Babichev, S.N. Shudaley *Vibration Mechanical Chemistry in Processes of Finishing – Hardening Working and Coatings of Machinery* (monograph) (DSTU Publishing House, Rostov-n/D, 2012)
2. V.V. Ivanov, *Vibrowave mechanical and chemical oxide coating*, 8, 8 p.p (2007)
3. J. Russell, *Mechanical plaiting*, (Book on Demand Ltd., 2013) ISBN13: 9785512398487
4. V.A. Lebedev, V.V. Ivanov, V.P. Fedorov *Materials Science and Engineering*, 124, 1-6, doi.10.1088/1757-899X/124/1/012160 (2016)

5. V.V. Ivanov, E.V. Marchenko., S.V. Samurgashev, A.A. Sarabashev, Galvanic world, № 1 **(32)** (2016)
6. V.V. Ivanov, N.S. Dontsov, A.V. Kirichek, *Key Engineering Materials*, **736**, 105-109, doi. 10.4028/www.scientific.net/KEM.736.105 (2017)
7. V.A. Lebedev, V.V. Ivanov, S.Y. Stine and I.V. Davydova, Uprochnyayushchiye tekhnologii i pokrytiya, **8,** 19-23 (2015)

Evaluation of cutting temperature of a tool with a curved cutting edge

Denis Moiseev[1,*], *Anatoliy* Ryzhkin[1] and *Gennadiy* Volokitin[2]

[1] Don State Technical University, «Metal-cutting machines and tools» department, 344000 Rostov-on-Don, Russian Federation
[2] Don State Technical University, «Mathematics» department, 344000 Rostov-on-Don, Russian Federation

Abstract. A model for solving the problem of heat distribution along the front surface of a tool with a shaped profile is considered. Analytical dependencies for estimating the temperature of the tool of elliptic profile for turning with radial feed in Cartesian and polar coordinate systems are obtained for the model under consideration. As a result, the obtained dependences can be applied to similar thermal calculations of instruments with other forms of plane curves both in turning with radial feed and in milling conditions. The calculation of a flat temperature field on the front surface of a shaped tool has been carried out and it has revealed areas with the highest temperatures at the contact area and the maximum wear values.

1 Introduction

Thermal state of the tool, working on forced modes, affects the rate of its wear. The levels of the temperatures while cutting can reach critical values, therefore the problem of evaluation of heat flows and temperatures using calculation methods is actual. Dependencies for calculation of the cutting temperatures for instruments with a straight cutting edge are known [1-3], but those dependencies can be used for the problem in a linear formulation, when it is impossible to determine the temperature distribution at any point on the front surface of the tool.

A shaped cutter of an elliptic profile is presented in the article as an object of the research, which can be used both for modeling of process of radial shaped turning and for milling.

2 Problem formulation and solution methods

The problem of distribution of the temperature field is solved on the front surface of the convex part of the cutter of the elliptical profile (region 2-3-4 in Fig. 1, a), because the cutting conditions will be practically identical for a concave part (regions 1-2 and 4-5) of the tool in view of the fact that the convex and the concave parts are the generating curves of the same elliptic profile.

The origin of coordinates for a convex part of the elliptic cutter with semiaxis a and b is located in the centre of the tool (system $X_1O_1Y_1$). For this case the equation of a convex part of ellipse in polar coordinates r and θ (Fig. 1, b) in the system $X_1O_1Y_1$ has the following form:

$$\begin{cases} x_1 = ra\cos\theta \\ y_1 = rb\sin\theta \\ \dfrac{\pi}{2} \le \theta \le \dfrac{3\pi}{2}, \quad 0 < r \le 1 \end{cases} \tag{1}$$

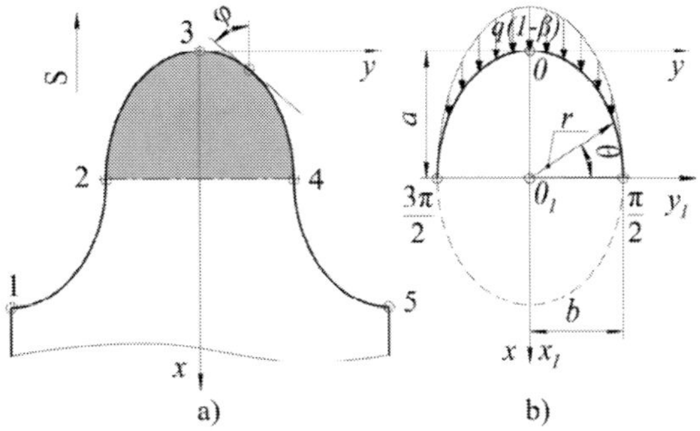

Fig. 1. The profile of the tool (a) and the scheme of heat flows (b)

For the XOY system, the ellipse equation will have the form:

$$\begin{cases} x = a\left(1 + r\cos\theta\right) \\ y = rb\sin\theta \end{cases} \tag{2}$$

The transformation formulas from polar coordinates to Cartesian x, y have the form:

* Corresponding author: denisey2003@mail.ru

$$\begin{cases} r^2 = \dfrac{(x-a)^2}{a^2} + \dfrac{y^2}{b^2},\ 0 \le r \le 1 \\ \theta = arctg\left(\dfrac{a-x}{y}\right) + \dfrac{\pi}{2},\ npu\ \dfrac{\pi}{2} \le \theta \le \pi \\ \theta = arcctg\left(\dfrac{x-a}{y}\right) + \pi,\ npu\ \pi \le \theta \le \dfrac{3\pi}{2} \end{cases} \quad (3)$$

We assume that the heat flow with q_0 density acts on the tool during cutting (Fig. 1, b). It propagates along the cutting blade unevenly and assumes the maximum value on the top of the tool (point 3 in Fig. 1, a), and approaching the points 2 and 4 it decreases and becomes equal to zero at these points. Let then the heat flow along the cutting blade vary according to the dependence:

$$f(\theta) = \frac{1}{\pi^2}\left(-4\theta^2 + 8\pi\theta - 3\pi^2\right), \quad (4)$$

from which the heat flow density function along the cutting blade takes the form:

$$q(\theta) = (1-\beta)\cdot q_0 \cdot f(\theta) \quad (5)$$

$f(\theta)$ is a normalized heat flow distribution function at the boundary; $(1-\beta)$ is a coefficient that takes into account the proportion of heat flowing into the tool.

This type of heat flow can be explained by the inconstancy of cutting geometry: rear angle of the tool takes maximum value on the top 3 (Fig.1,a), which provides the best cutting conditions; when approaching points 2 and 4 it decreases and becomes equal to *0*, i.e. the cutting process becomes impossible.

The stationary heat equation for the conditions, described above, has the following form:

$$\Delta T - m_1 T = 0, \quad (6)$$

where Δ is the Laplace operator; $m_1 > 0$ - coefficient that takes into account the heat exchange of the instrument with the environment.

In the polar coordinate system with the origin in the center of the tool, the differential heat equation has the form:

$$\begin{gathered} \frac{1}{r}\frac{\partial}{\partial r}\left(r\frac{\partial T}{\partial r}\right) + \frac{1}{r^2}\frac{\partial^2 T}{\partial \theta^2} - m_1 T = 0, \\ T = T(x,y) = T(r,\theta) \end{gathered} \quad (7)$$

with the following boundary conditions:

$$\begin{gathered} \left.\frac{\partial T}{\partial r}\right|_{\substack{r=1 \\ \frac{\pi}{2}\le\theta\le\frac{3\pi}{2}}} = \frac{q(\theta)}{\lambda_1} \\ T\left(r,\frac{\pi}{2}\right) = T\left(r,\frac{3\pi}{2}\right) = 0,\ T(0,\theta) < \infty \end{gathered}, \quad (8)$$

λ_1 –coefficient of thermal conductivity of the tool material.

The solution of the boundary value problem (7), (8) has been found using the method of separation of variables:

$$T = Y(r)\Phi(\theta) \quad (9)$$

Substituting (9) into (7), we obtain [4]:

$$\frac{r^2 Y'' + rY' - m_1 r^2 Y}{Y} = -\frac{\Phi''}{\Phi} = v^2, \quad (10)$$

where $v = const$.

The dependence (10) has 2 solutions:

$$\Phi'' + v^2\Phi = 0,\ \Phi \ne 0 \quad (11)$$

$$r^2 Y'' + rY' - \left(m_1 r^2 + v^2\right)Y = 0,\ Y \ne 0 \quad (12)$$

The variables are also separated in the boundary conditions. It follows from (8):

$$\Phi\left(\frac{\pi}{2}\right) = \Phi\left(\frac{3\pi}{2}\right) = 0 \quad (13)$$

The solution of the boundary-value problem (11), (13) has the following form:

$$\begin{gathered} \Phi_v(\theta) = \Phi_n(\theta) = A_n \cos n\theta + B_n \sin n\theta \\ (v = n,\ n \in Z) \end{gathered} \quad (14)$$

Using the substitution $\rho = \sqrt{m_1} r$ and the equality $v = n$, we obtain from (12) an ordinary differential equation:

$$\rho^2 Y'' + \rho Y' - \left(\rho^2 + n^2\right)Y = 0,\ n \in Z \quad (15)$$

Dependence (15) is a Bessel equation of diffusion type, which defines the modified functions: the Bessel function of the imaginary argument and the modified Hankel function (MacDonald function) [5]. If the argument *n* is also valid, then both functions take real values.

Since the solution of the problem is limited when $\rho = 0$, due to the singularities at zero of the functions of MacDonald we get:

$$Y_n(\rho) = C_n I_n(\rho) \quad (16)$$

Consequently:

$$T_n(r,\theta) = Y_n(\rho)\Phi_n(\theta) = \\ = I_n(\sqrt{m_1}r)(A_n\cos n\theta + B_n\sin n\theta) \quad (17)$$

Using the principle of superposition, the solution of the boundary value problem is:

$$T = \sum_{n=0}^{\infty} I_n\left(\sqrt{m_1}r\right)\left(A_n\cos n\theta + B_n\sin n\theta\right) \quad (18)$$

The coefficients A_n, B_n in the expression (18) can be determined from the boundary condition (8).

Using the expansion of expression (4) in the Fourier series and the boundary conditions (8), we obtain formulas for the coefficients A_n $(B_n = 0)$ and as a result, we obtain a solution of the boundary value problem for the thermal field of the tool in polar coordinates [6].

$$T = \frac{(1-\beta)q_0}{3\lambda_1\sqrt{m_1}I_0'\left(\sqrt{m_1}\right)}I_0\left(\sqrt{m_1}r\right) + \\ +\sum_{k=1}^{\infty}\left[I_k\left(\sqrt{m_1}r\right)\left(\frac{(1-\beta)q_0}{\lambda_1\pi^2\sqrt{m_1}}\right)\left(\frac{a_k}{I_k'\left(\sqrt{m_1}\right)}\right)\cos k\theta\right] \quad (19)$$

Using the dependencies (3) and (19) we obtain the final expression for the tool temperature field in Cartesian coordinates:

$$T = \frac{(1-\beta)q_0}{\lambda_1\sqrt{m_1}}\left\{\frac{I_0\left[\sqrt{m_1\cdot\left(\frac{(x-a)^2}{a^2}+\frac{y^2}{b^2}\right)}\right]}{3I_0'\left[\sqrt{m_1}\right]} + \right. \\ +\sum_{k=1}^{\infty}\frac{\left[m_1\left(\frac{(x-a)^2}{a^2}+\frac{y^2}{b^2}\right)\right]^{\frac{1}{2}}\left(\frac{-\pi\cos\frac{\pi k}{2}}{k^2}-\frac{2\sin\frac{\pi k}{2}}{k^3}\right)}{I_k'\left[\sqrt{m_1}\right]} \cdot \\ \left.\cdot\frac{8I_k}{\pi^3}\cdot\cos\left[k\left(arctg\frac{a-x}{y}+\frac{\pi}{2}\right)\right]\right\} \quad (20)$$

I_k is a Bessel function of an imaginary argument of the k-th order.

To calculate the temperature of the tool according to the equations (19) and (20) we must have the number of original data:

- heat flow $q_0 = 0,252\cdot\sigma\cdot\frac{V}{K_a}$ [7-8].
- the distribution coefficient of the heat flow β.

According to the solution presented in [1], the distribution coefficient of heat flow is equal to:

$$\beta = \frac{\lambda_2}{\lambda_2 + \lambda_1\sqrt{m_1}\cdot\sqrt{1,5a_2\tau}}, \quad (21)$$

where: λ_2 – the thermal conductivity coefficient of the material of the workpiece; σ – the tensile strength of the workpiece material; a_2 – the thermal diffusivity coefficient of the workpiece material; τ–the contact time of the chip with the front surface of the tool equal to [1]:

$$\tau = \frac{l_k}{V_c} = \frac{2S\cdot\sin\varphi\cdot K_a\left[K_a(1-tg\gamma)+\frac{1}{\cos\gamma}\right]}{V} \quad (22)$$

in dependence (22): l_k – length of the chip contact with the front surface of the tool; S – feed value; K_a – chip shrinkage ratio; φ – the main angle in plan of the tool; γ – front angle of the tool; V – cutting speed.

The heat flow distribution coefficient is a function of the variables K_a, γ, φ which take into consideration the inconstancy of the tool geometry (this approach is considered in [2]). Therefore, it is necessary to know the values of these parameters at each point of the cutting blade.

The functional dependences of the parameters entering into (21) - (22) are the following:

- for this case we assume the value of the front angle γ to be constant and equal to 0°;
- the dependence of the chip shrinkage coefficient K_a for the processing of steel 45 with a carbide cutter of grade T15K6 has been obtained experimentally in the following form:

$$K_a = 0,462\cdot V^{0,414}\cdot e^{-0,433V}\cdot S^{-0,238}\cdot\varphi^{-0,09}, \quad (23)$$

where cutting speed V has the dimension m/s, feed S - m/rev., and the angle in plan φ - rad.

- the value of the angle in the plan φ can be determined from the equation of the ellipse in the XOY coordinate system (Fig. 1,a), using the derivative of the function $y = \frac{b}{a}\sqrt{2ax-x^2}$ [6].

$$\varphi = arctg(y') = arctg\frac{b(a-x)}{a^2\sqrt{1-\frac{(x-a)^2}{a^2}}} \quad (24)$$

Thus, on the basis of dependences (21) - (23) it can be concluded that the heat flow distribution coefficient in this case is a function dependent on the main angle in plan; therefore, in calculating the heat leaving the tool, the position of the point on the cutting tool blade should be considered.

3 Example of calculation of the thermal distribution of the tool

The calculation of the thermal field of the cutter of the elliptical profile while turning with a radial feed of steel 45 with a carbide T15K6 cutter is presented below:

Initial data:

Cutting modes: $S = 3\cdot10^{-4}$ m/rev., $V = 3$ m/s

Parameters of the ellipse: a=4,5·10^{-3}m, b=5·10^{-3}m

Tool (hard alloy T15K6): $\lambda_1 = 27,2$ W/m.·gr., $m_1 = 40\,\mathrm{m}^{-1}$

Workpiece (steel 45): $\sigma = 598$ MPa, $a_2 = 8\cdot10^{-6}$ m²/s., $\lambda_2 = 40,1$ W/m·gr.

Given the coordinate values within the front surface of the tool ($x \in [0;a]$ and $y \in \left[-b\sqrt{1-\frac{(x-a)^2}{a^2}};\, b\sqrt{1-\frac{(x-a)^2}{a^2}}\right]$), we obtain the temperature distribution (Fig.2).

Fig. 2 shows the distribution of temperatures along the front surface of the tool in the form of level lines.

Fig. 3 shows the distribution of temperatures along the front surface of the tool in sections *A-A*, *B-B*, ... *E-E*, cutting the tool by planes parallel to the *OX* axis with an increment of $\Delta = 1\cdot10^{-3}$ m.

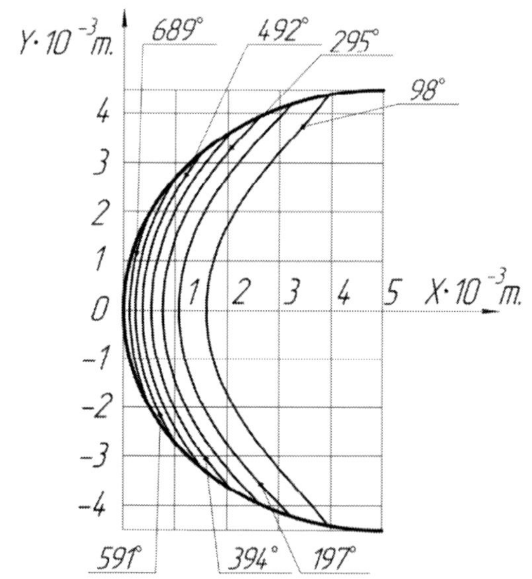

Fig.2. Temperature field on the front surface of the tool

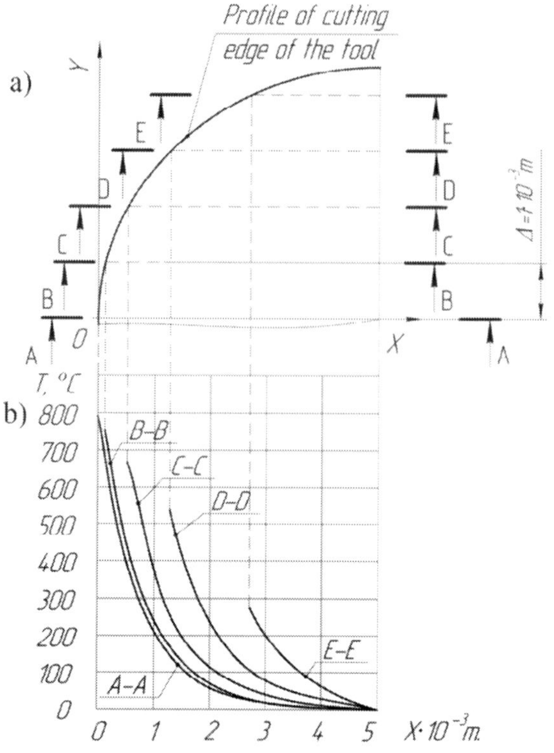

Fig.3. The tool profile (a) and the corresponding temperature distribution (b) in sections parallel to the tool axis

As we can see from the obtained diagrams, the temperature assumes minimum values far from the cutting edge and increases as it approaches it. The maximum temperature value, equal to 788 degrees, is observed on the top of the tool, where the angle in plan and the rear angle take the maximum values (point with coordinates $x = 0, y = 0$).

4 Conclusions

The solution of the problem of heat propagation on the working surface of the tool allows to evaluate the distribution pattern and the level of emerging temperatures on the front surface of the instrument, to identify areas of possible maximum wear and to calculate the wear rate at any point of the blade, and to predict the tool's durability.

References

1. A.A. Ryzhkin, K.G. Shuchev, M.M. Klimov, *Obrabotka materialov rezaniem* (Fenix, Rostov-on-Don, 2008) [in Russian].
2. A.A. Ryzhkin, D.V. Moiseev, Vestnik DSTU. Temperature evaluation under curved-edge tool cutting, 29-36 (2014) [in Russian].

3. A.A. Ryzhkin, *Teplofizicheskie protsessy pri iznashivanii instrumental'nykh rezhushhikh materialov* (Izd. tsentr DGTU, Rostov-on-Don, 2005) [in Russian].
4. A.N. Tikhonov, A.A. Samarskiy, *Uravneniya matematicheskoy fiziki* (Nauka, Moscow, 1977) [in Russian].
5. E. Yanke, F. Emde, F. Lesh, *Spetsialnyie funktsii* (Nauka, Moscow,1964) [in Russian].
6. I.N. Bronshtein, K.A. Semendyaev, *Spravochnik dlya inzhenerov i uchaschihsya vtuzov* (Nauka, Moscow, 1986) [in Russian]
7. A.N. Reznikov, *Teplofizika rezaniya* (Mashinostroyenie, Moscow,1969) [in Russian].
8. A.N. Reznikov, *Teplofizika protsessov mekhanicheskoj obrabotki materialov* (Mashinostroyenie, Moscow, 1981) [in Russian].

Improving the efficiency of the milling process by reducing the oscillation of the technological system

Alexander M. Kozlov[1,*], *Evgeny V.* Kiryuschenko[2], and *Andrey A.* Kozlov[1]

[1]Lipetsk State Technical University, 398600, 30, Moskovskaya Str., Lipetsk, Russia
[2]PJ-SCo. Novolipetsky Metallurgical Complex, 398040 2, Metallurgists' Square, Lipetsk, Russia

Abstract. The paper reports the dynamics of the technological system during the milling of complex parts, and the investigation of possibly efficiency increase of the cutting process by creating conditions for system self-adjustment, what will cause the reduction of the common level of its oscillation.

1 Introduction

The modern technology of multiaxis milling allows manufacturing parts with various forms, but the productivity of such a process has a number of limitations. Such limitations may be requirements to accuracy assurance in a form, parameters of worked surface roughness, and in some cases, - rigidity of a tool and a technological system on the whole [1, 2].

The impossibility to increase rigidity considerably causes oscillations when cutting which affects in a negative way upon quality and productivity of processing.

As a rule there is an approach to the problem of milling efficiency increase from the standpoint of vibrations decrease. A basic problem at the development of methods for milling efficiency increase is an adequate presentation of basic regularities in the excitation of forced oscillations and self-oscillations as factors defining machining accuracy [3, 4, 5].

2 System investigation

A technological system is considered as a mechanical oscillation system with the contour of some masses connected between each other with elastic and dissipative elements (Fig. 1).

The vector equation of the motion of the i-th mass:

$$m_i \cdot \overline{a}_i = \overline{F}_i + \overline{P}_i - \overline{F}_{i+1} \qquad (1)$$

P_i – external force influence upon the i-th mass, H; F_i – force causing elastic damping on the i-th unit, H; m – mass of the unit, kg; a – unit acceleration, mm/s^2.

$$F_i = C_i \cdot (X'_{i-1} - X'_i) + h_i \cdot (v_{i+1} - v_i)$$

$Fu_i = -m_i \cdot a_i$ – inertial force on the i-th unit, H;

X'_i – vibrodisplacement of the i-th mass, mm; v_i, – vibrovelocity of the i-th mass, mm/s. [6]

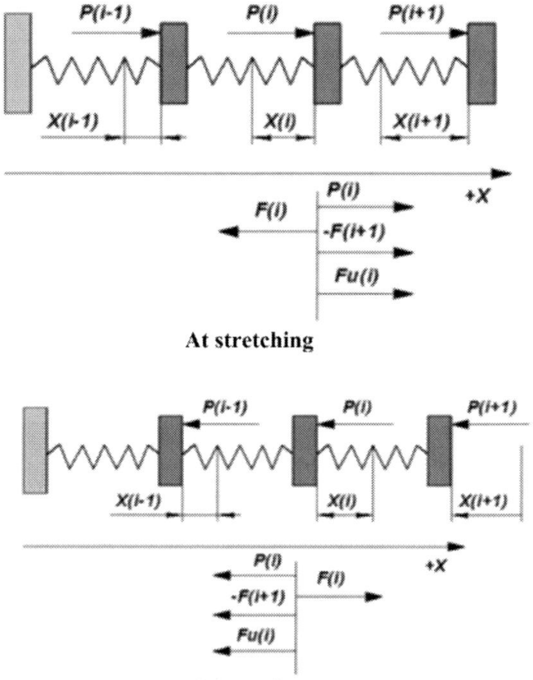

Fig. 1. Design scheme of an elastic system contour.

Vibration velocity is a velocity of an oscillating mass displacement. From the standpoint of vibro-diagnostics it is a universal parameter which allows comparing different oscillating processes. With its aid it is possible to determine a correlation of an oscillating process with cutting forces changes.

It is well-known [7, 8, 9] that when milling the primary oscillations are formed with secondary resonant

* kam-48@yandex.ru

forced oscillations. In such a way, the vibrations of the technological system are a sum of some oscillating processes. Hence, on the basis of the analysis of experimental data there is obtained a dependence describing changes in vibro-velocity of a milling machine spindle:

$$v = A_1 \cdot e^{-0,1S} \cos(f_1 \cdot S + \varphi_{11}) + A_2 \cdot e^{-0,1n} \sin(f_2 \cdot n + \varphi_{12}) + v_0 \quad (2)$$

v – vibrovelocity in the system, mm/s; A_1 and A_2 – amplitudes of corresponding oscillating processes, mm; S – feed, mm/min; n – rotation frequency, rpm; f – frequency of corresponding oscillations, Hz; φ – difference of oscillation phases in the corresponding direction, rad; v_0 – vibrovelocity out of a cutting area, mm/s.

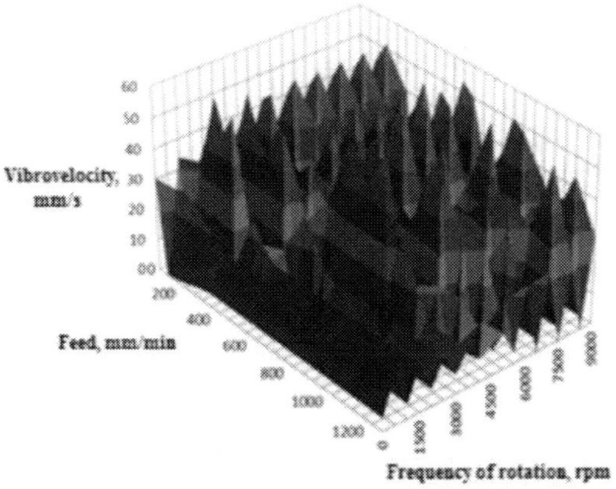

Fig. 2. Machining parameter impact upon vibrovelocity.

The peculiarity of a milling process is an interrupted character of cutting which may be characterized by lag time. Lag time corresponds to the value of phase difference this is the same time interval, but measured in a phase (id est, in the angles of the phase):

$$\varphi_{21x} = \omega_x \cdot \tau_{21} = \omega_x \cdot (k + i) \cdot T_x = (k + i) \cdot 2\pi \quad (3)$$

φ_{21x} – difference of phase oscillations in the system on X axis during two adjacent teeth motion through the given point in the area, rad;

ω_x – angular frequency of oscillations on X axis, rad/s.

If we consider a trace on the surface of cutting as a driving force with the frequency p, then the phase of forced oscillations of the system must lag behind the phase of a disturbing force for the value:

$$2\pi \cdot i_p = -\arctan\frac{2 \cdot \varepsilon \cdot \gamma}{1 - \gamma^2}$$

ε – dimensionless factor of attenuation; γ – ratio of the frequency of the driving force p to the own frequency f_x.

During the analysis of vibrovelocity changes (Fig.2) on the basis of equations (2) and (3) there is observed a soft excitation of regenerative oscillations. It appears from this that the system steady at cutting on a trace but with a small coefficient of rigidity may become unsteady at cutting along a trace when a rigidity coefficient will exceed a certain value. At the equality of own and driving frequencies the phase displacement does not depend upon ε may be equal only to $-\pi/2$. Strictly speaking, this value of the phase displacement corresponds to conditions of resonance occurrence [10].

In such a way, the increase of oscillations level at disturbances on a trail is determined, in the first place, by resonance and near-resonance phenomena caused by a coincidence of the own frequency of the system with the frequency of the wave on the surface of cutting. The results obtained have also shown that the parameter affecting the level of oscillations is a phase displacement at the initial moment of trail cutting. At that a minimum level of oscillations occurs at the initial displacement equal to $\pi/2$, and a maximum one – at $-\pi/2$. At that, if the initial displacement of phases differs from $-\pi/2$, then a part of oscillations energy is consumed for the system tuning, and oscillations are attenuated rather quickly. If the initial displacement of phases is equal to $-\pi/2$ or close to this value, then the system is set up immediately to the essential displacement of phases and in the system arise the established oscillations. In consequence of the existence of friction in the system such oscillations are damped gradually and damping time depends upon the initial displacement of phases.

3 Oscillations damping

It is possible to eliminate resonant oscillations with large amplitudes, if a damping device is introduced into the structure of metal-cutting equipment. In this case a damping device is an additional body with the mass of m_2 connected with the object through a spring with the rigidity of c_2 and a damper with the coefficient of viscous resistance, h.

In the presence of a damper a complete oscillations damping becomes impossible as it affects in addition a technological system. But a damper allows obtaining the limited amplitude of system oscillations in all the frequency range at an efficient selection of damper parameters (Fig. 3). This phenomenon may be called a self-tuning of the oscillation system [11, 12].

Differential equations of the system with a damper become:

$$m_1\ddot{x}_1 + h\dot{x}_1 + (c_1 + c_2)x_1 - h\dot{x}_2 - c_2x_2 = F_i$$

$$m_2\ddot{x}_2 - h\dot{x}_1 - c_2x_1 + h\dot{x}_2 + c_2x_2 = 0$$

m_1 – system mass, kg; m_2 – damper mass, kg; c_1 – system rigidity, c_2 – damper rigidity, h – coefficient of viscous resistance.

Let us add to these equations some dimensionless parameters: β, η, δ.

$$\beta = \frac{m_2}{m_1}, \quad \delta = \frac{\sqrt{c_2/m_2}}{\omega_0}, \quad \eta = \frac{h}{2m_2\omega_0},$$

ω_0 – eigen frequency of machine oscillations, 1/s; n_2 – partial frequency of the damper, 1/s.

Then, after transforms we obtain an expression for the dimensionless amplitude of damper oscillations:

$$A_1 = \sqrt{\frac{(\delta^2 - \gamma^2)^2 + 4\eta^2\gamma^2}{[(1-\gamma^2)(\delta^2-\gamma^2) - \beta\delta^2\gamma^2]^2 + 4\eta^2\gamma^2(1-\gamma^2-\beta\gamma^2)^2}}$$

The expression obtained allows calculating the amplitude of oscillations for a damping device for the achievement by a technological system a self-tuning state.

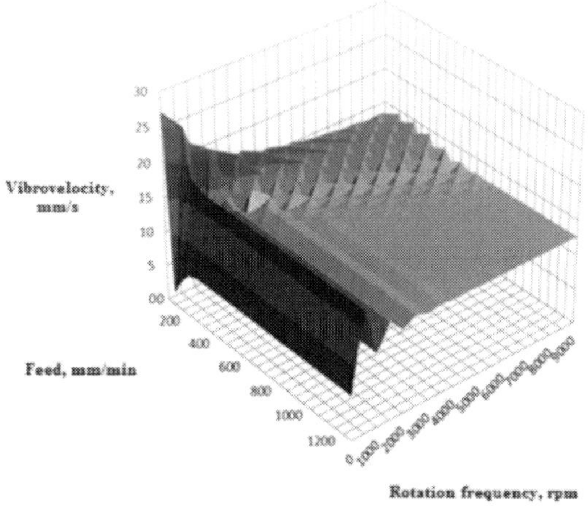

Fig. 3. Theoretical limitation of values of the spindle vibrovelocity in milling equipment at damping.

The expression obtained allows calculating the amplitude of oscillations for a damping device for the achievement by a technological system a self-tuning state.

In such a way, the more power for tuning is consumed, the less it becomes for oscillations excitations. To reduce a regenerative effect it is possible by means of the system tuning from resonance with a trail. If a wave period on the surface of cutting does not coincide with the period of eigen oscillations, then resonance will not appear and the level of oscillations will be considerably lesser.

4 Conclusion

In such a way, at the phase displacement for -π/2 occurs a system self-tuning as a result of which a part of oscillation processes is consumed and a general level of oscillations will drop. The more energy is consumed for tuning, the lesser it is for oscillations excitation. It results from this that one of the efficient ways for regenerative oscillations damping may be the formation of such conditions for its operation at which it should be necessary for it to carry out constantly such a retuning. It may be achieved with the aid of a special damping device located on bearings of machine spiral gears. At that if the period of the wave on the surface of cutting does not coincide with the period of eigen oscillations, then resonance will not appear and the level of oscillations will be much lesser.

References

1. R. A. Radulescu, General Cutting Process Model for High Speed Machining Dynamic and Thermal Considerations: PhD Thesis, University of Illinois at Urbana-Champaign, p. 257 (1993)
2. V.M. Svinin, Metal Machining, 3(28), 28–30 (2005)
3. A.M. Kozlov, E.V. Kiryushchenko, Reference Book. Engineering Journal, 1, 30–34 (2013)
4. A.M. Kozlov, E.V. Kiryushchenko, Fundamental and Applied Problems in Engineering and Technology, 3, 65–73 (2012)
5. S. Sastry, S. G. Kapoor, R. E. DeVor, G.E. Dullerud, ASME J. Eng. Indus., 123, 753–756 (2001)
6. A.M. Kozlov, E.V. Kiryushchenko, S.F. Kuznetsov, Fundamental and Applied Problems in Engineering and Technology, 5, 91-97 (2014)
7. Yu.I. Gorodetsky, S.N. Strebulyaev, Yu.E. Maiorova Bulletin of Scientific Technical Development, National Technological Group, 9(25), 91–95 (2005)
8. S.A. Vasin, *Prediction of Tool Vibration Resistance at Turning and Milling* (Mechanical Engineering Moscow, 2006), 384
9. A.M. Kozlov, G.E. Malyutin, Bulletin of Mechanical Engineering, 12, 71-75, 2014
10. A.M. Kozlov, E.V. Kiryushchenko, S.F. Kuznetsov *Current Problems in Mechanical Engineering: Proceedings of the VII-th Inter. Scientific-Tech. Conf.* (Tomsk Polytechnic University, Tomsk, 2013), 295–299
11. A.M. Kozlov, G.E. Malyutin, Science Intensive Technologies in Mechanical Engineering, 14-18 (2013)
12. A.M. Kozlov, E.V. Kiryushchenko, Current Fundamental and Applied Investigations, 1(4) 32-38, (2012)

Fuzzy controllers in the control system of a brushless electric motor using HIL technology

Gennady Kalach[1,*], *Nina* Kazachek[1], and *Georgi* Volosnyh[1]

[1]Moscow Technological University (MIREA), Moscow, Russian Federation

Abstract. This article proposes a method for creation of a control system for a brushless electric motor based on a fuzzy logic apparatus. The use of a fuzzy controller in this case can increase stability and improve the quality of the system under consideration, which was implemented in the Simulink environment using HIL technology. This technology increases the chances of successfully passing the test phase, considering the control system in prototype.

1 Introduction

Traditional automated control systems are built on the basis of linear object models according to a group of optimality criteria, for example, integral evaluations of transition process quality. Controllers built in this way are stable and optimal with regard to the models of real processors which serve as their foundations – control and regulation objects. However, methods of simplification and linearization which are often implemented in non-linear, dynamic, unclearly defined, objects do not deliver the expected results with regard to stable control and the quality of control over the real processor. Where objects' structures and functions are more complicated, it is even more difficult to use classic control methods [1-4, 8, 9].

In recent years, with increasing requirements for control systems, it becomes difficult to ensure high quality indicators (speed, overshoot) with reliable maintaining of stability conditions. In many cases, the "proven" over the years types of the control systems like PID-controllers are no longer cope with the task, and the next generation of controllers comes on the agenda – intellectual controllers [1].

Fuzzy controllers are used in many industry sectors [7]. By virtue of its fuzzy inference provides a parallel interpretation of knowledge through special high-performance hardware. In conjunction with the implementation of complex nonlinear transformations based on fuzzy logic, it can be concluded about the prospects of the creation of technology-based intellectual fuzzy logic of a new type of high-speed nonlinear controllers - fuzzy controllers for objects with a high degree of controllability, such as [10-12].

For the executive control subsystem in this work, a controller based on fuzzy logic was used instead of a traditional PID controller [5, 13, 14].

One of the more important stages of the development of embedded systems is the testing stage, which allows the developer to check the entire system's performance and reliability under non-laboratory conditions [6-7].

Hardware-in-the-loop (HIL) is a method which can be used to develop and test complicated real-time embedded systems. HIL provides an effective platform by adding a sophisticated control system for testing of the platform [16-17]. The sophisticated control system is present at the development and testing stage, because it adds a mathematical representation of all the linked dynamic systems. This mathematical representation is termed software modelling. During testing of an embedded system, it interacts with the software model.

2 Modelling control systems

As an example of HIL, let us consider a model with the following structure: Discrete time algorithm on a debugging board – control object on a computer, where a fuzzy PID controller serves as the algorithm, and a model of a brushless DC electric motor is the object [18].

An **STM32F4DISCOVERY** debugging board based on a **Cortex M4** core, connected to the computer through a **USB-UART** converter was used [15].

The process of modelling a control system is divided into two stages:

1) *Modelling the control algorithm on the debugging board*

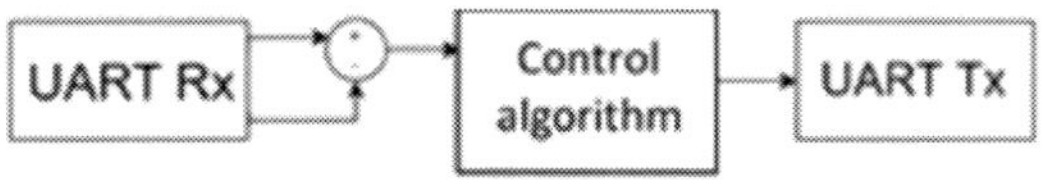

Fig. 1. Model of the control algorithm of a brushless DC electric motor on the debugging board.

In this model, the UART Rx unit receives two signals: the setpoint and object output signal. Any received errors are fed through the controller and the

* Corresponding author: kalach@mirea.ru

 DOI: 10.1051/matecconf/201713202001

UART Tx unit to the computer, upon which our object is realised.

The controller unit has the following structure (figure 2):

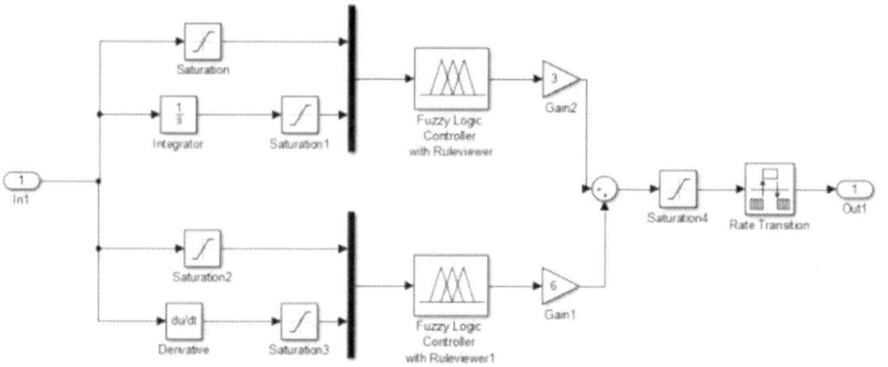

Fig. 2. Structure of the control algorithm unit.

Where the Saturation unit is needed to limit the controller's output, and the Rate Transition unit automatically harmonizes the remaining units' quantisation steps.

2) *Modelling the control object on the computer*

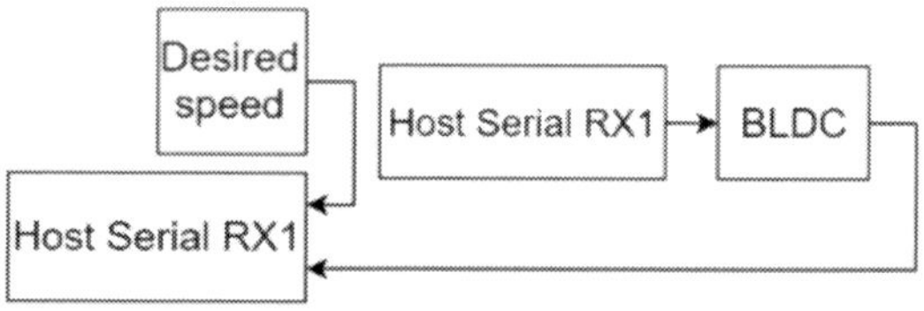

Fig. 3. Model of a brushless DC electric motor on the computer.

The BLDC unit has the following structure [19]:

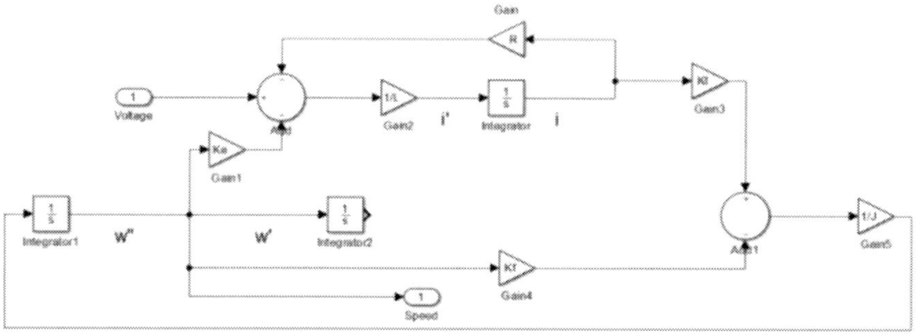

Fig. 4. Structure of the BLDC unit.

In this model, the Host Serial RX1 unit receives the control signal from the board and passes it to the control object. The Host Serial TX1 unit passes essential information to the board.

3 Tuning the fuzzy controller

The process of synthesizing a fuzzy PID controller has four steps [20].

1) *Synthesizing the P controller*

The P controller is synthesized for the object described by the following system of differential equations.

$$\begin{cases} u_e(t) = R_a i_a(t) + L_a \dfrac{di_a(t)}{dt} + e_b(t) \\ e_b(t) = k_e \omega_m(t) \\ T_m(t) = k_T i_a(t) \\ T_m = J_m \dfrac{d^2 Q_m}{dt^2} + B_m \dfrac{dQ_m}{dt} \end{cases} \qquad (1)$$

It is necessary to guarantee a transmission process with a low rise time, a steady-state error of $\Delta \leq 3\%$ and deregulation of $d \leq 10\%$.

a) First, a controller with a single gain factor is considered. Such a regulator does not influence the control process. We select the base gain factor kp=3, which gives the condition ymax=1.

b)A system with three rules is considered. The terms of the controller's output variable are arranged along the base scale according to figure 6, and the terms of the input scale according to figure 5. By method of trial and error, the value x1=0.1 is selected, which provides for a minimal rise time for the given overshoot.

c) Next, the system with three rules is considered again. We set the controller's gain factor to x2=0.6, which returns a steady-state error of 3%.

d) Next, the non-linear law of control must be obtained:

Table 1. Linguistic law of command.

	NB	NM	NS	Z	PS	PM	PB
e	-1	-0.66x2	-0.33x1	0	0.33x1	0.66x2	1
u	-1	-0.66	-0.33	0	0.33	0.66	1

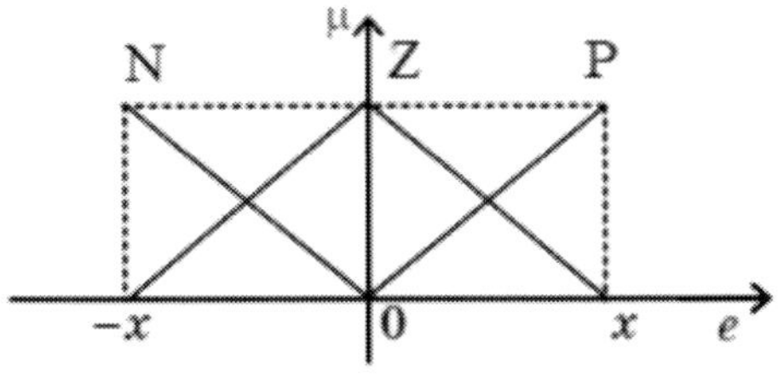

Fig. 5. Diagram of the controller input with three rules.

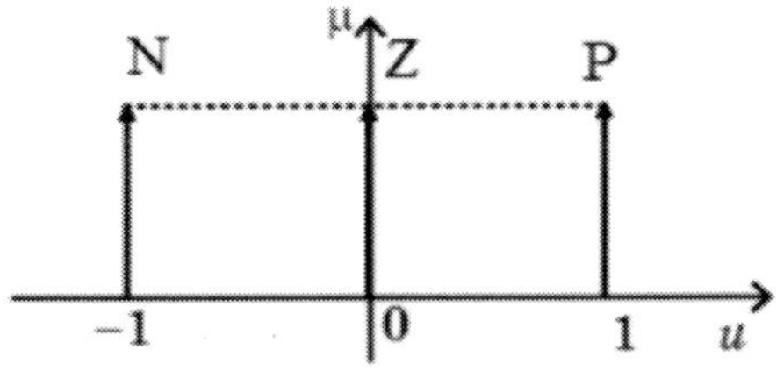

Fig.6. Diagram of the controller output with three rules.

2) *Synthesizing the PD controller*

The first stage is repeated, but the LP terms "derivative control errors" are arranged evenly (fig.7). The controlling rules are formed according to Table 2.

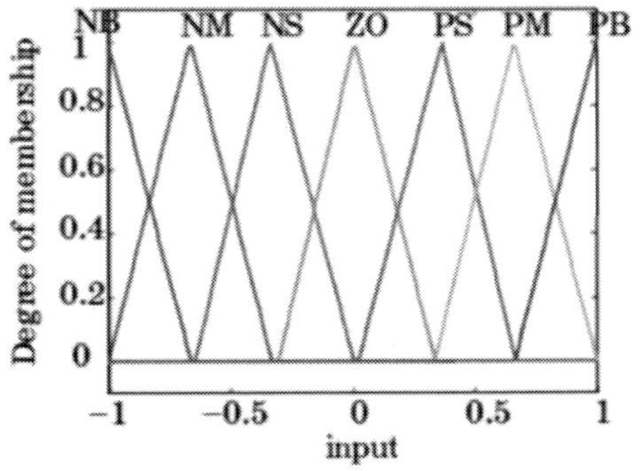

Fig 7. Linguistic description of derivative control errors.

3) *Synthesizing the PI controller*

A fuzzy controller of the PI type can be considered to have the same inputs as a PD controller. As such, the law of control showed in Table 2 is applicable. The essence consists of the method of forming the control signal, which can be used to minimize the system's static error.

Table 2. Linguistic law of command.

Rule table		e^*							
		NB	NM	NS	Z	PS	PM	PB	
Δe^*	NB	NB	NB	NB	NB	NM	Z	PB	u^*
	NM	NB	NB	NB	NM	NS	PS	PB	
	NS	NB	NB	NM	NS	Z	PM	PB	
	Z	NB	NM	NS	Z	PS	PM	PB	
	PS	NB	NM	Z	PS	PM	PB	PB	
	PM	NB	NS	PS	PM	PB	PB	PB	
	PB	NB	Z	PM	PB	PB	PB	PB	

4) *Synthesizing the PID controller*

The PID controller's equation can be expressed as follows:

$$u(t) = (\frac{1}{2}k_p e(t) + k_d \frac{de(t)}{dt}) + (\frac{1}{2}k_p e(t) + k_i \int e(t)dt) \quad (2)$$

It follows that a fuzzy PID controller can be represented as a PD and PI controller connected in parallel. As a result, we receive a synthesized fuzzy PID controller (fig.8).

4 Conclusions

As such, the demonstrated characteristics confirm the efficacy of using fuzzy PID controllers to control a brushless electric motor. The utility of HIL technology's utility lies in its ability to improve the quality of testing, decrease the influence of the human factor, and also reduce the costs of practical experiments

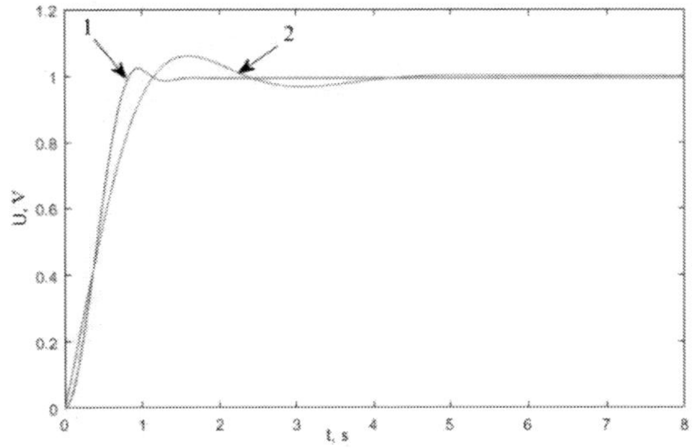

Fig. 8. 1 – The transformation process using a fuzzy controller
2 – the transformation process using a classic PID controller.

This work is supported by the Russian Science Foundation, according to research project No. 16-19-00052 of 2016.

References

1. I. M. Makarov, V. M. Lokhin, Intellectual Automatic Control Systems (2001)
2. I. M. Makarov, V. M. Lokhin, S. V. Manko, M. P. Romanov, M. S. Sitnikov, Information Tehnologies, no. 2, 2013, available at: http://aimagin.com/blog/using-hardware-in-the-loop-with-waijung/ (accessed 15.07.2017)
3. I. M. Makarov, V. M. Lokhin, S. V. Manko, M. P. Romanov, *Artificial Intelligence and Intelligent Control Systems* (Nauka, Moscow, 2006)
4. I. M. Makarov, V. M. Lokhin, *Artificial Intelligence and Complex Objects Control* (Edwin Mellen Press, New York, 2000)
5. A. Pegat, *Fuzzy Modelintg and Control* (Physica-Verl, New-York, 2001)
6. N. A. Kazachek, V. M. Lokhin, S. V. Manko, M. P. Romanov, *International Journal of Innovative Computing Information and Control*, **11**, 985-999, (2015)
7. N. A. Kazachek, Science Review, no. 20, 165-170 (2015)
8. K. Tanaka, T. Hori and H. O. Wang, *IEEE Transactions on Fuzzy Systems*, **11**, no. 4, 582-589 (2003).
9. H.K. Lam and F.H.F. Leung, *Stability Analysis of Fuzzy-Model-Based Control Systems: Linear-Matrix-Inequality Approach* (Springer, Heidelberg 2011).
10. V. R. Barseghyan, *International Conference Stability and Oscillations of Nonlinear Control Systems (Pyatnitskiy's Conference)*, 1 – 3 (2016)
11. Y. Hao, Z. Duan, W. Ren, *35th Chinese Control Conference,* 831 – 836 (2016)
12. Z. Lu, L. Zhang, L. Wang, *35th Chinese Control Conference)*, 8230 – 8235 (2016)
13. I. M. Makarov, V. M. Lokhin, S. V. Manko, M. P. Romanov, M. V. Kadochnikov, *Information technology S8*, 1-32 (2010)
14. V. M. Lokhin, M. P. Romanov, P. E. Tripolsky Russian technological journal, **1 (2)** 99-105 (2014)
15. STM32F4DISCOVERY, available at: http://www.st.com/en/evaluation-tools/stm32f4discovery.html (accessed 15.07.2017)
16. National Instruments (2016), Hardware-in-the-Loop (HIL) Test System Architectures, available at: http://www.ni.com/white-paper/10343/en/
17. MathWorks (2015), available at: https://www.mathworks.com/videos/hil-simulation-and-testing-with-simulink-real-time-and-speedgoat-target-hardware-116258.html (accessed 15.07.2017)
18. Aimaiging (2015), available at: http://aimagin.com/blog/using-hardware-in-the-loop-with-waijung/ (accessed 15.07.2017)
19. Matlab, https://matlab.ru/webinars/upravlenie-elektricheskim-privodom-v-simulink (accessed 15.07.2017)
20. M.V. Burakov, *Fuzzy controllers*. (GUAP, SPB. 2010)

A feasibility study for implementing advanced control of a heating system

Valerii Fomin[1,*]

[1]Saint Petersburg State University of Architecture and Construction, Saint Petersburg, Russian Federation

Abstract. The article justifies the necessity of implementation of a new technique of controlling heat consumption by a heating system. The formation of a heat carrier's parameters from the beginning of heat generation until supply to the consumer's premises has been discussed. An analysis of heat energy overconsumption in the period of increased external air temperature during automatic regulation in accordance with external air temperature is presented. The parameters not accounted for in the generation of a controlling signal under the current practices of regulating the heating of buildings.

1 Introduction

The enhanced requirements to reductions of specific heat losses for heating [1] are conducive to searching for ways [2, 3, 4, 5, 6, 7] to optimize heat consumption for heating systems of apartment residential and public buildings. The application of heat consumption system controllers in heat network connection assembles has opened new opportunities in improvement of regulation algorithms and individual approach to the generation of heat consumption conditions of each building. At the same time, the diversity of manufacturers of control automatics has created tools to increase efficiency of using heat energy when controlling heating systems. At the same time, the diversity of manufacturers of control automatics implies deep studies of the software functions of the equipment therewith complicating their efficient use by servicing personnel. The non-use of the automatic potential fully often does not justify the expectations of interested parties (consumers of heat energy) and delays further implementation of energy-saving technologies.

2 Formation of the heat carrier's temperature

Currently the mechanism of formation of the temperature on the premises of a building with central heat supply is as follows (Fig. 1). At the outlet of a power facility (heat power station, district boiler house) the parameters of a supplying heat carrier are formed: temperate 2.1 and pressure 2.2 (Fig. 1) in accordance with the applicable law [8] on the basis of average temperatures 1 (Fig. 1). The proposed methodologies of the outgoing parameters including [9] are above all conducive to the enhancement of operation characteristics of the heat installation itself, but not of the end consumers of the heat energy. As the heat carrier is transported, its parameters are reduced due to the influence on it of both its design features 3.1 (Fig. 1) [10] and weather conditions 3.2 (Fig. 1) [11]. Thus, at the boundary of the heat network and the heating unit the parameters can considerably differ de-pending on the distance from the power facility to the building. The heating unit equipped with control automatics must account for changes in the parameters occurring at the boundary of the connection to the heat network, and generate the characteristics of the heat carrier: temperature 5.1 and pressure 5.2 (Fig.1) to the consumer's premises, the consumption of the heat carrier taking account of the current exterior air temperature 4 (Fig. 1). Having various versions of controllers offered by manufacturers [12, 13] available, the most widespread practical way of the heat carrier's temperature formation is according to the supplying heat carrier's temperature. When an external air temperature changes, a signal enters a controller that verifies, in accordance with the control temperature schedule, the correspondence of the temperature of the supplying heat-carrier of the heating system and sends a command to the electric drive of the control valve that makes an opening or a closing, striving to bring the sensors' readings to correspond to the schedule. At that, the controller does not generate a base of parameters of individual characteristics of heat consumption of the facility to pick up an individual temperature schedule, but depends on the right adjustment of the temperature schedule, i.e. the competence and responsibility of the adjustment personnel.

* Corresponding author: sienef@yandex.ru

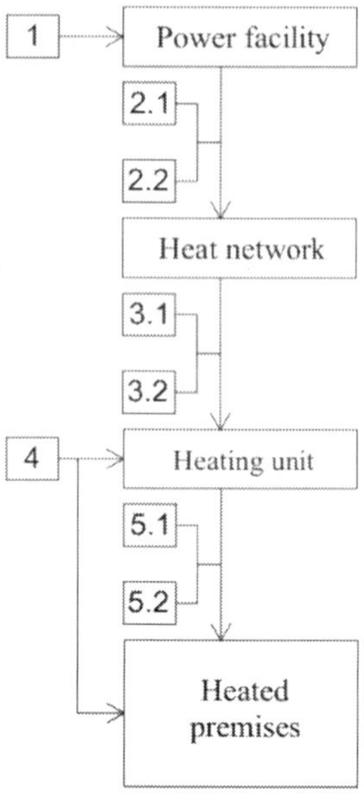

Fig. 1. Formation of the heat carrier's temperature.

Thus, the current features of the equipment of the control systems are unable to ensure comfortable temperatures [14, 15] on the premises of a building due to incomplete accounting for of disturbing factors when determining the quantity of heat for heating

3 A practical example of overconsumption of heat energy in automatic heating control

The obviousness of implementing adaptive control systems and inclusion in the process of generating quantities of heat required for creating comfortable temperatures is confirmed by an analysis of the heat consumption of the building equipped with temperature control automatics. These heat consumptions indicate heat energy overconsumptions in the period of exterior air temperature rise. be centred and should be numbered with the number on the right-hand side.

To determine the heating characteristic of a building, we will use the formula (1):

$$q_v \cdot V_{\text{bld}} = \frac{Q}{t_{\text{int}} - t_{ext}} \qquad (1)$$

where: q_v is specific heating characteristic of a building, kCal/m^3·°C;

V_{bld} is volume of the building, m^3;

Q is heat losses of the building, GCal/h;

t_{int} is internal air temperature, °C;

t_{ext} is external air temperature, °C;

To enhance visibility of the dynamics of changes in the heat consumption on the graphs, the volume of the building is withdrawn from the formula.

As one of the examples let us consider a new brick monolithic building of the volume 63,854 m^3 located in Saint Petersburg and commissioned in 2009. The heating power plant is equipped with an automated individual heating unit with ECL 300 controller (Danfloss). The connection of the heating system is independent with an option of control by risings mains (balancing valves) and thermostatic heads on heatsinks.

Table 1 shows the actual (a heat calculator of the heat energy meter, HEM) and calculated data showing the construction and thermotechnical characteristics of the building.

Table 1. Heating system parameters.

Date	Actual data of HEM					Calculated data		
	T_{ext}, °C	G, t/day	T_1, °C	T_2, °C	Q_{act}, GCal/day	Q_{calc}, GCal/day	$q_v \cdot V_{bld}$, kCal·day/°C	η, %
01.01.17	1,2	142,60	72,8	47,6	5,40	5,36	11,35	1
02.01.17	-2,1	174,26	76,0	49,9	6,60	6,30	11,89	5
03.01.17	-7,8	183,43	79,9	51,9	7,60	7,92	10,99	-4
04.01.17	-14,5	166,26	86,6	53,6	8,60	9,80	10,10	-14
05.01.17	-18,1	156,98	92,1	54,9	9,57	10,83	10,20	-13
06.01.17	-18,6	148,10	92,7	54,6	9,68	10,97	10,20	-13
07.01.17	-18,4	146,27	92,8	54,1	9,72	10,92	10,28	-12
08.01.17	-9,4	143,08	93,3	54,7	9,81	8,35	13,47	15
09.01.17	-1,1	73,48	93,1	51,5	7,63	5,99	14,41	21
10.01.17	0,8	38,90	93,5	50,8	6,50	5,47	13,40	16
11.01.17	-4,3	82,40	91,5	51,8	7,62	6,92	12,54	9
12.01.17	-3,7	87,75	91,8	51,9	7,75	6,75	13,07	13
13.01.17	-2,2	73,45	91,2	52,2	7,25	6,32	13,01	13
14.01.17	0,0	57,87	89,4	50,6	6,40	5,70	12,68	11
15.01.17	-0,9	64,84	87,5	50,9	6,48	5,95	12,33	8
16.01.17	-2,8	77,37	87,2	51,3	6,73	6,50	11,76	3
17.01.17	-4,0	99,13	86,8	51,6	7,12	6,81	11,89	4
18.01.17	-1,9	91,50	87,5	50,9	6,91	6,22	12,60	10
19.01.17	0,9	53,61	88,2	50,5	6,05	5,42	12,56	10
20.01.17	-0,5	68,31	87,3	50,8	6,37	5,83	12,34	8
21.01.17	-3,5	96,29	86,9	51,6	7,08	6,69	12,03	6
22.01.17	-1,4	79,30	87,5	51,5	6,65	6,08	12,39	9

where: T_{ext} is average daily external air temperature, °C;

G is daily consumption of the heat carrier for heating, t/day;

T_1 is average daily temperature of the heat carrier in the supplying pipeline, °C;
T_2 is average daily temperature of the heat carrier in the return pipeline, °C;
Q_{act} is actual daily heat consumption, °C;
Q_{calc} is calculated daily heat consumption, °C;
$q_v \cdot V_{bld}$ is specific heating characteristic of a building for the total volume of the building, kCal·day/°C;
η is deviation of actual heat consumption from calculated heat consumption $(Q_{act}-Q_{calc})/Q_{act} \cdot 100$, characterizing the overheating and underheating of the building, %
On graph 1 one can see the region of overheating, the region Q_{calc} located above Q_{act} and characterizing the "excess" of heat upon an increase in external air temperature.

Graph 1. Heating system parameters.

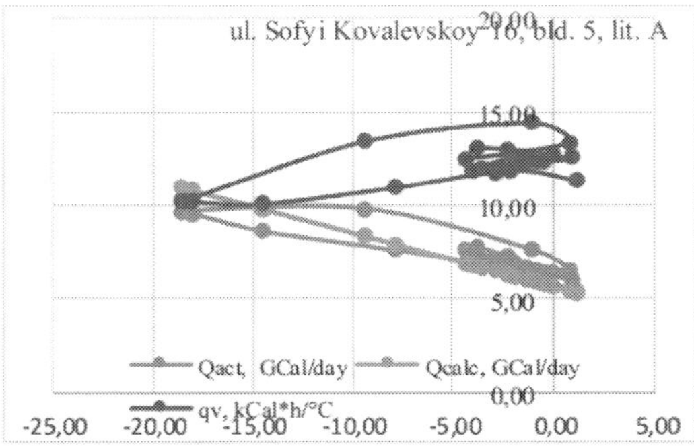

This residential building has the potential to reduce heat consumption due to the fact that the specific heating load of the building grows with an increase in the external air temperature, and the temperature in the apartments did not drop was in the range of 21÷28 °C.
On the basis of the table of the average external air temperature t (°C) and the heat consumption Q (GCal) on the heating system per month the graphs have been built.

Table 2. Average temperature and heat consumption.

Heating season	October	November	December	January	February	March	April
2013-2014	7,3	4,4	0,9	-7	0	2,2	6,5
	117,71	144,07	182,59	229,40	255,25	146,29	135,14
2014-2015	5,2	0,8	-1	-2,7	-0,6	2,6	5,1
	97,58	177,42	197,10	237,49	207,34	135,14	114,86
2015-2016	5,6	3,1	2,1	-11,2	0	1	6,3
	69,67	127,46	134,73	249,88	177,00	165,34	134,56
2016-2017	5	-1,8	-1,2	-3,9	-3,5	1,3	2,8
	110,92	187,12	206,37	212,40	199,86	157,35	126,12

Graph 2. Average temperature.

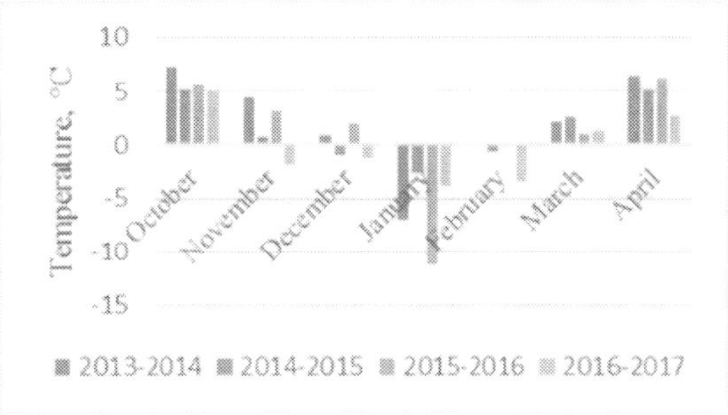

Graph 3. Heat consumption.

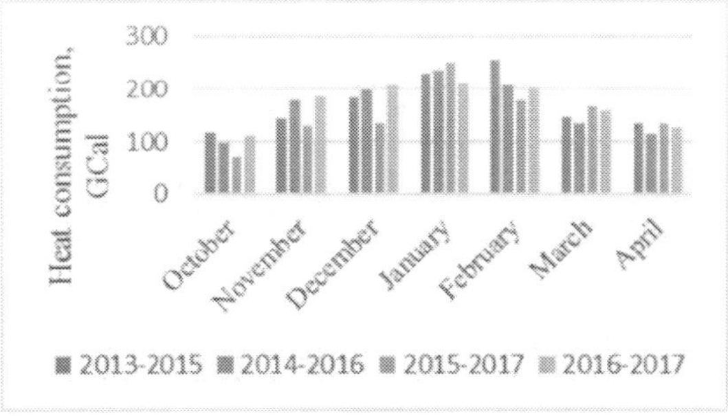

A conclusion can be drawn from the given schedules that there is no direct dependency of the building's heat consumption on external air temperature, which shows the insufficiency of this criterion in the efficient control of the heating system. Thus, the absolute conventionality is confirmed, taken when making an analysis of an energy-saving event on the criterion of external air temperature due to the imperfection as the only disturbing parameter.

4 Parameters not accounted for

Thus, to maintain the required comfort on the premises intended for various uses, one has to identify the factors (Fig. 2) not accounted for in the control process, but influencing the quantities of heat energy when the building is heated. The parameters influencing the heat consumption can be conditionally divided into disturbing and static ones.

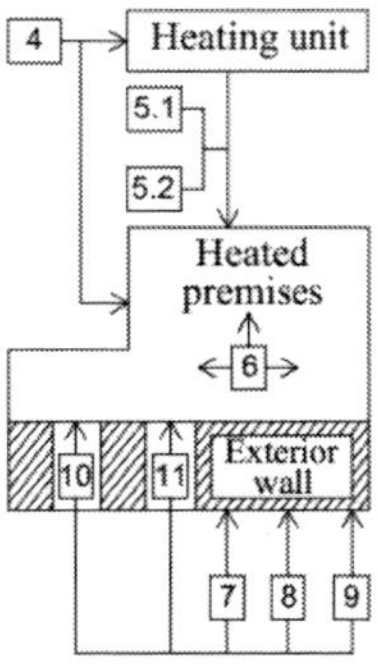

Fig. 2. Parameters not accounted for.

4.1. Disturbing parameters

The disturbing parameters are in their turn divided into interior and exterior ones.

4.1.1 Interior parameters

The interior disturbing parameters depend on the intended use of the building and its operating conditions. They include, in the first turn, heat or cold inputs [16] arising in the process of operation of the equipment or activities of people 6 (Fig. 2).

4.1.2 Exterior parameters

The wind effects 7 (Fig. 2) on enclosing structures consist of wind velocity and direction with respect to the building facades and are able to considerable increase the heat losses [17]. In the transition period of the heating season the effects of solar radiation increase 8 (Fig. 2) depending on the position of the building in respect of the cardinal directions and the glazing of the facades [18]. The humidity of the enclosing structures 9 (Fig. 2) arising both when precipitation falls and due to the location of the building on the coastal territory of seas and oceans, is able to reduce the thermotechnical characteristics both in current time and having effects when they are operated [19].

4.2 Statistical parameters

The statistical parameters are characterized by individual architectural 10 (Fig. 2) and thermotechnical features 11 (Fig. 2) of the building. The degree of influence of the statistical parameters increases with the increase in the effects of the exterior disturbing parameters [20].

4.2.1 Architecture of the building

The architectural features of the building, namely the altitude of the building and its compactness are determining ones when the wind or its directions act.

4.2.2 Thermotechnics of the enclosing structures

The high thermotechnical characteristics of the enclosing structures of the building with high heat-retaining capacity increase the efficiency of using heat energy with positive fluctuations of the exterior air temperature.

5 The proposed control method

The construction of the suggested control model is based on creating an adaptive advanced control system. The purpose of the control consists in searching for an individual approach to the control of each heated building with generation of a base of parameters serving as elements for constant adjustment (adaptation) to changes in disturbing factors taking into account the static characteristics of the building. The determining factor in this system is advanced generation of a controlling signal based on the full range of weather conditions, not on the one, albeit basic, parameter for control, i.e. the exterior air temperature. The issue of advanced control is discussed in the article [21] offering short-term forecast based on daily fluctuations (day/night) of the temperatures. The advantage in implementing this algorithm is attained by using the existing controller installed in a heating unit. The next step in implementing the advanced control of the heating system might be the creation of a system on the basis of local forecast of weather conditions [22], with the use of the practical value of heat accumulation of the building β [23] with frequent fluctuations of the exterior air temperature [24].

6 Conclusion

The presented practical results of heat consumption of the heating system of a modern residential building evidence the imperfection of local automatic control. The implementation of the advanced control system will make it possible to select an individual control schedule and to increase the efficiency of heat consumption when heating residential and public buildings.

References

1. Order of the Ministry of Construction of Russia dated 06.06.2016 No.399/order, On approving the Rules for determining the class of power efficiency of apartment buildings
2. Ye.L. Prokopchuk, Patent No.73509 Russian Federation, MPK G05D23/19
3. D.A. Shnaider, V.F. Postaushkin, L.S. Kazarinov, M.V. Shishkin, Patent No.2196274 Russian Federation, MPK F24D19/10
4. M.N. Masov, Patent No.2348061 Russian Federation, MPK G05D23/00
5. I.A. Kashmanov, Patent No.2492392 Russian Federation, MPK F24D3/00
6. A.N. Potapenko, A.V. Belousov, Ye.A. Potapenko, S.V. Kostrikov, Patent No.2247422 Russian Federation, MPK G05D23/19
7. T.A. Datsyuk, Yu.P. Ivlev, V.A. Modern problems of science and education, no.5 (2014)
8. Rules for operation of power stations and networks of the Russian Federation, RD 34.20.501-95
9. R.Yu Rozhkov, Heat supply news, no.2, 24-28 (2013)
10. S.Yu. Andreyev, Bulletin of the National Technical University Kharkov Polytechnic Institute. Series: Informatics and Stimulation, 9-12 (2005)
11. A.N. Ganzha, V.N. Podkopay, Power savings. Power engineering. Power Audit, No.8, 8-13 (2013)
12. Operation manual ECL Comfort 210/310, Danfoss Denmark

13. Operation manual TPM232M, OVEN Russian Federation
14. GOST 30494-2011, Residential and public buildings. Indoor climate parameters, 2013
15. L. Bankhidi, *Heat indoor climate* (Stroyizdat, Moscow, 1981)
16. Yu.A. Kuvshinov, *Power savings in indoor climate systems of buildings* (Publishing house of the Association of construction institutions of higher education, Moscow, 2010)
17. D.V. Drozd, Yu.V. Yelistratova, A.S. Seminenkov, Modern science-intensive technologies, No.8, 37-39 (2013)
18. A.I. Kruglova, *Climate and enclosing structures* (Stroyizdat, Moscow, 1970)
19. Yu.N. Leontyeva, Forecasting the dynamics of hygrothermal state of enclosing structures of residential buildings of Saint Petersburg and increasing their power efficiency: thesis of cand.of techn.sc. Saint Petersburg (2002) [in Russian]
20. O.D. Samarin, *Thermal physics. Power savings. Power efficiency* (ASV, Moscow, 2014)
21. Ye.L. Prokopchuk, A synthesis of a system of advanced control of heat supply to heat a building: thesis of cand. of techn.sc. Dzerzhinsk, (2009) [in Russian]
22. A.A. Fevralev, Yu.S. Prikhodko, Bulletin of South Ural State University. Construction and architecture, **16**, No.2, 48-51 (2016)
23. T.G. Manukovskaya, Design of a system supplying heat power to industrial facilities taking account of heat-retaining capacities of buildings: thesis of cand.of techn.sc. Voronezh, (2012) [in Russian]
24. N.V. Kobyshev, Quality of heat supply of cities, 110-127 (2004)

MATEC Web of Conferences 132, 02003 (2017) DOI: 10.1051/matecconf/201713202003
DTS-2017

Development of a mathematical model of a biomechatronic finger prosthesis as part of a multifunctional prosthetic hand

Aleksey Lysenko[1,*], *Denis* Khashev[1], and *Ekaterina* Uryanskaya[2]

[1]Don State Technical University, Robotics resource centre, 344000 Rostov-on-Don, Russia
[2]Don State Technical University, "Robotics and Mechatronics" department, 344000 Rostov-on-Don, Russia

Abstract. A mathematical model of a biomechanical finger prosthesis is proposed, which makes it possible to study its dynamics under the condition of an increased degree of mobility of the structure under consideration. Increased degree of mobility avoids the disadvantages of creating a multifunctional biomechanical hand prosthetic.

1 Introduction

The development of a fully functioning artificial prosthesis of a hand operating at speeds close to normal human responses and with a force comparable to a physiologically human-like one with real-time control is an urgent task of prosthetics of the upper limbs. Unfortunately, the modern component base and interface technologies are still far from realizing this task. A modern prosthesis can be considered as an instrument, not a replacement of a limb [1,2,3].

The person during the capture makes a new mechanism of the hand, depending on the purpose of the movement, on the nature of the object (size, mass, shape, consistency). The basis for a variety of movements are six types of capture: hook, interdigital, planar, plucked, cylindrical, ball [4,5,6]. Having considered a technological process standard for the hand, conditionally it consisting of three phases: compression of the object; holding the object; release of the subject [7,8,9]. When carrying out these actions with a natural hand, it is noticed that the greatest efforts are observed in the first phase. Based on the experience of performing typical actions with the use of the hand, the average value of F = 50 N tension on the joint between the 2 and 3 phalanges of the finger was obtained. This requirement corresponds to a lifting by one finger of a cargo of 5 kg weight sufficient to perform the basic actions of the natural hand.

2 Solution of design problems

In the framework of this paper, the structure of the mechatronic module is proposed (Fig. 1).

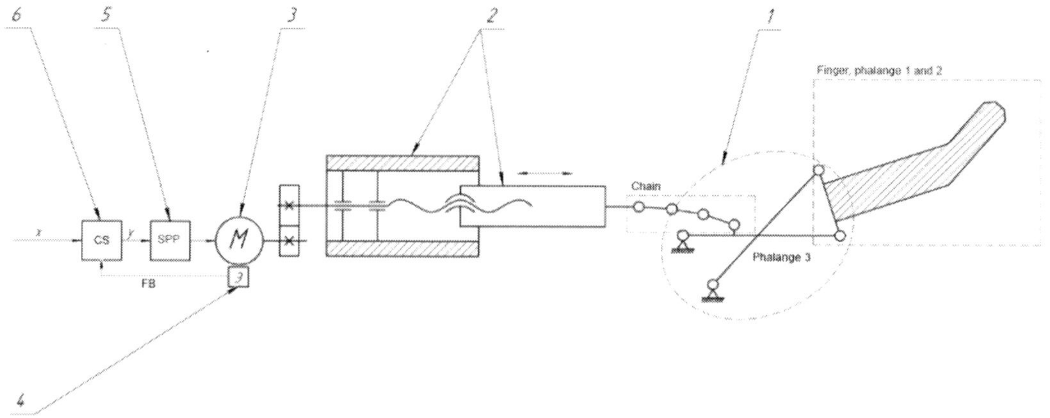

Fig. 1. Structural scheme of the mechatronic module.
1- The lever mechanism of the finger; 2 - Transmission of the screw-nut of sliding; 3 - Micromotor "M"; 4 - the Sensor of speed of rotation of a shaft of the engine "E"; 5 - Driver of the engine (block "SPP" - power semiconductor converters); 6 - SPP control pulse shaping system, and microprocessor control system; 7 - Miointerface (signals "x" from the sensors of the miopotential).

Corresponding author: aflysenko@gmail.com

In the kinetostatic calculation, the method of dividing the kinematic chain of the mechanism into Assurian groups [10, 11] is used, which are statically determinable. A calculation of the kinematic chain mobility degree is made according to the Somov-Malyshev formula for spatial mechanisms, the links of which can move in different mutually perpendicular planes [11]:

$$W = 6n - kp_i \qquad (1)$$

where n is the number of moving links;
kp_i - the number of 5, 4, 3, 2, 1 class kinematic pairs.

In the considered kinematic scheme, all kinematic pairs are rotational, of which p_3: 4, p_4: 2. Substituting the values in formula (1), the degree of mobility is W = 16. Often, in similar models, the finger is presented in a simplified form without one kinematic pair of the third class. In this case, p_3: 3, p_4: 2, and the degree of mobility is W = 13. Thus, the number of possible independent displacements of the model links is significantly reduced, which for the kinematic chain as a whole means a loss of mobility capabilities of 18.75%. The model of the lever mechanism of the finger (Figure 2), compiled in accordance with the chosen design, is considered. Knowing the lengths of the links and their types, a model of the kinematic scheme in Simulink is compiled using the SimMechanics package [12,13].

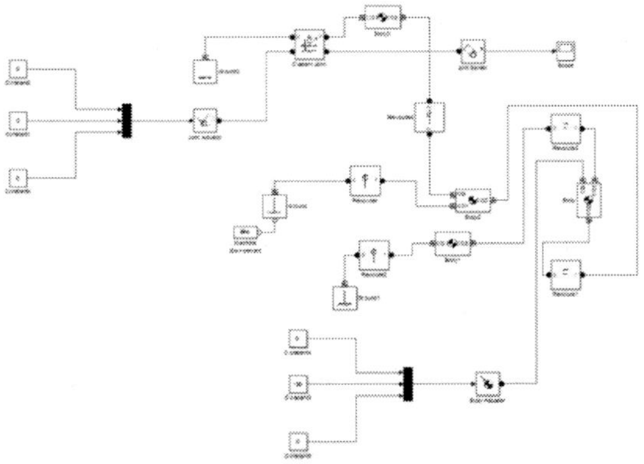

Fig. 2. Model of the lever mechanism.

The forces in the mechanical transmissions of a pair of screw-nut and gears are calculated in order to determine the required mechanical power of the engine. Based on the calculations in Table 1, a P = 0.14 W motor with the smallest size and equipped with a reducer was chosen.

The analysis of the continuous-mechanical system and the synthesis of digital regulators for the control object are performed. A block diagram of the finger control system is considered, where in Fig. 3 the main parts of the model are marked: Speed regulator blocks, the task of the input action, the transfer function of the regulator and the block for limiting its output; Current controller blocks, transfer function and feedback factors; Blocks PWT, the generator of a sawtooth signal, an ideal key and the smoothing filter; Blocks DC motor, load problem, transfer functions of the engine.

Table 1. Calculation of engine parameters.

Parameter	Formula	Value
Transmission ratio of screw-nut transmission.	$u = \frac{\pi D}{S}$	22
Circular force on the pinion of the screw, N	$F_{окр1} = \frac{F_{треб}}{u}$	5,27
Circular force on the gear wheel of the engine, N	$\frac{F_{окр1}}{F_{окр2}} = u$	3,19
Torque on the engine, Nm	$M_{треб} = F_{окр2} * \frac{D}{2},$	0,011
Engine speed, rpm	$n = \frac{V}{S},$	120
Engine power, W	$P = \frac{Mn}{9.55},$	0.14

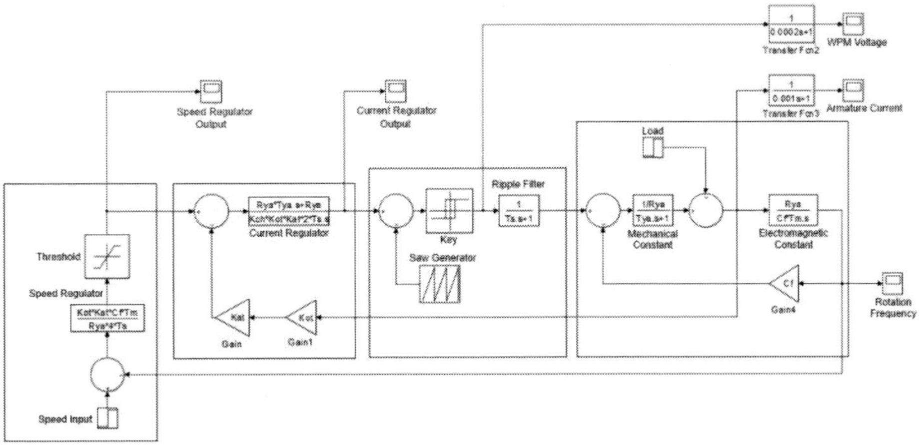

Fig. 3. Structural diagram of the control system.

The numerical values of the transfer functions of the continuous current and speed regulators, as well as the transfer functions of the selected DC motor are calculated. In order to do so the current feedback coefficient was found using the formulas from Table 2 [14,15].

Table 2. Value of transfer units and their coefficients.

Formula	Value
$K_{\text{от}} = \dfrac{U_{\text{огр}}}{I_{\text{н}} * K_{\text{ат}}}$	13.7
$U_{\text{огр}} = U_{\text{шп}}$	2^{16}
$K_{\text{ат}}$	1.986e + 04
$U_{\text{ш}}$	1.05
$K_{\text{ш}} = \dfrac{U_{\text{ш}}}{U_{\text{шп}}}$	1.602e − 05
$cF = \dfrac{U_{\text{н}} - R * I_{\text{н}}}{\omega_{\text{н}}}$	0.0024
$K_{\text{осн}} = \dfrac{N_{\text{дн}} T}{2 * \pi}$	0.0398
$W_{\text{рт}}(p) = \dfrac{R * (T_a * p + 1)}{K_{\text{ш}} * K_{\text{от}} * K_{\text{ат}} * 2 * T_s * p}$	$W_{\text{рт}}(p) = \dfrac{5.7 * ((1.1578e-05) * p + 1)}{(1.602e-05) * 13.75 * (1.986e+04) * 2 * 0.005 * p}$
$W_{\text{рс}}(p) = \dfrac{K_{\text{от}} * K_{\text{ат}} * cF * T_m}{R * 4 * T_s}$	$W_{\text{рс}}(p) = \dfrac{13.75 * (1.986e+04) * 0.0024 * 0.012}{5.7 * 4 * 0.005}$

Calculation of the numerical data of the current loop is performed for a filter with a time constant Ts = 0.005sec. The filter is only necessary for the synthesis of the current loop of a continuous subordinate control system.

In the course of the study, the regulator was reduced to a discrete form. As a result, during the modeling of electromechanical mechanical system of the prosthesis it was revealed that when the period of sampling of the regulator changes, the simulated model loses its stability when the sampling period is more than 0.00001 seconds.

3 Conclusion

Having examined the graphs of the simulation results in Fig. 4, with given input impacts of angular velocity of 500 rad/s, at the time t = 0.4 sec, and at the time t = 1.4 sec the load applied to the motor shaft is 0.3 Nm, it is concluded that the mathematical model of the control system is synthesized correctly, the output values of the regulators do not exceed the permissible limits, the control time is close to the constant time of the motor itself.

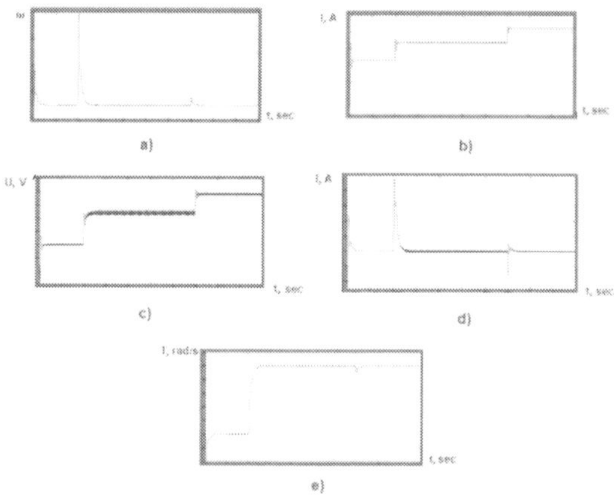

Fig. 4. The graphs of the control system investigation: (a) the output of the speed controller, the discrete code 0...65535, (b) the current controller output, the discrete code 0...65535, (c) the average PWT voltage, (d) the armature current, (e) the shaft speed , rad/s

References

1. D. Craig, Introduction to Robotics. Mechanics and management, Publishing Institute of Computer Research, p.564 (2013)
2. R. H. Taylor, *Proc. IEEE, Spec. iss. "Medical Robotics"*, **94**, No.9, 1652-1664 (2006)
3. G.V. Savrasov, Biomed. Radioelectronics, **10**, 42-46 (2007)
4. Bone-Surgery, available at: http://bone-surgery.ru/view/kist_kak_organ/
5. M.C. Carrozza, G. Cappiello, S. Micera, B. B. Edin, L. Beccai, C. Cipriani, Biol Cybern **95**, 629-644 (2006)
6. Grabcad, available at: https://grabcad.com/library/bionic-hand-concept-1
7. Enciklopediya Tehniki, available at: http://enciklopediya-tehniki.ru/promyshlennost-na-r/rychazhnyy-mehanizm.html
8. B.I. Glotov, Power calculation of plane latch mechanism. Methodical instructions for the course project on the theory of mechanisms and machines for students, Barnaul (2000)
9. V.D. Britsky, G.B. Zamoruev, M.A. Nozdrin, Study of the lever mechanism. Tutorial, SPb: SPbSU ITMO, p.30 (2007)
10. I.I. Artobolevsky, The theory of machines and mechanisms. Textbook for universities, 4th ed., p.640 (1988)
11. I.I. Borisov, S.S. Reznikov, *The Week of Science of SPbSPU. Materials of the sci-pract. conf. with int. particip.* Pt 1., 85-88 (2014)
12. I.I. Borisov, S.V. Krivosheev, S.S. Reznikov, *Collection of reports theses of the young scientists congress*, **2**, 351-352 (2014)

13. I.I. Borisov, S.V. Krivosheev, S.S. Reznikov, Almanac of scientific works of young scientists at the University ITMO, 55-58 (2014)
14. V.S. Shcherbakov, M.S. Korytov, A.A. *Ruppel Modeling and visualization of movement of mechanical systems in Matlab*, (SibAD, Omsk, 2007)
15. P.N. Kuninin, The study of the microprocessor control device with pulse-width converter by means of Matlab: Laboratory practical.

Dynamic monitoring of the driving wheels' pre-sliding of the railway transport

Mikhail Chuveyko[1,*], *Sergey* Nosachev[1]

[1]Don State Technical University, Automation of Production Processes Department, 344000, Gagarin sq. 1, Rostov-on-Don, Russia

Abstract. The proposed method of dynamic monitoring of the driving wheels' state of the railway transport makes it possible to detect in a timely manner the processes accompanying the transition of the wheel from the adhesion to slipping mode. This in turn helps prevent a loss of adhesion traction that have a wide range of negative consequences. The method is based on the spectral analysis of the vibration processes of the driving wheels of the railway transport.

1 Introduction

The task of identifying and preventing loss of traction appeared in the same period when the first self-propelled wheeled vehicles were created. Due to the continuous increase in the power of the railway transport drives, the likelihood of loss of traction increases, and therefore this problem is most actual in railway transport. Despite the deceptive simplicity, the problem of adhesion the wheel with the road surface is one of the oldest and most complex problems of wheeled vehicles [1, 2].

2 The model description

To solve this problem, consider the properties of the subsystem "wheel-rail". In contrast to the classical simplified concepts of the transformation of rotational motion into translational motion, the speed of rolling of the wheel is determined not only by its angular frequency and radius, but also by the deformation processes occurring in the wheel. This phenomenon is known as elastic sliding or "creep". It was considered in more detail in [3-8]. This phenomenon leads to the fact that the actual speed of the wheel during the action of the external load and in the absence of loss of adhesion is less than its ideal value (determined by the formula ωR). The difference between the actual and ideal velocities determines the speed of elastic sliding. Experiments show that the speed of elastic sliding depends on many factors, including external tangential F_τ and normal F_n loads on the wheel [9-12]. A mathematical model describing the dynamic processes taking place in the subsystem "wheel-rail" was considered in [6]. Let's consider it in more detail.

In constructing the model, the wheel was considered as an elastic-dissipative rim interacting with the wheel axis by means of an elastic-dissipative coupling formed by the disk. The concept of building a model is based on the use of the principle of flows and accumulators, which makes it possible to greatly simplify the analysis of the dynamics of the system. Simplification is achieved by using the division of the simulated wheel into small stationary (relative to the rail) sectors. Thus, the variables characterizing the state of the wheel in individual sectors are "tied" not to specific moving wheel points, but to the geometric position of these sectors relative to the contact spot. While its movement, the material forming the rim of the wheel rotates around the wheel axis. This movement leads to the fact that through the boundaries of each such stationary sector passes a certain amount of mass of the rim of the wheel. To describe the velocity of the material passing through the sector boundary, we will use the value of the "material flow" S [kg/s]. In the absence of deformation of the rim, the total mass of the incoming (with a positive sign) and the outgoing (with a negative sign) material will be zero. The flow in the absence of deformations and determined only by the angular frequency of rotation of the wheel axis will be denoted as $\bar{S}$. However, in the presence of deformations, the mass of material can accumulate within the sector. To numerically characterize the accumulation phenomenon, we introduce the notion of the coefficient of compaction of material τ, which characterizes the increase in the density of the material in each individual sector. Deformations are a consequence of forces acting on the elements of the wheel rim. In the model, forces are distinguished: the elastic-dissipative interaction of the rim of the wheel and the disk, the elastic-dissipative interaction of individual elements of the wheel rim with each other and the forces formed in the zone of tribo-contact of the wheel and rail.

To characterize the force influence of the disk on the wheel rim element, the coefficients of the specific tangential elasticity k_1 and the specific tangential dissipation h_1 of the disk are introduced. The coefficient k_1 characterizes the force reaction acting from the disk side to the element of the rim of the wheel of unit length

* Corresponding author: chuveyko@icloud.com

when it is displaced relative to the undeformed state per unit length. This coefficient is determined by the design of the wheel and the material shear module. To describe the amount of shift of an individual element of the wheel rim with respect to its undeformed state, the so-called tension value N [kg] is introduced. It determines the mass of the rim material that must pass through a fixed cross-section to neutralize the deformation. Thus, each section of the wheel is characterized by its value N.

To characterize the force effect of individual elements of the wheel rim, one introduces the coefficients of the specific longitudinal elasticity k_2 and the specific longitudinal dissipation h_2. The coefficient k_2 characterizes the force response of the sector of the rim of the wheel of a unit length when it is stretched per unit length. This coefficient is determined by the design of the rim of the wheel and the Young's modulus for the material. The coefficients h_1 and h_2 characterize the dissipative component of the reaction forces and determine the values of the forces arising in response to the change in the strain quantities.

Dynamic connection, formed by tribological processes of the contact zone of the wheel and rail, is the cumulative result of the interaction of individual elements of the wheel and rail. At the same time, various elements participating in the interaction can be in different modes of operation. According to modern ideas about the process of friction [13-16], we can conditionally distinguish two modes of operation: "adhesion mode" and " sliding mode":

The "adhesion mode" is characterized by the presence of physico-mechanical bonds between the elements of the interacting surfaces. These links prevent the relative motion of the surfaces, which mathematically can be expressed by means of the force function F in the state coordinates. Studies have shown that the form of the force function is identical to Hooke's law: $F = k_3\Delta x$.

"Sliding mode" is characterized by the absence of permanent physico-mechanical bonds between the contacting surfaces, while the surfaces are in constant motion relative to each other. The process of formation and destruction of new bonds leads to the appearance of an interaction force, which is called the sliding friction force. The magnitude of this force is determined by the formula: $F = k_4F_n$. This expression includes the value of the coefficient of friction k_4. The magnitude of the coefficient of friction is determined by the properties of the interacting surfaces, as well as by the speed of relative slip. As the relative slip speed decreases, the intensity of bond formation increases, while the intensity of their destruction decreases, and with a decrease in speed below the threshold value this process leads to the formation of stable bonds and the transition to the meshing regime. It is empirically found that the effect of velocity can be well approximated by the expression:

$$k(v) = sign(v)\left(1 - \xi\left(1 - e^{-\chi v}\right)\right) \quad (1)$$

Taking all this into account, the mathematical model describing the dynamics of a moving wheel has the form:

$$\begin{cases} \frac{dS}{dt} = k_2\frac{\partial}{\partial l}\frac{1}{\tau} - h_2\frac{\partial}{\partial l}\left(\frac{1}{\tau^2}\frac{d\tau}{dt}\right) - \\ \quad -\frac{1}{H\bar{Q}}\left(k_1N + h_1\frac{dN}{dt}\right) + f(v) \\ \frac{dN}{dt} = S - \bar{S} \\ \frac{d\tau}{dt} = -\frac{1}{H\bar{Q}}\frac{\partial S}{\partial l} \\ \frac{dx}{dt} = v \\ \frac{dv}{dt} = \frac{1}{m}\left(\int_D f(v)dl - F_n\right) \end{cases} \quad (2)$$

where: H is the area of the wheel rim; $\bar{Q}$ is the initial density of the rim material; $f(v)$ is the specific force of the contact interaction; D is the extent of the contact spot; x is horizontal coordinate of the wheel center; v is translational velocity of the wheel center; m is the mass of the wheel; F_n is the force of external action.

3 The method description

Studies of this model have shown that as the "wheel-rail" system approaches the pre-sliding state, the dynamic properties of the contact change. This is due to changes occurring in the contact spot.

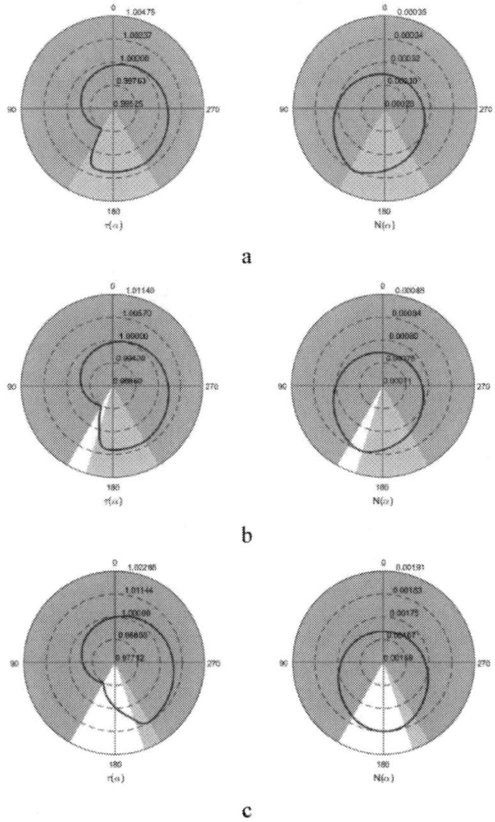

Fig. 1. The distribution of the coefficient τ and N along the surface of the wheel with variations in the tangential load: a - 2kN, b - 5kN, c - 11kN.

 DOI: 10.1051/matecconf/201713202004

In Figure 1 shows the diagrams of the dependence of the coefficient of compaction of the material τ and the value of the tension N on the angle of the sector on the wheel. The Figure 1 shows the results of experiments for three different values of the tangential load and the following color scheme is applied: dark gray - the area of the wheel, which is outside direct contact with the rail; light gray - the zone of the wheel, which is in an elastically-engaged interaction with the rail; white - the area of the wheel, which is in the sliding mode along the surface of the rail.

The diagrams in Fig. 1 it is easy to see that as the system approaches the pre-sliding state, a substantial change in the picture of the deformation distribution is observed, and, in view of the fundamental dependence of the elastic sliding effect on deformation processes, it is natural to expect a change in the properties of the system conjugate to this effect. As the system approaches the pre-sliding state, a change in the parameters of small oscillations of the wheel speed in response to external perturbations should be observed.

Considering the linearized model of the "wheel-rail" system (1) about some steady-state motion, it is possible to introduce transfer functions characterizing the influence of the external conditions (F_n, F_τ and ω) on the translational speed of the wheel:

$$\begin{pmatrix} v(p) \\ x_2(p) \end{pmatrix} = \begin{pmatrix} W_{v\tau} & W_{vn} & W_{v\omega} \\ W_{x\tau} & W_{xn} & W_{xw} \end{pmatrix} \begin{pmatrix} F_\tau(p) \\ F_n(p) \\ \omega(p) \end{pmatrix} \quad (3)$$

where: $F_n(p)$, $F_\tau(p)$, $\omega(p)$ are the Laplace transform corresponding to the deviations of the tangential force, normal force and angular frequency from its stationary values; W_{ij} is transfer functions by the corresponding channels.

Of special interest from the point of view of diagnosis is the transfer function $W_{v\tau}$, which describes the influence of the normal component of the external force on the speed of the wheel. Without going into all the properties of the given transfer function, we note the key property: in the low-frequency range (tens of hertz), this transfer function can be considered as a gain k. In addition, the value of this coefficient essentially depends on the stationary regime of motion. As the system approaches the pre-sliding state, k increases significantly (see Figure 2).

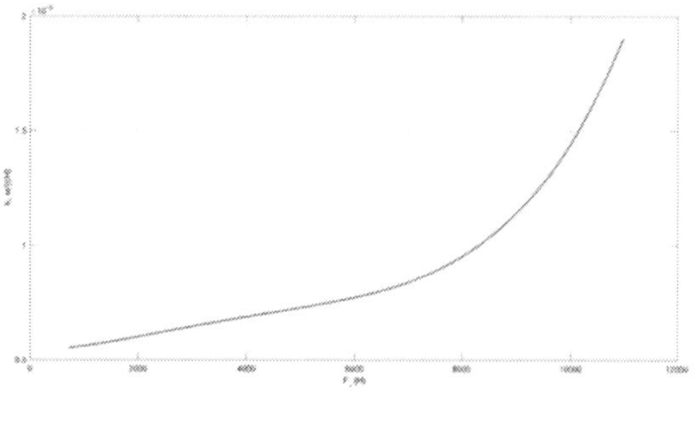

Fig. 2. Dependence of the coefficient of influence k on the tangential load on the wheel

Thus, this value can be used as a diagnostic criterion for the pre-sliding state. For its identification, it is proposed to use the spectral analysis of the vertical and horizontal oscillations of the wheel.

Consider a simplified "wheel-rail" system (see Figure 3).

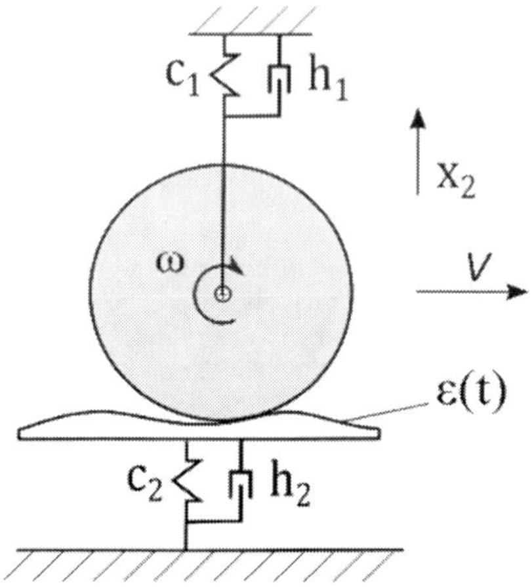

Fig. 3. The "wheel-rail" system.

Here ε characterizes the unevenness of the railway track and acts as a source of vertical disturbances. The properties of ε are considered in detail in [17-20]. Without going into detail on this issue, we note that ε can be considered as white noise. The mathematical model describing the vertical oscillations in the given system will have the form:

$$m\ddot{x}_2 + (h_1 + h_2)\dot{x}_2 + (c_1 + c_2)x_2 = c_2\varepsilon(t) + h_2\dot{\varepsilon}(t)$$

The mathematical model describing the horizontal oscillations in the operator form will have the form:

$$\tilde{v}(p) = W_{vn}(p)(c_1x_2(p) + ph_1x_2(p))$$

It is easy to see that the transfer function of such a system along the $\varepsilon \rightarrow \tilde{v}$ channel can be written as follows:

$$W_{\varepsilon\, v}(p) = W_{v\tau}\,\frac{(c_2+h_2p)(c_1+h_1p)}{mp^2+(h_1+h_2)p+(c_1+c_2)}, \quad (4)$$

but along the $\varepsilon \rightarrow x_2$ channel:

$$W_{\varepsilon\, x}(p) = \frac{c_2+h_2p}{mp^2+(h_1+h_2)p+(c_1+c_2)}. \quad (5)$$

The results of numerical simulation show that the main intensity of the wheel oscillations is concentrated in a narrow frequency range determined by the system parameters (ω_{max}= 51 Hz for typical railroad and wheel parameters). See Figure 4.

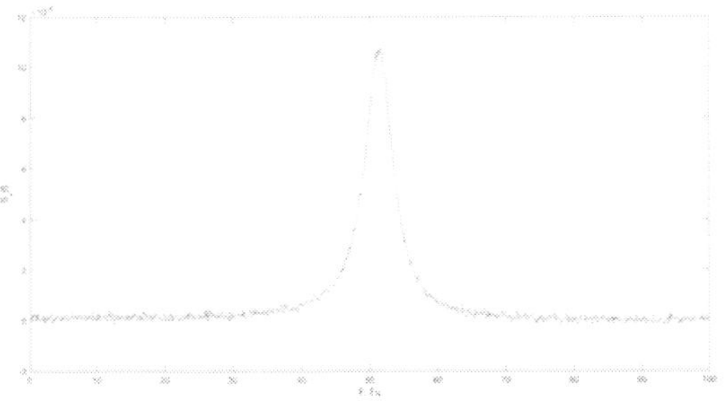

a

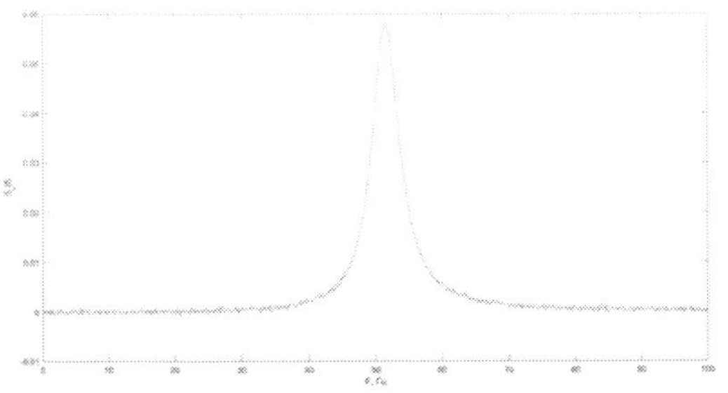

b

Fig. 4. The spectral density of oscillations: a – vertical, b - horizontal.

Based on the results of the simulation, it can be concluded that the main intensity of the wheel oscillations is concentrated in the frequency range f = 51 Hz and is determined mainly by the properties of track stiffness and dissipation, and not by the elastic-dissipative properties of the suspension. In addition, it can be noted that the relationship between the vertical oscillations of the centre of the wheel and horizontal velocity fluctuations is mainly at the frequency of vertical vibrations.

Also, based on the results of the simulation, it can be clearly seen that the spectral density of velocity oscillations at a frequency f = 51 Hz depends significantly on the value of the tangential load F_τ and increases significantly as it approaches the pre-sliding state.

The value of the spectral density at the considered frequency is practically independent of the frequency properties of $W_{v\tau}$ and is mainly determined by the influence coefficient k. Indeed, analysing the frequency characteristics of $W_{v\tau}$, it is easy to see that in the given frequency range, $W_{v\tau} \approx k$. Using this relation, as well as formulas (4) and (5), one can obtain an expression for the spectral densities of the observed signals:

$$S_v(\omega) = k^2 \left| \frac{(c_2 + h_2 j\omega)(c_1 + h, j\omega)}{m(j\omega)^2 + (h_1 + h_2)j\omega + (c_1 + c_2)} \right| S_\varepsilon(\omega)$$

$$S_x(\omega) = \left| \frac{c_2 + h_2 j\omega}{m(j\omega)^2 + (h_1 + h_2)j\omega + (c_1 + c_2)} \right| S_\varepsilon(\omega)$$

The ratio of the relations obtained is invariant to an external unobservable disturbance:

$$\frac{S_v(\omega)}{S_x(\omega)} = k^2(c_1^2 + h_1^2\omega^2)$$

and allows the identification of the parameter k with known suspension parameters (c_1 and h_1) at a specified frequency ω_{max}:

$$k = \sqrt{\frac{S_v(\omega_{max})}{S_x(\omega_{max})(c_1^2 + h_1^2\omega_{max}^2)}}. \quad (6)$$

As the value of the parameter k is related to the state of the contact and increases as it approaches the pre-sliding state, this value can be used as a diagnostic criterion for the pre-sliding state. Thus, the task of dynamic diagnostics can be solved by observing the power ratio of the vertical and horizontal oscillations of the wheel to detect the critical value k_{max} at which a loss of adhesion traction occurs.

4 Conclusion

A mathematical model of the "wheel-rail" system is proposed, which considers the deformation processes occurring in the wheel. Numerical simulation of the system made it possible to find out that the loss of adhesion traction occurs gradually as the tangential load increases. In this case, structural changes occur in the contact leading to a change in the deformation diagram of the wheel. This influences the effect of so-called deformation sliding, which in turn affects the relationship of small wheel vibrations in the horizontal and vertical directions. Formula (6) allows, by observing the ratio of spectral densities, to identify the coefficient linking these oscillations k. The information on the magnitude of this coefficient makes it possible to determine the degree of proximity of the system to the pre-sliding state.

Practical implementation of this method does not require the use of complex, expensive technical solutions, and, therefore, may be of practical interest.

References

1. R.V. Chernov, *Modern anti-skid protection on electric locomotives* (The Urals Electromech. Institute of Engineering, Transp., 1977)
2. V.V. Deev, G.A. Ilyin, GS Afonin. *Draft of trains* – (Transp., Moscow, 1987)
3. V.K. Garg, R.V. Dukkipati, *Dynamics of rolling stock* (Transp., Moscow, 1988)
4. F.T. Barweli, R.G. Woolacott, Proc. Instit. of Mechanical Engineers, **177**, 145 (1963)
5. A.Ya. Kogan, Vniizht bulletin. **5**, 33 (2004)
6. M.V. Chuveyko, S.V. Nosachev, Bulletin of the Don state tech. univ., **78**, 127 (2014)
7. M.V. Chuveyko, S.V. Nosachev, Modern trends of development of science and tech., **2-3**, 112 (2016)

8. N. Matsui, K. Yokose, Railway Technical Research Institute, **3** (1966)
9. J.J. Kalker, Vehicle System Dynamics, **5**, 317 – 358 (1979)
10. J.J. Kalker, In The General Problem of Rolling Contact. Transactions of American Society of Mechanical Engineers, **40**, 77-92 (1980)
11. F.T. Barweli, Woolacott R.G. Proceedings Institution of Mechanical Engineers, **177**, 145 – 160 (1963)
12. G.S. Itami, *The study of Friction - Creep Phenomenon of adhesion between Steel Wheels and Rail* (General Motors Institute, 1968)
13. K. Johnson, *Mechanics of contact interaction.* (Moscow, The World, 1989)
14. A.Yu. Ishlinsky, I.V. Kragelsky, Technical Physics, **45**, 276 - 282 (1944)
15. L. Bureau, Comptes Rendus de l'Académie des Sciences, **2**, 695-796 (2001)
16. V.L. Veits, M.S. Bundur, V.E. Hitrik, Journal of Friction and Wear, **6**, 653 – 660 (1985)
17. M.F.Verigo, A.Ya. Kogan, *Interaction of the track and rolling stock.* (Transp., Moscow, 1986)
18. S.V. Vershinsky, V.N. Danilov, I.I. Chelnokov, *The dynamics of the car.* (Transp., Moscow, 1978)
19. C. Esveld, A. Kok, The Railways of the World, **12** (2000)
20. V.F. Ushkalov, L.M. Reznikov, L.P. Boyarintseva, Vniizht bulletin, **6** (1986)

Building embedded quasi-time-optimal controller for two-wheeled self-balancing robot

Hai N. Phan[1,*] and *Chiem X.* Nguyen[2]

[1]Department of Software Engineering, Le Quy Don Technical University, Hanoi, Vietnam
[2]Automation & Engineering Department, Le Quy Don Technical University, Hanoi, Vietnam

Abstract. In this paper, two-wheeled self-balancing robot problem is introduced, and a quasi-time-optimal approach is applied to synthesize the control law. The result is compared with that of other methods. The process for designing and building the testting model, implementing the synthesized control law is also described.

1 Introduction

Two-wheeled self-balancing robot is a robot with two coaxial wheels. The idea of building such a robot, proposed firstly by Seway Inc., is based on two pillars, one being the way human-being keeping balance on two feet and the other being the movement of two-wheeled vehicle. Nowadays the robot has been developed to commercial product. Fig. 1 shows a typical two-wheeled self-balancing robot.

Fig. 1. Two-wheeled self-balancing robot.

Two-wheeled self-balancing robot has been extensively used as a means of communication thanks to its advantages, such as: small size, environmental protection, simplicity in controlling, flexibility in various topographies, simplicity in transport, installing and.

Qualitatively, the principle for keeping balance in two-wheeled robots is as follows. A robot in the equilibrium position is titled with an angle φ when subjected to a force. For the robot to regain its equilibrium, a force F is needed to apply to it so that its center of gravity is located in the middle of two wheels as shown in Fig. 2.

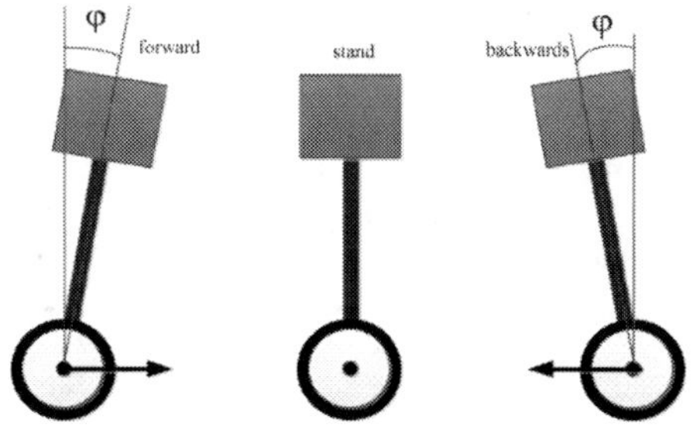

Fig. 2. The qualitatively balancing principle.

2 Mathematical model of the robot

The mathematical model of two-wheeled self-balancing robot is described in equations set (1) [1]:

$$\begin{bmatrix}\dot{x}\\ \ddot{x}\\ \dot{\phi}\\ \ddot{\phi}\end{bmatrix}=\begin{bmatrix}0 & 1 & 0 & 0\\ 0 & \frac{k_m k_e(M_p lr-I_p-M_p l^2)}{Rr^2\alpha} & \frac{M_p^2 gl^2}{\alpha} & 0\\ 0 & 0 & 0 & 1\\ 0 & \frac{2k_m k_e(r\beta-M_p l)}{Rr^2\alpha} & \frac{M_p^2 gl^2\beta}{\alpha} & 0\end{bmatrix}\begin{bmatrix}x\\ \dot{x}\\ \phi\\ \dot{\phi}\end{bmatrix}+\begin{bmatrix}0\\ \frac{2k_m(I_p+M_p l^2-M_p lr)}{Rr\alpha}\\ 0\\ \frac{2k_m(M_p l-r\beta)}{Rr\alpha}\end{bmatrix}U_a \quad (1)$$

Where: x - displacement (m); $\dot{x}$- velocity (m/s); $\ddot{x}$ - acceleration (m/s^2); φ - angular displacement (rad); $\dot{\varphi}$ - angular velocity(rad/s); $\ddot{\varphi}$ - angular acceleration(rad/s^2); U_a - voltage (V); k_m - rotation moment constant (Nm/A); k_e - reactive power constant (Vs/rad); R - nominal resistance (Ω); l - distance between the weight center of the wheels and the weight center of the robot (m); g - gravity acceleration (m/s^2); M_p - the chassis volume (kg); r - wheel radius (m); I_p - moment of inertia of the chassis (Nm); I_ω - moment of inertia of the wheeles (Nm); M_w - the wheel weight (kg). In addition:

* Corresponding author: phanhai@mail.ru

$$\beta = (2M_{\omega} + \frac{2I_{\omega}}{r^2} + M_p)$$

$$\alpha = (I_p\beta + 2M_p l^2 (M_{\omega} + \frac{I_{\omega}}{r^2}))$$

The problem of synthesizing control law for two-wheeled self-balancing robot is to find the force F via the voltage U_a that guarantees the robot to move from the start position $(x(0), \varphi(0))$ to the end position $(x_{sp}, 0)$ and stand still there.

Controlling a self-balancing robot is equivalent to the problem of controlling an inverted pendulum which is a typical system for simulation and implementation of controlling methods [2-7]. The optimal controls for the system were proposed in [2] and [3], in which the law using the signum function was used. This also means that the time optimal problem was not addressed. In this paper, we proposed a method for synthesizing quasi – time optimal nonlinear control law to overcome the above-mentioned problem, and implement it in an embedded board.

3 Using the quasi-time optimization approach

The quasi-time optimization approach for synthesizing control law is developed by Neydorf R. A's group [21, 22]. By using this approach, the problem of synthesizing control law for a wide spectrum of nonlinear objects can be solved. It is reported that the synthesized control law has several advantages, such as: quasi-optimal minimum-time, asymptotic stability and robustness.

To simplify model (1), we introduce the following variables:

$$a = \frac{2km.ke(M_p lr - I_p - M_p l^2)}{Rr^2\alpha};$$

$$b = \frac{M_p^2 g l^2}{\alpha}; \quad c = \frac{2km(I_p + M_p l^2 - M_p lr)}{Rr\alpha};$$

$$a_1 = \frac{2km.ke(r\beta - M_p l)}{Rr^2\alpha}; \quad b_1 = \frac{M_p g l\beta}{\alpha};$$

$$c_1 = \frac{2km(M_p l - r\beta)}{Rr\alpha}; \quad u = U_a;$$

It can be seen that $c_1 a = a_1 c$, or $a - a_1 c/c_1 = 0$.

Substituting these values into equation set (1), we get:

$$\begin{cases} \dot{x} = \dot{x} \\ \ddot{x} = a\dot{x} + b\varphi + cu \\ \dot{\varphi} = \dot{\varphi} \\ \ddot{\varphi} = a_1\dot{x} + b_1\varphi + c_1 u \end{cases} \quad (2)$$

Changing state variables as follows:

$$y_1 = x - \frac{c}{c_1}\varphi; \quad y_2 = \dot{x} - \frac{c}{c_1}\dot{\varphi};$$
$$y_3 = \varphi; \quad y_4 = \dot{x} \quad (3)$$

It is easy to verify that Jacobian matrix of transformation (3) is not degraded:

$$\det\left[\frac{\partial y}{\partial X}\right] = \begin{vmatrix} 1 & 0 & -c/c1 & 0 \\ 0 & 1 & 0 & -c/c1 \\ 0 & 0 & 1 & 0 \\ 0 & 1 & 0 & 0 \end{vmatrix} = \frac{c}{c_1} \neq 0$$

Where: $X = (x \quad \dot{x} \quad \varphi \quad \dot{\varphi})^T$.

Equations by $Y = (y_1, y_2, y_3, y_4)^T$:

$$\begin{cases} \dot{y}_1 = y_2 \\ \dot{y}_2 = (b - \frac{b_1 c}{c_1})y_3 + (a - \frac{a_1 c}{c_1})y_4 \\ \dot{y}_3 = -\frac{c_1}{c}y_2 + \frac{c_1}{c}y_4 \\ \dot{y}_4 = by_3 + ay_4 + cu \end{cases} \quad (4)$$

Equations (4) is in a controllable Jordan form [22]. Virtual system (4), which has asymptotic stability and quasi - time optimum, is used to synthesize the control law u with $z_1 = y_1$; u is found to introduce the system (4) into the virtual system (5) [22]. The obtained control law is a differentiable function of state variables, of which the formula is very long and therefore not given here.

$$\begin{cases} \dot{z}_1 = -v\frac{z_1 - x_{sp}}{\sqrt{(z_1 - x_{sp})^2 + \varepsilon_1^2}} + z_2 \\ \dot{z}_2 = -\frac{z_2}{\varepsilon_2} + z_3 \\ \dot{z}_3 = -\frac{z_3}{\varepsilon_3} + z_4 \\ \dot{z}_4 = -\frac{z_4}{\varepsilon_4} \end{cases} \quad (5)$$

Where $h_0(z, x_{sp})$ is quasi – time optimal function ($h_0(z, x_{sp}) \approx sign(z - x_{sp})$) [21].

Fig. 3 shows the simulation result of robot movement using the synthesized control law. The values set for simulation are: $k_m = 0.022$, $R = 1.7$, $r = 0.035$, $M_p = 1$, $M_w = 0.03$, $I_p = 0.0012$, $I_w = 0.000016$, $l = 1$, $g = 9.81$, $v = 1$, $\varepsilon_1 = \varepsilon_2 = \varepsilon_3 = \varepsilon_4 = 0.5$ $m_t = 2$, $m_p = 0.2$, $y(0) = x(0) = 0$, $x_{sp} = 1$, $\dot{x}(0) = 0$, $\varphi(0) = 0$, $\dot{\varphi}(0) = 0$.

Fig. 3 demonstrates the synthesized quasi-time optimal control. As compared to fuzzy control shown in Fig. 4, the synthesized quasi-time optimal control has several advantages. First, signal u is smooth. Second, there is no oscillation of control variable x. And finally, the speed of control is faster.

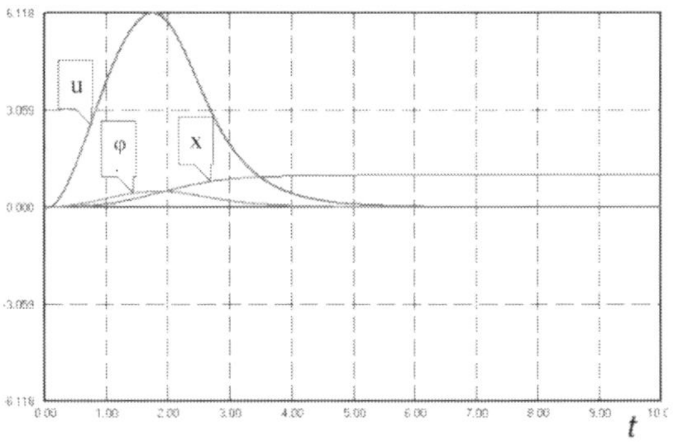

Fig. 3. Transition process with quasi-time-optimal control law.

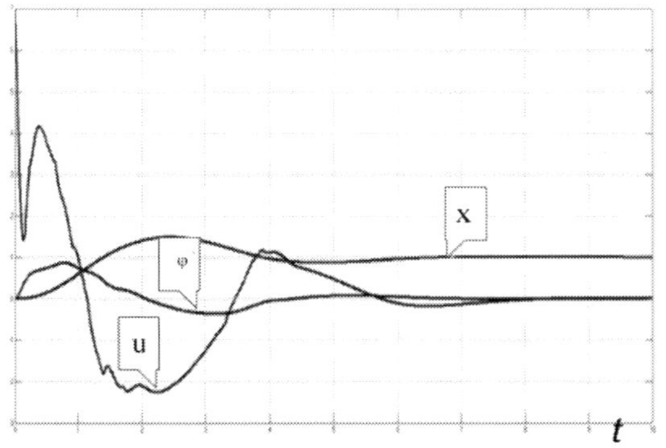

Fig. 4. Transition process with fuzzy control law

4 Designing and building embedded control system

To prove the feasibility of the simulation result, the testing model was build using Arduino microcontroller with a 16MHz crystal frequency to implement the control law as shown in Fig. 5.

Fig. 5. Two-wheeled self-balancing robot model

The most challenging issue in controlling a self-balancing robot is to keep the angle stable to guarantee the balance of the robot. The testing model is designed to keep the angle stable.

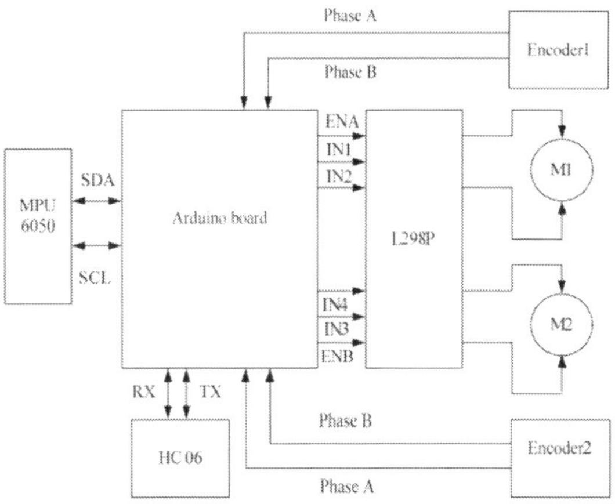

Fig.6. The schematic of two-wheeled self-balancing robot hardware

Where: M1 and M2 are the motor controlling the two wheels, L298P is the power block, and MPU 6050 is the sensing block.

The most difficult problem in implementing quasi - time optimal control law in embedded system is that the control law is quite complex and cumbersome because it contains 4th order derivative of the quasi – time optimal function. This results to long time for computation, making it difficult for a system to meet the real-time requirement. To simplify the calculation process, the intermediate functions, which are the high order derivative of sub-optimal function, were used in the embedded program. Equations (6) and (7) are the quasi – time optimal function (5) and its 1st order derivative. The high order derivatives are not shown here due to the fact that they are very cumbersome.

```
float h0(float z, float zs){
    return ((z-zs)/sqrt((z-zs)*(z-zs)+f1*f1));   (6)
}
float h1(float z, float zs){
    return (f1*f1/sqrt(((z-zs)*(z-zs)+
    f1*f1)*((z-zs)*(z-zs)+                        (7)
    f1*f1)*((z-zs)*(z-zs)+f1*f1)));
}
```

The angular displacement for balancing the robot is shown in Fig. 7. In general, the system has stabilized well but is still oscillating due to the following reasons. First, the actual model was built manually with relatively high measurement error. The actual model slips as the system is stable at zero angles. Second, the epsilons in the law are approximate parameters, which are not optimized since it is very hard to estimate them accurately in the simulation. Third, the measurement of feed-back values is with relatively high error. Fourth, time parameter is discretized in the embedded system. Improving the quality of the control part is the goal of our future work.

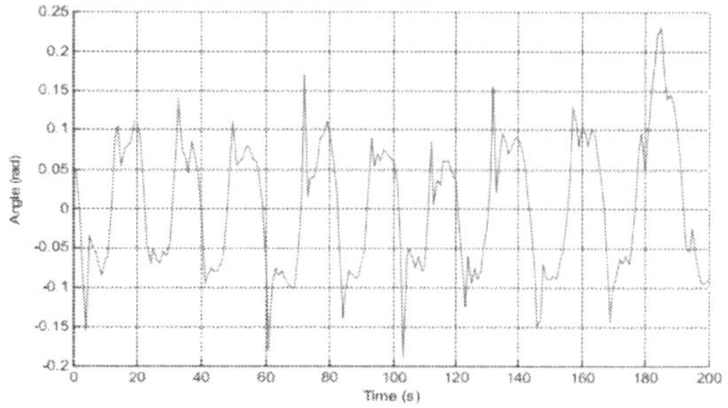

Fig. 7. The change of balance angle with time

5 Conclusions

In this paper, the problem designing and building two-wheeled self-balancing robot has been addressed comprehensively. First, the problem of synthesizing control law was addressed theoretically. By using the quasi – time optimal approach, the problem of synthesizing control low was solved. Then the robot and embedded board were designed to implement the synthesized control law.

References

1. Y.C. Ong, M.B. Shukri, Z. Abidin, *Proceedings of the IEEE 7th. Research and Development* (*SCOReD 2006*) (2006)
2. L.B. Prasad, B. Tyagi, H.O. Gupta, *International Journal of Automation and Computing*, **11**(6), 661-670 (2014)
3. S.C. Lin, C.C. Tsai, H.C. Huang, *Proceedings of the 2009 IEEE International Conference on Systems, Man, and Cybernetics*, 1965-1970 (2009)
4. S.C. Lin, C.C. Tsai, W.L. Luo, *In proc. 33rd Annu. IEEE IECON*, 868-873 (2007)
5. H. Ohara, T. Murakami, *IEEE Trans. Ind. Electron*, **55**, 1277-1285 (2008)
6. K.E. Vinodh, J. Jovitha, *Procedia Engineering*, **64**, 169-178 (2013).
7. R. Mellah, F. Lahouazi, S. Djennoune, S. Guermah, R. Toumi, *International Journal of Control, Automation and Systems*, **1** (2012)
8. O. Boubaker, *International Journal of Advanced Robotic Systems*, 1-9 (2013)
9. R.P.M. Chan, K.A. Stol, and C.R. Halkyard, *Annual Reviews in Control*, 89-103 (2013)
10. A. Filipescu and V. Minzu, *Proc. Int. Conf. System Theory, Control, and Computing*, 1-6 (2011)
11. E.J. Hwang, H.S. Kang, C.H. Hyun, and M. Park, *International Journal of Advanced Robotic Systems*, **10**, 1-8 (2013)
12. B.S. Park, S.J. Yoo, J.B. Park, and Y.H. Choi, *IEEE Transactions on Control Systems Technology*, 207-214 (2009)
13. R. Neydorf, *SAE Technical Paper*, 2015-01-2481, 8, doi: 10.4271.2015-01-2481 (2015)
14. R. Neydorf, Y. Sigida, V. Voloshin and Y. Chen, *SAE Technical Paper* 2013-01-2111, 10, doi:10.4271/2013-01-2111 (2013)
15. A. Filipescu and V. Minzu, *Proc. Int. Conf. System Theory, Control, and Computing*, 1-6 (2011)
16. J. Wu, Y. Liang, and Z. Wang, *6th international forum on strategic technology (IFOST)*, 1031–1035 (2011)
17. I. Baturone, F.J. Moreno-Velo, V. Blanco, and J. Ferruz, *IEEE Trans. Ind. Electron.*, **55**, no. 2, 928-936, (2008)
18. F. Grasser, A. D. Arrigo, and S. Colombi, *IEEE Trans. Ind. Electron.*, **49**, no. 1, 107-114 (2002)
19. Xiaojun, X.Z. Gao, and Xianlin, *In proc. 2004 IEEE International Conference on Systems*, Man and Cybernetics, **3**, 2286 – 2291 (2004)
20. P.V. Kokotovic, *Control Systems Magazine, IEEE*, **12** (3), 7–17 (2008)
21. R.A. Neudorf, Vestnik DSTU. Management and diagnostics in dynamic systems, 13-21 (1999) [in Russian]
22. N.H. Fan, Metody raspoznavanija upravljaemyh zhordanovyh form dinam., Cand. tech. sciences. Diss., Rostov-on-Don, 193 (2008) [in Russian]

The influence of external mechanical impact on the dynamics of an electromagnetic actuator

Irina Bolshenko[1], *Denis* Batishchev[1,*], and *Yuri* Klimenko[1]

[1]Platov South-Russian State Polytechnic University (NPI), Research Institute of Electromechanics, 346428 Novocherkassk Prosveshcheniya St. 132, Russian Federation

Abstract. A special case of studying the external mechanical acting factors - impact - influence on the electromagnet actuator dynamic characteristics is considered. An algorithm is proposed and software tools for its implementation are determined. The results of mathematical simulation are obtained, and their analysis is performed.

1 Introduction

The problems of construction of driving electromagnetic mechanisms with the specified noise and vibration levels occurred rather long ago [1]. They are especially acute in the automobile and shipbuilding industries, in the design of power plants mechatronic modules of autonomous transport systems including submarine and super marine vehicles. Initially their solution was reduced to the study and development of different methods for electromechanical transducers vibration and noise reducing. With the requirements strengthening in general and with the special low-noise and low-vibration electromechanical converters emergence occurrence, it became evident that in many cases the issues of providing requirements in terms of vibration and noise characteristics (VNC) must be solved at the devices design stage. The results of the research and ways to ensure the VNC were summarized in VNIIEM at the development of a low noise electrical machine design system [1]. However, despite existing progress in the field of vibration and noise research, for the driving electromagnetic mechanisms there is no possibility to predict their levels with the acceptable accuracy at the design stage. To the new autonomous transport systems being created, the high requirements not only on the level of comfort, but also on the possibility of detection due to the increased level of noise are applied. The devices' noise and vibration form the underwater noise of Navy ships and submarines worsening their furtively. The most "noisy" are the electromagnet actuators (EMA) used as valve actuators in control circuits, hydraulic and pneumatic control systems switchgears, fuel supply and air-and-gas exchange devices of internal combustion engines.

The previously conducted researches [2-5] aimed at studying the possibilities to create EMA resistant to mechanical impact factors (IMF) have shown that in order to obtain the specified dynamic properties that ensure the stable operation of the electromagnetic mechanism under vibration conditions, it is reasonable to use the integrated approach based on the definition and implementation of the required control law taking into account the configuration, geometric relationships and features of the electromagnet design.

In the works presented before the main emphasis was made on the research of external vibration loads effect on the EMA dynamic characteristics. In the present work the emphasis is made on the research of the various impact mechanical action parameters influence on the electromagnetic drive response dynamics at the design work stage. Among the external factors affecting the EMA it is necessary to note the mechanical effects, impact being the most dangerous of which. The most possible reasons of impacts actions on the products may be collisions, explosions, supersonic pressure, accidental falls, etc. The named effects can occur during the EMA exploitations at airplanes, ships, rockets, vehicles, as well as during their transportation.

2 Research method

To conduct the research, the approach that has proved itself to be successful in performing studies about the vibration effect on the electromagnetic drive characteristics was used [2]. At the development of the model the *FEMM 4.2* software package [6] in conjunction with the *LTspice IV* package [7] was used. The choice of this approach is due to the possibility of free distribution and usage of the named software for commercial purposes. For example, to conduct such a research with the widely used *ANSYS* software it is necessary to use the *ANSYS Maxwell* ($ 39,664.22) and *ANSYS Simplorer Advanced* ($ 31,133.49) [8] which is not reasonable within the framework of the research performed.

As an object of research the device for the back-run choice line electrical interlocking being developed for

* Corresponding author: batishchevd@gmail.com

 DOI: 10.1051/matecconf/201713202006

the concept cars of the VAZ-2116, 2117 and 2118 type platforms was adopted (Figure 1) [9, 10].

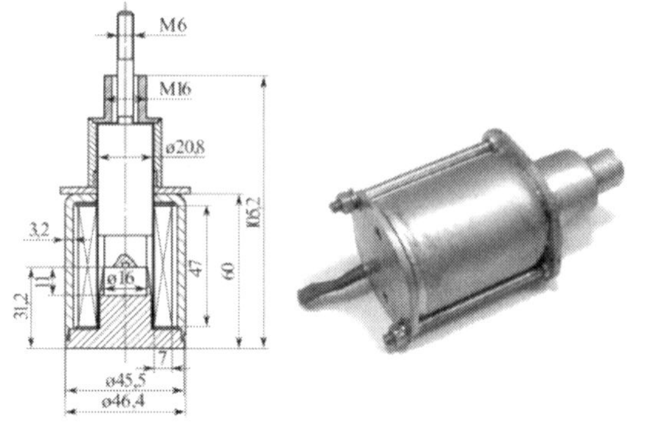

Fig. 1. Gear Shift Lock Solenoid.

Based on the solution of the field problem (Figure 2) families of characteristics $\Psi = f(i, x)$ and $P_{em} = f(i, x)$ were formed for further use in the *LTspice* program (Figure 2). The families of characteristics were calculated in the automatic mode using scripts written in the *Lua* programming language.

In the case of impact load simulation, the model of the impact load source was represented as a functional voltage source *Bimp* (Figure 3d). At this the main parameters of the impact pulse (Figure 3a-c) – the shock acceleration peak value (A), the impact pulse duration (D) and the impact pulse form – the law of the acceleration variation over time were specified in accordance with GOST 28213-89 [11], supposing that the result of the impact loads effect on the product is determined by its dynamic behavior that are characterized by its mass, toughness and natural oscillations frequency.

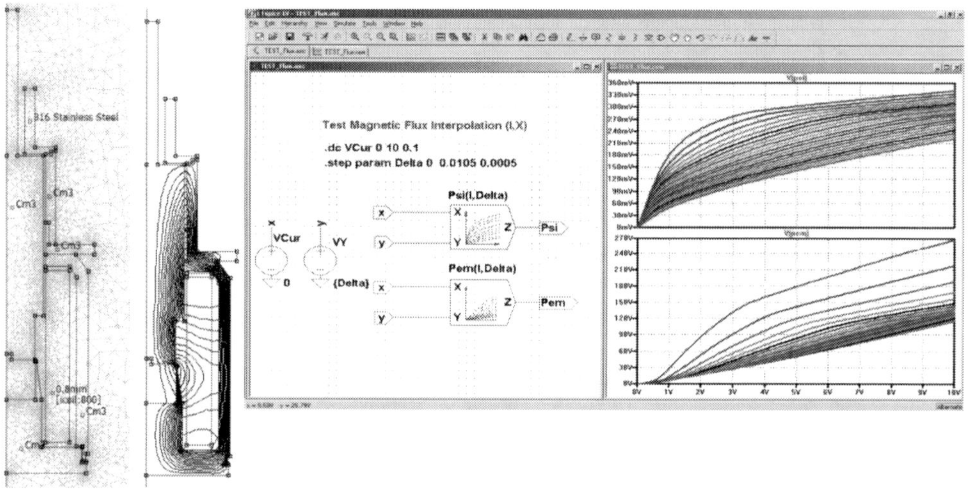

Fig. 2. The results of the field calculation for the EMA and the family of the drive characteristics $\Psi=f(i, \delta)$ and $P_{em}=f(i, \delta)$.

The impact pulse of the half-sine shape is used the most often in the tests (Figure 3a).

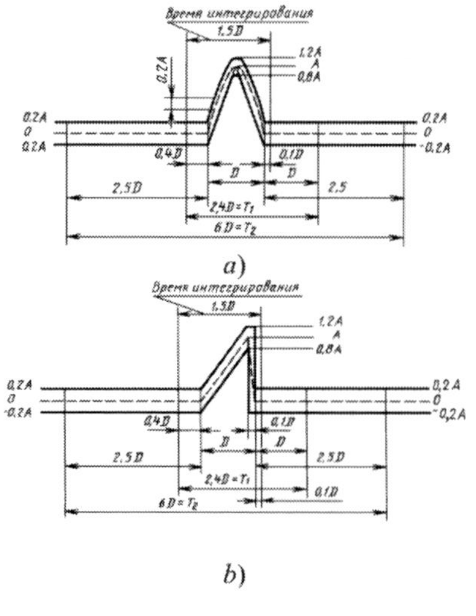

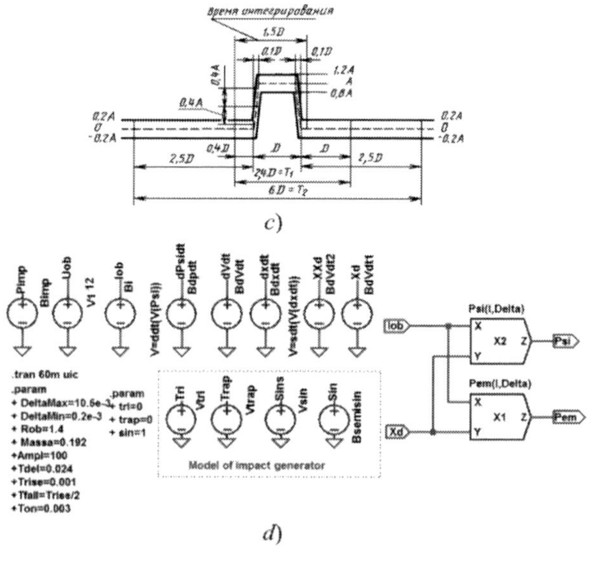

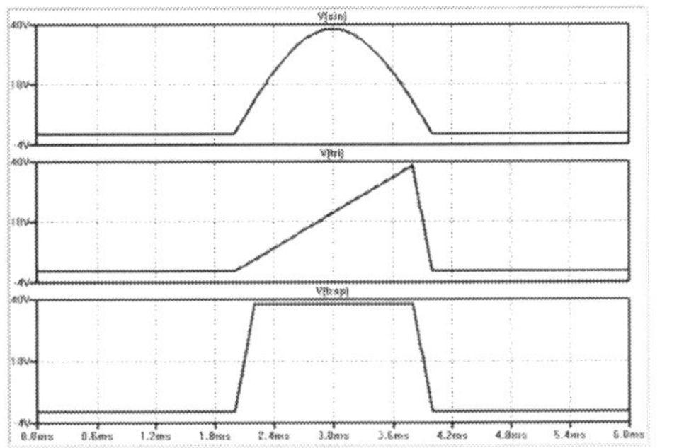

e)

Fig. 3. The forms of the impact pulses (*a-c*) and their representation in the *LTspice* program (*d, e*): a) the half-sine pulse; b) the saw tooth pulse with the peak at the end; c) trapezoidal impulse.

According to GOST 17516.1-90 [12] the articles installed in mobile railless complete products belong to the mechanical design group M30. This imposes the following requirements in terms of resistance to the IMF and in particular to impacts: the shock acceleration peak value of single shot impacts is 200m s^{-2} (20g), the duration of shock acceleration peak value of single-shot impacts is 2ms-15ms. Such impact level is high [12, 13].

3 Results and discussion

The results of the single shock impact on the dynamic performance simulation (speed *V(dxdt)*, motion *V(xd)* and winding current *V(iob)*) in the direction of external actions coinciding with the direction of armature movement and directed in the opposite direction are shown in graphs (Figure 4, Figure 5) .

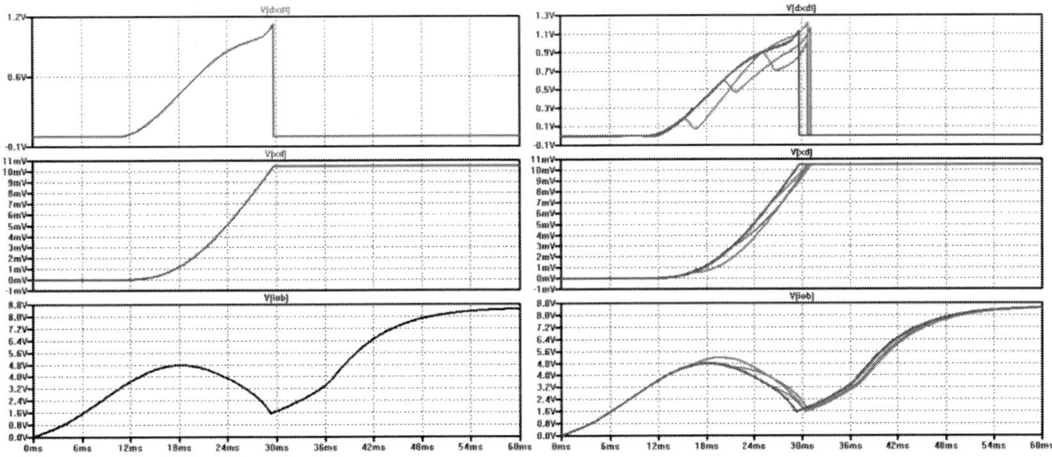

Fig. 4. The results of the single shock impact, the half-sine pulse with the duration of 2 ms at the variation of the shock impact action from 0 to 30 ms (the characteristics without the EMA are represented in the left side).

The results of the single stroke impact simulation, the half-sine pulse with the duration from 2 to 14 ms are shown in Figure 5, the moment of the impulse generation at the beginning of the electromagnet armature moving, 11 ms (left graph) and at the moment corresponding to half of the armature movement path, 24 ms (right graph).

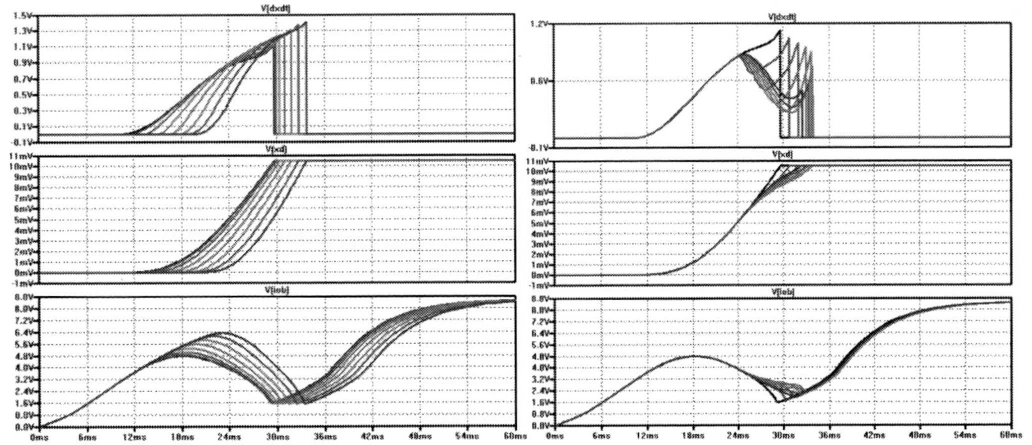

Fig. 5. The results of the single stroke impact simulation of the half-sine shape at the pulse duration variation.

4 Conclusion

Basing on the results of the presented studies it can be seen that the impact load influence significantly affects both the EMA response time and the terminal speed, i.e. on the drive VNC. It is possible to ensure the stability and required parameters of the VNC EMA either due to the rational shape of the working gaps bearing surfaces and the controlled conductivity of the workless gaps [14, 15] or due to the external influence on the control signal –the coils voltage supply forcing or applying special circuits for controlling their power [16-20].

Within the framework of the present article a small part of the research associated with the consideration of the impact load influence on the EMA dynamic characteristics is presented. More complete information will be provided in subsequent articles and in generalized STS.

The research was performed funded with the Russian Science Foundation grant (project № 17-79-10055).

References

1. V.A. Voronkin, V.Ya Gecha, E.A. Gorodetsky, V.V. Evlanov, A.B. Zakharenko, B.I. Zubenkov, A.I. Kaplin, V.I. Ledovskoi, I.V. Sklyarova, L.P. Smirnova, M.H. Shapiro, The electromechanics issues. Works of VNIIEM, **103** (2006)
2. D.V. Batishchev, *Methods and algorithms for electromagnetic drives designing with account of the external vibration effects* (dissertation in support of candidature for the technical degree, Novocherkassk, 2011)
3. Y.I. Klimenko, D.V. Batishchev, A.V. Pavlenko, V.P. Grinchenkov, Russ. Electr. Eng. **10**, 588-593 (2015)
4. A.V. Pavlenko, D.V. Batishchev, A.A. Gummel, I.A. Bolshenko, Russian Electromechanics, **3**, 17-27 (2017)
5. I.A. Bolshenko, *The electromagnetic drive of the gas-distributing mechanism valve of the internal combustion engine* (dissertation in support of candidature for the technical degree, Novocherkassk, 2015)
6. D.C. Meeker, *Finite Element Method Magnetics, Version* 4.2, available at: *http://www.femm.info/* (free)
7. *LTspice Simulation system description*, available at: *http://LTspice.linear-tech.com/software/scad3.pdf* (free)
8. *General Services Administration* (*GSA*), *Price List Current Through Modification #PS*-0081, (2017)
9. D.V. Batishchev, V.V. Medvedev, University News. North-Cauc. Region. Techn. Scien., Spec.Iss., 54-56 (2005)
10. *The power interlocking of the back-run choice line. Technical requirements for the cobbled unit* (Scientific and Technical Center of JSC AVTOVAZ, 2002)
11. GOST 28213-89 *The basic methods of testing for external factors influence. Part 2. Trials. The Ea test and the Manual: Single Stroke* (1989)
12. GOST 17516.1-90 *The electrotechnical products. General requirements regarding the mechanical external factors resistance* (1990)
13. GOST 30631-99 *General requirements for Machines, instruments and other industrial products in terms of resistance to mechanical external influencing factors during operation* (1999)
14. A.V. Pavlenko, D.V. Batishchev, University News. North-Cauc. Region. Techn. Scien., Spec.Iss., 24 (2003)
15. D.V. Batishchev, University News. North-Cauc. Region. Techn. Scien., Spec.Iss., 11-13 (2009)
16. A.G. Butkovsky, Yu.A. Cherkashin, *The optimal control of the electromechanical DC devices* (1972)
17. Yu.A. Cherkashin, *The optimal control of the electromechanical devices* (1969)
18. A.V. Pavlenko, V.P. Grinchenkov, D.V. Batishchev, A.A. Gummel, I.A. Pavlenko, E. Kallenbach, *The electromagnet drives for the mechatronic devices and vehicle systems. Design and optimization* (Mechatronics, automation, management, Taganrog, 2007)
19. A.V. Pavlenko, A.A. Gummel, I.A. Bolshenko, Proc. Eng., **129**, 571-577 (2015)
20. A.V. Pavlenko, A.A. Gummel', D.V. Batishchev, E. Baumbach, Russ. Electr. Eng. **4**, 189-193 (2016)

Algorithm for increasing the accuracy of three-axis MEMS orientation sensors

Ksenia Goryanina[1,*], *Oleg* Katin[1], and *Miroslav* Vernezi[1]

[1]Don State Technical University, Rostov-on-Don, Russia

Abstract. The authenticity of the readings of accelerometers and magnetometers is of great importance when using sensors in orientation systems. However, traditionally determined methods of calibration cannot always provide the necessary work and stability of indications. The method of calibration of MEMS accelerometers and magnetometers in the original setting of problems of identification of sensor parameters is considered in the article. The stochastic approach allows for the passage of a calibration from without an exact binding to a specific orientation of the accelerometer, since this does not require additional sensors to determine the angle of rotation. This allows us to use the algorithm for sensors, already fixed on the equipment. The experimental measurements were carried out using a three-axis MEMS sensor MPU-9265. The proposed method makes it possible to completely eliminate the displacement of zero, compensate for the non-unit scale of the axes, and also to determine the orientation of the sensitive axes of the magnetometer. The standard deviation of the experimental data from the reference cloud is less than 1%.

1 Introduction

In solving scientific-practical problems of synthesis and analysis of signals obtained from navigation devices, great progress has been made. The relevance of the topic is determined by the need to create hardware-software complex of calibration [1, 2] of three-axial MEMS accelerometers and magnetometers in order to reduce the error in determining the angular coordinates.

The practical importance of the work is determined, on the one hand, by the high need for such calibration systems (the use of MEMS accelerometers and magnetometers currently amounts to tens and hundreds of thousands), and, on the other hand, in reducing the cost of mass-produced sensors and, consequently, elimination of presale calibration of sensors. The development of a fairly simple and accurate method and complex for calibration of orientation sensors will certainly be in demand.

In general, the resolution of the problem of the authenticity of the data obtained depends on the accuracy of the calibration of the navigation sensors and the algorithm for processing their signals. Traditional calibration methods don't always provide the accuracy and stability of the readings required for operation [3–8]. The proposed approach providing a statistically significant estimate is different from most of the proposed calibration methods.

The solution of the calibration problem realizes the identification of the sensitivity ellipsoid over the cloud of points, which, unlike deterministic methods, makes it possible to determine not only the zero shift and statistically validate the gain factors, but also to determine the orientation of the sensitive axes.

2 Error Analysis

For the correct solution of the problem of calibration of orientation sensors, it is first necessary to perform an analysis and determine the main errors. The main source of errors in MEMS accelerometers can be considered sensitive part sensor [9]. Errors in the readings of the three-axial accelerometer arise due to three factors [10 – 12] associated with the structure of the board:

– presence of a constant zero offset;

– own noise;

– non collinearity of the axes of the object's coordinate system and the coordinate system of the accelerometer.

The main sources of MEMS magnetometers errors are:

– the presence of magnetic inclination due to the specificity of the earth's magnetic field;

– the presence of artificial fields around the sensor, created, for example, motor magnets, batteries, power wires, etc. In foreign literature, such errors are called Hard Iron [13 – 16];

– distortions caused by the presence of ferromagnetic materials near the sensor (Soft Iron) [13 - 16].

Thus, the errors of the MEMS orientation sensors are the displacement of the coordinate system of the data cloud, the non-unit scale of the axes, and the rotation of the ellipsoid of data around the origin.

* Corresponding author: gorianina.k@yandex.ru

3 Description of the method

When posing the problem of calibration of orientation sensors in stochastic interpretation, the sensor response is represented as a point in three-dimensional space [17]. Due to the influence of errors, the points form a cloud of data in the form of an ellipsoid with a shifted center, rotated relative to the origin.

Thus, the problem of calibration first of all becomes the determination of coefficients in the equation for the transformation of an ellipsoid into a sphere [18]:

$$\begin{pmatrix} a_x \\ a_y \\ a_z \end{pmatrix} = \left(M_{xyz}\right)_{3\times3} \begin{pmatrix} 1/k_x & 0 & 0 \\ 0 & 1/k_y & 0 \\ 0 & 0 & 1/k_z \end{pmatrix} \begin{pmatrix} r_x - a_{x0} \\ r_y - a_{y0} \\ r_z - a_{z0} \end{pmatrix} \quad (1)$$

To eliminate the displacement error of zero, it is necessary to center data [19]:

$$\bar{X}_a = \frac{1}{n}\sum_{i=1}^{n} X_{ia}$$

$$X = X_{ia} - \bar{X}_a \quad (2)$$

where are: *Xa* – the average component; *Xia* – numerical components of points of a cloud of data; *X* – the matrix of centered data.

Also, when calibrating the magnetometer, it is necessary to take into account the effect of magnetic inclination, which depends on the coordinates of the location of the experiment. To eliminate this influence, the value of the angle, called magnetic declination, is added or subtracted to the readings of the sensor.

To eliminate the error of non-unit scale axes, the problem is to identify the semi-axes of the ellipsoid, which can be defined as the problem of minimizing the functional J(Ω) with respect to the parameters *a, b* and *c*:

$$J = \sum_{i=1}^{n}\left(\frac{(x_i)^2}{a^2} + \frac{(y_i)^2}{b^2} + \frac{(z_i)^2}{c^2} - 1\right) \xrightarrow[a,b,c]{} \min \quad (3)$$

The calculation of the parameter estimation vector is carried out using the method of least squares according to the formula (4):

$$\vartheta = \left[X^T * X\right]^{-1} * \left(X^T * Y\right), \quad (4)$$

$$X = \begin{bmatrix} x_1^2 & y_1^2 & z_1^2 \\ x_2^2 & y_2^2 & z_2^2 \\ \vdots & \vdots & \vdots \\ x_n^2 & y_n^2 & z_n^2 \end{bmatrix}, \quad Y = \begin{bmatrix} Y_1 \\ Y_2 \\ \vdots \\ Y_n \end{bmatrix}, \quad \vartheta = \begin{bmatrix} 1/a^2 \\ 1/b^2 \\ 1/c^2 \end{bmatrix}$$

where are: *X* – explanatory variables matrix; *Y* – experimental results vector; ϑ – parameters estimates vector.

Results of simulation modeling are presented in the paper [17]. The error is 2 – 3%, which provides the necessary accuracy.

4 Calibration features of the measuring unit

The complexity of calibration of the measuring unit, consisting of the MEMS accelerometer and magnetometer, makes it necessary to set the sensor axes to a level with the horizon, which in practice is almost impossible. Also, in fact, the exact direction of the magnetic field vector is not known in connection with the features of the Earth's magnetic field [20]. As a result, the axes of the accelerometer and the magnetometer are mismatched, and the vector of the Earth's magnetic field does not coincide with the axis of the magnetometer "Fig. 1".

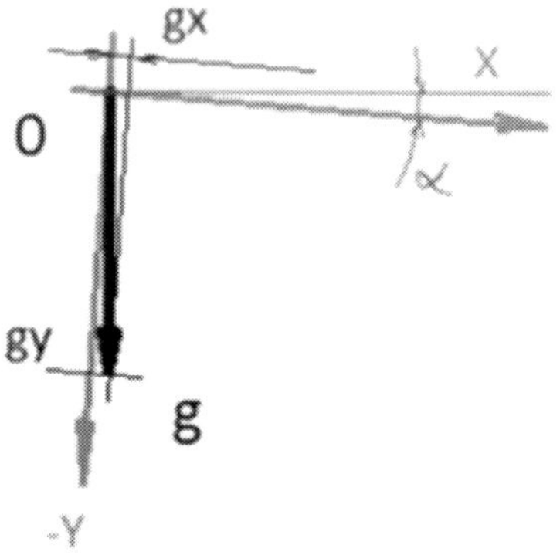

Fig. 1. Location of the sensor axes

The solution of this problem can be the initial calibration of the accelerometer and the further centering of the magnetometer data. In this case, the actual direction of the magnetic field vector is calculated depending on the exact coordinates of the measurement site. Compensating the angle between the true vector of the Earth's magnetic field and the axis of the magnetometer, we can bring the data cloud of the sensor to the desired value.

5 Experimental results

Experimental measurements were carried out in the operating temperature range using a three-axis accelerometer sensor, a magnetometer and a gyro MPU-9265.

In the case of the miniature microcircuit MPU-9265 two crystals are united: on one placed three-axis gyroscope and a three-axis sensor, and on the second - a three-axis magnetometer AK8963.

Data from the sensors are digitized by 16-bit ADCs, processed and transmitted to the external microcontroller via the I2C / SPI bus. The GY-9250 module is characterized by low power consumption, low cost and high performance.

MPU-9265 work in robotics, in 3D mouse designs, in gesture recognition devices, in navigation devices, etc.

Figures 2, 3 show the measurement results.

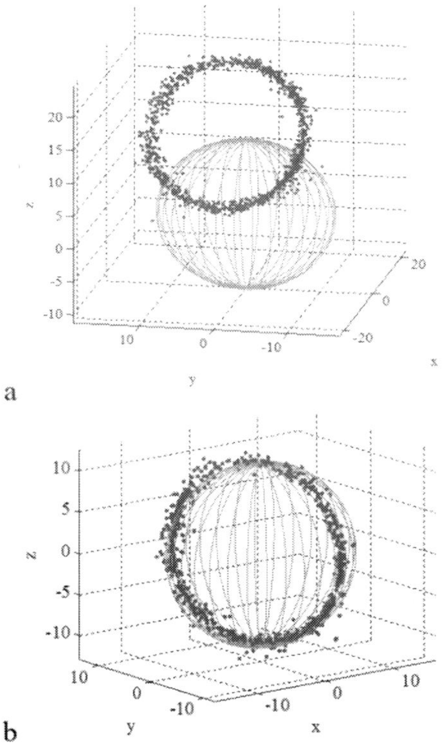

Fig. 2. Data cloud of the MEMS accelerometer on the sensitivity ellipsoid: a – uncalibrated; b – calibrated

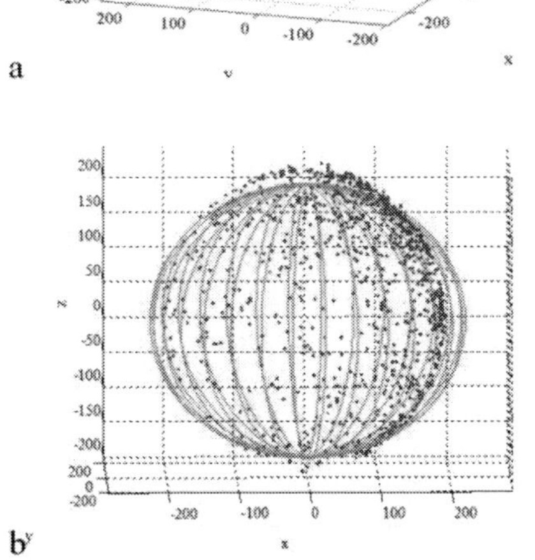

Fig. 3. Data cloud of the MEMS magnetometer on the sensitivity ellipsoid: a – uncalibrated; b – calibrated

The root-mean-square deviation along the axes is less than 1%, which satisfies the necessary requirements for accuracy of operation.

6 Conclusions

The proposed method allows you to completely get rid of the error of zero displacement, and also compensate for the non-unit scale of the sensor axes and the error in the angular orientation of the magnetometer. The main advantage of the method of calculating the estimation vector with the help of OLS is the ability to perform calculations without reference to the coordinate system. This is more important when calibrating the magnetometer, when you can't take into account the angle of inclination and perform calibration independently, without a gyroscope or an accelerometer. This allows the use of an algorithm for sensors already installed on the equipment.

References

1. Yu.V. Bolotin, V.P. Golikov, S.V. Larionov, A.V. Trebukhov, *Gyroscopy and navigation*, 3 (2008)
2. A.V. Tyuvin, Trudy MAI, № 71, (2013) [in Russian]
3. I.S. Khlopov, *Radar and radio navigation systems*, 2 (2015)
4. S.L. Lakoza, V.V. Meleshko, *Radiooptics*, 1 (2015)
5. P.A. Ivanov. *Development and research of test methods for micromechanical inertial modules*, Cand. tech. sciences. Diss., St. Petersburg, 146 (2011) [in Russian]
6. J.C. Hung, J.R. Thacher, H.V. White. *Proceedings of the IEEE Aerospace and Electronics Conference NAECON*, 1 (1989)
7. W. A. Feng. *Journal of Information & Computational Science*, 1551 – 1558 (2013)
8. A.V. Sobetsky, *Proceedings of 15th International Scientific-Practical Conference of Students, Post-graduates and Young Scientists "Youth and Modern Information Technology"*, 345-346, **1** (2016) available at: http://earchive.tpu.ru/handle/11683/36934
9. V.Benevicius, V. Ostasevicius, R. Gaidys, Sensors 13(9) (2013)
10. I. Skog, P. Handel. *XVIIIIMEKO World Congress, Metrology for a Sustainable Development* (Rio de Janeiro, Brazil, 2006)
11. Tao Wang Ba, *Dynamic method of investigating the sinuses of the triad of micromechanical accelerometers*, Cand. tech. sciences. Diss., St. Petersburg (2015) [in Russian]
12. H. N. Myint, Science and Education of Bauman MSTU (2014) available at: http://technomag.bmstu.ru/doc/691573.html

13. P. Veclak, P. Ripka, A. Platil et al., *Sensors and Actuators A: Physical*, 129 (2006)
14. Calibrating an eCompass in the Presence of Hard- and Soft-Iron Interference (2015)
15. VN-100 Hard and Soft Iron Calibration, Overview and Calibration Method (2012)
16. MEMS Industry Group (2011), available at: https://memsblog.wordpress.com/2011/03/22/hard-and-soft-iron-magnetic-compensation-explained/
17. A.D. Lukyanov, K.I. Goryanina, D.T. Fam, *Electronics and electrical engineering*, 2 (Russia, 2016)
18. J. Ganssle, DigiKey Electronics (2012), available at: http://www.digikey.com/en/articles/techzone/2012/jul/a-designers-guide-to-mems-sensors
19. Correlation, covariance and deviation (2015), available at: https://habrahabr.ru/post/263907/
20. Bol'shoy informatsionnyy arkhiv, available at: http://big-archive.ru/geography/the_role_of_water_in_the_formation_of_the_Earth_crust/66.php

Nonlinear orientation effect in liquid crystals to create a linear displacement sensor

Olga Denisova[1,*], *Roman* Abramishvili[2]

[1]Department of Physics, Ufa State Petroleum Technological University, Kosmonavtov Str.,1, Ufa, 450062, Republic of Bashkortostan, Russia
[2]Department of Management and Service in Technical Systems, Ufa State Petroleum Technological University, Kosmonavtov Str.,1, Ufa, 450062, Republic of Bashkortostan, Russia

Abstract. The results of experimental studies of thin layers of nematic liquid crystal under the influence of low frequency periodic oscillation (frequency about 100 Hz) are presented in this work. Before reaching the critical impact of the liquid crystal director varies symmetrically with respect to the normal to the cell. Further, when the oscillation amplitude is 0,8 microns, the liquid crystal molecules tilt relative to the normal to the cell while continuing to oscillate. The results of this research can be used in the development of liquid crystal sensors for diagnostics of mechanical displacements.

1 Introduction

Despite the fact that liquid crystals were discovered more than 100 years ago, they still cause great interest in the scientific community. Liquid crystals (LC) are a state of matter intermediate between a liquid and a solid. This state has a many ranges of practical applications. In the words of P. de Gennes, "Liquid crystals are beautiful and mysterious. I admire both". First of all, the liquid crystals have accelerated the revolution in the art of visual presentation devices (displays), and it is of interest to those involved in basic research. The second, the liquid - crystalline state inherent to many biologically active systems, including the human body, and therefore is one of the most important areas of biological research. But, unfortunately, until now, many of the fundamental problems remain unresolved.

Nematics and cholesterics are very sensitive to external fields. The first effects (due to the influence of the magnetic field) were found in the 30's and 40's of last century, by a group of Soviet scientists headed by V.K. Frediriks and V.N. Tsvetkov. Recently, interest increased to liquid crystals as nanomaterials. The size of the liquid crystal molecules is from a thickness of 5 Å and 30 Å in length to 200 Å in thickness and a length of 3000 Å, for example, tobacco mosaic virus. Biological membranes - a thin (~ 80 Å) from the leaves of lipids and proteins. They play a key role in many life processes, but their structure is little known. Most physical experimentation (for example, NMR) may not be carried on a separate membrane because it contains too little substance. However, one can create a model system of water and lipids or even of lipid, protein and water, having a lamellar (layered) structure. It is assumed that each individual layer is the analogue membrane.

The aim of the present scientific work was to the discover and investigate the stability of orientation, which is observed in the mechanical vibrations. The effect has a threshold of occurrence. It occurs due to violation of the spatial symmetry of the LC - system. LC procedure affects many of the physical properties of matter, and therefore the liquid crystals can be investigated by various experimental methods: ultrasonic, nuclear magnetic resonance, electron paramagnetic resonance, methods of linear and non-linear optics. A new method has been proposed and developed for the detection and experimental study of the effects mentioned above.

The influences of external factors (acoustic, electric, magnetic, temperature fields) to the liquid crystals accompany by a variety of effects and structural changes in a wide spectral range $10^{-1} \leq \omega \leq 10^{8}$ Hz. According to numerous visual observations, there is three stages of changing the structure of oriented sample: homogeneous, spatially periodic and non-uniform deformation.

A comparison of the characteristics of the different acousto-optic effects in liquid crystals in the presence of a temperature gradient, electric or magnetic fields, and without them, gives reason to conclude that the basis of the effects caused by the periodic shear deformation is the same physical mechanism.

Obviously, the presence of vibrations is characteristic of mechanisms containing moving elements. If the permissible readings are exceeded, an accident or equipment failure can occur [1-3]. Recently, non-destructive testing methods have been actively introduced, which provide timely detection of problems in equipment. Vibration is one of the most accessible and giving the necessary information parameters, both for a

* Corresponding author: denisovaolga@bk.ru

comprehensive assessment of the technical condition of equipment, and for individual units and aggregates [4 - 7]. Timely detection of technical problems, including vibration parameters, reduces the costs of repair and maintenance of equipment. They make it possible to solve such technical and economic problems as monitoring the current state of units and assemblies, preventing accidents and failures, as a consequence of preventive repairs while reducing indicators, bringing equipment to repair in accordance with its technical condition, which will increase the inter-repair interval.

To solve similar problems in some branches of science and technology, increased sensitivity and speed are required, and this requires new scientific and technical solutions.

Now there is a large number of instruments and systems for determining vibration parameters, which are based on various physical effects and phenomena [8-20].

The authors propose to create a highly effective system for determining the vibration parameters in nodes and mechanisms based on the acousto-optic effect in liquid crystals of the nematic type described in the article.

The development of this system would give an increase in sensitivity (up to 0,1 μm) and an increase in speed due to a pronounced orientation effect in liquid crystals (the relaxation time is 10^{-6} s).

2 Method

The experimental data and theoretical development for the high frequency range (1 MHz < ω < 100 MHz) are mainly reflected in the available scientific papers. The low frequency range (20 Hz < ω < 20 kHz) external influences are rather poorly investigated. Thus, the current study is the instabilities that occur when exposed to the liquid crystal of low-frequency periodic shift and their practical use.

Rayleigh (Strutt) John William, L.D. Landau and E.M. Lifschitz (the course of theoretical Physics [6]) considered the motion, which occurs in a viscous fluid with fluctuations immersed in it of solids, which has a number of unique features. The quickly damped viscous shear wave occurs in an isotropic liquid under oscillatory motion with frequency ω and amplitude a, which penetrates to a depth δ.

To a nematic liquid crystal (NLC) with an initial homeotropic orientation of the director, the oscillating body performs periodic oscillations with a frequency ω and a maximum amplitude a. In the linear approximation of the equations of motion of the director n and the molecular centers of inertia can be written as [7]:

$$K_{33}\frac{\partial^2\theta}{\partial z^2}-\eta_4\frac{\partial\theta}{\partial t}+\eta_1\frac{\partial\upsilon_x}{\partial z}=0,$$

$$\rho\frac{\partial\upsilon_x}{\partial t}+\eta_1\frac{\partial^2\theta}{\partial z\partial t}-\eta_3\frac{\partial^2\upsilon_x}{\partial z^2}=0, \qquad (1)$$

where η_1, η_3, η_4 - coefficient of viscosity.

The solution can be written in a general form:

$$\upsilon\approx\upsilon_o\exp\left[-\frac{\sqrt{2}}{2}q(z-\delta)\right]\cos\left[\frac{\sqrt{2}}{2}q(z-\delta)\right]\cos\omega t, \qquad (2)$$

$$\theta\approx\frac{\rho\upsilon_o}{\eta_1 q}\exp\left[-\frac{\sqrt{2}}{2}q(z-\delta)\right]\sin\left[\frac{\sqrt{2}}{2}q(z-\delta)\right]\cos\omega t,$$

$$\sigma_{zy}\approx(\alpha_4+\alpha_6)\frac{\partial\upsilon_y}{\partial z}+\alpha_3\frac{\partial\theta}{\partial t},$$

at $\theta|_{z=0}=0$ $\quad -<\sigma_{zy}u>\approx\frac{u_o^2}{2}\sqrt{\frac{\omega(\alpha_4+\alpha_6)\rho}{2}}$.

If the boundary conditions for θ weak, everything is determined by the condition of rapid energy dissipation. It is determined by the thermal conductivity $\chi_\perp << \chi_\parallel$, and that is why in the bed set the transition from n_z to n_x, and whose laws will be examined below.

The movement of the oscillating body (movable plate) in the NLC, leading to heating, will change the order parameter to the threshold of thermal convection. It will lead to instabilities of the liquid crystal orientation.

The object of study of this scientific work was a nematic liquid crystal n - type - methoxybenzylidene - n - butylaniline with a temperature range of mesophase existence of 18 - 42 °C. The main method of study was the method of interferometry birefringence based on a polarization microscope [8, 9]. The cell consists of a double sandwich (a massive two glass plates, where between was placed a third glass plate thickness of the cover glass, less than the distance between the massive plates). The record of the cover glass was free to oscillate in its (horizontal) plane. To create a homeotropic orientation of the LC molecules, chromium metal was sputtered on the substrate.

For registration and initial processing of signals studied experimental setup was assembled, where the basic elements of the recording were a polarization optical microscope Amplival Pol U console and spectrophotometric. To create an acoustic contact, membrane of the vibrator was connected by a waveguide. Measuring of the amplitude of the vibrations of thin cell substrate is held by inductive method [9].

According to the experimental procedure of sample preparation and introduction of the acoustic wave in the sample, we have a two layer system on the LC - film thickness h_c. Effect of acoustic wave on such a system gives rise to birefringence. Moreover, the polarizer and analyzer are crossed and form 45° with the direction of the shift. Constant intensity I_c as well as alternating signal components $I(t)$ in time, having a frequency range of harmonic excitation ω are recorded.

For the intensity of the optical signal, according to the Malus law:

DOI: 10.1051/matecconf/201713202008

$$I = I_o\{[\sin^2\frac{\pi hb}{\lambda}<\sin^2\theta_c>+\frac{\pi hb}{\lambda}[2<\sin\theta_c\theta_t>+$$
$$+<\theta_t^2>]\sin 2\frac{\pi hb}{\lambda}<\sin^2\theta_c>+\left[\frac{\pi hb}{\lambda}\right]^2[2<\sin\theta_c\theta_t>+$$
$$+<\theta_t^2>]^2\cdot\cos^2\frac{\pi hb}{\lambda}<\sin^2\theta_c>\}, \qquad (3)$$

where λ is the length of the incident light wave, the constant $b = 0.23$, θ_c is the constant of the component of the angle of inclination of the molecules of the liquid crystal (director), θ_t is the variable component of the slope angle of the LC molecules.
Components of the optical signal:

$$I_c = I_o\{\sin^2\frac{\pi hb}{\lambda}<\sin^2\theta_c>+\frac{\pi hb}{\lambda}\frac{\theta_o^2}{2}\sin 2\frac{\pi hb}{\lambda}\cdot$$
$$\cdot<\sin^2\theta_c>+2\left[\frac{\pi hb}{\lambda}\right]^2<\sin\theta_c\theta_o>^2\cos^2\frac{\pi hb}{\lambda}\cdot$$
$$\cdot<\sin^2\theta_c>+\frac{3}{8}\left[\frac{\pi hb}{\lambda}\right]^2(\theta_o^2)^2\cos^2\frac{\pi hb}{\lambda}<\sin^2\theta_c>\}, \qquad (4)$$

$$I_{1\omega} = I_o\{2\frac{\pi hb}{\lambda}<\sin\theta_c\theta_o>\sin 2\frac{\pi hb}{\lambda}<\sin^2\theta_c>+3\left[\frac{\pi hb}{\lambda}\right]^2\cdot$$
$$\cdot<\sin\theta_c\theta_o>\theta_o^2\cos^2\frac{\pi hb}{\lambda}<\sin^2\theta_c>\}\sin\omega t, \qquad (5)$$

$$I_{2\omega} = I_o\{-\frac{\pi hb}{\lambda}\frac{\theta_o^2}{2}\sin 2\frac{\pi hb}{\lambda}<\sin^2\theta_c>-2\left[\frac{\pi hb}{\lambda}\right]^2\cdot$$
$$\cdot<\sin\theta_c\theta_o>^2\cos^2\frac{\pi hb}{\lambda}<\sin^2\theta_c>-\left[\frac{\pi hb}{\lambda}\frac{\theta_o^2}{2}\right]^2\cdot$$
$$\cdot\cos^2\frac{\pi hb}{\lambda}<\sin^2\theta_c>\}\cos 2\omega t. \qquad (6)$$

Indignation weakly damps on the thickness of the cell, and captures all of the liquid crystal layers in this frequency range.

According to formula (5), the signal recorded at the excitation frequency ω, occurs only when there is a stationary tilt θ_s director and the normal to the surface, i.e. in violation of symmetry LC - the system relative to the normal to the layer. The appearance of the AC signal on the disturbance frequency determines a value of the threshold of an acoustic wave amplitude in the NLC, leading to the formation of a stationary director tilt in one direction or another deflection of the vibrating plate.

3 Result and discussion

According to the experimental data, the behaviors of the variable optical component signal I_c (a) and the second harmonic $I_{2\omega}(a)$ (Figure 1), the mean square amplitude of the sine components constant director tilt $<\sin^2\theta_c>(a)$ and the variable part $<\theta_o^2>(a)$ are calculated (Figure 2).

a)

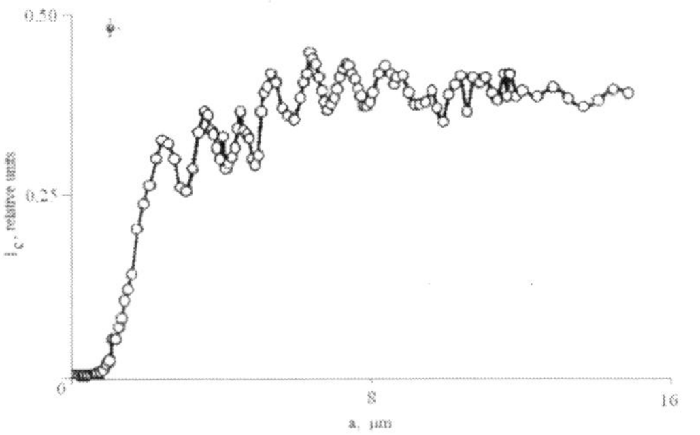

b)

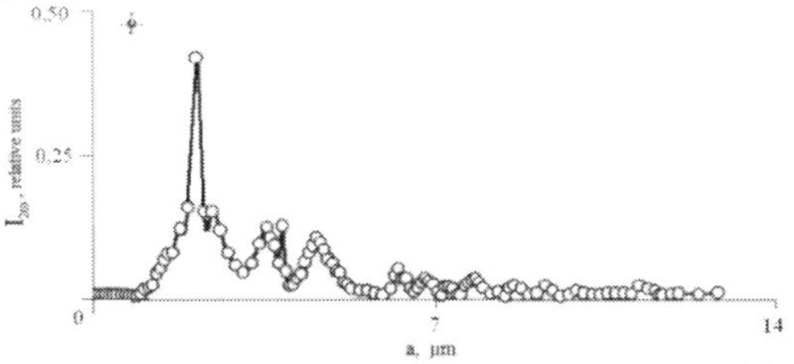

Fig. 1. Amplitude dependences of a) constant component of the optical signal I_c, b) second harmonic optical signal $I_{2\omega}$.

a)

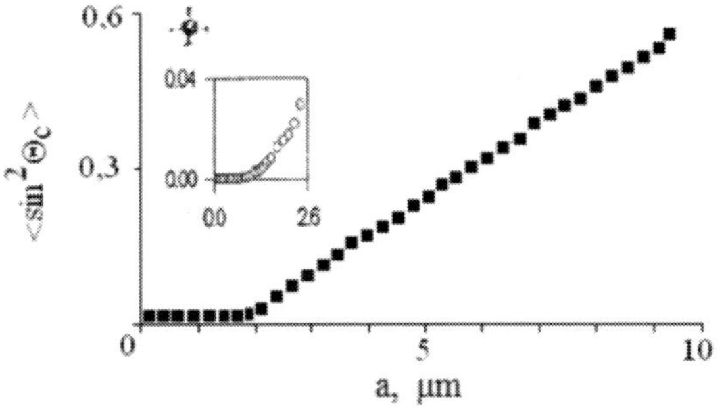

b)

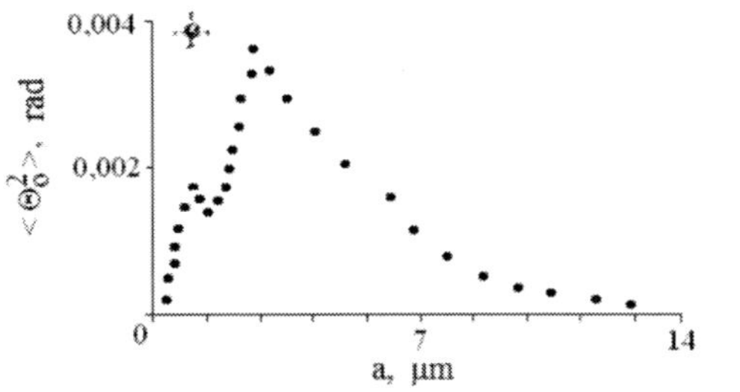

Fig. 2. Amplitude dependences of a) mean square of the sine of the angle of inclination of the stationary $<\sin^2\theta_c>$, b) variable part $<\theta_o^2>$ of director of the tilt angle.

As we see (Figure 2), the function $<\sin^2\theta_c>(a)$ increases monotonically, tending to the maximum value - to 1, and $<\theta_o^2>(a)$ reaches a maximum at a certain value of amplitude and, then begins to subside. The increase of the stationary tilt of the director leads to a reduction of the dynamic susceptibility LC - layer.

The question of the definition of this type of orientation approach in shear wave field is reduced to the study of the frequency dependence of the threshold value an education inpatient director tilt. If $a_n(\omega)$ = const, then we can talk about the acoustic analogue of the Fredericks transition [7]. This behavior is confirmed by the data [10]. The observed pattern of behavior homeotropic NLC layer at periodic low-frequency shifts were predicted in theoretical studies [11, 12].

Orientation instability of this type leads to permanent changes in time orientation of the molecules. In the case where the wavelength of a viscous layer of considerably greater thickness, it is proportional to the threshold of instability LC thickness layer and does not depend on the strain rate. We have identified a threshold determined to depending delete on the temperature and the thickness of the LC layer. We can say that the threshold forming the stationary director tilt in the temperature range of $18<T<45$°C and a thickness of $20 < h_c < 125$ microns remains constant ($a_n \approx 0{,}8$ microns) (Figure 3) [13 - 20].

We are invited a new method for controlling the vibrations of the nodes of mechanisms and aggregates. The uniqueness of the method lies in the application of the acousto-optic effect in liquid crystals (LC). The LCD cell serves as an indicator, the light transmission, which depends on the amplitude of the oscillations. Its essence lies in the fact that a monochromatic beam of light, passing through a polarizer, falls on a cell filled with a liquid crystal.

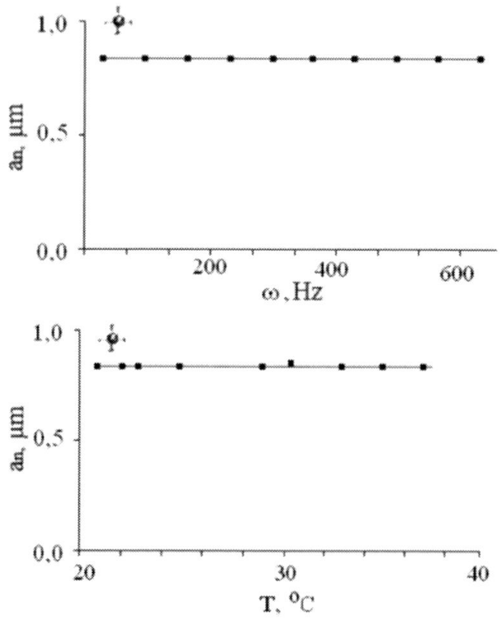

a) b)

Fig. 3. Dependence of the threshold shift amplitude of an effect the formation of a stationary angle of the director of a) the frequency of exposure, b) temperature.

The cell is designed in such a way that the liquid crystal is between two massive transparent plates separated by spacers. In addition, there is one more thin plate between the plates that can make shear vibrations in the plane of the cell. This plate is connected by a waveguide to a mechanism, the vibration on which it is necessary to control. During its operation, oscillations are transmitted along the waveguide to a moving plate. From it, shear perturbations propagate in a liquid crystal. When the threshold value of the vibration amplitude is reached, initially homeotropically oriented liquid crystal molecules deviate from the normal to the cell. As a result, the light beam incident on the cell passes and is fixed by the photodetector. Further, the processed signal is output to the operator's monitor in a form convenient for analysis.

It should be noted that the occurrence of this acousto-optic effect is of a threshold nature. The threshold value of the amplitude of the action does not depend on the frequency, temperature, thickness of the LC layer. Dependence of the deviation angle of the director on the amplitude of vibration has a linear relationship, which simplifies the calibration of the sensor. The use of a cell on liquid crystals as an indicator will increase the accuracy and sensitivity of the vibration monitoring method. Since liquid crystals are systems more sensitive than solid crystals. The relaxation time of the orientational effects in liquid crystals is $\sim 10^{-6}$ s.

4 Conclusion

In summary, the work carried out showed the spread of transverse vibration of a viscoelastic wave in NLC under the influence of low-frequency periodic shift. It leads to instability of the orientation of nematic liquid crystals. The effect of the formation of the stationary tilt of the director in a homeotropically oriented layers of nematic liquid crystals. It is shown that the amplitude of the critical independence of the orientation of the transition from the frequency shift, the thickness of the crystal temperature. The results presented research can be used in the development of acousto-optic sensors for diagnostics of mechanical vibrations. From the practical point of view, this method is of interest for use in the oil and gas industry and the pulp and aircraft industry.

Reference

1. V.N. Chelomei (prev.), M.D. Genkina (Ed.) *Vibrations in technology: Handbook. In the 6th edition of Ed. Advice*, **5** (Mashinostroenie, Moscow, 1981)
2. A.I. Yashura. *The system of maintenance and repair of general industrial equipment: a directory* (Izd-vo NTs ENAS, Moscow, 2006)
3. F.M. Dimentberg. *Flexural vibrations of rotating shafts* (Izd. Akad. Nauk SSSR, Moscow, 1959).
4. V.P. Sivakov, V.N. Mikushina, E.N. Stepanova. Modern problems of science and education, **6** (2014) [in Russian]
5. A.I. Morenko, D.V. Tarakanov, Sensors and systems, **3** (2001)

6. L.D. Landau, E.M. Lifshitz. *Theoretical physics*, **6** (Gos.isd. fis-mat.lit., Moscow, 1986)
7. De Gennes P., *Physics of Liquid Crystals* (Mir, Moscow, 1977)
8. O.A. Denisova, Electrical and information complex and systems, **9** (2), 107-113 (2013)
9. O.A. Denisova, The nonequilibrium structural changes of liquid crystals in electric fields and currents acoustic. Scientific publication (UGAES, Ufa, 2012)
10. G.N. Belova, E.I. Remizova, Acoustic magazine, **3**, 289-295 (1985)
11. I.A. Chaban, Acoustic magazine, **1**, 132-134 (1985)
12. I.A. Chaban, Acoustic magazine, **24**, 260-270 (1978)
13. O.A. Denisova. Electronic scientific journal "Oil and gas business", **5** (2011)
14. O.A. Denisova, O.A. Skaldin, Letters on materials, **6** (3), 168-172 (2016)
15. O.A. Baimakova , O.A. Skaldin, A.N. Chuvyrov, Mol. Cryst. Liq. Cryst., **265** (1995)
16. O.A. Denisova, A.N. Chuvyrov, Liq. Cryst. and their Appl., **1** (2012)
17. O.A. Denisova, A.N. Chuvyrov. Liq. Cryst. and their Appl., **2** (2011)
18. O.A. Denisova, A.N. Chuvyrov. Liq. Cryst. and their Appl., **3** (2011)
19. V.V. Leonov, O.A. Denisova, Electrotechnical and Information Systems and Systems, **2** (2015)
20. O.A. Denisova, A.N. Chuvyrov, Liq. Cryst. and their Appl., **2** (44) (2013)

Industrial Control System of Roll Stamping

Aleksandr Pilipenko [1,*], *Anastasia* Pilipenko [1], and *Olga* Pilipenko [1]

[1]Federal State Budgetary Educational Institution of Higher Education «Orel State University named after I.S. Turgenev», Department of Automated Control Systems and Cybernetics, Orel, Russia

Abstract. The article describes approaches to the automation of roll stamping. Authors deeply cover aspects of design of the hardware and PID controller. In the article authors represent algorithms of the existing automated system and examples of use of machine vision for solving of application-oriented tasks.

1 Introduction

Now a priority task for the industrial enterprises is the release of competitive production with the minimum expenses. Production of products and details by methods of form changing operations is characterized by profitability of consumption of metal, an opportunity to receive details of high precision with high technical characteristics. Rather new development of these methods is the technology of plastic processing of metals with complex local loading of a billet which received the name "roll stamping" [1,2,3]. Roll stamping combines in a single technological process operations of volume stamping and local deforming by not drive or drive rolls. The technique of roll stamping is applied in practice to receive solid and hollow details, thin-walled products of small sizes with high accuracy and quality of manufacture with the technological force much smaller, than for traditional methods of volume stamping. Roll stamping found its classical application in the field of production of axisymmetric cylindrical details of high accuracy including multistage ones.

The diagram of process of roll stamping is carried out as follows (Figure 1): forming of billet 1 is made in the roller matrix including the rotating stop 2 and form-building rollers 5, punch 3 and besides punch 3 and stop 2 rotate together with billet 1 with angular speed ω1, and rotation of rollers 5 is carried out with angular speed ω2, connected with ω1 through a ratio of diameters of rollers 5 and detail, and rotation of billet 1 is carried out at the expense of forces of contact friction with rollers 5. On a line item of roll stamping two operations are executed at the same time. The first one is piercing or forming of billet 1 by punch 3 with force P, the second one is rolling of billet 1 on an outside circuit by rollers 5. Punch 3 makes the working course down to complete shaping of billet 1.Use a two-column format, and set the spacing between the columns at 8 mm. Do not add any page numbers.

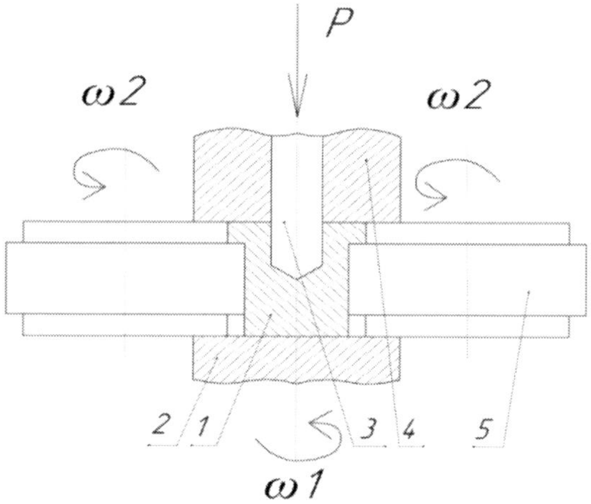

Fig. 1. Scheme of process of roll stamping.

The classical equipment for roll stamping is the hydraulic press and die tooling. A large number of scientific and technical works is devoted to automation of hydraulic press. However, automated process control systems of roll stamping are not practically mentioned anywhere. It should be taken into account that technological processes of roll stamping last from several hours to several days, demand obligatory participation of the operator for control of accuracy, speed of rotation of rolls or a billet, subcontrol. [3] Taking all this into account, it is possible to assume that development of an automated control system for non-stationary technological process with intellectual algorithms of computer modeling and machine vision is a new relevant task.

2 Conception

In technological processes of roll stamping it is necessary not only to give to a product the required geometry and to create a certain microstructure of material, but also to provide stable repeatability of result

* Corresponding author: a@pilipenko.info

of operation of the equipment when unsteady, long duration and danger of technological processes of deformation. This problem is solved by the created modern system of the automated management of the lower level, capable to analyze big amounts of information from sensors in the confidant to real time mode. Use of machine vision for identification of reject during its emergence expands possibilities of system. The following requirements to the developed system of automated management are formulated [4,5]:

- processing of information from sensors, visualization of the processed information;
- creation of computer imitating model of process of roll stamping;
- analysis of operation of the equipment on the basis of comparison of information from sensors and from computer model;
- recommendations about selection of optimum working parameters;
- self-diagnostics;
- convenient display of all information for the operator;
- a possibility of work in the automatic mode;
- a possibility of the analysis of emergence of reject during processing.

2.1 Design

According to the created requirements the specification of the equipment of management system of technological process of forming, executed on the hydropress equipment was developed. And authors tried to make system most unified and suitable for any hydropress equipment therefore NICompactRIO was selected as a controller [6]. It is a worthy modular industrial controller with the modern icore processor having onboard necessary quantity of analog and digital inputs and outputs and also the buses Ethernet, RS-485 and RS-285, and working with the standard TCP/IP, Modbus and Can protocols. The panel of manual control set on a hydraulic press is connected to the industrial controller via the bus RS-485. Via the bus Ethernet remote system management through a Web server is provided. [7] Stepping motors with reducers for control of throttles and also the frequency transformers for management of engine pump and the equipment engine which rotates rolls are connected to digital outputs. Sensors of relocation, pressure, density of pressure fluid, etc. are connected to analog inputs of the device.

The generalized structure of management system is provided in figure 2. In an automatic mode of operation the PC controller sets starting values $x(t) = p0(t), x0(t), v0(t)$, influencing rotating speed of the engine of the hydraulic pump ω and the expenditure of the liquid passing through the throttle q. The sensors set on a hydraulic press, measuring parameters on the unit of a self-adjustment which, comparing the data obtained experimentally and by computer simulation calculates the adjusting coefficient k_{adj}., using the logical law of control: if $|p_{exp}| < 0{,}2 * |p_{mat}|$, then $k_{adj} = p_{exp} * k_1$; otherwise $k_{adj} = p_{exp} * k_2$; where $k_1 < k_2$. Regulator creates controlling influence $U(t)$ and transfers it to used controller.

The novelty of management system consists in the developed principle of combined control integrating the principle of control on a deviation and the logical non-linear law of control.

At the same time by the principle of control on a deviation the system is described by transfer function of open-ended system and the equation of closing: $x(t) = g(t) - y(t) Woc(p)$. The algorithm of system operation is in the aspiration to reduce an error $x(t)$ to zero.

One of advantages of such a diagram is that feedback leads to reduction of an error irrespective of factors having caused it (changes of parameters of an adjustable object or external conditions).

Non-linear laws of control can have other forms which are implemented by means of not functional but more or less difficult logic devices. We will call them logical non-linear algorithms. In a system logical non-linear algorithm is applied to save controlling impacts on an object (and also saving of power consumption for needs of control).

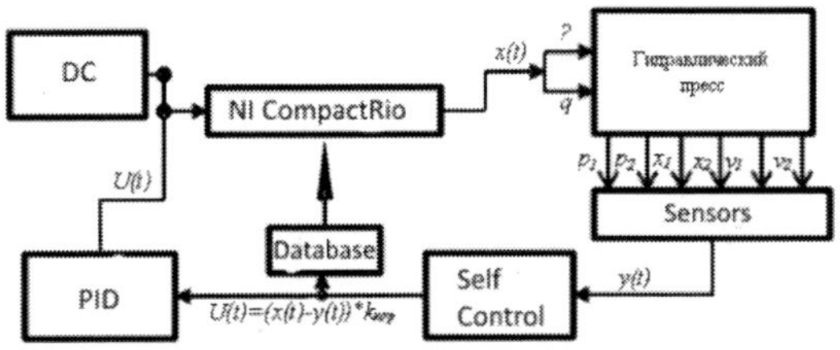

Fig. 2. The generalized structure of a control system.

2.2 Modelling and regulation

Subject to regulation in this system is the valve of a throttle and the frequency converter which regulates a hydraulic pump engine speed. It should be taken into account that there is a computer model which specifies desirable movement of a traverse depending on the current pressure [8, 9, 10, 11]. The condition of an object is defined by: 1) pressure; 2) movement. The operator model of a control system is presented in figure 3.

Having carried out synthesis of system of regulation by a throttle and system of regulation by frequency converters complex regulation of pressure in a general view is received:

$$Wop = \frac{k * e^{\tau_{BO} * P}}{T_{BO} * p + 1} + \frac{k_{пч-д}}{T_m p + 1} * \frac{k_н}{T_н p + 1}$$

$$Wop = \frac{k_{пч-д} k_н T_{BO} * p + k_{пч-д} k_н + k * e^{\tau_{BO} * P} T_m p^2 T_н p + k * e^{\tau_{BO} * P} T_m p + k * e^{\tau_{BO} * P} p T_н p + k * e^{\tau_{BO} * P}}{(T_{BO} * p + 1) * (T_m * p + 1) * (T_н * p + 1)}$$

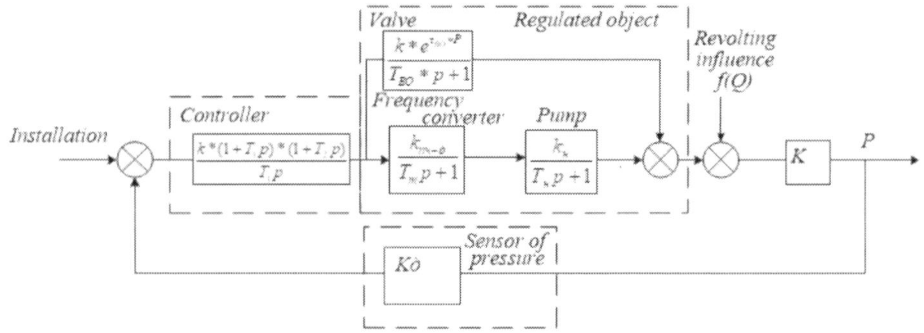

Fig. 3. Operator model of an automation object. A valve; a regulated object; the revolting influence; a controller; a setpoint; a frequency converter; a pump; a sensor of pressure.

So, we will admit: *Pressure* (*t*) – current pressure in the pressure head trunk (a present situation of time **t**), *Relocation* (*t*) – current relocation. Also we have a Step (*t*) – current step of a throttle, *Frequency* (*t*) – rotating speed of the engine of a hydraulic pump and *Pressure* (*t*+1) – new value of engine speed which depends on temperature. It is necessary to receive as a result *Step* (*t*+1) – new position of the throttle and *Frequency* (*t+1*) – new frequency of the engine:

Step (*t*+1) = *Function {Step (t), Pressure (t), Relocation (t), Pressure* (*t*+1)*}*

Frequency (*t*+1) = *Function {Step* (*t*+1)*, Frequency* (*t*)*, Pressure* (*t*)*, Relocation* (*t*)*, Pressure* (*t*+1)}

The first function gives the current value of a step which depends on all above-mentioned, and the second gives the current frequency rate.

It is necessary to know value of the current error (discrepancy), and it can be calculated as the difference of estimated pressures in the future and current ones:

Error (*t*) = *Pressure* (*t+1*) — *Pressure* (*t*)

In case pressure accurately corresponds to required one the error will be equal to zero. This error shows how strong influence shall be. The formula for the following step can be written as follows now:

Step (*t*+1) = *Function* {*Step* (*t*), *Error* (*t*)}

Now it is possible to put value of the previous step outside brackets:

Step (*t*+1) = *Step* (*t*) + *Function* {*Error* (*t*)}

On each step of regulator operation current position of the stepping motor is set. The stepping motor is operated by offsets therefore the current step is not so important. As a result it is possible to pass to such a record:

Change of Step (*t*+1) = *Function* {*Error* (*t*)}

It is a little more difficult as far as frequency is concerned since it does not make a sense to control directly two devices. As a result control of the frequency transformer (frequency change) begins only when throttle is in limit position, thus limits of control extend and accuracy increases. Function of determination of limit value of the stepping motor looks as follows:

Limit SM (*t*) = *the Step* (*t*+1) - *the Step* (*t*)

In case step accurately corresponds to required one, the limit of SM will be equal to zero.

Function for frequency looks as follows:

Frequency (*t+1*) = *Limit SM* (*t*) *Frequency* (*t*) + *Function* {*Error* (*t*) }

The developed mathematical apparatus is set in the system of computer simulation (figure 4).

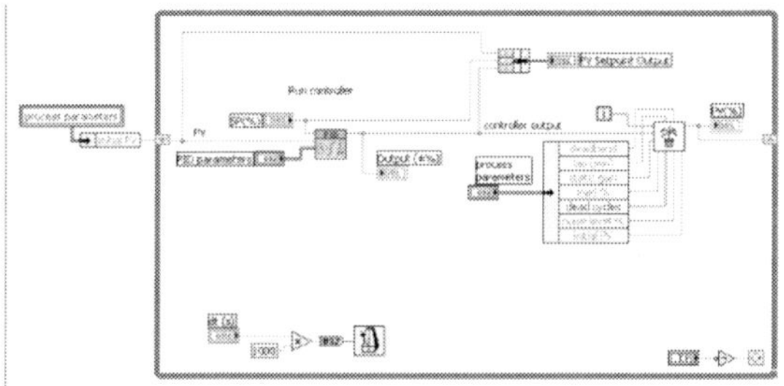

Fig. 4. The flowchart of system of regulation with use of mathematical model

At the same time, the PID controller used in the flowchart in figure 4 has its own flowchart shown in figure 5.

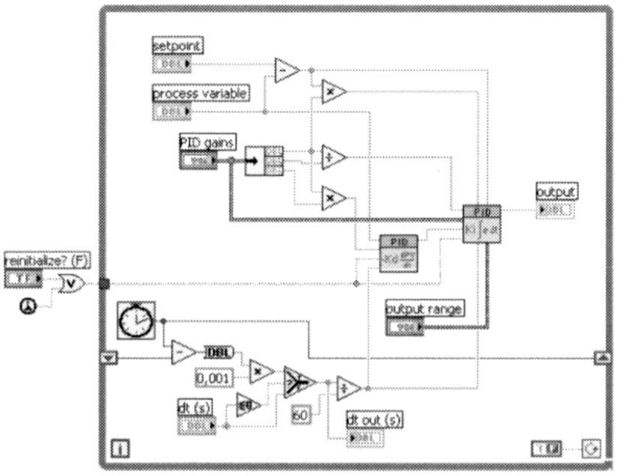

Fig. 5. Flowchart of the PID controller

2.3 Algorithmization and programming

The management system algorithm for the PC-controller is provided in figure 6. In it the following sequence of actions is executed:

1 – Beginning ; 2 – Input of the name of test, comment and N-quantity of the required cycles; 3 - Operation mode choice: Automatic control; 4 – regulator: the frequency transformer of the drive of a hydraulic pump and to move a traverse to the maximum position; 5 – regulation by throttle the frequency transformer, to move a traverse to the maximum position; 6 – Procedure of initialization of the data collecting device; 7 – Start of a subsystem of machine vision for the analysis of appearance of spoilage; 8 – Start of the subprogram of computer simulation and PID-regulation; 9 – 13 – Material data entry: L, d, h; 14 – Choice of material and its parameters; 15 – 19 – Type of material, density, diagram of resistance of deformation, etc.; 20 – Choice of parameters of equipment operation; 21-25 – Parameters of regulation, pressure, speed, etc.; 26 – Data collection from ADC [12]; 27 – Data filtering; 28 – Data handling and conversion to C; 29 – Solution of differential equations with the Runge–Kutta 2 method [13, 14]; 30 – Output of diagrams and data to the screen; 31 – Subprogram of comparing of data; 32 – i=N?; 33 – Do you want to change N of cycles and to continue work?; 34 – Detection of an error of system operation or defect of a detail. 35 – End of the program.

The code of the subprogram of detection of defects with use of machine vision [15, 16] is provided in figure 7.

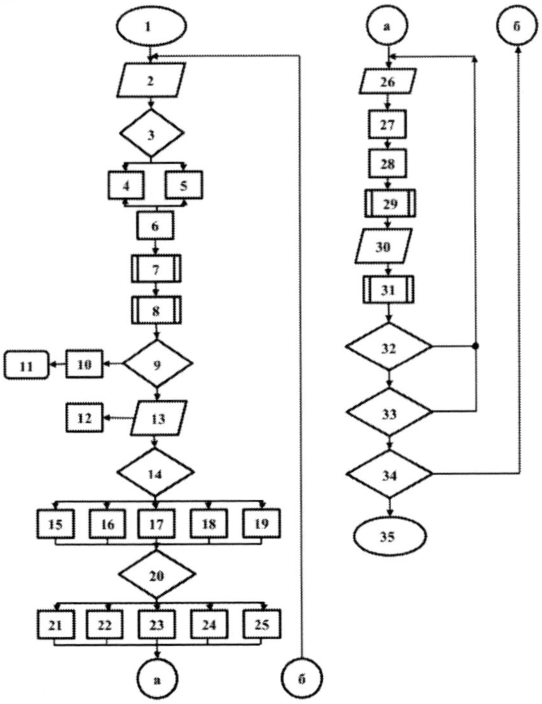

Fig. 6. An algorithm of the adaptive system of control for the PC controller

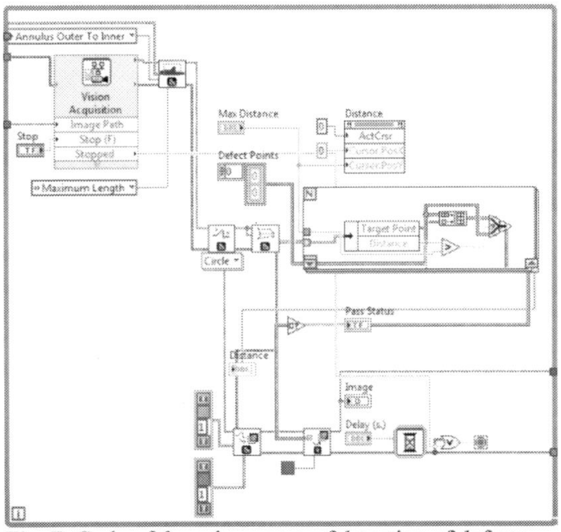

Fig. 7. Code of the subprogram of detection of defects

3 Conclusion

This work contains the solution of the task of development of an automated control system for the hydropress equipment on the basis of the adaptive algorithms constructed on mathematical models of dynamics of the hydropress equipment with the equipment for roll stamping. The created hardware and software system [17, 18] underwent approbation in the conditions of real manufacturing enterprise.

It is defined that achievement of stability of management system requires control on the PID controller [19, 20]. The proportional component in the PID controller works at the time of appearance of a mismatch. When traverse begins to move, power begins to fall down and when reaches the minimum point the device is switched off. The effect of influence is revealed with delay. An object is also influenced by the environment: traverse mass, compressibility of liquid and pipeline. For compensating of "external" influence the integral component is added to a circuit. As a result of such approach the integral becomes stable therefore value of the given power becomes constant. For compensating of influence of time delays between influence and response of system the differential component is added to the system.

References

1. S. Radchenko, Razrabotka novyh tekhnologicheskih processov valkovoj shtampovki tonkostennyh izdelij i metodov ih proektirovaniya, Thesis for a degree of Doctor of Technical sciences, 308 (2003) [in Russian]
2. V. Golenkov, S. Radchenko, D. Dorokhov, High technologies in mechanical engineering, no. 3, 31-37 (2011) [in Russian]
3. S. Teslenko, Casting and metallurgy, no. 3 (66) (2012) [in Russian]

4. A. Pilipenko, Information systems and technologies, 4/66,115-119 (2011) [in Russian]
5. A. Pilipenko, Fundamental and application-oriented problems of technique and technology, no. 6-2 (296), 47-52 (2012) [in Russian]
6. M. Rosol, A. Pilat, A. Turnau, *Computer Science and Information Technology (IMCSIT),* Proceedings of the 2010 International Multiconference on. – IEEE, 825-830 (2010)
7. J. Travis, J. Kring, *Lab VIEW for everyone: graphical programming made easy and fun* (Prentice-Hall, 2007).
8. A. Pilipenko, Certificate of registration of the computer program, No. 2011611940, 3.03.2011 (2011) [in Russian]
9. A. Pilipenko, Certificate of registration of the computer program No. 2011610002, 11.01.2011 (2011) [in Russian]
10. A. Pilipenko Certificate of registration of the computer program No. 2011614723, 16.06.2011 (2011) [in Russian]
11. A. Pilipenko Fundamental and application-oriented problems of technique and technology, No. 6-2 (296), 58-64 (2012) [in Russian]
12. D. Budanov, D. Morozov, M. Pilipko, Radio engineering and electronics, **62**, No. 4, 399-410 (2017) [in Russian]
13. E. Kamke *Spravochnik po obyknovennym differencial'nym uravneniyam* (Publishing house "Science", Principal edition of physical and mathematical literature, Moscow, 1971), p. 576. [in Russian]
14. A. Kim, S. Polnikov, Modern innovations, No. 8, 5-7. (2016) [in Russian]
15. A. Pilipenko, Industrial ICS and controllers, No. 5, 39-40 (2009) [in Russian]
16. A. Pilipenko, Industrial ICS and controllers, No. 4, 63-69 (2013) [in Russian]
17. Pilipenko A. Certificate of registration of the computer program No. 2011611941, 3.03.2011. (2011) [in Russian]
18. Pilipenko A. Certificate of registration of the computer program. No. 2011611939, 3.03.2011. (2011) [in Russian]
19. Alexandrov A., Palenov M. Automatic equipment and telemechanics, No. 2, 16-30. (2014) [in Russian]
20. Chertkov A., Tormashev D., Saburov S. Bulletin of the State university of sea and river fleet named after Admiral S.O. Makarov, No. 5 (27) (2014) [in Russian]

Automation of a dynamic system with computer simulation of heat losses

Aleksandr V. Pilipenko[1,*], *Sergey P. Petrov*[1], and *Olga V.* Pilipenko[1]

[1]Federal State Budgetary Educational Institution of Higher Education «Orel State University named after I.S. Turgenev», Department of Automated Control Systems and Cybernetics, Orel, Russia

Abstract. The article describes approaches to automation of a non-linear system, the stability of which essentially depends on the impact of external factors and requires compliance with high safety standards. The authors model the processes occurring in the system, substantiate the selection of the equipment. In the paper, the authors take into account (with accepted admissions) the heat losses in the pipes.

1 Introduction

Recently, heating of apartment buildings has been realized using stand-alone or roof-mounted gas boiler houses, the uninterrupted operation of which is ensured by the fulfillment of a number of requirements described in the normative documents [1,2] and the rules for the installation of electrical stations. One of the important requirements in the design of modern boiler houses is the lack of maintenance personnel. At the same time, an increasingly important role is played by a two-level automation system of the boiler house, which consists of a remote control system for boiler and local automation installed on boilers and control panels.

Two-level automation of important and dangerous objects can reduce labour costs and increase energy efficiency, but it does not improve reliability and safety of the system, which makes it difficult to transfer the system to automatic mode.

According to the Federal Target Program "Reducing Risks and Mitigating the Consequences of Natural Disasters in the Russian Federation until 2020," the reduction in the risks of emergencies of all types and scales and their negative consequences will be achieved through the implementation of the following key areas:

- systemic research and improvement of organizational management bases in the field of increasing safety and security of critical facilities against threats of anthropogenic nature;
- systemic studies aimed at creating scientific basis for determining the patterns of occurrence of emergencies and management of these risks;
- development of infrastructure for information situation analysis of emergency situations.

In this regard, the development of a three-level automated boiler management system with an auxiliary industrial safety management system is an urgent scientific and technical task.

2 Conception

The basic two-level system provides cheap automation of technological equipment and technological processes due to PID TPM regulators of OVEN company, additional technological devices of control, data collection, signaling and indication [3,4]. This connection allows manual controlling of each boiler separately, if necessary, but also programmatic controlling of all boilers from the control panels (local and remote), by connecting regulators, a modem, a dashboard and a monitor to an industrial controller [5].

Thus, the automation shields of hot-water boilers perform the following functions [6]:

- exclusion of pulsation during ignition of the boiler due to delay;
- control of gas burners;
- control of the recirculation pump;
- diagnostics of the equipment status of the switchboard and sensors;
- heating the boiler during the initial start-up;
- blocking the operation of the boiler in case of emergency high values of the sensors;
- keeping a log of boiler failures;
- filtration of operation of discrete sensors;
- issuance of alarm signals of the boiler to the general boiler board of automation.

The control panel of general boiler equipment provides the following functions [7,8]:

- PID control to maintain the set temperature of the heating circuit;
- calculation of the heating circuit settings by comparing the temperature sensor data and the temperature mode models;
- automatic and manual switching to night / day mode;
- automatic control of the pumps taking into account the worked time;

* Corresponding author: a@pilipenko.info

- prohibition of starting boilers with pumps disconnected;
- light and sound indication of emergency parameters of the boiler room;
- diagnostics of the state of the shield equipment and temperature sensors;
- issuing alarms to the controller's panel via GSM-communication.

The auxiliary control system (subsystem of the third level) of industrial safety is built on a more efficient NI CompactRIO controller and works in parallel with the main two-level system in real time, and also contains the software modules realizing [9]:

- computer models of the state of boilers (dangerous and non-hazardous);
- models of situations exceeding the level of security in both statics and dynamics;
- algorithms for processing sensor data for modeling;
- decision-making algorithms;
- algorithms for working with the knowledge base.

The auxiliary system performs the following functional:

– continuous automatic monitoring of the water temperature at the boiler inlet;
– continuous automatic monitoring of the water temperature at the outlet from the boiler;
– continuous automatic control of water pressure at the boiler inlet;
– continuous automatic control of water pressure at the boiler outlet;
– continuous automatic monitoring of water pressure at the output of charging, main pumps and hot water pumping stations;
– continuous automatic control of the operation of executive mechanisms and devices;
– continuous automatic control of the operation of regulators and PLC-100;
– continuous automatic monitoring of the concentration of carbon monoxide in the boiler room;
– continuous automatic monitoring of the concentration of carbon monoxide in the boiler room;
– continuous automatic control of the concentration of carbon dioxide;
– collection and backup of all data, including data passing through the rs485 buses connected to the OVEN PLC-100.

2.1 Designing

As described above, the system consists of three sub-systems, with standard requirements for reducing costs and increasing reliability. The lower level of automation is represented by a combination of the OVEN TPM 132 PID controller, two-channel OVEN TPM 202 meter, frequency converters, chokes/valves, sensors and the OVEN IP320 control panel [10, 11]. The general scheme for connecting devices to the cabinet is shown in Figure 1, where: 1-5 are temperature sensors; 6 and 7 are solenoid valves; 8 and 9 are frequency converters and motors; 10 and 11 are high-pressure sensors; RS is data bus line RS-485.

The second level of automation should provide local and remote dispatching, maintaining an archive, changing the regimes common to all boilers (summer/winter, day/night), etc. Thus, it was decided to use OVEN PLK-100, TPM 148, MT6070ip operator panel, data transmission module via SHMR-16U radio channel, analog input module MB110-8A. Moreover, during the commissioning of the automated system, such an algorithm for checking the data transmission module over the radio channel was developed, which not only shows the level of the signal, but also the passage of packets along the path:

- install a jumper RX-TX on the maintained ShMP-16;
- connect a main one to the computer via RS485 or RS232;
- in the PLConnect program, select "main", set the mode of the port, set the "echo mode" and the desired packet p-p (for your case, 60-100 bytes);
- start the exchange and see the quality of the packets passing (about 100% with good connection).

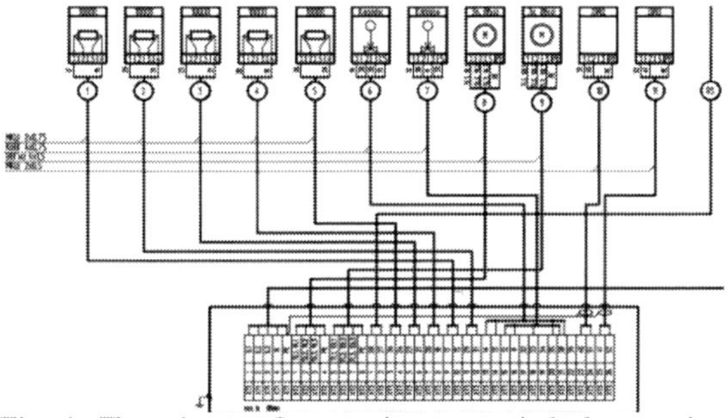

Fig. 1. The scheme of connecting external devices to the control panel.

The auxiliary system (a subsystem of the third level) of industrial safety requires computing power to operate in real or near real time mode, because it produces a solution of differential equations, which are models of the behavior of the system or regulatory object [12]. This system is built on a modular industrial controller NI CompactRIO with analog and digital input and output modules, which allows connecting all sensors and actuators (frequency converters and relays) directly.

2.2 Modelling

Types of models used in the algorithms of this automated system can be divided into two:

- mathematical models described by differential equations;
- behavioral models based on sets of settings, embedded in the structure of multiple choice Case.

The use of mathematical equations can be seen in the example of using the heat loss model in the system to calculate the correction coefficients of temperature measurements.

When performing the parametric synthesis of the control system of temperature regime (CSTR), an important step is to determine the boundary conditions within which the synthesis is performed [13, 16].

The basis of the synthesis is the description of the control system by one of the known operators, for example, by differential equations or transfer functions. In the latter case, an operator method is used, which makes it possible to determine the generalized transfer function of the control system (CS) and its characteristic equation [14, 15].

Developed in accordance with the new approach [15,17], the CSTR includes a number of subsystems, the main role of which belongs to the subsystem for controlling the final consumer of thermal energy, since it is here that the dynamics of the transient process is determined under the influence of disturbing factors, in particular, from the change in the outdoor air temperature [13, 18, 19]. The dynamic characteristics of the automation object are key when determining the boundary conditions and performing the parametric synthesis of the CSTR. The operator model of the automation object is shown in Figure 2.

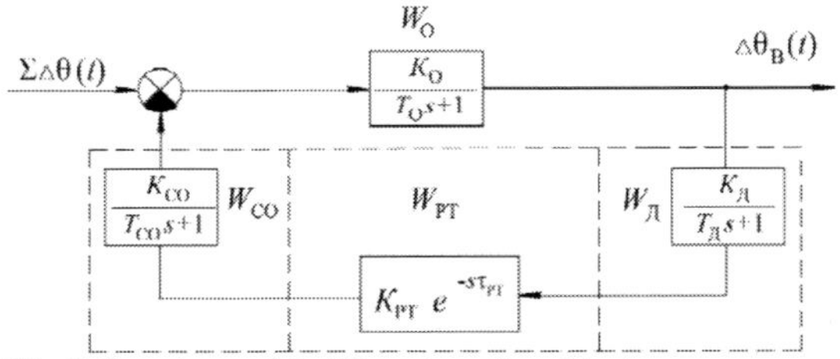

Fig. 2. Operator model of the automation object.

The expression of the transfer function is received according to the operator model of the automation object:

$$W(s)=\frac{\dfrac{K_O}{T_O\cdot s+1}}{1+\dfrac{K_O\cdot K_{\text{Д}}\cdot K_{PT}\cdot K_{CO}}{(T_O\cdot s+1)\cdot(T_{\text{Д}}\cdot s+1)\cdot(T_{CO}\cdot s+1)}e^{-s\tau_{PT}}} \qquad (1)$$

where K_O, K_{CO}, $K_{\text{Д}}$, K_{PT} are the coefficients of transmission of the automation object, the heating system (warming), sensors, temperature controller; T_O, T_{CO}, $T_{\text{Д}}$ are time constants of the automation object, heating system (warming), sensors; τ_{PT} is time delay temperature regulator; s is Laplace.

In the stability area, we consider stability at the boundary of aperiodicity, when all the roots of the characteristic equation are negative real ($\alpha < 0$, $\omega = 0$), and on the stability boundary, when the roots of the characteristic equation are complex with negative real part ($\omega > 0$, $\alpha = 0$) [15, 20].

Let us determine the dynamic characteristics of the object of automation, which ensure the stable operation of the CSTR on the stability boundary. The investigation of the control subsystem is performed for a system in which there is a delay associated with the time of movement of the regulating organ. To investigate the behavior of the CSTR with delay at the stability boundary, the characteristic equation is considered for a purely imaginary root value, i.e. when $s = j\omega$, where $0 \le \omega \le \infty$.

When the s operator is replaced by a complex variable $j\omega$, the dependences $K_O = f_5(T_O)$, $K_O = f_6(T_{CO})$, $K_O = f_7(T_{\text{Д}})$, $K_O = f_8(\tau_{PT})$ on the stability boundary are obtained [1]:

$$T_O = \frac{\left(T_{\text{Д}}T_{CO}\omega - \dfrac{1}{\omega}\right)tg\,\omega\tau_{PT}T_{\text{Д}} - T_{CO}}{1 - T_{\text{Д}}T_{CO}\omega^2 - \left(T_{\text{Д}} + T_{CO}\right)\omega\, tg\,\omega\tau_{PT}}; \qquad (2)$$

$$T_{CO} = \frac{\left(T_{\text{Д}}T_O\omega - \dfrac{1}{\omega}\right)tg\,\omega\tau_{PT}T_{\text{Д}} - T_O}{1 - T_{\text{Д}}T_O\omega^2 - \left(T_{\text{Д}} + T_O\right)\omega\, tg\,\omega\tau_{PT}}; \qquad (3)$$

$$T_{\text{Д}} = \frac{\left(T_O T_{O\Pi}\omega - \dfrac{1}{\omega}\right)tg\,\omega\tau_{PT}T_O - T_{O\Pi}}{1 - T_O T_{O\Pi}\omega^2 - \left(T_O + T_{O\Pi}\right)\omega\, tg\,\omega\tau_{PT}}; \qquad (4)$$

$$\tau_{PT} = \frac{1}{\omega}arcrc\frac{\left(T_O + T_{\text{Д}} + T_{CO}\right)\omega - T_O T_{\text{Д}}T_{CO}\cdot\omega^3}{\left(T_O T_{CO} + T_{\text{Д}}T_{CO} + T_{\text{Д}}T_O\right)\omega^2 - 1}. \qquad (5)$$

The notation for the coefficients entering into formulas (2 ... 5) is given in formula (1). The dependences $K_O = f_5(T_O)$, $K_O = f_6(T_{CO})$, $K_O = f_7(T_{\text{Д}})$, $K_O = f_8(\tau_{PT})$, obtained at the stability boundary, were investigated, and it was established that the maximum transmission coefficient of the CSTR at the stability boundary is much larger than at the aperiodicity boundary. Moreover, for $K_O = f_5(T_O)$ and $K_O = f_6(T_{CO})$, this excess is correspondingly 5 and 10 times, respectively. For $K_O = f_7(T_{\text{Д}})$ and $K_O = f_8(\tau_{PT})$, this excess is 30 and 10 times.

2.3 Algorithmization and programming

All three described subsystems work on algorithms of two types: an algorithm for keeping the minimum and maximum values of set points; an algorithm for keeping the difference in the results of computer simulation and the results of the real operation of the system within the given hysteresis. In the first case, 4 set points are set for each sensor, which determine the inclusion of the regulating bodies, and in the second case a sophisticated algorithm for analyzing deviations from the computer model based on trained neural networks is used. The MS SQL database is connected in the industrial safety management subsystem. The block diagram of the subprogram responsible for interaction with the database using the NI LabVIEW Data Base Connectivity Toolkit is shown in Figure 3.

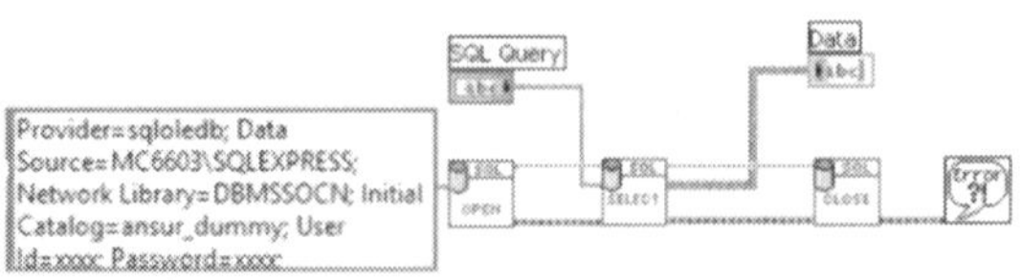

Fig. 3. Block diagram of the subprogramme responsible for interaction with the database.

The requirement for a system to work in close to real time causes problems with processing super-large data sets. In this connection, the following algorithms are used to process data from sensors: an algorithm for monitoring the smallest and largest values of the indicators; an algorithm for controlling the difference between the largest and the smallest values of indicators; an algorithm for controlling the arithmetic mean; an algorithm for controlling emissions.

The algorithm for controlling the smallest and largest values of indicators has the following form:

1. Obtain the M set of accounting objects of type t.
2. Select the accounting object from the set M.
3. For the selected accounting object, compare the current value of the indicator m with the minimum and maximum allowable values for this indicator (max and min).
4. If the value of the indicator m> max or m <min, then place the accounting object in the set P_n. Otherwise, place the accounting object in the set P_m.
5. If the set M is non-empty, go to step 2.
6. Output a list of accounting objects from the set P_n, which failed to pass the control of the smallest and largest values of the indicators. Output a list of accounting objects from the set P_m, which successfully passed the control of the smallest and largest values of the indicators.

The algorithm for controlling the difference between the largest and the smallest values of the indicators is the following:

1. Obtain the M set of accounting objects of type t.
2. Select the accounting object from the set M.
3. For the indicator r of the selected accounting object, find the maximum and minimum values (s_{max} and s_{min}).
4. Compare the difference module of the found values with the allowable difference between the largest and the smallest values for the exponent r.
5. If $| s_{max} - s_{min} | >$ int, then place the accounting object in the P_n set. Otherwise, place the accounting object in P_r.
6. If the set M is non-empty, go to step 2.
7. Output a list of accounting objects from the set P_n, which failed to control the difference between the largest and the smallest values for the exponent r. Output a list of accounting objects from the set P_r, which successfully passed the control of the difference between the largest and the smallest values for the indicator r.

The algorithm for controlling the arithmetic mean looks like this:

1. Obtain the M set of accounting objects of type t.
2. Select the accounting object from the set M.
3. For the indicator a of the selected accounting object, find the arithmetic mean of its values.
4. Compare the found arithmetic mean with the minimum and maximum admissible values of the arithmetic mean for the indicator a (a_{min} and a_{max}).
5. If avr (k, S)> amax or avr (k, S) <amin, then place the accounting object in the set P_n. Otherwise, place the accounting object in the set P_a.
6. If the set M is non-empty, go to step 2.
7. Output a list of accounting objects from the set P_n, which failed to control the arithmetic mean of the indicator a. Output a list of accounting objects from the set Pa, which have successfully passed the control of the arithmetic mean of the indicator a.

The algorithm for controlling emissions – the values of indicators that differ sharply from the basic number of values - looks like this:

1. Obtain the M set of accounting objects of type t.
2. Select the accounting object from the set M.
3. For the indicator g of the selected accounting object, find the arithmetic mean of its values except for the current one.
4. Compare the difference between the average arithmetic mean and the current value of g with the allowable emission value for the g indicator.
5. If | avr (h, S) - sh + 1 |> out, then place the accounting object in the set P_n. Otherwise, place the accounting object in the set P_g.
6. If the set M is non-empty, go to step 2.
7. Output a list of accounting objects from the set P_n, which failed to control the emissions for the indicator g. Output a list of accounting objects from the set of P_g, which successfully passed the emission control for the indicator g.

When using models of fuzzy criteria for checking data, the difference between their algorithms and those described above will be that for accounting objects the value of the membership function will be calculated, which will divide accounting objects into subsets assigned by the function.

Despite a large number of subsystems for data processing and regulation, the operator interface at all levels remained simple and convenient. A snapshot of the window form of the dispatcher control panel installed in the boiler house is shown in Figure 4, and Figure 5 shows the interface of the dispatch center operator from the second level system. The interface of the industrial safety system was not developed; the system launches and runs in fully automated mode.

Fig. 4. The interface of the dispatching control panel, the system of the lower one.

 DOI: 10.1051/matecconf/201713202010

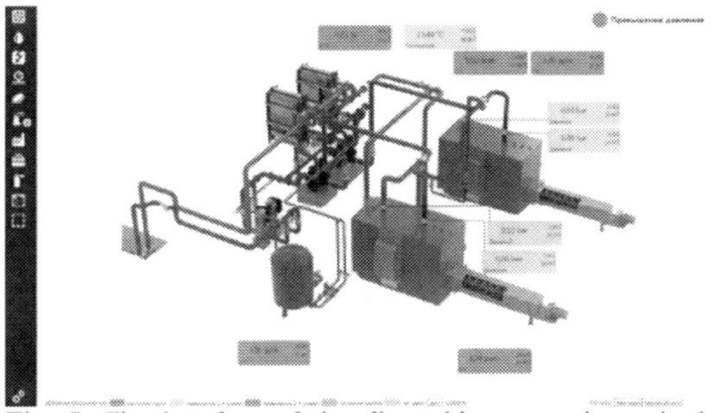

Fig. 5. The interface of the dispatching control panel, the system of the lower one.

3 Conclusion

The developed software and hardware complex was installed in the boiler house of the city of Orel. Complex automation took a month of work, but at the same time, a complex shutdown of the boiler house occurred only once per one working day. The carried out complex of works confirms that the development of application systems with improved safety and reliability characteristics is not only an important technical task, but also requires the carrying out of research works.

References

1. E. L. Palej, *Normativnye trebovanija i prakticheskie rekomendacii pri proektirovanii kotel'nyh* (Piter, SPb, 2014) [in Russian]
2. Svod pravil SP 89.13330.2012 Kotel'nye ustanovki. Aktualizirovannaja redakcija SNiP II-35-76 (Analitik, Moscow, 2012) [in Russian]
3. A. V. Shhegol'kov, M. A. Mishin, Polzunovskij vestnik, no. 1, 257-265 (2011) [in Russian]
4. A. G. Aleksandrov, M. V. Automatics and telemechanics, no. 2, 16-30 (2014) [in Russian]
5. T. A. Kramina, I. F. Safin, G. A. Zaripova, *Proceedings of Kazan state University of architecture and construction*, no. 1, 50-54. (2014) [in Russian]
6. V. Polonskij, G. Titov, A. Polonskij, *Avtonomnoe teplosnabzhenie*, 152 (Izdatel'stvo Associacii stroitel'nyh vuzov, Moscow, 2007) [in Russian]
7. D.V. Paap, *Novye tehnologii v teplosnabzhenii i stroitel'stve*, no. 8, 293 (2010) [in Russian]
8. Ju. F. Davydov, *Vestnik Kazahstansko-Amerikanskogo Svobodnogo Universiteta*, Issue 6, Ust'-Kamenogorsk,178 (2011) [in Russian]
9. V. A. Nesterov et. al., *Vestnik Voronezhskogo gosudarstvennogo universiteta*. Serija: Sistemnyj analiz i informacionnye *tehnologii*, no.1, 77-83. (2014) [in Russian]
10. O. V. Shishov *Avtomatizacija i proizvodstvo*, no.1, 36-37 (2014) [in Russian]
11. O. V. Shishov, *Jelementy sistem avtomatizacii: kontrollery, operatornye paneli, moduli udalennogo dostupa* (Direkt-Media, Moscow – Berlin, 185, 2015) [in Russian]
12. A. A. Gorjunkova, Ju. I. Okuneva, *Prioritetnye napravlenija razvitija nauki i tehnologij*, 124 (2015) [in Russian]
13. S.P. Petrov, *Analiz i sintez sistemy upravlenija processom peredachi teplovoj jenergii v ko-generacionnoj sisteme centralizovannogo tep-losnabzhenija. Monografija* (Orlik and K Publishing House, Orel, 314, 2009) [in Russian]
14. K.V. Podmaster'ev, Datchiki i sistemy, no. 6, 14-18. (2008) [in Russian]
15. S.P. Petrov, *Avtomatizacija kogeneracionnyh sistem teplosnabzhenija s raspredelennymi pikovymi nagruzkami. Monografija* (Mashinostroenie, Moscow: 1, 304, 2007) [in Russian]
16. A.V. Pilipenko, Informacionnye sistemy i tehnologii, no. 4/66, 115-119. (2011) [in Russian]
17. K.V. Podmaster'ev, Kontrol'. Diagnostika, no. 5, 45-57 (2010) [in Russian]
18. A.V. Pilipenko, Fundamental'nye i prikladnye problemy tehniki i tehnologii, no. 6-2(296), 58-64. (2012) [in Russian]
19. K.V. Podmaster'ev, Mir Izmerenij, no. 1, 36-41. (2010) [in Russian]
20. S.P. Petrov, Mehatronika, avtomatizacija, upravlenie, no. 5, 46 -53. (2006) [in Russian]

Analysis of the influence of positive feedback on the quality of the control system

Irina Tsybriy[1,*], *Vitaliy* Ignatenko [2]

[1]Department of Instrument-making, Don State Technical University, Russia
[2]Department of Instrument-making, Don State Technical University, Russia

Abstract. The paper proposed a matrix mathematical apparatus of synthesis optimal drive control system, providing the change of the angular acceleration of the output shaft for a given law. An example is given of constructing a matrix model for optimal control of a high-torque direct current drive with thyristor control.

1 Introduction

The conditions for high-performance equipment require the use of drives built on the principle of optimal control [1 - 7] by the criterion of maximum speed. The control criterion can be formulated as follows [8,9] "... the optimum speed is achieved in those cases when the selected executive motor for a given typical control action develops the maximum possible torque in the required direction during the entire time of the transient process."

As you know [3], the drive with the maximum speed has a step-like control action, performed in two stages:

- the first one accelerates the engine with the maximum possible acceleration and the engine speed changes by a value $\Delta\Omega$;
- on the second - braking of the engine with the maximum possible deceleration and speed change by the value $\Delta\Omega$.

Depending on the parameters of the drive, the value of the control time t and the set value of the movement S, the following modes of playback of this path are possible:

- when acceleration is limited; acceleration is limited by the maximum torque of the motor, the static torque of the load and the moment of inertia of the drive;
- under the restriction of acceleration and the derivative of acceleration; the derivative of acceleration is limited by the maximum rate of change of the torque in the function of time;
- when speed and acceleration are limited; speed is limited by the maximum engine speed for a given load value;
- with the limitation of speed, acceleration and derivative of acceleration.

The limitation of the rate of change of the motor torque in time is of considerable importance only at a low moment of inertia of the drive. Let us consider the possibility of reproducing a given trajectory of motion with an acceleration limitation.

To analyze the behavior of a system with viscous friction and a nonlinear dependence of the moment of resistance on the angle of rotation of the motor shaft, we consider an electric direct current drive with a high-torque electric motor with a control system built on the principle of subordinate regulation.

2 Statement of the problem of optimal control

For example, such a system can always be represented by a system of three linear differential equations of the third order, starting from the form of the transfer function [10]:

$$W(p)=\frac{1}{K_C}\;\frac{1+8\tau p}{64p^3\tau^3+32p^2\tau^2+8p\tau+1}=\frac{\omega_1(p)}{\theta_U(p)} \quad (1)$$

where: τ - time constant of the pulse-phase control system of the thyristor rectifier; K_C - coefficient of speed feedback; ω_1 - frequency of rotation of the motor shaft; θ_U - voltage at the motor armature (control signal).

The control problem is formulated as follows: drive control must ensure that the angular acceleration of the output shaft is changed according to a given law $\varepsilon=\varepsilon(t)$. The mechanical part of the drive can be represented by a two-mass model.

We denote the law of change of the control action as $\theta(t)$.

The control criterion can be written in the form:

$$J_U=\int_0^{\tau}[\theta(t)-\varphi_2]^2\,dt\Rightarrow\min \quad (2)$$

where: φ_2 - angle of rotation of the actuator, reduced to the shaft of the electric motor.

Obviously, the input action can be specified either by the angle of rotation of the motor shaft as a function of time, or by the angular acceleration as a function of time.

* Corresponding author: irconst@mail.ru

The control criterion in matrix form [11 - 15]:

$$J = \int_0^{\infty} \left\{ \mathbf{X}^T \mathbf{Q} \mathbf{X} + \mathbf{U}^T \mathbf{R} \mathbf{U} - 2\mathbf{X}^T \mathbf{P} \mathbf{U} \right\} dt \Rightarrow \min \quad (3)$$

where: **X** - n-dimensional state vector of the control object; **U** - n-dimensional vector of the state of control actions.

The choice of matrices **P** and **Q** is determined by the type of the optimization criterion, but the transition to the matrix form of recording with the previous formulation of the problem is impossible, so we will carry out its correction.

We consider that the required law of change of control action is the law of change of voltage on the armature of the motor. In this formulation, the system can be described in state variables.

In our example, the equations of a system consisting of a controlled thyristor rectifier, a DC motor, and a mechanical part replaced by an equivalent two-mass model have the following form:

$$\begin{aligned} &\dot{\omega}_1 = y \\ &\dot{y} = z \\ &\dot{z} = \theta(t)a + bz + cy + d\omega_1 \\ &\dot{M}_{12} = q(\omega_1 - \omega_2) \\ &\dot{\omega}_2 = (M_{12} - M_C(\varphi_2) - \beta\omega_2)s \\ &\dot{\varphi}_2 = \omega_2 \end{aligned} \quad (4)$$

where: $a = \frac{K_C}{64\tau^3}$; $q = \frac{1}{J_1}$; $b = \frac{1}{2\tau}$; $s = \frac{1}{J_2}$; $c = \frac{1}{8\tau^2}$; $r = 8\tau$; $d = \frac{1}{64\tau^3}$; β - coefficient of viscous friction in the actuator; Mc - the moment of resistance, $M_{12}(\varphi_2)$ - dependence of the moment of resistance on the angle of rotation of the actuator, reduced to the shaft of the electric motor, J_1, J_2 - moments of inertia reduced to the motor shaft.

Then the vector of control actions will take equal:

$$\mathbf{U}(t) = \begin{pmatrix} \varepsilon(t) \\ -M_C(\varphi_2) \end{pmatrix} \quad (5)$$

3 Solution of the problem by dynamic programming method

To solve the problem, we use the method of dynamic programming, which is widely used to solve problems of optimal control [16 - 21]. It is based on the principle: any terminal part of the optimal trajectory is the optimal trajectory. This means that the optimal control is determined by the state of the system at a given moment in time and does not depend on the states at the previous moments of time.

Consider this method in application to an object described in the space of state variables. In our case, the initial state of the object $\mathbf{X}(0)$, the domain of admissible controls $\Psi(\mathbf{U})$ and the optimality criterion are given.

We assume that the upper limit of integration τ is given. We do not impose constraints on the state vector and its final value of constraints. The argument in the formulas $F(\mathbf{X},\mathbf{U},t)$ and $\mathbf{U}(t)$, means that they can be non-stationary.

In this formulation, we can formulate the following problem: it is necessary to find in the class of admissible controls a control $\mathbf{U}(t)$ under which the functional reaches a minimum.

The task is related to the type of tasks with a free right end and a fixed control time: $J = \min$, when: $\mathbf{U}(t) \in \Psi(t)$; $t \in [t_0, T]$.

We write the Bellman equation:

$$-\frac{\partial S}{\partial t} = \min \left\{ F(\mathbf{X},\mathbf{U},t) + \sum_{k=1}^{n} \frac{\partial S}{\partial x_k} \cdot f_k \right\} \quad (6)$$

and perform the following transformations:

$$-\frac{\partial S}{\partial t} = \min \left\{ \mathbf{X}^\mathbf{T}\mathbf{Q}\mathbf{X} + \mathbf{U}^\mathbf{T}\mathbf{R}\mathbf{U} - 2\mathbf{X}^\mathbf{T} \cdot \mathbf{U} + \frac{\partial S}{\partial \mathbf{X}} \cdot (\mathbf{A}\mathbf{X} + \mathbf{B}\mathbf{U}) \right\} -$$

$$-\frac{\partial}{\partial \mathbf{U}} \left\{ \mathbf{X}^\mathbf{T}\mathbf{Q}\mathbf{X} + \mathbf{U}^\mathbf{T}\mathbf{R}\mathbf{U} - 2\mathbf{X}^\mathbf{T}\mathbf{P}\mathbf{U} + \frac{\partial S}{\partial \mathbf{X}} \cdot (\mathbf{A}\mathbf{X} + \mathbf{B}\mathbf{U}) \right\} = 0 \quad (7)$$

where **A** - own matrix of the object.

Further we receive:

$$2\mathbf{U}^\mathbf{T} \cdot \mathbf{R} - 2 \cdot \mathbf{X}^\mathbf{T} \cdot \mathbf{P} + \frac{\partial S}{\partial \mathbf{X}} \cdot \mathbf{B} = 0 . \quad (8)$$

From the last equality we find the expression for the control vector. For further reasoning we need expressions for **U** and $\mathbf{U}^\mathbf{T}$:

$$\mathbf{U}^\mathbf{T} = \left(\mathbf{X}^\mathbf{T} \cdot \mathbf{P} - \frac{1}{2} \cdot \frac{\partial S}{\partial \mathbf{X}} \cdot \mathbf{B} \right) \cdot \mathbf{R}^{-1} ; \quad (9)$$

$$\mathbf{R}^\mathbf{T} \cdot \mathbf{U} = \left(\mathbf{X}^\mathbf{T} \cdot \mathbf{P} - \frac{\partial S}{\partial \mathbf{X}} \cdot \mathbf{B} \right)^T = \mathbf{P}^\mathbf{T} \cdot \mathbf{X} - \mathbf{B}^\mathbf{T} \cdot \left(\frac{\partial S}{\partial \mathbf{X}} \right)^T ; \quad 10)$$

$$\mathbf{U} = \left(\mathbf{R}^\mathbf{T} \right)^{-1} \cdot \left(\mathbf{P}^\mathbf{T} \cdot \mathbf{X} - \mathbf{B}^\mathbf{T} \cdot \left(\frac{\partial S}{\partial \mathbf{X}} \right)^T \right) \quad (11)$$

The resulting expressions and substitute into the Bellman equation, while we seek its solution in the form:

$$S(\mathbf{X}(t)) = \mathbf{X}^\mathbf{T}(t) \cdot \mathbf{M}(t) \cdot \mathbf{X}(t), \quad (12)$$

where: $\mathbf{M}(t)$ is an unknown symmetric nonstationary matrix of dimension $[n \times n]$; n - the order of the system under study.

The expressions needed for further transformations are written in the form:

$$\frac{\partial S}{\partial \mathbf{X}(t)} = 2 \cdot \mathbf{X}^\mathbf{T}(t) \cdot \mathbf{M}(t) ; \quad (13)$$

$$\frac{\partial S}{\partial t} = \mathbf{X}^\mathbf{T}(t) \cdot \dot{\mathbf{M}}(t) \cdot \mathbf{X}(t) ; \quad (14)$$

DOI: 10.1051/matecconf/201713202011

$$\mathbf{U}^{\mathrm{T}} = \mathbf{X}^{\mathrm{T}} \cdot \mathbf{P} - \mathbf{X}^{\mathrm{T}} \cdot \mathbf{M} \cdot \mathbf{B} = \mathbf{X}^{\mathrm{T}} \cdot (\mathbf{P} - \mathbf{M} \cdot \mathbf{B}); \quad (15)$$

$$\mathbf{U} = (\mathbf{P} - \mathbf{M} \cdot \mathbf{B})^{\mathrm{T}} \cdot \mathbf{X}. \quad (16)$$

After carrying out the appropriate substitutions, we obtain the Bellman equation in the form:

$$\mathbf{X}^{\mathrm{T}}\dot{\mathbf{M}}\mathbf{X} = \mathbf{X}^{\mathrm{T}}\mathbf{Q}\mathbf{X} + \mathbf{X}^{\mathrm{T}}(\mathbf{P} - \mathbf{M}\mathbf{B})(\mathbf{P} - \mathbf{M}\mathbf{B})^{\mathrm{T}}\mathbf{X} -$$

$$- 2\mathbf{X}^{\mathrm{T}}\mathbf{P}(\mathbf{P} - \mathbf{M}\mathbf{B})^{\mathrm{T}}\mathbf{X} + 2\mathbf{X}^{\mathrm{T}}\mathbf{M}\mathbf{A}\mathbf{X} + 2\mathbf{X}^{\mathrm{T}}\mathbf{M}\mathbf{B}(\mathbf{P} - \mathbf{M}\mathbf{B})^{\mathrm{T}} \quad (17)$$

From the last equality it is obviously possible to obtain an equation for determining the matrix **M**:

$$\dot{\mathbf{M}} = \mathbf{Q} + 2\mathbf{M}\mathbf{A} + (\mathbf{P} - \mathbf{M}\mathbf{B})(\mathbf{P} - \mathbf{M}\mathbf{B})^{\mathrm{T}} - 2\mathbf{P}(\mathbf{P} - \mathbf{M}\mathbf{B})^{\mathrm{T}} +$$

$$+ 2\mathbf{M}\mathbf{B}(\mathbf{P} - \mathbf{M}\mathbf{B})^{\mathrm{T}} \quad (18)$$

Conducting the transformations of the last expression, we have:

$$\dot{\mathbf{M}} = \mathbf{Q} + 2\mathbf{M}\mathbf{A} + (-2\mathbf{P} + 2\mathbf{M}\mathbf{B} + (\mathbf{P} - \mathbf{M}\mathbf{B})) \cdot (\mathbf{P} - \mathbf{M}\mathbf{B})^{\mathrm{T}} \quad (19)$$

$$\dot{\mathbf{M}} = \mathbf{Q} + 2\mathbf{M}\mathbf{A} + (\mathbf{M}\mathbf{B} - \mathbf{P}) \cdot (\mathbf{P} - \mathbf{M}\mathbf{B})^{\mathrm{T}} \quad (20)$$

Thus, we have obtained a nonlinear quadratic equation with respect to the matrix $\mathbf{M}(t)$.

Since the solution is sought in the quadratic form:

$$S(\mathbf{X}(t)) = \mathbf{X}^{\mathrm{T}}(t)\mathbf{M}(t)\mathbf{X}(t), \quad (21)$$

the solution of this equation, for $t = \tau$, must satisfy the condition $\mathbf{M}(\tau) = \mathbf{Q}$.

The object under study is stationary, which means that the matrix **M** is constant. Then the equation under study has the form:

$$\mathbf{Q} + 2\mathbf{M}\mathbf{A} - (\mathbf{P} - \mathbf{M}\mathbf{B})(\mathbf{P} - \mathbf{M}\mathbf{B})^{\mathrm{T}} = 0. \quad (22)$$

We perform fairly simple transformations:

$$(\mathbf{P} - \mathbf{M}\mathbf{B})(\mathbf{P} - \mathbf{M}\mathbf{B})^{\mathrm{T}} = \mathbf{P}\mathbf{P}^{\mathrm{T}} - \mathbf{M}\mathbf{B}\mathbf{P}^{\mathrm{T}} -$$

$$- \mathbf{P}\mathbf{B}^{\mathrm{T}}\mathbf{M}^{\mathrm{T}} + \mathbf{M}\mathbf{B}\mathbf{B}^{\mathrm{T}}\mathbf{M}^{\mathrm{T}}; \quad (23)$$

$$(\mathbf{P} - \mathbf{M}\mathbf{B})(\mathbf{P} - \mathbf{M}\mathbf{B})^{\mathrm{T}} = \mathbf{P}\mathbf{P}^{\mathrm{T}} - \mathbf{M}\mathbf{B}\mathbf{P}^{\mathrm{T}} -$$

$$- \mathbf{M}(\mathbf{P}\mathbf{B}^{\mathrm{T}})^{\mathrm{T}} + \mathbf{M}\mathbf{B}\mathbf{B}^{\mathrm{T}}\mathbf{M}^{\mathrm{T}}; \quad (24)$$

$$(\mathbf{P} - \mathbf{M}\mathbf{B})(\mathbf{P} - \mathbf{M}\mathbf{B})^{\mathrm{T}} = \mathbf{P}\mathbf{P}^{\mathrm{T}} - 2\mathbf{M}\mathbf{B}\mathbf{P}^{\mathrm{T}} +$$

$$+ \mathbf{M}\mathbf{B}\mathbf{B}^{\mathrm{T}}\mathbf{M}^{\mathrm{T}}. \quad (25)$$

The final equation can be written as:

$$\mathbf{Q} + \mathbf{P}\mathbf{P}^{\mathrm{T}} - 2\mathbf{M}\left(\mathbf{A} - \mathbf{B}\mathbf{P}^{\mathrm{T}}\right) + \mathbf{M}\mathbf{B}\mathbf{B}^{\mathrm{T}}\mathbf{M}^{\mathrm{T}} = 0 \quad (26)$$

4 Conclusion

Thus, the matrix equation is obtained, the solution of which will allow to determine the matrix and will make it possible to synthesize the control law according to the specified criterion.

References

1. F. Vasilyev, *Optimization methods* (Moscow, Factorial Press, 2002)
2. Fond, S., Optimal feedback control policies for stochastic-parameter systems, Journal of Optimization Theory and Applications, **30(3)**, 501-512 (1980)
3. A. Alexandrov *Optimal and adaptive systems*, (Moscow, Higher School, 1990)
4. N. Polovinchuk, V. Trofimenko, N. Rudenko, S. Ivanov Optimal control of a stochastic dynamical system under terminal constraints, Dual technologies, **4**, 40-43 (2013)
5. E.M. Galeev, Zelikin M.I., Konyagin, S.V. Osmolovsky, *Optimum control* (Moscow, MCNMO, 2008)
6. V. Komissarchik *Methods of optimization and optimal control* (Tver state technical University, 2002)
7. D. White, *Optimality and Efficiency* (Wiley, New York, 1982)
8. V. Surkov, *Analytical design of optimal controllers by the criteria of accuracy, speed, energy saving* (Tula State University, 2005)
9. D. Sue, A. Meyer, *Modern theory of automatic control and its application* (Moscow, Mechanical engineering, 1972)
10. V. Miroshnichenko, *Research of dynamic characteristics of metal-cutting machine components at the design stage* (Rostov-on-Don, DSTU Publishing Center, 2006)
11. V. Miroshnichenko, *Problems of Synthesis of Modal Regulators of Machine Drives*, (Rostov-on-Don, DSTU Publishing Center, 2009)
12. A. Balakrishnan *Applied functional analysis* (Moscow, Nauka, 1980)
13. F. Gontmakher, *Matrix Theory* (Moscow, Nauka, 1988)
14. R. Bellman, Introduction to the theory of matrices (Moscow, Mir, 1969)
15. A. Egorov On the matrix Riccati equations, Differ. equations, **1**, (1995)
16. R. Bellman, On the theory of dynamic programming, Proceedings of the National Academy of Sciences, **38**, 716-719 (1952)
17. R. Bellman, The theory of dynamic programming, Bulletin of the American Mathematical Society, **60**, 503-515 (1954)
18. Y. Maksimov, E. Filipovskaya *Algorithms for solving nonlinear programming problems* (Moscow: MEPhI, 1982)
19. M. Sniedovich, *Dynamic Programming* (Marcel Dekker, New York, 1992)
20. M. Puterman, *Dynamic Programming and its Application* (Academic Press, Ney York, 1978)
21. T. Abdul-Razaq, C. Potts Dynamic programming state-space relaxation for single-machine scheduling, Journal of the Operational Research Society, **39(2)**, 141-152 (1988)

The analitical modeling of the motion process of the inverted pendulum system

Pavel Obukhov[1,*], *Yaroslav* Mogilin[2]

[1]Department of automation of production processes, Don State Technical University, Russia
[2]Department of automation of production processes, Don State Technical University, Russia

Abstract. In the article the conceptual description of the nonlinear mathematical model of the inverted pendulum is presented. The Lagrange's equations of the second kind are applying to build of the model, the generalized coordinate are selected and the system of the differential equations of the fourth odder is formed.

1 Introduction

The problem of the building of the systems of the automatic control process of the object movement type the inverted pendulum is urgent and very complex. These systems widespread in practice, when the large objects with a high center of the gravity are moving. The mathematical modeling play important role in the synthesis of the such control system by which it is possible to abstract from the real variables and to efficient explore the problem. The quality of the system control functioning ultimately depends of the accuracy of the construction of the control object model.

2 Main part

The parameters of the inverted pendulum necessary to identify for the conceptual description of the control object. The transport complex which consist of the platform of mass m2 which is moving on rails or surface (horizontal) and the vertical rod of mass m1 and length L are attached to the platform by using of the electromechanically device (without signal control it works such as hinge) is presented on the figure 1. The rod is deviate from the balance position at the angel φ by the disturbance (wind flow G=Fв/L) and the force of gravity P acts at the center of the gravity $P_{ЦТ} = \frac{2L}{3}$. The thrust force Fu applied to the platform for the motion in the horizontal direction are used. The moment turns the rod at the certain angel for its return to the balance position.

With compiling the mathematical model (MM) of the control object (CO) as a control actions are taken the force Fu that causes the platform to move at the horizontally direction and the moment Mu are rotated the rod at the certain angel. As the disturbance we will take the wind which acts on the all surface along the length of the rod is considered in the following as a stream.

Some assumptions to obtain the MM OC: the parameters of the OC are concentrated (the rod and the platform are uncompressible objects); the platform moves on rails without friction and only at the horizontally direction; the rod rotates on the hinge mechanism without friction; the cylinder is forming a shape of the rod, the moment of the inertia of the cylinder are considered.

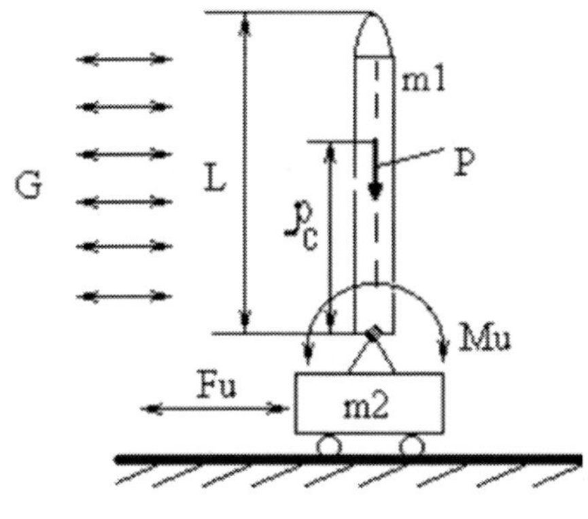

Fig. 1. The transport complex

The Lagrange's equations of the second kind are applying to build of the MM, for this the generalized coordinates are selected, in this case it may be the motion of the platform at the horizontally direction X and the deviate angel of the rod at the balance position φ (1),

$$\begin{cases} q_1 = X \\ q_2 = \varphi \end{cases} \quad (1)$$

then δX - the variation of the displacement for a given deviation, $\delta\varphi$ - the variations of the rotate angel of the rod (see figure 2).

* Corresponding author: pobuhov@spark-mail.ru

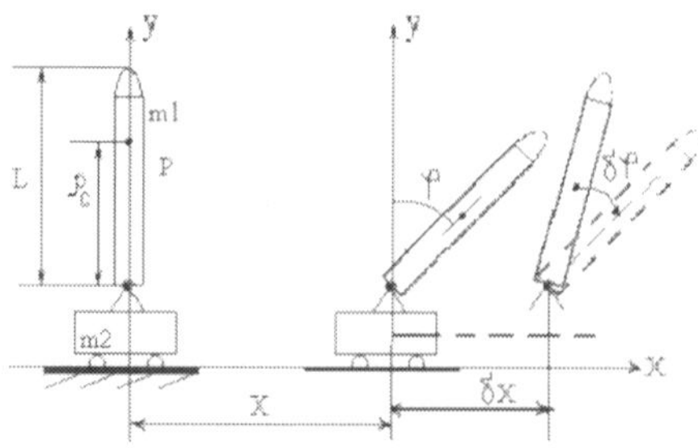

Fig. 2. The choice of the generalized coordinates

The equations system with the generalized coordinates chosen takes the form:

$$\begin{cases} \dfrac{d}{dt}\left(\dfrac{\partial T}{\partial \dot{X}}\right) - \dfrac{\partial T}{\partial X} = Q_X - \dfrac{\partial \Pi}{\partial X} \\ \dfrac{d}{dt}\left(\dfrac{\partial T}{\partial \dot{\varphi}}\right) - \dfrac{\partial T}{\partial \varphi} = Q_\varphi - \dfrac{\partial \Pi}{\partial \varphi} \end{cases} \quad (2)$$

where T – kinematic energy; Π - potential energy; Q_X, Q_φ - generalized forces on displacement X and angel of the deviation φ.

The expressions of the kinematic and potential energy of the control object and their derivatives and also expressions for the generalized forces are found by the formulas:

$$\frac{\partial T}{\partial X} = 0, \text{ т.к. } T = T(\varphi, \dot{X}, \dot{\varphi}) \quad (3)$$

$$\frac{\partial T}{\partial \varphi} = -\frac{2m_1}{3} \cdot L \cdot \dot{X} \cdot \dot{\varphi} \cdot \sin\varphi \quad (4)$$

$$\frac{\partial T}{\partial \dot{\varphi}} = \dot{\varphi} \cdot \frac{m_1 \cdot L^2 \cdot 7}{9} + \frac{2m_1 \cdot L}{3} \cdot \dot{X} \cdot \cos\varphi \quad (5)$$

$$\frac{\partial T}{\partial \dot{X}} = \dot{X} \cdot (m_1 + m_2) + \frac{2m_1 \cdot L}{3} \cdot \dot{\varphi} \cdot \cos\varphi \quad (6)$$

$$\frac{d}{dt}\left(\dot{X} \cdot (m_1 + m_2) + \frac{2m_1 \cdot L}{3} \cdot \dot{\varphi} \cdot \cos\varphi\right) = \\ = \ddot{X} \cdot (m_1 + m_2) + \frac{2m_1 \cdot L}{3} \cdot \left(\ddot{\varphi} \cdot \cos\varphi - \dot{\varphi}^2 \cdot \sin\varphi\right) \quad (7)$$

$$\frac{d}{dt}\left(\dot{\varphi} \cdot \frac{m_1 \cdot L^2 \cdot 7}{9} + \frac{2m_1 \cdot L}{3} \cdot \dot{X} \cdot \cos\varphi\right) = \\ = \ddot{\varphi} \cdot \frac{m_1 \cdot L^2 \cdot 7}{9} + \frac{2m_1 \cdot L}{3} \cdot \left(\ddot{X} \cdot \cos\varphi - \dot{X} \cdot \dot{\varphi} \cdot \sin\varphi\right) \quad (8)$$

$$\begin{cases} Q_X = G \cdot L \cdot \cos\varphi + F_U \\ Q_\varphi = G \cdot \dfrac{2L^2}{3} \cdot \cos^2\varphi + M_U \end{cases}; \quad \Pi = m_1 \cdot g \cdot \rho_C \cdot \cos\varphi;$$

$$\begin{cases} \partial \Pi / \partial \varphi = -m_1 \cdot g \cdot \rho_C \cdot \sin\varphi \\ \partial \Pi / \partial X = 0 \end{cases} \quad (9)$$

The expression is substituted into the formula (2) and obtain the system of equations:

$$\begin{cases} \ddot{X} \cdot (m_1 + m_2) + \dfrac{2m_1 \cdot L}{3} \cdot \left(\ddot{\varphi} \cdot \cos\varphi - \dot{\varphi}^2 \cdot \sin\varphi\right) = G \cdot L \cdot \cos\varphi + F_U \\ \ddot{\varphi} \dfrac{m_1 \cdot L^2 \cdot 7}{9} + \dfrac{2m_1 \cdot L}{3} \cdot \left(\ddot{X} \cdot \cos\varphi - \dot{X} \cdot \dot{\varphi} \cdot \sin\varphi\right) + \dfrac{2m_1}{3} \cdot L \cdot \dot{X} \cdot \dot{\varphi} \cdot \sin\varphi = \dfrac{G2 \cdot L^2}{3} \cdot \cos^2\varphi + M_U + \\ + m_1 \cdot g \cdot \dfrac{2L}{3} \cdot \sin\varphi \end{cases}$$

(10)

The system of The Lagrange's equations of the second kind descriptions the control object of study from the point of view the dynamic of the mechanical system and it is her equivalent of the mathematical model based on the assumptions made in the problem formulation.
For further research of the control object the system is converted to the form of Cauchy

$$\begin{cases} \dot{X} = V \\ \dot{\varphi} = \omega \\ \dot{\omega} = \dfrac{\dfrac{2m_1 \cdot L \cdot \omega^2 \cdot \sin\varphi}{3 \cdot (m_1 + m_2)} + \dfrac{G L \cdot \cos\varphi}{(m_1 + m_2)} + \dfrac{F_U}{(m_1 + m_2)} \dfrac{G L \cdot \cos\varphi}{m_1} - g \cdot tg\varphi \dfrac{3 \cdot M_U}{2m_1 \cdot L \cdot \cos\varphi}}{\left(\dfrac{2L \cdot \cos\varphi}{3 \cdot \left(1 + \dfrac{m_2}{m_1}\right)} \dfrac{7 \cdot L}{6 \cdot \cos\varphi}\right)} \\ \dot{V} = \dfrac{tg\varphi\, \omega^2 + \dfrac{G3}{2m_1} + \dfrac{3 \cdot F_U}{2m_1 \cdot L \cdot \cos\varphi} \dfrac{G6 \cdot \cos^2\varphi}{m_1 \cdot 7} \dfrac{6\, g \cdot \sin\varphi}{L \cdot 7} \dfrac{M_U \cdot 9}{m_1 \cdot L^2 \cdot 7}}{\left(\dfrac{\left(1 + \dfrac{m_2}{m_1}\right) \cdot 3}{L \cdot \cos\varphi} \dfrac{\cos\varphi\, 6}{L \cdot 7}\right)} \end{cases} \quad (11)$$

3 Conclusions

Thus, the mathematical model of the inverted pendulum is the system of the equations of the first order, which contains nonlinearity of the type of the product. The next stage of the research is the task of the analyze of the mathematical model of this object and synthesis of the control system.

References

1. A. A. Kolesnikov, Synergetic control of mechanical system "inverted pendulum on a cart" nonlinear coordinate transformation, TRTU, (2001)
2. Yu. G. Martynenko, A. M. Formalsky, Reports of Academy of Sciences, **439**, N6 (2011)
3. E.I. Butikov, Komp'yuternyye instrumenty v obrazovanii, № 5, 40 – 51 (2010)

Simulation of physical processes in devices with non-magnetic, milk-dispersed secondary part

Maksim Minkin[1,*], *Denis* Kuimov[1], and *Andrej* Mozgovoj[2]

[1]DSTU, Department of automation of production processes, 344000 Rostov-on-Don, Russia
[2]DSTU, Department of safety and environmental protection, 344000 Rostov-on-Don, Russia

Abstract. At present, most machine-building enterprises focus on the problem of saving energy resources. The proposed system of air regeneration in industrial premises will allow to keep the air in the room at a comfortable temperature without removing it to the atmosphere. The second problem solved by this system is the prevention of pollutants entering the atmosphere. At the same time, the efficiency of this system is not inferior to analogs, and many of them even surpass. Energy costs are reduced by at least 30% because there is no need to use the system of regeneration of working elements. However, the development of this system requires the solution of a number of problems. One of the main problems is the control of the motion of particles in the corona discharge region. A number of differently directed forces act on a particle of a finely dispersed substance, which actually form the equation of motion of this particle. However, when composing the equation of particle motion, it is necessary to take into account the space charge in the outer region of the corona. With a fairly complex geometry of the core, it is impossible to take into account the influence of a volumetric charge analytically, and, consequently, the use of computer modeling tools is required. Another factor that needs to be taken into account is the appearance of an aerodynamic force arising from such a poorly understood phenomenon as the "ion wind" that appears in the presence of a corona discharge. The solution of the described problem will allow to create a system of air regeneration in industrial premises, allowing recirculation of polluted air without diversion into the atmosphere, which will reduce the cost of energy.

1 Formulation of the problem. Construction of a mathematical model

In general, the energy-saving air regeneration system in industrial premises is built on the basis of a standard system of shop ventilation. At the design stage, or in an already existing system, it is proposed to build in several additional elements. After the modernization of the system, its main working element is the electromechanical device for suppressing dust flows (electrostatic shutter). The scheme of the energy-saving air regeneration system in production facilities (Figure 1).

The problem of developing devices for blocking dust emissions requires a lot of attention, since the development of many industries, combined with conservation and rational use of its resources, is often closely associated with the problem of suppressing dust-containing flows. The urgency of its solution increases with the increase of the territorial concentration of industrial enterprises and the increase in their individual capacities.

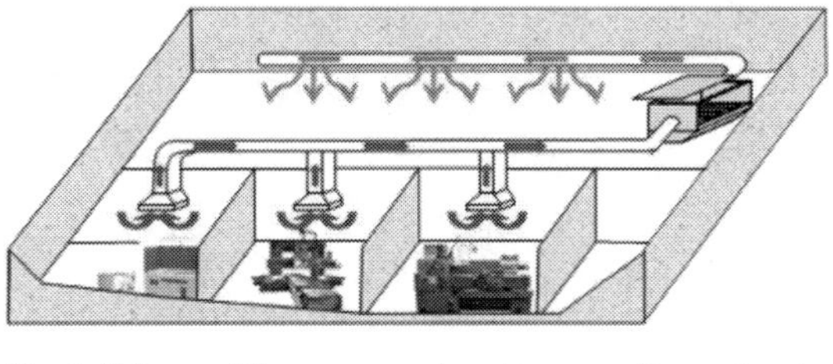

Fig. 1. Scheme of the energy-saving air regeneration system in production premises.

Under corona discharge conditions, the force exerted by the electric field acting on the particles of the fine-dispersed fraction is formed under the influence of the space charge, which affects the magnitude of this force [1,2]. Just take into account the aerodynamic effort, called the "ion wind," which in these devices, according to earlier experimental studies, is about 15-20% of the force from the side of the electric field. The distribution of fine-dispersed particles by physical properties and dimensions is also an important factor, since it affects the dynamics of particle acquisition and aerodynamic properties. The system of equations for the physical processes of particle motion in an inhomogeneous electric field, in general,

* Corresponding author: msi_58@mail.ru

with an arbitrary moving part (aerosol cloud) is constructed on the basis of a mathematical model developed earlier by the leader of this project; however, in contrast to it, one can take into account the influence of the force acting on the part of the "ion wind ". The numerical solution of this system will allow predicting the behavior of a fine particle fraction, and based on the forecast, to develop an efficient system of air regeneration in industrial premises without air venting to the atmosphere.

To solve the problem, it is proposed to use computer simulation tools based on the finite element method. The computer model is based on a mathematical model describing physical processes in the core of an electromechanical device for blocking dust emissions. This mathematical model is constructed on the basis of the following regularities. The following forces act on a particle in an inhomogeneous electric field and under the influence of external perturbations: 1. Gravity; 2. The force of the electric bullet on a charged particle; 3. The force due to the uneven distribution of the electric field strength; 4. The strength of the resistance of the medium to the motion of particles; 5. Aerodynamic force generated by the "ion wind".

The system of equations for the physical processes of particle motion in an inhomogeneous electric field, in general, with an arbitrary moving part (aerosol cloud) is built on the basis of the mathematical model developed earlier by the project manager [1-5], but, unlike it, acting on the part of "ion wind" and includes the following equations:

$$\frac{d^2\mu}{dx^2} + \frac{d^2\mu}{dy^2} = -\frac{\rho}{\varepsilon_0} \quad (1)$$

$$\rho = \frac{dq(t)}{dv} \quad (2)$$

$$F_1 = mg \quad (3)$$

$$F_2 = q(t)E \quad (4)$$

$$F_3 = \rho\frac{\pi r^2}{4} \quad (5)$$

$$F_4 = \rho_{i.wind}\frac{\pi r^2}{4} \quad (6)$$

$$m\frac{dV}{dt} = \sum F \quad (7)$$

$$q(t) = q_m\frac{en_0kt}{4\varepsilon_0 + en_0kt} \quad (8)$$

$$q_m = 4\pi\varepsilon_0(1 + 2\frac{\varepsilon-1}{\varepsilon+2})r^2E \quad (9)$$

In this system of equations: φ - is the scalar electric potential in the computational domain; ρ - is the volumetric charge density; q (t) - is the charge in the analyzed volume of the calculated region; v - is the volume of the investigated space; m - is the mass of the dust particle; qm - is the limiting charge of the dust particle; p - air pressure in the dust-air mixture flow; p.vetra - air stream pressure created by the ion wind; r - radius of the dust particle.

Based on the analysis of the charging process of a particle as it moves in the core of the device, a preliminary evaluation of the operation of the electrostatic shutter can be made. In view of the fact that the physical properties of the finely dispersed fraction and the characteristics of the technological system have a great influence on the patterns of processes occurring in the interelectrode space of the dust-blocking devices, it is necessary to consider the operation of the devices based on the results given the characteristics of the finely dispersed fraction, and also the influence of the force acting with side of the "ion wind".

The numerical solution of this system will allow predicting the behavior of a particle of a finely dispersed fraction.

The process of the program for modeling physical processes begins with entering the initial data from the display screen:

1) the total number of nodes in the calculated area of a given section of the device;

2) the estimated number of nodes in the region of a given section;

3) geometric parameters of the electrostatic shutter, gate height, gate width;

As a result of the work of the above-described program, both the characteristics of the physical processes occurring in the electrostatic shutter and the characteristics of the processes affecting the behavior of the fine-particle fraction are obtained. A characteristic difference of this model of calculation of physical processes is the fact that, unlike all previously developed models, in the considered, it becomes possible to realize the modeling of the process of acquiring a charge by a particle of a finely dispersed fraction and a picture of the forces acting on the particle. This was possible, thanks to the introduction of special subprograms. In all previous studies, only the characteristics of the electric field were calculated, which did not give a complete picture of the processes taking place in electrostatic gates.

Based on the results of the simulation of electrical processes using the above computer model, it is determined:

1) characteristics of the electric field in each finite element of a given section of the device;

2) the dynamics of the acquisition of charge by a particle of a finely dispersed fraction;

3) the force acting on the particle of the finely dispersed fraction from the side of the electric field;

In the framework of this work, a simulation of physical processes for several designs of dust-blocking devices was conducted, the results of the research were used in developing devices for specific operating conditions.

The study of physical processes is shown on the example of two variants of the electrostatic shutter design: "blind-grid", "comb-grid". The constructions of these devices are presented

The initial data, for all studies, were divided into the following groups.

1) Immutable technical data determined by the technological process:

- dust fraction - with diameter from 40 to 160 microns;

- the temperature of the dust-air mixture, taking into account the reduction when entering an unheated zone - 28 ° C;

2) Immutable technical data determined by the parameters of the power supply network and the design features of the device under consideration:
- power is supplied from a single-phase network with a voltage U = 220 V / 30kV with a frequency f = 50 Hz / 0 Hz (reverse polarity);
- cleaning of working parts - air.

In the considered versions of dust emission blocking devices, a number of physical processes have a direct effect on the efficiency of this type of device.

The quality of the device is the higher, the greater is the longitudinal component of the electric field at the entrance to the interelectrode space. The required configuration of the electric field lines at the entrance to the interelectrode space is achieved by a rational choice of the geometry of the shaping electrodes [6-9].

To the designer of the gate structure, it is necessary to achieve this direction and magnitude of the force vector acting on the side of the electric field in the active zone of the device, in which an efficient ejection of fine particles from the interelectrode space towards the total flow of the dust-air mixture occurs.

The blocking of dust emissions occurs in three stages:

1. charging of the fine-dispersed fraction of the dust-air mixture;
2. inhibition of the dust fraction;
3. return of the dust fraction to the dust source or to a specially designated place for collecting the finely dispersed fraction.

To determine the design parameters of the device, under which its effective operation is ensured, it is necessary to use mathematical modeling of physical processes in the device.

It should be noted that in the study of physical processes in the "comb-grid" type of organization, modeling is performed proceeding from the assumption of a plane-parallel picture of the field in the interelectrode space [10-13]. Such an assumption becomes possible in the case when the distance between the electrodes belonging to the same comb is small enough to introduce significant distortions in the picture of the processes occurring.

For shutter-type "blind-mesh", in view of the design features, it makes sense to consider processes exclusively in the interelectrode window between the corona electrodes.

The simulation of physical processes in was performed with the change:

1) the interelectrode distance between the corona electrodes;

2) the diameter of a particle of a finely dispersed substance;

3) positions of the corona system with respect to the shaping electrode;

4) the values of the supply voltage from 30kV to 50kV.

The obtained results of modeling the electric field made it possible to formulate the following recommendations for the construction of electrostatic gates.

1) the best results on the magnitude and nature of the variation of the longitudinal component of the electric field strength are obtained in the case when the points of the tip of the corona electrode and the points of intersection of the perpendiculars from the tip to the surface of the forming electrode lie at the vertices of an equilateral triangle;

2) on the particle of the fine-dispersed fraction, the effect from the side of the electric field balances the force of pressure from the side of the air flow, the faster, the more its geometric dimensions;

3) the presence of an additional shaping electrode at the entrance to the interelectrode space leads to an increase in the time of passage through the dust particles of the outer region of the corona, which increases the charge acquired by the particle;

4) the change in the interelectrode distance makes it possible to reduce the magnitude of the "dip" of the electric field strength along the symmetry axis between the corona electrodes;

5) the optimal distance between the corona and shaping electrodes should be selected from the calculation that the particle will travel from the entrance to the active zone of the electrostatic shutter to the guaranteed stop line in a time approximately equal to 0.1 seconds.

2 Experimental research

After obtaining theoretical results, necessary experimental studies were carried out to verify their adequacy. The abovementioned methodological approaches to conducting research to solve the tasks set in the project are implemented on the basis of a specially designed for this purpose experimental laboratory complex. To obtain a dust mixture, an experimental sample of the vortex layer activator is used, which allows processing of the material in a wide range. During the experimental research, the equipment of the Inter-departmental Resource Center for Collective Use (DMSTU) of the DSTU is included in the federal list of the Central Control Center of the Russian Federation, in particular, the centrifuge (CPS Disk Centrifuge DC24000) for sedimentation analysis of particulate matter of the aerosol mixture.The experimental data and the establishment of analytical dependencies using the methods of mathematical statistics, experiment planning, etc.

In the electrostatic gate, the shaping electrode is made in the form of a grid of non-magnetic material stretched perpendicular to the gas-air mixture flow and is connected to the positive pole of the high-voltage voltage source. The corona electrode is made of metal and consists of a series of angled plates at an equal distance from each other parallel to the shaping electrode and is connected to the negative pole of the power source. Such a design is very effective in view of the fact that an increase in the area of the corona electrodes, an increase in the residence time of the particle in the zone of the corona electrode, and also a partial mechanical removal of the dust fraction.

In the devices under consideration, the fine particle moves in the interelectrode space due to excess air pressure. Then the force acting on the particle from the air side will be determined as the product of the excess air pressure inside the hopper or the container at the cross section of the particle. This force is counteracted by the complex of forces that act on it from the side of the electric field in the interelectrode space. Moreover, as the particle moves to the outlet, its charge increases and the braking force increases, while the force on the airflow side can be considered unchanged. The shutter will be considered to have performed its functions if the path traveled by the particle to a stop is less than the distance from the entry point of the particle to the outer corona discharge area to the hopper outlet [14-17].

The main physical processes ensuring the functioning of the shutter are: 1) the creation of a region of drift of electric charges in the space between the corona and forming electrodes; 2) adsorption of electric charges by finely dispersed particles; 3) deceleration of charged particles in an electric field; 4) redirection of the motion of charged dust particles. Thus, three stages can be distinguished in the working area of the device (Figure 2): 1) the region of particle charge; 2) area of inhibition; 3) the area of removal of particles from the working area.

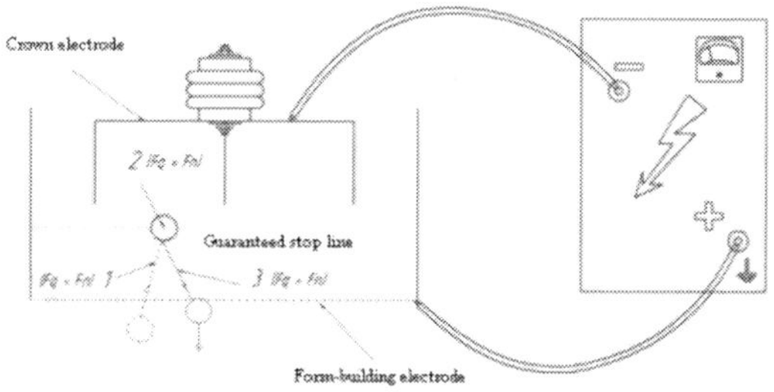

Fig. 2. Three stages of blocking dust emissions: 1) charging of the finely divided fraction of the dust-air mixture; 3) inhibition of the dust fraction; 2) return of the dust fraction to the source of dust or to a specially designated place for collecting the finely dispersed fraction.

To date, a large number of different devices for blocking dust emissions into the atmosphere have been developed, manufactured and used. They differ in a number of parameters, but most devices are air filtration systems, whereby filtration is carried out by depositing dust particles on the structural elements [18]. This method of cleaning affects the dimensions of the cleaning device and energy consumption for regeneration elements. The device, called an electrostatic shutter, performs a non-contact method of blocking a finely dispersed fraction in which the trapping and deposition of the dust fraction on the structural elements is not the main but a concomitant process [19-20].

The closest competitors of the electrostatic shutter are bag filters and electrostatic precipitators. The functional capabilities of the device proposed for examination are superior to known analogs (bag filters) or have similar characteristics (electrostatic precipitators) and provide a high efficiency of suppression of dust streams (degree of suppression of the dust fraction at least 95%) with significantly reduced energy costs, which leads to the economization of valuable material and increase the efficiency of transportation of raw materials within the technological system. The selection of effective geometric parameters allows to reduce the size of devices for blocking dust emissions by 3-4 times and to reduce energy consumption by 5-10 times (up to 20 times). A high degree of efficiency of suppression of dust flows allows using these devices on an industrial scale (Fig. 3).

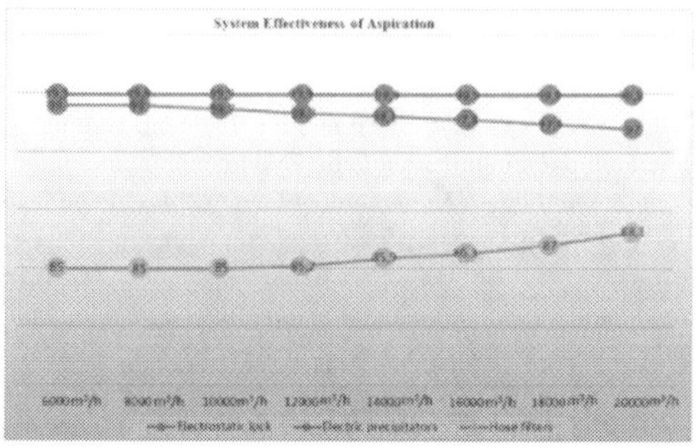

Fig. 3. The effectiveness of the aspiration system.

The essential advantage of an electrostatic shutter is its cost. On average, the cost of the electrostatic shutter is 1.5 times lower than the cost of a bag filter and 2 times lower than the cost of an electrostatic precipitator manufactured domestically. Thus, the total costs for the acquisition and operation of an electrostatic shutter are 15 to 20 times lower than the available analogs for the same cleaning efficiency (Figure 4).

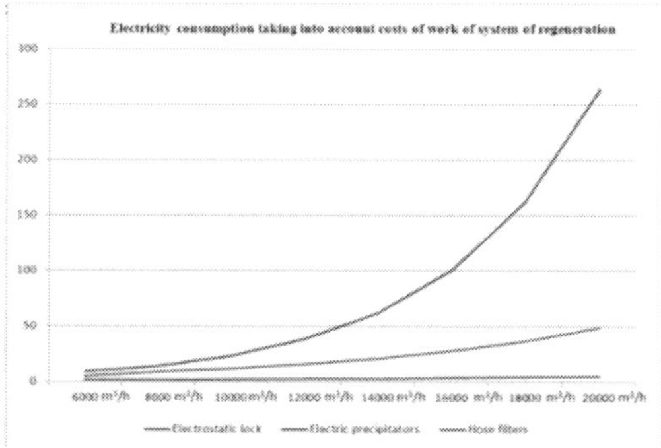

Fig. 4. Electricity consumption, taking into account the costs of the regeneration system.

On the basis of the conducted researches it is possible to assert that this device is effective and competitive in relation to analogues.

3 Alternative applications of electromechanical dust suppression devices

As an alternative application, one can consider the use of electromechanical devices as corona separation devices. Obviously, the underlying principles of these devices allow to broaden their scope of application provided that the method of controlling the behavior of the fine fraction in the active device area and use the ring separation electromechanical devices to select the fine fraction with the desired properties. At the present time, the most effective way of separating the finely dispersed fraction is electrical separation. A fairly large number of types of devices are known to divide the flow of a finely dispersed fraction.

Electrical separation is used to enrich granular bulk materials with a particle size of 3-0.05 mm. Depending on the method of creating on the charge particles and its transfer in the process of electrical separation, there are: electrostatic, corona, dielectric, triboadhesive separation

Under corona discharge conditions, the force exerted by the electric field acting on the particles of the fine-dispersed fraction is formed under the influence of the space charge, which affects the magnitude of this force. We should also take into account the aerodynamic effort called "ion wind", which in these devices, according to earlier conducted experimental studies is about 15-20% of the force, from the side of the electric field. The distribution of fine-dispersed particles by physical properties and dimensions is also an important factor, since it affects the dynamics of particle acquisition and aerodynamic properties.

The system of equations for the physical processes of particle motion in an inhomogeneous electric field, in general, with an arbitrary moving part (aerosol cloud) is built on the basis of the mathematical model developed earlier by the project manager, but, unlike it, acting on the part of "ion wind" and includes the following equations:

$m\frac{dV}{dt} = \sum F$ - the equation of motion of the particle (1)

$F_2 = q(t)E$- Coulomb force acting on a charged particle (2)

$F_1 = mg$-the particle force of a dust particle (3)

$F_3 = \rho\frac{\pi r^2}{4}$ - the force of excess air pressure (4)

$F_4 = \frac{I_k \times h_{м.э.}}{k_{ионов}}\frac{\pi r^2}{4}$ -the force created by the "ion" wind (5)

In this system of equations: φ is the scalar electric potential in the computational domain, ρ is the volumetric charge density, q (t) is the charge in the analyzed volume of the calculated region, v is the volume of the investigated space, m is the mass of the dust particle, qm is the limiting charge of the dust particle, p - air pressure in the dust-air mixture flow, p.vetra - air stream pressure created by the ion wind, r - radius of the dust particle.
After substituting expressions (2-5) in (1), we obtain the following expression for the balance of forces:

$$m\frac{dV}{dt} = mg + q(t)E + \left(\frac{I_k \times h_{м.э.}}{k_{ионов}} - \rho\right)\frac{\pi r^2}{4} \qquad (10)$$

The numerical solution of the given equation will allow to predict the behavior of a particle of a finely dispersed fraction. After receiving theoretical results, to verify their adequacy, it is necessary to conduct experimental studies.

Further studies of the physical characteristics of the electrostatic shutter are related to the study of the efficiency of work with a change in the particle size of the finely dispersed fraction passing through the active zone. Investigation of the possibility of the shutter performing its functions was carried out at the moment the dust particle moved along the trajectory passing through the region of the least impact from the electric field. The corresponding trajectory of motion is found on the basis of the analysis of the characteristics of the magnitude of the force acting on a particle of matter along the guaranteed stop line. In this construction, this trajectory will coincide with the line of symmetry between the corona electrodes. The force acting on the dust fraction, as the core passes through the pre-selected trajectory, is shown in Fig. The investigations were carried out for particles with a diameter from 40 μm to 160 μm. The calculated stopping point is indicated by the symbol (•).

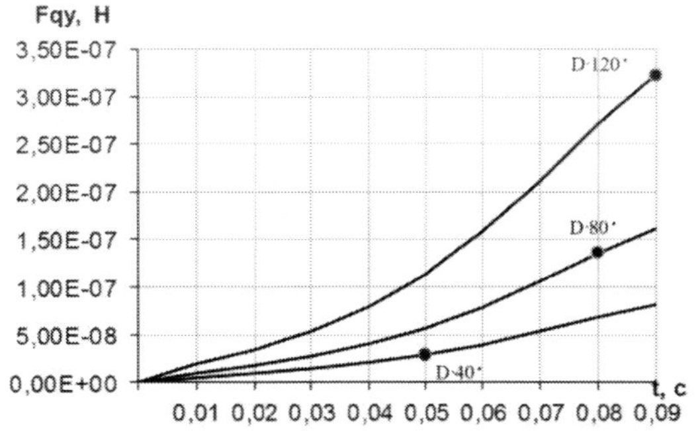

Fig. 5. Dynamics of the increase in force from the side of the electric field as the particle acquires a charge of a diameter from 40 μm to 160 μm as the core passes through the electrostatic shutter.

The study was carried out under the worst conditions for charging the particles. The analysis of the results obtained made it possible to conclude that the excess air pressure acting on the dust particle is balanced by the action from the electric field the faster the larger the geometric dimensions of the finely dispersed fraction. Modeling of physical processes in this design was carried out for a reverse polarity supply voltage of 30 kV and a distance between the corona electrodes of 100 mm.

Conclusion

The field of application of the electrostatic shutter is quite extensive, because the bulk of the pollutants are gaseous substances and finely dispersed materials that can travel for long distances and accumulate. The most demanded ESZ will be in the technological processes in the cement industry, metallurgy, brick and slate plants, for cleaning the air flow in the main electric locomotives and wagons, cleaning the smoke from solid fractions, cleaning the air in centralized air intakes of ventilation systems. For example, in the cement industry, the ESP can be used: for installation on the exhaust holes of clinker silos; when loading in road tankers; for installation on exhaust openings of cement silos; on the nodes of the transfer from the bucket conveyor; for post-treatment of the exhaust of the spike cooler; for loading wagons.

DOI: 10.1051/matecconf/201713202013

References

1. M.S. Minkin, Electromechanical devices of blocking of dust emissions, Cand. tech. sciences. Diss., Novocherkassk, 144 (2013) [in Russian]
2. M.S. Minkin, G.I. Volodin, News of higher educational institutions. Electromecanics, no. 5, 33-38 (2013)
3. G.I. Volodin, V.I. Nadtoka, I.I. Nadtoka, V.I. Hludeev, P.F. Sidorenko, V.P. Breslavets, A.Y. Boasters, M.S. Minkin, Patent No.2438794, Russian Federation, MPK B03C3/00
4. G. I. Volodin, V. I. Nadtoka, I.I. Nadtoka, V.P. Breslavets, Nis Ya.Z., P.F. Sidorenko, M. S. Minkin, Patent No.107077, Russian Federation
5. M. S. Minkin, News of higher educational institutions. Electromecanics, no. 5, 61-66 (2010)
6. E. Karakas, A. Begum, M. Laroussi, A Positive, *In Proc.IEEE Transactions on Plasma Science*, **36**(4), 950-951 (2008)
7. D. Lacoste, D. Pai, C. Laux, *In Proc. 42nd AIAA Aerospace Sciences Meeting and Exhibit* (2004)
8. R. Na, G. Fan, *International Conference on Materials for Renewable Energy & Environment* 2014-2017 (2011)
9. R. Chengyan, W. Jue, Y. Ping, X. Rong, W. Tao, *IEEE 11th International Conference on the Properties and Applications of Dielectric Materials*,784-787 (2015)
10. V.L. Bychkov, V.A. Chernikov, S.A. Volkov, D.V. Bychkov, A.A. Kostiuk, J. *IEEE Transactions on Plasma Science*, no. 11, 2640-2641 (2011)
11. T. Johnson, S. Jakobsson, B. Wettervik, B. Andersson, A. Mark, F. Edelvik, Journal of Electrostatics, 27-36 (2015)
12. Y. Zhang, L. Liu, J. Ouyang, Journal of Electrostatics, no. 1, 76-81 (2014)
13. A. Iuga, R. Morar, A. Samuila, I. Cuglesan, M. Mihailescu, L. Dascalescu, J. *IEEE Trans, on Ind. Appl*, no.3 (1999)
14. R.Morar, A.Iuga, I.Cuglesan, O.Muntean, L.Dascalescu, J. *IEEE Trans, on Ind. Appl*, no.1 (1999)
15. G. Kazkaz, J. *IEEE Trans, on Ind. Appl*, no. 4 (1998)
16. A. Iuga, V. Neamtu, I. Suarasan, R. Morar, L. Dascalescu, J. *IEEE Trans, on Ind. Appl*, no. 2, 290-326 (1998)
17. L. Dascalescu, D. Rafiroiu, A. Samuila, R. Tobazeon, J. *IEEE Trans, on Ind. Appl*, no. 1 (1998)
18. L. Daskalescu, R. Morar, A. Iuga, V. Neamtu, I. Suarasan, J. Magn. and Electrical Separ. No. 2, 91-106 (1993)
19. N. Szedenic, I. Kiss, *8 Int. Conf. on El.st.*, 88-92 (1997)
20. M.S. Minkin, A.D. Lukyanov, D.N. Kuimov S, Materials Science Forum, 625-629 (2016)

Technological processes intensification in devices with magneto-fluidized bed

Ivan Egorov[1,*], *Nikolay* Egorov[2]

[1]Don State Technical University, Rostov-on-Don, Russian Federation
[2]Southern Federal University, Rostov-on-Don, Russian Federation

Abstract. The paper shows the efficiency of application of magneto-fluidized bed formed of ferromagnetic materials in press-forms dosed filling and milling devices. For powders without natural fluidity, dosing device is situated between electromagnets creating constant magnetic field with horizontal induction lines and alternating gradient magnetic field with vertical induction lines and higher induction gradient in the lowest part of chamber with powder. This configuration allows development of uniform efflux speed of ferromagnetic powders from 2 mm diameter opening. For example, mass efflux speed of strontium ferrite powder with 1 μm average particle size in magneto-fluidized bed reached 181.1 mg/s and the dosing time of 2000 mg of this powder was 11.1 s. For practical usage of magneto-fluidized bed in the milling device, beater mill is situated between electromagnet poles in the way that induction lines of constant and alternating gradient magnetic fields were mutually perpendicular and parallel to the plane of rotating beaters. Milling of particulate strontium ferrite with 1558.5 μm average particle size during 120 minutes in magneto-fluidized bed allows increasing milling degree in 16.2 times by comparison with processing without electromagnetic effect.

1 Introduction

Powder metallurgy uses various particulate metals and alloys as raw materials. Particle size decrease increases interparticle interaction leading to changing of particulate system technological characteristics [1]. Considerable theoretical and experimental material about influence of magnetic fields on rheological state of magnetic fluids [2-4] and particulate materials in fluidized state [5-7] was accumulated. There is a wide range of devices that fluidize particulate medium by means of hydration, aeration, ultrasonic, electrodynamic and electromagnetic effects [8-11]. In case of ferromagnetic materials electromagnetic effect on particulate medium (uniform alternating magnetic field, rotating magnetic field, superposition of alternating gradient and constant magnetic fields [12-15]) is the most effective way of obtaining dynamically stable weighted state of particulate system – magneto fluidized bed. Using of magneto-fluidized bed can largely intensify technological processes of mixing, milling, dosing, drying etc. [16-19].

For practical use of magneto-fluidized bed, ferromagnetic powder is placed in working chamber situated between poles of electromagnets forming constant and alternating gradient magnetic fields. Disposition of working chamber, poles of electromagnets and their configuration depend on required character of particle motion in the process.

2 Problem

Ferromagnetic powders don't have natural fluidity (don't flow through an opening in bunker to press-form) as their particles form aggregates and they stuck in the opening. Magneto-fluidized bed allows to destroy particles aggregates and greatly reduce interparticular friction. This makes possible flowing of powder through bunker opening.

In case of dry milling particle aggregates are also a problem as large part of energy goes to aggregate destruction and small particles that surround large ones protect them from milling effect. This problem can be partially solved by powder fluidization as it destructs particle aggregates.

The goal of the work is in development of determination principles of the most effective regimes and required configuration of alternating gradient and constant magnetic fields for intensification of technological processes of dozed press-form filling and milling in rheological controlled fluidized bed.

3 Experiment

In alternating gradient and constant magnetic fields depending on regimes of electromagnetic effect on ferromagnetic powder, take place processes of aggregate destruction as well as formation of secondary aggregates.

* Corresponding author: egorovivan1@gmail.com

Direction of particle motion in magneto-fluidized bed depends on electromagnets poles configuration, values of constant field induction and alternating field gradient [20] and is conditioned by requirements to the device. Fig. 1 presents diagram of device for development of ferromagnetic powders stimulated fluidity. Poles of direct current electromagnet 1 form magnetic field with horizontal induction lines. Poles of alternating current electromagnet 2 form magnetic field with vertical induction lines with higher induction gradient below working chamber 3. For development of high gradient, we sharpened one of alternating current electromagnet poles. Working chamber contains wire mesh 4 to prevent clogging of efflux opening with large particles. To develop stimulated fluidity through efflux opening 5 with diameter 2 mm in the receiving chamber 6, the magneto fluidized bed is formed by the effect of magnetic fields. In order to minimize arched effects, we used working chamber with flat bottom. Higher gradient of alternating magnetic field below efflux opening makes ferromagnetic particles flow to the receiving chamber.

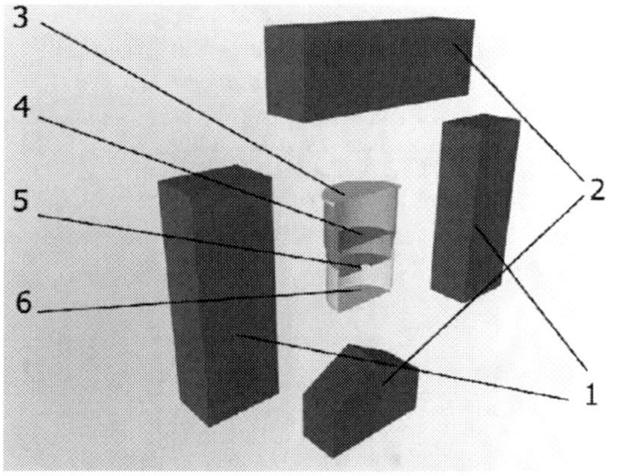

Fig. 1. Diagram of device for ferromagnetic powders stimulated fluidity development: 1 – direct current electromagnet poles, 2 – alternating current electromagnet poles, 3 – working chamber, 4 – wire mesh, 5 – efflux opening, 6 – receiving chamber

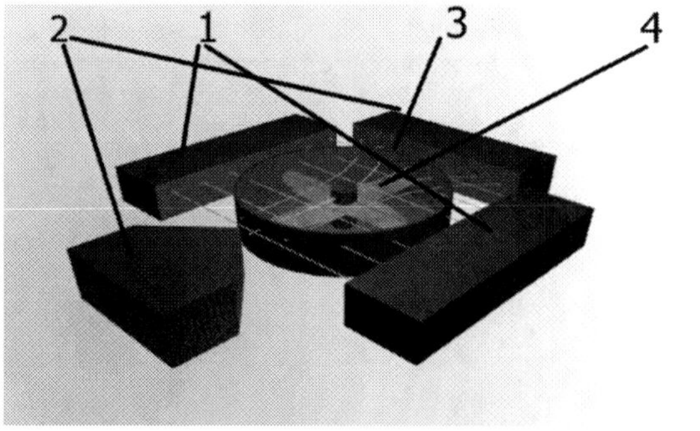

Fig. 2. Diagram of ferromagnetic material milling device: 1 – direct current electromagnet poles, 2 – alternating current electromagnet poles, 3 – beater mill, 4 - beaters

Fig. 2 presents diagram of device for milling particulate ferromagnetic materials. Between the poles of direct current electromagnet 1 and alternating current is located beater mill 3 with nonmagnetic body. Induction lines of constant and alternating gradient magnetic fields are mutually perpendicular and parallel to the plane of rotating with 15000 rpm beaters 4. Dynamics of magneto-fluidized bed of particles and aggregates in beater mill determinates the character of impact interactions of these elements with rotating beaters and between themselves. Large particles are destructed by rotating beaters, smaller ones have less probability to be hit by beaters but are restlessly hit by other particles, that leads to self-milling. An important condition for development of magneto-fluidized bed during milling of particulate ferromagnetic material is the intense particle motion providing intensification of the technological process.

At given configuration of magnetic fields for making choice of the most effective electromagnetic effect regimes in each case of practical realization one need to pick up the ratio of constant magnetic field induction and alternating field induction gradient providing rheological features of magneto-fluidized bed.

4 Results and discussion

Significant demands at continuous processes of dosing, press-form filling and fabrication of permanent magnets are made to provision of constant efflux speed of particulate material from opening, because dosing error influences quality of final product [21].

Fig. 3 presents experimental dependencies of mass efflux speed of barium ferrite powder with 1 μm average particle size (a) and of strontium ferrite with 1 μm (b) and 50 μm (c) average particle size through opening with 2 mm diameter from constant magnetic field induction at fixed values of alternating magnetic field induction gradient. Presented dependencies show that ferrite powders mass efflux speed grows with increase of induction gradient at fixed constant magnetic field induction. This can be explained by considering that in gradient alternating magnetic field the force proportional to the induction gradient acts on particles of ferromagnetic material. In case of barium ferrite powder with 1 μm average particle size (Fig. 3a) in the field with induction gradient 250 mT/m mass efflux speed increases to 34.3 mg/s when increasing constant magnetic field induction (B_c) to 6.8 mT. Further B_c increase leads to mass efflux speed decrease as high constant magnetic field begin to hold particles in the area of its induction lines. Induction gradient increase to 593 mT/m at B_c=15.7 mT makes mass efflux speed increase to 96.6 mg/s. In case of strontium ferrite with 1 μm average particle size in field with induction gradients of 250 and 593 mT/m (Fig. 3b) maximal values of efflux speed are 97.4 and 181.1 mg/s, correspondingly. For the same induction gradient values in case of strontium ferrite 50 μm average particle size (Fig. 3c) values of corresponding speeds are 83.4 and 143.4 mg/s.

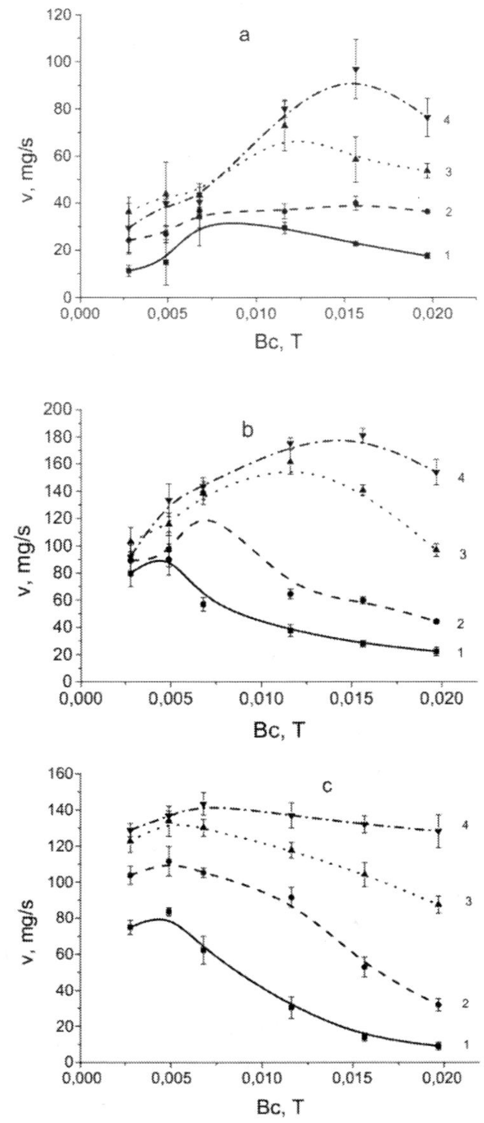

Fig. 3. Dependencies of mass efflux speed of barium ferrite (a) and strontium ferrite (b, c) powders through 2 mm diameter opening from constant magnetic field induction at various alternating field induction gradient values: 1 - 250 mT/m, 2 - 390 mT/m, 3 - 525 mT/m, 4 - 593 mT/m

Obtained results show that to fabricate permanent magnet with mass of 2000 mg from barium ferrite and strontium ferrite powders with 1 μm average particle size in regime of B_c=15.7 mT and induction gradient 390 mT/m dosing time will be 50.0 and 33.3 seconds correspondingly. In case of induction gradient increase to 593 mT/m, dosing time will decrease to 20.0 and 11.1 seconds correspondingly.

Main drawbacks at powder charge preparation for pressing in existing ferrite production are: long milling duration, large energy consumption and poor reproducibility of particle size distribution that leads to variance of permanent magnet parameters [22].

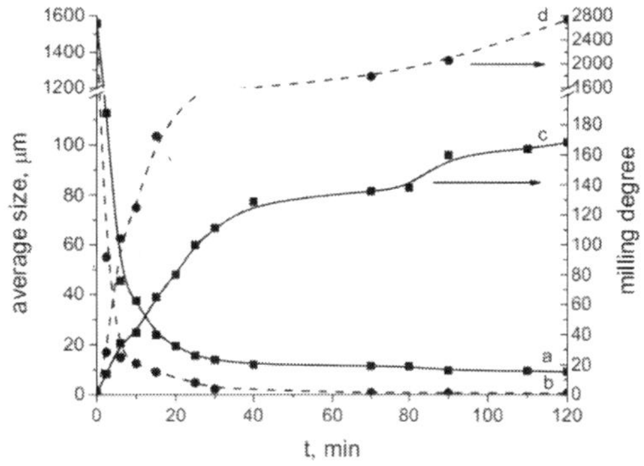

Fig. 4. Dependence of average particle size and milling degree of strontium ferrite powder from milling time without electromagnetic effect (a, c) and in magneto fluidized bed (b, d)

Fig. 4 presents experimental dependencies of average particle size and milling degree (ration of initial average particle size to current particle size) from milling time in beater mill without electromagnetic effect and in magneto fluidized bed, formed in constant magnetic field with induction of 15.3 mT and in alternating one with induction gradient of 90 mT/m. Particulate strontium ferrite with average particle size 1558.5 μm, maximal size of 3100 μm, median of 1477 μm was milled. In case of processing initial particulate material in mill during 120 minutes without electromagnetic effect, average particle size decreased to 9.26 μm and in magneto-fluidized bed to 0.57 μm, milling degree increased in 16.2 times, that witness of milling intensification in magneto-fluidized bed.

The increase of milling efficiency is archived through holding particulate medium in rotating beaters area and self-milling of particles as a result of collisions of particles with each other in alternating gradient magnetic field.

On early milling stages beaters make almost all job, but the less particle size becomes the less effective they become. Rotating beaters don't mix powder in the chmber enough, some of it lies on the bottom slightly involved in the milling process. Magneto-fluidized bed develop additional mizing that involves all the powder into milling process.

Calculations show that value of particle kinetic energy obtained in magneto-fluidized bed is not enough for particle destruction in the interparticle collision act. However, despite insufficient speeds of particle motion in magneto-fluidized bed, intensive self-milling takes place there. Multiple interparticle collisions cause fatigue destruction of powder particles. Processing of strontium ferrite particulate material with 81.3 μm average particle size only in magneto-fluidized bed formed in constant magnetic field with induction of 15.4 mT and alternating magnetic field with frequency 50 Hz and induction gradient of 90.0 mT/m, allowed to decrease dispersed composition of obtained product. Quality of particles smaller than 5.5 μm contained in initial powder increased in 31 times [23] after 70 minutes of processing. In this way, intensification of

DOI: 10.1051/matecconf/201713203001

milling process in magneto fluidized bed is conditioned by self-milling and mixing of particulate material.

Milling in magneto-fluidized bed require additional energy consumption on power supply of electromagnets. However, getting strontium ferrite powder with 9 μm average particle size from initial 1558.5 μm requires 130 minutes and the 15000 J of energy. In case of milling in magneto fluidized bed for getting the same average particle size requires 15 minutes and the energy constitutes only 3126 J.

Presented results show that application of magneto-fluidized bed to milling of particulate ferromagnetic materials not only intensifies the process but also decreases its energy consumption.

5 Conclusion

The paper experimentally substantiates effectiveness of application of magneto-fluidized bed of particulate ferrite materials in devices of dosed press-form filling and milling.

Application of magneto-fluidized bed to press-forms filling process allows developing steady efflux of ferromagnetic powders without natural fluidity. Electromagnets configuration required for the process is horizontal constant magnetic field and vertical alternating magnetic field with high gradient below efflux opening. High and steady efflux speed is archieved at constant magnetic field induction about 15 mT (for 1 μm barium and strontium ferrite powders) and alternating field induction gradient about 600 mT/m. This allow to develop automated dosing devices for press-forms filling and futher pressing of powder charge into permanent magnet blank.

Application of magneto-fluidized bed to the process of particulate ferrite milling in beater mill allows dry milling process to average powder particle size less than a micrometer. This is unattainable at the same mill without electromagnetic effect. Development of magneto-fluidized bed holds powder particles around rotating beaters making all particles involved in the milling process. Restless interparticle collisions, though small, but develop fatigue stresses that cause particle fatigue destruction (like in vibrating mill).

Analysis of obtained results shows that the choice of magnetic fields configuration optimal for specific process and determination of most rational regimes to develop predetermined rheological state of particulate system in magneto-fluidized bed allow to greatly intensify technological processes.

The work was supported by grant of RFFI № 16-38-00365.

References

1. J.M. Valverde, Springer Netherlands, **136** (2013)
2. J. De Vicente, D.J. Klingenberg, R. Hidalgo-Alvarez, Soft Matter 7 3701 (2001)
3. H. See, Rheologica Acta **42**, 86 (2003)
4. T. Gadipelly, S. Rajender, Physica B: Condensed Matter **448**, 346 (2014)
5. J.Y. Hristov, *5th European Rheology Conference*, (Portoroz, Slovenia, 1998)
6. M.J. Espin, M.A.S. Quintanilla, J.M. Valverde, Chem. Eng. J. **277**, 269 (2015)
7. M.J. Espin, J.M. Valverde, M.A.S. Quintanilla, J. of Fluid Mech. **732**, 282 (2013)
8. I.B. Uriev, N.V. Mikhaylov, P.A. Rebinder, Rep. of AS of USSR **184**, 192 (1969) [in Russian]
9. I.M. Kirko, M.V. Filippov, J. Tech. Phys. **30**, 1081 (1960) [in Russian]
10. P.A. Kouzov, *Basics of analysis of disperse compound of industrial dusts and milled materials*, **264** (1987) [in Russian]
11. O.A. Myazdrikov, *Electrodynamic fluidizing of particulate systems*, **160** (1984) [in Russian]
12. A.P. Lupanov, A.Yu. Vinarov, A.P. Kanavin, V.I. Panteleev, J. Tech. Phys. **77**, 134 (2007) [in Russian]
13. Yu. A. Buevich, S.V. Syutkin, V.V. Tetyuhin, Magn. Hydrodyn. **4,** 3 (1984) [in Russian]
14. M.K. Bologa, I.F. Marta, Magn. Hydrodyn. **3**, 103 (1988) [in Russian]
15. I.N. Egorov, S.I. Egorova, Russ. J. Non_Ferr. Met. **55**, 371 (2014) [in Russian]
16. I.N. Egorov, S.I. Egorova, A.I. Chernyi, Pat 022640 (EA, 2016)
17. I.N. Egorov, S.I. Egorova, Yu.M. Vernigorov, Pat 2306180 (RF, 2007)
18. T. Chou, G. Nishizawa, M. Hatakeyama, C. Ishizaka, Pat 7416613 (US, 2008)
19. S. Okumura, A. Nakamura, Pat. 2003/0209842 (US, 2003)
20. I.N. Egorov, S.I. Egorova, G.F. Lemeshko, *Perspective directions of developing finishing methods of workpiece processing* (Symp. Eng. Techn., Rostov-on-Don, 2016) [in Russian].
21. A.M. Stryzhankov, P.F. Pool, Pat 1253738 (USSR, 1986).
22. D.D. Logvinenko, O.P. Shelyakov, *Intensification of technological processes in devices with vortex layer* (1976) [in Russian].
23. N.Ya. Egorov, I.N. Egorov, V.P. Kryzhanovsky, A.I. Chernyi, Int. sc. Inst. "EDUCATIO", **2**, 41 (2015) [in Russian].

Investigation of the influence of contact friction forces in tests on the strength of compressing wall materials

Nataliya Lukinova[1,*], *Andrey* Matrosov[1], *Irina* Serebryanaya[2], *Yuliya* Terekhina[2], *Dariya* Nizhnik[1]

[1]Don State Technical University, Rostov-on-Don, Russia
[2]Academy of Architecture and Construction, Don State Technical University, Rostov-on-Don, Russia

Abstract. The results of the experiment are presented, which estimate the effect of the support friction in standard tests on axial compression on the output strength characteristics of the hand-formed brick. The effect of surface parameters on the compressive strength was established. The optimal shape and dimensions of the samples, as well as the method of preparation of their surfaces, were determined.

1 Introduction

Axial compression testing is the main type of strength control of most building materials. The results of these tests are laid in the designs of buildings and structures, from which it follows that the errors of these experiments should be reduced to a minimum. When analyzing the results of tests of building materials on compressive strength, it is assumed that these values were obtained under conditions of a linear stress state, in reality, a volumetric stress state is realized in tests, which is confirmed by the nature of the destruction of the samples [1-2]. The volumetric stress state is due to the presence of frictional contact forces between the sample and the press plates, which in turn leads to an overestimation of the strength parameters [3-4]. This explains the practical relevance and relevance of the chosen topic, since a reliable estimate of compressive strength is a necessary condition for correct calculation of buildings and structures. An analysis of the results of experiments aimed at investigating the effect of frictional contact forces in compressive strength tests will make it possible to develop an optimal test method that ensures that the results obtained are close to the true value of the measurand [5].

2 Methods of research

The tests were carried out on tested certified equipment. In order to measure the destructive load, a hydraulic press P-125 equipped with a torsion dynamometer was used. For the production of samples representing the halves of the product, the PRG-1-50K press was used, which is a hydrostatic force multiplier with a manual drive.

The bearing face (bed) of ceramic bricks always has significant deviations from flatness, which does not ensure the uniformity of the distribution of load on the entire plane of the sample when tested for compressive strength [6]. Therefore, in the preparation of samples for testing, the surfaces are leveled, which in the structure and, accordingly, during testing, are located perpendicular to the direction of the axis of the compressive load.

According to the current legislation of the Russian Federation in accordance with GOST 530-2012 "Brick and ceramic stone. General specifications" preparation of bearing faces of brick samples for acceptance testing is performed by grinding. The deviation from the flatness of the bearing surfaces of the test specimens shall not exceed 0.1 mm for every 100 mm of length. The nonparallelism of the support surfaces of the test specimens (the height difference measured by the four vertical edges) should not be more than 2 mm [7]. It should be noted that the surface of the hand-formed brick, due to the special features of the production, is characterized by great heterogeneity, is characterized by the presence of grooves, cracks, corrugations, bugholes and spallings that are not defects, and serving as necessary means for implementing the decorative function of the product and imitating the ancient brick laying [8-9]. Thus, it is logical to assume that as a result of grinding, the contact surface of the tested samples will be substantially changed, which in turn will inevitably lead to distortion of the measurement results and overestimation of the strength characteristics of the batch. It is also permissible to apply other methods of leveling the reference surfaces of samples when carrying out acceptance testing, provided there is a correlation between the results obtained by different methods, as well as the availability of verification of information that is the basis for such a connection [10].

Regarding the European legislation, in accordance with EN 772-1: 2014 "Methods of test for masonry units Part 1. Determination of compressive strength" of the surface of the test specimen is made to conform to the requirements for flatness and straightness. The deviation from the flatness of the loaded surfaces of the sample

* Corresponding author: lukinova-n@ya.ru

(article or fragment) should not exceed 0.1 mm for every 100 mm of length. The deviation from the straightness of the upper surface of the test specimen should not exceed 1 mm for every 100 mm of length. However, with a strong change in the contact surfaces due to grinding, they are leveled using a cement-sand mortar [11-12]. In our opinion, this method also has a significant drawback, since it leads to the appearance of a mechanical interaction of the product with a grout based on the penetration of the grout in bugholes with the formation of a locking engagement. And this is the same as in the case of grinding will lead to an additional increase in strength.

Other methods for preparing a brick surface for compressive strength tests are known, in particular in the Netherlands, in accordance with NEN 6790: 1991, two whole bricks are used in the tests, using felt liners between them and the support surfaces of the felt [13]. In the USA, brick tests are performed in accordance with ASTM C67-03a. According to the specified standard, it is possible to use two methods of surface preparation: leveling with gypsum mortar or sulfur concrete solution [14]. In India, however, there is no standard for test methods for bricks, but traditionally one whole brick is used for testing, placed between two plywood sheets [15].

Based on the analysis of the domestic and foreign regulatory framework, the selection of methods for preparing the surfaces for the experiment was made.

For grinding the samples, a belt grinder MAKITA 9404 was used.

To prepare the surface of the remaining samples for testing, the following materials were used:

• a cement-sand mortar consisting of equal parts of cement grade 400 and sand sieved through a sieve with mesh No. 1,25 (V / C = 0,40), layer thickness 5 mm;

• gypsum mortar consisting of gypsum grade G-16, layer thickness 5 mm, water-gypsum ratio 0.35;

• 5mm technical thin-wool felt in two layers;

• plywood birch 9mm.

The shapes and sizes of the samples, which are acceptable in accordance with different standards, are also unequal. In accordance with Russian legislation, until 2012, when determining the compressive strength, it was allowed to use two halves of bricks, in particular those formed after the bending test, stacked on opposite surfaces with each other [16-17]. In accordance with the amendments introduced in 2012, it is allowed to use only whole bricks, laid on top of each other. According to the European standard, a sample consisting of one whole product is used. An exception is made only if, after grinding, the height of the sample is less than 40 mm or the ratio of height to width is less than 0.4, a composite sample is made by stacking the products one on top of another without using a solution, a connecting material or a separation layer. And, finally, under American law, the test samples represent single halves of bricks. Samples should be obtained by any method that allows the separation of bricks into two equal parts without cracks and destruction.

Based on the results of the analysis of the domestic and foreign regulatory framework, a sample of methods for preparing samples for the experiment was made.

3 Experimental results

Based on the tests performed, a summary table of results was compiled (Table 1).

Table 1. Results of the determination of compressive strength.

Alignment method	Number of products for one sample	R_1, MPa	R_2, MPa	R_3, MPa	R_4, MPa	R_5, MPa	Average value R, MPa
Without alignment	two whole	11,957	13,252	16,622	12,959	12,697	13,497
	two halves	18,296	19,362	15,636	16,064	17,261	17,323
	one whole	18,781	16,027	14,766	17,179	20,998	17,550
	one half	20,983	19,625	21,202	15,01	22,516	19,867
Alignment with cement-sand mortar	two whole	14,691	12,971	15,682	14,459	13,121	14,184
	two halves	18,289	19,321	16,136	17,064	18,261	17,814
	one whole	19,396	19,652	18,841	15,42	17,188	18,099
Grinding alignment	two whole	15,284	15,771	17,391	16,069	15,859	16,074
	one whole	18,187	18,962	19,776	18,179	20,998	19,220
Aligning with gypsum	one whole	20,166	19,991	21,533	16,011	21,215	19,783
Felt Alignment	two whole	10,809	10,073	10,268	7,862	9,573	9,717
	two halves	10,882	11,276	12,372	11,47	11,409	11,481
Plywood Alignment	one whole	16,129	18,944	15,139	16,961	19,933	17,421

Based on the obtained results, it was established that in addition to the method of leveling the sample surface before the test, the size and shape of the sample exert a large influence on the compressive strength value. So, it is evident that, other things being equal, the samples consisting of two whole products show the least strength, regardless of the method of alignment. Further strength increases with decreasing dimensions of the supporting surfaces and reaches its peak on a sample consisting of one half of the product. This is clearly demonstrated in the diagram shown in Fig. 1. For its compilation, alignment methods were used, according to which the test results with the largest number of sample types were used. As for the stability of the results, the smallest range of data is characteristic of samples consisting of two halves of the product stacked on top of each other. This is due to the fact that this type of specimen in shape is most closely approximated to the shape of the cube.

DOI: 10.1051/matecconf/201713203002

It is also worth noting that for all types of leveling, the average strength of samples consisting of two halves is almost equal to the strength of samples consisting of a single whole product, but the latter is less stable.

Thus, as the dimensions of the sample and the areas of the contact surfaces increase, the resistance decreases. The decrease in strength in large samples with the usual method of testing them should be explained by the diminishing influence of the support friction.

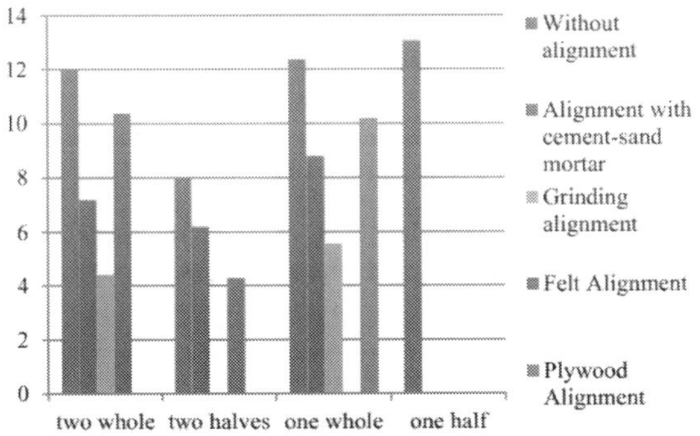

Fig. 1. Dependence of strength value on sample size and surface leveling method

In general, the lower the sample, the stronger the effect of the reference friction on the value of the time resistance to compression. Conversely, as the height of the sample increases, the transverse expansion of the latter becomes more free and the material reaches its ultimate strength under a lower load [18-19].

With regard to the method of leveling the surface, the samples aligned with the pads made of technical felt, regardless of shape and size, show less strength, compared with other methods of surface preparation, but are characterized by a smaller spread of values. When leveling with felt, often the sample splits along the entire length (Fig. 2). In other cases, des6rtuctions are presented by the scattering of the faces of the samples.

a - the sample before the test; b - sample after the test

Fig. 2. The nature of the destruction of a sample lined with felt pads

As can be seen from the figure, the through crack passes through the entire sample exactly in the middle. Such results are associated with a decrease in the influence of frictional force, since the felt is smoother than the untreated surface of the hand-formed brick or the cement solution. Cracks are formed in the sample, parallel to the direction of compression (perpendicular to the bed). In this case, friction no longer prevents the development of transverse deformations of the end faces of the sample, and they increase until a break in the brick occurs [20-21]. Accordingly, destruction from the break occurs at a much lower compressive load than shear failure. In other cases, friction delays transverse deformations of the sample (transverse expansion) and does not allow them to develop up to the limiting value corresponding to the rupture.

Along with this, the increased strength parameters of the samples, leveled by means of a cement-sand mortar, are caused by the mechanical interaction of the solution with the surface of the sample, based on the penetration of the first into the cavities with the formation of a locking engagement.

The high strength of the samples, leveled by means of grinding, is due to a significant change in their contact surface, which also makes it difficult to form an objective quality assessment.

4 Results of the numerical experiment

To verify the conclusions of the full-scale experiment, finite element modeling of the stressed-deformed state of the hand-formed brick was carried out using the ANSYS 17.2 software package (Fig. 3).

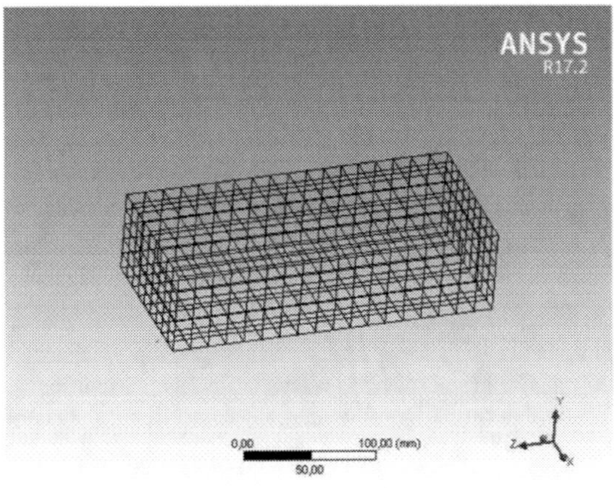

Fig. 3. Finite Element Model

Modeling was carried out taking into account various ways of preparing the surface of samples for testing, as well as various types of samples in shape and size.

The results of the numerical experiment very well agree with the field experiment, which shows the correctness of mechanical model selection.

5 Conclusion

On the basis of the conducted research, the effect of contact frictional forces on the output characteristics of the strength of the hand-formed brick was established. It is shown that with standard tests of bricks in accordance with GOST 530-2012 a systematic error arises due to the appearance of a three-dimensional stress state. This is confirmed by the nature of the destruction of the samples. The volumetric stress state is due to the presence of frictional contact forces between the sample and the press plates, which in turn leads to an overestimation of the strength parameters. According to the results of the experiment, the most optimal method of preparing surfaces for testing is the use of linings from technical felts. The tests themselves are recommended to be carried out on samples consisting of two halves of the product laid on top of each other.

The results of the numerical experiment agree well with the results of the field experiment, which indicates that the model constructed in ANSYS 17.2 reliably describes the real picture and can be used as a basis for further research.

References

1. I.A. Serebryanaya, J.V. Terehina, N.A. Lukinova, *International Conference PHENMA 2016*, Surabaya, Indonesia, 237 (2016)
2. S.S Kirilenko, Herald of Civil Engineers, **2(23)**, 48-50 (2010)
3. V.Z. Vasiliev, S.S Kirilenko, *III International Scientific Conference "Actual Problems of Mechanics and Mechanical Engineering"*, Almaty, 142-145 (2009)
4. V.V. Pangaev, Proceedings of high schools. Building, **10**, 99-106 (2004)
5. N.A. Lukinova, I.A. Serebryanaya, Young researcher of the Don, **2(2)**, 1-3 (2016)
6. N.A. Lukinova, A.A. Matrosov, D.A. Nizhnik, I.A. Serebryanaya, J.V. Terehina, *V International Scientific and Practical Conference "Innovative Technologies in Science and Education"*, Divnomorskoe, 179-181 (2017)
7. GOST 8462-85 Wall materials. Methods for determination of compressive and flexural strengths (1985)
8. V.A. Khatlamadgiyan, J.V. Terehina, A.V. Kotlyar, *XX International Scientific and Practical Conference "Modern technology and technology"*, Rostov-on-Don, 113-114 (2014)
9. J.V. Terehina, V.D. Kotlyar, Yu.I. Nebegco, *XVII International Scientific and Practical Conference "Modern Technologies and Technologies"*, Rostov-on-Don, 165-166 (2014)
10. GOST 530-2012 Bricks and stone, ceramic. General specifications (2012)
11. EN 771-1:2009 Specification for masonry units Part 1. Clay masonry units (2009)
12. EN 772-1:2014 Methods of test for masonry units Part 1. Determination of compressive strength (2014)
13. NEN 6790:1991 Masonry structures - Basic requirements and determination methods (1991)
14. ASTM C67-03a Standard Test Methods for Sampling and Testing Brick and Structural Clay Tile (2003)
15. N.A. Lukinova, I.A. Serebryanaya, J.V. Terehina, *International scientific and practical conference "Modern condition and prospects of development of engineering and environmental systems, building technologies, materials and quality in construction"*, Rostov-on-Don, 396-397 (2015)
16. GOST 530-2007 Bricks and stone, ceramic. General specifications (2007)
17. N.A. Lukinova, I.A. Serebryanaya, J.V. Terehina, *International Scientific and Practical Conference "Construction and Architecture - 2016"*, Rostov-on-Don, 256-259 (2016)
18. A.A. Matrosov, I.A. Serebryanaya, N.A. Lukinova, T.N. Brovko, *XII all-russian school-seminar "Mathematical modeling and biomechanics in the modern university"*, Divnomorskoe, 92-93 (2017)
19. L.M. Abramov, A.V. Orehov, I. L. Abramov, Industrial and civil construction, 1, 58 (2012)
20. A.A. Mungin, Modern technologies – transport, **2**, 64-67 (2013)
21. A.A. Mungin, *International Scientific and Practical Conference "Education and Science: Current State and Prospects for Development"*, Tambov, 86-90 (2013)

Metallic materials for medical use

Anatoly Illarionov[1], *Sergey* Belikov[1], *Stella* Grib[1,*], and *Artem* Yurovskikh[1]

[1]Ural Federal University named after the first President of Russia B.N. Yeltsin, Russian Federation

Abstract. This article provides a brief overview of the metallic materials used as implants in orthopedics, the alloying system and a complex of the physical-mechanical properties for metallic materials certified for medical use, as well as the advantages and drawbacks of using metallic materials as implants. Approaches to improve the quality of an implant made of metallic materials are noted.

1 introduction

According to the Hench and Polak classification [1], metals and their alloys refer to inert biomaterials of the "first generation". It is necessary to take into account the conventionality of this classification, since the materials of the "second" (biologically active and biodegradable polymers and ceramics [2]) and the "third" (materials aimed to achieve certain cellular responses on the molecular level, for example, composite materials [2]) are not intended to replace previously created materials; their purpose is to create qualitatively new, improved methods of therapy. The use of more expensive biomaterials and the complication of their production, as in the case of composites, significantly increase financial costs, which in turn limit the commercial availability of the most advanced medical technologies. Therefore, the use of financially feasible solutions using traditional biomaterials, in particular metal, remains relevant even now, especially in the field of hard tissue restoration.

2 Metallic materials in orthopedics

"Biomaterial" usually means any material intended to partially or completely replace and perform the function of an organ or tissue of a living organism [3]. Therefore, two main applications of biomaterials are distinguished in orthopedics: the replacement (augmentation) of defects in bone tissue, for example, in the treatment of fractures; replacement and reconstruction of the musculoskeletal system, such as joints, ligaments, intervertebral disks, resected bone fragments. In the latter case, metal materials have found wide application in medicine as bone fixing devices and prosthetic studs [4].

The common biomaterials used in orthopedic surgery can be divided into two groups [5]: metals and non-metals (polymers, ceramics, etc.). One of the important characteristics for materials used as implants are high strength and resistance to fatigue failure, and the modulus of elasticity close to the modulus of elasticity of the human bone. Polymers, despite their lightness, manufacturability, flexibility and low modulus, are highly prone to creep, which can lead to a 20% weakening of the initial retaining force of the polymer screws and, as a consequence, the mobility of the implant [6]. In case of alternating loads, polymers often fails at stress levels even below the fatigue limit (σ_{-1}) of bone tissue (see Table 1). The main disadvantage of ceramics is their low tensile and flexural strength, as a result, they shows brittle fracture under cyclic loads. Usually, metal materials have high strength characteristics (yield strength $\sigma_{0.2}$ and temporary failure resistance (ultimate strength) σ_B) and high fatigue resistance (σ_{-1}) in comparison with ceramics and polymers (see Table 1), and therefore, at present time more than 60% of all implants are made of metallic materials [7]. Metal materials such as stainless steel, cobalt-based alloys, titanium and its alloys are used in orthopedic surgery nowadays. Some of the physical and mechanical properties that can be obtained by means of heat and mechanical treatment for certified for medical use metallic materials are given in Table 1 compared to the properties of bone, polymers and ceramics.

A number of advantages of titanium alloys can be noted compared to other metallic materials. According to their strength characteristics, titanium alloys are not inferior to cobalt-based alloys and stainless steel, while their specific strength (σ_B/ρ) can exceed them. Thus, according to Table 1, the maximum specific strength is 17.3 km for stainless steel, 23.2 km for cobalt-based alloys and 29.2 km for titanium-based alloys. Moreover, titanium-based alloys exhibit a much lower elastic modulus E (50-121 GPa) compared to other metal alloys, such as stainless steel (190-230 GPa) and cobalt-based alloys (200-541 GPa), indicating a better mechanical compatibility of titanium alloys with a bone with less than 30 GPa elastic modulus.

It should be noted that the most obvious drawbacks of both stainless steels and (Co, Mo, Cr)

* Corresponding author: s.v.grib@urfu.ru

Table 1. Some of the physical and mechanical properties of bone, nonmetallic and metallic materials used in orthopedics

Material (alloying system)	**Standards**		**Properties**				
			Physical		Mechanical[1]		
	ASTM	ISO	E, GPa	ρ, kg/m³	$\sigma_{0.2}$, MPa	σ_B, MPa	σ_{-1}, MPa (10^7 cycles)
Bone tissue							
Cortical	-	-	15-23,8[8]	1800-2000[9]	114-129[8]	68-157[8]	30-78[10,11]
Trabecular	-	-	0,17[8]	100-1500[9]		3,85[8]	
Polymers[10]							
UHMWPE	-	-	0,5-1,3	930-950	20-30	30-40	13-20
PMMA	-	-	1,8-3,3	1190	35-70	38-80	19-39
Ceramics[10, 12, 13]							
Al_2O_3	-	-	366-380	3990	-	310-350 3790-4500*	-
ZrO_2	-	-	150-201	5680	-	200-500 2000-7500*	-
Metals and alloys approved for medical use [2, 3] [4, 7, 10, 12, 14-16]							
Stainless steel							
Fe-18Cr-14Ni-2.5Mo	F138	5832-1	190-230	7800	170-1213	465-1351	180-820
Cobalt alloys							
Co-28Cr-6Mo	F75 F799 F1537	5832-4	200-541	8900	241-2000	430-2068	200-1220
Co-20Cr-15W-10Ni--1.5Mn	F90	5832-5					
Co-35Ni-20Cr-10Mo	F562	5832-6					
Titanium alloys							
Commercially Pure Ti (Grade 1-4)	F67	5832-2	50-121	4500	480-1060	240-1312	300-689
Ti-6Al-4V	F136 F1472	5832-3					
Ti-6Al-7Nb	F1295	5832-11					
Ti-15Mo-5Zr-3Al	-	5832-14					
Ti-13Nb-13Zr	F1713	-					
Ti-12Mo-6Zr-2Fe	F1813	-					
Ti-15Mo	F2066	-					
Ti-45Nb	AMS4982	-					
Ti-55Ni	F2063	-					

NOTE 1 – tensile properties unless otherwise specified
2 – the intervals of the physical and mechanical properties correspond to values can be obtained by means of various thermomechanical treatment
3 – density (ρ) is given for pure metals (Fe, Co, Ti)
* – compression properties
E – the modulus of elasticity; $\sigma_{0.2}$ – the yield strength; σ_B – the ultimate strength; σ_{-1} – the fatigue limit

alloys and a number of titanium alloys used in medicine are, firstly, alloying chemical elements (Al, V, Ni, Co, Cr) that causing adverse reactions from the living organism; secondly, the existing discrepancy between the implant and the surrounding bone tissue due to the difference in the elastic modulus of the bone and the metal implant [8, 16].

Research into the development of metal medical materials have been aimed at solving these problems in the last two decades. One of the approaches is the development of new titanium alloys that do not contain chemical elements causing toxic and allergic reactions of the organism, based on molecular-orbital calculations of electron structures, followed by the creation of special structural-phase states by means of heat and deformation treatments, which allow obtaining lower values of elastic modulus (90…42 GPa) [8, 17-20]. However, numerous studies [8, 17-20] have shown that the reduction of the elastic modulus by the methods of alloying and structure formation is limited to a value of at least 40 GPa, which is still too high for the bone implant (see Table 2).

Further reduction of the elastic modulus while maintaining acceptable strength characteristics clearly should be associated with the switchover to the use of porous (mesh, foam-like) product structures obtained by powder metallurgy methods [21, 22] or layer-by-layer laser melting / selective laser melting ("3D printing") [23] instead of solid materials. Porous titanium and its alloys ensure the flow of body fluids and bone tissue ingrowth at a pore size of 200-500 μm [24], whereas a reduction in size to 100 μm makes osteoblast ingrowth impossible [25]. In addition, the size, shape, pores fraction and their distribution in the

Table 2. Compositions, processing and elastic modulus of new β-type Ti alloys [8]

Alloy composition, wt. %	Production methods / processing	E, GPa
Ti-29Nb-13Ta-4Mo	Melting/Solutionized / Aged	50-80
Ti-29Nb-13Ta-6Sn		65-70
Ti-29Nb-13Ta-4.6Sn		55-78
Ti-29Nb-13Ta-2Sn		45-48
Ti-30Nb-10Ta-5Zr	Sintering / Hot-forging and swaging / Heat treatment	66.9
Ti-35Nb-4Sn	Melting / Cold rolling / Heat treatment	42-55
Ti-30Zr-3Cr-3Mo	Solution treated/Cold rolling	66/78
Ti-12Mo-3Nb	Melting / Solution treated	105
Ti-12Mo-5Ta	Annealed	74
Ti-50Ta	Solution treated / Aged	77/88/93
Ti-50Ta	Solution treated	88
Ti-30Zr-(5, 6, 7)Mo		75/63/66
Ti-30Zr-(5, 6, 7)Mo	Cold rolling	59/61/73
Ti-36Nb-2.2Ta-3.7Zr-0.3O (at. %)	High pressure torsion	43-65
Ti-31Fe-9Sn	Cast	147
Ti-39.3Nb-13.3Zr-10.7Ta		71
Ti-25Nb-11Sn	Swaged	53
Ti-12Mo-5Zr	Solution treated	64
Ti-25Nb-2Mo-4Sn	Cold rolling / Aging	65

material have a significant effect on the complex of physical and mechanical properties. Thus, obtained in [24] titanium with a porosity of 78% had a compressive strength of 35 MPa with elastic modulus of 5.3 GPa, which is close to the characteristics of the bone. On the other hand, the introduction of macro- and micropores into material reduces the level of strength characteristics. Therefore, it is essential to take into account many factors, from the pore size to the microstructure of the alloy, when creating a porous titanium material for orthopedics [8].

Various methods are developed to predict mechanical performance of porous materials. Analytical models like proposed by Gibson and Ashby [26] used idealized conditions or simplified assumptions, while the finite element models are able to consider more realistic structures like material with random pores distribution [27].

Due to titanium high melting point and and extreme chemical affinity with atmospheric gases, methods of solid-state powder metallurgy were most widely used to create porous Ti-based materials (while the liquid state techniques are common in production of porous aluminium, zinc and magnesium). Some of the most recently used are partial (loose) sintering of metal powders [28], space holder method [20], spark plasma sintering (porous Ti with yield strength 27.2–94.2 MPa and elastic modulus 6.2–36.1 GPa was derived in [29]), microwave sintering (Ti-6Al-4V/multiwall carbon nanotubes composite with porosity of approximately 25%, yield strength of 145.48 ± 27.28 MPa and elastic modulus of 10.87 ± 2.46 GPa obtained in [30]), combustion synthesis [31], selective laser melting (porous Ti-10Mo alloy with compressive yield strength of 95.59 MPa and an elastic modulus of 4.89 GPa was created in [23]). "3D printing" techniques seems to be most promising among them, since they allow the formation of desired porous structure along with exact implant shape according to the CAD model based on tomography data.

3 Conclusion

Thereby, it can be concluded that among the metal materials for orthopedic use, titanium-based alloys are of particular interest due to their high specific strength and biocompatibility; and the modeling and creation of porous titanium implants to replace bone defects (augments) is a promising direction in the development of new metallic materials for medical purposes, as this
allows us to bring the physical-mechanical characteristics of the augment closer to those of the bone tissue.

We hereby acknowledge the support of the Ministry of Science and Education of the Russian Federation, in accordance to the decree of the Government of April 9, 2010, №218, project number 03.G25.31.0234.

References

1. L.L. Hench, J.M. Polak, Science, **295**, 1014-1017 (2002)
2. S.K. Bhullar, N.L. Lala, S. Ramkrishna, Rev. Adv. Mater. Sci., **40**, 303-314 (2015)
3. F. Mahyudin, L. Widhiyanto, H. Hermawan, *Advanced Structured Materials*, **58**, 161-181 (2016)

4. M. Navarro, A. Michiardi, O. Castaño, J.A. Planell, J. R. Soc. Interface, **5**, 1137–1158 (2008)
5. F.A. Rodriguez-Gonzalez, *Biomaterials in Orthopedic Surgery* (ASM International, Materials Park, OH, 2009).
6. L.E. Claes. Clinical Materials, **10**, 41-46 (1992)
7. S.R. Golish, W.M. Mihalko, J Bone Joint Surg Am., **93**(2), 206-212 (2011)
8. Y. Li, C. Yang, H. Zhao, S.Qu, X. Li, Y. Li., Materials, **7(3)**, 1709-1800 (2014)
9. I.A. Khlusov, V.P. Pichugin, M.A. Ryabtseva, *Basics of biomechanics of biocompatible materials and biological tissues* (Tomsk Polytechnic University 2007)
10. B.D. Ratner, A.S. Hoffman, F.J. Schoen, J.E. Lemons, *Biomaterials science*: *an introduction to materials in medicine* (Boston: Elsevier, 2013).
11. J.D. Currey, Proc. Inst. Mech. Eng., H. **212**(6), 399-412 (1998)
12. N.R. Patel, P.P. Gohil, Int J Emerging Technol Adv Eng. **2**(4), 91-101 (2012)
13. Z. Sheikh, S. Najeeb, Z. Khurshid, V. Verma, H. Rashid, M. Glogauer, Materials, **8**(9), 5744-5794 (2015).
14. D.C. Hansen, The Electrochemical Society Interface, **17**(2), 31-14 (2008).
15. H. Hermawan, D. Ramdan, J. R. P. Djuansjah, *Biomedical Engineering - From Theory to Applications* (R. Fazel-Rezai, Ed., chapter 17, InTech, Rijeka, Croatia, 2011).
16. M.Yu. Kollerov, V.S. Spektor, S.V. Skvortsova, A.M. Mamonov, D.E. Gusev, G.V. Gurtovaya, Titanium, **2**, 42-53 (2015).
17. M. Abdel-Hady, K. Hinoshita, M. Morinaga, Scripta Mater, **55,** 476-480 (2006)
18. M. Abdel-Hady Gepreel, M. Niinomi, Journal of the mechanical behavior of biomedical materials, **20**, 407-415 (2013)
19. P. Laheurte, F. Prima, A. Eberhardt, Journal of the mechanical behavior of biomedical materials, **3**, 566-573 (2010)
20. M. Niinomi, Y. Liu, M. Nakai, H. Liu, H. Li. Regen Biomater, **3**(3), 173-185 (2016)
21. H. Nakajima, T. Ikeda, S.K. Hyun, Advanced Engineering Materials, **6**(6), 377-384 (2004)
22. C.E Wen, M. Mabuchi, Y. Yamada, K. Shimojima, Y. Chino, T. Asahina, Journal of Materials Science: Materials in Medicine, 13 (4), 397-401 (2002)
23. F.X. Xie, X.B. He, S.L. Cao, X. Lu, X.H. Qu., Corros. Sci., **67**, 217-224 (2013)
24. C.E. Wen, M. Mabuchi, Y. Yamada, K. Shimojima, Y. Chino, T. Asahina, Scr. Mater. **45**(10), 1147-1153 (2001)
25. W. Xue, B.V. Krishna, A. Bandyopadhyay, S. Bose., Acta Biomater, **3**(6), 1007-1018 (2007)
26. L.J. Gibson, M.F. Ashby, *Cellular Solids: Structure and Properties* (Cambridge University Press: Cambridge, UK, 1997)
27. A.P. Roberts, E.J. Garboczi, Acta Mater, **49**, 189-197 (2001)
28. I.H. Oh, N. Nomura, N. Masahashi, S. Hanada, Scr. Mater, **49**, 1197-1202 (2003)
29. F.M. Zhang, E. Otterstein, E. Burkel. Adv. Eng. Mater, **12**, 863-872 (2010)
30. C.Y. Tang, C.T. Wong, L.N. Zhang, M.T. Choy, T.W. Chow, K.C. Chan, T.M. Yue, Q. Chen, J. Alloy, Compd., **557**, 67-72 (2013)
31. B.Y. Li, L.J. Rong, Y.Y. Li, V.E. Gjunter, Intermetallics, **8**, 881-884 (2000)

Morphological features and mechanics of destruction of materials with different structures under impact drop cyclic loading

Valery N. Varavka[1,*], *Oleg V.* Kudryakov[1], *Igor Yu.* Zabiyaka[1], and *Antonina A.* Shapovalova[1]

[1]Don State Technical University, Material Science Department, 344000 Gagarin sq. 1, Rostov-on-Don, Russia

Abstract. The process of erosion destruction of steels with austenitic, sorbitol and martensitic structure under the influence of high-speed liquid-drop collisions is studied. The characteristics of the morphological features and mechanisms of the process of surface degradation of steels with different structures are given. Their classification criterion is proposed on the basis of the diagrams of limiting states. Based on the theory of Paris-Erdogan, the computational and analytical model of the fatigue fracture of martensitic steel has been developed.

1 Introduction

The impact of a water drop on a metal surface includes several components: mechanical, cavitational (hydrodynamic and bubble) and corrosive [1-5]. The prolonged cyclic action of droplets causes fatigue failure of the metal, which takes place in several stages [6, 7]. The first of these, incubation, is characterized by the fact that there are no signs of destruction and loss of mass at this stage. Therefore, the duration of the incubation period is usually taken as an anti-erosion property of the material under the conditions of droplet impact. In practice, the conditions for droplet erosion are formed during condensation in vapor-gas environments and lead to progressive deterioration of the blades of steam-gas turbine and compressor equipment. This problem is very relevant and is being actively studied [8-14].

To date, the physical nature of the initiation of wear during impact-dropping effects is not fully understood. In a number of works [3, 15-19] we have studied the features of this phenomenon in such *ductile* metallic materials as single-phase titanium alloys, steel with an austenitic or sorbitol structure.

The beginning of wear in this case is determined by the development of subsurface fatigue porosity (Fig. 1a), which appears in the metal in the impact zone by the percolation mechanism [20, 21], followed by pore fusion and the formation of primary surface erosional wear craters (Fig. 1b). In this case, the craters have a distinctive morphology with a very branched wall surface. Morphological features of such wear and its calculation and analytical model, developed on the basis of the principles of the fatigue theory of Goodman, are most thoroughly represented in the work [3]. In *hardened* metal materials, the percolation mechanism

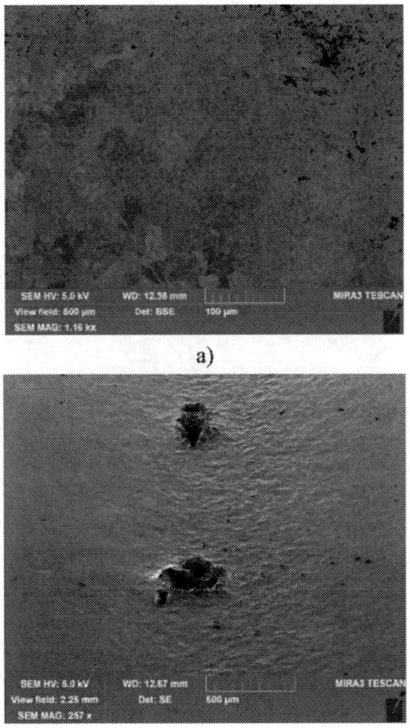

a)

b)

Fig. 1. Development of wear during the droplet erosion of samples of austenitic steel 08Cr18Ni10Ti (SEM): a – formation of fatigue porosity at the collision site (cross section of the sample in depth); b – formation of the first erosion craters (erosion track on the sample surface)

* Corresponding author: varavkavn@gmail.com

cannot be realized, since the size of the plasticity zones is too small. In this case, the fatigue porosity is not formed, and the morphology of the primary wear craters is determined by the mechanism of nucleation, growth, and branching of brittle fracture fatigue cracks. Distinctive features of such a mechanism for the nucleation of erosion craters are the relatively smooth surface of their walls (Fig. 2a) and the presence of cracks in the cross section of the craters (Fig. 2b). On the basis of these experimental data, the main task of the present work was to find regularities and create a computational and analytical model of erosion wear of such materials. At the same time, the problem of criterial evaluation of materials was solved, with the help of which they could be categorized as plastic or hardened in order to be described within the corresponding mechanical model of fatigue failure.

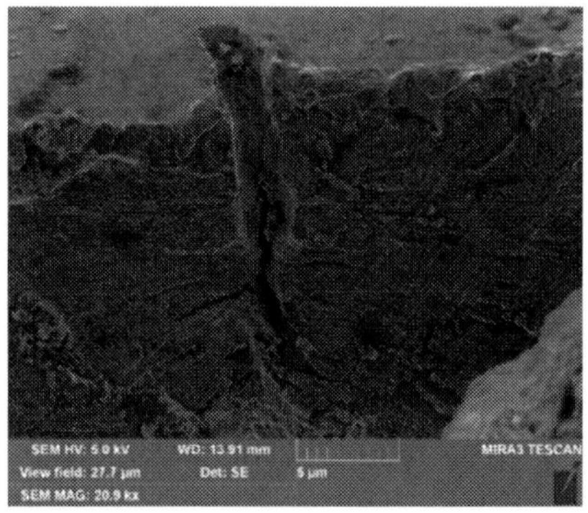

a)

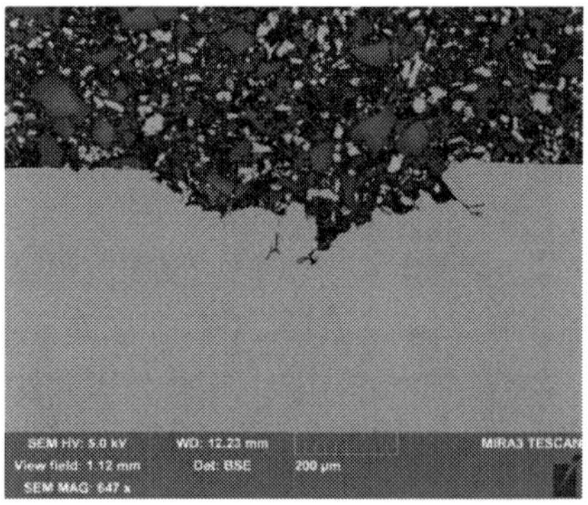

b)

Fig. 2. Erosion craters on the surface of a sample of martensitic steel 20Cr13 (SEM): a – the wall of the primary crater with a branched sharp crack; b – cross section of the crater with fatigue cracks

2 Limiting state diagrams

Analysis of experimental data obtained from the results of bench erosion testing of steel samples with different structures makes it possible to recommend their *limiting state diagrams* as a classification criterion for ductile and durable materials. Such diagrams of austenitic and martensitic steels are presented in Fig. 3 and Fig. 4, respectively. The linear graph represents the limiting, and the nonlinear – the critical stress. Austenitic steel does not have points of intersection on the diagram (Fig. 3), which indicates the absence of cracks and allows it to be categorized as ductile materials with percolation fatigue failure mechanisms under conditions of droplet impingement erosion wear. The points of intersection of the graphs in the martensitic steel diagram (Fig. 4) are the initial l_0 and critical l_C fracture dimensions by their physical entity. For $l < l_0$, the crack does not spontaneously grow, and there is no danger of brittle failure.

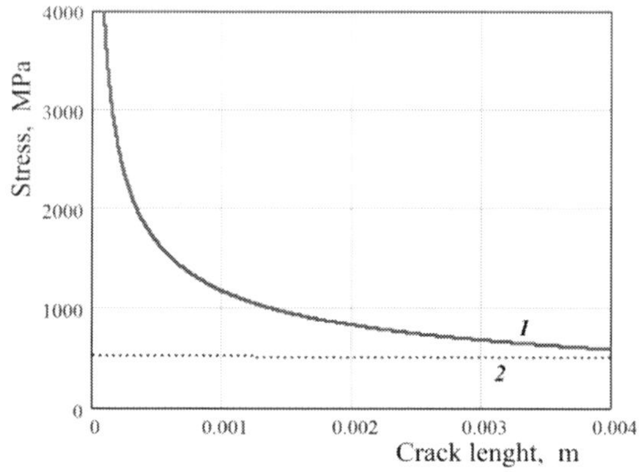

Fig. 3. The diagram of the limiting states of austenitic steel 08Cr18Ni10Ti.

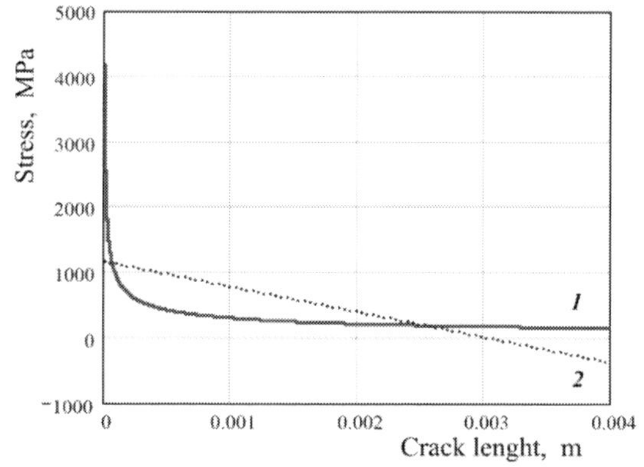

Fig. 4. The diagram of the limiting states of steel 20Cr13 with a martensitic structure

At $l \geq l_C$, crack growth is irreversible and autocatalytic is accelerated, and destruction (formation of an erosion wear crater) is developing catastrophically fast. The values of l_0 and l_C in the framework of the proposed calculation and analytical model are found by solving equations (1) for ultimate σ_p and critical σ_k stresses for a material specified by such properties as hardness *HV*, tensile strength σ_B, fracture toughness K_{IC} taking into account the geometric parameters of the crack *Y*. The factor *Y* characterizes the type of crack and can be determined from directories, for example [22]. In this case, depending on the choice of the crack type, the values of Y can differ substantially and they must be coordinated with the experiment. For example, for steel 20Cr13 with a martensitic structure, we have: $l_C = 68$ μm – for the projection of the crack on the surface of the sample and $l_C = 7$ μm – for the projection of the crack along the sample section (depth).

$$\sigma_p(l) = HV - \frac{\sigma_B^3}{4 \cdot K_{1C}^2} \cdot l \; ; \quad \sigma_k(l) = \frac{K_{1C}}{Y \cdot \sqrt{\pi \cdot l}} \qquad (1)$$

In the interval between l_0 and l_C, the rate of crack growth (per one loading cycle, i.e, one dropping impact) is determined in accordance with Paris equation (2) of the Paris-Erdogan fatigue theory [23]. To use the methodical apparatus of this theory, it is necessary to experimentally determine the parameters of the Paris equation m and A, which are the empirical constants of the material. Thus, for example, bench tests performed by us on samples of steel 20Cr13 with a martensitic structure for the conditions of droplet impingement erosion give the values of the parameters of the Paris equation: $m = 2.5$ and $A = 3.403 \cdot 10^{-9}$ m / cycle (m / impact).

$$\frac{dl}{dN} = A \cdot \Delta K^m \qquad (2)$$

where the value of ΔK is determined by the expression:

$$\Delta K = K_{\max} - K_{\min} = \Delta\sigma \cdot \sqrt{\pi \cdot l} \qquad (3)$$

taking into account the following conditions:
$K_{\max} = \sigma_{\max}\sqrt{\pi \cdot l}$ and $K_{\min} = \sigma_{\min}\sqrt{\pi \cdot l}$ for $\sigma_{min} > 0$. Here K is the stress concentration coefficient for the corresponding maximum σ_{max} and minimum σ_{min} stresses of a single cycle.

3 Single loading cycle

It was shown in [3] that when a liquid drop hits a metal surface, the stress σ is distributed over the surface in time τ in accordance with the graph in Fig. 5. The value σ is expressed in relative units $\rho_0 \cdot c_0 \cdot V_0$, called the hydraulic shock margin. Time τ also has a relative scale, where ρ_0 – is the density of the liquid, c_0 – is the velocity of sound in the liquid under normal conditions, V_0 is the impact velocity, and R_0 is the radius of the drop. Fig. 5 also shows the stress values adopted in the theory of fatigue, in particular in the Paris–Erdogan theory (maximal σ_{max}, minimum σ_{min}, mean σ_m, etc.). However, they need to be refined for use in the Paris equation (2).

4 Multi-cycle loading

Fig. 5 shows the distribution of stresses for a single impact, and the Paris equation is used for multi-cycle loading. In a single impact with an increase in the time $\sigma_{min} \to 0$, the single loading cycle becomes repeating-stress cycle. And the time between impacts falling into one point exceeds several times the time of a single impact.

With a repeating-stress cycle in the formula (3) in accordance with Fig. 5 it should be taken that

$$\Delta\sigma = \sigma_{max} = 0{,}528 \cdot \rho_0 \cdot c_0 \cdot V_0. \qquad (4)$$

By the solution of the Paris equation (2) for the conditions of droplet impingement erosion, an acceptable number of cycles (droplet impacts) N_C that determines the duration of the incubation period of erosive wear of this material can be found:

$$N_C = \int_0^{N_C} dN = \int_{l_0}^{l_C} \frac{da}{A \cdot (\Delta K)^m} \qquad (5)$$

Strictly speaking, the value of N_C obtained by integration is not the value of the incubation period of droplet impingement erosion m_0, since it does not take into account the number of N_3 cycles necessary to form a crack of size l_0. Therefore, for application of expression (5), correlations of the N_C value with the bench test data are necessary, taking into account that $m_0 = N_3 + N_C$.

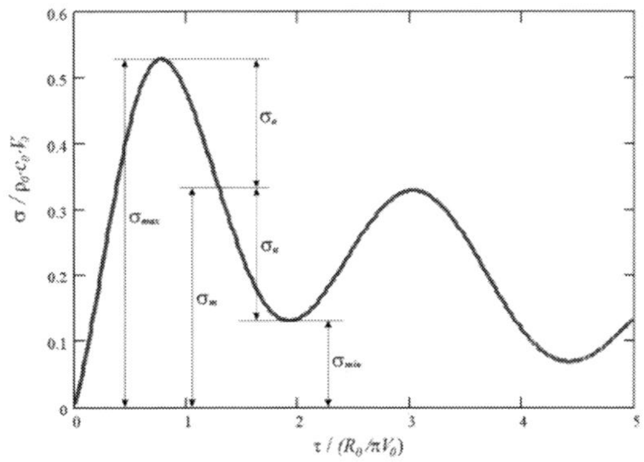

Fig. 5. Calculated parameters of the loading cycle for a drip collision

Without going into the details of this correlation, which are fulfilled in the development of the computational and analytical model, we present the result of the use of the Paris-Erdogan theory (2) + (3) + (5), taking into account the features of the droplet impingement erosion mechanism (4) using the example of steel 20Cr13 with a martensitic structure, presented in Fig. 6. Dependences in Fig. 6 can be considered as the *applied criterion of erosion failure*, showing the minimum value of the impact velocity at which wear in steel is initiated. If the graph of the function in Fig. 6 is below the threshold line, the distance over which the crack propagates in one loading cycle lies in the region of elastic deformations and elastically relaxes after the load is removed. In this case, the crack will not grow. For example, it follows from Fig. 6 that if there are already cracks in the martensite of steel 20Cr13 of length $l_0 = 1.5$ μm (line 2), then they will develop at the impact velocities V_0 exceeding 150 m/s.

DOI: 10.1051/matecconf/201713203004

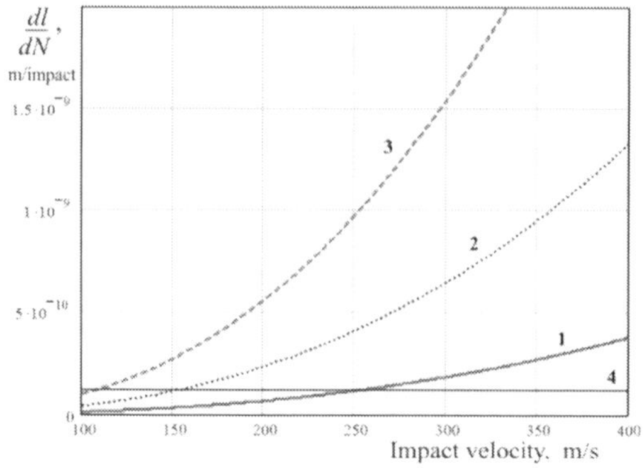

Fig. 6. The dependence of the rate of crack growth in steel 20Cr13 with a martensitic structure on the impact velocity for different values of the initial crack length: 1- 0.55 μm; 2 - 1.5 μm; 3 - 3.0 μm; 4 - threshold value of growth of a sharp crack

5 Conclusion

Thus, the criterion of the minimum impact velocity at which wear is initiated can be considered as universal for any material. A similar criterion was proposed in [3] on the basis of the Goodman fatigue model for *plastic* metal alloys, and in this paper it was used for *reinforced* (hardened) metal alloys based on the Paris-Erdogan model. This provides a single measuring space for materials, the surface degradation of which during droplet impingement erosion occurs in fundamentally different mechanisms – by developing brittle cracks in hardened metal materials and by forming fatigue porosity by the mechanism of percolation in ductile metal materials.

The reported study was funded by RFBR in the framework of the research project no. 15-08-06181.

References

1. A. Chizhov, A. Shmidt, Tech. Phys., **45**, 1529 (2000)
2. K. Haller, Y. Ventikos, D. Poulikakos, P. Monkewitz, J. Appl. Phys., **92**, 2821 (2002)
3. V. Varavka, O. Kudryakov, J. Friction and Wear, **36** (1), 71 (2015)
4. V. Varavka, O. Kudryakov, J. Friction and Wear, **36** (2), 153 (2015)
5. O. Kudryakov, V. Varavka, V. Irkha, I. Moiseeva, Vestnik Don St. Tech. Univ., **17** (2), 56 (2017)
6. V. Varavka, O. Kudryakov, Al. Mednikov, V. Irkha, Izv. Vyssh. Uchebn. Zaved. Sev. Kavkaz. Reg. Tekhn. Nauki, 6, 92-98 (2011)
7. V. Ryzhenkov, G. Kachalin, A. Mednikov, Al. Mednikov, O. Kudryakov, V. Varavka, Nadezhn. Bezopasn. Energ., **16** (1), 67 (2012)
8. S. Hattori, M. Takinami, Wear, **269**, 310 (2010)
9. V. Ryzhenkov, A. Lebedeva, Al. Mednikov, Therm. Eng., **58** (9), 713 (2011)
10. A. Mednikov, V. Ryzhenkov, L. Seleznev, A. Lebedeva, Therm. Eng. **59** (5), 414 (2012)
11. E. Tobin, T. Young, D. Raps, O. Rohr, Wear, **271**, 2625 (2011)
12. S. Sandeep, Int. J. Mech. Eng. & Rob. Res., **1** (3), 214 (2012)
13. N. Kamkar, F. Bridier, P. Bocher, P. Jedrzejowski, Wear of Materials, **301** (1-2), 442 (2013)
14. A.S. Vasiliev, S.S. Volkov, A.A. Belov, S.Yu. Litvinchuk, S.M. Aizikovich, Int. J. Eng. Sci., **112**, 7 (2017)
15. V. Varavka, O. Kudryakov, Vestnik Don St. Tech. Univ., **11** (8), 1376 (2011)
16. O. Kudryakov, V. Varavka, Materialoved. (5), 36-43 (2012)
17. O. Kudryakov, V. Varavka, Materialoved. (6), 14-19 (2012)
18. V. Varavka, O. Kudryakov, Izv. Vyssh. Uchebn. Zaved. Sev.Kavkaz. Reg. Tekhn. Nauki, (3), 45-50 (2012)
19. O. Kudryakov, V. Varavka, Uprochn. Tekhnol., Pokr., (10), 40-47 (2012)
20. M. Shtremel', *Durability of Alloys. Ch. 2. Deformation*, Moscow: Mos. Inst. Stali & Splavov, 527 (1997)
21. M. Ashby, D. Jones, *Engineering Materials*, Pergamon, 142 (1989)
22. Y. Ito, Y. Murakami, N. Hasebe, R. Yuki, M. Toya and others, *Handbook on stress intensity factors*, Pergamon, 90-107 (1987; Mir, 1990)
23. P. Paris, F. Erdogan, Trans. ASME, J. Basic Engng., **15**, 528 (1963)

On accounting for tempering effect in unsteady creep model for metals

Ivan Volkov[1], *Leonid* Igumnov[1,*], *Aleksandr* Ryabov[1], and *Denis* Shishulin[1]

[1]National Research Lobachevsky State University of Nizhni Novgorod, Research Institute for Mechanics, 603950 Nizhny Novgorod, Russian Federation

Abstract. A mathematical model describing processes of nonstationary creep of metals under multiaxial stress state is developed. A phenomenon of reverse creep modeling is considered in more details. The results of numerical modeling of the reverse creep process in variety of structural steels are presented. Obtained numerical results are compared with available results of physical experiments.

1 Introduction

The study of the laws of inelastic deformation of structural materials under uniaxial and multiaxial stressed state is important for the development of the fundamentals of the mathematical theory of creep. With a considerable number of experimental studies of the creep process of structural alloys [1], experiments on the study of the phenomenon of reverse creep (the phenomenon of decreasing deformations in time after removing the load) of metals is comparatively small. As a result, this effect is not taken into account in most cases during the calculation of structural elements creep nowadays.

At the same time, in order to verify the physical reliability of the defining equations of the creep theory, calculations and comparison of the obtained numerical results with the data available in the literature on a wide range of experiments are required [2-16]. The results of such researches can also be useful in determining (specifying) the material parameters of the scalar functions and functionals involved in defining equations of nonstationary creep. Moreover, fast unloading of elements and components of load-bearing structures can lead to stress loading of the opposite sign, which in some cases is unacceptable (deformation of reverse creep is commensurable with elastic deformation).

A mathematical model of the nonstationary creep of metals under a multiaxial stressed state was developed in [17-19]. In [19], reliability estimation was carried out and the boundaries of applicability of the developed defining relations of nonstationary creep under multiaxial stress state were determined. Applicability boundaries of the developed defining relations of nonstationary creep under multiaxial stress state were determined as well as reliability estimation was done in [19].

This model is modified to describe the phenomenon of inverse creep and an estimate of its reliability by comparing the obtained numerical results with the experimental data available in the literature on the inverse creep of structural steels is given in this paper.

2 The defining relations of nonstationary creep

General considerations of the nonstationary creep equations [17-19] are:

- Initially isotropic media are considered.
- The strain and strain rates tensors are sums of the "instantaneous" and " time-dependent" components. The "instantaneous" component consists of elastic strains that do not depend on the loading history and determined by the final state of the process, and plastic strains depending on the loading history. The time component (creep strains) describes the time dependence of deformation processes at low loading speeds.
- The equipotential surface evolution is determined by changes in its radius C_c and displacements of its centre ρ_{ij}^c.
- The change in the body volume is determined only by the elastic strains: $e_{ii}^p = e_{ii}^c = 0$.
- Deformation processes characterized by small strains are considered.

The strain and strain rate tensors e_{ij} and $\dot{e}_{ij}$ are sums of the elastic strains e_{ij}^e, $\dot{e}_{ij}^{\prime e}$, plastic strains e_{ij}^p, $\dot{e}_{ij}^{\prime p}$ and creep strains e_{ij}^c, $\dot{e}_{ij}^{\prime c}$, i.e.:

$$e_{ij} = e_{ij}^e + e_{ij}^p + e_{ij}^c\,;\ \dot{e}_{ij} = \dot{e}_{ij}^e + \dot{e}_{ij}^p + \dot{e}_{ij}^c\,,\ i,j = 1,2,3.$$

The relationship between the stress tensor and elastic strain tensor is based on thermoelasticity equations:

$$\sigma = 3K\left[e - \alpha T)\right];\ \dot{\sigma} = 3K\left[\dot{e} - \dot{\alpha}T - \alpha\dot{T})\right] + \frac{\dot{K}}{K}\sigma\,,$$

* Corresponding author: igumnov@mech.unn.ru

$$\sigma'_{ij} = 2Ge'^e_{ij},\ \dot{\sigma}_{ij} = 2G\dot{e}'^e_{ij} + \frac{\dot{G}}{G}\sigma'_{ij},\ e'^e_{ij} = e'_{ij} - e^c_{ij} \quad (1)$$

where σ, e are spherical and σ'_{ij}, e'_{ij} are deviatoric components of the corresponding stress σ_{ij} and strain e_{ij} tensors; $G(T)$ is the shear modulus, $K(T)$ is the volumetric compression modulus, $\alpha(T)$ is the coefficient of linear temperature expansion of the material.

To describe the creep processes, we introduce in the stress space the equipotential creep surfaces F_c having the same centre ρ^c_{ij} and different radii C_c, determined by the current state of stress:

$$F^{(i)}_c = S^c_{ij}S^c_{ij} - C^2_c = 0,\ S^c_{ij} = \sigma'_{ij} - \rho^c_{ij},\ i = 0,1,2,... \quad (2)$$

In accordance with the associativity law

$$\dot{e}^c_{ij} = \lambda_c \frac{\partial F^{(i)}_c}{\partial S^c_{ij}} = \lambda_c S^c_{ij}, \quad (3)$$

where λ_c corresponds to the current surface $F^{(i)}_c$, defining current stress state S^c_{ij}.

The surface with radius $\bar{C}_c$ may be outlined among these equipotential surfaces as corresponding to zero creep rate:

$$F^{(0)}_c = \bar{S}^c_{ij}\bar{S}^c_{ij} - \bar{C}^2_c = 0,\ \bar{S}^c_{ij} = \bar{\sigma}'_{ij} - \rho^c_{ij}, \quad (4)$$

where $\bar{S}^c_{ij}$ and $\bar{\sigma}'_{ij}$ are set of strained states that correspond (with a certain tolerance) to the zero creep rate.

Consider that

$$\bar{C}_c = \bar{C}_c(\chi_c, T);\ \dot{\chi}_c = \left(\frac{2}{3}\dot{e}^c_{ij}\dot{e}^c_{ij}\right)^{1/2};\ \chi^c = \int_0^t \dot{\chi}^c dt,$$

$$\lambda_c = \lambda_c(\psi_c, T);\ \psi_c = \left[\frac{\left(S^c_{ij}S^c_{ij}\right)^{1/2} - \bar{C}_c}{C_c}\right], \quad (5)$$

where $\bar{C}_c$ and λ_c are experimentally determined functions of temperature T.

The evolution equation for the coordinates of the centre of the creep surface has the following form [17-19]:

$$\dot{\rho}^c_{ij} = g^c_1\dot{e}^c_{ij} - g^c_2\rho^c_{ij}\dot{\chi}_c, \quad (6)$$

where g^c_1 and $g^c_2 > 0$ are experimentally determined material parameters.

By specifying (3) the law of gradientality can be represented as

$$\dot{e}^c_{ij} = \lambda_c(\psi_c, T)S^c_{ij} = \lambda_c\psi_c S^c_{ij} = S^c_{ij}\lambda_c\left(\frac{\sqrt{S^c_{ij}S^c_{ij}} - \bar{C}_c}{C_c}\right). \quad (7)$$

The intensity of the creep strain rate tensor has the following form:

$$\dot{e}^c_u = \lambda_c\left(\frac{\sqrt{S^c_{ij}S^c_{ij}} - \bar{C}_c}{C_c}\right)S^c_u. \quad (8)$$

By taking into account (8) we are getting the following expression for the length of the creep strain path:

$$\dot{\chi}_c = \sqrt{\frac{2}{3}}\dot{e}^c_u = \sqrt{\frac{2}{3}}\lambda_c\left(\sqrt{S^c_{ij}S^c_{ij}} - \bar{C}_c\right), \quad (9)$$

and

$$\lambda_c = \frac{\dot{\chi}_c}{\sqrt{\frac{2}{3}}\left(\sqrt{S^c_{ij}S^c_{ij}} - \bar{C}_c\right)}. \quad (10)$$

The dependence of χ_c on the time t with $S^c_u = const$ in the case of multi-axis straining over the ray trajectory is shown in Fig. 1.

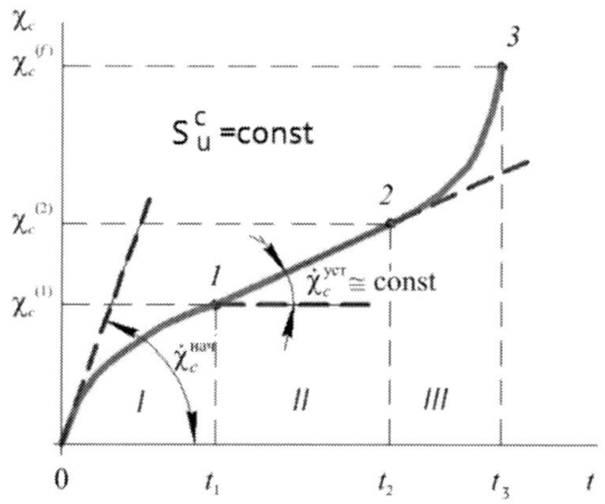

Fig. 1. The dependence of χ_c on the time t in case of multi-axis straining.

The curve $\chi_c \sim t$ (Fig. 1) can be conventionally divided into three segments:

I. segment of unsteady creep ($0 - \chi^{(1)}_c$) where the creep-induced strain rate $\dot{\chi}_c$ decreases.

II. segment of steady creep ($\chi^{(1)}_c - \chi^{(2)}_c$) where the creep-induced strain rate is approximately constant $\dot{\chi}_c \cong const$.

III. segment of unsteady creep ($\chi_c > \chi^{(2)}_c$) where the creep-induced strain rate $\dot{\chi}_c$ drastically increases (followed by fracture).

The lengths of these segments depend on the value of $S^c_u = const$.

In the case of uniaxial stress state of the specimen we have

$$\dot{\chi}_c = \dot{e}_{11}^c,\ \bar{C}_c = \sqrt{\frac{2}{3}}\bar{\sigma}_c$$

$$\left(\sqrt{S_{ij}^c S_{ij}^c} - \bar{C}_c\right) = \sqrt{\frac{2}{3}}\left(\sigma'_{11} - \frac{3}{2}\rho_{11}^c - \bar{\sigma}_c\right); \quad (11)$$

$$\dot{e}_{11}^c = \frac{2}{3}\lambda_c\left(\sigma'_{11} - \frac{3}{2}\rho_{11}^c - \bar{\sigma}_c\right); \quad (12)$$

$$\dot{\chi}_c = \dot{e}_{11}^c\ \chi_c^{(1)} = e_{11}^{c(1)};\ \chi_c^{(2)} = e_{11}^{c(2)};\ \chi_c^{(3)} = e_{11}^{c(3)}, \quad (13)$$

where $\bar{\sigma}_c = \bar{\sigma}_c\left(e_{11}^c, T\right)$ is the creep limit of the material corresponding to zero creep rate. In this case the diagram $\chi_c \sim t$ is equivalent to the diagram $e_{11}^c \sim t$. λ_c may be obtained using (13) on the second segment ($e_{11}^{c(1)} \le e_{11}^c \le e_{11}^{c(2)}$):

$$\lambda_c^{II} = \frac{3}{2}\frac{\dot{e}_{11}^{fat}}{\left(\sigma'_{11} - \frac{3}{2}\rho_{11}^c - \bar{\sigma}_c\right)}. \quad (14)$$

On the segment ($0 \le e_{11}^c \le e_{11}^{c(1)}$)

$$\dot{e}_{11}^c = \begin{cases} \dot{e}_{11}^{c(init)}, \text{when } e_{11}^c = 0, \\ \dot{e}_{11}^{c(fat)}, \text{when } e_{11}^c = e_{11}^{c(1)}. \end{cases}$$

Taking into account these equations we may represent λ_c^I on the first segment of creep curve as:

$$\lambda_c^I = \lambda_c^{(0)}\left(1 - \frac{e_{11}^c}{e_{11}^{c(1)}}\right) + \lambda_c^{(1)}\frac{e_{11}^c}{e_{11}^{c(1)}}, \quad (15)$$

where

$$\lambda_c^{(0)} = \frac{3}{2}\frac{\dot{e}_{11}^{c(init)}}{(\sigma'_{11} - \bar{\sigma}_c)}, \lambda_c^{(1)} = \frac{3}{2}\frac{\dot{e}_{11}^{c(fat)}}{\left(\sigma'_{11} - \frac{3}{2}\rho_{11}^c - \bar{\sigma}_c\right)} \quad (16)$$

are the values of λ_c in points "0" and "1", respectively.

On third segment ($e_{11}^{c(2)} < e_{11}^c \le e_{11}^{c(3)}$) prior to fracture:

$$\lambda_c^{III} = \lambda_c^{II} / (1-\omega)^{r_c}, \quad (17)$$

where ω is a measure of damage and r_c is a material parameter [17].

Extending the resultant relations to the non-one-dimensional case, we obtain:

$$\lambda_c = \begin{cases} \lambda_c^I, 0 \le \chi_c \le \chi_c^{(1)} \\ \lambda_c^{II}, \chi_c^{(1)} \le \chi_c \le \chi_c^{(2)} \\ \lambda_c^{III}, \chi_c^{(2)} \le \chi_c \le \chi_c^{(3)} \end{cases}, \quad (18)$$

Equations (1)-(18) describe unsteady and steady-state segments of the creep curve at different stress levels and the main effects of the creep process under alternating stress.

To describe the phenomenon of inverse creep, the law of gradientality can be represented in the form:

$$\dot{e}_{ij}^c = \begin{cases} \lambda_c^{(+)}\left(\sqrt{S_{ij}^c S_{ij}^c} - \bar{C}_c^{(+)}\right) n_{ij}^c, \sqrt{S_{ij}^c S_{ij}^c} > \bar{C}_c^{(+)} \\ 0, \quad : \bar{C}_c^{(-)} \le \sqrt{S_{ij}^c S_{ij}^c} < \bar{C}_c^{(+)} \\ \lambda_c^{(-)}\left(\sqrt{S_{ij}^c S_{ij}^c} - \bar{C}_c^{(-)}\right) n_{ij}^c, \sqrt{S_{ij}^c S_{ij}^c} < \bar{C}_c^{(-)} \end{cases}, \quad (19)$$

where $n_{ij}^c = S_{ij}^c / \sqrt{S_{ij}^c S_{ij}^c}$, $\lambda_c^{(+)}$ obtained using equation (19) and $\lambda_c^{(-)} = const$ determined from the reverse creep experiments.

It is always a compromise between the correspondence of the model to the actual behaviour of the material under operational conditions, the possibility and efficiency of the model in solving applied problems on a computer, the possibility of experimentally determining the material parameters of the model and the accuracy of these parameters. It is necessary to find this compromise while developing models of deformation and fracture of materials for a wide range of thermomechanical loads with a wide range of applicability. In case it is not possible to accurately determine the material parameters the model can be useless in practice [17]. That is the reason why this variant of the defining relations of nonstationary thermal creep under the complex stress state (1)-(18) was formulated as a system of "nested" models.

In the absence of the required experimental data or due to the unimportant effect of these or other physical effects, the particular relationships of unsteady creep under a complex stress state may be obtained by omitting corresponding material parameters.

3 Results of numerical calculations

The results of experimental studies of reverse creep conducted by McWethie and Waugh on a number of structural steels are presented in [20]. In the first example, we calculated a creep of a specimen made of steel with content of 0,35% C (steel 35) at the temperature $T = 450^o C$ and stress $\sigma_{11} = 52\text{MPa}$ within $t = 9070\text{h}$. In this case a value of strain is $e_{11} = 0{,}587\%$. After that the specimen was quickly unloaded to the value $\sigma_{11} = 7{,}7\text{MPa}$, elastic strain decreased to $e_{11} = 0{,}029\%$. Later the inverse creep strain of value $e_{11}^c \approx 0{,}019\%$ ($e_{11}^c \approx 0{,}9 e_{11}^e$) was observed after $t_1 = 700\text{h}$.

The analysis for straining process of the specimen using the model of nonstationary creep was performed with 35 steel material parameters shown in Table 1.

The test results and their comparison with the obtained numerical data are shown in Fig. 2, which shows the creep curve (the solid line indicates the numerical results obtained by authors, and the markers indicate the corresponding experimental data). Qualitative and quantitative coincidence of experimental and calculated data observed.

Table 1. Physico-mechanical characteristics and parameters of the nonstationary creep model for steel 35 and ATV

	Steel 35 at $T = 450^oC$	Steel ATV at $T = 538^oC$
K, MPa	128400	128400
G, MPa	59270	59270
$\bar{C}_c^{(+)}$, MPa	15	70
$\bar{C}_c^{(-)}$, MPa	1,1	11,6
g_1, MPa	2800	5000
g_2	100	200
$\lambda_c^{(0)}$, $(\text{MPa}\cdot\text{h})^{-1}$	$4{,}13\cdot10^{-8}$	$7\cdot10^{-8}$
$\lambda_c^{(1)}$, $(\text{MPa}\cdot\text{h})^{-1}$	$4{,}13\cdot10^{-8}$	$7\cdot10^{-8}$
$\lambda_c^{(-)}$, $(\text{MPa}\cdot\text{h})^{-1}$	$1\cdot10^{-5}$	$6\cdot10^{-6}$

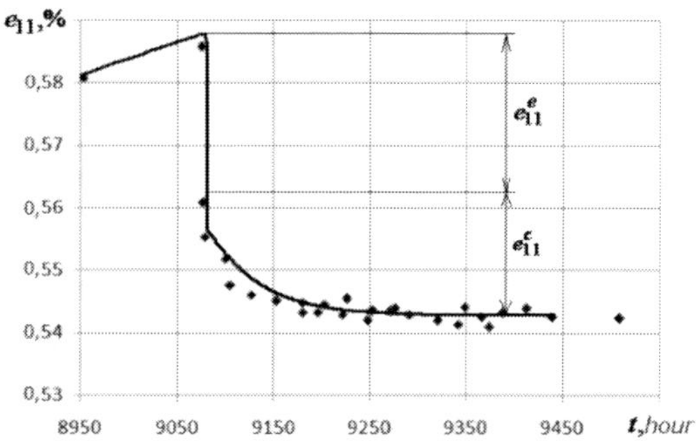

Fig. 2. Reverse creep curve of 35 steel at the temperature $T = 450^oC$.

In the next example the steel ATV specimen was tested at the temperature $T = 538^oC$ and stress $\sigma_{11} = 138\text{MPa}$ within $t = 8060\text{h}$. After that the specimen was quickly unloaded to the value $\sigma_{11} = 7{,}7\text{MPa}$, elastic strain decreased to $e_{11} = 0{,}0921\%$. Later the inverse creep strain of value $e_{11}^c \approx 0{,}08\%$ was observed after $t_1 = 2100\text{h}$. ($e_{11}^c \approx 0{,}9e_{11}^e$ the same with previous example).

The analysis of the straining process of specimen using the model of nonstationary creep was performed with ATV steel material parameters shown in Table 1.

The test results and their comparison with the obtained numerical data are shown in Fig. 3, which shows the creep curve (the solid line indicates the numerical results obtained by authors, and the markers indicate the corresponding experimental data). Qualitative and quantitative coincidence of experimental and calculated data observed here as well.

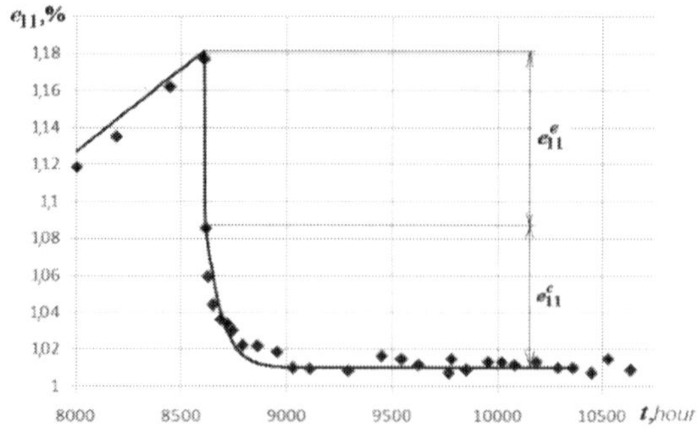

Fig. 3. Reverse creep curve of ATV steel at the temperature $T = 538^oC$.

4 Conclusions

A mathematical model describing the processes of nonstationary creep of structural materials (metals and their alloys) under complex loading is developed.

The results of numerical simulation and experimental data of reverse creep processes of structural steels are compared. It is shown that the developed variant of the defining relations of nonstationary creep allows to describe the processes of reverse creep of metals with required for engineering calculations accuracy.

The research was supported by the Russian Science Foundation, grant No.16-19-10237.

References

1. A.M. Lokoshchenko, *Polzuchest' i dlitel'naia prochnost' metallov* (Fizmatlit, Moscow, 2016) [in Russian]
2. I.A. Volkov, L.A. Igumnov, Yu.G. Korotkih, D.A. Kazakov, A.A. Emel'yanov, I.S. Tarasov, M.A. Guseva, Problemy prochnosti i plastichnosti, **78**, №2, 188-207 (2016) [in Russian]
3. A.G. Kazancev, Problemy prochnosti, №5, 25-31 (1985) [in Russian]
4. Yu. N. Rabotnov, *Polzuchest' ehlementov konstrukcij,* (Nauka, Moscow, 1966) [in Russian]
5. D.A. Gohfel'd, O.S. Sadakov, *Plastichnost' i polzuchest' ehlementov konstrukcij pri povtornyh nagruzheniyah* (Mashinostroenie, Moscow, 1984) [in Russian]
6. V.P. Degtyarev, *Plastichnost' i polzuchest' mashinostroitel'nyh konstrukcij.* (Mashinostroenie, Moscow, Mashinostroenie, 1967) [in Russian]
7. N.N. Malinin, *Prikladnaya teoriya plastichnosti i polzuchesti.* (Mashinostroenie, Moscow, 1968) [in Russian]
8. A.M. Lokoshchenko, *Polzuchest' i dlitel'naya prochnost' metallov* (Fizmatlit, Moscow, 2016) [in Russian]
9. Dzh. Bojl, Dzh. Spens, *Analiz napryazhenij pri polzuchesti.* (Mir, Moscow, 1986) [in Russian]

10. Ohasi Ono Kavai, Teoreticheskie osnovy inzhenernyh raschetov, ASME, **104**, №3, 1-8 (1982) [in Russian]
11. A. Miller, Teoreticheskie osnovy inzhenernyh raschetov, ASME, №2, 1-20 (1976) [in Russian]
12. Yu. N. Shevchenko, R.G. Terekhov, *Fizicheskie uravneniya termovyazkoplastichnosti,* (Naukova Dumka, Kiev, 1982) [in Russian]
13. P. Pezhina, *Osnovnye voprosy vyazkoplastichnosti,* (Mir, Moscow, 1968) [in Russian]
14. J.L. Chaboche, Inter. J. of Plasticity, **5**, No. 3, 247-302 (1989) [in Russian]
15. N.N. Malinin, G.M. Khadjinsky, Inter. J. of Mechanical Science, **14**, 235-246 (1972) [in Russian]
16. R.D. Krieg, J.C. Swearengen, R.W. Rohde, *A physically-based internal variable model for rate-dependent plasticity. Inelastic behavior of pressure vessel and piping components* ASME/CSME, PVP-PB-028, 15-28 (1978) [in Russian]
17. I.A. Volkov, Yu.G. Korotkikh, *Uravneniia sostoianiia viazkouprugoplasticheskikh sred s povrezhdeniiami* (Fizmalit, Moscow, 2008) [in Russian]
18. I.A. Volkov, L.A. Igumnov, Iu.G. Korotkikh, *Prikladnaia teoriia viazkoplastichnosti* (UNN, Nizhny Novgorod, 2015) [in Russian]
19. I.A. Volkov, L.A. Igumnov, D.A. Kazakov, D.N. Shishulin, I.V. Smetanin, Problems of strength and plasticity, **78**, 436 (2016)
20. A. Nadai, *Plastichnost' i razrushenie tverdykh tel* (Izd-vo inost. lit., Moscow, 1969) [in Russian]

Definition of the strain-stress distribution of porous glass in the retarded cooling temperature range

Irina Grushko[1,*]

[1]SRSPU (NPI), Department of Industrial, Civil Engineering, Geotechnics and Foundation Engineering, 346428 Novocherkassk, Russia

Abstract. The estimation of the strain-stress distribution (SSD) of porous glass (foamed slag glass, FSG) is assessed by annealing temperature curves according to the given values of the thermomechanical and thermophysical properties of porous glass, which are in correlation with the properties data of the host glass and its structure. When calculating cooling processes (cooling rate) of porous glass products, the A.N. Dauvalter's formula, which takes into account only the stresses arising from the safe product cooling, but does not take into account those that remained there to the cooling start point, is usually used. The cooling rate in the interval of the annealing zone itself should be sufficiently low so that residual stresses, arising after they pass it, have small values. Since methods, that make it possible to determine the residual stresses that appear in the porous glass after passing through the initial annealing zone, are currently poorly developed, numerical simulation methods should be used to determine the porous glass SSD under the influence of thermal loads. Numerical study of the strain-stress distribution of porous glass allowing for thermal loads in the annealing temperature range was carried out in the *Ansys Workbench* software package.

1 Introduction

In industrial and civil engineering thermal insulation materials based on glass and designed to perform effectively the functions of heat protection and the heat accumulation and cold in buildings are of great importance [1]. In conjunction with properties and thermophysical characteristics, porous glass (foam glass) is considered to be promising in use [2]. The powder method of production is the most prevalent in industrial production. It includes four main stages: heating of the foaming mixture (charge), foaming, quenching, annealing. The annealing of the material is aimed to stabilize the structure and to reduce the appearing stresses to acceptable values. It is necessary to note two factors that occur when a temperature changes in a porous glass during its production. When the ambient temperature that surrounds the sample of the porous glass changes, the heat is gradually transferred along the layers of material due to the limited thermal conductivity of the material, so that there will be the temperature difference between the outer and inner layers in the sample. This heat transfer process requires a certain time. The glass matrix of the porous glass changes its linear dimensions when the temperature changes (it expands upon heating and shrinks upon cooling), and, consequently, the layers of material tend to accept the temperature-determined boundaries, which leads to the formation of stresses capable of breaking the sample.

There, in the glass matrix, may appear stresses of the second kind associated with the formation of a crystalline phase. These include the stresses, the appearance of which depends on the presence of the crystallization process in the glass. The resulting crystalline phase has a thermal expansion coefficient different from glass (TCLE). Due to this, there will be a difference between the compression values of the crystal and the glass matrix, which will lead to stresses at the contact boundary. Also, when it comes to improper cooling of the glass, there may occur residual stresses, which reduce the strength of the finished product and subsequently lead to destruction, even after a long time.

Both the structure relaxation process and the stress relaxation process in the glass are well studied and provide the basis for the development of stress analysis methods [3, 4]. The calculation of stress relaxation is based on the traditional approach to the solution of viscoelastic problems – using the Boltzmann superposition principle, the validity of the application is confirmed by experimental data [5, 6]. The finite element method is defined as the most promising for mathematical modeling of heat exchange processes and stress analysis for the desired material at the technological stage of annealing [7].

2 Object description

The object (Figure 1) is a sample of a porous glass with dimensions 30 × 30 × 30 mm, a perforation coefficient of 0.2589, an average pore diameter of 3 mm, placed on a 100 × 100 × 3 mm support, 12X18H9T support material [8]. The gas in the pores is CO_2.

* Corresponding author: grushkois@gmail.com

DOI: 10.1051/matecconf/201713203006

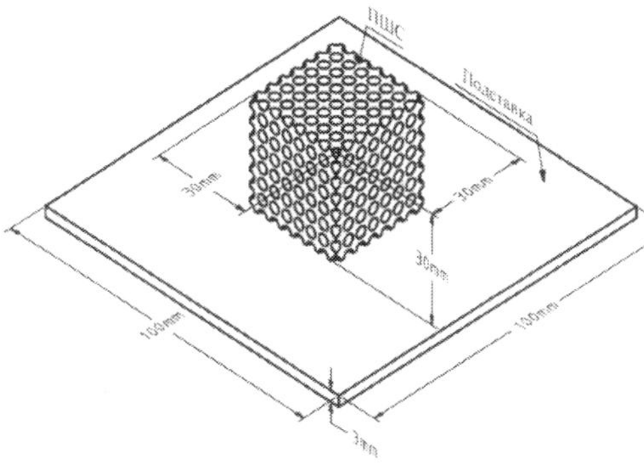

Fig. 1. Study subject

The algorithm for analyzing the strain-stress distribution of a porous glass sample consists in solving the system of equations [9] by the finite element method in the software complex *Ansys* [10] using the technique proposed in [11], allowing for the relaxation of material properties:

$$T_{fi} = T - \int_0^t \exp\left[-\int_{t'}^{t} \frac{dt''}{\tau_i(t'')}\right] \frac{dT(t')}{dt'} dt',$$

$$T_{fi} = \sum_{i=1}^{n} g_{i\tau} T_{fi},$$

where $\sum_{i=1}^{n} g_{i\tau} = 1$,

$$\lg \tau_i = \left(A + \frac{B}{T_{fi} - T_0} - \lg K_i\right) \frac{T_{f\tau}}{T} - \lg \tau_0 \left(\frac{T_{f\tau}}{T} - 1\right),$$

$$P = P_i(T_{fp}) + \int_{T_{fp}}^{T} \left(\frac{\partial P}{\partial T'}\right) dT',$$

where T – temperature, T_f – structural temperature, t – time, τ – relaxation time, g_i – weighted coefficients, n – amount of internal parameters characterizing the structure of substance, A, B и T_0 – constants in the Fulcher-Tamman equation; P – property; P_i – property of substance in the equilibrium (liquid) state;

$$K_i = \frac{\eta}{\tau_i},$$

here η – viscosity.

Mathematical modeling of the heat transfer process in the study object is based on a numerical solution of the initial boundary value problem:

$$\rho c \frac{\partial T}{\partial \tau} = \frac{\partial}{\partial x}\left(\lambda_x \frac{\partial T(x,y,z,\tau)}{\partial x}\right) + \frac{\partial}{\partial y}\left(\lambda_y \frac{\partial T(x,y,z,\tau)}{\partial y}\right) + \frac{\partial}{\partial z}\left(\lambda_z \frac{\partial T(x,y,z,\tau)}{\partial z}\right),$$

where T (t, x, y, z) – desired temperature distribution function; λ_x, λ_y, λ_z – coefficients of thermal conductivity in the direction of the x, y, z axes; ρ – density; c – specific ambient heat.

The stationary heat equation for an isotropic medium has the form:

$$div(\lambda \cdot gradT) = 0,$$

where T – desired temperature distribution function; λ – coefficient of thermal conductivity.

In accordance with the peculiarities of the technological mode of the annealing stage, the decrease in the ambient temperature is carried out linearly with a low speed (2-3 °C / min).

In light of this, the stresses are found from the system of equations of the following form [12]:

$$\sigma_{ik} = 2G\left[\varepsilon_{ik} + \frac{\mu}{1+\mu}\frac{S}{2G}\delta_{ik} - \alpha T \delta_{ik}\right],$$

$$\varepsilon_{ik} = \varepsilon_{ki} = \frac{1}{2}\left(\frac{\partial U_k}{\partial x_i} + \frac{\partial U_i}{\partial x_k}\right),\ i, k = 1, 2, 3,$$

$$\sum \frac{\partial \sigma_{ki}}{\partial x_k} = 0,$$

where σ_{ki} – stress, (σ_{ii} – normal stresses, σ_{ki} ($i{\neq}k$) – tangential stresses; ε_{ik} – strains; U_i – displacement; $S = \sum \sigma_{ii}$ – sum of the normal stresses, T – ambient temperature, G – shear modulus, $2G = \frac{E}{(1+\mu)}$, E – the Young's modulus; μ – the Poisson's ratio; α – thermal-expansion coefficient; $\delta_{ik} = \begin{cases} 1, i = k, \\ 0, i \neq k, \end{cases}$ – Kronecker symbol.

The main stages in solving this problem are – building a model, setting the initial data, solving and processing the results. It should be noted that the results of calculating the thermal state are the basis for solving the strength problem.

To create geometric models, CAD systems can be used. In this paper we implemented a comprehensive approach to creating a solid model – import of the geometry of a porous glass sample from the *PC "GMSH"* [13] with subsequent refinement in *PC ANSYS* (fig. 1).

The process of creating a model of a porous glass sample is automated – the geometry of the sample is formed depending on its size and porosity. The location of the pores is structured – it can have a hexagonal or ordinary arrangement of pores (depends on the thickness of the matrix walls that defines the skeleton of the porous glass); the pore size can be constant with the volume of the sample, or to have random nature. In the presented calculation, pore material is described by a certain number of spherical cavities of small diameter (in the region of 3 mm) and a constant composition possessing gas properties.

Work with materials is carried out by the Engineering Data module used to organize and store material data, describe material properties and input parameters of mathematical models using the additional interface of the Workbench. When calculating the temperature fields in the Transient Thermal module, the physical properties of materials such as thermal conductivity, heat capacity, etc. are available in the Toolbox window. It is necessary for thermal analysis to set the thermal conductivity in the material properties. The material can be anisotropic or isotropic (orthotropic). For an anisotropic material, the value of the thermal conductivity coefficient does not depend on the direction, i.e. is constant along the *X*, *Y*, and *Z* axes, and for the orthotropic material three thermal conductivity coefficients are specified in the *X*, *Y*, and *Z* directions.

When performing static strength calculations in the Static Structural module in the Toolbox window the mechanical properties of materials and models are available to describe the behavior of materials under the action of loads (plasticity, creep models, etc.).

To verify the efficiency of the proposed approach to the evaluation of heat transfer in a porous glass, numerical studies of thermal processes in the target material were carried out, the main task of which was to reveal the distribution of the temperature field with the volume of the sample with a change (decrease) in the temperature at the outer boundary of the calculating region (the thermal chamber wall) according to the annealing process. The calculations used real thermophysical and radiative properties of carbon dioxide, a matrix of porous glass and a steel support [14, 15, 16, 17]. The values of the thermal conductivity of carbon dioxide at the upper and lower boundary curves, at the critical point and on isobars from 30 to 200 kg / cm^2 to a temperature of 1000 °C are presented as temperature dependence. The thermal conductivity and the specific heat of the porous glass matrix were determined by the laser flash method [18] in the temperature range 25-700 °C in the TC-9000H thermal conductivity measuring device at the collective use center "Technologies and Materials of the Belgorod State University", Belgorod. The remaining physical characteristics of the porous glass matrix were chosen in accordance with its chemical composition [8].

The results of the temperature and the strain-stress distribution analyses are made at the calculated points (fig. 2).

The change in temperature over the cross section of the sample (along the lines, fig. 2) for the time point 12000 (the termination of annealing) is shown in fig. 3. The time point corresponds to the temperature at the boundary of the calculated region (furnace wall) of 25 °C. The results of stress analyses are shown in fig. 4.

3 Conclusion

The nature of the change in the temperature field in a sample of porous glass corresponds to the assumptions presented in the literature [19, 20, 21]. The created mathematical model adequately describes the thermal processes and stresses arising in the porous glass at the annealing stage and can be used to create optimal (for energy-consuming) technological processes for the production of porous glass.

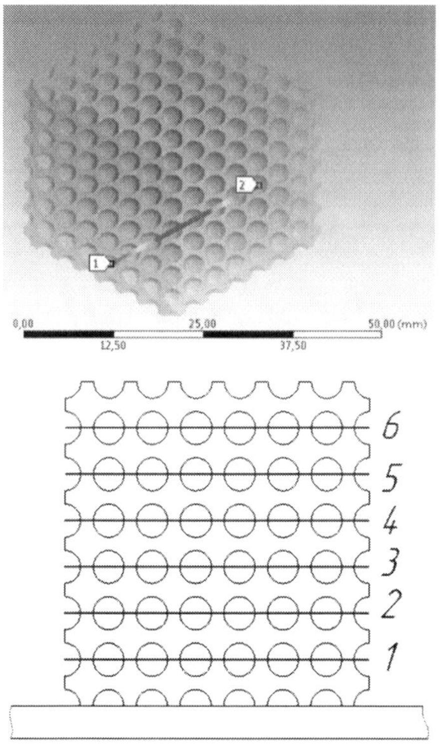

Fig. 2. Location of points and lines of temperature measurement (for example, line 1).

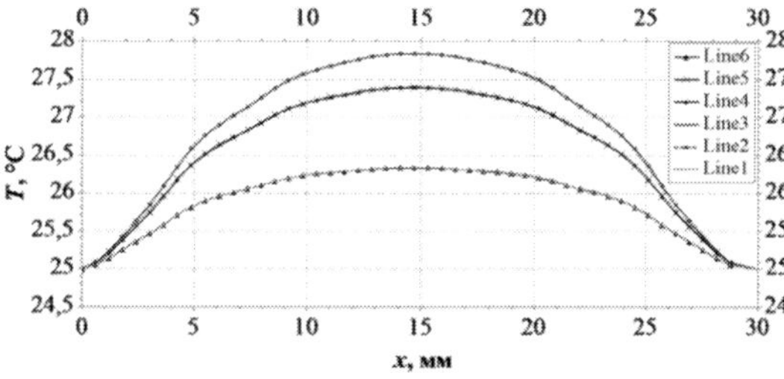

Fig. 3. The change in temperature in the sample at a time point of 12000 s.

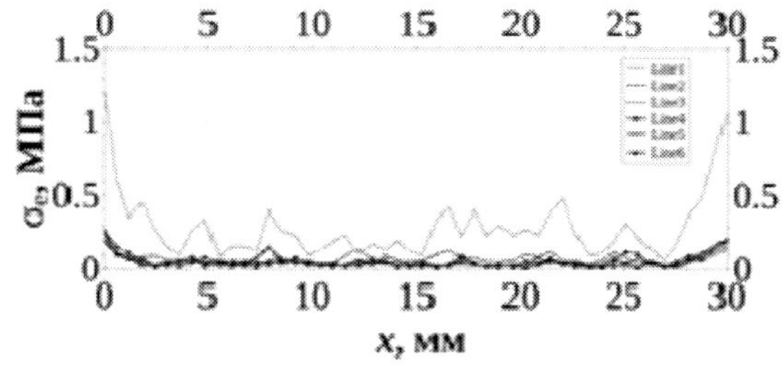

Fig. 4. The voltage magnitude in the sample at time point of 12000 s.

The article is prepared on the results of a study carried out with the financial support of the Russian Foundation for Basic Research as part of the scientific project No. 16-33-60177 mol_a_dk.

References

1. N.I. Minko, O.V. Puchka, M.N. Stepanova, S.S. Vaysera, *Teploizolyatsionnye materialy. Penosteklo.* (Monografiya), Belgorod (2016) [in Russian]
2. M.N. Stepanova, O.V. Puchka, *Naukoemkie tekhnologii i innovatsii, sbornik dokladov mezhdunarodnoy nauchno-prakticheskoy konferentsii*, 385-389 (2016) [in Russian]
3. O.V. Mazurin, Y.L. Belousov *Otzhig i zakalka stekla*, 11-21 (1984) [in Russian].
4. E.H. Lee, T.G. Rogers, *J. Am. Ceram. Soc.* **48 [9]** 480-487 (1965)
5. S.M. Rekhson, Y.K. Startsev, *Fizika i himiya stekla,* **2**, 140-1486 (1977) [in Russian]
6. V.P. Klyuev, A.S. Totesh, *Metody i apparatura dlya kontrolya vyazkosti stekla,* (VNIIEHSM, Moscow, 1975) [in Russian]
7. I.S. Grushko, *Steklo i keramika*, **10**, 3-9 (2016) [in Russian]
8. I.S. Grushko *Razrabotka teploizolyatsionnogo steklo- i steklokristallicheskogo kompozita stroitelnogo naznacheniya na osnove zoloshlakovyh othodov,* Cand. tech. sciences. Diss., South-Russian State Polytechnic University (NPI), St. Petersburg (2015) [in Russian]
9. O.V. Mazurin, *Steklovanie*, (Nauka, Leningrad, 1986)
10. N.N. Fedorova, S.A. Valger, M.N. Danilov, Y.V. Zaharova, *Osnovy raboty v ANSYS 17,* (DMK Press, Moscow, 2017)
11. O.Y. Smetannikov, N.A. Trufanov *Vychislitelnaya mekhanika sploshnyh sred,* **1**, 92-107 (2008) [in Russian]
12. E. Melan, G. Parkus, *Termouprugie napryazheniya vyzvannye statsionarnymi temperaturnymi polyami* (Fizatgiz, Moscow, 1958) [in Russian].
13. *GMSH Reference Manual. Gmsh* 3.0., available at: *http://gmsh.info/doc/texinfo/gmsh.html* (2017) [in Russian]
14. GSSSD 101-86 *Dioksid ugleroda. Koehffitsienty vyazkosti, teploprovodnosti i chislo Prandtlya razrezhennogo gaza v diapazone temperatur 150...2000 K, Tablitsy spravochnyh standartnyh dannyh* (Izd-vo standartov, Moscow, 1986) [in Russian].
15. N.V. Tsederberg, *Teploprovodnost gazov i zhidkostey* (Gosenergoizdat, Moscow – Leningrad, 1963) [in Russian].
16. L.N. Latyev, V.A. Petrov, V.Y. Chekhovskiy, E.N. Shestakov *Izluchatelnye svoystva tverdyh materialov: spravochnik* (Energiya, Moscow, 1974) [in Russian]
17. A. Misnar, *Teploprovodnost tverdyh tel, zhidkostey, gazov i ih kompozitsiy*, (Mir, Moscow, 1968) [in Russian]
18. V.G. Baranov, A.V. Tenishev, A.V. Lunev, S.A. Pokrovskiy, A.V. Hlunov, *Yadernaya fizika i inzhiniring*, **2(4)**, 291-302 (2011) [in Russian].
19. I.I. Kitaygorodskiy, *Penosteklo: uchebnoe posobie*, Stroyizdat ,Moscow, 1958) [in Russian].
20. B.K. Demidovich, *Penosteklo* (Nauka i tekhnika, Minsk, 1975) [in Russian].
21. S.V. Alekseev *Sovershenstvovanie protsessa otzhiga vysokoporistyh materialov na osnove stekla* (Belgorod Shukhov State Technological University, Belgorod, 2002) [in Russian]

Developing of complex for hot plastic deformation modeling of steel type 20-30CrNiMoV for heavy forging

Vladimir Dub[1,*], *Alexandr* Churyomov[1], *Igor* Soloviev[2], *Anton* Popkov[3]

[1]The National University of Science and Technology MISiS, Russian Federation,
[2]Ural Federal University named after the first President of Russia B.N.Yeltsin», Russian Federation,
[3]United Heavy Machinery Plants, Russian Federation.

Abstract. Production of heavy forging of bars weighing more then 235 tons for such products as rotors made of steel type 20-30CrNiMoV is a critical independent work, failure to perform which entails high costs related to repeated production (in case of defective product) and untimely launch of production plants. One of the frequent causes of a defective product is the impossibility of ultrasonic testing in the barrel-gate zones on the rotor workpiece, which is due to the microstructure of the metal, namely the grain size. Determining the stages of deformation process wich causes such defects in structure is the main goal of this work.

1 Formatting the title, authors and affiliations

Decrease of defective products at all processing stages of metallurgical production is a priority task for researchers and production engineers. As is known, the grain size under deformation depends on the following parameters: degree of deformation ε, rate of deformation $\dot{\varepsilon}$ s^{-1}, temperature *T*. These physical parameters in turn are associated with the process parameters, such as tool operation time on the workpiece, *T* of metal on the surface. Prediction of the structure change at different degrees of deformation and at a certain temperature is described on the basis of the theory of plastic deformation and can be modeled for specific technological processes [1,2].

In turn, during deformation, recrystallization always occurs, which can be of three types: static, metadynamic and dynamic. One of the ways to optimize production processes at the stage of hot plastic deformation (HPD) of ingots is to model the structure using mathematical models and the finite elements method.

There are several phenomenological models that establish the connection between the stressed and deformed state of the medium [3-20]:

- Johnson-Cook model;
- Steinberg-Cochran-Guinan- Lund model;
- Zerilli-Armstrong model;
- Mechanical threshold stress model (MTS);
- Preston-Tonks-Wallace model.

At the present time in order to model the evolution of structure during the HPD process, the microstructure characteristics relationship models (volume fraction of recrystallized grains and grain size) with HPD parameters based on equations of Avrami-Kolmogorov type [3-8] are most widely used. They show a high predictive capability, but require a lot of experimental data in order to determine unknown constants.

Kinetics of dynamic recrystallization can be described by the following set of equations (1-4):

$$X_{DRX} = 1-\exp\left[-\beta_d \cdot \left(\frac{\varepsilon - a_{10}\cdot\varepsilon_p}{\varepsilon_{0.5}}\right)^{k_d}\right] \tag{1}$$

$$\varepsilon_{0.5} = a_5 \cdot d_0^{h_5} \cdot \varepsilon^{n_5} \cdot \dot{\varepsilon}^{m_5} \cdot \exp\left(\frac{Q_5}{R\cdot T}\right) + c_5 \tag{2}$$

$$\varepsilon_p = a_1 \cdot d_0^{n_1} \cdot \dot{\varepsilon}^{m_1} \exp\left(\frac{Q_1}{R\cdot T}\right) + c_1 \tag{3}$$

$$\varepsilon_c = a_2 \cdot \varepsilon_p \tag{4}$$

where d_0 – initial grain size, µm; $\varepsilon_{0.5}$ – degree of deformation at 50% of the volume of recrystallized grains; a_1, n_1, m_1, Q_1, c_1, a_2, a_5, h_5, n_5, m_5, Q_5, c_5, β_d, a_{10}, k_d – rheological parameters of the material, determined on the basis of the results of experimental studies.

To determine the size of dynamically recrystallized grains, the polynomial empirical equation with temperature dependence according to the Arrhenius law was used (5):

$$d_{DRX} = a_8 \cdot d_0^{h_8} \cdot \varepsilon^{n_8} \cdot \dot{\varepsilon}^{m_8} \cdot \exp\left(\frac{Q_8}{R\cdot T}\right) + c_8 \tag{5}$$

where a_8, h_8, n_8, m_8, Q_8, c_8 are the material rheological parameters determined on the basis of the results of experimental studies (if $d_{DRX} > d_0$, then $d_{DRX} = d_0$).

Grain growth during the annealing of a non-deformed metal, or in the intervals between deformations, can be described by the following equation (6):

* Corresponding author: vladimir.dvchem@gmail.com

$$d_{gg} = \left[d_0^m + a_9 t \exp\left(-\frac{Q_9}{RT} \right) \right]^{\frac{1}{m}} \tag{6}$$

where a_9, Q_9, m – are the material parameters determined on the basis of the results of experimental studies.

To determine the constant rheological parameters for steel type 20-30CrNiMoV included in the model, steel grade 26HN3M2FA was chosen and experimtens with the use of the Gleeble 3800 complex were carried out: мelting of 12 kg of ingots followed by forging into bars and annealing in order to obtain a structure with grains of different sizes (50, 200, 1000 μm), simulating different zones of the ingot. An experiments matrix was developed to study the effect of the rate, degree and temperature of deformation on the size of austenite grain to determine the rheological parameters included in the phenological model.

Curves of deformation according to which the coefficients describing the kinetics of dynamic recrystallization for steel 26HN3M2FA were determined, are shown in Fig. 1. Kinetic parameters were determined using the method of minimizing the error between calculated and experimental values of the flow stress in the section of the deformation curve from the maximum value to the value corresponding to the steady-state stage of deformation.

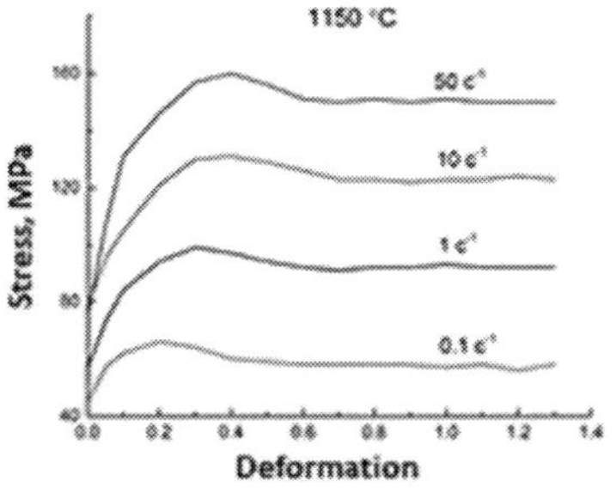

Fig. 1. Curves of deformation of dynamic recrystallization

Deformation curves for determination of dynamic recrystallization kinetics for 26HN3M2FA steel.

Unknown coefficients were determined on the basis of the results of microstructural studies of grain size after different deformation modes. Figure 2 shows typical microstructure of 26HN3M2FA steel with dynamically recrystallized structure.

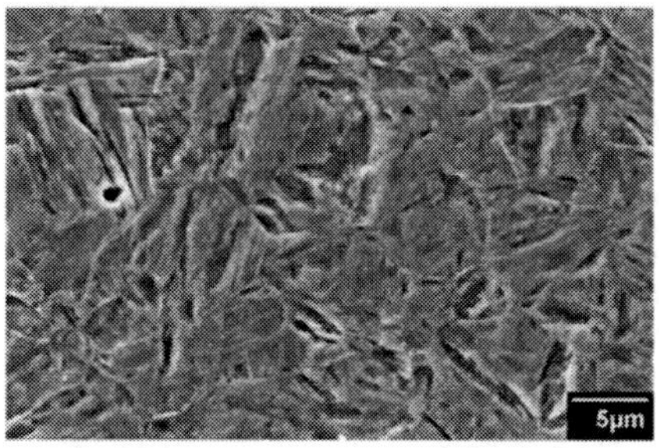

Fig. 2. Microstructure of the 26HN3M2FA steel (after deformation in the mode T=1150C°, $\dot{\varepsilon}$=50 c-1,ε=1,3, τ=0 c.

Comparison of experimental and calculated grain sizes for 26HN3M2FA steel is shown in Fig. 3. Average calculation error for the 26HN3M2FA steel was 5,2%

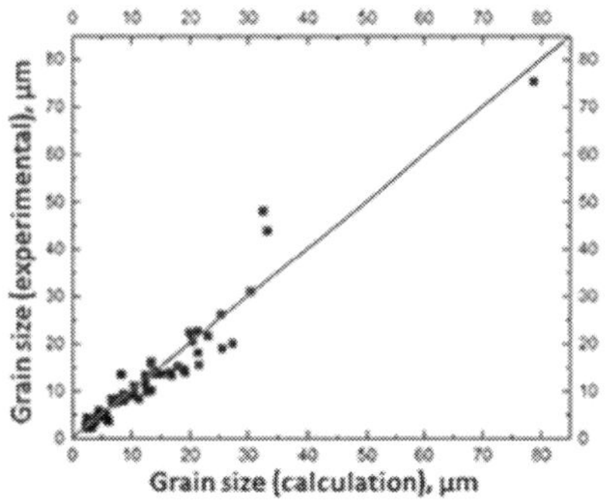

Fig. 3. Comparison of experimental and calculated grain sizes for 26HN3M2FA steel

2 Comparison of calculated and experimental values of dynamically recrystallized grains for 26HN3M2FA steels

More than 540 studies were carried out in order to determine the rheological parameters of 26HN3M2FA steel. As a result, a model that describes the deformation of 26XH3M2ФA steel was developed (7-8):

$$d_{DRX} = 67352.5 \times d_0^{-0.051} \times \varepsilon^{-0,672} \times \dot{\varepsilon}^{-0,075} \times exp\left(\frac{94000}{R \times T}\right) \tag{7}$$

$$d = d_{DRX} X_{DRX} + (1 - X_{DRX}) d_0 \tag{8}$$

For modeling of the actual structure formation in the process of industrial HPD, obtained mathematical dependences (values of the coefficients) and rheological properties of materials are downloaded to the program that allows to calculate deformation processes using the finite element method. The model (7-8) is applied to determine grain in each element In the environment of finite element modeling, the initial geometry of the workpiece and deforming tool, the initial and boundary conditions, including the initial distribution of grain sizes are set. After that, the process conditions in the form of time dependencies of the workpiece position relative to the tool, its speed and the surface temperature of the workpiece are set.

Result of the calculation is the dependence of characteristics of the grain structure microstructure at the final moment of forming at each point of the workpiece on the HPD parameters (set by the corresponding time dependencies) (9):

$$d = f(F_{GPD}(\tau), x_i, y_i, z_i) \tag{9}$$

This model was used to study the distribution of grain for a rotor workpiece of bar with mass 235 tons in order to identify the least deformable zones.

Distribution in the initial ingot was modeled using the DeForm software package (40,000 items were studied).

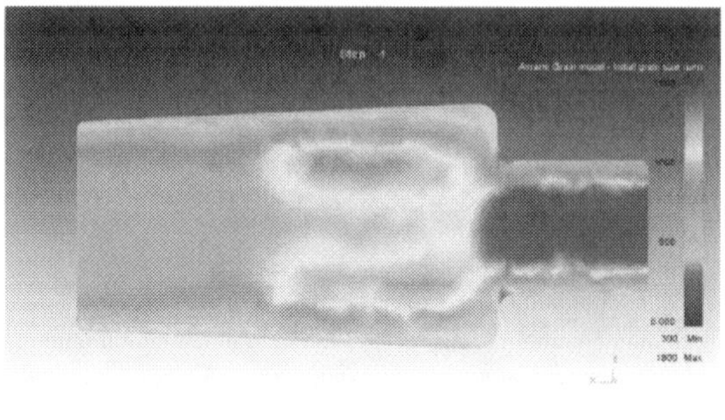

Fig. 4. Stage of hot deformation model of rotor forging

Further, the entire forging process was modeled taking into account the actual deformation conditions with measurement of geometric pattern, time and temperature. As a result, the most critical stage of forging the rotor workpiece was determined: deformation into circle.

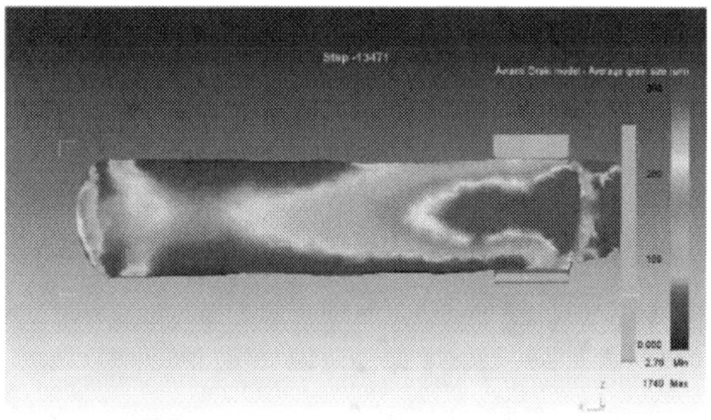

Fig. 5. Stage of hot deformation model of rotor forging

And the final distribution of grain after heating:

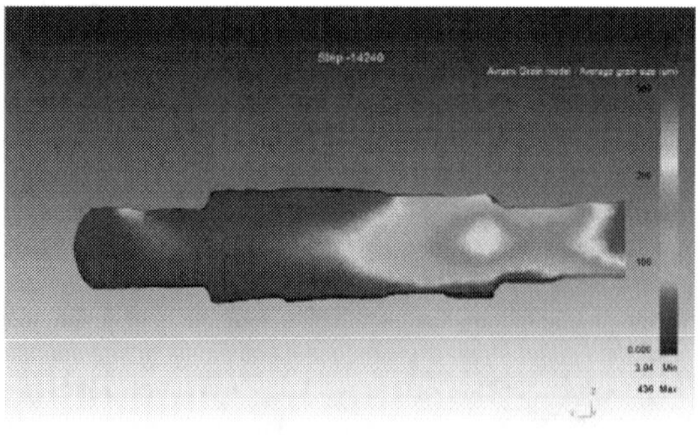

Fig. 6. Final stage of hot deformation model

It can be seen that the most problematic zone is «barrel-gate» with a grain size of more than 170 μm. These results are fundamental for predicting the technology of rotor deformation and adjusting the factory technology.

3 Conclusion

1. A phenomenological model describing the process of deformation for unique heavy forging (more than 230 tons) for power machine made of 20-30CrNiMoV type steel was developed.

2. On the basis of large array of experimental data, the unique constant parameters of a steel of the selected type, which determine the rheological properties of the material for the following thermal-deformation modes, were determined for the first time: T=1230-900°C; degree of deformation 0,17-1,3; rate of deformation from 0.1 to 50 c^{-1}

3. Numerical modeling of the rotor workpiece forging process from an ingot of 235 tons was performed, and occurrence of critical zones with large grains (more than 170 μm) and the stage of formation of these zones was predicted for the first time.

This study was sponsored by the Ministry of Education and Science of the Russian Federation in frame of the Federal Targeted Program on Research and Development in Priority Areas of Development of the Russian Science and Technology Sector for 2014–2020 (Application ID: 2015-14-579-0173-366; unique ID of applied research: RFMEFI57815X0114).

References

1. R.W. Hertzberg *Deformation and fracture mechanics of engineering materials*, 786 (1996)
2. A.V. Dub, V. A. Durynin, V. S. Dub, [et al.] *Proc. Sci.-Pract. Conf. Innovative Materials and Technologies for Nuclear and Other Types of Power Plants and Heavy Machinery Construction*, Moscow, p. 8, (2011)
3. A.M. Elwazri, E. Essadiqi, S. Yue, ISIJ International, **44**, N. 4, 744-752 (2004)
4. G. Shen, S.L Semiatin, and R. Shivpuri, Metallurgical and Materials Transactions A, **26A**, 1795-1803, (1995)
5. A.J. Brand, K. Karhausen, R. Kopp., Materials Science and Technology. **12**. 963-969 (1996).
6. X. Lei, J. Huang, X. Jin, S. Chen, X. Zhao, Materials Letters, **181**, 240-243 (2016)
7. F. Yin, L. Hua, H. Mao, X. Han, D. Qian, R. Zhang, Materials & Design, **55**, 560-573 (2014)
8. Z. Jina, K. Lia, X. Wua, H. Dong, Proceedings 2S, 460-465 (2015)
9. S. Kim, Y. Lee, D. Lee, Y. Yoo., Mater. Sci. Eng. A. **355**, 384-393 (2003)
10. K.R. Kinsman, H.I. Aaronson. *Transformation and Hardenability in Steels,* Climax Molybdenum Co., Ann. Arbour (Michigan, USA, 1967)
11. M.H. Jacobs, D.W. Pashley. The Mechanism of Phase Transformation in Crystalline Solids (Institute of Metals, London) Monograph no. **33**, 43 p. (1969)

12. Kwak D., Micro-Magnetic Examination of Thermal Fatigued Heat-Resistant Steels, PhD dissertation (2004)
13. M. Hojny, M. Głowacki and Z. Malinowski, High Temperature Materials and Processes, **28**, 4, 245-252 (2009)
14. M. Głowacki and M. Hojny, Inverse Problems in Science and Engineering, **17**, 2, 159-174 (2009)
15. P.J. Wray, Metall. Trans. A, 15A, 2041-2058 (1984)
16. Y. Saito et al., Kawasaki Steel Techn. Rep., No.9, 12-21 (1984)
17. Y.W. Chen, R.L. Tobler, B.J. Fills, K.J. Coakley, NIST Technical Note 1500-6, Materials Reliability Series (1999)
18. N. Hatta, J. Kokado, S. Kikuchi, H. Takuda, Steel Res., 56, 575-582 (1985)
19. S. Sakai, T. Sakai, K. Takeishi, Trans ISIJ, 17, 718-725 (1977)
20. D.C. Ludwigson, Metall. Trans., 2, 2825-2828 (1971)

Quenching temperature influence on elastic and hardness behavior in a biocompatible Ti-based alloy

Anatoliy Illarionov[1,*], *Maksim* Karabanalov[1], *Andrey* Korelin[1], and *Stepan* Novokreschenov[1]

[1]Ural Federal University, 19 Mira St., 620002 Ekaterinburg, Russia

Abstract. Elasticity modulus and hardness changes in deformed semifinished rods manufactured of biocompatible Ti-6Al-7Nb alloy under quenching from the temperatures in a range of 500-1050°C have been studied by means of indentation. The elasticity modulus and Vickers hardness have been shown to vary from 95 to 115 GPa and from 330 to 520 HV respectively depending on the quenching temperature. Relationship between the elastic and hardness behavior and the alloy phase composition has been found.

1 Introduction

The Ti-6Al-7Nb alloy [1] was purposely developed as a biocompatible material for implants to replace the Ti-6Al-4VEli alloy being widely used in medical practice. The main β-stabilizing element in Ti-6Al-7Nb is niobium because it is more biologically inert and exhibits better nontoxicity compared to vanadium [2]. Nowadays the Ti-6Al-7Nb alloy for medical implants is manufactured by means of traditional metallurgy methods, e.g., severe plastic deformation [3, 4], powder metallurgy processes [5, 6], and selective laser melting [7,8], with processing parameters greatly influencing the structure and mechanical properties of semifinished material. It should be noted that the Ti-6Al-7Nb alloy in its equilibrium state possesses rather high elasticity modulus of 105 – 110 GPa [9, 10] compared to most other Ti-based medical alloys [11, 12] because of considerable aluminum content, and this somewhat reduces the alloy biocompatibility for implant application. Quenching from different temperatures corresponding to two-phase (α + β)-field was shown by using the VT16 Ti-based medical alloy [13] to result in considerable elasticity modulus variation due to obtaining various phase composition. As for the Ti-6Al-7Nb alloy, its phase composition also changes with the quenching temperature corresponding to two-phase (α + β)-field [14], but the influence of those changes on the alloy elasticity modulus and hardness has not been studied yet. It is within the context of the above that the aim of this research is to study the influence of quenching temperature on the deformed semifinished Ti-6Al-7Nb alloy elasticity modulus and hardness, with the quenching being carried out from two-phase (α + β)- and single-phase β-fields, and to consider the relationship between the elastic and hardness behavior and the phase composition of the alloy quenched.

2 Materials and Methods

Hot-rolled rods (20 mm in diameter) supplied by VSMPO-AVISMA Corporation were manufactured of the Ti-6Al-7Nb alloy with the composition of Ti-6,28Al-7,24Nb-0,19O–0,19Fe-0,015C-0,009N-0,001H (wt.%) [14] and subjected to the deformation in two-phase (α + β)-field.

The temperature of complete polymorphous α + β → β transformation (T_{pt}) determined by means of test quenching was 990±5 °C. The heat treatment procedure of the Ti-6Al-7Nb alloy specimens consisted of heating up to the temperatures in a range of 500-1050°C with the interval of 50°C, holding during 1 hour, and water quenching.

The elasticity modulus and hardness were measured by means of indentation using MHTX equipment from CSM instruments at a load of 9N, with calculations made according to Oliver-Pharr technique [15]. X-ray diffraction analysis was carried out using a DRON 3M X-ray diffractometer and a Neophot 2 light microscope.

3 Results and Discussion

The Ti-6Al-7Nb alloy structure analysis in the initial hot rolled alloy state has revealed that the structure mostly consists of the bright primary α-phase precipitates of rather complex shape which is sometimes similar to the globular one (Fig. 1). Space around the primary α-phase particles is occupied by the matrix turned into β-phase, with the increased matrix etchability making it darker compared to the primary α-phase. There are the second

* Corresponding author: illarionovag@mail.ru

 DOI: 10.1051/matecconf/201713203008

α-phase precipitates (dispersed particles) within the above matrix (Fig. 1).

It has been found by means of X-ray diffraction analysis that the alloy structure in its hot rolled state includes only α- and β-phases with the following parameters: $a_α$=0.29298 nm, $c_α$=0.46748 nm, c/a=1.5956, $a_β$=0.32577 nm. The c/a parameter obtained for the α-phase is higher than that of pure titanium (1.587 [16]), with this fact being characteristic for the aluminium doped alloys [17]. The β-phase lattice constant is rather high, and this can be observed when the high-temperature β-solid solution does not break down completely during its cooling from its last treatment temperature. The values of elasticity modulus (105 GPa) and hardness (345 HV) in the hot rolled state are typical for the alloy [10, 18].

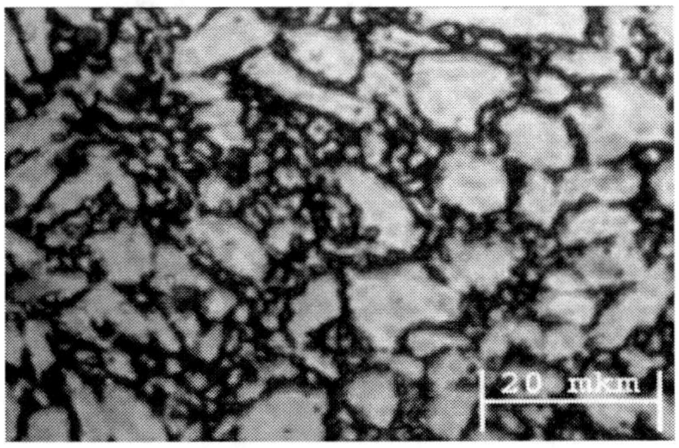

Fig. 1. Structure of as-hot rolled Ti-6Al-7Nb alloy

Figure 2 shows the quenching temperature dependences of the Ti-6Al-7Nb alloy hardness and elasticity modulus measured by means of indentation. These temperature dependences reveal their minimum after the quenching from 750°C. The alloy hardness and elasticity modulus decrease from 430 to 330 HV and from 105 to 95 GPa respectively with increasing quenching temperature from 500 to 750°C, whereas the alloy hardness and elasticity modulus grow up to 500-520 HV and 115 GPa respectively at the higher quenching temperatures in a range of 800-1000°C. The alloy hardness values obtained after quenching are always higher (Fig. 2a), whereas the alloy elasticity modulus values are higher only after quenching from the temperatures in a range of 950-1050°C (Fig. 2b) compared to the above values in the initial hot-deformed alloy state.

The quenching temperature dependences of the Ti-6Al-7Nb alloy hardness and elasticity modulus (Fig. 2) are consistent with the changes of the alloy phase composition (Table 1, [14]) and the lattice constant of the metastable phases observed under quenching (Table 2, Fig. 3).

Table 1. Ti-6Al-7Nb alloy phase composition after quenching from different temperatures [14]

Quenching temperature, °C	500-700	750	800-900	950	1000-1050
Phase composition	α+β	α+α''+ β	α+α''	α+α'	α'

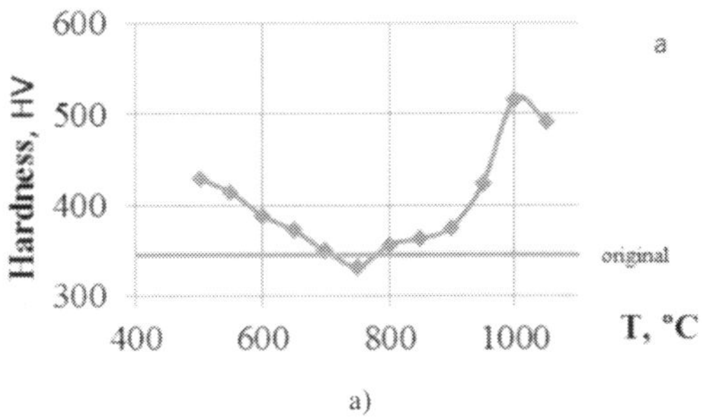

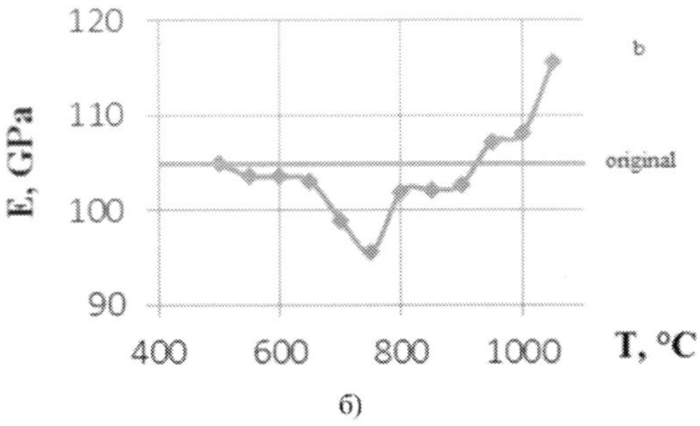

Fig. 2. Change of (a) hardness and (b) elasticity modulus in Ti-6Al-7Nb alloy with the quenching temperature.

Table 2. β-phase lattice constant after quenching

Tq,°C	500	550	600	650	700
$a_β$, nm	0.32435	0.32414	0.32465	0.32477	0.32508

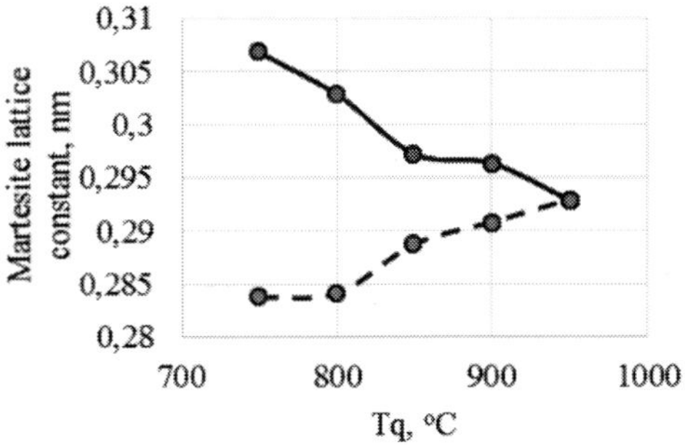

Fig. 3. Change of martensite lattice constant in Ti-6Al-7Nb alloy with the quenching temperature ($a_{α''}$ – solid line, $(b/\sqrt{3})_{α''}$ – dashed line).

The Ti-6Al-7Nb alloy structure includes α- and β-phases under the quenching at the temperatures in a range of 500-700 °C, with the β-phase lattice constant in the temperature range mentioned above being lower than that in the initial hot-deformed alloy state. This β-phase lattice constant reduction is obviously connected with the breakdown in the initial β-solid solution during the holding in the quenching temperature range mentioned above. The breakdown is accompanied by the second α-phase precipitation which contributes to obtaining higher alloy hardness values compared to those in the initial alloy state. With increasing heating temperature before the quenching in the temperature range mentioned above the second phase precipitates are coarsening and their volume fraction is decreasing (the fact is proved by the β-phase lattice constant growth), the hardening effect is reducing reaching its minimum at the temperature of 700°C. At the quenching temperature of 750°C the metastable β-phase solid solution being observed during the heating before quenching does not break down but partly undergoes the martensitic β → α''-transformation during the cooling. In this case the fact that the second α-phase precipitations do not exist and the α''-martensite which is rich enough in alloying elements but is known [17] not to possess high hardness is observed leads to obtaining minimal hardness values over the whole quenching temperature range studied. The further quenching temperature increase up to 900°C results in the complete martensitic β → α''-transformation of the β-phase solid solution during quenching, but according to [14] the forming α''-martensite is less rich in β-stabilizing elements (niobium, iron), exhibits less rhomboidity (being characterized by the difference in $a_{\alpha''}$ and $(b/\sqrt{3})_{\alpha''}$ values, see Fig.3) but higher hardness. This contributes to some growth of the alloy hardness values compared to those in its hot rolled state. At the quenching temperature of 950°C and higher the β-phase solid solution undergoes the martensitic β → α'-transformation. The α'-martensite is shown in [17] to possess the best hardness behavior compared to β-, α''-, and α-phases, it is this fact which leads to the considerable hardness growth compared to the alloy hardness in its hot rolled state with increasing quenching temperature up to T_{pt}. When quenching from the temperatures higher than T_{pt} the alloy hardness somewhat decreases because the β-grain size was shown by microstructure to grow from 600 μm at T_q = 1000°C to 900 μm at T_q = 1050°C, and the forming α'-martensite needles coarsen correspondingly. The similar hardness dependence on quenching temperature along with the martensite formation was observed in the VT18U alloy [19] which contains about 6% of Al as well as the Ti-6Al-7Nb alloy does but less amount of Nb (1.13%).

Unlike hardness the Ti-6Al-7Nb alloy elasticity modulus value after the quenching from 500°C (105 GPa) coincides with that of the hot rolled semifinished material. This is due to fact that the elasticity modulus in not influenced by the alloy structural features (e.g., existence of the second phase precipitates) [20], but it is connected with the alloy phase composition which remains the same after the quenching from 500°C and in the hot rolled alloy state. Meanwhile, the alloy elasticity modulus value tends to reduce with increasing quenching temperature up to 750°C which is due to the β-phase solid solution volume fraction growth observed in [14] under quenching. (Note that the β-phase solid solution exhibits low elasticity modulus.) The further quenching temperature increase up to 1050°C results in the fact that the alloy elasticity modulus value tends, on the contrary, to grow (Fig.2b). This is connected with the following: the α''-martensite forming under the quenching from the temperatures lower than 950°C is becoming less alloyed by β-stabilizing elements (as shown above, Fig.3) with increasing quenching temperature but, at the same time, the α''-martensite elasticity modulus was established in the VT16 alloy [13] to be growing. However, it should be noted that the alloy elasticity modulus value is less than that in the initial hot rolled alloy state because the elasticity modulus of the α-phase being the main part of the initial hot rolled alloy structure is higher than that of the α''-martensite forming under the quenching. After the quenching from the temperature of 950°C and higher the α'-martensite is observed in the alloy structure (Table 1), it exhibits higher elasticity modulus [19] compared to that of α-phase and, therefore, the alloy elasticity modulus values obtained are higher than those in the initial hot rolled alloy state and increase up to T_{pt} due to reducing primary α-phase volume fraction and growing α'-martensite volume fraction [14].

On the whole, the Ti-6Al-7Nb alloy elasticity modulus values vary from 95 to 115 GPa over the quenching temperature range studied. The minimal value of 95 GPa is 10-15 GPa less than that of the alloy in its equilibrium state and obtained after the quenching from 750°C which is close to this alloy critical temperature (T_c), the quenching from any temperature above T_c leads to the β-phase solid solution martensitic transformation [21]. The maximum elasticity modulus value of 115 GPa is 5-10 GPa higher than that of the alloy in its equilibrium state and obtained after the quenching from above T_{pt}, from β-phase field with T_q = 1050°C, with the 100% α'-martensite structure being observed. It follows from the above that to obtain the Ti-6Al-7Nb alloy elasticity modulus values about 95 GPa it is necessary to use heat treatment including the quenching from the temperatures close to T_c.

4 Conclusions

Thus, the following conclusions can be made based on the research carried out:

1. The as-hot rolled Ti-6Al-7Nb alloy is a two-phase α + β-alloy consisting predominantly of the primary α-phase of complex, near to globular morphology which is imbedded into the matrix turned into β-phase with the second α-phase precipitates. The alloy possesses its typical values of hardness (345 HV) and elasticity modulus (105 GPa).
2. The temperature dependences of the alloy hardness and elasticity modulus within the temperature range of 500-1050 °C have been found to show their minimum at 750 °C which is close to this alloy critical temperature. The ranges of the alloy Vickers hardness and elasticity modulus variations under quenching are 330-520 HV and 95-115 GPa respectively.
3. It has been shown that the alloy hardness variation under quenching depends on both its structural (whether or not the second phase precipitates exist) and phase transformations, whereas the alloy elasticity modulus variation is mostly influenced by its phase composition changes.

We hereby acknowledge the support of the Ministry of Science and Education of the Russian Federation in accordance to the decree of the Government of April 9, 2010, №218, project number 02.G25.31.0234.

 DOI: 10.1051/matecconf/201713203008

References

1. M. Semlitsch, F. Staub, H. Weber, Biomed. Technik, (30), 334-339 (1985)
2. A. Biesiekierski, J. Wang, Mohamed Abdel-Hady Gepreel, C. Wena. Acta Biomater, 8, 1661 (2012)
3. P. Barbosa, S. Button, Proceedings of the Institution of Mechanical Engineers Part L: Journal of Materials: Design and Applications, 214, 23-31, (2000)
4. V. Polyakova, I. Semenova, R. Valiev, Materials Science Forum, **667-669**, 943-948 (2011)
5. V. Henriques, C. Da Silva, Key Engineering Materials, **189-191**, 443-448 (2001)
6. L. Bolzoni, E. Ruiz-Navas, E. Gordo, Journal of the Mechanical Behavior of Biomedical Materials, 67, 110-116 (2017)
7. T. Sercombe, N. Jones, R. Day, A. Kop, Rapid Prototyping Journal, 14, 300-304 (2008)
8. E. Chlebus, B. Kuźnicka, T. Kurzynowski, B. Dybała, Materials Characterization, 62, 488-495 (2011)
9. M. Niinomi, Materials Science and Engineering A, 243, 231-236 (1998)
10. R. Boyer, G. Welsch and E. Collings *Materials Properties Handbook*: *Titanium Alloys*. Ed. by, *ASM International*, (Materials Park, OH, USA, 1994)
11. M. Geetha, A. Singh, R. Asokamani, A. Gogia, Progress in Materials Science, 54, 397-425 (2009)
12. C.-W. Lin, C.-P. Ju, J.-H. Lin, Materials Transactions, 45, 3028-3032 (2004)
13. A. Popov, A. Illarionov, S. Stepanov, O. Ivasishin, Physics of Metals and Metallography, 115, 517-522 (2014)
14. A. Illarionov, M. Karabanalov, S. Stepanov, Metal Science and Heat Treatment, 52, 481-486 (2011)
15. W. Oliver, G. Pharr, Journal Materials Research, 7, 1564-1583 (1992)
16. U. Zwicker, *Titanium and its alloys* (Metallurgija, Moscow, 1979)
17. A. Popov, A. Illarionov, N. Rossina, S. Grib, Metal science and heat treatment of titanium alloys. Structure and properties, UrFU (2012)
18. D. Gadeev, A. Illarionov, S. Demakov, Metal Science and Heat Treatment, 57, 463-468 (2015)
19. L. Bolzoni, E. Ruiz-Navas, E. Neubauer, E. Gordo, Journal of the Mechanical Behavior of Biomedical Materials, 9, 91-99 (2012)
20. B. Livshits, Physical properties of metals and alloys, Textbook, 2nd ed., Ext. and rev. (1980)
21. N.F. Anoshkin, G.A. Bochvar and others, *Metallography of titanium alloys* (Metallurgia, Moscow, 1980)

Inhomogeneity of deformed state during compression testing of titanium implant

Yury Loginov[1], *Stepan* Stepanov[1,*], and *Catherine* Khanykova[1]

[1]Ural Federal University, 620002, 19 Mira str., Ekaterinburg, Russia

Abstract. This study examines the inhomogeneity of deformed state during compression testing of porous titanium implant. The theoretical part of the article includes numerical simulation of deformation of a prismatic titanium sample compressed with absolutely rigid plates. The porosity was provided by the circular shape pores with titanium struts between them. To solve the problem by means of the finite element analysis, the boundary conditions were set using the ABAQUS software. The fields of strain, stresses and displacements were determined. Presumable place of fracture coincide with the highest values of strain which are localized in vertical struts. Physical modeling of the implant compression was performed at the second part of the study on the testing machine with video recording. It was shown that in the real process the localization of failure corresponds to the calculated data.

1 Introduction

The implant industry takes an ever-increasing living space in the field of bioengineering. It`s achievements allow for prolonging an active life of a human as well as "repairing" individual body parts [1]. Titanium alloys are often utilized as a material for such invasions due to their chemical inertness and high strength to weight ratio [2-4]. Manufacturing of titanium implants employs methods of powder metallurgy used to obtain the desired architecture (the ratio of the shape and the size of pores in materials) [5, 6]. The predetermined porosity reduces the weight of implant, which allows for achieving the weight of natural bone tissue. A greater shock absorbing of joints, when impact and static loads take place, is provided by a smaller modulus of elasticity. The better intergrowth of the living tissues through pore space is ensured, thus integration with this organism is achieved.

Originally, porous structures of metals were obtained using the powder sintering. Sintering mode provided the required porosity. However, this did not allow for designing the architecture of the material at each point. The development of additive manufacturing enabled to define the structure of the material at macro level by using the software for controlling the process of a metal medium growing [7]. At the same time, the structural and phase state of the alloys used can be set [8, 9].

Although the modulus of elasticity of widely used titanium alloys can be adjusted by changing phase composition [10, 11, 12], this method regulates the elastic modulus in a very limited range. A more effective technique is to change the porosity of the material [13]. In addition, the effect of the porous space architecture can be added. Thus, a whole class of so-called cellular materials origination appeared [14].

An additional elastic effect can be provided by a gas located within the pores due to the appearance of throttling effect through the pores with open porosity and compressive resistance effect for closed porosity [16]. The architecture of the volume of a porous medium is also significant. For example, the tortuosity of pores is introduced to account for the resistance to throttling in the mechanics of liquids and gases transiting in the porous media. Another lever for controlling the properties of porous materials is to modify the shape of the pores by methods of metal forming [16, 17], including isostatic treatment in a hydrostatic or gas-static version.

The compression testing of porous implants has some peculiarities, since their elastic and strength characteristics are determined both by the properties of the material from which they were made and by the shape, as well as by the size and arrangement of the pores.

The purpose of the work is to study the behavior of the material in the compression test of additive manufactured implant.

2 Simulation

This study employs mathematical modeling based on the finite element analysis (FEA) to describe a porous/cellular material with the simple circular cells uniformly distributed in the metallic medium.

A planar case is considered to simplify the stress-strain state. A schematic representation of such a structure is shown in Fig. 1. The architecture of the material in the plane is shown. It consists of the metal 1 and the circular pores 2 formed therein. The selected fragment 3 was utilized for calculations introduces the radius of curvature R (as half the pore diameter) and the

* Corresponding author: s.i.stepanov@urfu.ru

thickness of the strut δ. In general, such a construction is regarded as a unit cell. The implant consists of a number of such cells. FEA simulation of the compression loading was performed using the ABAQUS software with the number of finite elements not exceeding 1000. The properties of the Ti-6Al-4V most frequently used alloy for implant production were set. It is characterized by the high strength and reasonable ductility. The strength properties are given by the dependence of the stress plastic flow σ_s on the relative reduction $\varepsilon_{\%}$, expressed in percent. It is approximated by the formula $\sigma_s = a + b * \varepsilon_{\%}^c$, where a = 980 MPa; b = 10.78 MPa; $c = 0.76$ – empirical coefficients.

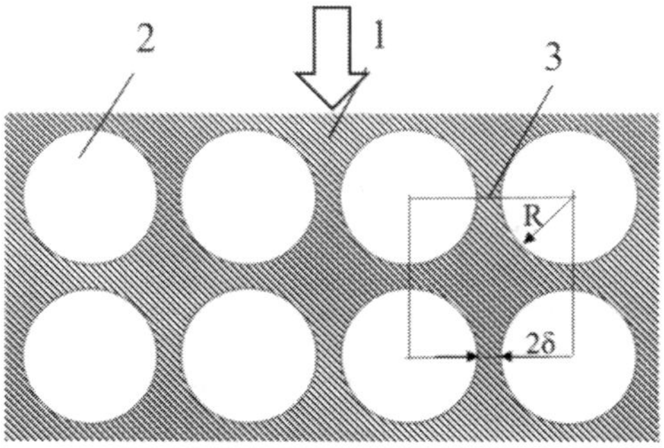

Fig. 1. Planar architecture of the porous titanium sample: 1 – metal; 2 – pore; 3 – a segment for simulation characterized by the radius R and the strut width δ.

The configuration of the unit cell for calculations is shown in Fig. 2 a. The load is applied in the XY plane in the vertical direction along the Y axis. The contour of the deformed body consists of five segments *A*, *B*, *C*, *D*, *E*. The Y axis is considered as an axis of vertical symmetry; therefore, on the boundary *E*, the horizontal displacements u_y on this axis are equal to zero. The X axis is considered as a horizontal axis of symmetry, therefore the vertical displacements u_y are also equal to zero on the boundary *A*. At the boundary *D*, the vertical displacement transmitted by the deformation tool is specified. The boundary *C* is free, thus no displacements take place there.

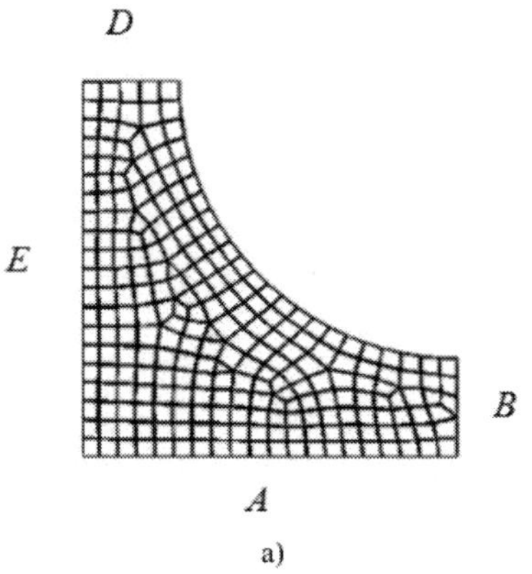

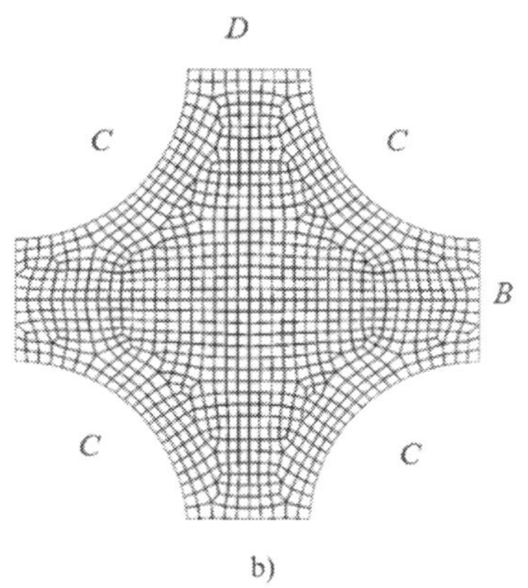

Fig. 2. Boundary conditions in the graphic form (a) and the configuration of the deformation zone non-loaded state with allowance for the symmetry conditions (b).

In the basic solution, the ratio R/δ = 75/25 = 3 was adopted. The cell height was assumed equal to the $H_0 = R + \delta$, then the relative reduction was u_y / H_0. The area of the square occupied by the unit cell is $H_0 * H_0$, the pore area in this square is $\pi R^2/4$, then the porosity can be estimated as the ratio of the area of the pore to the area of the square, expressed as a percentage: $P=100*\pi R^2/4/(R+\delta)^2$. For the relations used in the study and for the flat formulation of the problem, we obtain the porosity $P = 44\%$.

3 Simulation results and discussion

Fig. 3 shows the vertical and horizontal displacement fields U2 and U1, respectively. The largest vertical displacement U2 (Fig. 3a) are localized near the contact surface. Displacements along the central vertical are higher than in the periphery. The horizontal displacements U1 have an inverse order of distribution: the largest in the periphery and the smallest in the center.

The distribution of equivalent plastic strain PEEQ and of the intensity of stresses *S* for $u_y/H_0 = 3/(7{,}5+2{,}5/2) = 0{,}34$ is presented in Fig. 4.

A field of equivalent plastic strain PEEQ in the scalar form is distributed unevenly in volume (Fig. 3).

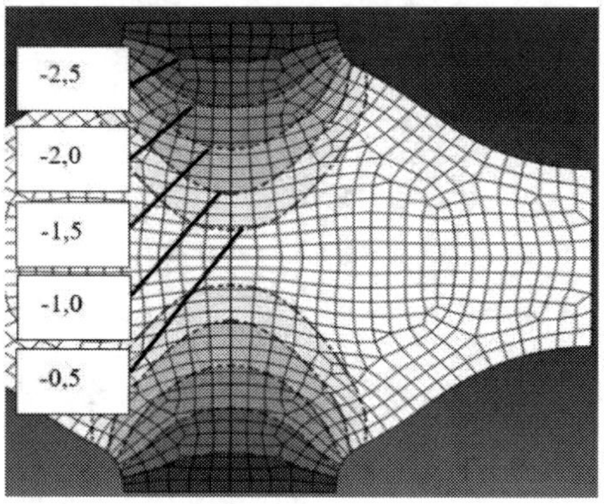

a)

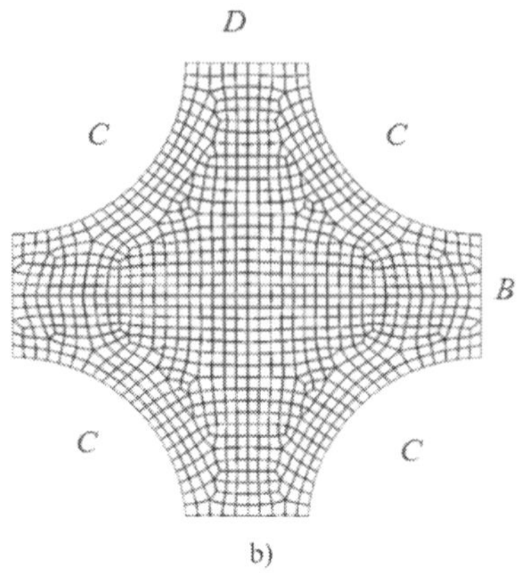

Fig. 3. Field of vertical U2 (a) and horizontal U1 (b) displacements (mm)

A larger value of PEEQ is acquired for the metal layers located on the free surface near the contact with the deformation tool and in the center of the unit cell. In addition, the point of maximum deformation is displaced from the center of the strut by a half of the thickness. The stress intensity is localized at the same point. In general, the stress intensity decreases from the center to the periphery. This is explained with the increase in the cross-section area through which the load is transmitted. Note, that both equivalent stresses and deformations are maximal in vertical struts.

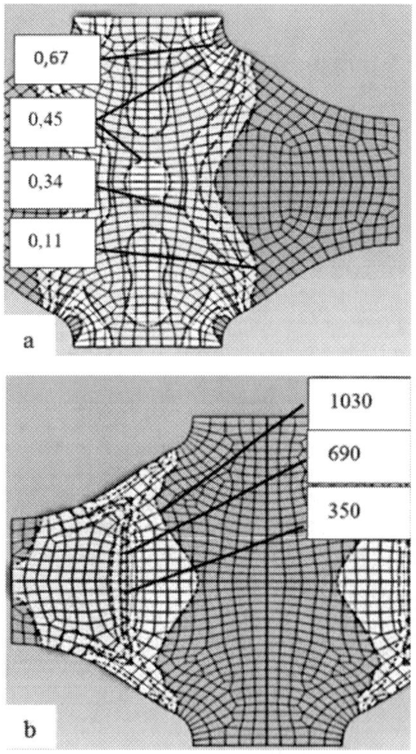

Fig. 4. Distribution of equivalent plastic strain PEEQ (a) and of the intensity of stress S (b) through the volume of the unit cell

Thus, the most loaded points were in the same place both for the deformed and for the stressed state. To assess the probability of failure, it is necessary to analyze the distribution of the mean (hydrostatic stresses), as shown in Fig. 5.

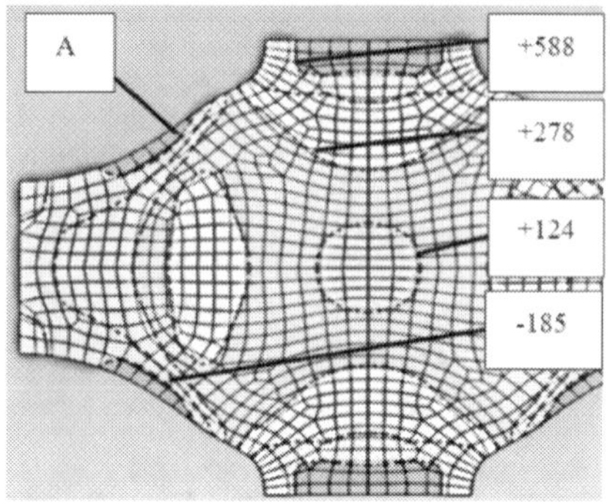

Fig. 5. Distribution of the pressure (mean stress taken with the opposite sign, [MPa]) through the volume of the unit cell

The cell volume is characterized by the occurrence of both compressive and tensile mean stresses as can be seen from the Fig. 5. The average stresses are close to zero in the center of the cell and in the periphery. A high compressive stresses act on the upper boundary, which is explained with the very type of the applied load. The mean stress decreases from the center to the periphery by analogy with the stress intensity (in absolute value). The appearance of a zone of tensile stresses on a segment of the cell radius (designated as region A) is of special interest.

Note, that the solution is based on the hypothesis that the material remains in a plastic state and does not become brittle. If it turns out that the ductility of the material decrease, then, instead of reaching the level of large plastic deformations, a brittle fracture will occur.

The simulation of the distribution of the stress-strain state parameters allows for estimation of the dangerous sections from the point of fracture of the whole structure. Such places in this case are the areas with the increased value of strain that are the horizontal bridges between the cells. In addition, the regions with tensile stresses appear near the vertical struts which can also lead to cracks formation and further fraction of the material. A general recommendation to improve strength of the total structure is to increase the thickness of the struts. At the same time, it should be remembered that the modulus of elasticity of the system will simultaneously increase.

4 Experimental procedures

The implant of the titanium alloy was manufactured by direct metal laser sintering using EOSINT 280 with a 200W pulse laser power. The powder of the (α + β) – titanium alloy Ti-6Al-4V was utilized, the composition and properties of which correspond to ISO 5832-3-2014 for metallic implants for surgery.

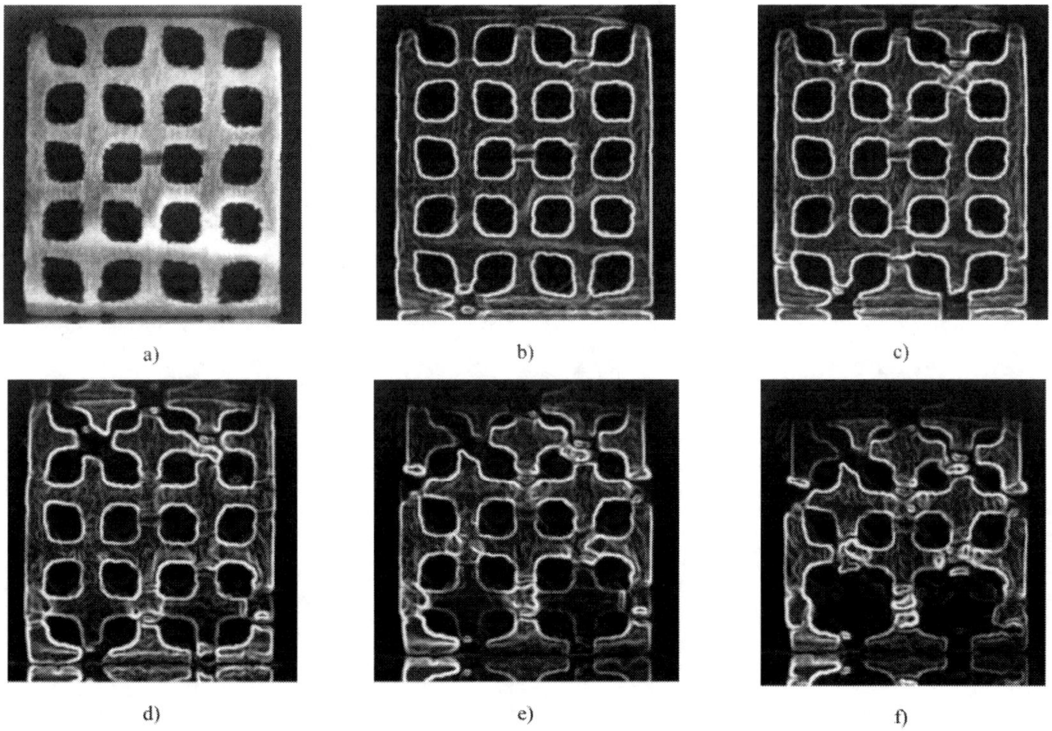

Fig. 6. Cell/pore configuration changing for strains of (%): 0(a); 0,7(b); 3,5(c); 9,2(d); 14,8(e); 27,1(f) during compression test

ISO 13314 specifies a test method for compressive properties of porous and cellular metals.

Rectangular prism shape samples with the width of 9.2 mm, height of 11 mm and the length of 11,02 mm were sectioned from the implants. The samples were tested in compression using an Instron 3380 tensile testing machine at a traverse speed of 1.8 mm/min. Video recording was performed during compression testing along with the logging of strain and loading by the monitoring system of the testing machine.

5 Results and discussion

When the specimen is compressed (Fig. 1), the shape of circular pores in the cross section becomes elliptical, elongated in the orthogonal direction with respect to the compression direction [18, 19]. However, such a scenario is optional, since the pore contour can provide additional resistance to deformation. What is obvious is that the B size (the width of the sample) under the compression would increase for an incompressible medium, but this is not obvious for a sample with pores.

A number of films typical of specific compression stages and corresponding to a certain relative strain (expressed in percent) were excerpted from the video record. The strain value depends on the method of measurement. If the sample were to be extracted every time to estimate this value, then the plastic component of the parameter would be calculated. But in this case the measurements were carried out immediately under load, so the represented relative reduction accounts for the sum of both the elastic and the plastic strain components. This sample collection of frames is shown in Fig. 6. All the drawings, except the first one, were modified by applying the function of increasing the contrast of the cell boundaries. This function provides a better distinguishing of the contours of these boundaries.

The initial pore configuration is not a circumferential, since the secant plane did not intersect the channels at an expected angle (Fig. 6a) when sectioning the samples. Even at a relative strain $\varepsilon = 0.7\%$ (Fig. 6 b), fracture was already initiated, however, it was localized in contact zones.

The fracture of the struts of the upper and lower rows of cells occurred at a relative reduction of 3.5% (Fig. 6 c). Further deformation at $\varepsilon = 9.2\%$ led to the spalling of the material within the upper and lower rows of cells. This process continues at $\varepsilon = 14.8\%$, but the central region of the sample remains intact. Its fracture is observed only at $\varepsilon = 27.1\%$.

Thus, the fracture of the sample occurs stepwise with the transition from the proximity rows to the central ones. This can be explained with the fact that frictional stress is applied on the metal near the contact surface, which, according to the principle of equivalent stresses calculation, must be added to the general stress value following the certain rules of material resistance. The procedure of such an experiment is described in [20] in more detail. A similar type of cellular structure failure was observed in [21].

Comparative analysis of the experimental results with the simulation data given in the first part of the study shows the following. The fracture is initially localized in the places where the greatest strains are accumulated and the maximum equivalent stresses occur, i.e. in vertical struts between the cells. In the second stage, the fracture moves into the region of horizontal struts. In the third stage, the failure process passes to a set of struts with a complete spalling of the material upon deformation.

The stress-strain curve is shown in Fig. 7. When the stress plateau responsible for struts bending considered, then the local extrema are observed in this interval.

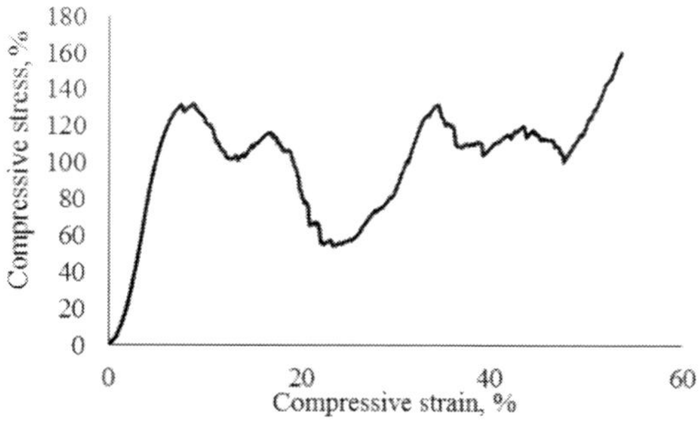

Fig. 7. Compression stress-strain curve for porous implant of titanium alloy

Moreover, one of the minima turns out to be especially deep (at a residual reduction of 21%). Two neighboring rows (4 and 5) were destroyed at once during this time interval as can be seen from Fig. 6. The increase in load at the end of the test is associated with the end of the fracture of the struts between the cells. The fractured material densified and formed its own cross-sectional area, as a result, greater stresses are required to transfer the material to the plastic state.

6 Conclusions

The boundary conditions were set and the boundary-value problem for the titanium implant loading with the cellular architecture was solved using FEA. Data on the stress and strain distribution were obtained. The struts between the individual cells were shown to be dangerous from the point of fracture. To obtain implants with a higher loading capacity, the thickness of the struts should be increased, however, excluding the high values of the modulus of elasticity. The sample of a porous titanium alloy intended for the manufacturing of implants was tested in compression. It was found that the change in cellular structure of the implant under loading occurred stepwise. At first, the peripheral contact regions of the metal were involved in deformation and subsequent fracture. Onward, the deformation extends into the material. The change of cell profile was nonmonotonic during the test due to the localization of stresses and strains within the vertical struts of the cellular structure of the implant. The noted nonmonotonicity led to inhomogenous deformation behavior, which was accomponied with the local extrema in the stress-strain curve.

We hereby acknowledge the support of the Ministry of Science and Education of the Russian Federation, in accordance to the decree of the Government of April 9, 2010, №218, project number 03.G25.31.0234.

References

1. X. Wang, S. et al., Biomaterials, **83**, 127-141 (2016)
2. F.S.L. et al., Acta Biomaterialia, **53**, 572-584 (2017)
3. R. Wauthle et al., Acta Biomater, **14**, 217-225 (2015)
4. K. Moiduddin et al., Biocybernetics and biomedical engineering, **36**, 719-730 (2016)
5. A. Yánez, A. Herrera, O. Martel, D. Monopoli, H. Afonso, Mater. Sci. and Eng.: C., **68**, 445-448 (2016)
6. G. Ryan, A. Pandit, DP Apatsidis. Biomaterials 27: 2651-70 (2006)
7. D. Herzog, V. Seyda, E. Wycisk, C. Emmelmann. Acta Materialia, **117,** 371-392 (2016)
8. S. Zhang, C. Li, W. Hou, S. Zhao, S. Li. J. of Mat. Sci. & Tech., **32**, 1098-1104 (2016)
9. L. Thijs, F. Verhaeghe, T. Craeghs, J. V. Humbeeck, J.-P. Kruth. Acta Mat., **58**, 3303-3312 (2010)
10. F. Wang, S. Williams, P. Colegrove, A. A. Antonysamy., Met. and Mat. Trans.: A., **44A**, 968 (2013)
11. B. Piotrowski, A.A. Baptista, E. Patoor, P. Bravetti, A. Eberhardt, P. Laheurte. Mat. Sci. and Eng. C., **38,** 151-160 (2014)
12. M. Niinomi, L. Yi, H. Liu, H. Li. Regenerative Biomat., 1-13 (2016)
13. H. Shen, H. Li, L.C. Brinson. Mechanics of Materials, **40,** 708-720 (2008)
14. L.J. Gibson, M.F. Ashby. *Cellular Solids: Structure and Properties* II ed., Cambridge University Press (1997)
15. Yu.N. Loginov. Izvestiya Vysshikh Uchebnykh Zavedenij. Chernaya Metallurgiya., **11**, 24-28 (2002)
16. B.I. Kamenetskij, A.L. Sokolov, Yu.N. Loginov. Kuznechno-Shtampovochnoe Proizvodstvo., **6**, 21-23 (1996)
17. A.A. Emel'yanov, I.P. Konakova, Yu.N. Loginov, A.A. Popov. Izvestia Akademii nauk SSSR. Metally., **1**, 133-137 (1995)
18. A.A. Ershov, Y.N. Loginov, S.L. Demakov., Metallurgist., **8**, 659-663 (2015)
19. Yu.N. Loginov. Steel in Translation., **31**, 63-68 (2001)
20. Yu.N. Loginov, A.A. Popov, S.I. Stepanov, E.Yu. Kovalev., Titan., **2** (2017)
21. A. Öchsner, K. Lamprecht., Mech. Research Com., **30**, 573-579 (2003)

MATEC Web of Conferences 132, 03010 (2017) DOI: 10.1051/matecconf/201713203010
DTS-2017

Numerical-analytical study on effect of three-phase poroelastic medium model parameters on dynamic displacements and porous pressures response

Leonid Igumnov[1], *Aleksandr* Belov[1,*], *Igor* Vorobtsov[1] and *Andrey* Petrov[1]

[1]National Research Lobachevsky State University of Nizhni Novgorod, Research Institute for Mechanics, 603950 Nizhny Novgorod, Russia

Abstract. The problem of loading on a one-dimensional partially saturated poroelastic rod is considered. A mathematical model of a partially saturated porous medium based on the Bio model was used. Solid skeleton displacements and pore pressures of fillers are chosen as basic functions defining wave process. A mathematical model of the boundary-value problem of the dynamic theory of poroelasticity is considrered in Laplace domain. The solution in time domain is obtained using the stepping method of numerical inversion of the Laplace transform. Dynamic displacement responses and pore pressures are presented for different values of the parameters of the poroelastic material model.

1 Introduction

Porous materials are widely used in nature and engineering. An example of such materials may be a water saturated soils, rocks, biological tissues, foam metals, etc. The mechanics of porous media is important in many domains of science, such as geotechnics, geomechanics, engineering geology, biomechanics, engineering and materials science.

Research of wave processes in saturated porous media was started by Frenkel [1] and Biot [2, 3]. The proposed approach was later developed by other authors for the case of partially saturated porous media containing several fluid fillers. A model describing the dynamic behavior of a three-phase porous medium is presented in Beskos [4], Beskos et al. [5, 6] and Vgenopoulou et al. [7] researches. Garg and Nayfeh [8], Santos [9, 10], Tuncay and Corapcioglu [11, 12], Wei and Muraleetharan [13], Lo [14], Lu [15] studies on the theory of porous media with two fillers belong conceptually to the general approach with different details in implementation. A general system of equations for a multiphase porous medium was proposed by Nikolaevskiy [16].

Applying a model of a fluid saturated porous material (even with simplifications) results in significantly more complicated computational scheme of the boundary value problem in comparison with elastic or viscoelastic formulations when considering wave processes. The dynamic of the filler fundamentally changes the form of the wave patterns, which can be predicted with the help of advanced computational methods.

Nevertheless, analytical solutions are important since allow to identify mechanisms of interaction between the solid and fluid phases, conduct a posteriori evaluation of numerical results, and investigate transient processes in a porous medium subjected to dynamic loading.

Review of available analytical solutions of wave propagation problems in saturated porous media can be found in Schanz article [17]. Li and Schanz [18] presented an analytical solution for a one-dimensional partially saturated poroelastic rod.

2 Problem formulation

A set of fully coupled governing differential equations of a porous medium saturated by two compressible fluids (water and air) subjected to dynamic loadings is considered. In this formulation the solid skeleton displacements u_i, water pressure p^w and air pressure p^a are presumed to be independent variables [18]. The final differential equations in Laplace domain yield:

$$\begin{aligned} &G\bar{u}_{i,jj} + (K + \frac{G}{3})\bar{u}_{j,ij} - (\rho - \beta S_w \rho_w - \gamma S_a \rho_a) s^2 \bar{u}_i - \\ &-(\alpha - \beta) S_w \bar{p}^w_{,i} + (\alpha - \gamma) S_a \bar{p}^a_{,i} = -\bar{F}_i, \end{aligned} \quad (1)$$

$$\begin{aligned} &-(\alpha - \beta) S_w s \bar{u}_{i,i} - (\zeta - S_{aa} S_w + S_u) s \bar{p}^a + \\ &+ \frac{\beta S_w}{\rho_w s} \bar{p}^w_{,ii} - (\zeta S_{ww} S_w + \frac{\varphi}{K_w} S_w - S_u) s \bar{p}^w = -\bar{I}^w, \end{aligned} \quad (2)$$

$$\begin{aligned} &-(\alpha - \gamma) S_a s \bar{u}_{i,i} - (\zeta S_{ww} + S_u) s \bar{p}^w + \frac{\gamma S_a}{\rho_a s} \bar{p}^a_{,ii} - \\ &-(\zeta S_{aa} + \frac{\phi}{K_a} S_a - S_u) s \bar{p}^a = -\bar{I}^a, \; x \in \Omega, \end{aligned} \quad (3)$$

* Corresponding author: belov_a2@mech.unn.ru

where K_w and K_a are bulk moduli of the fluid, $\bar{F}_i, \bar{I}^w, \bar{I}^a$ are bulk body forces. The porosity φ is defined as

$$\varphi = \frac{V_{void}}{V} \quad (4)$$

where V_{void} is volume of interconnected pores in the specimen, V is total volume of the material. The bulk density is denoted by

$$\rho = (1-\varphi)\rho_s + \varphi S_w\rho_w + \varphi S_a\rho_a , \quad (5)$$

where ρ_s is the density of the solid, ρ_w is water density, ρ_a is air density. The saturation degrees are defined as the ratios of the volume occupied by the fluid V_w or V_a to the void volume, i.e. it holds

$$S_w = \frac{V_w}{V_{void}},\ S_a = \frac{V_a}{V_{void}},\ S_w + S_a = 1. \quad (6)$$

The following abbreviations

$$\zeta = \frac{\alpha - \varphi}{K_s}, \quad (7)$$

$$S_{ww} = S_w - \vartheta(S_w - S_{rw}), \quad (8)$$

$$S_{aa} = S_a + \vartheta(S_w - S_{rw}), \quad (9)$$

$$S_u = -\frac{\vartheta(S_{ra} - S_{rw})}{p^d}\left(\frac{S_w - S_{rw}}{S_{ra} - S_{rw}}\right)^{\frac{\vartheta+1}{\vartheta}}, \quad (10)$$

are introduced, where S_{rw} is the residual water saturation and S_{ra} is air entry saturation. The symbol p^d is non-wetting fluid entry pressure, ϑ skeleton grain size distribution coefficient while the value of ϑ lies between 0.2 and 3, normally. The symbols β and γ are Laplace parameter dependent variables and expressed as

$$\beta = \frac{\kappa_w\phi\rho_w s}{\varphi S_w + \kappa_w\rho_w s},\ \gamma = \frac{\kappa_a\varphi\rho_a s}{\phi S_a + \kappa_a\rho_a s} \quad (11)$$

where κ_w and κ_a the phase permeability of the water and the air are given by $\kappa_w = K_{rw}k/\eta_w$ and $\kappa_a = K_{ra}k/\eta_a$ respectively. K_{rw} and K_{ra} denotes the relative fluid phase permeability, k denotes the intrinsic fluid permeability, η_w and η_a are viscosity of the fluid. To evaluate relative phase permeability following equations are used

$$K_{rw} = S_e^{(2+3\vartheta)/\vartheta},\ K_{ra} = (1-S_e)^2\left[1 - S_e^{(2+\vartheta)/\vartheta}\right]. \quad (12)$$

S_e denotes the effective water saturation degree given by

$$S_e = \begin{cases} 0 & S_w \le S_{rw} \\ \frac{S_w - S_{rw}}{S_{ra} - S_{rw}} & S_{rw} < S_w < S_{ra} \\ 1 & S_w \ge S_{ra} \end{cases} \quad (13)$$

For the extreme case $S_w = 0$, the equations turn out to be simple a elasticity problem, for the extreme case $S_w = 1$ – saturated poroelacticity problem.

3 Analytical solution

Consider the problem of loading as a Heaviside function in time on a one-dimensional partially saturated poroelastic rod (1-d column) of length l. The scheme of the problem is presented in fig. 1.

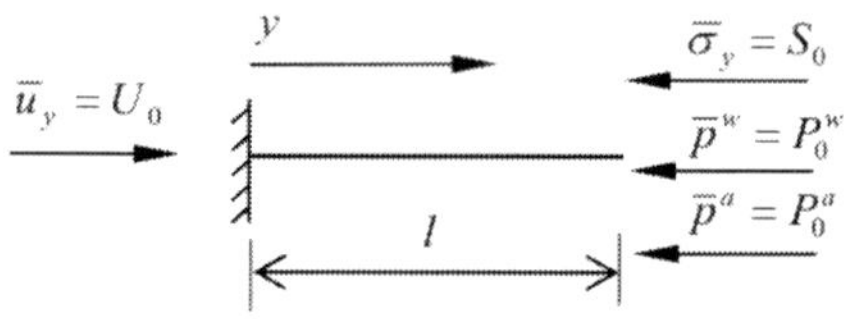

Fig.1 A 1-D column under dynamic loads.

Differential equations of one-dimensional propagation of poroelastic waves in Laplace domain has the following form:

$$(K + \frac{G}{3})\bar{u}_{y,yy} - (\rho - \beta S_w\rho_w - \gamma S_a\rho_a)s^2\bar{u}_y - \\ -(\alpha-\beta)S_w\bar{p}^w_{,y} + (\alpha-\gamma)S_a\bar{p}^a_{,y} = 0, \quad (14)$$

$$-(\alpha-\beta)S_w s\bar{u}_{y,y} - (\zeta S_{aa}S_w + S_u)s\bar{p}^a + \\ + \frac{\beta S_w}{\rho_w s}\bar{p}^w_{,yy} - (\zeta S_{ww}S_w + \frac{\varphi}{K_w}S_w - S_u)s\bar{p}^w = 0, \quad (15)$$

$$-(\alpha-\gamma)S_a s\bar{u}_{y,y} - (\zeta S_{ww} + S_u)s\bar{p}^w + \\ + \frac{\gamma S_a}{\rho_a s}\bar{p}^a_{,yy} - (\zeta S_{aa} + \frac{\phi}{K_a}S_a - S_u)s\bar{p}^a = 0. \quad (16)$$

Boundary conditions in Laplace domain in addition to equations (14)-(16):

$$\bar{u}_y\big|_{y=0} = 0,\ \bar{\sigma}_y\big|_{y=l} = 0, \quad (17)$$

$$\bar{p}^w\big|_{y=l} = P_0,\ \bar{q}^w\big|_{y=l} = 0, \quad (18)$$

$$\bar{p}^a\big|_{y=l} = 0,\ \bar{q}^a\big|_{y=l} = 0. \quad (19)$$

Exponential substitution

$$\bar{u}(y) = Ue^{\lambda sy},\ \bar{p}^w = U^w e^{\lambda sy},\ \bar{p}^a = U^a e^{\lambda sy} \quad (20)$$

allow to have the following analytical solution:

$$\bar{u}_y = \frac{P_0}{Ms}\sum_{i=1}^{3}\left(t_i\frac{e^{\lambda_i s(y-l)} - e^{-\lambda_i s(y+l)}}{1+e^{-2\lambda_i sl}}\right) \quad (21)$$

$$\bar{p}^w = \frac{P_0}{M}\sum_{i=1}^{3}\left(a_i t_i\frac{e^{\lambda_i s(y-l)} + e^{-\lambda_i s(y+l)}}{1+e^{-2\lambda_i sl}}\right) \quad (22)$$

$$\bar{p}^a = \frac{P_0}{M}\sum_{i=1}^{3}\left(b_i t_i\frac{e^{\lambda_i s(y-l)} + e^{-\lambda_i s(y+l)}}{1+e^{-2\lambda_i sl}}\right) \quad (23)$$

Coefficients in equations (21)-(23) may be found in paper [18].

4 Method for numerically inverting Laplace transform

The stepping method of numerical inversion of the Laplace transform used to obtain the solution in time domain. The method, based on the operational calculus theorem, allows to calculate the required original function using quadrature formula [19]:

$$y(0) = 0, \quad y(n\Delta t) = \sum_{k=1}^{n}\omega_k(\Delta t), \quad n = 1,...N. \quad (24)$$

Weights ω_k are determined using the Laplace image of the inverse function $\bar{f}(s)$ and the linear multistep method

$$\omega_n(\Delta t) = \frac{R^{-n}}{L}\sum_{l=0}^{L-1}\bar{f}(s)se^{-in\phi}, \quad (25)$$

$$s = \frac{\vartheta(z)}{\Delta t}, \quad z = Re^{i\phi}, \quad \phi = 2\pi\tfrac{l}{L}, \quad (26)$$

$$\vartheta(z) = 3/2 - 2z + z^2/2, \quad (27)$$

where N is number of time steps, L is number of nodes for numerically integrating on argument ϕ.

5 Numerical results

The following values of the boundary value problem parameters used for calculations: $l = 10m$, $P_0 = 1N/m^2$. Water-saturated sand was used as a porous material of the rod with following parameters:

$K = 1.02\cdot10^9 N/m^2$,
$G = 1.44\cdot10^9 N/m^2$,
$\phi = 0.23$,
$\rho_s = 2650kg/m^3$,
$\rho_w = 997kg/m^3$,
$\rho_a = 1.10kg/m^3$,
$K_s = 3.55\cdot10^{10} N/m^2$,
$K_w = 2.25\cdot10^9 N/m^2$,
$K_a = 1.10\cdot10^5 N/m^2$,
$\kappa = 2.5\cdot10^{-12} m^2$,
$\eta_w = 1.0\cdot10^{-3} Ns/m^2$,
$\eta_a = 1.8\cdot10^{-5} Ns/m^2$,
$S_{rw} = 0$, $S_{ra} = 1$, $\theta = 1.5$.

The solution in time domain found with values $\Delta t = 5\cdot10^{-5}$, $N = L = 2000$, $R = 0.997$.

Displacement and pore water pressure are presented on fig. 2-4.

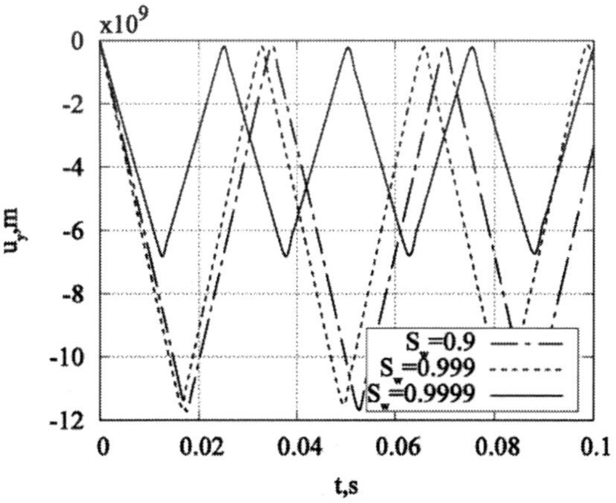

Fig. 2. Displacement versus time for different water saturation

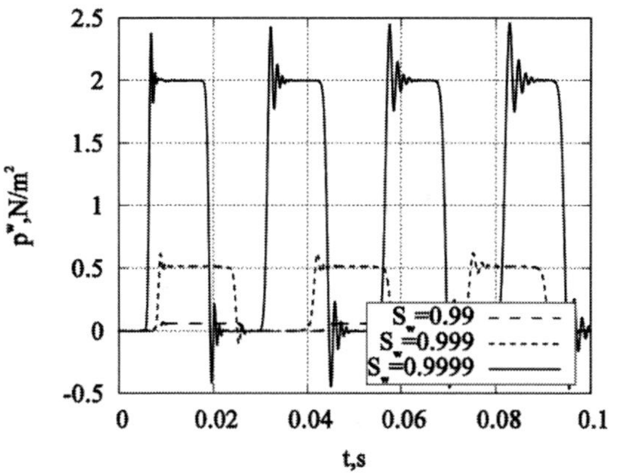

Fig. 3. Pore water pressure versus time for different water saturation

With an increase of the saturation degree from 0.99 to 0.9999, the amplitude of displacements decreases and velocity of the longitudinal wave increases (fig. 2). An increase in the amplitude is observed on the dynamic responses of pore pressures (fig. 3-4). The results of calculations are in qualitative agreement with the results presented in papers [18, 20]. Although these results were obtained for the case of loading $\bar{\sigma}_y = 1N/m^2$, similar effect of the saturation degree observed.

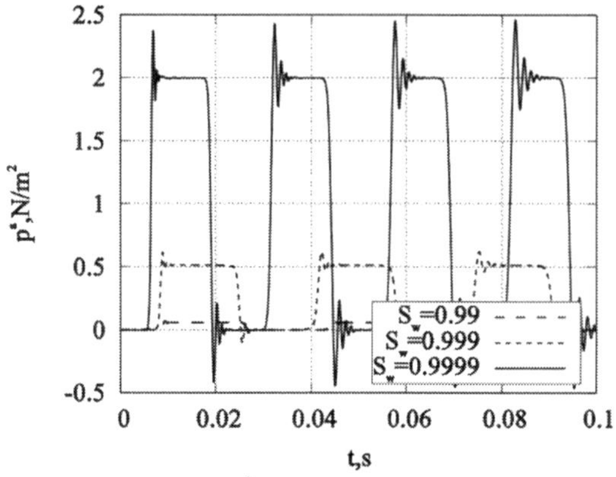

Fig. 4. Pore air pressure versus time for different water saturation

6 Conclusions

The analytical solution for the problem of one-dimensional partially saturated poroelastic rod dynamic loading is considered. The results of calculations obtained using the stepping method of numerical inversion of the Laplace transform are presented. The effect of the saturation degree on the dynamic responses of displacement and pore pressures is demonstrated. A qualitative coincidence of the results with the results of other authors observed.

The work is financially supported by the Russian Foundation for Basic Research (RFBR) under grant 17-08-01262.

References

1. J. Frenkel, Journal of Physics, **8**, 230 (1944)
2. M.A. Biot, J. Acoust. Soc. Am., **28**, 168 (1956)
3. M.A. Biot, J. Acoust. Soc. Am., **28**, 179 (1956)
4. D.E. Beskos, J. Eng. Mech., **115**, 982 (1989)
5. D.E. Beskos, C.P. Providakis, H.S. Woo, J. Eng. Mech., **115**, 1017 (1989)
6. D.E. Beskos, I. Vgenopoulou, C.P. Providakis, J. Eng. Mech., **115**, 996 (1989)
7. I. Vgenopoulou, D.E. Beskos, I. Vardoulakis, Acta Mech., **85**, 115 (1990)
8. S.K. Garg S.K., A.H. Nayfeh, J. Appl. Phys., **60**, 3045 (1986)
9. J.E. Santos, J.M. Corberó, J.Jr. Douglas, J. Acoust. Soc. Am., **87**, 1428 (1990)
10. J.E. Santos, J.Jr. Douglas, J. Corberó, O.A. Lovera, J. Acoust. Soc. Am., **87**, 1439 (1990)
11. K. Tuncay, M.Y. Corapcioglu, J. Geophys. Res., **101**, 25149 (1996)
12. K. Tuncay, M.Y. Corapcioglu, J. Appl. Mech., **64**, 313 (1997)
13. C. Wei, K.K. Muraleetharan, Int. J. Eng. Sci, **40**, 1807 (2002).
14. W.-C. Lo, G. Sposito, E. Majer, Water. Resour. Res., **41**, 140(2005).
15. J.-F. Lu, A. Hanyga, D.-S. Jeng, J. Appl. Geophys., **62**, 89 (2007).
16. V.N. Nikolaevskij, Mechanics of porous and fractured media. Singapore, World Scientific, 1990, 472 p.
17. M. Schanz, Appl. Mech. Rev, **62(3)** (2009).
18. P. Li, M. Schanz, Geophys. J. Int., **184**, 1341 (2011).
19. L.A. Igumnov, Y.Y. Rataushko, S.Y. Litvinchuk, A.A. Ipatov, *Advanced Materials - Studies and Applications* (Nova Science Publishers, 2015).
20. L.A. Igumnov, A.N. Petrov, I.V. Vorobtsov, Key Eng. Mater., **685**, 276 (2016).

 DOI: 10.1051/matecconf/201713203011

Solution of a dual integral equation arising in the contact problems of elasticity theory with the full Fourier series as the right-hand side

Andrey Vasiliev[1,2*], *Sergei* Volkov[1,2]

[1]Research Institute of Mechanics, National Research Lobachevsky State University of Nizhni Novgorod, 603950, Nizhni Novgorod, Russia
[2]Research and Education Center "Materials", Don State Technical University, 344000, Rostov-on-Don, Russia

Abstract. A class of dual integral equations is analyzed which arises in solution of a wide range of plane and antiplane contact problems of elasticity theory for a half-plane with functionally graded coating. In particular, a similar equation arises in solution of the contact problem on indentation in the presence of tangential stresses on a surface. The solution of the dual integral equation is sought in the form of a sum of even and odd functions. It makes possible to reduce the problem to independent solution of two dual integral equations over odd and even functions. Kernel transform of these equations is approximated by a product of fractional quadratic functions. The solution of dual integral equations is constructed in approximated analytical form by the bilateral asymptotic method. The expressions obtained are asymptotically exact for small and large values of a characteristic geometrical parameter.

1 Introduction

Contact mechanics attracts the researchers from all over the world for more than a century. For solution of contact problems, as well as the other mixed boundary value problems, methods based on complex analysis, the theory of potential functions and the theory of dual and singular integral equations are widely used [**Ошибка! Источник ссылки не найден.**, 1].

From the last decade of the previous century functionally graded (continuously inhomogeneous) and layered coatings are actively studied [3–28]. Deformation of such materials is investigated caused by only mechanical loading [3–7], simultaneous mechanical and thermal loading [8–10], electromechanical loading [11, 12], etc. Other important research directions are fracture of such materials [13], deformation of stiff structures (beams and plates) made from such materials [14, 15] as well as bending of beams and plates lying on such materials [16, 17], etc. Functionally graded and layered composite coatings are used for protection from wear, erosion, oxidation forced by high temperature, cracks formations, etc.

Contact problems for functionally graded or layered coatings are usually reduced to the solution of singular integral equations or dual integral equations, which requires development of special methods. Among methods of solution these equations the following should be noticed: regular [18] and singular [19] asymptotic methods, orthogonal polynomials [20], collocation [21] and bilateral asymptotic [22] methods. The advantages of the bilateral asymptotic method in comparison with the others consist in following: the solution is constructed in the approximated analytical form and has high accuracy for coatings of any thickness; the method is effective for arbitrary variation of properties in depth of the coating.

The present paper addresses the solution of one class of dual integral equations arising in the solution of wide range of plane and antiplane contact problems, see for example [5, 17, 24, 25]. Feature of the paper is that the right-hand side of the integral equation is represented by the full Fourier series that can be used to model non-symmetric contact. The solution of this dual integral equation is of interest for the tribology, in particular, in modeling of lubricated contact of solids with functionally graded or layered coatings [25–28].

2 Solution of dual integral equation

Let us consider the dual integral equation:

$$\begin{aligned} &\int_{-\infty}^{\infty} Q(\alpha)|\alpha|^{-1} L(\alpha\lambda) e^{-ix\alpha} d\alpha = 2\pi f(x),\ |x| \le 1 \\ &\int_{-\infty}^{\infty} Q(\alpha) e^{-ix\alpha} d\alpha = 0,\ |x| > 1 \end{aligned} \tag{1}$$

Here $Q(\alpha)$ is the Fourier image of function $q(x)$ to be obtained, λ is real parameter (usually, describing geometrical properties), $L(\alpha\lambda)$ is the kernel transform of integral equation possessing:

* Corresponding author: andre.vasiliev@gmail.com

$$L(\alpha) = A + B|\alpha| + C\alpha^2 + O(|\alpha|^3), \quad \alpha \to 0$$

$$L(\alpha) = 1 + D|\alpha|^{-1} + E\alpha^{-2} + O(|\alpha|^{-3}), \quad \alpha \to \infty$$

Function $f(x)$ can be represented by the full Fourier series:

$$f(x) = a_0 + \sum_{n=1}^{\infty} (a_n \cos(n\pi x) + b_n \sin(n\pi x))$$

Let us represent $q(x)$ in the form of sum of even $q_+(x)$ and odd $q_-(x)$ functions. Then for Fourier images the following equation is satisfied:

$$Q(\alpha) = Q_+(\alpha) + iQ_-(\alpha) \tag{2}$$

Substitution of (2) into (1) makes it possible to split the dual integral equation into two independent dual integral equations:

$$\begin{aligned} &\int_{-\infty}^{\infty} \frac{Q_-(\alpha)}{|\alpha|} L(\alpha\lambda)\sin(\alpha x)d\alpha = \\ &= -2\pi \sum_{n=1}^{\infty} b_n \sin(n\pi x), \; |x| \le 1 \\ &\int_{-\infty}^{\infty} Q_-(\alpha)\sin(\alpha x)d\alpha = 0, \; |x| > 1 \end{aligned} \tag{3}$$

$$\begin{aligned} &\int_{-\infty}^{\infty} \frac{Q_+(\alpha)}{|\alpha|} L(\alpha\lambda)\cos(\alpha x)d\alpha = \\ &= 2\pi \sum_{n=0}^{\infty} a_n \cos(n\pi x), \; |x| \le 1 \\ &\int_{-\infty}^{\infty} Q_+(\alpha)\cos(\alpha x)d\alpha = 0, \; |x| > 1 \end{aligned} \tag{4}$$

Equations (3) and (4) can be solved using the bilateral asymptotic method [22]. For this purpose, let us approximate the kernel transform by the ratio of two polynomials in the variable $\alpha^2\lambda^2$

$$\begin{aligned} &L(\alpha\lambda) \approx L_N(\alpha\lambda) = \frac{P_1(\alpha^2\lambda^2)}{P_2(\alpha^2\lambda^2)} \\ &P_1(\alpha) = \prod_{i=1}^{N} (\alpha + A_i^2), \; P_2(\alpha) = \prod_{i=1}^{N} (\alpha + B_i^2) \end{aligned} \tag{5}$$

Algorithms for calculation of parameters A_i, B_i (i=1,…,N) were described earlier [30]

Let us introduce the notation:

$$p(x) = (2\pi)^{-1} \int_{-\infty}^{\infty} Q_-(\alpha)|\alpha|^{-1} \sin(x\alpha)d\alpha \tag{6}$$

and rewrite first equation in (3) in the form of differential equation using (5):

$$P_1(-\lambda^2 D^2)p(x) = P_2(-\lambda^2 D^2)\left(-\sum_{n=1}^{\infty} b_n \sin(n\pi x)\right), \; D = \frac{d}{dx}$$

Solution of this equation has the form:

$$\begin{aligned} p(x) = &\sum_{m=1}^{N} \left(C_m \operatorname{ch}(A_m\lambda^{-1}x) + D_m \operatorname{sh}(A_m\lambda^{-1}x)\right) - \\ &- \sum_{n=1}^{\infty} b_n L_N^{-1}(n\pi\lambda)\sin(n\pi x) \end{aligned} \tag{7}$$

From the oddity of the function $p(x)$ it follows that $Cm=0$ (m=1,…,N). Using (6) and second equation in (3) following dual equation is obtained:

$$\begin{cases} \int_0^{\infty} Q_-(\alpha)\alpha^{-1} \sin(x\alpha)d\alpha = \pi p(x), & |x| \le 1 \\ \int_0^{\infty} Q_-(\alpha)\sin(x\alpha)d\alpha = 0, & |x| > 1 \end{cases} \tag{8}$$

where $p(x)$ is described in (7). Let us differentiate first equation in (8) with respect to x and apply an operator $\int_0^t \frac{dx}{\sqrt{t^2 - x^2}}$, while operator $\int_t^{\infty} \frac{dx}{\sqrt{x^2 - t^2}}$ is applied to the second equation

$$\begin{cases} \int_0^{\infty} Q_-(\alpha)J_0(t\alpha)d\alpha = 2\int_0^t \frac{p'(x)dx}{\sqrt{t^2 - x^2}}, & t \le 1 \\ \int_0^{\infty} Q_-(\alpha)J_0(t\alpha)d\alpha = 0, & t > 1 \end{cases} \tag{9}$$

Inverting the Hankel transformation in (9) yields

$$Q_-(\alpha) = \pi\left(\sum_{m=1}^{N} \tilde{D}_m \Psi_-(\alpha, A_m\lambda^{-1}) - \sum_{n=1}^{\infty} \tilde{b}_n \Psi_-(\alpha, i\pi n)\right) \tag{10}$$

where $\tilde{D}_m = D_m A_m \lambda^{-1}$ and $\tilde{b}_n = \pi b_n n L_N^{-1}(n\pi\lambda)$,

$$\Psi_{\pm}(x, A) = x\frac{J_1(x)I_0(A)\{A, x\} \mp \{x, A\}J_0(x)I_1(A)}{x^2 - A^2}$$

Let us invert Fourier transformation in (10):

$$\begin{aligned} q_-(x) = &\frac{1}{2\pi}\int_{-\infty}^{\infty} Q_-(\alpha)e^{-i\alpha x}d\alpha = \\ &\frac{x}{\sqrt{1 - x^2}}\left[\sum_{n=1}^{\infty} \tilde{b}_n Z_-(x, i\pi n) - \sum_{m=1}^{N} \tilde{D}_m Z_-(x, A_m\lambda^{-1})\right] \end{aligned} \tag{11}$$

where

$$\begin{aligned} &Z_{\{+,-\}}(x, A) = I_{\{1,0\}}(A) + \\ &+ A\frac{\sqrt{1 - x^2}}{\{-1, x\}}\int_x^1 \frac{tI_0(A)\{\operatorname{ch}, \operatorname{sh}\}(A(t - x)) - I_1(A)\{\operatorname{sh}, \operatorname{ch}\}(A(t - x))}{\sqrt{1 - t^2}}dt \end{aligned}$$

Coefficients D_m(m=1,…,N) are calculated from the system of linear algebraic equations below (n=1,…,N):

$$\sum_{m=1}^{N} \tilde{D}_m F_-(A_m\lambda^{-1}, B_n\lambda^{-1}) = \sum_{k=1}^{\infty} \tilde{b}_k F_-(i\pi k, B_n\lambda^{-1}) \tag{12}$$

$$F_{\pm}(a_+,a_-)=\frac{a_{\pm}I_0(a_+)\mathrm{K}_1(a_-)+a_{\mp}I_1(a_+)\mathrm{K}_0(a_-)}{a_+^2-a_-^2}$$

Solution of dual integral equation (4) was obtained earlier [17, 23, 24]:

$$Q_+(\alpha)=\widetilde{P}J_0(\alpha)-\pi\sum_{m=1}^{N}\widetilde{C}_m\Psi_+\left(\alpha,\frac{A_m}{\lambda}\right)- \\ -i\pi\sum_{n=1}^{\infty}\widetilde{a}_n\Psi_+(\alpha,i\pi n) \quad (13)$$

$$q_+(x)=\left(\frac{\widetilde{P}}{\pi}+\sum_{m=1}^{N}\widetilde{C}_m Z_+\left(\frac{A_m}{\lambda},x\right)\right)\frac{1}{\sqrt{1-x^2}}+ \\ +i\sum_{n=1}^{\infty}\widetilde{a}_n Z_+(i\pi n,x)\frac{1}{\sqrt{1-x^2}} \quad (14)$$

where $\widetilde{C}_m=C_mA_m\lambda^{-1}$ and $\widetilde{a}_n=\pi a_n n L_N^{-1}(n\pi\lambda)$. Coefficients $C_m(m=1,\ldots,N)$ are calculated from the system of linear algebraic equations below ($n=1,\ldots,N$):

$$\sum_{m=1}^{N}\widetilde{C}_n F_+\left(\frac{A_m}{\lambda},\frac{B_n}{\lambda}\right)=\frac{\widetilde{P}\lambda}{\pi B_n}\mathrm{K}_0\left(\frac{b_n}{\lambda}\right)- \\ -i\sum_{k=1}^{\infty}a_kF_+\left(i\pi k,\frac{b_n}{\lambda}\right), \quad (15)$$

Thus, solution of the original dual integral equation has the form:

$$q(x)=q_+(x)+q_-(x). \quad (16)$$

To determine unknown constant $\widetilde{P}$ an additional condition is required. For the indentation problem such condition is equilibrium of the punch: $P=\int_{-1}^{1}q(x)dx$, where P is the dimensionless indentation force.

Expressions (11)-(15) describe the approximated analytical solution of the dual integral equation (1). It was shown [22] that if one approximates the kernel transform in (1) by expression (5) then the obtained solution is asymptotically exact for $\lambda\to 0$ and $\lambda\to\infty$. Let us rewrite (1) in the operator form:

$$\Pi_N(p)+\Sigma(p)=f,$$

where $\Pi_N(p)=f$ is the approximated equation. Therefore, the discrepancy $\Sigma(p)$ satisfies the following inequalities:

$$\max_{x\in(-1,1)}\left|\Sigma(p)\sqrt{1-x^2}\right|\le\lambda M_0,\ \lambda\to 0\ (\lambda<\lambda_0),$$

$$\max_{x\in(-1,1)}\left|\Sigma(q)\sqrt{1-x^2}\right|\le\frac{M_\infty}{\lambda},\ \lambda\to\infty\ (\lambda>\lambda_\infty),$$

where λ_0, λ_∞, M_0 and M_∞ are some constants independent from λ. It is obvious that the solution's precision for the intermediate values of λ depends on the error of approximation of the kernel transform by expression (5). In the example of torsion problem it was shown [22] that the error of the solution is of the same order of magnitude as the error of the approximation of the kernel transform.

3 Conclusion

The results obtained can be used in solution of plane and antiplane contact problems of elasticity theory for solids with coatings (layered or functionally graded). In particular, indentation of an elastic half-plane with a functionally graded coating by a rigid punch with complicated geometry can be reduced to the dual integral equation (1). Full Fourier series in the right-hand side makes it possible to consider a punch with non-symmetric profile. Using Fourier series one can model additional non-symmetric loading acting in a finite region (region may differ from the contact area). Dual integral equation (1) can be also used in solution of frictional contact problems (for instance, caused by lubricant flow [26, 27, 29]). In this case an iterative process can be organized to calculate contact pressure.

Work was supported by the Russian Foundation for Basic Research (RFBR) (grant numbers 16-07-00958-a, 15-07-05820-a) and the President of the Russian Federation (grant number MK-5342.2016.1).

References

1. L.A. Galin, *Kontaktnyye zadachi teorii uprugosti*, (Gostekhizdat, Moscow, 1953)
2. N.I. Muskhelishvili, *Nekotorye osnovnye zadachi matematicheskoj teorii uprugosti* (Nauka, Moscow, 1966)
3. A.E. Giannakopoulos, S Suresh, Int. J. Solids Struct. **34** (1997)
4. M.A. Guler, F. Erdogan, Int. J. Solids Struct., **41** (2004)
5. L.-L. Ke, Y.-S. Wang, Int. J. Solids Struct., **43** (2006)
6. Yu. Tokovyy, C.-C. Ma, J. Appl. Mech., **83** (2016)
7. A.S. Vasiliev, M.V. Swain, S.M. Aizikovich, E.V. Sadyrin, Arch. Appl. Mech., **86** (2016)
8. J. Liu, L.-L. Ke, Y.-S. Wang, J. Yang, F. Alam, Int. J. Mech. Sci. **63** (2012)
9. Yu. Tokovyy, C.-C. Ma, Int. J. Eng. Sci. **92** (2015)
10. L.I. Krenev, Y.V. Tokovyy, S.M. Aizikovich, N.M. Seleznev, S.V. Gorokhov, J. Thermal Stresses, **107** (2016)
11. J. Su, L.-L. Ke, Y.-S. Wang, Int. J. Solids Struct., **90** (2016)
12. A.S. Vasiliev, S.S. Volkov, S.M. Aizikovich, Acta Mech., **227** (2016)
13. G.-Y. Huang, Y.-S. Wang, D. Gross, Eur. J. Mech. Solids., **21** (2002)
14. H. Altenbach, V.A. Eremeyev, Arch. Appl. Mech., **78** (2008)
15. H. Altenbach, V.A. Eremeyev, Computers Mater. Continua, **9** (2009)

16. S.S. Volkov, A.N. Litvinenko, S.M. Aizikovich, Y.-C. Wang, A.S. Vasiliev, Struct. Eng. Mech., **59** (2016)
17. S.M. Aizikovich, B.I. Mitrin, N.M. Seleznev, Y.-C. Wang, S.S. Volkov, Struct. Eng. Mech., **58** (2016)
18. I.I. Vorovich, Iu.A. Ustinov, J. Appl. Math. Mech., **23** (1959)
19. V.M. Aleksandrov, I.I. Vorovich, J. Appl. Math. Mech., **28** (1964)
20. G.Ia Popov, J. Appl. Math. Mech., **33** (1969)
21. A.I. Kalandija, *Matematicheskie metody dvumernoj uprugosti* (Nauka, Moscow, 1973) [in Russian]
22. S.M. Aizikovich, J. Appl. Math. Mech., **54** (1990)
23. S.M. Aizikovich, S.S. Volkov, B.I. Mitrin, J. Appl. Math. Mech., **81**, 6 (2017)
24. A.S. Vasiliev, S.S. Volkov, S.M. Aizikovich, B.I. Mitrin, Z. Angew. Math. Phys., **68** (2017)
25. I.I. Kudish, S.S. Volkov, A.S. Vasiliev, S.M. Aizikovich, J. Tribol., **138** (2016) http://dx.doi.org/10.1115/1.4030956
26. I.I. Kudish, S.S. Volkov, A.S. Vasiliev, S.M. Aizikovich, J. Tribol., **138** (2016)
27. I.I. Kudish, S.S. Volkov, A.S. Vasiliev, S.M. Aizikovich, J. Tribol., **139** (2017)
28. I.I. Kudish, S.S. Volkov, A.S. Vasiliev, S.M. Aizikovich, Math. Mech. Solids., (2017)
29. I.I. Kudish, S.S. Volkov, A.S. Vasiliev, S.M. Aizikovich, Math. Mech. Solids., (2017)
30. S.M. Aizikovich, A.S. Vasiliev, J. Appl. Math. Mech. **77** (2013)

On the problem of shear of a functionally graded half-space by a punch

Vladimir Zelentsov[1,*]

[1]Don State Technical University, Research and Education Center "Materials", 344000 Rostov-on-Don, Russia

Abstract. Contact problem on shear of a functionally graded half-space by a strip bunch was considered. Shear modulus of the half-space is exponentially increasing by depth. The contact problem was reduced to a convolution integral equation of the first kind. Solution of the integral equation was constructed by asymptotic methods over the characteristic parameter of the problem. Dependence of the obtained problem solution and its main characteristics on the shear modulus of the half-space was analysed. Expressions for the main characteristics of the problem are given.

1 Introduction

From the mid-20th century functionally graded inhomogeneity of materials is accounted for in analysis of plates, foundations [1, 2] etc. In modern times, FGMs, apart from the civil engineering, are widely used in industry, especially in microelectronics [3, 4], where traditional materials like alumina, silicon etc. are oxidized to form an insulator layer. Oxides can have elastic modulus two times or more as the bulk material, their thickness can present a sufficiently large amount of the component thickness. Therefore, analysis of stress and strain, electrostatic or magnetostatic state of such materials should take into account variation of elastic moduli, electric and magnetic parameters in volume of the material. Common analytical approach consist in solution of elasticity problems using reduction to an integral equation. Various methods were developed for solution of these equations, including orthogonal polynomials method [5], collocation method [6], bilateral asymptotic method [7]. These approaches are widely used nowadays to get the solution of the contact problems for FGMs [8–15]. However, improvement of solutions' accuracy needs development of new analytical approaches to FGM mechanics modelling. In the present work, an elastic contact problem is considered on a pure shear of a functionally graded half-space by a punch. Solution of the problem is reduced to the integral equation. Approximated analytic solution is constructed by asymptotic methods. Dependence of main mechanical characteristics of the problem on the half-space shear modulus parameters is analysed.

2 Problem statement

The rigid strip punch with width $2a$ ($|x| \le a$, $y = 0$) shears by the amount ε ($w(x,0) = \varepsilon$, $|x| \le a$) the surface of the elastic half-space made of functionally graded material with shear modulus dependent on depth $\mu(y) = \mu e^{2dy}$ ($0 < y < \infty$), where $\mu(0) = \mu$ is shear modulus on the half-space surface ($y = 0$), d is the gradient parameter ($d > 0$). Contact stresses $\varphi(x) = -\sigma_{yz}(x,0)$ arising under the punch are to be determined, along with displacements $w(x,0)$ of the half-space surface outside of the contact region ($a < |x| < \infty$). It is assumed that displacements of the half-space $w(x,y)$ decay at infinity.

In the half-space, the pure shear deformation conditions are fulfilled: $\sigma_{xx} = \sigma_{xy} = \sigma_{yx} = \sigma_{yy} = \sigma_{zz} = 0$, where σ_{xx}, σ_{yy}, σ_{zz} are normal stresses, σ_{xy}, σ_{yx} are tangential stresses [16].

In the considered case, elasticity equations are simplified to the equation

$$\frac{\partial \sigma_{zx}}{\partial x} + \frac{\partial \sigma_{zy}}{\partial y} = 0 \quad (1)$$

where σ_{zx}, σ_{zy} are tangential stresses, which can be expressed in terms of strains

$$\sigma_{zx} = \mu(y)\frac{\partial w}{\partial x}, \quad \sigma_{zy} = \mu(y)\frac{\partial w}{\partial y} \quad (2)$$

* Corresponding author: vbzelen@gmail.com

DOI: 10.1051/matecconf/201713203012

where $w(x,y)$ are displacements of the half-space along the z axis.

Substituting (2) to the equation (1), differential equation of elasticity in terms of displacements is obtained:

$$\frac{\partial^2 w}{\partial y^2}+\frac{\mu'(y)}{\mu(y)}\frac{\partial w}{\partial y}+\frac{\partial^2 w}{\partial x^2}=0 \tag{3}$$

Taking into account that $\mu(y)=\mu e^{2dy}$, the differential equation (3) with variable coefficient dependent on y turns into the differential equation with constant coefficients

$$\frac{\partial^2 w}{\partial y^2}+2d\frac{\partial w}{\partial y}+\frac{\partial^2 w}{\partial x^2}=0 \tag{4}$$

Mixed boundary conditions of the problem have the form

$$y=0 \quad w(x,0)=\varepsilon \qquad |x|\le a \tag{5}$$

$$\sigma_{yz}(x,0)=\begin{cases}-\varphi(x) & |x|\le a\\ 0 & a<|x|<\infty\end{cases} \tag{6}$$

where $\varphi(x)$ are unknown contact stresses under the punch.

At infinity displacements $w(x,y)$ turns to zero, i.e. $\lim\limits_{\sqrt{x^2+y^2}\to\infty} w(x,y)=0$.

3 Integral equation of the problem

Using the Fourier integral transform

$$w^F(\alpha,y)=\int_{-\infty}^{\infty} w(x,y)e^{i\alpha x}\mathrm{d}x,$$
$$w(x,y)=\frac{1}{2\pi}\int_{-\infty}^{\infty} w^F(\alpha,y)e^{-i\alpha x}\mathrm{d}\alpha \tag{7}$$

where F index above w stands for the Fourier transform of the $w(x,y)$ function, the differential equation (4) is reduced to an ordinary differential equation with constant coefficients

$$\frac{\mathrm{d}^2 w^F}{\mathrm{d}y^2}+2d\frac{\mathrm{d}w^F}{\mathrm{d}y}-\alpha^2 w^F(\alpha,y)=0 \tag{8}$$

The general solution of this equation decaying at infinity ($y\to\infty$) has the form

$$w^F(\alpha,y)=C(\alpha)e^{\lambda_- y},\ \lambda_-=-d-\sqrt{d^2+\alpha^2},\ d>0 \tag{9}$$

where $C(\alpha)$ is the unknown constant determined from the boundary condition (6) after applying the Fourier transformation (7) to it:

$$C(\alpha)=-\mu^{-1}(0)\varphi^F(\alpha)\exp(-\lambda_- y)/\lambda_-$$

where $\varphi^F(\alpha)$ is the image of $\varphi(x)$.

Determining $w(x,y)$ using the inverse Fourier transform in (9) and satisfying mixed boundary conditions (5), (6), the integral equation of the problem was obtained

$$\int_{-a}^{a}\varphi(\xi)k(\xi-x)\mathrm{d}\xi=2\pi\mu(0)\varepsilon \qquad |x|\le a \tag{10}$$

$$k(x)=\int_{-\infty}^{\infty}K(\alpha)e^{i\alpha t}\mathrm{d}\alpha,\ K(\alpha)=\left(d+\sqrt{d^2+\alpha^2}\right)^{-1} \tag{11}$$

The integral equation (10), (11) was then represented in dimensionless formulation

$$\int_{-1}^{1}\varphi(\xi)k\left(\frac{\xi-x}{\lambda}\right)\mathrm{d}\xi=2\pi\frac{\mu(0)}{a}\varepsilon \quad |x|\le 1,\ \lambda=\frac{1}{da} \tag{12}$$

$$k(t)=\int_{-\infty}^{\infty}K(\alpha)e^{i\alpha t}\mathrm{d}\alpha,\quad K(\alpha)=\left(1+\sqrt{1+\alpha^2}\right)^{-1} \tag{13}$$

being the Fourier convolution integral equation of the fist kind with differential kernel. The $K(\alpha)$ function is the Fourier image of the integral equation kernel $k(t)$ and is even, multivalued in the complex plane with algebraic branch points $\alpha=\pm i$. $K(\alpha)$ express the following asymptotic properties

$$K(\alpha)=|\alpha|^{-1}+\mathrm{O}(\alpha^{-2}) \quad \text{at } |\alpha|\to\infty$$
$$K(\alpha)=K(0)+\mathrm{O}(\alpha^2) \quad \text{at } |\alpha|\to 0 \tag{14}$$

where $K(0)=1/2$. Calculation of (13) gives the following representation for the kernel $k(t)$

$$k(t)=2\left(K_0(t)-\int_t^{\infty}\mathrm{d}\tau\int_\tau^{\infty}K_0(\xi)\mathrm{d}\xi\right)e^t \tag{15}$$

where $K_0(t)$ is the Macdonald function. At small t the following estimation takes place

$$k(t)=-\ln|t|+\mathrm{O}(1) \qquad \text{at } t\to 0 \tag{16}$$

At the fulfillment of the above stated properties of the kernel $k(t)$ the integral equation (12) has the unique solution [17].

4 Asymptotic solution of the integral equation

The most effective analytic methods of the integral equation (12) solution are asymptotic methods with respect to the dimensionless parameter $\lambda = da$ for its small ($0 < \lambda < 1$) or large values ($\lambda > 2$) [17].

At small values of λ the solution of the integral equation (12) is constructed in the form of the Neumann series with zeroth term represented as

$$\varphi(x) = \varphi_+\left((1+x)\lambda^{-1}\right) + \varphi_-\left((1-x)\lambda^{-1}\right) - \varphi_\infty(x\lambda^{-1}) \quad (17)$$

where functions $\varphi_\pm(x)$ are determined from the Wineher-Hopf integral equation on the half-axis [18]

$$\int_0^\infty \varphi_\pm(\xi)k(\xi - x)\mathrm{d}\xi = 2\pi\frac{\mu(0)}{a\lambda}\varepsilon \qquad 0 < x < \infty \quad (18)$$

while function $\varphi_\infty(x)$ is determined from the convolution integral equation on the infinite interval

$$\int_{-\infty}^\infty \varphi_\infty(\xi)k(\xi - x)\mathrm{d}\xi = 2\pi\frac{\mu(0)}{a\lambda}\varepsilon \quad -\infty < x < \infty \quad (19)$$

Solution of the integral equation (19) is constructed using the Fourier integral transform (7) and has the form

$$\varphi_\infty(x) = \frac{\mu(0)}{a\lambda}\frac{\varepsilon}{K(0)} \qquad -\infty < x < \infty \qquad (20)$$

Solution of the integral equations (18) on the half-axis is constructed by the Wiener-Hopf method. On the first stage, equation (18) is extended to the whole axis x

$$\int_0^\infty \varphi_\pm(\xi)k(\xi - x)\mathrm{d}\xi = \begin{cases} 2\pi\dfrac{\mu(0)}{a\lambda}\varepsilon & 0 < x < \infty \\ 2\pi\dfrac{\mu(0)}{a\lambda}w_\mp(x) & -\infty < x < 0 \end{cases} \quad (21)$$

where $w_\mp(x)$ are given by operators

$$w_\mp(x) = \int_{-\infty}^0 \varphi_\pm(\xi)k(\xi - x)\mathrm{d}\xi \qquad -\infty < x < 0$$

which present displacements of the half-space free surface outside the contact region.

On the first stage of the Wiener-Hopf method, the Fourier transform (7) is applied to solve integral equation on the whole axis (21), performing transition of the latter to functional equations in transforms $\varphi_\pm^F(\alpha)$ and $w_\mp^F(\alpha)$

$$\varphi_\pm^F(\alpha)K(\alpha) = \frac{\mu(0)}{\lambda a}\varepsilon\left(-\frac{1}{i\alpha}\right) + \frac{\mu(0)}{\lambda a}w_\mp^F(\alpha)$$
$$\eta_- < \mathrm{Im}(\alpha) < \eta_+ \qquad (22)$$

where

$$\varphi_\pm^F(\alpha) = \int_0^\infty \varphi(\xi)e^{i\alpha\xi}\mathrm{d}\xi\,, \quad w_\mp^F(\alpha) = \int_{-\infty}^0 w_\mp(\xi)e^{i\alpha\xi}\mathrm{d}\xi$$

are unknown transforms to be determined from (22). Assuming that factorization of $K(\alpha)$ is known, i.e.

$$K(\alpha) = K_+(\alpha)K_-(\alpha) \qquad (23)$$

where $K_+(\alpha)$ is regular at $\mathrm{Im}(\alpha) > \eta_-$, $K_-(\alpha)$ is regular in half-plane $\mathrm{Im}(\alpha) < \eta_+$ ($\eta_+ > \eta_-$). Substituting (23) to (22), left- and right-hand sides of the obtained equation was divided by $K_-(\alpha)$. The function $f(\alpha) = \left(i\alpha K_-(\alpha)\right)^{-1}$ was then represented in the form of composition

$$f(\alpha) = \frac{1}{i\alpha K_-(\alpha)} = f_+(\alpha) + f_-(\alpha)$$

where $f_+(\alpha)$ is regular in the upper half-plane $\mathrm{Im}(\alpha) > \eta_-$, while $f_-(\alpha)$ is regular in the lower half-pane $\mathrm{Im}(\alpha) < \eta_+$, that in in the considered case was fulfilled by elementary means

$$f_+(\alpha) = \frac{1}{K_-(0)i\alpha}\,, \quad f_-(\alpha) = \frac{K_-(0) - K_-(\alpha)}{i\alpha K_-(0)K_-(\alpha)}$$

After this, the functional equation is represented in the form of equality of two functions in the complex plane

$$\begin{gathered}\varphi_\pm^F(\alpha)K_+(\alpha) + \frac{\mu(0)}{a\lambda}f_+(\alpha) = \\ \frac{\mu(0)}{a\lambda}\varepsilon f_-(\alpha) + \frac{\mu(0)}{a\lambda}\frac{w_\mp^F(\alpha)}{K_-(\alpha)} = F(\alpha) \\ \eta_- < \mathrm{Im}(\alpha) < \eta_+\end{gathered} \qquad (24)$$

where left-hand side of (24) is analytical extension to the upper half-plane ($\mathrm{Im}(\alpha) > \eta_-$), and right-hand side to the lower ($\mathrm{Im}(\alpha) < \eta_+$) half-plane of a some $F(\alpha)$. As analytical extensions of $F(\alpha)$ to the upper and lower half-planes decay at infinity $|\alpha| \to \infty$ it follows that $F(\alpha) = 0$ in the complex plane. From (24) two equations follow, from which it was determined that

$$\begin{aligned}\varphi_\pm^F(\alpha) &= -\frac{\mu(0)}{a\lambda}\frac{\varepsilon}{K_-(0)}\frac{1}{i\alpha K_+(\alpha)} \\ w_\mp^F(\alpha) &= \varepsilon\left(\frac{1}{i\alpha} - \frac{1}{K_-(0)}\frac{K_-(\alpha)}{i\alpha}\right)\end{aligned} \qquad (25)$$

Inverting the Fourier transform (7) of equations (25), solution of the integral equations (21) were obtained

$$\varphi_{\pm}(x) = -\frac{\mu(0)}{a\lambda}\frac{\varepsilon}{K_-(0)}\frac{1}{2\pi}\int_{-\infty}^{\infty}\frac{e^{-i\alpha x}}{i\alpha K_+(\alpha)}d\alpha \quad x > 0 \quad (26)$$

$$w_{\mp}(x) = \frac{\varepsilon}{2\pi}\int_{-\infty}^{\infty}\left(\frac{1}{i\alpha} - \frac{1}{K_-(0)}\frac{K_-(\alpha)}{i\alpha}\right)e^{-i\alpha x}d\alpha \quad x < 0 \quad (27)$$

To calculate integrals in (26), (27) the factorization of the function $K(\alpha)$ from (13), i.e. representation in the form (23), is needed. To obtain compact analytic solution factorization of (23) should be performed by elementary means, which is no case for the function $K(\alpha)$ (13). So the function $K(\alpha)$ is replaced by its approximation

$$K(\alpha) = \left(\sqrt{(K^{-1}(0))^2 + \alpha^2}\right)^{-1} \quad (28)$$

which is close to the exact function on the real axis (which is the axis of integration in (12)) and accomplishes to asymptotic properties (14), (15). Error of such approximation along the real axis do not exceed 13%. Factorization (23) of the approximation (28) is achieved by elementary means in the form

$$K_{\pm}(\alpha) = \left(K^{-1}(0) \mp i\alpha\right)^{-1/2} \quad (29)$$

where $K_+(\alpha)$ is regular in the upper half-plane $\mathrm{Im}(\alpha) > \eta_-$, $K_-(\alpha)$ is regular in the lower half-plane $\mathrm{Im}(\alpha) < \eta_+$ of the complex plane $\alpha = \xi + i\eta$, where $\eta_{\pm} = \pm K^{-1}(0)$. The function $K_+(\alpha)$ is analytical extension of $K(\alpha)$ from the strip $|\mathrm{Im}(\alpha)| < K^{-1}(0)$ to the upper half-plane, $K_-(\alpha)$ to the lower half-plane. Substituting $K_+(\alpha)$ from (29) and calculating integral in (26) it was obtained that

$$\varphi_{\pm}(x) = \frac{\mu(0)}{a\lambda}\frac{\varepsilon}{K_-(0)}\left(\frac{1}{\sqrt{\pi x}}\exp\left(-K^{-1}(0)x\right) + \sqrt{K^{-1}(0)}\,\mathrm{erf}\left(\sqrt{K^{-1}(0)x}\right)\right) \quad x > 0 \quad (30)$$

Substituting $K_-(\alpha)$ from (29) and calculating integral in (27) it was obtained that

$$w_{\mp}(x) = \varepsilon\,\mathrm{erfc}\left(\sqrt{K^{-1}(0)(-x)}\right) \quad x < 0 \quad (31)$$

where $\mathrm{erf}(x)$ is the error function [19], $\mathrm{erfc}(x) = 1 - \mathrm{erf}(x)$.

Substituting $\varphi_{\pm}(x)$ from (30) and $\varphi_{\infty}(x)$ from (20) to (17), solution of the integral equation (12) was obtained, together with the problem solution. Functions $w_{\mp}(1 \pm x/\lambda)$ in (31) give displacements of stress-free surface of the half-space outside of the contact region to the left

$$w_-\left(\frac{1+x}{\lambda}\right) = \varepsilon\,\mathrm{erfc}\left(2\left(-\frac{1+x}{\lambda}\right)\right) \quad \text{at } x < -1$$

and to the right

$$w_+\left(\frac{1-x}{\lambda}\right) = \varepsilon\,\mathrm{erfc}\left(\sqrt{2\left(-\frac{1-x}{\lambda}\right)}\right) \quad \text{at } x > 1$$

from the punch.

Punch shear force T, which is an integral characteristic of the problem, can be obtained by integration of (17) over interval $[-1, 1]$

$$T = T_+ + T_- - T_{\infty},$$

$$T_{\pm} = \int_{-1}^{1}\varphi_{\pm}\left(\frac{1 \pm x}{\lambda}\right)dx,\ T_{\infty} = \int_{-1}^{1}\varphi_{\infty}\left(\frac{x}{\lambda}\right)dx \quad (32)$$

Taking solution of the integral equation (12) in multiplicative form [17]

$$\varphi(x) = \varphi_+\left(\frac{1+x}{\lambda}\right)\varphi_-\left(\frac{1-x}{\lambda}\right)\Big/\varphi_{\infty}\left(\frac{x}{\lambda}\right) \quad (33)$$

and integrating it over interval $[-1, 1]$, one can obtain the formula, convenient for analysis, which is asymptotically equivalent to (32):

$$T = 2\mu(0)\varepsilon(1+\gamma),\quad \gamma = 2/\lambda = 2da \quad (34)$$

The obtained formulas allow to deduct other important problem's characteristics, such as displacements $w(x,y)$ and stresses $\sigma_{yz}(x,y)$ in the half-space and point out their dependence on shear modulus $\mu(y)$

$$w(x,y) = -\frac{a}{\mu(0)}\frac{1}{\pi}\int_{-1}^{1}\varphi(\xi)\theta\left(\frac{\xi - x}{\lambda}, dy\right)d\xi \quad y > 0 \quad (35)$$

$$\sigma_{yz}(x,y) = -\frac{da}{\pi}\sqrt{\frac{\mu(y)}{\mu(0)}}\ln\sqrt{\frac{\mu(y)}{\mu(0)}} \times \int_{-1}^{1}\varphi(\xi)\kappa\left(\frac{\xi - x}{\lambda}, dy\right)d\xi \quad y > 0 \quad (36)$$

where

$$\theta(u,v) = \int_0^v \sqrt{\frac{\mu(0)}{\mu(\eta/d)}}\ln\sqrt{\frac{\mu(0)}{\mu(\eta/d)}}\kappa(u,\eta)d\eta$$

$$\kappa(u,v) = \left(u^2 + v^2\right)^{-1/2}K_1\left(\sqrt{u^2 + v^2}\right)$$

and $K_1(x)$ is a cylinder function [19].

From (36) it follows that shear stresses $\sigma_{yz}(x,y)$ increase, whilst displacements $w(x,y)$ decrease with depth coordinate y proportionally to $\sqrt{\mu(y)}$.

To construct solution of the integral equation (12) of the problem at large values of λ (λ > 2) it is sufficient to use results described in [20].

The financial support from Ministry of Education and Science of Russia through the Governmental Assignment no. 9.1481.2017/4.6 is acknowledged.

References

1. G.Ia. Popov, Izvestija vuzov. Serija: Stroitel'stvo i arhitektura (11-12), 11-19 (1959)
2. S.M. Aizikovich, Mech. Sol. (5), 74-80 (1978)
3. Y. Miyamoto et al., *Functionally Graded Materials: Design, Processing and Applications* (Springer, New York, 2013)
4. Y. Shiraki, N. Usami (eds), *Silicon-Germanium (SiGe) Nanostructures: Production, Properties and Applications in Electronics* (Woodhead Publishing, Cambridge, 2011)
5. G.Ia. Popov, J. Appl. Math. Mech. **33**(3), 503-517 (1969)
6. A.I. Kalandia, *Mathematical Methods of Two-Dimensional Elasticity*, (Nauka, Moscow, 1973)
7. S.M. Aizikovich, J. Appl. Math. Mech. **54**(5), 719-724 (1990)
8. J. Su, L.L. Ke, Y.S. Wang, Int. J. Sol. Struct. **45**(59), 45-59 (2016)
9. J. Ma, S. El-Borgi, L.-L. Ke, Y.-S. Wang, J. Therm. Stress. **39**(3), 245-277 (2016)
10. M.A. Guler, F. Erdogan, Int. J. Sol. Struct. **41**(14), 3865-3889 (2004)
11. P. Chen, S. Chen, Acta Mech. **223**(3), 563-577 (2012)
12. H.J. Choi, G.H. Paulino, J. Mech. Phys. Sol. **56**(4), 1673-1692 (2008)
13. S. Aizikovich, A. Vasiliev, N. Seleznev, Int. J. Eng. Sci. **48**(10), 936-942 (2010)
14. A.S. Vasiliev, S.S. Volkov, S.M. Aizikovich, Acta Mech. **227**(1), 263-273 (2016)
15. I.I. Kudish, S.S. Volkov, A.S. Vasiliev, S.M. Aizikovich, J. Trib. 138(2), 21504 (2016)
16. V.M. Aleksandrov, E.V. Kovalenko, *Continuum Mechanics Problems with Mixed Boundary Conditions* (Nauka, Moscow, 1986)
17. I.I. Vorovich, V.M. Aleksandrov, V.A. Babeshko, *Non-classical Mixed Problems of Elasticity Theory* (Nauka, Moscow, 1974)
18. B. Noble, *Methods Based on the Wiener–Hopf Technique for the Solution of Partial Differential Equations* (Pergamon Press, London, 1598)
19. H. Bateman, A. Erdélyi, *Table of Integral Transforms*, Vol. 1. (McGraw-Hill Book Company, New York, 1954)
20. V.M. Aleksandrov, A.V. Belokon', J. Appl. Math. Mech. **31**(4), 718-724 (1967)

Mechanism of pitting corrosion protection of metals and alloys in new-generation water treatment plants

Vladimir Grachev[1,*], *Igor* Nechaev[2], *Andrei E.* Rozen[3], *and Andrei A.* Rozen[3]

[1] Rosatom State Corporation, 24 Bolshaya Ordynka Str., 119017 Moscow, Russian Federation
[2] «Karpov Institute of Physical Chemistry», 10-bld. 1 Vorontsovo Pole str., 105064 Moscow, Russian Federation
[3] Penza State University, 40 Krasnaya St., 440026 Penza, Russian Federation

Abstract. In this article authors set out a principle of pitting corrosion protection, suggested a new class of multilayer materials with high corrosion resistance. They substantiated the choice of the layers for the multilayer material designed for exploitation in oxidizing and non-oxidizing environment. The sphere of application of the multilyer materials was defined.

1 Introduction

Currently, annual direct metal losses resulting from corrosion make up 12% of the total weight of metal in stock in Russia, which corresponds to the loss of up to 30% of annually produced metal1. Global corrosion losses account for 4-6% of the national income in developed countries [2]. Especially strongly this process is expressed in water-supply systems. Which historically, for a long while, have been principally made of steel pipes. At an average service life of water conduits and distribution network over 30 years, most part of metal conduits are in the sad state. Indirect losses associated with the structural failure, process shutdown and depressurization of equipment cause even greater economic damage. This leads to the necessity of improving the methods of protection against corrosion and developing new corrosion resistant materials and products.

Modern chemical, petrochemical and oil refining industries, as well as communal service branches, are characterized by engineering units of high capacity and processing medium that is highly corrosive. This leads to the use of equipment with an increased metal content. According to the estimates made by Japanese experts [3] and data provided by the NACE International (International Association of Corrosion Engineers) [4], more than 60% of the engineering units' failures occur due to the influence of temperature and corrosive environment, and the annual cost of corrosion is estimated at $ 2.2-2.5 trillion. In particular, in 2014, they amounted to about $445.0 billion in the USA, $ 95.0 billion – in Germany, more than $ 50.0 billion – in Russia. This corresponds to the loss of up to 30% of the annually produced metal.

2 Matherials and methods

When designing a new unit, it is important to use materials with an improved corrosion resistance, e.g. multilayer materials with an internal protector [5, 8, 9]. This solution is particularly true for the industrial water treatment systems as well as for water carriage systems of the big industrial units. Namely in such systems, especially difficult environment for pipelines and metal facilities service, is observed. At the same time, there has not been any accomplished technical solution that would make it possible to use the internal protector in order to increase the guaranteed technical lifetime of the products operating in the conditions of an increased corrosion attack.

Pitting corrosion is typical for metals susceptible to passivation due to the presence of strong oxidants or external anodic protection. The main condition for the pitting occurrence is an electrochemical potential shift towards more a positive one than a certain critical value, which is known as a pitting potential. Simultaneous presence of activating anions of fluorine, chlorine, bromine, iodine, and of such oxidizing agents as oxygen in the solution favors pitting formation. Superficially, active haloid ions are adsorbed on the metal surface and displace oxygen that ensures a passive state. Oxide film destruction takes place and the surface activation occurs. Pitting formation may happen at the locations with a distorted structure, non-metallic inclusions, and reduced concentration of chromium and other alloying elements in steel. The surface of pitting becomes anodic, and the rest of the passive metal surface becomes cathodic.

The difference in the activation potential and the steady-state (equilibrium) potential is regarded as a characteristic of resistance to pitting corrosion. The larger the difference, the more resistant to pitting corrosion the metal is.

* Corresponding author: secretary@nifhi.ru

DOI: 10.1051/matecconf/201713203013

In this regard, technical solution for producing a multilayer material with the internal protector is of a certain interest.

The corrosion process is transformed due to the multilayer structure. Pitting corrosion of the outer layer shifts to the contact corrosion of the specific sacrificial layer (Fig.1).

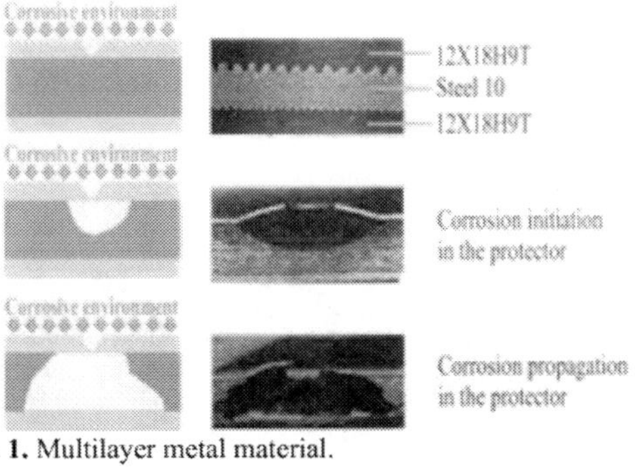

Fig. 1. Multilayer metal material.

The key point of the suggested technology is as follows. Instead of mono- or bimetal, a multilayer material with at least three layers is used. Such multilayer composite can be adapted for the one-way or two-way contact with the corrosive environment or media of a different composition. Composition of the multilayer material's layers depends on the composition of media and on the electrochemical potentials of the constituent metals. Applying a protector between the protected layers is principally new.

We have considered two cases:

• When a multilayer material contacts with the processing medium containing aqueous solutions of alkalis, acid salts or acids, whose anions are not oxidizing agents;

• When a multilayer material contacts with the processing medium containing aqueous solutions of alkalis, acid salts or acids, whose anions are oxidizing agents. It allows of further discussion of the first and second type multilayer composite

2.1 Multilayer composite of the first type

For the first (outer) layer contacting the processing non-oxidizing medium, a material that has a sufficiently high corrosion resistance and passivity in such environment should be selected.

The material of the first layer is characterized by the anode A curve and the cathode K curve (Fig.2). There is an area of passivity on the anode curve. When the layer contacts with the corrosive environment that does not contain oxidants, a fixed potential E_1 is set on the layer.

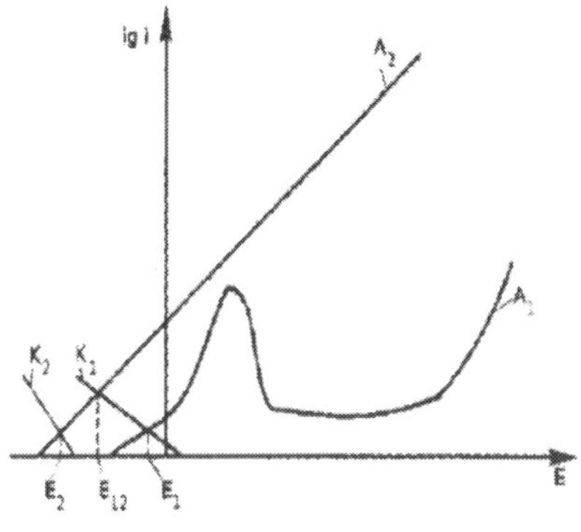

Fig. 2. Schematic arrangement of the metal polarization curves in the multilayer material: K_1, K_2 – cathodic curves of the outer and inner layers; A_1, A_2 – anodic curves, respectively.

As a result of the medium action, corrosion spots in the form of pitting appear in the outer layer. These corrosion spots deepen over time and penetrate into the second layer. The material of the second layer is selected so that the value of its steady-state electrochemical potential E2 in contact with the processing medium is lower than the steady-state electrochemical potential of the first metal layer. The state of the second layer material – the protector – is characterized by the anode A2 and cathode K2 curves.

When pitting reaches the second layer metal, the steady-state potential E12 is set due to the contact potential difference of the first and second metal layers. When this happens, the second layer metal becomes the anode, and the first layer metal becomes the cathode. The second layer becomes the protector, i.e. a sacrificial electrode, and gradually dissolves. Anodic dissolution reaction can proceed until the formation of a sizable cavity – a lens – in the protector. On the first layer material, depending on the composition of the medium, the evolution of hydrogen, the reduction of oxygen or other electrochemical reactions take place.

Anode:

$$Fe \rightarrow Fe+2 + 2e \qquad (1)$$

Cathode:

$$2H+ + 2e \rightarrow H2 \qquad (2)$$

$$Fe+3 + e \rightarrow Fe+2 \qquad (3)$$

The Composition of the third layer is similar to the first one. When the depth of the cavity in the protector becomes equal to its thickness, the third layer also becomes the cathode – the same as the first layer. The corrosion rate of the second layer may increase, and then corrosion may take place until the complete dissolution of the protector. If the reaction products are insoluble materials, they can slag individual pitting sites and reduce the rate of destruction of the three-layer material as a whole.

Fig.3 represents the corrosion dynamics in the multilayer material in the medium containing aqueous

 DOI: 10.1051/matecconf/201713203013

solutions of alkalis, acid salts or acids, whose anions are not oxidizing agents.

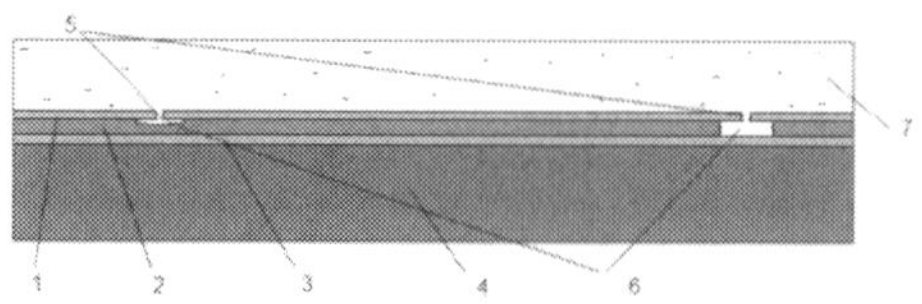

Fig. 3. Corrosion dynamics in the multilayer material in the medium that doesn't contain oxidizing agents: 1 – outer layer in contact with the processing medium; 2 – protector; 3 – third layer; 4 – base layer; 5 – pitting in the outer layer; 6 – lens in the protector; 7 – corrosive environment

2.2 Multilayer composite of the second type

For the first (outer) layer contacting the processing oxidizing medium, a material that has a sufficiently high corrosion resistance and passivity in such environment should be selected

The second layer's steady-state electrochemical potential should be higher than the first layer electrochemical potential; moreover, the material of the second layer should have lower hydrogen overvoltage than the material of the outer layer. The first layer material is characterized by the anode A1 and cathode K1 curves (Fig.4). There is an area of passivity on the anode curve.

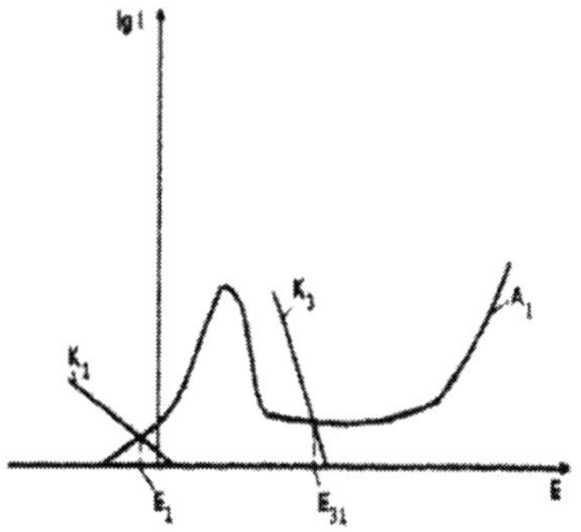

Fig. 4. Schematic arrangement of the metal polarization curves in the multilayer material: K_1 – cathodic curve of the outer layer; A_1 – anodic curve of the outer layer; K_3 – cathodic curve of the inner layer.

While in operation, pitting is formed in the outer layer and penetrates up to the second layer in the course of time. The contact potential difference between the layers appears. The second layer metal retains its passive state. Anodic dissolution of the first layer metal is inhibited due to the formation of the reaction products that are slightly soluble. The potential of the first layer metal shifts to more positive values of E31 potentials; it leads to additional passivation and, as a result, stops the growth of pitting.

In this case, the corrosion potential of the outer layer steel maintains a stable positive value corresponding to the passive state. While in operation, the passive film on the outer layer may dissolve due to the chemical reaction.

Fig.5 represents the corrosion dynamics in the multilayer material in the medium containing aqueous solutions of alkalis, acid salts or acids, whose anions are oxidizing agents.

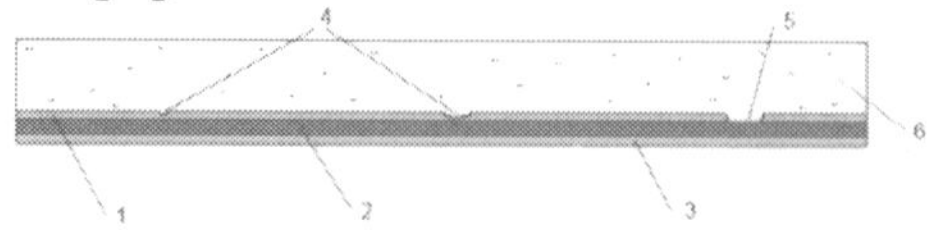

Fig. 5. Corrosion dynamics in the multilayer material in the medium that contain oxidizing agents: 1 – outer layer in contact with the processing medium; 2 – protector; 3 – third layer; 4 – pitting in the outer layer; 5 – dissolution of the outer layer; 6 – corrosive environment.

Composition of the third layer material is identical to the material of the first outer layer.

Table 1 shows structural metals and alloys in the order of increasing Ecor corrosion potential in sea water [6, 10, 11]. Data in the table indicate that the grades 10, 15, 20 low-carbon steels have a more negative potential compared with the 12X17, 12X18H9, 10X17H13M2T high-alloy steels in the passive state. It means that low-carbon steels may be used as a protector in combination with the corrosion-resistant high-alloy steels in the media containing aqueous solutions of alkalis, acid salts or acids, whose anions are not oxidizing agents

According to the Fig. 5, the 12X17, 12X18H9, 10X17H13M2T high-alloy steels in active state have a more negative potential compared with copper, titanium, and Hastelloy. It means that high-alloy steels can be used as a protector in combination with the indicated metals and alloys in media containing aqueous solutions of alkalis, acid salts or acids, whose anions are oxidizing agents.

Metal	E_{cor}, B	Metal	E_{cor}, B
Magnesium	-1.45	Brass (40 Zn)	-0.20
Magnesium alloy(6 A1, 3 Zn, 0.5 Mn)	-1.20	Bronze (6 In)	-0.20
Zinc	-0.80	Nickel (active state)	-0.12
Aluminum alloy (10 Mg)	-0.74	Brass (30 Zn)	-0.10
Aluminum alloy (10 Zn)	-0.70	Bronze (5-10 A1)	-0.10
Aluminum	-0.53	Copper	-0.08
Duralumin	-0.50	Corrosion-resistant steel 20X13 (passive state)	+0.03
Iron	-0.50	Nickel (passive state)	+0.05
Carbon steel	-0.40	Corrosion-resistant steel 12X17 (passive state)	+0.10
Gray cast-iron	-0.36	Corrosion-resistant steel 12X18H9 (passive state)	+0.17
Corrosion-resistant steel 12X13 and 12X17 (active state)	-0.32	Hastelloy (Ni, 20 Mo, 18 Cr, 6 W,7 Fe)	+0.17
Corrosion-resistant steel 12X18H9 (active state)	-0.30	Corrosion-resistant steel 10X17H13M2T (passive state)	+0.20
Corrosion-resistant steel 10X17H13M2T (active state)	-0.30	Silver	+0.12... +0.20
Lead	-0.30	Titanium	+0.12... +0.20

Fig.5. Ecor potential values in sea water

Any multilayer materials under development should have at least three layers [12,14]. A three-layer material should be composed by matching materials for each layer so that the corrosion of the product is minimal, whereas the protective ability is maximal. As the base layer, a layer of structural steel of the required thickness can be welded to the multilayer material in order to achieve high structural strength.

Materials containing two or more layers of different composition may be obtained by various technologies. These include cast sheathing, hot pack rolling, cold sheathing, explosive welding, and weld deposition [7, 13, 14, 15, 18]. One of the most popular and highly efficient methods of industrial production of multilayer materials is explosive welding.

Explosive welding is a high-velocity welding under pressure, wherein a non-detachable joint is formed as a result of the welded elements'collision.

The source of energy is explosives. Upon explosive initiation, a detonation front (shockwave) spreads, and an expansion of gaseous products occurs. There is a jump in pressure, which enables high acceleration of the element colliding with the stationary element. In the collision area, pressure can reach 10...20 GPa. There is a joint plastic deformation of the metals in a relatively thin layer that is up to 2 mm deep. The result is a firm non-detachable joint.

Explosive welding has advantages due to short duration and high pressure, which is achieved in the detonation of the explosives [17, 19, 20]. These advantages can be pointed out as follows:

1. It is possible to obtain compounds of similar and dissimilar metals and alloys, including those with differing physical and mechanical properties (melting point, coefficient of linear thermal expansion, strength, ductility).
2. Chemical composition of the base and cladding layers doesn't change during the explosive welding, and mechanical properties vary in a narrow band of width of 3...6 mm that is adjacent to the joint line.
3. Explosive welding is carried out in the solid phase; this allows joining materials that form brittle intermetallic compounds.
4. It is possible to obtain compounds with the area of 20m2 and more, the thickness of the base layer that is limited only by the sheet or product gauge, and the cladding layer thickness from fractions of a millimeter to several millimeters.
5. It is possible to obtain multilayer materials in a single welding cycle.
6. There is no need for complicated and expensive equipment.
7. The cost of consumables and energy consumption is relatively low.
8. It is possible to produce flat blanks and cylindrical billets and to clad curved surfaces.

According to the Skolkovo Innovation Center, the global market of produced corrosion-resistant multilayer material in various industries amounts to more than $900 million. Of these, 33% is energy industry, 25% – oil refining industry, 21% – chemical industry, 14% – nuclear industry, and 7% – shipbuilding

3 Conclusion

Depending on the corrosive environment, the equipment's operational life is increased by a factor of 5-15 as compared with austenitic stainless steels. A multilayer metal material consists of minimum number of expensive components. It is possible to carry out cost-neutral and safe external diagnostics of corrosion, including ultrasonic testing.

The obtained multilayer metal material is a completely new product. There are no direct equivalents. The main areas of application are chemical, oil and gas industries and nuclear power. The obtained material is in the same segment with tantalum and platinum by performance characteristics, and it is at the same segment with conventional chromium-nickel stainless steels by price range.

The present research was supported by the Ministry of Education and Science of the Russian Federation (Grant of the 20th of October, 2014, № 14.576.21.0053; The Applied Research Unique Identifier RFMEFI57614X0053)

References

1. E.N. Kablov, Science and Life, **11**, 16-21, (2012)
2. V.N. Ivanovsky, Engineering practice, **6**, 4-14, (2010)
3. ESPEC Technology Report. Test Navi Report No. 21 (Issue 78)
4. NACE International. *Corrosion Resistant Coatings*, (2011)
5. Multilayer material with enhanced corrosion resistance (embodiments) and method for its production (embodiments). EAPO patent 016878 dd 30.08.2012.
6. N. D. Tomashov, G. P. Chernova, Metallurgiya, Moscow (1986)
7. A. G. Kobelev et al., Intermet Engineering, Moscow, (2002)
8. Yu. A. Konon, L. B. Pervukhin, A. D. Chudnovskii, Mashinostroenie, Moscow, (1987)
9. A. E. Rozen, I. S. Los, Yu. P. Perelygin, L. B. Pervukhin, Yu. A. Gordopolov, G. V. Kiriy, P. I. Abramov, S. G. Usatyi, D. B. Kryukov, O. L. Pervukhina, I. V. Denisov, A. A. Rozen. Eurasian Patent 016878, (2008)
10. B. Wallén, S. Henrikson, Werkstoffe und Korrosion, **40**, 602, (1989).
11. A. Mollica et al, Corrosion, **45**, 1, 48-56, (1989)
12. A. E. Rozen, L. I. Reformatskaya, E. V. Kuznetsov, I. S. Los, P. A. Abramov, I. V. Denisov, A. V. Khorin, A. A. Rozen, Explosive Production of New Material: Science, Technology, Business, and Innovations, **9**, (2010).
13. S. Yu. Kireyev, I. S. Los, A. E. Rozen, Yu. P. Perelygin, G. V. Kozlov, D. B. Kryukov, A. A. Rozen, Explosive Production of New Materials: Science, Technology, Business, and Innovations, **68**, (2012).
14. I. S. Los, A. E. Rozen, Yu. P. Perelygin, S. G. Usatyi, A. V. Khorin, Resource-Saving Technologies to Repair, Restore, and Reinforce Parts of Machines,

Mechanisms, Equipment, Tools, and Accessories from Nano- to Macrosize, **1**, 465–468, (2010).

15. Yu. P. Perelygin, S. Yu. Kireyev, I. S. Los, A. E. Rozen, Prot. Met. Phys. Chem. Surf., **60** (7), 856–859 (2014).
16. H. Bisadi, M. Khaleghi Meubodim J. Mech. Eng. Sci., **225**, (2011).
17. F. Findik, Mater Des, **32**, 1081–1093, (2011)
18. S. Saravanan, K. Raghukandan, Materials and Manufacturing Processes, (2013).
19. J. Song, A. Kostka, M. Veehmayer, D. Raabe, Mat Sci Eng A, (2011).
20. H. Paul, M. Faryna, M. Prazmowski, R. Banski, Arch. Metall. Mater., **56**, 463-474 (2011).

Comparative Observation of Silver Nano and Microstructures Deposited from Aerosol and Fog

Anna Zheltova[1,*], *Yurii* Biryukov[2], *Sergey* Ryndya[3], *Alexander* Smolyanskiy[2,4]

[1]Lomonosov Moscow State University, Faculty of Bioengineering and Bioinformatics, 1-bld. 73, bx. 433 Leninskie Gory, GSP-1, 119234 Moscow Russian Federation
[2]«Karpov Institute of Physical Chemistry», 10-bld. 1 Vorontsovo Pole str., 105064 Moscow, Russian Federation
3National Research Nuclear University "MePhi", 31 Kashirskoe shosse, 115409, Moscow, Russian Federation
4Dmitry Mendeleev University of Chemical Technology of Russia, 9 Miusskaya sq., 125047 Moscow, Russian Federation

Abstract. A comparative study of the structure and fractal properties of arrays of the silver nano-/micro-particles deposited on the silicon substrate both from the aerosol and fog showed that the form of the silver individual particles and nano-/microstructures greatly depends on the deposition conditions. By passing an aerosol through isopropyl alcohol, the formation of fractal aggregates of the silver nano-/micro-particles both in the air and in alcohol was observed. Deposition of the silver nano-/micro-particles in the atmosphere of the saturated isopropyl alcohol vapours led to formation of fog. Micro-droplets of the silver colloidal solution were deposited on the substrate. The further evaporation of alcohol created the silver nano/microstructures in the form of annular layers. It was found that the concerned annular layers contained silver particles of the same shape in the form of a Crescent (or Janus-nano-/microparticles). The nature of discovered effects is discussed.

1 Introduction

It is known [1], that anomalous optical properties of the metal nano-/microparticle (NMP) arrays can be determined by the interaction of light with the surface plasmon polaritons and localized plasmons. The structure and form of the metallic NMP are the important factors determining the efficiency of the concerned "plasmon-photon" interactions. Theoretical analysis [2] [3] [4] showed that the sickle shaped metallic NMP can enhance the interaction of localized plasmon with light photons, compared to the NMP having the spherical shape. However, the proposed method of the sickle metallic NMP synthesis by means of the colloidal microlithography method [5] is multi-stage, and requires complicated hardware design. So the studies aim to simplify the method of the sickle metallic NMP synthesis.

The disadvantage of the newly proposed method of the metallic nano-/microstructures (NMS) layers synthesis by the "dry" aerosol deposition [6] consists in the random ballistic mechanism of coatings formation on the substrate surface. These coatings contain both the NMP of different sizes and the NMS of chain and dendritic shapes that can lead to significant scatter of the optical and other properties of the resulting porous metal films [7]. Therefore it is desirable to develop a method of the metallic NMS formation in the controlled conditions which could provide an opportunity for receiving the NMS of the same shape and, preferably, containing the arrays of the metallic NMP having the same geometric shape. Adherence to conditions allowed achieving the stability and reproducibility of the optical properties of the metallic NMS.

A promising approach to solve this problem may be the creation of the metallic NMS precipitates formed during the evaporation of the micro-droplets from the metallic NMP colloidal solutions in various solvents [8], [9].

In the given case, depending on the conditions of the evaporation process [10], the metallic NMP deposits can exist either in the form of solid spots (this situation is considered in the "coffee stains" model [8]), or form of ring structure.

According to the research [8] [9] [10], the concerned method of the metallic NMS preparation can have many practical applications such as production of plasmon resonance sensors, catalytic surfaces, etc. The different forms of the received metallic NMS are due to the occurrence of the sedimentation processes in the highly non-equilibrium conditions which lead to the development of the self-organization. In that case the shape of the NMS deposits is determined by the deviation degree of the evaporation conditions from the equilibrium [10]. Usually the colloidal NMP solutions can be considered as the micro-emulsions which can be formed and stabilized by the surfactants. Then it is necessary to wash off the NMP sludge from the surfactant and emulsifier admixtures that can result in the deformation and the NMP damage. This problem can

* Corresponding author: annettezv@gmail.com

be solved if the deposition of the aerosol metallic particles on the silicone surface has been carried out in the atmosphere of the solute saturated vapors. Metallic NMP is the center where the condensation of the solute molecules and micro-droplet formation is taking place. Finally, the initial aerosol of the metallic particles transforms into fog [11]. After that, the evaporation of micro-droplets deposited on the silicone surface forms the metallic NMS, which do not contain the surfactant and emulsifier admixtures.

The aim of present study consisted in:

a) the experimental confirmation of the possibility to form the ring metallic NMS synthesized by the evaporation of micro-droplets deposited on the substrate surface from the fog formed as a result of the interaction between the aerosol NMP and the solvent molecules;

b) the study of the structure and shape of metal NMP contained in the NMS, which formed either after the micro-droplet deposition and evaporation, or the passing of aerosol NMP through the solvent media;

c) conducting the fractal analysis of images that show the metal NMP shape and the structure of the metal NMS synthesized in both modes, obtained by transmission and scanning electron microscopy (TEM and SEM, respectively), and atomic force microscopy (AFM).

2 Experimental

To obtain the colloidal solutions of silver NMP in isopropyl alcohol (IPA) without the admixture of the emulsifier and surfactant molecules the self-made experimental facility [6] has been used (Fig.1(a)). The scheme of the concerned facility includes an aerosol spark generator, a diffusion aerosol spectrometer (DAS), the ultrasonic generator of the liquid and the aerosol flow stabilizer. After mixing the silver aerosol and the IPA saturated vapour the condensation processes were developed in the volume of the aerosol flow stabilizer and the micro-droplets of the colloidal solution appeared. Then these micro-droplets were deposited on the surfaces of three silicon wafers located at the output of the aerosol flow stabilizer. IPA vapours were obtained using the ultrasonic generator. For that purpose an IPA media with the chemical grade "pure for analysis" (State Standard 9805-84) was placed in ultrasonic bath before switching on the ultrasonic generator. Experiments were carried out at room temperature and at the atmospheric pressure. The IPA saturated vapour pressure was ~25.3 mm Hg [12], however not measured or fixed specifically. The silicon plates with dimensions 10 × 10 mm and the thickness 0.5 mm (State Standard 8.592-2009) were used in research. The duration of the separate experiment did not exceed one hour.

The silver NMP aerosol was created with the help of the spark discharge between two silver electrodes and an air stream passed at the speed of 1 cm/s through the working chamber of the aerosol spark generator [6]. The frequency of the spark breakdown was 50 min−1. Then the silver aerosol was transported to the ultrasonic generator and to the DAS measuring system, which determined the size distribution and concentration of the silver NMP in the aerosol (Fig. 1(b)).

During the experiment the composition of the silver aerosol particles was controlled every minute. It has been established that the largest part of the aerosol composition was about 45 - 50 nm (Fig. 1(b)-1). The total number of the silver aerosol particles received in the volume of the working chamber during the experiment was about ~5.2 × 106. As it is shown in Fig. 1(b)-2, the particle size distribution can be satisfactorily approximated by lognormal function [11], [13].

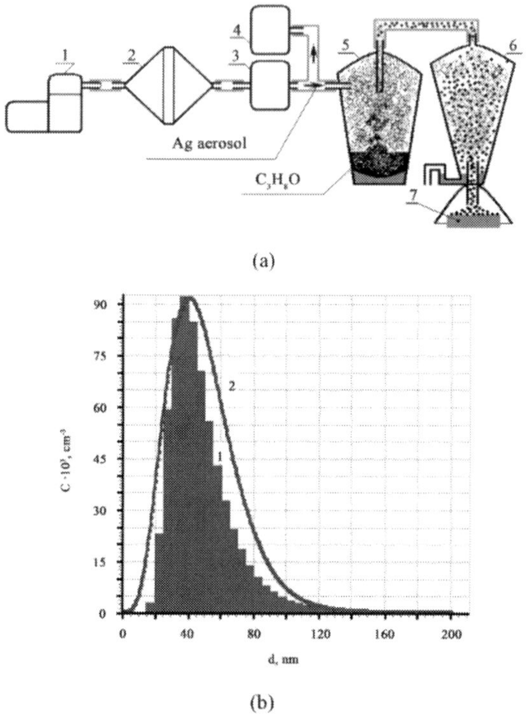

Fig. 1. (a) Technological scheme of the silver NMP micro-droplets deposition: 1—micro-pump; 2—high efficiency particulate air filter; 3—aerosol spark generator; 4—diffusion aerosol spectrometer; 5—ultrasonic generator; 6—aerosol flow stabilizer; 7—silicon substrate. Black dots represent aerosol particles of silver; (b) Histogram (I) and the aerosol silver particle size distribution function (II) obtained by the approximation of experimental data by the lognormal distribution.

The colloidal solutions of silver NMP were also prepared by passing the aerosol through the IPA medium. Before carrying out the TEM and SEM measurements the colloidal solutions of silver NMP in IPA were treated in an ultrasonic bath for 10 minutes. Then the sample of solution (about ~0.1 ml) was deposited on the copper mesh modified by carbon. Investigation of sludge formed on the silicon surface and carbon substrate after evaporation of the colloidal solution of micro-drops was conducted using TEM and SEM electronic microscopes FEI Tecna G2 F20 S-Twin TMP and Quanta-3D FEG (Fig. 2 and Fig. 3 respectively). TE-microscope FEI Tecna G2 F20 S-Twin

TMP worked at an accelerating voltage of 200 kV in bright field. The device resolution was 0.14 nm (line resolution).

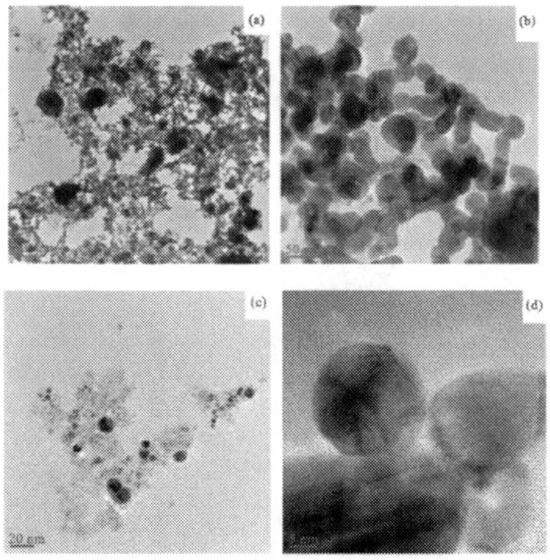

Fig. 2. Images of the obtained silver nano-/microstructures on the copper-carbon substrate at different magnifications: 9900× (a); 38,000× (b); 97,000× (c); and 450,000× (d).

It should be noted that the study of sludge which formed after evaporation of the micro-drops received by the second method was carried out during half an hour after it had been placed on a copper-carbon substrate. When the micro-drops were received using the first method, the TEM- and SEM-measurements were performed after the storage of the precipitates in the air and at the room temperature within two weeks.

AFM investigation of the silicon surface roughness after deposition of the silver NMP arrays was carried out by Multimode 8 scanning probe microscope, manufactured by Bruker (USA).The concerned array of AFM data was further processed using Image J 1.49 [14] software with the help of the Fraclac plug-in [15].

3 Results and Discussion

3.1 Silver Nano-/Microstructures Deposited from Aerosol

As follows from the TEM images obtained in the first case the silver aerosol passage through IPA media resulted in intense NMP coagulation and formation of the different NMS types (Fig. 2). In that case the formation of the numerous silver NMS resulted in the appearance of the porous silver film [6], that is an array of the randomly linked chains consisting of several (at least 10) silver NMP with sizes in the range of 10 - 20 to 200 nm (Fig. 2(a)-(c)). The facets of silver nano/microcrystals can be observed on the NMP surface (Fig. 2(d)).

It is known [16], that the chain structure of NMS occurs as a result of the silver NMP coagulation in the air. Possibly, coagulation can take place in the IPA medium. The secondary structure observed in Fig. 2(a) and Fig. 2(b), consists of chaotically intertwined chains containing the numerous silver NMP's. Such structure occurs both at the stage of the storage in alcohol, and the evaporation of nano-/micro-drops. The emergence and growth of the secondary structures (or fractal aggregates [17]) is a result of the aggregation of the numerous silver particles in chains, and it can be described in terms of different models of aggregation [18]. The estimation of the fractal dimension for layers of the silver NMS located in the regions with high and low NMP concentrations (Fig. 2(a) and Fig. 2(b)) gives the values of the surface fractal dimension D = 1.85 and 1.99 which correspond to the case of Brownian motion and/or the motion of NMS's on a linear trajectory in the model of the three-dimensional "cluster–cluster association" [18]. Therefore, at the formation of the silver NMS layers which is described within of the "cluster–cluster association" model, the different mechanisms of silver particles fusion and structures can be implemented.

3.2. Silver Nano-/Microstructures Deposited from Fog

The observed ring shape of the silver NMS, synthesized by deposition from the fog, repeats the profile of evaporated micro-droplet (Fig. 3). At the same time it was found that the silver NMS's with sizes from 0.5 to 20 nm are uniformly distributed on the substrate surface that is consistent with the ideas of the "coffee stains" model [8],[9],[10].

Considering the arrays of the silver NMS, which are formed by the micro-droplets of colloidal silver solution in isopropyl alcohol on the silicon plate surfaces after deposition from fog (Fig. 3(a) and Fig. 3(b)).

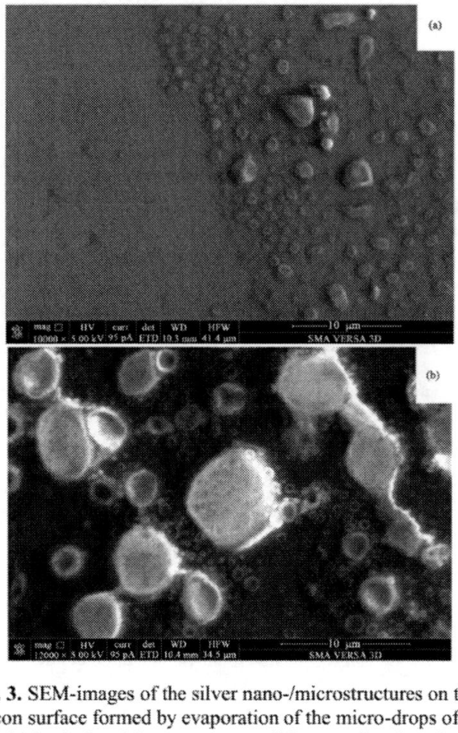

Fig. 3. SEM-images of the silver nano-/microstructures on the silicon surface formed by evaporation of the micro-drops of colloidal solution. Measurement conditions are listed on the insert.

Despite the significant time interval between the deposition of the micro-droplets on the substrate and SEM-measurements, the border of drops can be clearly observed on the silicon surface. This may be due to the sorption of the alcohol molecules by the silicon substrate. The numerous silver NMSs with the size from <0.1 to 3 μm can be observed on the SEM images also. It should be noted that a pronounced ring structure is characteristic for silver NMSs with the size from <0.1 to 1 μm.

According to [9] [10] [19], capillarity leads to one micron particle dynamics in evaporated drop. In this case, mass transfer processes can move the silver NMP either to the drop surface, or in the opposite direction depending on the ratio of the speed of the drop shape relaxation and the evaporation rate. The formation of the ring structure (or self-organization) may be considered a result of the mass transfer processes.

Unfortunately, due to the high level of the electron field emission the origin of which is associated with the adsorbed solvent molecules, it was impossible to carry out the SEM-studies of the ring silver NMS at the higher magnifications. Further investigation of the structure and distribution of silver NMPs in the ring boundary layer was carried out by AFM (Figure 4). As follows from the AFM-images (Figure 4(a) and Figure 4(b)), the size of the observed numerous silver NMP is about from 0.01 to 0.5 μm. All the silver particles considered have the same Crescent shape as well as the same orientation. All the observed silver NMPs are oriented by thickened side towards the outer border of the ring, while the Crescent is oriented towards the center of the vaporized micro-droplet.

It is possible to result the following arguments in favour of the fact that with the help of AFM method exactly the silver particles have been registered. First, in the AFM profiles the silver NMP revealed itself as the fluctuations of the spatial distribution of the surface height with the magnitude from 30 to 35 nm (Fig. 4(a) and Fig. 4(c)), that is much higher than the surface roughness of the silicon substrate (2.5 - 5.0 nm). Second, the maximum height of the discovered sickle silver NMP is 44 nm that coincides with the maximum of the size distribution (40 - 45 nm) in the aerosol containing the silver NMP which has been determined previously (Fig. 1(b)).

The most part of the sickle silver NMP's, shown in the Fig. 4, has a two-or three-layer structure (Fig. 5(a)). At least two aerosol particles participated in the formation of the considered silver NMP, which originally had a spherical shape (Fig. 2(d)). Possibly, the deformation of the considered two-or three- particle clusters tends towards its boundaries away from the center. Probably the mass-transfer processes have been moving in that direction during the evaporation of the IPA molecules.

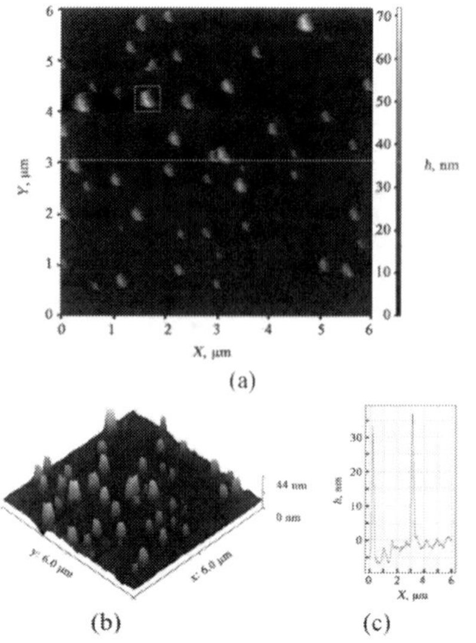

Fig. 4. 2D- (a) and 3D-images (b) of the silver NMP arrays on the surface of the silicon support recorded by AFM-method in the field of the ring structure appeared in the place of the vaporized micro-droplets; (c) the AFM-profile of the silicon surface containing the silver nano-and micro-particles (shown by the white line on the image (a)). An enlarged image of the silver particle, selected by white rectangular frame, is shown in the Fig. 5(a).

An assessment of geometric characteristics of the Crescent-shaped silver NMP has been done. For this purpose, the shape of the particles has been represented in the form of a Crescent (Fig. 5(b)) and the geometric parameters of the silver NMP have been calculated using the following formulas [20] [21]:

$$\begin{cases} h_1 = h_2 + \Delta h; s_i = \alpha_i r_i; r_i = h_i + d_i = \dfrac{h_i}{2} + \dfrac{c^2}{8h_i} \\ c = \dfrac{2r_i \sin\alpha_i}{2} = 2\sqrt{h_i\left(2r_i - h_i\right)}; \alpha_i = \dfrac{2\arccos d_i}{r_i} \\ A_i = \dfrac{1}{2} r_i^2\left(\alpha_i - \sin\alpha_i\right) \end{cases} \quad (1)$$

In the system of the Equation (1) i = 1, 2; hi = h1, h2 is the height of the circle segments, the arc lengths s1, s2 are part of these segments, correspondingly, limiting the considered particle with external and internal sides; Δh is the thickness of the Crescent; d1, d2 are the distances from the centres of circles O1 and O2 to the middle of the chord AB, which have the lengths c; r1, r2 are the radii of the circles with centres O1 and O2; α1, α2 are the angles AO1B and AO2B; Ai is the area of segments. The area of the Crescent is determined as the difference of the squares of segments.

According to the calculations, the radii of the circles O1 and O2 are practically identical (Table 1). Consequently, the Crescent-shaped silver particle is formed by two deformed aerosol NMP (Fig. 5(a)). On the other hand, the estimated values of the circle radii O1 and O2 are 4 - 5 times higher than the most probable silver NMP size in the initial aerosol (Fig. 1(b)). Probably, the obtained difference is due to the presence of a layer of the adsorbed isopropyl alcohol molecules, i.e. the cluster "silver NMP–isopropyl alcohol molecules" can be seen as a direct micelle. Possibly, the evaporation of micro-droplets is accompanied by the coagulation of silver particles existing in the solution in the form of direct micelles. Further, clusters containing two or three NMP's, which surrounded by a layer of the IPA molecules, have been created.

Table 1. The geometrical characteristics of the Crescent modelling the silver sickle nano-/micro-particles.

№	Name, designation	Value	
		Arc1	Arc2
1*	The thickness of the Crescent in the center of the figure Δh, μm	0,18	
2*	The length of the chord c, μm	0,36	
3*	The height of the segment in the center h, μm	0,29	0,11
4	The radius of the circle r, μm	0,201	0,205
5	The length of the arc s, μm	0,82	0,44
6	Angle α, degree	232,6 9	125,72
7	The area of the segment Ai, μm2	0,10	0,03
8	The area of the Crescent (particles), μm2	0,07	

Note: (*) The values are determined by measuring the Fig. 5(a).

The processes of spinodal decomposition [22] of the considered clusters and the formation in the cluster volume of the two-phase system containing the layer of the deformed Crescent shape silver NMPs have been developed during further drying and storage of the sample. Subsequent evaporation of the IPA molecules leads to the formation of the observed silver Crescent-shaped NMP. It is known [23], that this phenomenon is often observed in the study of the metastable heterogeneous colloidal systems as a result of the complex thermo- and mass -transfer processes [24] and it is called "the syneresis".

According to [25] [26] [27] the driving force the effect of syneresis is concentration gradient of the alcohol molecules in the cluster and on the substrate surface during the evaporation process. The value of the occurring stress concentration is sufficient for cluster splitting, and deformation of the original spherical silver NMP's.

The value of the fractal dimension for silver Crescent-shaped NMP (Fig. 5(a)) is about of D ~1.69, that is significantly lower than the established similar values for the silver NMP fractal dimension obtained by passing the silver aerosol though the IPA medium [18]. This value of the fractal dimension D corresponds to the model of the "cluster–cluster association in the regime of Brownian motion model" [18].

Thus, experimentally synthesized silver NMS with sickle NMP's was obtained by means of controlled evaporation of the colloidal solutions droplets deposited from the fog on the surface of the silicone support. It is possible that the arrays of the silver Crescent-shaped NMP's may be associated with the processes of silver nanoparticles self-organization, which develop during the micro-drops evaporation. In conclusion, it should be noted that the detected silver Crescent shaped NMP belongs to the recently discovered kind of specific asymmetric nanoparticles–the so called Janus silver nanoparticles [25],[26] [27].

The present research was supported by the Ministry of Education and Science of the Russian Federation (Grant of the 20th of October, 2014, № 14.576.21.0053; The Applied Research Unique Identifier RFMEFI57614X0053).

References

1. V.V. Klimov, Nanoplasmonics, 480 (2009)
2. A. Aubry, Y.L. Dang, S.A. Maier, J.B. Pendry, A Physical Review, 82B (2010)
3. N. Bocchio, M. Alvarez, F. Gaul, H. Rochholz, K. Koynov, M. Kreiter, Molecular Plasmonics. International Symposium, Abstracts. Jena (2007)
4. V. Krapek, A.L. Koh, L. Brinek, M. Hrton, O. Tomanec, R. Kalousek, S.A. Maier, T. Sikola, Optics Express , 23, 11855 (2015)
5. H. Rochholz, , N. Bocchio, M. Kreiter, New Journal of Physics , 9, 53, (2007) https://doi.org/10.1088/1367-2630/9/3/053
6. A.S. Smolyanskii, V.A. Zagaynov, Yu.G. Biryukov, E.P. Magomedbekov, O.G. Stepanova, L.I. Trakhtenberg, Composites and Nanostructures, 6, 137-147 (2014)
7. S.V. Gaponenko, N.N. Rosanov, E.L. Ivchenko, A.V. Fedorov, A.V. Baranov, A.M. Bonch-Bruevich, T.A. Vartanyan, S.G. Przybelski, Nedra, 326 (2005)
8. M. Layani, M. Gruchko, O. Milo, I. Balberg, D. Azulay, S. Magdassi, ACS Nano, 3, 3537-3542 (2009)
9. V.V. Vysotskaya, O.J. Uryupin, I.N. Senchikhin, V.I. Roldugin, Colloidal Journal , 75, 161-169 (2013)
10. I.V. Vodolazsky, Yu.Yu. Tarasevich, O.P. Isakova, Nonlinear World , 8, 142-150 (2010)
11. P. Reist, *Aerosols* (Mir, Moscow, 1997)
12. S.S. Kutateladze, V.M. Borishanskii, *Handbook of Heat Transfer* (2012)
13. K.W. Lee, H. Chen, Aerosol Science and Technology, 3, 327-334 (1984)
14. T. Ferreira, W. Rasband, ImageJ User Guide-IJ 1.46r. (2012)
15. A.L. Karperien, Fraclac for ImageJ, 1-36, (2013)
16. A.E. Aloyan, V.O. Arutyunyan, A.A. Lushnikov, V.A. Zagaynov, Journal of Aerosol Science, 28, 67-85 (1997)
17. A.V. Kolesnichenko, M.Y. Marov, *Modeling of Processes of Formation of Dust Fractal Clusters as the Basis of Loose Preplanetesimals in the Solar Deplanatum the Cloud* (Preprint IPM, Moscow, 2014)

18. B.M. Smirnov, Uspekhi Fizicheskih Nauk, 149, 177-219 (1986)
19. I.P. Suzdalev, Nanostructures and Nanomaterials (2006)
20. G. Korn, and T. Korn, Mathematics Handbook (for Science Officers and Engineers (1977)
21. Online Calculator. http://planetcalc.ru
22. A.A. Rempel, A.A. Valeeva, Materials and Methods of Nanotechnology, (2015)
23. D.A. Fridrikhsberg, Course of Colloid Chemistry, (1995)
24. J.E. Geguzin, *Physics of Sintering* (Nauka, Moscow, 1984)
25. A.Y. Olenin, T.R. Nizamov, G.V. Lisichkin, Russian Nanotechnology, 9, 19-24 (2014)
26. Y. Song, S.W. Chen, Chemistry—An Asian Journal, 9, 418-430 (2014)
27. M. Lattuadaa, T.A. Hattonb, Nano Today, 6, 86-308, (2011)

Fractal microgeometry of original and hydgrogen-charged graphene multilayer films surface: research by the low-temperature sorption of nitrogen method

Svetlana Vasilyeva[1], *Anna* Zheltova[2], *Alexey* Merkushkin[3], *and Alexander* Smolyanskiy[1,3]

[1]«Karpov Institute of Physical Chemistry», 105064 Moscow, Russian Federation.
[2]Lomonosov Moscow State University, Faculty of Bioengineering and Bioinformatics, 119234 Moscow Russian Federation
[3]Dmitry Mendeleev University of Chemical Technology of Russia, 125047 Moscow, Russian Federation

Abstract. Changes in the surface microgeometry of multilayer graphene films, as a result of hydrogen adsorption (1% mass.), have been examined by the method of low-temperature fractal sorption of nitrogen. It is established that hydrogen atmosphere treatment leads to the significant development of graphene surface and lowers the first layer heat adsorption from 289 down to 260J/kg. There have been found out as well the overall mesopores concentration growth and pore size distribution function broadening. At that, the surface fractal dimension value has grown from 2,52 up to 2,67, which is to give us an evidence of the surface chaotization as a result of hydrogen withdrawal into the atmosphere and the graphene surface erosion.

1 Introduction

The use of water purification plants based on the ozone electrochemical generators necessitates the utilization of hydrogen evolved on the electrolyzer cathode [1]. The multilayer graphene films are considered to be an advanced sorbent for the hydrogen absorption [2]. It is commonly known [2, 3, 10, 11] that the hydrogen absorption on graphene is a complex physicochemical process, which, in particular, leads to some substantial changes in the film surface microgeometry. However, the conclusions [3], being based on the results of roughness measurement of the original and hydrogen-charged graphene films surface (hereinafter – G and GH respectively) obtained by the atomic force microscopy (AFM) method, require an additional experimental proof which is achievable through the sorption methods [4, 5] use. An advantage of such sorption methods is a possibility of an overall analysis of the surface of the test sample, whereas AFM-measurement provides only local surface analysis.

On the other hand [4], the polymolecular adsorption theory of Brunauer–Emmett–Teller (BET) is based on the presumption that the surface the adsorbate molecules are deposited on, is to be perfectly smooth. For the surface roughness recordkeeping, the fractal sorption approach [5] is applied.

Therefore, the use of the approach in question [5] is essential for the estimation of the G surface changes caused by the hydrogen adsorption.

The purpose of the present research was to examine the G multilayer films surface, before and after treatment in the hydrogen atmosphere, using the low-temperature nitrogen sorption method; to analyze the obtained sorption data using the BET sorption isotherm and the fractal sorption equations [5].

2 Experimental technique

To prepare the graphene oxide using the method [6], 2 g of graphite powder, 105 ml of sulphuric acid and 1 g $NaNO3$ were being stirred during 30 minutes at the ambient temperature. Then, 8 g $KMnO4$ were gradually being added into the reaction mixture during 2 hours, keeping the reaction mixture stirred in the ice bath.

After the stirring within one hour at 35°C, 200 ml of distilled water were been added slowly to the reaction mixture. Then, the temperature of the reaction mixture was gradually raised up to 98°C and the reaction went on within other 15 minutes. After the cooldown of the mixture to the ambient temperature, 150 ml of distilled water containing 6 ml of hydrogen peroxide was added thereto. After the process termination the reaction mixture was filtered, rinsed and redispersed in the 5% HCl solution. Then, the graphene oxide powder was being washed with water until the neutral pH was reached and then was dehydrated in vacuum at 60°C. The derived product was exposed to a flash drying at 1000°C during 12 minutes under control.

The produced G hydrogenation was carried out in the special device under various pressure conditions (up to 1 MPa (~0,987 atm.) at 77 K (Fig. 1).

It was found out that hydrogen sorption on G could be adequately described by the Freundlich sorption isotherm equation [7]:

* Corresponding author: vasileva@nifhi.ru

$$lnA=lnK+ 1/n lnP \quad (1)$$

where the regression parameters lnK = -6,294 ± 0,157, 1/n = 0,696 ± 0,019, the correlation factor value is 0,994 (Fig. 1).

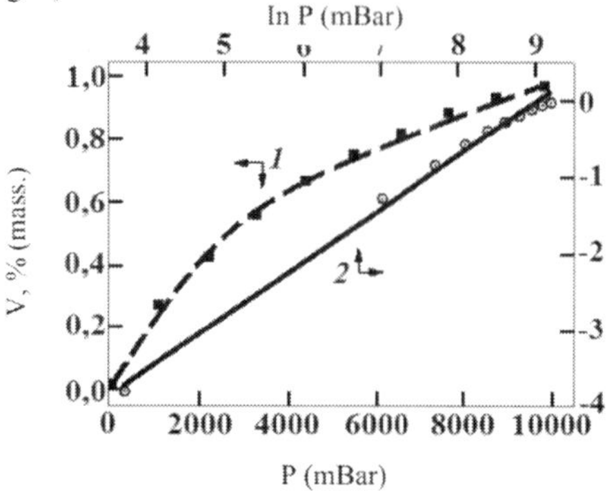

Fig. 1. Parameters of hydrogen sorption on graphene surface during hydrogenation process.

The Freundlich sorption isotherm is observable in case of adsorbates uptake in heterogeneous systems featured by the presence of the interactive active centers (AC) sorption heat exponential distribution [4, 7]. The calculated parameter values of the Freundlich pattern – the invaiables K~1,85•10-3 (% (mass.) H2/g)1-n(% (mass.) H2/g)n as well as power index n~1,437, - feature the graphene hydrogen absorption capacity and the sorption intensity. In the present research the G samples with the highest hydrogen absorption (~1 % (mass.)) were used.

The examination of low-temperature nitrogen sorption on G and GH samples was carried out at 77 K on the Quantachrome QuadraWin plant. The sorption experiment results interpretation in the context of the BET sorption isotherm equation was carried out using the proper software for the device – there were determined the values of the BET equation invaiables – C- constant as well as monolayer capacity Vm, S – surface area of G and GH (Table 1).

Table 1. Comparison of nitrogen sorption parameters of normal and hydrogenated graphene

Name	G	GH
C	53,52	15,17
Q_1, kJ/kg	289,02	260,19
S, m^2/g	342,8	1201,3
V_m, cm^3/g	0,098	0,345
B	-0,48 ± 0,01	-0,33 ± 0,01
Δ	-0,4477	0,0065
D_s	2,52	2,67

3 Results and Discussion

The hydrogenization of G leads to the substantial nitrogen sorption parameters change (Table 1): it was found triple C-constant derate whereas S value grows up by a factor of 4 circa. It is known [4, 12, 13, 14, 17], that C-constant features an interaction between the adsorbate and G, GH – surfaces as well as nitrogen molecules interaction:

$$C=exp[((Q1-\lambda))/RT] \quad (2)$$

where Q1 – adsorption heat in the first layer, which is determined by the nitrogen molecules and G, GH-surface interaction, kJ/kg; λ – the nitrogen condensation heat (198 kJ/kg [8]); T – temperature, K; R – universal gas constant for nitrogen (297 J/(kg•K) [8]). As it follows from the Table 1, the hydrogen adsorption leads to the Q1 value decrease which proves the change of type of the nitrogen molecules and G interaction over adsorbed hydrogen.

Let us discuss the surface development effect in the context of the fractal sorption approach [5, 15, 16, 18]. The point of the given approach is the Ds - surface fractal dimension determination:

$$ln(V/V\_m)=A+Bln[ln(p\_s/p)] \quad (3)$$

where V – adsorbtion, cm3/g; A, B – regression factors; p – the nitrogen vapor pressure current value, mbar; ps – the nitrogen saturated vapor pressure (1940,44 mBar at 77 K [8])

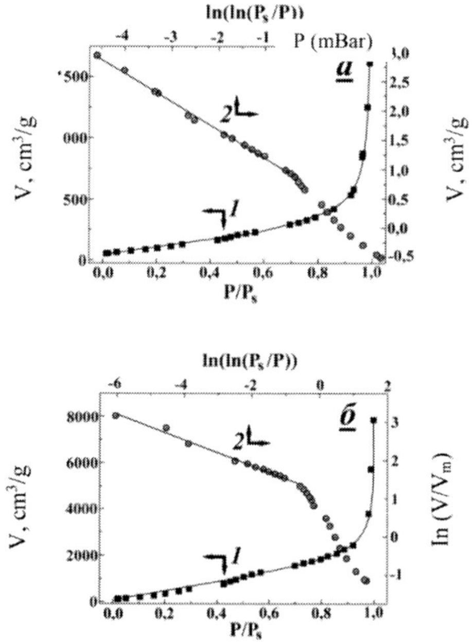

Fig. 2 (a, b). Determination of regression factor value for normal and hydrogenated graphene.

It is necessary to point out that 2 < Ds < 3, moreover, Ds=2 corresponds to an xy-plane in Euclidean space and Ds=3 – to a perfectly chaotic surface. The value B for G,

GH can be determined from the tilting angle of the straight-line portion of ln(V/Vm) or ln[ln(ps/p)] functional connection(Fig. 2 (a, b), curve 2; Table1).

In accordance with [5], the second regression factor depends on Ds and on the adsorption mechanism (Van der Waals forces interaction or capillary condensation). In case of Van der Waals interaction mechanism

$$B=(D_s-3)/3 \qquad (4)$$

In case of capillary condensation mechanism:

$$B=D_s-3 \qquad (5)$$

In case of minor coverage of surface the adsorbate – adsorbent interaction mechanism choice is to be done with the following criteria:

$$\delta=3(1+B)-2 \qquad (6)$$

The results of the parameter δ calculations are shown in the Table 1. In accordance with [5], for G, the negative δ value says for capillary condensation mechanism prevalence by the nitrogen adsorption on graphene. In this case the G surface fractal dimension can be determined according to the formula (5) (Table 1).

For GH $\delta \geq 0$ this fact means that the hydrogen modification of the graphene surface can lead to the nitrogen-GH surface interaction Van der Waals mechanism realization. The surface fractal dimension value (2,0069) calculated according to the formula (4) is slightly different from the critical value Ds=2, which corresponds to the smooth surface. In view of this one can presume that the graphene surface hydrogenization can lead to a significant reduction of surface roughness.

However, this conclusion is in controversy with the abovementioned evidence of the hydrogen-charged graphene surface area S increase as well as with the results of the pore size distribution function calculation (PDF) for G and GH, shown in the Fig. 3.

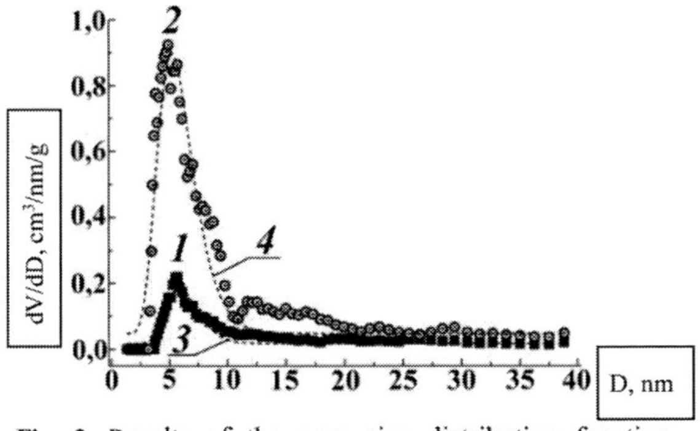

Fig. 3. Results of the pore size distribution function calculation for normal and hydrogenated graphene.

As it follows from the Fig. 3, hydrogenization leads to the significant PDF extension as well as to the pore concentration growth in the highest PDF (5 – 7 nm) by the factor of 5 in comparison with the original graphene PDF. Furthermore, PDF of GH contains two new wide peaks of low intensity at ~10 and~15 nm. These peaks appearance may display the coalescence processes for the pores of the basic highest size (5 – 7 nm). It is necessary to notice as well that in both cases the PDF form could be adequately described using the lognormal distribution method (Fig. 3, curves 3, 4).

Therefore, the aggregate of the presented facts, namely, increase of S and significant broadening of PDF in case of GH, is in controversy with the conclusion about the surface roughness reduction trend. Furthermore, determination of the δ-parameter has been done inaccurately, - taking into consideration the lower variation range limit in the B-parameter value, inserting into (6) the condition $\delta < 0$ which means the interaction mechanism is determined as capillary condensation. Then the surface fractal dimension calculation according to the equation (5) leads to the value of Ds=2,67, that says for the hydrogen-charged graphene surface development.

4 Conclusions

The ambiguousness in the determination of the nitrogen molecules – hydrogen-charged graphene surface interaction revealed in the present research, may prove the fact that the hydrogen sorption process is the complex one and comes with graphene surface erosion due to dehydrogenation [9,19, 20]. It can be presumed that formation of new nano-/micropores on the hydrogenated graphene surface takes place as the result both of such erosion and hydrogen evolution. The subsequent nitrogen adsorption allows fixing the changes on the surface that have taken place. The Van der Waals interaction realization during the nitrogen molecules sorption on the hydrogen-charged graphene surface can be explained by the fact that nano- and micropores of the nitrogen molecules occurring on the adsorbent surface are full of hydrogen molecules which excludes any possibility of capillary condensation mechanism realization. Reduction of the sorption heat value in the first layer (table 1) can as well favor such hypothesis.

The further experimentation of the original and hydrogen-charged graphene surface using scanning electron microscopy (SEM) and AFM in order to prove (or disprove) the results shown in the present article, is evident. The results of multifractal analysis of SEM- and AFM-pictures of the G-and GH-surfaces will be stated in the following article.

The present research was supported by the Ministry of Education and Science of the Russian Federation (Grant of the 20th of October, 2014, № 14.576.21.0053; The Applied Research Unique Identifier RFMEFI57614X0053). The authors hereof would like to thank Mr. Zahra Gohari Bajestani (Sabancı Üniversitesi, İstanbul, Turkey) for the original and hydrogen-charged graphene samples production.

References

1. G.Ph. Potapova, M.I. Ikim, S.A. Korneychuk, A.S. Smolyanskiy, Composites and nanostructures **5**, 23 – 32 (2013)

2. Yu.S. Nechaev, T.N. Veziroglu, Int. J. Chem. **7**, 207 – 212 (2015)
3. S. Goler, C. Coletti, V. Tozzini, V. Piazza, T. Mashoff, F. Beltram, V. Pellegrini, S. Heun, J. Phys. Chem. 117, 11506 –11513 (2013)
4. S. Greg, K. Sing, *Adsorption, specific surface, porosity*, 306 (1984)
5. M.K. Wu, Aerosol Sci. and Technol. **25**, 392-398, (1996)
6. W.S. Hummers, R.E. Offeman, J. Amer. Chem. Soc. 80, 1339 (1958)
7. H.M.F. Freundlich, J. Phys. Chem. A57, 385 (1906)
8. M.B. Vargaftik, *Reference book for thermal-physical properties of gases and liquids*, 720, (1972)
9. A. Thangaraja, S.M. Shinde, K. Golap, R. Papon, S. Sharma, R. Vishwakarma, K.P. Sharma, M. Tanemura Appl. Phys. Lett., **106**, 253106 (2015)
10. C. Wonbong, *Graphene: synthesis and applications*, 35-36, (2012).
11. D. Dreyer, S. Park, C. Bielawski, R.S. Ruoff, Chem. Soc. Rev., **39**, 228-240, (2010)
12. Z. M. Ao, Q. Jiang, R. Q. Zhang, T. T. Tan, S. Li, J. Appl. Phys., (2009)
13. O. Leenaerts, B. Partoens, and F. M. Peeters, Phys. Rev. B 80, 245422, (2009)
14. P. Ruffieux et al., Phys. Rev. B 66, 245416 (2002).
15. S. Ross and J. Olivier, *On Physical Adsorption*, (1964)
16. W. Choi, I. Lahiri, R. Seelaboyina, Y.S. Kang, Crit. Rev. Solid State Mater. Sci., **35**, 52, (2010)
17. S. Gilje, S. Han, M. Wang, K. L. Wang, R. B. Kaner, Nano Lett. 7, 3394 (2007)
18. M.R. Falvo, G.J. Clary, M. Taylor II, V. Chi, F.P. Brooks Jr., S. Washburn, R. Superfinem, Nature, 389, 582, (1997)
19. S. Bae et al., Nat. Nanotechnol. **5**, 574–578, (2010)
20. A. W. Tsen et al., Science 336, 1143–1146 (2012)

The electromechanical converter in the systems of desulfurisation of crude oil

Denis Kuimov[1,*], *Maxim* Minkin[1]

[1]Don State Technical University, Chair "Production Process Automation", 344000, Gagarin square, Rostov-on-Don, Russia

Abstract. In article authors have investigated a question of a possibility of application of hydrodynamic cavitation processing of crude oil for the purpose of decrease in content of sulphurous compounds. The electromechanical converter with a secondary discrete part, the being device exciting in the processed material the cavitation and shock field by means of heavy traffic of big set of ferromagnetic elements under the influence of external magnetic field is presented. Features of initiation of hydrodynamic cavitation on the example of the movement of a single ferromagnetic element are considered. Results of the mathematical description of process of desulphurization of crude oil are presented. On the basis of economic efficiency of the presented technological process, authors have offered options of introduction of technology on production objects.

1 Introduction

The modern oil branch of the Russian Federation is characterized by increase in a share of high-sulphurous grades of the extracted oil in the general structure of reserves of hydrocarbonic raw materials. As of 2016 the share high-sulphurous (1,68%) and sulphurous grades (0,61-1,68%) has made 69%, and in the future will only increase, taking into account exhaustion of fields of "easy" grades.

Widespread ways of processing of raw materials for the purpose of minimization of content of sulphurous compounds are: hydrotreating, the adsorptive cleaning and sonocatalitic desulphurization. All listed methods have the proved efficiency in oil desulphurization processes, and have already found the application at many oil refineries of various power. Nevertheless, they aren't deprived of shortcomings among which: high power consumption, irrevocable losses of expensive catalysts, complexity of system.

According to many scientists both foreign, and domestic, a perspective method of processing of oil for the purpose of change of rheological properties and change of fractional structure is hydrodynamic cavitation processing.

Within the previous researches [1,2], authors of work have studied the phenomenon of hydrodynamic cavitation in devices with a secondary discrete part, the new configuration of the electromechanical converter is offered and the possibility of its application in the systems of primary oil refining for the purpose of decrease in content of sulphurous compounds is investigated.

The purpose of this research is:

1. Theoretical assessment of overall performance of the device;
2. Economic assessment of the offered technology.

2 Main part

2.1 Technological process

The electromechanical transformer with a secondary discrete part represents the cylindrical multiphase inductor with the concentrated winding, on the place of a rotor the working camera from non-magnetic material is located. As a working organ the ferromagnetic elements located in the working camera and making under the influence of an outside magnetic field heavy traffic from one active pole to another are applied. Movement of ferromagnetic elements with high speeds generates hydrodynamic cavitation in processed liquid material.

The phenomenon of hydrodynamic cavitation [3, 4, 5] arises at the time of approximation of liquid pressure to pressure of saturated steam of liquid raw materials. In case of the close values in liquid processed material abundant separation of the bubbles filled with gas is watched. The considerable effect of cavitation processes is connected to high concentration of the energy emitted in the course of collapse in the processed environment thereby initiating processes of corrupting of sulphurous compounds. This phenomenon at the time of achievement by a bubble of the minimum radius preceding collapse is explained by the small volume of substance.

In case of collapse of a cavitation bubble kinetic energy [6,7] which turns into activation energy of different responses is emitted:

* Corresponding author: muzhik1010@yandex.ru

$$W_{cin} = 2\pi\rho U^2 R^4 \int_R \frac{dr}{r^2} = 2\pi\rho U^2 R^3 \quad (1)$$

where, U - the speed of a spherical wall; R - instant radius of the slamming sphere.

Speed of a spherical wall is determined by the following formula:

$$U = \sqrt{\frac{2}{3}\frac{P}{\rho}\left(\frac{R_0^3}{R^3} - 1\right)}, \quad (2)$$

where, R_0 - the initial radius of a bubble.

We will write down the total energy emitted at collapse of a set of cavitation cavities through number of cavitation cavities on a formula:

$$N = k_c \cdot V_l \ , \quad (3)$$

where, k_c - concentration of cavitation cavities which for oil accepts value 10^{11} 1/m^3, V_l - liquid volume with the developed cavitation, m^3.

The way passed by a single ferromagnetic element on the studied trajectory will be accepted to liquid volume with the developed cavitation.

Considering expressions (1-3), we will write down a formula of the energy which is marked out at collapse of cavities at supercavitation:

$$W_{cin} = \frac{4}{3} k_c V_l \pi P \left(\frac{R_0^3}{R^3} - 1\right) R^3, \quad (4)$$

Having considered effect of cavitation and the energy emitted at collapse of cavitation cavities, further for a theoretical research of efficiency of work of the electromechanical converter with a secondary discrete part it is necessary to investigate fractional composition of the processed raw materials.

The analysis of fractional composition of liquid hydrocarbons of various fields has shown that distribution of sulfur has difficult character. For a theoretical research sulphidic oil of the Arlansky field has been chosen. For the purpose of simplification of calculations, all variety of sulphurous compounds has been divided into three groups with the closest molar mass and energy of communication (tab. 1).

Table 1. Sulphurous compounds

№	Name	Share from the general contents CC, %	Average molar weight, g/mol	Average energy of a rupture of communication, kJ/mol
1	Thiophene	76,17	44	82
2	Sulphide	23,6	50	80
3	Mercaptan	0,23	80	360

Within model of process of desulfurization it is supposed that reaction of a rupture of sulphurous compounds is carried out in one stage on C-S communication, at achievement by cavitation of necessary value of energy of a rupture of communication, in the set volume, at the same time calculation was carried out in two steps. At the first stage the volume of the liquid hydrocarbonic raw materials subjected to processing as a result of hydrodynamic cavitation paid off. The way passed by ferromagnetic elements with a speed called further critical at which the condition of emergence of the supercavitation field is satisfied was accepted to volume.

$$V = N_{el}\pi r^2 L_{mean} 15 k_{tr} + N_{el}\pi r^2 L_c 15 k_{tr} \quad (5)$$

where N_{el} - quantity of the ferromagnetic working elements participating in processing of liquid hydrocarbonic raw materials, piece; r - radius of a single ferromagnetic element in the widest part (a supercavitating profile), m; L_{mean} - average length of a run of an element from one pole to another, m; k_{tr} – a share of the way passed with cavitation.

The value 15, determines the number of switching for one motion cycle of ferromagnetic elements by a circle along boring of the inductor and is equal to quantity of phases of the inductor. In the second item, the value 2 defines the number of switching for which ferromagnetic elements pass average distance on a path lengthways center of the working camera.

The analysis [8] of movement of a one ferromagnetic element, showed that ferromagnetic elements, in case of relocation from one active pole to another, pass through stages of acceleration and braking. As a result, the analysis of process of movement of a one ferromagnetic element, allowed to obtain some averaged data showing a share of the way passed by such element with a speed below the critical, expressed by coefficient path k_{tr} and equal 0,86.

The amount of the energy emitted in the course of collapse of cavitation cavities is defined with expressions (4) and (5).

At the second stage computation of the molar mass of the sulphurous compounds subjected to cavitation processing was carried out. The amount of substance of sulphurous compounds, is calculated in volume of the liquid hydrocarbonic raw materials subjected to cavitation processing taking into account initial percentage and average molar mass for each group of the compounds provided to tab. 1. on the following expression:

$$\upsilon = m / M \ , \quad (6)$$

where, m – a lot of sulphurous compounds; M – average molar mass of group of sulphurous compounds, g/mol.

Taking into account told the algorithm of calculation of number of the broken-off sulphurous compounds in the course of cavitation processing shown on rice is offered. In fig. 2 the dependence of the broken-off sulphurous compounds on processing time in the electromechanical converter with a secondary discrete part is presented. It is possible to note good removal of sulphurous compounds in the first 20 sec. of processing. Further removal is slowed down in view of continuous hashing and decrease in concentration of sulphurous compounds.

Results of theoretical researches have shown efficiency of the choice as the dominating power influence hydrodynamic cavitation, unlike earlier applied energy of impact, allows to realize effective processing of liquid materials, including oil.

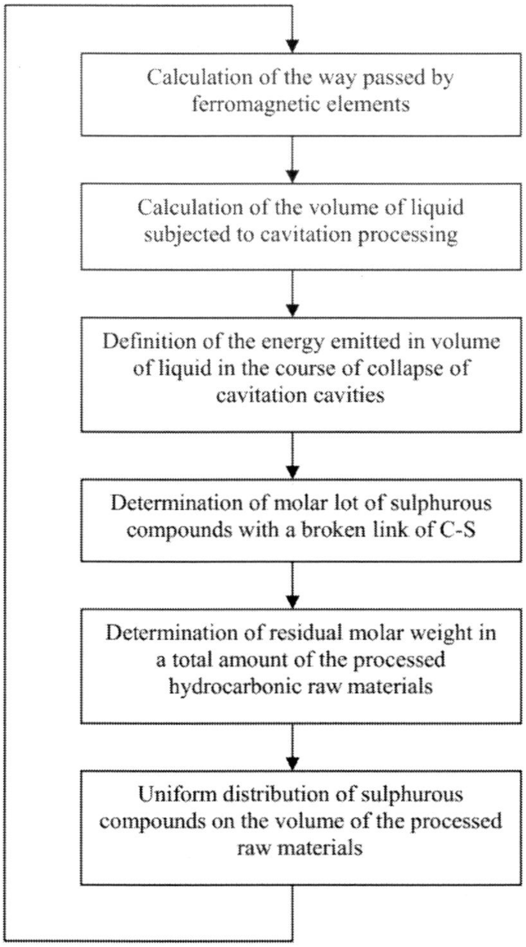

Fig. 1. Algorithm of Calculation of Number of the Broken-off Sulphurous compounds in the course of cavitation processing.

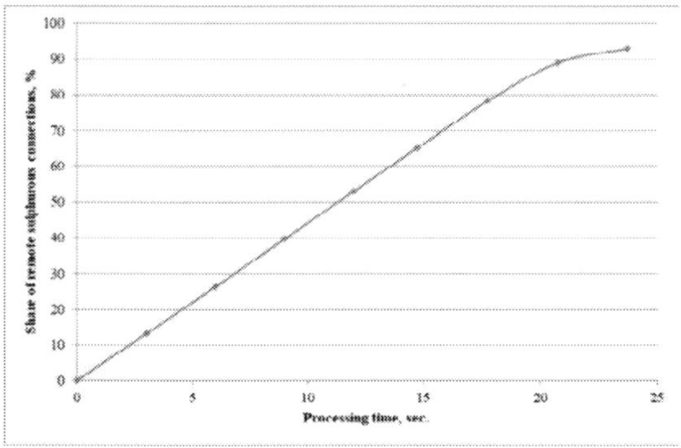

Fig. 2. Desulfurization process the electromechanical converter with a secondary discrete part

Nevertheless, the efficiency and expediency of use of this or that device has to be estimated and compared to the technologies and methods which are the most extended today not only from the point of view of efficiency, but also profitability.
The economic appeal of process of desulfurization of crude oil is connected by the electromechanical converter with a secondary discrete part, first, with lack of need of use of expensive reagents, secondly, with soft conditions of carrying out, moderate temperature and pressure, thirdly, with simplicity of processing equipment.

3 Assessment of economic efficiency

Relying on results of a theoretical research of process of desulfurization by the electromechanical converter with a secondary discrete part, approximate calculation of cost of processing of 1 ton of initial raw materials has been carried out with a productivity of system 10t/day and also the comparative analysis with methods of desulfurization of crude oil most widespread and perspective today is carried out.

Certainly, within economic assessment of the presented earlier technological process, the essential value has such concept as an electric power expense as any mechanical or electromechanical device demands her consumption. For this reason accounting of these expenses is essential when accounting expenses on operation of technical devices within the offered technological process.

The lack of need of use of reagents and catalysts reduces the main expenses of production to ensuring energy for implementation of process of movement of a secondary discrete part on the set trajectory.

Thus, it is possible to conclude that the sum of all material inputs on the electric power necessary for processing of 1 ton of crude oil, was respectively, 1078.82 roubles.

Table 2. Costs of the electric power

Name of the equipment	Power of the equipment, kW	The required time for processing of 1 ton of initial raw materials, hour	Cost is one kW/hour of the electric power, rub.	Total material inputs, rub.
Electromechanical converter	4,9	2,45	1,16	55,7
Capacitor bank	90	2,45	1,16	1023,12
General costs of processing of 1 ton of initial raw materials, rub				1078,82

Process most widespread today removal of undesirable impurity of sulphur and sulphurous compounds is hydrotreating process. It agrees [9,10], the cost of processing of 1 ton of liquid hydrocarbonic raw materials by method of hydrotreating is 4398,9 rub. It allows to draw conclusions on economic efficiency of the introduced method of cavitation processing and alleged cost cutting for 75,4%.

4 Conclusion

Achievement of 90% of remote sulphurous compounds allows to speak about efficiency of the offered technology of desulphurization of oil and the accompanying device for realization of technology – the electromechanical converter with a secondary discrete part. Possibilities of the electromechanical converter with a secondary discrete part allow to use it not only to mini-oil refinery, but also at large oil refineries, for the purpose of improvement of quality of output products and decrease in economic expenses. At the same time, It should be noted that on the basis of the presented calculations, use of the presented device in the offered configuration is inexpedient for processing the mercaptan types of oil in which in quantity, dominating or commensurable with sulphides and thiophenes, to contain mercaptans, having high values binding energy. Further researches have to be devoted to removal the mercaptan compounds.

References

1. D.N. Kuimov, M.S. Minkin, A.D. Lukyanov, Materials Science Forum, **870**, 671-676 (2016)
2. M.S. Minkin, D.N. Kuimov, M.N. Kulinich, *Dynamics of the technical Collection of Works XII of the International Scientific and Technical Conference systems*, 392-396 (2016)
3. E. A. Denisyuk, S.P. Shevelyov, The NGIEI Bulletin, **1(2)** (2011)
4. A.V. Stavitskaya, Development of methods of oxidizing modifying oil and products of her processing abstract. Wuchang. step. Cand. Tech.Sci., Moscow (2015)
5. V.A. Koryagin, *Szhiganiye vodotoplivnykh emul'siy i snizheniye vrednykh vybrosov* (monografiya) (Nedra, SPb., 1995), 304p. [in Russian]
6. O.V. Lavrinenko, E.I. Savina, G.V. Leonov, The Polzunovsky messenger, 59-63 (2007)
7. R. Knepp, J. Daley, F.Hemmit, *Cavitation* (Mir, Moscow, 1974) [in Russian]
8. D.N. Kuimov, News of higher educational institutions. Electromecanics, **5 (547)**, 11-15 (2016)
9. E.S. Gridneva, Desulfuration of oil products under the influence of ultrasound: abstract. Wuchang. step. Cand.Tech.Sci., Moscow (2010)
10. A.V. Fkopyan, Oxidizing desulphurization of hydrocarbonic raw materials peroxide hydrogen in the presence of salts of transitional metals: abstract. Wuchang. step. Cand. Chem.Sci., Moscow (2015)

Maxwell Homogenization Scheme in Micromechanics: an Overview

Igor Sevostianov[1]

[1]Department of Mechanical and Aerospace Engineering, New Mexico State University, Las Cruces, NM 88003, USA France

Abstract. The paper discuss the classical Maxwell's (1873) homogenization method and the problems appearing in its practical implications. The method equates the far fields produced by a set of inhomogeneities and by a fictitious domain with unknown effective properties. It is re-written in terms of the property contribution tensors. The shape of the effective inclusion substantially affects the overall elastic properties. The choice of this shape in the case of anisotropic composite is a non-trivial problem. Another problem is related to the possible appearance of artificial singularities in the process of homogenization. We illustrate the method by several examples.

1 Introduction

The Maxwell scheme appears to be one of the best homogenization techniques in terms of its applicability to cases of anisotropic multiphase composites and accuracy. In his original work [1], Maxwell considered the effective conductivity of a matrix of conductivity k_0 containing spherical inhomogeneities of conductivity k_1 and volume fraction φ. He considered a sufficiently large spherical domain possessing the yet unknown effective conductivity k_{eff} embedded in the background material. He calculated the far-field asymptotics of the perturbation, by the inhomogeneities, of the externally applied electric field, in two different ways: (1) as a sum of far-fields generated by the small spheres treated as non-interacting ones, and (2) as the far-field generated by the mentioned large sphere (Fig. 1). Equating the two yields the effective conductivity in the form

$$k_{eff} = k_0 \frac{1+2\varphi\Psi}{1-\varphi\Psi} \tag{1}$$

where $\Psi \equiv (k_1 - k_0)/(k_1 + 2k_0)$. This relation is non-linear in concentration of inhomogeneities. Comparing it with results for a periodic arrangement, Lord Rayleigh [2] found that Maxwell's formula remains sufficiently accurate up to volume fractions as high as 0.4. As discussed in [3], the formula (1) has been misinterpreted by several later authors who linearized it with respect to φ – the operation that substantially worsened its accuracy [4, 5]:

$$k_{eff} = k_0(1+3\Psi\varphi) \tag{2}$$

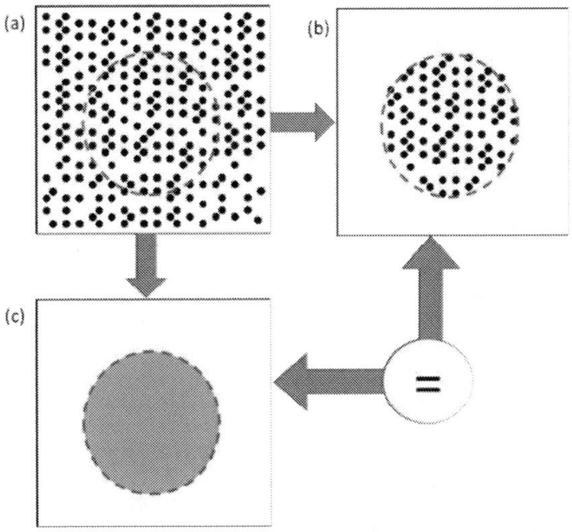

Fig. 1. Scheme of Maxwell's homogenization method. Effective properties of a composite (a) are calculated by equating effects produced by set of inhomogeneities embedded in the matrix material (b) and by fictitious domain having yet unknown effective properties (c).

This formula is sometimes erroneously identified with the original Maxwell's method (see, for example, [6]). Moreover, in contrast with the original Maxwell's result, formula (2) violates Hashin-Shtrikman bounds for the effective conductivity [3]. As further remark on the linearized formula (5.8.2), we note that Jeffrey [4], seeking to improve its accuracy, considered series expansion of k_{eff} in powers of φ and calculated the coefficient at the φ^2-term:

* Corresponding author: igor@nmsu.edu

$$\begin{aligned} k_{eff}/k_0 &= (1+3\Psi\varphi)+ \\ &+\varphi^2\left(3\Psi^2+\frac{3\Psi^3}{4}+\frac{9\Psi^3}{16}\frac{\alpha+2}{2\alpha+3}+\frac{3\Psi^4}{64}+\ldots\right)+o(\varphi^2) \end{aligned} \tag{3}$$

where $\alpha = k_1/k_0$. Fig. 2 compares Maxwell's original result (1) with the linearized version (2) and experimental data presented in [7] for (thermal) conductivity of epoxy containing particles of (a) silica ($\alpha = k_1/k_0 = 7.69$), (b) silica coated aluminum nitride (SCAN) ($\alpha = 1128$), and (c) alumina ($\alpha = 185$). The insets show shapes of the ceramic particles. The plots also show the result (3) (the coefficients at φ^2-term were taken from the plot in Jeffrey's paper as 1.89, 4.51, and 4.35 for silica, SCAN and alumina particles respectively) and, also, the Hashin-Shtrikman bounds for conductivity. Note that adding the quadratic term to the linearized version moves it closer to the *original* (non-linearized) Maxwell's formula but still violates the bounds.

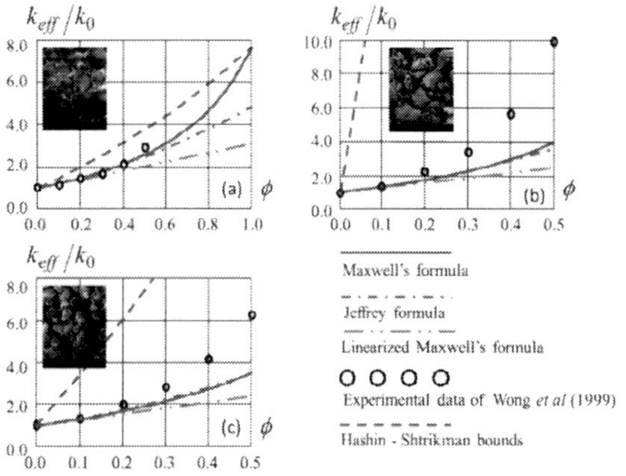

Fig. 2. Effective thermal conductivity of epoxy matrix filled with (a) silica, (b) SCAN (silica coated aluminum nitride) and (c) alumina particles: predictions by Maxwell's formula (1), its linearized variant (2), Jeffrey formula (3) are compared with experimental data of Wong and Bollampally (1999) and with Hashin–Shtrikman bounds for thermal conductivity. Insets show shape of particles of each material used for reinforcement.

Figure 2 also illustrates the importance of appropriate modeling of shapes of inhomogeneities. Indeed, the Maxwell's result for a composite with spherical silica particles coincides with experimental data up to the volume fraction of inhomogeneities 0.5. For SCAN and alumina particles having rather irregular shapes, modeling them as spherical ones leads to acceptable agreement with data at the volume fraction up to only 0.2.

Recently, Maxwell's scheme started to attract increasing attention. We mention the works [8, 9], where the scheme was formulated for the elastic properties of a material containing either random or aligned ellipsoidal inhomogeneities of identical aspect ratios. In [10], Maxwell' scheme was applied to finding the elastic, electric and poroelastic properties of composite containing identical parallel ellipsoidal inhomogeneities. Following [11], we apply Maxwell's approach to effective elastic properties of composites containing non-spherical particles of generally diverse shapes and of arbitrary orientation distribution. To this end, we reformulate Maxwell's assumptions using property contribution tensors. These tensors – that, as discussed throughout the present book, are fundamental quantities in the problem of effective properties – are particularly useful for the Maxwell's scheme. Indeed, the far fields entering the scheme are directly expressed in terms of these tensors (Section 3.2): the far-field asymptotics of the elastic field generated by an inhomogeneity determines its contribution to the effective elastic properties and vice versa.

2 Maxwell scheme in terms of property contribution tensors

According to Maxwell's idea, we evaluate far-field perturbations due to inhomogeneities in two different ways and equate the results. Firstly, we evaluate this field as the one generated by a homogenized region Ω possessing the (yet unknown) effective properties. As discussed in [12], this field can be expressed in terms of the compliance contribution tensor $\boldsymbol{H}^{eff}$ of the domain Ω possessing the unknown effective properties. Secondly, we consider the sum of far fields generated by all the individual inhomogeneities within Ω (treated as non-interacting ones); these fields are expressed in terms of $\boldsymbol{H}$-tensors of the said inhomogeneities. Equating the two quantities yields:.

$$\frac{V_\Omega}{V}\boldsymbol{H}^{eff} = \frac{1}{V}\sum_i V_i \boldsymbol{H}^i \tag{4}$$

The right hand side of the equation is known, however, the left hand side, besides being dependent on the effective properties, depends also on the shape of Ω. Explicit analytic expressions for $\boldsymbol{H}^{eff}$ are available for the ellipsoidal shapes only [13], and in this case equation (4) can be explicitly solved for the effective properties. In the following, we consider domains Ω of the ellipsoidal shape, assuming that it provides sufficient flexibility. Then equation (4) can be specified as follows [13]:

$$\frac{V}{V_\Omega}\left[\left(\boldsymbol{S}^{eff}-\boldsymbol{S}^0\right)^{-1}+\boldsymbol{Q}_\Omega\right] = V\left[\sum_i V_i \boldsymbol{H}^{(i)}\right]^{-1} \tag{5}$$

where $\boldsymbol{Q}_\Omega$ is fourth-rank Hill tensor [14, 15] that depends on the specific ellipsoidal shape of domain Ω. Inverting (5) yields the effective compliance:

$$\left(\boldsymbol{S}^{eff}-\boldsymbol{S}^0\right)^{-1} = \left[(1/V_\Omega)\sum_i V_i \boldsymbol{H}^{(i)}\right]^{-1} - \boldsymbol{Q}_\Omega \tag{6}$$

In this form, Maxwell's covers mixtures of inhomogeneities of diverse shapes. Note that the non-interaction approximation is recovered if the first term in the brackets is much larger than the second one; in other words, tensor Q_Ω is the quantity that describes the difference between Maxwell's scheme and the NIA. Alternatively, the Maxwell scheme can be formulated in terms of stiffness contribution tensors N.

We note certain logical inconsistency of the scheme: in ascribing to homogenized domain Ω the effective properties, interactions between inhomogeneities within Ω are assumed to be taken into account (otherwise, the scheme would have coincided with the NIA), whereas in considering the far fields generated by inhomogeneities, the interactions are disregarded. In other words, the left-hand part of the equation (4) is sensitive to the shape of Ω whereas the right-hand part is not. This is not fully consistent: the interaction effect experienced by an inhomogeneity can be characterized by placing it into the field $\sigma^\infty + \Delta\sigma$ where $\Delta\sigma$ is the correction due to interaction with neighbors within Ω, and the corresponding correction to the far field generated by this inhomogeneity is of the first order in $\Delta\sigma$. This discrepancy leads to sensitivity of the predictions to the shape of Ω (indeed, as shown in [16, 17], incorporating pair interactions into the Maxwell scheme eliminates the sensitivity of the prediction to the shape of Ω). The question arises on the optimal (from the viewpoint of accuracy of the scheme) choice of the shape. In the isotropic cases (fully random orientations, or spherical inhomogeneities), the spherical shape of Ω appears to be appropriate since there are no preferred directions. This choice may, possibly, be reasonable in cases of overall anisotropy as well (this assumes ascribing anisotropic properties, of the same symmetry as the effective ones, to the spherical domain Ω), provided the type of the overall anisotropy can be specified a priori.

3 The choice of shape of the effective inhomogeneity

As discussed above, the shape of the effective inhomogeneity is one of the key parameters in Maxwell homogenization scheme. This shape has to obey the following two requirements:

- The shape of Ω should be ellipsoidal. Otherwise, one cannot proceed from formula (4) to (5) and thus cannot separate the two effects, of the domain shape and of the elastic contrast;
- The shape of Ω should properly reflect the shapes of individual inhomogeneities, their orientations and properties as discussed in the text to follow.

We now specify the requirement B using certain known results: the one for isotropic composites [8, 18] and the one for a composite with parallel ellipsoidal inhomogeneities [9]. We utilize these results in conjunction with our observations on connections between the tensors entering expression (6), to come up with general recommendation regarding the shape of Ω (of the effective inhomogeneity in Maxwell scheme), for an arbitrary composite material (that may be anisotropic and may contain several sets of inhomogeneities).

3.1 Isotropic distribution of identical inhomogeneities

Since the only isotropic ellipsoidal shape is a sphere, the effective inhomogeneity has to be spherical so that

$$Q_\Omega = 3K_0\alpha_K\left(\frac{1}{3}II\right) + 2\mu_0\alpha_G\left(J - \frac{1}{3}II\right) \qquad (7)$$

where $J_{ijkl} = \left(\delta_{ik}\delta_{lj} + \delta_{il}\delta_{kj}\right)/2$, $II_{ijkl} = \delta_{ij}\delta_{kl}$ and

$$\alpha_K = \frac{2}{3}\frac{1-2\nu_0}{1-\nu_0} \quad \text{and} \quad \alpha_\mu = \frac{1}{15}\frac{7-5\nu_0}{1-\nu_0} \qquad (8)$$

Note that the shape of individual inhomogeneities may be arbitrary, not necessarily ellipsoidal. For example, Giraud and Sevostianov [19] used Maxwell scheme to calculate the effective elastic properties of a material containing randomly oriented concave superspherical pores. In the case of randomly oriented spheroids [11, 20]

$$\frac{K_{eff}}{K_0} = \frac{1-\alpha_K\sum_i \varphi_i B_i}{1+(1-\alpha_K)\sum_i \varphi_i B_i};$$

$$\frac{G_{eff}}{G_0} = \frac{1-\alpha_G\sum_i \varphi_i C_i}{1+(1-\alpha_G)\sum_i \varphi_i C_i} \qquad (9)$$

Formulas (9) coincide with the ones given in [8, 18, 21].

3.2 Parallel identical ellipsoidal inhomogeneities

In this case,

$$\frac{1}{V}\sum_i V_i \mathbf{H}^i = \varphi\mathbf{H} \qquad (10)$$

where φ is the volume fraction of the inhomogeneities. McCartney [9] showed that the best choice of the shape of the effective inclusion is the shape of an individual inhomogeneity; thus $Q_\Omega = Q$, where Q is Hill's tensor of an individual inhomogeneity. Then the expression (6) for the effective compliance takes the form

$$\begin{aligned}\mathbf{S}^{eff} &= \mathbf{S}^0 + \varphi\left\{(S^1 - S^0)^{-1} + (1-\varphi)Q\right\}^{-1} = \\ &= \mathbf{S}^0 + \varphi\left\{(1-\varphi)H^{-1} + \varphi(S^1 - S^0)^{-1}\right\}^{-1}\end{aligned} \qquad (11)$$

Thus, interactions in Maxwell's scheme are reflected only via tensor Q_Ω that depends on the shape of the inhomogeneities and elastic constants of the matrix and inhomogeneities (in the case of multiphase materials it also depends on the partial volume fractions of the phases).

4 Maxwell scheme as one of effective field methods

Maxwell's scheme can be considered as a version of the effective field method. Indeed:

$$\begin{aligned}\langle \boldsymbol{\varepsilon} \rangle &= \mathbf{S}^{eff} : \boldsymbol{\sigma}^{\infty} = \mathbf{S}^{0} : \boldsymbol{\sigma}^{\infty} + \left[\left(\sum_i \varphi_i \mathbf{H}^{(i)} \right)^{-1} - \mathbf{Q}^{\Omega} \right]^{-1} : \boldsymbol{\sigma}^{\infty} = \\ &= \mathbf{S}^{0} : \boldsymbol{\sigma}^{\infty} + \sum_i \varphi_i \mathbf{H}^{(i)} : \underbrace{\left[\mathbf{J} - \mathbf{Q}^{\Omega} : \left(\sum_i \varphi_i \mathbf{H}^{(i)} \right) \right]^{-1} : \boldsymbol{\sigma}^{\infty}}_{\boldsymbol{\sigma}^{eff}}\end{aligned} \tag{12}$$

In the context of geophysical applications, Kuster and Toksöz [21] proposed a model for wave propagation in a media with randomly oriented spheroidal inhomogeneities that – as far as the effective elastic properties are concerned – coincides with Maxwell's model, with the domain Ω taken to be spherical. In the case of spherical inhomogeneities, the Kuster-Toksöz' scheme also coincides with the Mori-Tanaka-Benveniste one [22].

References

1. J.C. Maxwell, *A Treatise on Electricity and Magnetism* (Clarendon Press, Oxford, 1873)
2. Lord Rayleigh, Philosophical Magazine, 34, 481-502 (1892)
3. I. Sevostianov and M. Kachanov, M. International Journal of Engineering Science, 58, 124-128 (2012)
4. D.J. Jeffrey, Proceedings of the Royal Society of London, A 335, 355-367 (1973)
5. K.Z. Markov, *Heterogeneous Media: Micromechanics Modeling Methods and Simulations*, (eds. K.Z. Markov and L. Preziozi) (Birkhauser, Boston, 1-162, 2000).
6. G.W. Milton, *The Theory of Composites* (Cambridge University Press, Cambridge, 2002)
7. C.P. Wong and R.S. Bollampally, Journal of Applied Polymer Science, 3396–3403, **74** (1999)
8. L.N. McCartney and A. Kelly, Proceedings of the Royal Society of London, A464, 423-446 (2008)
9. L.N. McCartney, Philosophical Magazine, 90, 4175 – 4207 (2010)
10. V. Levin, S. Kanaun, and M. Markov, International Journal of Engineering Science, 75–8661 (2012)
11. I. Sevostianov and A. Giraud, International Journal of Engineering Science, 64, 23-36 (2013)
12. I. Sevostianov and M. Kachanov, International Journal of Solids and Structures, 48, 2340-2348 (2011)
13. I. Sevostianov and M. Kachanov, Journal of the Mechanics and Physics of Solids, 50, 253-282 (2002)
14. R. Hill, Journal of the Mechanics and Physics of Solids, 11, 357-372 (1963)
15. T.T. Wu. International Journal of Solids and Structures, 2, 1-8 (1966)
16. V.I. Kushch, I. Sevostianov, and A.S. Belyaev, Mechanics of Materials, **89**, 1–11 (2015)
17. V.I. Kushch and I. Sevostianov, International Journal of Engineering Science, **98**, 36-50 (2016)
18. L. Shen and S. Yi, International Journal of Solids and Structures **38**, 5789-5805 (2001)
19. A. Giraud and I. Sevostianov, International Journal of Rock Mechanics and Mining Sciences, **62**, 23-27 (2013).
20. I. Sevostianov, Mechanics of Materials, **75** 45-59 (2014)
21. G.T. Kuster and M.N. Toksöz, Geophysics, **39**, 587-606 (1974)
22. J. Berryman and P. Berge, Mechanics of Materials, **22** 149-164 (1996)

Increase robustness of the method of diagnostics and identification of high-precision positioning systems

Anton Lankin[1,*], *Valery* Grechikhin[1], *Michail* Lankin [1]

[1]SRSPU (NPI), Department of Information and Measuring Systems and Technologies, 346428 Novocherkassk, Russia

Abstract. Today, high-precision positioning systems are used in various fields of science and technology, providing fast and accurate control of movements. Considering the fact that such systems perform important tasks, diagnostics and timely detection of defects is an urgent task. To effectively diagnose and identify defects of such systems, a method based on the transition to the space of principal components was developed. Such a method will make it possible to represent each characteristic in the form of a point in a new orthogonal space. Despite the possibility of using such a method, a drawback was revealed, which is the low stability of this method to the gross errors in the measurement. In this paper, we propose an approach that provides a significant increase in the stability of the method for diagnosing and identifying high-precision positioning systems.

1 Introduction

For diagnostics and identification of high-precision positioning systems in [1, 2, 3] method of principal components (PC) is used, one of the disadvantages of which is the sensitivity to gross errors in measurement (misses). The presence of several such errors greatly distorts the results of the diagnostic, or even completely makes them unreliable. Authors [4], the methods of increasing the robustness of the PC method were investigated [12].

As a high-precision positioning system, proportional electromagnets were considered. To analyze the quality of proportional electromagnets in [5, 6] it is proposed to use the dynamic magnetization characteristic (DCM). In this regard, in our opinion, it will be effective to increase the robustness of the diagnostic method by using an approximated polynomial of the sixth degree [7, 13].

2 Experimental studies of increasing robustness

An experiment was performed to illustrate this approach. A group of thirty-five DCM was formed [8, 9, 10, 11]. The first (DCM1) is obtained by measuring the proportional electromagnet. The following five characteristics (DCM2- DCM6) were obtained as a result of the addition to the DCM1 points of the random error of current measurement of 1, 2, 3, 4 and 5%. Four characteristics (DCM7- DCM10) received additions to the coordinates of one point DCM1 gross error in 5%, 10%, 15% and 20%. The following eight characteristics (DCM11- DCM18) are obtained similarly to the previous four, but adding the same errors at three and five points. The characteristics DCM19- DCM35 is the result of approximation by a sixth-degree polynomial, characteristics beginning with DCM2 according to DCM18.

For clarity, we will represent the DCM1 forms without the introduced error (Figure 1), DCM6 with an error of 5% (Figure 2) and DCM10 with one point of rough error of 20% (Figure 3). And also their shape after the approximation: DCM23 (Figure 4) and DCM27 (Figure 5).

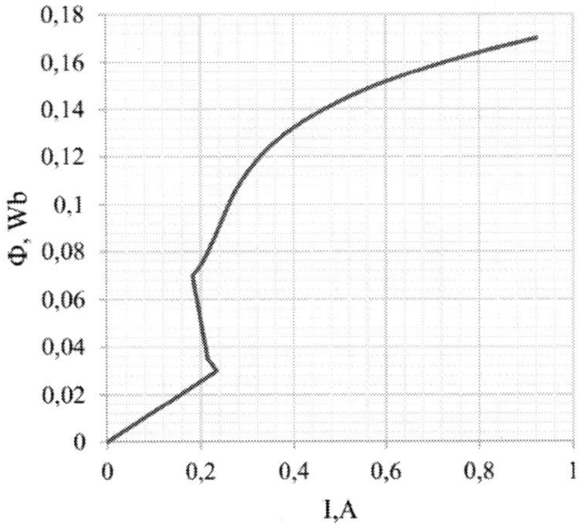

Fig. 1. DCM1 without the introduced error

* Corresponding author: lankinjohn@yandex.ru

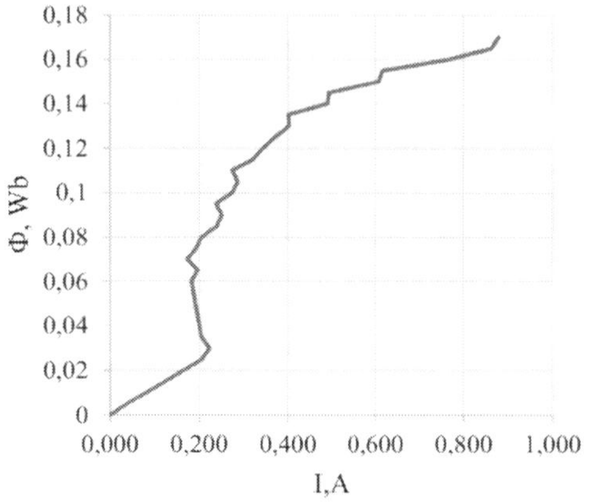

Fig. 2. DCM6 with a random error of 5%

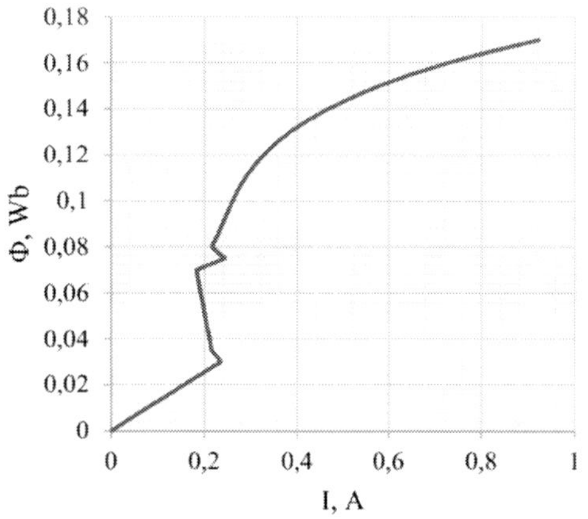

Fig. 3. DCM10 with a rough error introduced at one point 20%

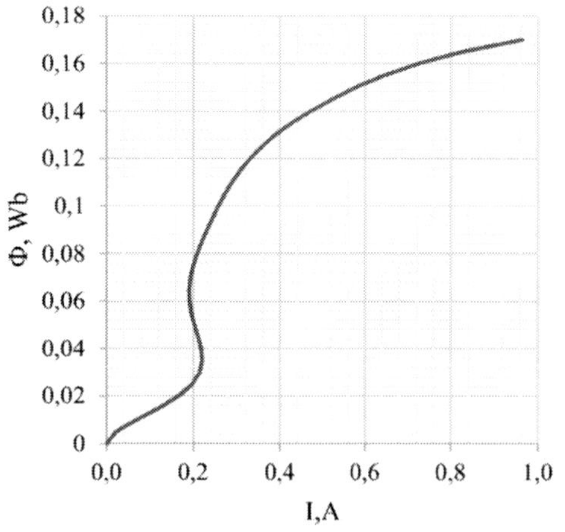

Fig. 4. Approximated DCM23 with an introduced error of 5%

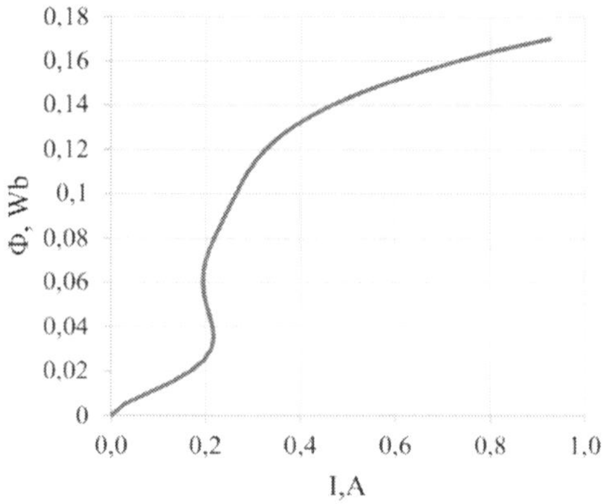

Fig. 5. Approximated DCM27 with a rough error introduced at one point 20%

When the various DCM are combined, it is precisely the approximated curves that are closest to the original DCM.

Consider how the DCM will behave in the course of projecting into the space of the PC [14, 15, 16]. The purpose of the experiment in the analysis of the location of the points of the corresponding DCM2- DCM35 in the PC space with respect to the point of the corresponding DCM1.

First we consider the influence of a random error. In Figure 7, the grey dot depicts DCM1 without error and DCM2- DCM6 with extrinsic errors, and the greater the error, the further the DCM from the original one.

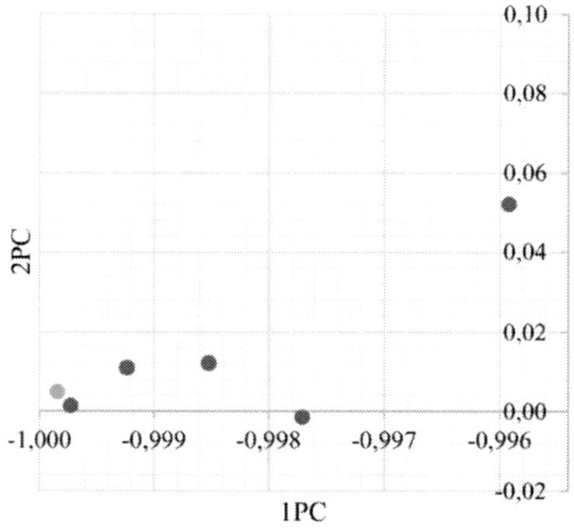

Fig. 7. Initial DCM1 and DCM2-DCM6 with errors from 1% to 5%

After approximating DCM containing random errors (Figure 8), the points corresponding to them in the PC space become much closer to each other than the original DCM1 (the distance for DCM with a maximum random error of 5% was reduced by a factor of 12).

DOI: 10.1051/matecconf/201713204001

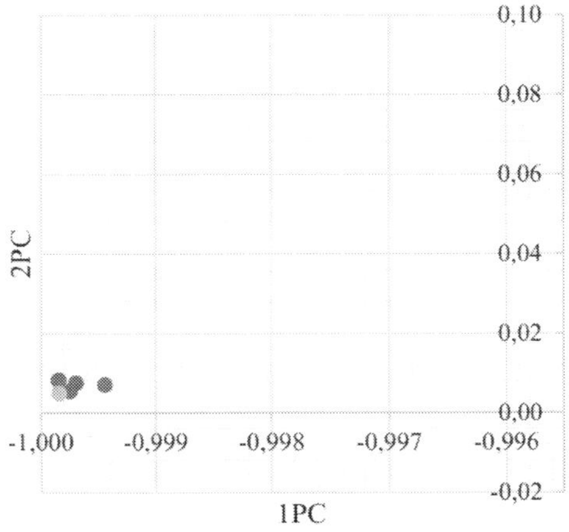

Fig. 8. Initial DCM1 and approximated DCM19-DCM23 with errors from 1% to 5%

Next, consider the effect of gross errors. The initial DCM1 and DCM with a rough error at one point are shown in Fig. 9. There are significant effects of deviation from the initial DCM1 for gross errors of 15% and 20%.

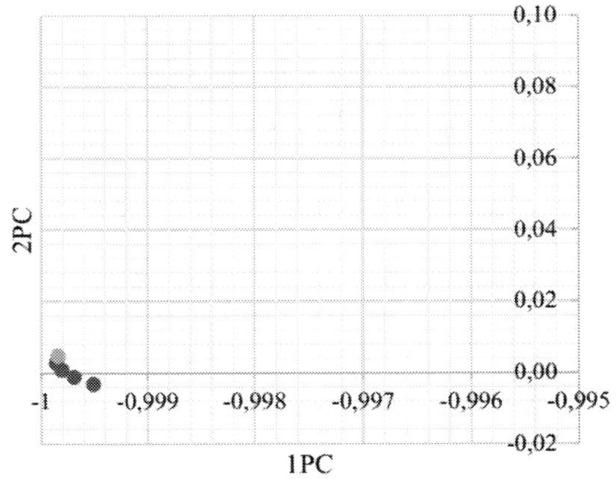

Fig. 9. Initial DCM1 and DCM7-DCM10 with a rough error at one point

After the approximation of the point in the space PC, as in the previous case, they become closer to each other (Figure 10).

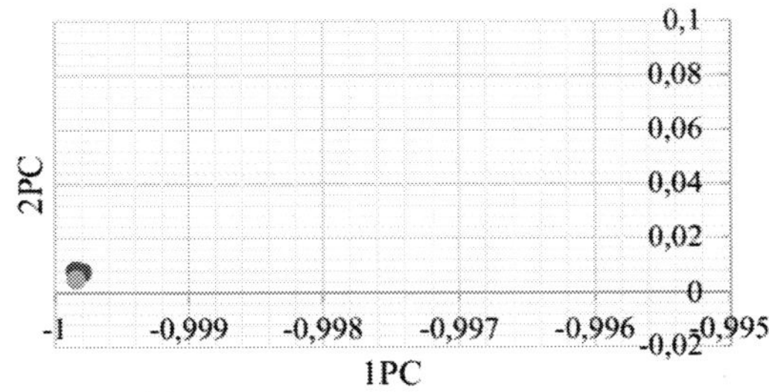

Fig. 10. Initial DCM1 and approximated DCM24-DCM27 with a rough error at one point

The introduction of a rough error at three and five points affects the location of DCM in the PC space much more strongly, as can be seen in Figures 11 and 12.

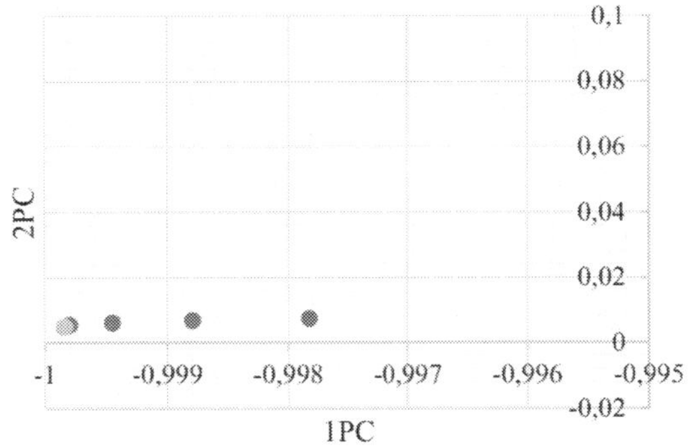

Fig. 11. Initial DCM1 and DCM11-DCM14 with a rough error at three points

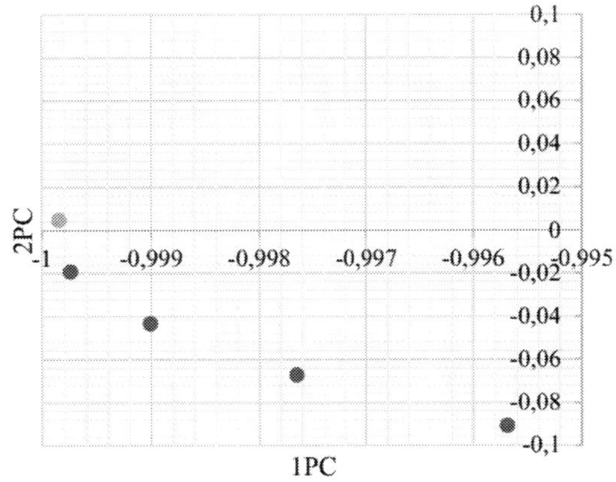

Fig. 12. Initial DCM1 and DCM15-DCM18 with a rough error at five points

After approximating DCM containing rough errors (Figs. 13 and 14), the points corresponding to them in the PC space become much closer to each other than the original DCM1 (the distance for DCM with a maximum gross error of 20% at five points was reduced by 25 times).

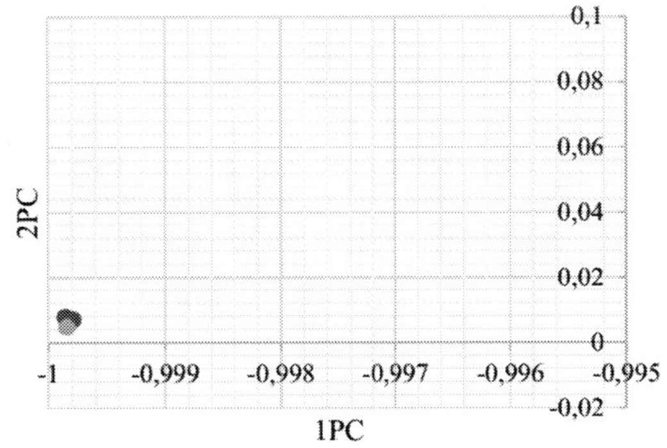

Fig. 13. Initial DCM1 and approximated DCM28-DCM31 with a gross error at three points

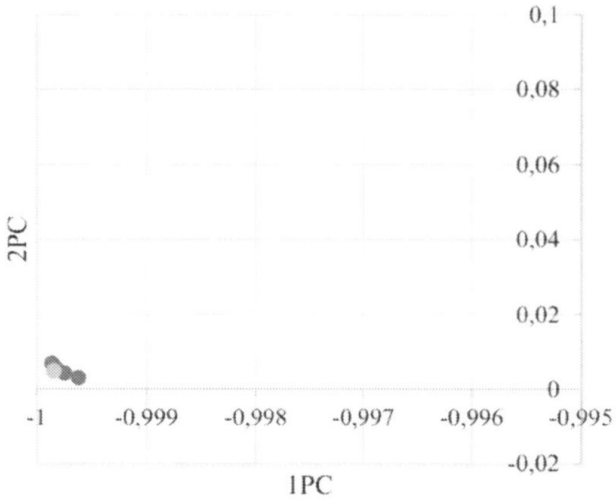

Fig. 14. Initial DCM1 and approximated DCM32-DCM35 with a gross error at five points.

This confirms the operability of the proposed approach for solving the problem of increasing the stability of the PC method in the presence of random and gross errors (misses).

3 Experimental studies of the influence of approximation

The use of polynomial approximation raises the question of its effect on the sensitivity of DCM to the defects of proportional electromagnets. In connection with this, an experiment was carried out to project into the space of PC DCM of various proportional electromagnets before and after the approximation [17, 18, 19].
For the comparison of DCM in the form of a series of points (magnetic flux, current) and polynomial approximation, DCM was selected for conditioning proportional electromagnets and a group of electromagnets with various defects (Table 3).

Table 3. Type of the DCM.

DCM number	Defect type
36-38	No
39-41	Incomplete operation
42-44	Stuck anchor in triggered position

For the nine DCM, the first two PC, describing 95% of the total variance (Table 4), are defined as a series of points and their corresponding characteristics in the form of a polynomial approximation.

Table 4 shows that there are minor deviations in the numerical values of DCM within ± 2.5% [20]. Let's construct the obtained values in the space of the first two PC (Figure 15).

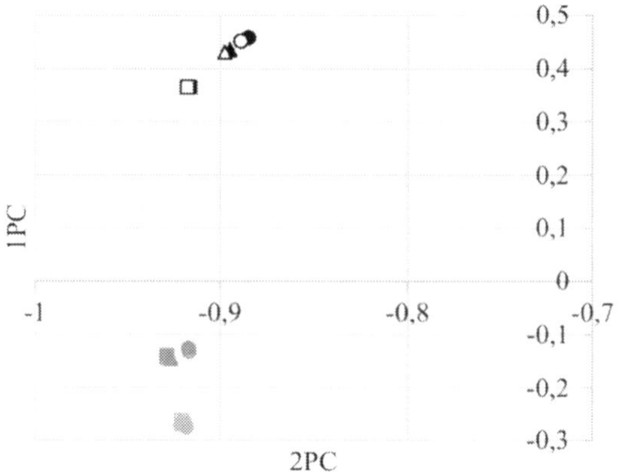

Fig. 15. Projections of DCM into the space of PC.

Table 4. The values of principal components.

Characteristic type / DCM number	DCM in the form of a series of points		Polynomial approximation of DCM	
	1 ГК	2 ГК	1 ГК	2 ГК
ДХН 36	-0,916	0,364	-0,918	0,364
ДХН 37	-0,895	0,435	-0,898	0,431
ДХН 38	-0,885	0,457	-0,889	0,451
ДХН 39	-0,920	-0,270	-0,921	-0,264
ДХН 40	-0,921	-0,269	-0,922	-0,262
ДХН 41	-0,918	-0,276	-0,919	-0,269
ДХН 42	-0,929	-0,140	-0,929	-0,144
ДХН 43	-0,917	-0,127	-0,917	-0,128
ДХН 44	-0,927	-0,144	-0,927	-0,146

4 Conclusion

Thus, the application of a proportional electromagnet to the DCM, the procedure for approximating a power polynomial of the sixth degree, makes it possible to increase the robustness of the diagnostic method and reduce the amount of measurement data without losing the sensitivity of DCM to the defects of proportional electromagnets.

The results of the work were supported by the project No. 2.7193.2017 / 8.9 "Development of the scientific basis for designing, identifying and diagnosing high-precision positioning systems using the inverse electrical engineering methodology" performed within the basic part of the state task.

References

1. A.M. Lankin, M.V. Lankin, M.Y. Lankina, IOP Conf. Ser.: MSE, **10**, 012074 (2017)
2. N.I. Gorbatenko, M.V. Lankin, Elektrotekhnika **8**, 55-58 (2004)

3. A. Lankin, D. Shaykhutdinov, V. Grechikhin, D. Schuchkin, N. Narakidze, N. Gorbatenko, Jour. of Eng. and Appl. Sci., **12(2)**, 363-373 (2017)
4. B.T. Polyak, M.V. Khlebnikov, Automation and Telemechanics, **3**, 130-148 (2017)
5. V.V. Grechikhin, M.V. Lankin, A.M Lankin, *Prom-Eng. WISTC*, 235-238 (2015)
6. N.I. Gorbatenko, A.M. Lankin, M.V. Lankin, *Procedia Engineering*, 129, 799-803 (2015)
7. A.M. Lankin, M.V. Lankin, M.Y. Lankina, *IOP Conf. Ser.: MSE,* **10**, 012075 (2017)
8. N.I. Gorbatenko, A.M. Lankin, M.V. Lankin, D.V. Shayhutdinov Inter. Jour. of Appl. Eng. Res., **10(3)**, 6509-6519 (2015)
9. A.M. Lankin, M.V. Lankin, N.I. Gorbatenko, D.V. Shayhutdinov MAS, **9(8)**, 247-261 (2015)
10. D.V. Shaikhutdinov, A.M. Lankin, N.D. Narakidze, V.V. Grechikhin, K.M. Shirokov, N.I. Gorbatenko Res. Jour. of Appl. Sci., **10(10)**, 555-557 (2015)
11. A.M. Lankin, M.V. Lankin, V.V. Grechikhin, D.V. Shaikhutdinov Res. Jour. of Appl. Sci., **10(10)**, 695-700 (2015)
12. N.I. Gorbatenko, A.M. Lankin, M.V. Lankin, *Procedia Engineering*, **150**, 1027-1031 (2016)
13. A.M. Lankin, M.V. Lankin, O.A. Naugolnov, *Procedia Engineering*, **150**, 1020-1026 (2016)
14. N.I. Gorbatenko, A.M. Lankin, M.V. Lankin, Met. Sci. and Heat Treat., 1-3 (2017)
15. N.I. Gorbatenko, A.M. Lankin, M.V. Lankin, Met. Sci. and Heat Treat., 4-6 (2017)
16. N.I. Gorbatenko, V.V. Grechikhin, M.V. Lankin, A.M. Lankin, Meas. Tech., **59(7)**, 747-751 (2016)
17. A.M. Lankin, M.V. Lankin, Nat. Ass. of Sch., **4-2 (9)**, 132-135 (2015)
18. A.M. Lankin, Mod. high tech., **9-3**, 420-424 (2016)
19. A.M. Lankin, Bas. res., **9-3**, 495-499 (2016)
20. M.V. Lankin, A.M. Lankin, *IOP Conf. Ser.: MSE*, **10**, 012031 (2016)

Mathematical modeling of wave processes and transport of bottom materials in coastal water areas taking into account coastal structures

Sofiya Protsenko[1,*], *Tatyana* Sukhinova[2]

[1]Don State Technical University, Rostov, Russia
[2]Southern Scientific Center of the Russian Academy of Sciences, Rostov, Russia

Abstract. The necessity of numerical modeling of the mechanisms of formation of the field of suspension and sediment flow determines the urgency of the problem under study. The process of developing shallow coastal zones, with constructive transformation and changes in coastal reliefs, is becoming increasingly important. Assessment of the hydrodynamic impact on shore protection structures and coastal structures installed at the bottom of a shallow water body is an important task at present. With a constructive transformation of the reliefs, one should take into account the dynamics of the formation of the coast investigate the formation of the profile of the bottom in the coastal zone of the reservoir under the influence of wave processes. For determining the dynamics and trend of phenomena occurring in the coastal waters and predicting possible interventions in the ecosystem, the need to build mathematical models of the transport processes of matter in the coastal waters, taking into account the presence of coastal structures, is actualized.

1 Introduction

There is a significant number of studies of dynamic processes occurring in the coastal zone of water bodies at present. Laboratory experiments performed by Russian and foreign scientists have revealed many regularities of formation of fields of a suspension under the influence of currents. The quantitative estimation and prediction of the distribution of suspended matter can be carried out using mathematical modeling, taking into account laboratory data obtained earlier.

Despite the large number of experimental and theoretical works aimed at the study of the specific dynamic processes of the coastal zone, the effectiveness of the proposed approaches are far from practical necessity. The difficulties of the research include the narrow scope of the models constructed on the basis of laboratory studies, the complexity of conducting experiments in a real sea, and, as a result, data limitations comprehensive full-scale measurements. Obtaining analytical solutions due to the nonlinearity and multidimensionality of the equations of Navier-Stokes equations underlying the considered problems, possible only in a very limited number of cases.

One of the most effective methods of investigation of real processes of hydrodynamics at the present time is the numerical simulation. For mathematical modeling of hydrodynamic processes in the reservoirs is an urgent problem of construction and practical use of computationally-efficient methods, the use of which would allow to obtain sufficiently accurate approximate numerical solution. Mathematical modeling of natural systems, including shallow ponds, in addition to, and in many cases allows to exclude expensive full-scale experiments with a real ecosystem. The use of mathematical models allows to predict the possible effects of changes in bottom topography due to natural climatic conditions or human intervention in the ecosystem of the pond. The development of information technology allows to significantly increase the scope and increase the complexity of the modeled task to expand the circle of users.

Wave processes are the main source of energy in the coastal zone of the reservoir, which significantly influences its dynamics; due to the fact that a successful field experiments with natural systems is difficult, and laboratory simulations, has a narrow specialization, mathematical modeling becomes an important tool for the analysis and management of coastal water systems; for the prediction of dynamic processes of the coastal zone there is a need of mathematical modeling of processes of movement of sediment in areas characterized by where the water is shallow under the influence of surface gravity waves.

There is a need to build models of three-dimensional wave processes and sediment transport, adapted for the simulation of hydrodynamic wave processes in the presence of various coastal structures, the scope of which is to build the velocity field and pressure of water environment, assessment of hydrodynamic impacts on coast protection works and coastal structures in the presence of surface waves, placed on the bottom of a shallow pond.

* Corresponding author: rab55555@rambler.ru

In accordance with the intended purpose, it is necessary to solve following tasks: to justify the importance of solving problems, to carry out a review of existing approaches to the problem; to analyze existing models of sediment transport in aquatic ecosystems, and to analyse the approaches to mathematical modeling of these processes in aquatic ecosystems, classifying describing their mathematical models; to develop a model of sediment transport in shallow waters, allowing to predict the dynamics of the bottom relief; to develop a set of interrelated models, three-dimensional wave processes and sediment transport, adapted for hydrodynamic modelling of wave processes in the presence of various shoreline structures and to assess the hydrodynamic impact on coast protection works and coastal structures in the presence of surface waves.

2 Mathematical model of sediment transport

To describe the dynamics of marine sediments, the equations that describe the reformation of the coastal zone of water bodies are applied in the work, where water and solid particles move in one direction. The equations of the process of sediment transport [1-5] can be written in the following form:

$$(1-\varepsilon)\frac{\partial H}{\partial t}+\frac{\partial Q_x}{\partial x}+\frac{\partial Q_y}{\partial y}=0, \qquad (1)$$

$$\vec{Q}=\begin{cases} A\varpi d|\vec{\psi}|^{\beta-1}\vec{\psi}, & |\tau|\geq\tau_{bc}, \\ 0, & |\tau|<\tau_{bc}; \end{cases} \quad \vec{\psi}=\frac{\vec{\tau}}{(\rho_1-\rho_0)gd},$$

where H – the depth of the bottom, measured from the unperturbed surface of the reservoir; ε – the porosity of the soil; $\vec{Q}=\{Q_x,Q_y\}$ – the sediment load, $|\vec{Q}|=Q$; x, y – the horizontal Cartesian coordinates; τ_b – the tangential stress at the bottom; τ_{bc} – the critical value of the tangential stress at which the sediment transfer starts, g – acceleration of gravity, ρ – the fluid density, A and β – dimensionless constants (A = 19.5, β = 3), ϖ – the wave frequency, d – the characteristic of precipitation.

The shear stress for an inclined bottom surface is

$$\vec{\tau}=\vec{\tau}_b-\alpha\sin S\vec{n}, \qquad (2)$$

where $S(x,y,t)$ – the acute angle between the vector of the normal to the surface of the bottom and the vector of the gravitational force at time t, $\vec{n}$ – the unit vector directed towards the depth gradient, $\alpha\sin S$ – the additional tangential stress at the bottom of the reservoir, caused by gravitational forces. In case of $|\vec{\tau}_b|=0$, $|\vec{\tau}|=\tau_{bc}$ we have the equality: $\tau_{bc}=\alpha\sin\varphi_0$, φ_0 – the angle of the natural slope of the soil in the water.

Thus, the system of equations for the Schilds parameter takes the following form:

$$\vec{\psi}_S=\frac{\vec{\tau}_b-\sin S\tau_{bc}\vec{n}/\sin\varphi_0}{(\rho_1-\rho_0)gd}, \quad \vec{n}tgS=gradH, \qquad (3)$$

where ψ_S – Childs' parameter for the sloping bottom.

Taking into account the restrictions on tangential stresses at the bottom of the computational domain, the system of equations (1)-(3) takes the form [3-5]:

$$(1-\varepsilon)\frac{\partial H}{\partial t}+div(k\vec{\tau}_b)= =div\left(k\left(\tau_{bc}/\sin\varphi_0\right)gradH\right), \qquad (4)$$

$$k=\frac{A\varpi d}{((\rho_1-\rho_0)gd)^{\beta}}\left|\vec{\tau}_b-\frac{\tau_{bc}}{\sin\varphi_0}gradH\right|^{\beta-1}\cdot h\left(\left|\vec{\tau}_b-\frac{\tau_{bc}}{\sin\varphi_0}gradH\right|-\tau_{bc}\right), \qquad (5)$$

where $h(x)=\begin{cases}1, & x\geq 0\\ 0, & x<0\end{cases}$ – the Heaviside's function.

The equation (4) is supplemented by the initial condition:

$$H(x,y,0)=H_0(x,y). \qquad (6)$$

At the boundary of the calculated region, there is no flow caused by the influence of gravitational forces:

$$H_0'(x,y)=0. \qquad (7)$$

The information about the field of the velocity vector at the bottom of the reservoir is necessary to calculate the tangential stresses in the model of sediment transport. The system of shallow water equations is used to describe the movement of water and environment [6-9].

On the basis of the constructed algorithms were developed software intended for modeling of sediment transport in shallow waters [2-3]. The developed software, used to calculate two-dimensional velocity vector of the flow of the water environment, takes into account the turbulent exchange, the geometry of the seabed, wind currents and friction on the bottom.

Fig. 1-2 shows the results of numerical experiments modeling the dynamics of changes in bottom topography. The size of the grid is 200x200, step of spatial variable 0.1 m, time step 0.01 s, speed wind 5 m/s and directed from left to right. Fig. 1 shows the initial bottom topography (bottom) and the position of the free surface (top).

The simulation of sediment transport has shown in about 20 minutes later, the geometry of the bottom begins to take on a conical structure. The velocity of the fluid flow the maximum in the peak part of the region (at minimum depths). A function of the elevation of the level takes positive values in the left part of the computational region (the windward side) and negative on the right. After 40 minutes there is the wavy structure of the function elevate the level not only in the left part, but in the region of peak depth. After 60 minutes – enhanced oscillation functions elevate the level of the left part and in the region of the peak depth. The ridges are formed at the bottom of the computational region.

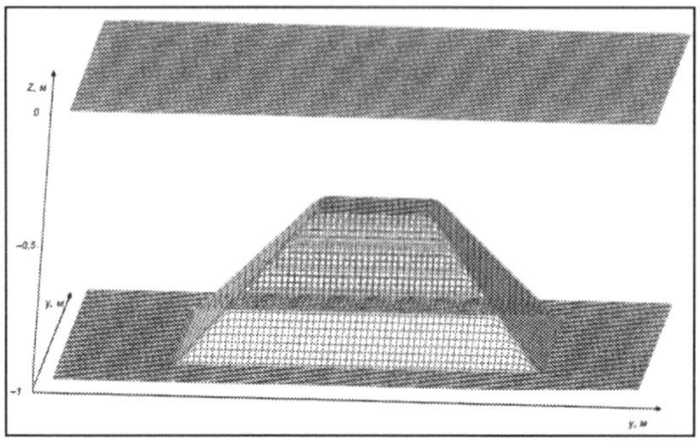

Fig. 1. The initial bottom topography and the position of the free surface

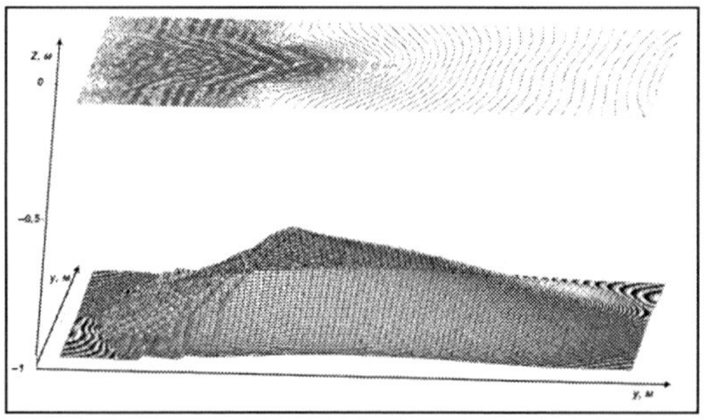

Fig. 2. The geometry of the bottom and the function of the elevation of the level after 80 minutes

After 80 minutes region with peak values of depth continues to shift to the left, there is an expansion region in the direction perpendicular to the movement of the wind (Fig. 2).

After 100 minutes region with peak values of depth continues to shift to the left, is the expansion of the area to the sides perpendicular to the motion of the wind, a function of the elevation of the level becomes highly oscillatory structure on the windward side, as well as the expansion of the area to the side. The results of the experiment allow us to analyse the dynamics of changes in the geometry of the bottom, the function of the elevation of the level of education wave patterns and sediment transport. This mathematical model and developed software allow to predict dynamics of change of the bottom topography, the appearance of a sea of ridges, their growth and transformation.

3 Mathematical model of wave hydrodynamics

For the mathematical description of the problem of wave dynamics of the fluid source equations are [1,2]:

– the Navier-Stokes equation

$$u'_t + uu'_x + vu'_y = -\frac{P'_x}{\rho} + (\mu u'_x)'_x + (\eta u'_y)'_y, \qquad (8)$$

$$v'_t + uv'_x + vv'_y = -\frac{P'_y}{\rho} + (\mu v'_x)'_x + (\eta v'_y)'_y + g; \qquad (9)$$

– the continuity equation for incompressible fluid:

$$u'_x + v'_y = 0. \qquad (10)$$

Equations (8)-(10) are considered with the following boundary conditions:

– at the bottom of the region:

$$P'_n(x,y,t) = 0, \quad V_n(x,y,t) = 0,$$

$$\rho\eta u'_y(x,y,t) = -\tau_x(t),$$

$$\rho\mu v'_x(x,y,t) = -\tau_y(t), \qquad (11)$$

– on the surface of the liquid:

$$u'_n(x,y,t) = 0, \qquad v(x,y,t) = P'_t / g\rho,$$

$$P'_n(x,y,t) = 0; \qquad (12)$$

– on a side border:

$$u'_n(x,y,t) = 0, \quad v'_n(x,y,t) = 0,$$

$$P'_n(x,y,t) = \frac{\rho\Pi}{\tau L}; \qquad (13)$$

where $V = (u,v)$ – the velocity vector of the aquatic environment; P – the pressure; μ, η – the turbulent exchange coefficients in the horizontal and vertical directions, respectively; g – acceleration of gravity; ρ – the fluid density; τ_x, τ_y – the components of the tangential stress at the bottom of the liquid; Π – the flux of the velocity vector through the lateral surface; L – the distance from the surface of the liquid to the bottom (the depth of the liquid taking into account the elevation of the level) at the lateral boundary. The axis Ox is combined with the surface of the unperturbed liquid and is directed toward the shore, the axis Oy – vertically down.

The tangential stress caused by bottom friction, according to the Van-Dorn law, is calculated by the formulas:

$$\tau_x = \rho C_p(|V|)u|V|, \qquad \tau_y = \rho C_p(|V|)v|V|, \qquad (14)$$

where $C_p(|V|) = \begin{cases} 0.0088, & |V| < 6{,}6 \text{ м/с}, \\ 0.0026, & |V| \geq 6{,}6 \text{ м/с} \end{cases}$ – the dimensionless coefficient.

The calculated area is inscribed in a rectangle. For numerical realization of the discrete mathematical model of the formulated wave hydrodynamics problem, a uniform grid is introduced:

$$w_h = \{t^n = n\tau, x_i = ih_x, y_j = jh_y;$$

$$n = \overline{0, N_t}, i = \overline{0, N_x}, j = \overline{0, N_y};$$

$$N_t \tau = T, N_x h_x = l_x, N_y h_y = l_y\},$$

where τ – time step, h_x, h_y – steps to space, N_t – the upper bound on the time, N_x, N_y – borders in space.

The calculated cells are rectangles, they can be filled, partially filled, or empty. The cell centers and nodes are $h_x/2$ and $h_y/2$ by coordinates x and y respectively. We denote by $o_{i,j}$ «fullness» of the cell (i,j). The degree of cell fullness is determined by the pressure of the liquid column inside this cell. If the average pressure at the nodes that belong to the vertices of the cell in question is greater than the pressure of the liquid column inside the cell, then the cell is considered to be full $(o_{i,j} = 1)$. In this case, the «fullness» of the cell can be calculated by the following formula::

$$o_{i,j} = \frac{P_{i,j} h(P_{i,j}) + P_{i-1,j} h(P_{i-1,j})}{\rho g h_y}, \quad (15)$$

The results of numerical experiments for calculating the movement of the aquatic environment in the coastal water area are shown in Fig. 3, which shows the change in the profile of the wave coming ashore in a different period of time, while the program displays the vectors of the velocity field, the palette shows the pressure field.

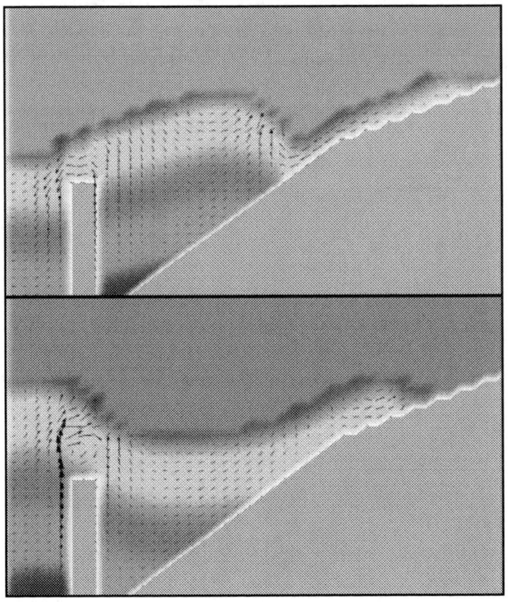

Fig. 3. The dynamics of the change in the profile of the elevation level of a liquid in the case of an artificial obstacle (breakwater)

4 Mathematical model of wave processes, taking into account the presence of above-water coastal structures

A three-dimensional hydrostatic model was used to construct a two-dimensional mathematical model of the motion of an aquatic environment. The initial equations of hydrodynamics are:

– Navier-Stokes euations system

$$u'_t + uu'_x + vu'_y + wu'_z = -\frac{1}{\rho}P'_x + (\mu u'_x)'_x + (\mu u'_y)'_y + (\eta u'_z)'_z,$$

$$v'_t + uv'_x + vv'_y + wv'_z = -\frac{1}{\rho}P'_y + (\mu v'_x)'_x + (\mu v'_y)'_y + (\eta v'_z)'_z, \quad (16)$$

– the continuity equation for incompressible fluid

$$u'_x + v'_y + w'_z = 0, \quad (17)$$

– the equation of hydrostatics

$$P = \rho g(z + \xi). \quad (18)$$

The system of equations (16)-(18) is considered with the following boundary conditions: at the bottom the condition of impenetrability and friction $\rho_v \eta u'_n = \tau_{x,b}(t)$, $\rho_v \eta v'_n = \tau_{y,b}(t)$, $V_n = 0$, on the surface sets level rise and wind stress $\rho\eta u'_n = -\tau_{x,p}(t)$, $\rho\eta v'_n = -\tau_{y,p}(t)$, $w = -\xi'_t$, on the lateral boundaries, the condition of sliding without friction $u'_n = 0$, $v'_n = 0$, $\xi'_n = 0$, where ξ – the raise function level (function of elevation), $V = \{u, v, w\}$ – the velocity vector of the aquatic environment, P – the pressure, μ, η – the turbulent exchange coefficients in the horizontal and vertical directions, respectively, g – the acceleration of gravity, ρ – the density of the fluid, τ_x, τ_y – the tangential stress at the bottom of the fluid, axis *Ox*, *Oy* have a horizontal direction.

The equation calculate the pressure in hydrostatic approximation in the presence of surface design, installed on the bottom of the reservoir [14]:

$$P = \rho g(z + \chi) + P_g = \rho g(z + \chi + P_g / \rho g).$$

The function of elevation in the case of a surface body on the surface is $\xi = \chi + P_g / \rho g$, where χ – the function describing the geometry of the bottom of the surface body.

The constructed program complex allows to specify the shape and intensity of the source of oscillations, as well as the geometry of the above-water object. The results of numerical experiments on the simulation of the propagation of wave hydrodynamic processes in the flow of an aqueous medium around a surface body, taking into account the geometries of the bottom of the object located in the liquid and the bottom of the reservoir, are shown in Fig. 5.

As an example of the practical use of a problem-oriented program complex, the problem of calculating the hydrodynamic effect of support waves on structures is solved. Dimensions of the above-water structure: width 5 m, length 10 m, immersion depth 20 cm. The structure is installed on the bottom of the reservoir with 6 supports. The selected modeling site has dimensions of 50 by 50 m and a depth of 1 m. The disturbance source is set at some distance from the above-water object.

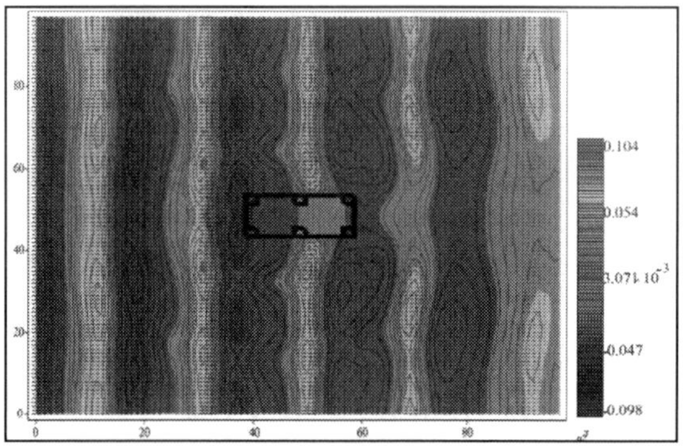

Fig. 4. Movement of the aquatic environment

At the initial time, the liquid is at rest. It is required to determine the subsequent movement of the aquatic environment in the presence of a surface object on the surface and the hydrodynamic forces on the supports of the structure. To solve the problem, a grid of 100x100 dimensions was used, the time step is 0,01 seconds.

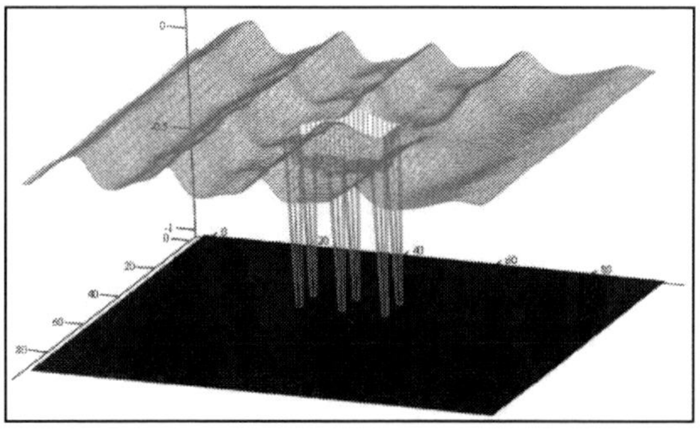

Fig. 5. The function of the elevation of the level when flowing over the water environment of a surface body having a support

Fig. 4, 5 shows that during the propagation of a plane wave that meets an obstacle in the form of a surface of the body, the reflection of wave oscillations from a fixed object, which in turn leads to a change of the wave profile. The source of vibrations distributed along the left border and has a sinusoidal form.

The results of numerical experiments on modeling the distribution of hydrodynamic wave processes and provide an opportunity to assess the impact of waves on structures having a support on the bottom of the pond. Fig. 6 shows the hydrodynamic force effects on the support surface of the structure installed on the bottom of the reservoir (Fig. 5, 6). The y-axis indicates the power loads in tons, the abscissa marks the time measured from the beginning of the oscillations in seconds.

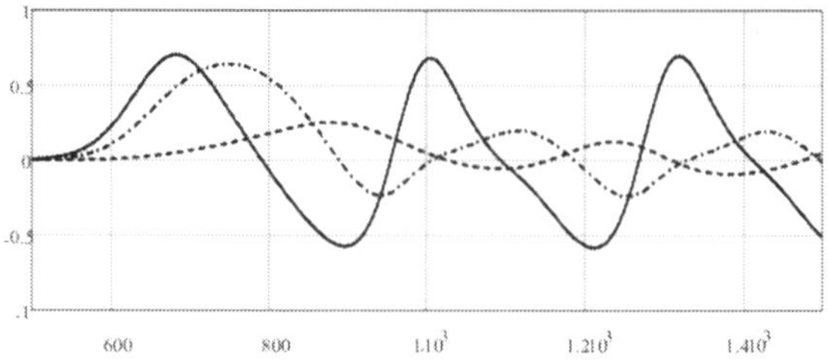

Fig. 6. Power loads on the supports of the above-water structure: 1 - on the front pair from the incoming wave, 2 - on the central pair, 3 - on the distant pair of supports

5 Conclusion

Models of sediment transport in coastal water systems are used in works related to protection from port and canal entry, construction of hydraulic structures, predictions of power impacts on coastal structures and bank protection structures.

The novelty of the scientific idea consists in the use of models of wave processes and transport of bottom materials in the coastal waters, taking into account the availability of coastal infrastructure facilities for studying hydrophysical processes in coastal water systems, as well as for constructing the velocity and pressure field of the aquatic environment, and assessing the hydrodynamic effect on shore protective structures and coastal structures in the presence of surface waves, installed on the bottom of a shallow water body.

Application of the research results is related to the planning of rational nature management: the construction of facilities and the use of specific sections of the shoreline, predicting possible reformation of the bottom.

The work was carried out with the financial support of the Program of Fundamental Research of the Presidium of the Russian Academy of Sciences I.33P, project No. 0256-2015-0078.

References

1. E. Alekseenko, B. Roux, A. Sukhinov, R. Kotarba, D. Fougere, Computers and Fluids, **77**, 24-35, (2013)
2. E. Alekseenko, B. Roux, A. Sukhinov, R. Kotarba, D. Fougere, Nonlinear Processes in Geophysics, **20** (2), 189-191 (2013)
3. N. Buzalo, P. Ermachenko, T. Bock, A. Bulgakov, A. Chistyakov, A. Sukhinov, E. Zhmenya, N. Zakharchenko, *Procedia Engineering*, **85**, 84-93 (2014)
4. I. O. Leontev, *Pribrezhnaya dinamika: volny, techeniya, potok osadkov*, (GEOS, Moscow, 2001) [in Russian]

5. A.V. Nikitina, A.I. Sukhinov, G.A. Ugolnitsky, A.B. Usov, A.E. Chistyakov, M.V. Puchkin, I.S. Semenov, Mathematical Models and Computer Simulations, **9** (1), 101-107 (2017)
6. E. A. Protsenko, Engineering journal of Don, **13**, No. 3, 23-31 (2010) [in Russian]
7. E. A. Protsenko, News of southern Federal University. Technical Sciences, **97**, No. 8, 71-75 (2009)
8. A.I. Sukhinov, A.E. Chistyakov, I.I. Levin, I.S. Semenov, A.V. Nikitina, A.A. Semenyakina, *5th International Conference on Informatics, Electronics and Vision,* 1128- 1133 (2016)
9. A. I. Sukhinov, News of TRTU, № 3 (58), 228-235 (2006) [in Russian]
10. A. I. Sukhinov, *Dvumernyye skhemy rasshchepleniya i nekotoryye ikh prilozheniya* (MAKS Press, Moscow, 2005) [in Russian]
11. A. I. Sukhinov, A. E. Chistyakov, E. A. Protsenko, Izvestiya SFedU. Engineering Sciences, №8 (121), 32-44 (2011)
12. A.I. Sukhinov, A.E. Chistyakov, Mathematical Models and Computer Simulations, **4**, 398-409 (2012)
13. A.I. Sukhinov, A.E. Chistyakov, A.V. Nikitina, A.A. Semenyakina, I. Korovin G. Schaefer, *2016 5th International Conference on Informatics, Electronics and Vision,* 1134-1139 (2016)
14. A.I. Sukhinov, A.E. Chistyakov, A.A. Semenyakina, A.V. Nikitina, Computer researches and modeling, **8**, No. 1, 151-168 (2016)
15. A.I. Sukhinov, A.E. Chistyakov, A.V. Shishenya, Mathematical Models and Computer Simulations, **6,** No. 3, 324-331 (2014)
16. A.I. Sukhinov, A.E. Chistyakov, E.F. Timofeeva, A.V. Shishenya, Mathematical Models and Computer Simulations, **5**, No. 2, 122-129 (2013)
17. A.I. Sukhinov, D.S. Khachunts, A.E. Chistyakov, Computational Mathematics and Mathematical Physics, **55**, No. 7, 1216-1231 (2015).
18. A.I. Sukhinov, A.V. Shishenya, Mathematical Models and Computer Simulations, **5**, N 3, 257-265 (2013)
19. V. S. Vasiliev, A. I. Sukhinov, Mathematical modelling, **15**, No. 10, 17-34 (2003)
20. E.V. Yakushev, A.I. Sukhinov, Yu.F. Lukashev, S.Yu. Fomin, V.G. Yakubenko, Oceanology, **43**, No 1, 39-47 (2003)

Investigation of nonlinear 2D bottom transportation dynamics in coastal zone on optimal curvilinear boundary adaptive grids

Alexander Sukhinov[1,*], *Alexander* Chistyakov[1], and *Valentina* Sidoryakina[2]

[1]Don State Technical University, Rostov-on-Don, Russian Federation
[2]Taganrog Institute after A.P. Chekhov (branch) RGEU (RINE), Taganrog, Russian Federation

Abstract. One of the practically important tasks of hydrophysics for sea coastal systems is the problem of modeling and forecasting bottom sediment transportation. A number of problems connected to ship safety traffic, water medium condition near the coastal line etc. depends on forecasting bottom deposit transportation under natural and technogenic influences. Coastal systems are characterized by a complicated form of coastline - the presence of long, narrow and curvilinear peninsulas and bays. Water currents and waves near the beach are strongly depend on complicated coastal line and in turn, exert on the bottom sediment transportation near the shore. The use of rectangular grids in the construction of discrete models leads to significant errors in both the specification of boundary conditions and in the modeling of hydrophysical processes in the coastal zone. In this paper, we consider the construction of a finite-element approximation of the initial-boundary value problem for the spatially two-dimensional linearized equation of sediment transportation using optimal boundary-adaptive grid. First, the linearization of a spatially two-dimensional nonlinear parabolic equation on the time grid is performed-when the coefficients of the equation that are nonlinearly dependent on the bottom relief function are set on the previous time layer, and the corresponding initial conditions are used on the first time layer. The algorithm for constructing the grid is based on the procedure for minimizing the generalized Dirichlet functional. On the constructed grid, finite element approximation using bilinear basis functions is performed, which completes the construction of a discrete model for the given problem. The using of curvilinear boundary adaptive grids leads to decreasing of total grid number in 5-20 times and respectively the total modeling time and/or it allows to improve modeling accuracy.

1 Introduction

One of the essential features of coastal and marine systems is complicated form of the coastline, which hardly influences on mathematical modelling of these systems. The use of rectangular, does not allow to take into account, with practical value accuracy, the presence narrow sea coastal bodies of complicated shape, such as bays, estuaries, straits, peninsulas etc. At the same time, the presence of these objects hardly determines the structure of coastal currents, especially for shallow water bodies of the South of Russia, similar to the Azov Sea and Taganrog bay. The development of optimal curvilinear grids, consistent with the complex shape of the shoreline (boundary-adaptive grids), as shown by the practice of numerical solution of problems of this class, can significantly improve the accuracy of the solution obtained and / or reduce the amount of computation. The paper deals with the construction of a finite-element approximation of the initial-boundary value problem for the spatially two-dimensional linearized equation of sediment transport on a boundary-adaptive grid. First, the linearization of a spatially two-dimensional nonlinear parabolic equation on the time grid is performed, when the coefficients of the equation that depend nonlinearly on the bottom depth function are set on the previous time layer. The rationale for this method of linearization (the correctness of the formulation of the problem) and the convergence of the solutions of the linearized chain of problems in the norm of the space L^1 with the step of the time grid tending to 0 was justified in the previous works of the authors [1-8]. The algorithm for constructing the grid is based on the procedure for minimizing the generalized Dirichlet functional. Previously, this approach has shown its effectiveness in the construction of minimal non degenerate grids for Z-shaped test regions, such as the "Maltese cross", etc., as well as in the numerical solution of problems of hydrodynamics of coastal systems, including in the Azov Sea [9]. On the constructed grid, in this paper, the finite element method is approximated using the bilinear basis functions of the linearized chain of sediment transport problems applied to the Azov Sea and the Taganrog Bay.

*Corresponding author: sukhinov@gmail.com

2 The formulation of the nonlinear initial-boundary value problem for two-dimensional models of sediment transport

Let us consider the equation of sediment transport according to the [10-15]:

$$(1-\varepsilon)\frac{\partial H}{\partial t}=div\left(k\frac{\tau_{bc}}{\sin\varphi_0}gradH\right)-div(k\boldsymbol{\tau}_b), \quad (1)$$

where $H=H(x,y,t)$ is the water depth; ε is the porosity of bottom materials; $\boldsymbol{\tau}_b$ is the vector of tangential stress at the water bottom; τ_{bc} is a critical value of the tangential stress; $\tau_{bc}=a\sin\varphi_0$, φ_0 is an angle of repose of soil in the water; $k=k(H,x,y,t)$ is the nonlinear coefficient, determined by the ratio:

$$k\equiv\frac{A\tilde{\omega}d}{((\rho_1-\rho_0)gd)^{\beta}}\left|\boldsymbol{\tau}_b-\frac{\tau_{bc}}{\sin\varphi_0}gradH\right|^{\beta-1}, \quad (2)$$

where ρ_1, ρ_0 are the density of bottom material and water particles respectively; g is the gravity acceleration; $\tilde{\omega}$ is the wave frequency; A and β are the dimensionless constants; d are characteristic dimensions of soil particles.

Let's $D\subset R^n$ is an area where a process occurs, and S is boundary with a piecewise smooth line. We considered a cylinder $\amalg_T=D\times(0,T)$ of height T with the base D as the definitional domain of the equation (1). Its border consists of the lateral surface $S\times[0,T]$ and two bases: the upper base $\bar{D}\times\{0\}$ and the lower base $\bar{D}\times\{T\}$. Further, for simplicity, the equation (1) was considered in a rectangular domain $D(x,y)=\{0<x<L_x, 0<y<L_y\}$.

We added the initial condition to the equation (1), assuming that the initial conditions of function belong to the corresponding class of smoothness:

$$\begin{gathered}H(x,y,0)=H_0(x,y), H_0(x,y)\in C^2(D)\cap C(\bar{D}),\\ grad_{(x,y)}H_0\in C(\bar{D}),(x,y)\in\bar{D}.\end{gathered} \quad (3)$$

The conditions on the border domain $\bar{D}$ is following:

$$\left|\boldsymbol{\tau}_b\right|\Big|_{y=0}=0, \quad (4)$$

$$H(L_x,y,t)=H_2(y,t), 0\le y\le L_y, \quad (5)$$

$$H(0,y,t)=H_1(y,t), 0\le y\le L_y, \quad (6)$$

$$H(x,0,t)=H_3(x), 0\le x\le L_x, \quad (7)$$

$$H(x,L_y',t)=H_4(x,t), 0\le x\le L_x, L_y'<L_y. \quad (8)$$

In addition to the boundary conditions (5)–(8), we assume the fulfillment of their smoothness conditions – the existence of continuous derivatives on the boundary of the domain $\bar{D}$:

$$grad_{(x,y)}H\in C(\bar{\amalg}_T)\cap C^1(\amalg_T). \quad (9)$$

We consider that there's a liquid layer of finite thickness always in this area and occur for the specified time interval at no drainage area, i.e.

$$H(x,y,t)\ge c_0\equiv const>0, 0\le x\le L_x, 0\le y\le L_y', 0\le t\le T. \quad (10)$$

The condition of non-degeneracy of the operator has the form:

$$k\ge k_0=const>0, \forall(x,y)\in\bar{D}, 0<t\le T. \quad (11)$$

The vector of tangential stress at the bottom is calculated using the unit vectors of the coordinate system:

$$\boldsymbol{\tau}_b=\bar{i}\tau_{bx}+\bar{j}\tau_{by},\ \tau_{bx}=\tau_{bx}(x,y,t),\ \tau_{by}=\tau_{by}(x,y,t). \quad (12)$$

3 Linearization of the sediment transportation equation on the time grid

We will assume that the time interval $[0,T]$ is divided into time layers by values $t_n, n=1,...,N$.

The values of the required function $H(x,y,t)$ and coefficient k of the differential equation (1) on the n-th time layer (i.e. at $t=t_n$) will be denoted by $H^n(x,y,t)$ and k^n respectively. These quantities are no longer time-dependent t, but remain functions of spatial coordinates, i.e.

$$H^n=H^n(x,y)=H^n(x,y,t_n), k^n=k^n(x,y)=k^n(x,y,t_n).$$

Further consideration of the problem is carried out on two time layers: the lower one $t=t_{n-1}$, on which the distribution of the unknown function $H(x,y,t)$ is known, (when $n=0$ the distribution is determined by the initial condition (3) $H^0(x,y)$), and the upper temporal layer $t=t_n$, on which the distribution of the desired function – $H^n(x,y)$.

Reasoning is similarly repeated for any layer $t_n, n=1,...,N$.

When $t=t_n$ equation (1) is written in the form:

$$(1-\varepsilon)\frac{\partial H^n}{\partial t}-div\left(k^{n-1}\frac{\tau_{bc}}{\sin\varphi_0}gradH^n\right)=-div(k^{n-1}\boldsymbol{\tau}_b), \quad (13)$$

and we supplement it with the initial conditions:

$$H^1(x,y,t_0)=H_0(x,y), H^n(x,y,t_{n-1})=H^{n-1}(x,y,t_{n-1}), \quad (x,y)\in\bar{D}, n=2,...,N. \tag{14}$$

Member of the species $div(k^{n-1}\tau_b)$ will be a known function of the right-hand side. The boundary conditions (3) - (11) are assumed to be satisfied for any $t=t_n$, $n=1,...,N$.

We also assume that the functions $H_0(x,y)$ and $H^1(x,y,t_0)$ from the initial conditions (14) are given by their values at the nodes of the spatial grid.

Consider the approximation of the differential equation (16) in time using the implicit scheme:

$$(1-\varepsilon)\frac{H^n-H^{n-1}}{\tau}=div\left(k^{n-1}\frac{\tau_{bc}}{\sin\varphi_0}gradH^n\right)-div(k^{n-1}\tau_b), \quad n=1,...,N, \tag{15}$$

where $\tau=t^n-t^{n-1}$.

When the time is sampled for the parameter τ a limitation is imposed on the type of the Courant criterion.

4 The construction of a discrete bottom sediment transportation model using the finite element method

4.1 The method of constructing an optimal boundary-adaptive grid (based on the example of the Sea of Azov and Taganrog Bay)

Then the optimal boundary-adaptive grid, constructed for the water area of the Azov Sea and Taganrog Bay, is used in the calculations. The grid is considered as a necessary auxiliary step in solving the problem of sediment transport using the finite element method.

To construct the grid, we used cartographic differential operators and the corresponding net functionals described in [16-18].

In the plane (x,y) a grid is under construction $\omega=\{(x_{i,j},y_{i,j}), i=\overline{1,N_1}, j=\overline{1,N_2}\}$ by the given coordinates of the boundary nodes. The grid area separates D on elementary subdomains in the form of quadrangles. Then we move from the variables x,y to variables ξ,η, given in the field Ξ, in which elementary subregions become "standard". On surface (ξ,η) a grid is under construction $\tilde{\omega}=\{(\xi_{i,j},\eta_{i,j}), i=\overline{1,N_1}, j=\overline{1,N_2}\}$, which is defined by functions $x=x(\xi,\eta)$, $y=y(\xi,\eta)$, displaying a rectangle $1\le\xi\le N_1, 1\le\eta\le N_2$ in the plane (ξ,η) on a curvilinear quadrangle in the plane (x,y).

For transformations

$$(\xi,\eta)\to(x,y): x=x(\xi,\eta), y=y(\xi,\eta), \tag{16}$$

$$(x,y)\to(\xi,\eta): \xi=\xi(x,y), \eta=\eta(x,y), \tag{17}$$

we require the difference from zero jacobians

$$J(\xi,\eta)=\frac{\partial x}{\partial\xi}\frac{\partial y}{\partial\eta}-\frac{\partial x}{\partial\eta}\frac{\partial y}{\partial\xi}, \tag{18}$$

$$\tilde{J}(x,y)=\frac{\partial\xi}{\partial x}\frac{\partial\eta}{\partial y}-\frac{\partial\xi}{\partial y}\frac{\partial\eta}{\partial x}. \tag{19}$$

For convenience of perception, it is advisable to represent the local consideration of grids by considering the corresponding cells $\omega,\tilde{\omega}$. Cell nodes from ω we set the coordinates $(x_{i,j},y_{i,j})$, $(x_{i,j+1},y_{i,j+1})$, $(x_{i+1,j+1},y_{i+1,j+1})$, $(x_{i+1,j},y_{i+1,j})$.

Bypassing nodes with coordinates $(x_{i,j},y_{i,j})$, $(x_{i,j+1},y_{i,j+1})$, $(x_{i+1,j+1},y_{i+1,j+1})$, $(x_{i+1,j},y_{i+1,j})$ should be performed counter-clockwise (when viewed along the direction of acceleration of free fall).

Continuous transformation $(x,y)\to(\xi,\eta)$ is defined on each concrete cell by the formulas

$$\begin{aligned} x=x(\xi,\eta)&=x_{i,j}+(x_{i+1,j}-x_{i,j})\xi+(x_{i,j+1}-x_{i,j})\eta+\\ &+(x_{i+1,j}-x_{i+1,j+1}-x_{i+1,j}+x_{i,j})\xi\eta,\\ y=y(\xi,\eta)&=y_{i,j}+(y_{i+1,j}-y_{i,j})\xi+(y_{i,j+1}-y_{i,j})\eta+\\ &+(y_{i+1,j}-y_{i+1,j+1}-y_{i+1,j}+y_{i,j})\xi\eta. \end{aligned} \tag{20}$$

The cells are convex quadrangles, where each of the angles is less than π, which ensures the positivity of the jacobian

$$\begin{aligned} J(\xi,\eta)=&\left(x_{i+1,j}-x_{i,j}+(x_{i+1,j}-x_{i+1,j+1}-x_{i+1,j}+x_{i,j})\eta\right)\cdot\\ &\cdot\left(y_{i+1,j+1}-y_{i,j}+(y_{i+1,j}-y_{i+1,j+1}-y_{i+1,j}+y_{i,j})\xi\right)-\\ &-\left(x_{i+1,j+1}-x_{i,j}+(x_{i+1,j}-x_{i+1,j+1}-x_{i+1,j}+x_{i,j})\xi\right)\cdot\\ &\cdot\left(y_{i+1,j}-y_{i,j}+(y_{i+1,j}-y_{i+1,j+1}-y_{i+1,j}+y_{i,j})\eta\right) \end{aligned}$$

and makes restrictions when choosing the shape of the cell.

By numbering the vertices of the cell from 1 to 4 clockwise from the node ij, we denote the doubled areas of the triangles $J_{1i,j}, J_{2i,j+1}, J_{3i+1,j+1}, J_{4i+1,j}$. For the node (i,j) we have $J_{i,j}=\frac{1}{4}(J_1+J_2+J_3+J_4)_{i,j}$.

It is obvious that the condition for the jacobian to be positive J, is equivalent to the positivity of the quantities J_1,J_2,J_3,J_4. At least one of these quantities will be zero on the boundary of the convex mesh constructed.

The procedure for constructing grids is connected with the consideration of the Dirichlet functional

$$I=\iint g dxdy=\iint \frac{g}{J} d\xi d\eta, \qquad (22)$$

$g=g_{11}+g_{22}$, $g_{11}=\left(\frac{\partial \xi}{\partial x}\right)^2+\left(\frac{\partial \xi}{\partial y}\right)^2$ and $g_{22}=\left(\frac{\partial \eta}{\partial x}\right)^2+\left(\frac{\partial \eta}{\partial y}\right)^2$ – components of the metric tensor of the coordinate transformation.

The Euler equations for the functional (22) can be written as a system of nonlinear equations with respect to the unknown functions $x=x(\xi,\eta)$, $y=y(\xi,\eta)$. A discrete analogue of a functional is the one obtained by summing over all four angles (triangles formed when a quadrilateral cell is partitioned by its diagonals), all grid cells, i.e.

$$\begin{aligned} I^h=\frac{1}{4}\sum_{i=1}^{N_1-1}\sum_{j=1}^{N_2-1}\Big\{\Big[&\left(x_{i,j+1}-x_{i,j}\right)^2+\left(y_{i,j+1}-y_{i,j}\right)^2+\\ &+\left(x_{i,j}-x_{i+1,j}\right)^2+\left(y_{i,j}-y_{i+1,j}\right)^2\Big]J^{-1}_{1i,j}+\Big[\left(x_{i+1,j+1}-x_{i,j+1}\right)^2+\\ &+\left(y_{i+1,j+1}-y_{i,j+1}\right)^2+\left(x_{i,j+1}-x_{i,j}\right)^2+\left(y_{i,j+1}-y_{i,j}\right)^2\Big]J^{-1}_{2i,j+1}+\\ &+\Big[\left(x_{i+1,j}-x_{i+1,j+1}\right)^2+\left(y_{i+1,j}-y_{i+1,j+1}\right)^2+\left(x_{i+1,j+1}-x_{i,j+1}\right)^2+\\ &+\left(y_{i+1,j+1}-y_{i,j+1}\right)^2\Big]J^{-1}_{3i+1,j+1}+\Big[\left(x_{i,j}-x_{i+1,j}\right)^2+\left(y_{i,j}-y_{i+1,j}\right)^2+\\ &+\left(x_{i+1,j}-x_{i+1,j+1}\right)^2+\left(y_{i+1,j}-y_{i+1,j+1}\right)^2\Big]J^{-1}_{4i+1,j}\Big\}. \end{aligned} \qquad (23)$$

Expression (23) can be obtained by replacing $x(\xi,\eta), y(\xi,\eta)$ in (22) by $x^h(\xi,\eta), y^h(\xi,\eta)$ approximations by quadrature formulas of integrals over square cells in the plane ξ,η.

Solution of the system

$$\frac{\partial I^h}{\partial x_{i,j}}=0,\ \frac{\partial I^h}{\partial y_{i,j}}=0 \qquad (24)$$

gives at least one solution - a convex grid. To find it, we minimize the function I^h.

When a grid is built on m - stepiteration, then $(m+1)$ - step it is determined from the solution of two linear equations written for each internal node [17]:

$$\begin{cases} \mu\dfrac{\partial J^{h(m)}}{\partial x_{i,j}}+\dfrac{\partial\left(\dfrac{\partial J^{h(m)}}{\partial x_{i,j}}\right)}{\partial x_{i,j}}\left(x^{(m+1)}_{i,j}-x^{(m)}_{i,j}\right)-\dfrac{\partial\left(\dfrac{\partial J^{h(m)}}{\partial x_{i,j}}\right)}{\partial y_{i,j}}\left(y^{(m+1)}_{i,j}-y^{(m)}_{i,j}\right)=0,\\ \mu\dfrac{\partial J^{h(m)}}{\partial x_{i,j}}+\dfrac{\partial\left(\dfrac{\partial J^{h(m)}}{\partial y_{i,j}}\right)}{\partial x_{i,j}}\left(x^{(m+1)}_{i,j}-x^{(m)}_{i,j}\right)-\dfrac{\partial\left(\dfrac{\partial J^{h(m)}}{\partial y_{i,j}}\right)}{\partial y_{i,j}}\left(y^{(m+1)}_{i,j}-y^{(m)}_{i,j}\right)=0, \end{cases}$$

where μ - iteration parameter, $\mu=1$.

We note that the necessary condition for a minimum of the functional is a consequence of the conformality of the transformation under consideration.

Minimization I^h on the set of convex meshes is realized by means of a quasi-Newtonian procedure in the computational domain (ξ,η). To do this, it is expedient to use the well-proven gradient methods, tested and described in [9] with respect to the water area of the Azov Sea and Taganrog Bay.

Change of the sign in the denominators of the integrands of the jacobian (22) - (23) functionals J, leads to discontinuities of the second kind, which is critical for second-order methods (quasi-newtonian procedure). To exit this situation near the degeneracy boundary($J=0$) coordinate transformation is performed J by

$$J'=\max(J,\varepsilon)=(J-\varepsilon)_+, \qquad (25)$$

ε - a sufficiently small number, $\varepsilon>0$.

It is important to note that when generating a grid, using the above procedure, the initial approximation in the form of a convex grid is specified. If this is not the case, then the initial grid-convex grid is determined by successively solving optimization problems.

Note that since cells with numbers containing $i=1,N_1$, $j=1,N_2$ are boundary, then for them the position of the nodes is not optimized. Therefore, further consideration of the problem is carried out at the nodes $i=\overline{2,N_1-1}$, $j=\overline{2,N_2-1}$.

We introduce the notation

$$f(J_k)=\frac{1+\left(1-\frac{J_k}{J'_k}\right)+\left(1-\frac{J_k}{J'_k}\right)^2}{J'_k}=\begin{cases}\dfrac{1}{J_k},\ J_k\ge\varepsilon,\\ \dfrac{1+\left(1-\frac{J_k}{\varepsilon}\right)+\left(1-\frac{J_k}{\varepsilon}\right)^2}{\varepsilon},J_k\le\varepsilon,\end{cases}\quad k=\overline{1,4}.$$

Taking into account the last equality, expression (23) can be represented in the following form

$$I^{h*}=\sum_{i=2}^{N_1-1}\sum_{j=2}^{N_2-1}\left[g_1 f(J_1)+g_2 f(J_2)+g_3 f(J_3)+g_4 f(J_4)\right]_{i,j}, \qquad (26)$$

where

$$J_{1i,j}=\left(x_{i+1,j}-x_{i,j}\right)\left(y_{i,j+1}-y_{i,j}\right)-\left(y_{i+1,j}-y_{i,j}\right)\left(x_{i,j+1}-x_{i,j}\right),$$

$$J_{2i,j}=\left(x_{i+1,j}-x_{i,j}\right)\left(y_{i,j}-y_{i,j-1}\right)-\left(y_{i+1,j}-y_{i,j}\right)\left(x_{i,j}-x_{i,j-1}\right),$$

$$J_{3i,j}=\left(x_{i,j}-x_{i-1,j}\right)\left(y_{i,j}-y_{i,j-1}\right)-\left(y_{i,j}-y_{i-1,j}\right)\left(x_{i,j}-x_{i,j-1}\right),$$

$$J_{4i,j}=\left(x_{i,j}-x_{i-1,j}\right)\left(y_{i,j+1}-y_{i,j}\right)-\left(y_{i,j}-y_{i-1,j}\right)\left(x_{i,j+1}-x_{i,j}\right),$$

$$g_{1i,j}=\frac{\left(x_{i,j+1}-x_{i,j}\right)^2+\left(y_{i,j+1}-y_{i,j}\right)^2+\left(x_{i,j}-x_{i+1,j}\right)^2+\left(y_{i,j}-y_{i+1,j}\right)^2}{4},$$

$$g_{2i,j}=\frac{\left(x_{i+1,j}-x_{i,j}\right)^2+\left(y_{i+1,j}-y_{i,j}\right)^2+\left(x_{i,j}-x_{i,j-1}\right)^2+\left(y_{i,j}-y_{i,j-1}\right)^2}{4},$$

$$g_{3i,j}=\frac{\left(x_{i,j-1}-x_{i,j}\right)^2+\left(y_{i,j-1}-y_{i,j}\right)^2+\left(x_{i,j}-x_{i-1,j}\right)^2+\left(y_{i,j}-y_{i-1,j}\right)^2}{4},$$

$$g_{4i,j}=\frac{\left(x_{i-1,j}-x_{i,j}\right)^2+\left(y_{i-1,j}-y_{i,j}\right)^2+\left(x_{i,j}-x_{i,j+1}\right)^2+\left(y_{i,j}-y_{i,j+1}\right)^2}{4}.$$

Solution of a system of algebraic equations of the form (24) for I^{h*} representing the necessary condition for the minimum of the functional I^{h*}, gives at least one solution (convex grid).

Further, inordertominimize I^{h*} they used the same quasi-Newtonian procedure as when minimizing the function I^h.

To determine the gradient components of a discrete functional I^{h*} represented locally in the point $(x_{i,j}, y_{i,j})$, the following derivatives were required:

$$\left(\frac{\partial J_1}{\partial x}\right)_{i,j}=-\left(y_{i,j+1}-y_{i,j}\right)+\left(y_{i+1,j}-y_{i,j}\right)=y_{i+1,j}-y_{i,j+1},$$

$$\left(\frac{\partial J_2}{\partial x}\right)_{i,j}=-\left(y_{i,j}-y_{i,j-1}\right)-\left(y_{i+1,j}-y_{i,j}\right)=y_{i,j}-y_{i+1,j},$$

$$\left(\frac{\partial J_3}{\partial x}\right)_{i,j}=\left(y_{i,j}-y_{i,j}\right)-\left(y_{i,j}-y_{i-1,j}\right)=y_{i-1,j}-y_{i,j-1},$$

$$\left(\frac{\partial J_4}{\partial x}\right)_{i,j}=\left(y_{i,j+1}-y_{i,j}\right)+\left(y_{i,j}-y_{i-1,j}\right)=y_{i,j+1}-y_{i-1,j},$$

$$\left(\frac{\partial g_1}{\partial x}\right)_{i,j}=\frac{1}{2}\left\{\left(-x_{i,j+1}+x_{i,j}\right)+\left(x_{i,j}-x_{i+1,j}\right)\right\},$$

$$\left(\frac{\partial g_2}{\partial x}\right)_{i,j}=\frac{1}{2}\left\{\left(-x_{i+1,j}+x_{i,j}\right)+\left(x_{i,j}-x_{i,j-1}\right)\right\},$$

$$\left(\frac{\partial g_3}{\partial x}\right)_{i,j}=\frac{1}{2}\left\{\left(-x_{i,j-1}+x_{i,j}\right)+\left(x_{i,j}-x_{i-1,j}\right)\right\},$$

$$\left(\frac{\partial g_4}{\partial x}\right)_{i,j}=\frac{1}{2}\left\{\left(-x_{i-1,j}+x_{i,j}\right)+\left(x_{i,j}-x_{i,j+1}\right)\right\},$$

$$\left(\frac{\partial J_1}{\partial y}\right)_{i,j}=-\left(x_{i+1,j}-x_{i,j}\right)+\left(x_{i,j+1}-x_{i,j}\right)=x_{i,j+1}-x_{i+1,j},$$

$$\left(\frac{\partial J_2}{\partial y}\right)_{i,j}=\left(x_{i+1,j}-x_{i,j}\right)+\left(x_{i,j}-x_{i,j-1}\right)=x_{i+1,j}-x_{i,j-1},$$

$$\left(\frac{\partial J_3}{\partial y}\right)_{i,j}=\left(x_{i+1,j}-x_{i,j}\right)-\left(x_{i+1,j}-x_{i+1,j-1}\right)=x_{i+1,j-1}-x_{i,j},$$

$$\left(\frac{\partial J_4}{\partial y}\right)_{i,j}=-\left(x_{i,j}-x_{i-1,j}\right)-\left(x_{i,j+1}-x_{i,j}\right)=x_{i-1,j}-x_{i,j},$$

$$\left(\frac{\partial g_1}{\partial y}\right)_{i,j}=\frac{1}{2}\left\{\left(-y_{i,j+1}+y_{i,j}\right)+\left(y_{i,j}-y_{i+1,j}\right)\right\},$$

$$\left(\frac{\partial g_2}{\partial y}\right)_{i,j}=\frac{1}{2}\left\{\left(-y_{i+1,j}+y_{i,j}\right)+\left(y_{i,j}-y_{i,j-1}\right)\right\},$$

$$\left(\frac{\partial g_3}{\partial y}\right)_{i,j}=\frac{1}{2}\left\{\left(-y_{i,j-1}+y_{i,j}\right)+\left(y_{i,j}-y_{i-1,j}\right)\right\},$$

$$\left(\frac{\partial g_4}{\partial y}\right)_{i,j}=\frac{1}{2}\left\{\left(-y_{i-1,j}+y_{i,j}\right)+\left(y_{i,j}-y_{i,j+1}\right)\right\},$$

$$(f_J')_k=\frac{1+2\left(1-\frac{J_k}{J_k'}\right)}{(J_k')^2}=\begin{cases}-\frac{1}{(J_k)^2}, J_k\geq\varepsilon,\\ -\frac{1+2\left(1-\frac{J_k}{\varepsilon}\right)}{\varepsilon^2}, J_k\leq\varepsilon, k=\overline{1,4}.\end{cases}$$

Taking into account the expressions obtained, we find the components of the gradient of the functional

$$I^{h*}=\left\{g_1 f(J_1)+g_2 f(J_2)+g_3 f(J_3)+g_4 f(J_4)\right\}_{ij},$$

соответствующие декартовым координатам $(x_{i,j}, y_{i,j})$ node (i,j):

$$\left(\frac{\partial I^{h*}}{\partial x}\right)_{i,j}=\left(\sum_{k=1}^{4}\frac{\partial g_k}{\partial x}f(J_k)+g_k\frac{\partial f(J_k)}{\partial J_k}\frac{\partial J_k}{\partial x}\right)_{i,j}, \quad (27)$$

$$\left(\frac{\partial I^{h*}}{\partial y}\right)_{i,j}=\left(\sum_{k=1}^{4}\frac{\partial g_k}{\partial y}f(J_k)+g_k\frac{\partial f(J_k)}{\partial J_k}\frac{\partial J_k}{\partial y}\right)_{i,j}. \quad (28)$$

Expressions (27) - (28) are determined by substituting in them the corresponding values of the derivatives from.

In Fig. 1 schematically represents the calculated grid (Azov Sea - 200 × 80, Taganrog Bay - 108 × 40 knots), the number of cells of which significantly exceeds the analogue of its predecessors.

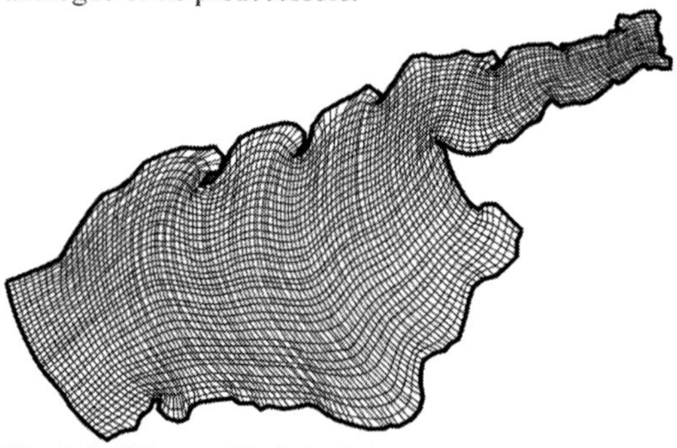

Fig. 1. Building a grid of physical area.

4.2 Approximation of the sediment transport equation using the finite element method

Equation (15) can be written in the form

$$-\left(\frac{\partial}{\partial x}\left(k^{n-1}\frac{\tau_{bc}}{\sin\varphi_0}\frac{\partial H^n}{\partial x}\right)-\frac{\partial}{\partial y}\left(k^{n-1}\frac{\tau_{bc}}{\sin\varphi_0}\frac{\partial H^n}{\partial y}\right)\right)+ \\ +\frac{(1-\varepsilon)}{\tau}H^n=\frac{(1-\varepsilon)}{\tau}H^{n-1}-\left(\frac{\partial}{\partial x}\left(k^{n-1}\tau_{bx}\right)+\frac{\partial}{\partial y}\left(k^{n-1}\tau_{by}\right)\right) \quad (29)$$

and note that in the new coordinates

$$H^n\equiv H^n\left(x(\xi,\eta),y(\xi,\eta)\right),\ H^{n-1}\equiv H^n\left(x(\xi,\eta),y(\xi,\eta)\right),$$
$$k^{n-1}\equiv k^{n-1}\left(x(\xi,\eta),y(\xi,\eta)\right).$$

Using the expressions for the derivatives

$$\frac{\partial H}{\partial x}=\frac{\partial H}{\partial\xi}\frac{\partial\xi}{\partial x}+\frac{\partial H}{\partial\eta}\frac{\partial\eta}{\partial x},\ \frac{\partial H}{\partial y}=\frac{\partial H}{\partial\xi}\frac{\partial\xi}{\partial y}+\frac{\partial H}{\partial\eta}\frac{\partial\eta}{\partial y}$$

for (29) we find

$$-\frac{\partial}{\partial\xi}\left\{k^{n-1}\frac{\tau_{bc}}{\sin\varphi_0}\left(g_{11}\frac{\partial H^n}{\partial\xi}+g_{12}\frac{\partial H^n}{\partial\eta}\right)\right\}-$$
$$-\frac{\partial}{\partial\eta}\left\{k^{n-1}\frac{\tau_{bc}}{\sin\varphi_0}\left(g_{22}\frac{\partial H^n}{\partial\eta}+g_{12}\frac{\partial H^n}{\partial\xi}\right)\right\}+\frac{(1-\varepsilon)}{\tau}H^n=$$
$$=\frac{(1-\varepsilon)}{\tau}H^{n-1}-\left(\frac{\partial}{\partial\xi}\left((k^{n-1}\tau_{bx})\frac{\partial\xi}{\partial x}+(k^{n-1}\tau_{by})\frac{\partial\xi}{\partial y}\right)+\right.$$
$$\left.+\frac{\partial}{\partial\eta}\left((k^{n-1}\tau_{bx})\frac{\partial\eta}{\partial x}+(k^{n-1}\tau_{by})\frac{\partial\eta}{\partial y}\right)\right). \quad (30)$$

We define the scalar product

$$(f,g)=\iint_D f(x,y)g(x,y)dxdy=$$
$$=\iint_\Xi f(x(\xi,\eta),y(\xi,\eta))g(x(\xi,\eta),y(\xi,\eta))Jd\xi d\eta \quad (31)$$

We multiply the equation (30) by the rule (31) by an arbitrary function $v(x(\xi,\eta),y(\xi,\eta))$. We get

$$-g_{11}\iint_\Xi\frac{\partial}{\partial\xi}\left(k^{n-1}\frac{\tau_{bc}}{\sin\varphi_0}\frac{\partial H^n}{\partial\xi}\right)vJd\xi d\eta-$$
$$-g_{22}\iint_\Xi\frac{\partial}{\partial\eta}\left(k^{n-1}\frac{\tau_{bc}}{\sin\varphi_0}\frac{\partial H^n}{\partial\eta}\right)vJd\xi d\eta-$$
$$-g_{12}\iint_\Xi\left(\frac{\partial}{\partial\xi}\left(k^{n-1}\frac{\tau_{bc}}{\sin\varphi_0}\frac{\partial H^n}{\partial\eta}\right)+\frac{\partial}{\partial\eta}\left(k^{n-1}\frac{\tau_{bc}}{\sin\varphi_0}\frac{\partial H^n}{\partial\xi}\right)\right)vJd\xi d\eta+$$
$$+\frac{(1-\varepsilon)}{\tau}\iint_\Xi H^n vJd\xi d\eta=\frac{(1-\varepsilon)}{\tau}\iint_\Xi H^{n-1}vJd\xi d\eta-$$
$$-\frac{\partial\xi}{\partial x}\iint_\Xi\left(\frac{\partial}{\partial\xi}(k^{n-1}\tau_{bx})\right)vJd\xi d\eta-\frac{\partial\xi}{\partial y}\iint_\Xi\frac{\partial}{\partial\xi}(k^{n-1}\tau_{by})vJd\xi d\eta-$$
$$\frac{\partial\eta}{\partial x}\iint_\Xi\left(\frac{\partial}{\partial\eta}(k^{n-1}\tau_{bx})\right)vJd\xi d\eta-\frac{\partial\eta}{\partial y}\iint_\Xi\frac{\partial}{\partial\eta}(k^{n-1}\tau_{by})vJd\xi d\eta.$$

Consider the approximation of functions defined on a polygonal domain, D using piecewise linear basis functions. Area D undergo a triangulation.

The set of basis functions φ_{ij} is defined in the standard way.

The required approximate solution $\tilde{H}^n$ of the problem under consideration can be represented by expanding in terms of the basis functions

$$\tilde{H}^n(x(\xi,\eta),y(\xi,\eta),t_{n-1})=\sum_{i=1}^{N_1-1}\sum_{j=1}^{N_2-1}z_{ij}\varphi_{ij}(x(\xi,\eta),y(\xi,\eta)). \quad (32)$$

To describe the computational procedures for finding finitely-elementary solutions of the initial-boundary problem of sediment transport with allowance for (32), it is convenient to write down its matrix analogue:

$$-Az+\frac{(1-\varepsilon)}{\tau}Bz=C, \quad (33)$$
$$z=(z_{ij}),$$

$$A=(A_{ij,kl}),\ A_{ij,kl}=g_{11}\iint_\Xi\frac{\partial}{\partial\xi}\left(k^{n-1}\frac{\tau_{bc}}{\sin\varphi_0}\frac{\partial\varphi_{kl}}{\partial\xi}\right)\varphi_{ij}Jd\xi d\eta+$$
$$+g_{22}\iint_\Xi\frac{\partial}{\partial\eta}\left(k^{n-1}\frac{\tau_{bc}}{\sin\varphi_0}\frac{\partial\varphi_{kl}}{\partial\eta}\right)\varphi_{ij}Jd\xi d\eta+$$
$$+g_{12}\iint_\Xi\left(\frac{\partial}{\partial\xi}\left(k^{n-1}\frac{\tau_{bc}}{\sin\varphi_0}\frac{\partial\varphi_{kl}}{\partial\eta}\right)+\frac{\partial}{\partial\eta}\left(k^{n-1}\frac{\tau_{bc}}{\sin\varphi_0}\frac{\partial\varphi_{kl}}{\partial\xi}\right)\right)\varphi_{ij}Jd\xi d\eta,$$
$$B=(B_{ij,kl}),\ B_{ij,kl}=\frac{(1-\varepsilon)}{\tau}\iint_\Xi\varphi_{kl}\varphi_{ij}Jd\xi d\eta\ ,$$
$$C=(C_{ij,kl}),\ C_{ij,kl}=\frac{(1-\varepsilon)}{\tau}\iint_\Xi H^{n-1}\varphi_{ij}Jd\xi d\eta-$$
$$-\frac{\partial\xi}{\partial x}\iint_\Xi\left(\frac{\partial}{\partial\xi}(k^{n-1}\tau_{bx})\right)\varphi_{ij}Jd\xi d\eta-\frac{\partial\xi}{\partial y}\iint_\Xi\frac{\partial}{\partial\xi}(k^{n-1}\tau_{by})\varphi_{ij}Jd\xi d\eta-$$
$$-\frac{\partial\eta}{\partial x}\iint_\Xi\left(\frac{\partial}{\partial\eta}(k^{n-1}\tau_{bx})\right)\varphi_{ij}Jd\xi d\eta-\frac{\partial\eta}{\partial y}\iint_\Xi\frac{\partial}{\partial\eta}(k^{n-1}\tau_{by})\varphi_{ij}Jd\xi d\eta,$$
$$i,k=\overline{1,N_1},\ j,l=\overline{1,N_2}.$$

The grid equation (33) is actually the SLAE for the unknown vector

$$\left(-A+\frac{(1-\varepsilon)}{\tau}\cdot B\right)\cdot z=C, \quad (34)$$

with a global stiffness matrix of the form $\left(-A+\frac{(1-\varepsilon)}{\tau}B\right)$ [19-20].

We note that after carrying out a sequential computational procedure for the compilation and solution of equations (34), we obtain a solution of the initial-boundary value problem on all time layers of interest to us $n=1,...,N$.

5 Conclusions

On the optimal boundary-adaptive grid constructed on the basis of minimization of the generalized Dirichlet functional, a discrete two-dimensional mathematical model for a linearized model of sediment transport in coastal systems that satisfies the basic conservation laws is constructed by the finite element method. A finite element approximation has been implemented in relation to the water area of the Azov Sea and the Taganrog Bay and, thus, a system of finite-element equations has been prepared for the subsequent numerical modeling of sediment transport processes.

This work was supported by the RNF (project code 17-11-01286).

References

1. A.I. Sukhinov, V.V. Sidoryakina, *The theory of operators, complex analysis and mathematical modeling. Abstracts of the XIII International Scientific Conference*, 184-185 (2016)

2. A.I. Sukhinov, V.V. Sidoryakina, Bulletin of the Taganrog State Pedagogical Institute, № 2, 270-274 (2016)
3. A.I. Sukhinov, V.V. Sidoryakina, Bulletin of the Taganrog State Pedagogical Institute, № 2, 245-249, (2017)
4. A.I. Sukhinov, V.V. Sidoryakina, A.A. Sukhinov, Bulletin of the Don State Technical University, **17**, № 1 (88), 5-17 (2017)
5. V.V. Sidoryakina, A.I. Sukhinov, Computational Mathematics and Mathematical Physics, **57**, № 6, 978-994 (2017)
6. V.V. Sidoryakina, A.I. Sukhinov, Zhurnal Vychislitel'noi Matematiki i Matematicheskoi Fiziki, **57**, № 6, 985–1002 (2017)
7. A.I. Sukhinov, A.E. Chistyakov, E.V. Alekseenko, Mathematical Models and Computer Simulations, **23**, №. 3, 3–21 (2011)
8. A.I. Sukhinov, A.A. Sukhinov, *Parallel Computational Fluid Dynamics, Mutidisciplinary Applications, Prcoceedings of Parallel CFD 2004 Conference*, 231-238 (2005)
9. V.S. Vasiliev, A.I. Sukhinov, Matem. Modelling, **15**, № 10, 17-34 (2003)
10. I.O. Leontyev, Coastal dynamics: waves, moving streams, deposits drifts (GEOS, Moscow, 2001)
11. A.I. Sukhinov, A.E. Chistyakov, E.A. Protsenko, News of SFedU. Technical science, № 8 (121), 32-44 (2011)
12. A.I. Sukhinov, A.E. Chistyakov, E.A. Protsenko, Matematicheskoe Modelirovanie, **25**, № 12, 65–82, (2013)
13. E. Alekseenko, B. Roux, A. Sukhinov, R. Kotarba, D. Fougere, Nonlinear Processes in Geophysics, **20**, № 2, 189-198 (2013)
14. A.I. Sukhinov, A.E. Chistyakov, E.A. Protsenko, Math. Models Comput. Simul., **6**, № 4, 351–363 (2014)
15. A.I. Sukhinov, A.E.Chistyakov, E.A. Protsenko, Numerical Methods and Programming, **15**, № 4, 610–620 (2014)
16. S.A.Ivanenko, Zhurnal Vychislitel'noi Matematiki i Matematicheskoi Fiziki, **28**, № 10, 1498-1506 (1988)
17. G.I. Marchuk, V.I. Agoshkov, *Introduction to projection-grid methods* (Science. The main edition of physics and mathematics, Moscow, 1981)
18. I.A. Vaseva, A.V. Kofanov, V.D. Liseikin, Yu.V. Likhanova, A.M. Kharitonchik, Vychislitel'noi Matematiki i Matematicheskoi Fiziki, **50**, № **1**, 99–117 (2010)
19. A.A. Samarskii, A.V. Gulin, *Numerical methods of mathematical physics*. 2-nd ed. (The scientific world, Moscow, 2003)
20. O.M. Belotserkovsky, *Numerical modeling at mechanics continuous environments* (Fizmatlit, Moscow, 2004)

On the developing of the mathematical model of all-wheel drive vehicle

Andrey Efimov[1], *Sergey* Kireev[1], and *Marina* Korchagina[1]

[1]Don State Technical University, Rostov-on-Don, Russia

Abstract. The mathematical model of all-wheel-drive vehicle is proposed, which makes it possible to investigate the economic indicators of the vehicle's movement for choosing the optimal type of central differential in a given range of road conditions.

1 Introduction

At vehicle movement, there are two factors that specify conflicting requirements for transfer box drive and axle drive. Let's describe them: each wheel rotating at different speeds and reduced adherence of one or more wheels [1-10]. These factors tend to be of time-varying nature. The nature of the studied processes is transient; it will be influenced by the inertia, stiffness and damping properties of the vehicle power train parts. All this makes the task to solve the task by the dynamics methods.

2 Main part

To assess the applicability of drive of a particular type at designed or already existing vehicle, a mathematical model of all-wheel-drive vehicle was developed to determine the main force and kinematic parameters of movement under the specified road conditions.

In analytical studies, the vehicle is represented as a multimass dynamic system, where $J_1, J_2, J_3, J_4, J_5, J_6, J_{xz}$ – are the rotational inertia of the flywheel and the moving mass of engine, the rotational inertia of clutch parts, gearbox and transfer case reduced to the carrier of the central differential, the rotational inertia of sun gear and ring gear and reduced parts, the rotational inertia of front and rear wheels, the rotational inertia of vehicle frame in the *xz*-plane with respect to the center of mass; m – vehicle mass. In the dynamic system, in addition to the drive circuit, the changes in soil reactions to the wheels, the stiffness and damping properties of the transmission parts, and the spin of the drive wheels are taken into account.

At the development of the mathematical model, the following assumptions are made: the frame of the machine is a solid body with a longitudinal axis of symmetry; through-the-thickness slip of the tires of the driving wheels is neglected; frictional forces in the transmission element are considered to be zero; assume the conditions for the adherence of the right and left wheels of the same bridge to be the same.

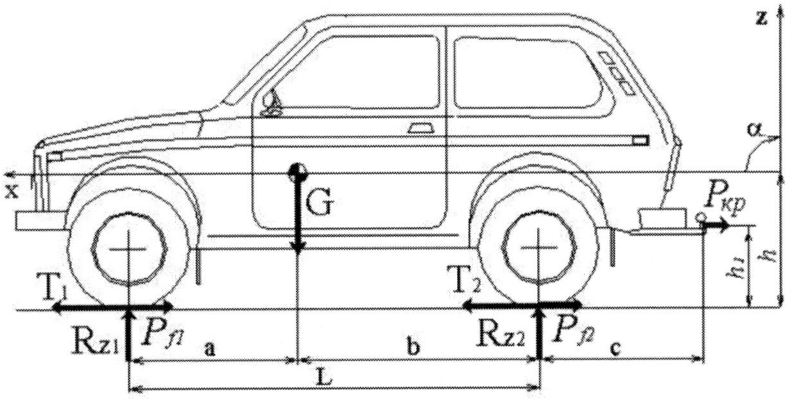

Fig. 1. Scheme of forces acting on the vehicle

* Corresponding author: efimovav2007@yandex.ru

Figure 2 shows the dynamic model of a wheeled vehicle with a differential transfer box drive,

Figure 1 shows the forces acting on the wheeled vehicle, where T_1, T_2 – tractive effort of front and rear wheels; P_{f1}, P_{f2} – power of rolling resistance of front and rear wheels respectively; R_{Z1}, R_{Z2} – vertical soil reactions to front and rear wheels respectively; P_h – hook power; G – vehicle gravity; a, b, L, h, h_1 – geometric dimensions of the vehicle.

where M_e – engine torque; $M\varphi$ – torque of clutch friction; $M_{\varphi1}$, $M_{\varphi2}$ – torques on the wheels from traction forces, $M_{\varphi1}=T_1r_{w1}$, $M_{\varphi2}=T_1r_{w2}$; r_{w1}, r_{w2} – dynamic radius of front and rear drive wheels; i_{12}, i_{35}, i_{46}, c_{12}, c_{35}, c_{46}, k_{12}, k_{35}, k_{46} – transmission ratios, coefficient of stiffness and damping coefficients of power train sections between first and second, third and fifth, fourth and sixths mass respectively; $c_{\chi1}, c_{\chi2}$, $k_{\chi1}, k_{\chi2}$ – extension rigidity and pitch damping coefficient of tires of front and rear wheels respectively; i_d – transmission ratio of differential,

$$i_d = \left.\frac{\omega_3}{\omega_4}\right|_{at\ \omega_2=0}.$$

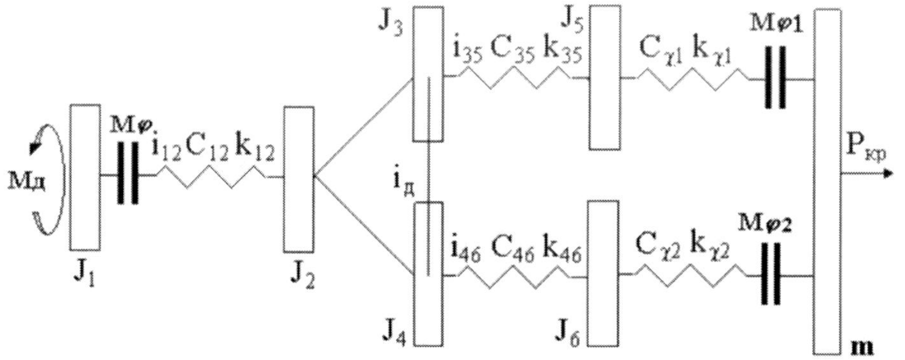

Fig. 2. Diagram of dynamic model of transmission with differential transfer box drive

Since ω_3 and ω_4 at stopped carrier ($\omega_2=0$) are oppositely directed for all epicyclic gears used as differential, then $i_d<0$, which explains the change of direction of torque M_{46} after its reduction (see Figure 3).

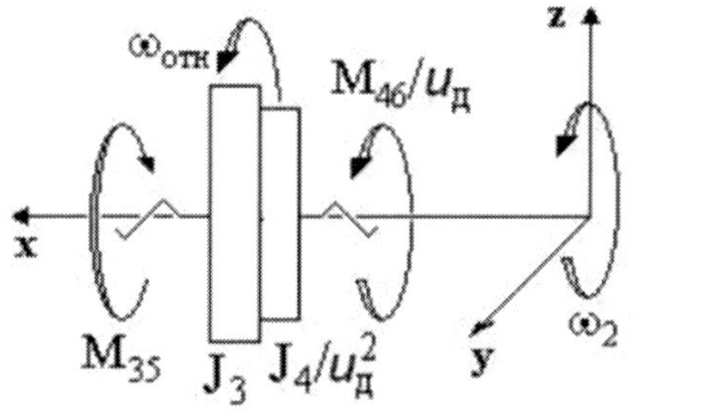

Fig. 3. Diagram of the dynamic model of the transmission section with a differential transfer box drive involved in the relative motion

Besides, $|i_d| = u_d \cdot u_d$ – transmission number of central differential.

The above configuration of the dynamic model of a vehicle with a differential drive is the most general case. Other types of drive can be described using the same model with some refinements.

The mathematical model of a wheeled vehicle equipped with a differential drive is represented as a system of differential equations:

$$\begin{cases} J_1\ddot{\varphi}_1 = M_{ea} - \dfrac{M_{12}}{i_{12}}; \\ (J_2+J_3+J_4)\ddot{\varphi}_2 = M_{12} - \dfrac{M_{35}}{i_{35}} - \dfrac{M_{46}}{i_{46}}; \\ J_5\ddot{\varphi}_5 = M_{35} - T_1r_{K1}; \\ J_6\ddot{\varphi}_6 = M_{46} - T_2r_{K2}; \\ \left(J_3 + \dfrac{J_4}{u_d^2}\right)\ddot{\varphi}_{rel} = \dfrac{M_{46}}{u_d i_{46}} - \dfrac{M_{35}}{i_{35}}; \\ m\ddot{x} = T_1 + T_2 - P_{f1} - P_{f2} - P_{KP}; \\ J_{XZ}\ddot{\alpha} = R_{Z1}a - R_{Z2}b + \\ +(T_1+T_2-P_{f1}-P_{f2})h - P_{KP}(h-h_1); \\ m\ddot{z} = R_{Z1} + R_{Z2} - G, \end{cases} \qquad (1)$$

For all-wheel-drive vehicle the following generalized coordinates are taken:

x, z - respectively horizontal and vertical movement of vehicle frame; $\varphi_1, \varphi_2, \varphi_5, \varphi_6, \varphi_{rel}$ - angular coordinates of the flywheel of the engine, carrier of central differential, front and rare wheels and rotation angle of sun gear of central differential against carrier; α - the angle of rotation of the vehicle frame around a horizontal axis passing through the center of mass.

The system of differential equations is supplemented by the equations of non-holonomic constraints [11]

* Corresponding author: efimovav2007@yandex.ru

$$\dot{\chi}_1 = r_{w1}\dot{\phi}_5 - \dot{x} - (\lambda_1 + k_{\delta 1})\dot{\phi}_5 T_1;$$
$$\dot{\chi}_2 = r_{w2}\dot{\phi}_6 - \dot{x} - (\lambda_2 + k_{\delta 2})\dot{\phi}_6 T_2, \tag{2}$$

where χ_1, χ_2 - longitudinal deformation of tires of driving wheels; λ_1, λ_2 - coefficients of tire elastic tangency; k_δ - coefficient of specific slip of the spinning driving wheel.

Consider the analytic dependence of the main parameters included in the system of differential equations (1).

The equation that defines the dependence of power on rate speed at a constant position of the throttle body, for gasoline engines, in accordance with [12] is as follows:

$$N_{ex} = N_e \frac{n_x}{n_N}\left(1 + \frac{n_x}{n_N} - \left(\frac{n_x}{n_N}\right)^2\right),$$

where N_e, n_N - nominal (maximum) rating power (kW) and corresponding rate speed of engine crankshaft (r/m) for this position of throttle shutter.

Let's use this formula to determine engine torque when the throttle shutter is not fully open. According to [12], torque of engine (N·m) is defined as:

$$M_e = 3*10^4 N_{ex}/(\pi n_x).$$

Angular velocity and shaft speed are related by the formula:

$$n = \frac{30}{\pi}\omega \cdot$$

Therefore,

$$M_e = \frac{N_e}{\omega_N}\left(1 + \frac{\omega}{\omega_N} - \left(\frac{\omega}{\omega_N}\right)^2\right), \tag{3}$$

where N_e, ω_N - maximum rating power (W) and corresponding angular velocity of engine crankshaft (r/m) when the throttle shutter is partially open, $N_e = k_N N_e^{\max}$,

where $N_e^{\max}$ - maximum rating power (W) when the throttle shutter is fully open;

k_N - power filling factor of engine (corresponds to a certain rate of opening of throttle shutter).

Based on the assumption of a linear relationship between ω_N and N_e,

$$\omega_N = (\omega_N^{MAX} - \omega_{xx})k_N + \omega_{xx},$$

where ω_N^{MAX} - angular velocity of engine crankshaft corresponding to the maximum engine power when the throttle shutter is fully open N_e^{MAX}; ω_{xx} - angular velocity at no-load.

Acceleration of the vehicle is carried out under unsteady operating conditions of the engine. According to [13], torque developed by the engine during acceleration is

$$M_{ea} = M_e\left(1 - \gamma\frac{d\omega}{dt}\right),$$

where γ – coefficient taking into account the decrease of torque developed by the engine due to the transient operation mode, γ=0,07-0,09.

Torques acting in the system between the corresponding masses:

$$M_{12} = c_{12}\left(\frac{\phi_1}{i_{12}} - \phi_2\right) + k_{12}\left(\frac{\dot{\phi}_1}{i_{12}} - \dot{\phi}_2\right);$$
$$M_{35} = c_{35}\left(\frac{\phi_3}{i_{35}} - \phi_5\right) + k_{35}\left(\frac{\dot{\phi}_3}{i_{35}} - \dot{\phi}_5\right); \tag{4}$$
$$M_{46} = c_{46}\left(\frac{\phi_4}{i_{46}} - \phi_6\right) + k_{46}\left(\frac{\dot{\phi}_4}{i_{46}} - \dot{\phi}_6\right),$$

where angle of rotation and speed of sun gear and ring gear of differential are expressed through the accepted generalized coordinates,

$$\varphi_3 = \varphi_2 + \varphi_{rel};$$
$$\varphi_4 = \varphi_2 - \frac{\varphi_{rel}}{u_d};$$
$$\dot{\varphi}_3 = \dot{\varphi}_2 + \dot{\varphi}_{rel}; \tag{5}$$
$$\dot{\varphi}_4 = \dot{\varphi}_2 - \frac{\dot{\varphi}_{rel}}{u_d}.$$

With locking differential, the relative motion of the differential is not possible, so the movement of a vehicle with a similar type of drive is described by a single dynamic model, the circuit of which is shown in Figure 4. The mathematical model of a vehicle with locking differential is described by the system of differential equations (1), supplemented by the condition:

$$\begin{cases} \varphi_{rel} = 0; \\ \dot{\varphi}_{rel} = 0; \\ \ddot{\varphi}_{rel} = 0; \end{cases}$$

Friction torque of the clutch M_Φ, according to [14]:

$$M_\Phi = \mu Q_{sp} r_f i, \tag{6}$$

where μ - coefficient of friction; i – number of friction pairs; r_f – friction radius; Q_{sp} - force of pressure springs;

$$Q_{sp} = \begin{cases} Q_{sp}^{\max}\dfrac{t}{t_{on}}, & at\ t \le t_{on}; \\ Q_{spP}^{\max}, & at\ t > t_{on}, \end{cases}$$

where $Q_{sp}^{\max}$ - maximum effort of pressing spring; t_{on} – clutch on-time.

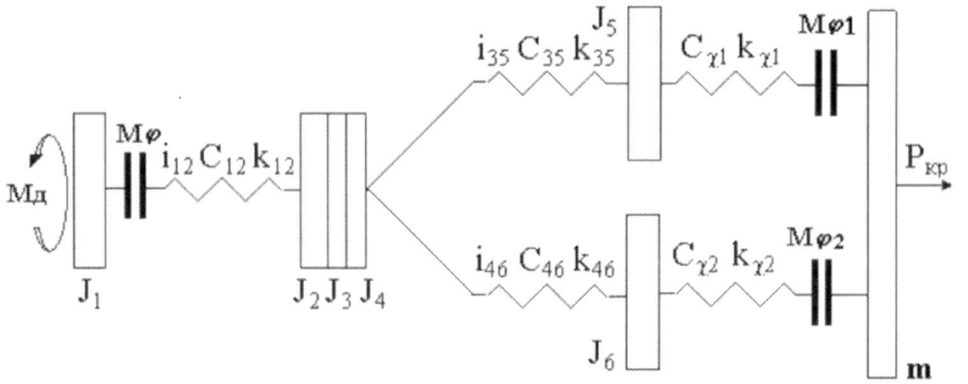

Fig. 4. Diagram of a dynamic transmission model without taking into account the relative motion in central differential drive

The frictional moment of M_ϕ clutch restricts the torque between the engine and M_{12} transmission.

Tangent traction force of the drive wheel, according to [11],

$$T = c_\chi \chi + k_\chi \dot{\chi}, \quad (7)$$

where c_χ, k_χ - longitudinal stiffness and coefficient of longitudinal damping of the tire of the corresponding wheel; χ - longitudinal deformation of tires of driving wheels.

Coefficient of specific spin by [11]

$$k_\delta = r_w \delta / T, \quad (8)$$

where T – total tractive effort of the wheels of this axle; r_w - axle wheels radius; δ - axle wheels spin. The spins of left and right wheels are equal. Consider their clutching conditions to be the same (see assumptions adopted in the mathematical model).

In accordance with [13], the tangential traction force of the wheel is calculated by the formula:

$$T = R_Z \phi_{adh}(1 - e^{-K|\delta|}) sign(V_T - V_{act}),$$

where T – tangential force of wheel traction; R_Z - vertical soil reaction to wheel; φ_{adh} - coefficient of wheel adherence with soil; K – empirical coefficient; V_T - theoretical wheel speed; V_{act} – actual vehicle speed.

From here, we express spin through tractive effort:

$$\delta = -\frac{1}{K}\ln\left(1 - \frac{|T|}{\varphi_{adh} R_Z}\right) sign(T), \quad (9)$$

The formula (9) is valid in the range of values δ = -1…1.

Vertical reaction to front and rear wheels

$$R_{Z1} = G_1 - C_1(z + \alpha a) - K_1(\dot{z} + \dot{\alpha} a);$$
$$R_{Z2} = G_2 - C_2(z - \alpha b) - K_2(\dot{z} - \dot{\alpha} b), \quad (10)$$

where G_1, G_2 - vehicle weight in statics falling on the front and rear wheels; C_1, C_2 – radial stiffness of tires and suspension of front and rear wheels; K_1, K_2 – coefficient of damping of tires and suspension of front and rear wheels.

Effort of resistance of front and rear wheels' rolling

$$P_{f1} = fR_{Z1};$$
$$P_{f2} = fR_{Z2}, \quad (11)$$

where f – rolling resistance coefficient, [15-18].

3 Conclusion

The solution of the system of equations (1) together with the analytic relations (2) - (11) can be carried out by the Runge-Kutta method [19-21]. Comparison of the calculated data with the experimental data has shown the adequacy of the mathematical model obtained. It allows carrying out further studies of the effect of the drive type on the spin efficiency of wheeled vehicle using computational methods, which has a number of advantages. As follows from the analysis of the results of theoretical studies, the following conclusions can be drawn:

1. The developed dynamic model of the vehicle is distinguished by a certain novelty and reflects the behavior of all links in the dynamic system.
2. The mathematical model developed in accordance with the dynamic model allows modeling the movement of a vehicle equipped with a differential, blocked, or differential drive with a limited gear ratio in a variety of road conditions and can be used for comparative analysis of vehicle spin losses equipped with different types of drive.

References

1. P. V. Aksenov, *Multi-axle cars* (Mechanical Engineering, Moscow, 1989)
2. A.V. Andreev, V.V. Vancevich, A.Kh. Lefarov, Differential Wheeled, Machines, Mechanical Engineering, 1987.

* Corresponding author: efimovav2007@yandex.ru

3. A. Yu. Barykin, Automobile industry, 9 (2004).
4. V.V. Vancevich, D.A. Dubovik, A.F. Andreev, V.I. Kabanov, V.G. Ermalenok, V.S. Voiteshonok, Automobile industry, 2 (2006).
5. A.V. Keller, G.D. Dragunov, Automotive industry, 1 (2004)
6. G.V. Novikov, V.V. Pogorelov, G.B. Shipilevsky, Automotive industry, 8 (2007).
7. I.V. Grinchenko, Investigation of the loading of transmissions of a multi-drive car while driving along a curvilinear trajectory, Diss. PhD. Moscow, (1976)
8. A.Kh. Lefarov, M.S. Vysotsky, V.V. Vantsevich, V.I. Kabanov, *Energy loading and reliability of differential mechanisms of transport-traction machines*, (Navuka i tehnika, Minsk, 1991)
9. B.M. Tverskov, Automotive industry, 3 (1998)
10. E.A. Chudakov, *Circulation of parasitic power in the mechanisms of a non-differential car* (Mashgiz, Moscow, 1950)
11. V.P. Tarasik, Tractors and agricultural machinery, 4 (1981)
12. A.I. Kolchin, V.P. Demidov, *Calculation of automobile and tractor engines: Guide for universities*, (Vysshaya shkola, Moscow, 2008)
13. A.I. Grishkevich, L.A. Moliboshko, O.S. Rukteshel, V.M. Belyaev, The use of computers in the design and calculation of vehicle, (1978)
14. I.P. Ksenevich, A.S. Solonsky, V.Ya. Rubenchik, L.I. Peskina, Tractors and agricultural machinery, 5 (1979).
15. A.A. Polungyan, *Design of all-wheel drive vehicles*, **1**, (Bauman University Publishing House, Moscow, 2008)
16. J. Mackerle, *Modern economical car*, Ed. A.R. Benedictova, (Mechanical Engineering, Moscow, 1987)
17. A. Sh. Khusainov, Operational properties of the car, (UlSTU, Ulyanovsk, 2011).
18. V.F. Platonov, Mechanical Engineering, 1989.
19. B.P. Demidovich, I.A. Maron, E.Z. Shuvalova *Numerical methods of analysis* (Fizmatgiz, Moscow, 1963)
20. N.S. Bakhvalov, N.P. Zhidkov, G.M. Kobelkov *Numerical methods* (Laboratory of Basic Knowledge, Moscow, 2001)
21. V.A. Il'ina, P.K. Silaev Numerical methods for theoretical physicists. Part 2, (Institute for Computer Research, Moscow-Izhevsk, 2004)

Development of structure of a subsystem of information security of the automated system

Yulia Bakhracheva[1,*]

[1]Department of Information Security, Institute of priority technologies, Volgograd State University, Prosp. Universitetsky, 100, 400062 Volgograd, Russian Federation

Abstract. Approach to synthesis of modular structure of a subsystem of information security of the automated process control system is offered. The task has been solved on the basis of a set-theoretic notation with elements of the graph theory. The graph model constructed thus will be used further during the setting and the decision of the task of synthesis of the optimum conceptual diagram of modular structure. Thus, we can conclude that graph theory is widely applied in the design of the information space of the organization and can be used at all stages – from the formation of the it strategy and building it infrastructure to optimize specific software tools subsystem of information security.

1 Introduction

The rapid development of the computer sphere of high technologies in the past two decades has led to the fact that information acquired particular financial, reputational, temporal and other expressions. In this context, data protection becomes a priority of the organizations. The state takes an active part in the process of formation of information security, as evidenced by the strengthening and toughening of requirements to protect confidential information (commercial secrets, personal data, banking secrecy, etc.), the adoption of new laws and regulations in this area, as well as guidance documents on the classification of information security based on the security requirements. One of the first and main stages of building a secure infrastructure of the organization analysis assessment of the security of the automated control system of technological processes.

Information security system organization is a complex system of processes of different nature. This creates problems associated with the peculiarities of such systems:

- the relationship of the processes, which does not allow to consider the system from an economic, social or technical side – all processes should be investigated together;

- lack of quantitative information and, consequently, the need to consider qualitative characteristics of the processes;

- non-stationary processes when the unknown nature of the changes of process parameters.

Such systems are called semi structured (semi). In such systems impossible with traditional mathematical approach to the analysis of processes to develop a comprehensive (i.e., covering various aspects of the system under study) solutions.

In mathematical modeling of complex systems, the researcher is faced with the necessity of finding a compromise between accuracy of simulation and the possibility of obtaining detailed information needed to build the model.

Recently for the modeling of complex poorly formalized systems use a cognitive approach [1]. This approach is based on cognitive aspects, which include the processes of perception, thinking, knowledge, explanations and understanding. A schematic, simplified description of the world picture that is related to the problem situation, often represented in the form of cognitive maps.

It can be visualized as a set of vertices, each of which corresponds to one factor or element of a simulated picture of the world, connected by arcs (the graph model).

The effectiveness of this method is determined by a qualified analyst, his knowledge and ability to determine the necessary to solve a set of basic factors and relations between factors. The construction of cognitive maps allows the analyst to familiarize himself with the problem, to better understand the role of individual factors and their interconnection.

Under the decision of any problem, studied in the framework of the graph model, as providing the required dynamics of the factors that constitute the essence of the problem. These factors (being "output" factors in the graph model) are called trust factors.

The control factors (input factors graph model) are those factors (impacts) by which the analyst can manage the processes in the model, applying certain control actions.

* Corresponding author: bakhracheva@volsu.ru

External factors the model's factors that are hanging vertices in the graph model. External factors correspond to factors in the external environment, the dynamics of which does not depend on substructuring system.

Thus, the dynamics of the external factors model is given by the analyst, not for the purpose of process control in the model, but only to reflect relevant influences of the environment on the simulated system.

Control task dynamics models is to select such set of control actions (departments), which compensate for undesirable effects for analysis of initial conditions and provide the desired dynamics of changes in the target factors.

2 Methods

In the industrial systems of automation there were high-quality and quantitative changes [1-4, 7, 8]. Insufficiency of the existing approaches to ensuring information security of automated process control systems, has led to emergence of multiple threats which realization leads to catastrophic consequences [5].

Special software of a subsystem of information security represent the difficult multilevel hierarchical system of the components of the automated system connected according to information and management [6].

Now introduction of a subsystem of support of modular structure of special software is necessary. It will allow to reduce the general time for creation of the project [9].

Key question is solution of the problem of rational distribution of the information calculating resources defined on a set of information and settlement tasks of each functional module so that the system of modules would provide the solution of objectives and would meet the criteria of quality shown to it [10]. The most general is definition of the module as functionally full components of special software of a subsystem of information security [11]. The module has three basic characteristics: the functions, logic, communications characterizing operations, their sequence and flows of information. The problem of modeling of processes of information security belongs to the class of problems of modeling of complex systems. Currently developed a wide range of mathematical models. These include models developed based on the methods of Queuing theory, graph theory, automata theory, theory of Petri nets and other [12 - 20].

It is offered to use a set-theoretic notation with elements of the graph theory.

3 Results and discussion

Let $P = \{p_1,...,p_j,...,p_J\}$ - a set of processing of data. $B=\{B_1,...,B_k,...B_K\}$ – special software of a subsystem of information security of a complex of the automation equipment of the relevant control centers; About = $\{ o_1..., o_t..., o_T \}$ - a set of the input, intermediate and output data (variables, objects) processed and transformed by processes of River. On sets of P and O ($P \bigcap O = \varnothing$) defined the relations of hierarchy of processes of $Hp \subseteq P \times B(P)$, hierarchy of the data $Ho \subseteq O \times B(O)$; $\Re = \{In, Out\}$ - the interaction relations: $In \subseteq B(O) \times P$ - the relations input objects of process - process; $Out \subseteq P \times B(O)$ - the relations process - output objects of process; $\Im \subseteq P \times B(P)$ - the relations of following of processes.

The relations of hierarchy of Hp (Ho) set compliance between separate processes (objects) and sets of the processes (objects) subordinated to them. They are represented in the form of the functional displays which areas of values are elements from bulean of B(P) (B(O)):

$$h_p: P \to B(P)\ ,\ \{p_j\} = h_p(p_i),\ p_j \in h_p(p_i), \tag{1}$$

$$h_o: O \to B(O)\ ,\ \{o_j\} = h_o(o_i),\ o_j \in h_o(o_i), \tag{2}$$

where $\{p_j\}$ - a set of the processes subordinated in model to process of p_i; $\{o_j\}$ - a set of the objects subordinated in model to o_i object.

Also the relations of interaction of $\Re$, setting input and output data of processes are in the same way described:

$$in: B(O) \to P\ ,\ \{o_i\} = in(p_i),\ o_i \in in(p_i), \tag{3}$$

$$out: P \to B(O)\ ,\ \{o_j\} = out(p_i),\ o_j \in out(p_i), \tag{4}$$

where $\{o_i\}$ - set input, and $\{o_j\}$ - output objects of process of p_i

The relations of following of $\Im$, the partial execution order of processes sets: $\Im: P \to B(P)$, $\{p_j\} = \Im(p_i)$, where $\{p_j\}$ - a set of processes which execution strictly precedes execution of process of p_i ($p_j \in \Im(p_i)$).

With model elements (processes and objects) communicate:

- set of names of processes of N_p and np function: $P \to N_p$;
- set of names of types of processes of T_p and t_p function: $P \to T_p$;
- set of names of objects of No and no function: $O \to N_o$;
- set of names of object types of T_o and to the function: $O \to T_o$;
- set of names of types of the relations of hierarchy of processes and objects of T_h and t_{hp} function: $P \to T_h$, t_{ho}: $O \to T_h$. The sets set thus and the relations form conceptual model of calculating processes.

This model can be delivered in compliance a processing graph of data of G= (V, U). Tops of such graph $V=\{v_i;\ i=1,R\}$ is processing of data from P, and edges $U=\{u_{ij}=(v_i,v_j)\}$ - a set of data from O_h, the general for the corresponding processes.

Let the set of all admissible subsets of in Boolean ($B(V)=\{M_q;\ q=1,2^R-1\}$; $M_q=\{v_i\}$)are given a subset of $M_f(C)$, whose elements satisfy the conditions M_q:

$$\bigcup M_q=V, \quad \text{где } M_q \in M_f(V), \quad q=1,Q, \tag{5}$$

$$M_q \bigcap M_{q'}=\varnothing, \text{ где } q \neq q',\ M_q, M_{q'} \in M_f(V). \tag{6}$$

A lot of $M_f(V)$ is an aggregated graph $G=(\Gamma_q,S)$ whose vertices are the subgraphs $\Gamma_q=(M_q,D_q)$, ($M_q=\{v_i\}$, $M_q \in M_f(V)$; $D_q=\{u_{ij}=(v_i,v_j),\ v_i,v_j \in M_q\}$ be a set of edges M_q), and S is the set of edges connecting the subgraphs Γ_q:

$$S = \bigcup_{q \neq q'} \{u_{ij}=(v_i,v_j),\ v_i \in M_q,\ v_j \in M_{q'}\}. \quad (7)$$

The subgraphs $\Gamma_q=(M_q,D_q)$ are called functional modules graph $\Gamma=(V,U)$ data processing.

The set of edges S of a graph G form an inter-module information system interface modules.

If $S = \varnothing$, then $G=(\Gamma_q,S)$ consists of separate, unrelated subgraphs (modules) process data.

If $\Gamma_q = \{v_i;\ i=1,R\}$ and $S = \varnothing$, then the graph G is no aggregation, which corresponds to the modular system the null level.

This graph model will be further useful in the formulation and solution of the problem of synthesis of optimal conceptual scheme of the modular structure. For this it is necessary to formally describe the conceptual scheme of the modular structure of special software of a subsystem of information security.

The conceptual diagram of the modular structure is a pair:

$$KS_m = \langle NM,SM \rangle, \quad (8)$$

where NM is the name of the modular structure; SM - schema modules, which is represented by a pair:

$$SM = \langle PSM,PFV \rangle, \quad (9)$$

PSM = {SMq; q=1,Q} - a lot of diagrams of modules, where SMq diagram of the q-th module:

$$SMq = \langle P, O, H, \Re, \Im \rangle \quad (10)$$

PFV={FVqq'; q,q'=1,Q} - multiple of the relations existing between the modules (Q - the number of modules), where: FVqq' - multiple inputs and outputs of the processes belonging to module q', called from module q:

$$(q \neq q'):\ FVqq' = \{\ o_i \mid o_i \in in(p_j)\ V\ o_i \in out(p_j),$$

$$p_j \in h_p(p_i);\ p_i \in SMq,\ p_j \in SMq'\}. \quad (13)$$

The optimal graph should meet the following requirements:

- save the reach, that is, each vertex of the graph must be reachable and have your route.

- possibilities of optimization in conditions of incomplete information. That is, all vertices are reachable.

Such a graph is built up gradually on the components with the test reach. Based on models of the strong reach and the reach of the base graph is constructed, and marked his ribs.

When optimizing the links of the graph according to the criteria of information compactness index and stratification of the removed edges, which have less weight.

The main advantages of this model can be called the ease of implementation, fast results of the assessment, the versatility for any degree of research risks. However, a significant drawback is the fact that, when considering large-scale information systems to construct and study the graph causes certain difficulties. Therefore, the most promising is the application of this model to small and medium-sized information systems. The most rational way of Preodolenie this restriction are the logical division of large networks into smaller segments and the implementation of this approach at the level of subsystems.

4 Conclusions

The graph model constructed thus will be used further during the setting and the decision of the task of synthesis of the optimum conceptual diagram of modular structure. Thus, we can conclude that graph theory is widely applied in the design of the information space of the organization and can be used at all stages – from building it infrastructure to optimize specific software tools subsystem of information security.

References

1. V. Diekert, M. Lohrey Theory of Computing Systems, **37**, 133-156 (2004)
2. V. Diekert, M. Lohrey Lecture Notes in Computer Science, **2285**, 501-512 (2002)
3. Y. Dimopoulos, A. Torres Theoretical Computer Science, **170**, 209-244 (1996)
4. M. Gagne-Portelance, D.G.C. McKeon Canadian Journal of Physics, **83**, 1051-1061 (2005)
5. S. Guerrini, Theoretical Computer Science, **227** (1999)
6. K. Li, Y. Pan, M. Hamdi, Parallel Computing, **26**, № 6, 723-735 (2000)
7. S. Liénart, P. Ruelle, O. Verhoeven, Nuclear Physics B, **592**, 479-511 (2001)
8. Yu. V. Matiyasevich, Mathematical Notes, **12**, № 6, 904-908 (1973)
9. P. A. Morris, Journal of the London Mathematical Society, **2**, № 1, 63 (1974)
10. S. B. Nadler, Journal of the London Mathematical Society, **47**, № 1, 167 (1993)
11. E. A. Palyutin, Journal of Mathematical Sciences, **167**, № 6, 825-840 (2010)
12. M. Polak, U. Romańczuk, V. Ustimenko, A. Wróblewska, Electronic Notes in Discrete Mathematics, **43**, 329-342 (2013)
13. A.V. Vasilyev, Yu.S. Bakhracheva, O. Kabore, Yu.O. Zelenskiy, Valve Cam Design Using Numerical Step-by-Step Method, Vestnik Volgogradskogo gosudarstvennogo universiteta. Seriya 10, Innovatsionnaya deyatelnost [Science Journal of Volgograd State University. Innovation Activity], **10**, № 1, 26-32 (2014)

14. A.V. Vasilyev, Y.S. Bakhracheva, *Procedia Engineering,* **150**, 312-317 (2016)
15. D.Y. Altimentova, N.I. Gdansky, Contemporary Problems of Social Work, **12**, 73-80 (2015)
16. Pyrus Helps Business Managers to Make Better Decisions Faster by Automating Workflows, available at: https://pyrus.com/en/workflows.
17. Top Banking Systems Software Products, available at: http://www.capterra.com/banking-systemssoftware/
18. A. Sirotskiy, The Technologies of Technosphere Safety. Scientific Internet Magazine, **50** (2013)
19. J. Berg, A. Mehta, Advances in Complex Systems, **4**, 309-319 (2001)
20. P. Grindrod, D. Higham, Proceedings of the Royal Society A*: Mathematical, Physical and Engineering Sciences,* **466**, 753-770 (2010)

 DOI: 10.1051/matecconf/201713204006

The analysis of passengers travel time to the distant from a central area transportation districts in the city of Rostov-on-Don

Anna Chernova[1,*], *Elena* Semchugova[2], and *Nikolai* Negrov[1]

[1]FSBEI of Higher Education «Don State Technical University» Academy of Construction and Architecture, Russia
[2]FSBEI of Higher Education «Don State Technical University» Academy of Construction and Architecture, Russia

Abstract. The paper describes one of the indicators of the passenger traffic quality – passengers travel time in the city of Rostov-on-Don. The studied quality index is analyzed in comparison with the standard value for urban passenger transport in Rostov-on-Don.The paper suggests the measures to improve the operation of urban passenger transport.

1 Introduction

The level of urban passenger transport has an impact on the life quality of urban residents and the development of the economy. Therefore, urban passenger transport services should be accessible to all segments of the population and meet their need for transportation. In accordance with GOST 51004-96 [1], evaluating the quality of passenger traffic, the total costs for the passengers delivery from the point of departure to the destination point, as well as the travel time, are taken into account.

2 The analysis of the passengers travel

For the analysis of the passengers travel time we used the Pikas software complex, which allows online monitoring of the operation of municipal public transport vehicles. The software complex Pikas allows to plan the efficient route of passengers traffic. This software provides the following information: city routes considering mode of transport; directions of the chosen route, as well as the schedule of public transport vehicles for all stops of urban passenger traffic with possible changings; departure time on the route (i.e, when the passenger arrives at the next stop of the route, if he leaves this stop at the chosen time), etc.[2,3,4,5,6,7]

For the analysis we calculated the travel time from the city center to the transportation districts of the city, in the forward and backward directions. An example is shown in Fig.1. The example shows passengers travelling to a distant from a central area district with a changing and walking to the stopping point. From the analyzed route alternatives, routes with the least travel time were selected, considering a number of changings. For all directions, the studies were carried out with a change in the planning parameters for passengers travelling, on the basis of which operational calculations were made. The analysis was carried out according to the following parameters [8-21]:

- departure time;
- options for the use of various modes of transport (tram, trolley, bus, fixed-route taxi);
- time for changing;
- distance of the pedestrian walking; average walking speed.

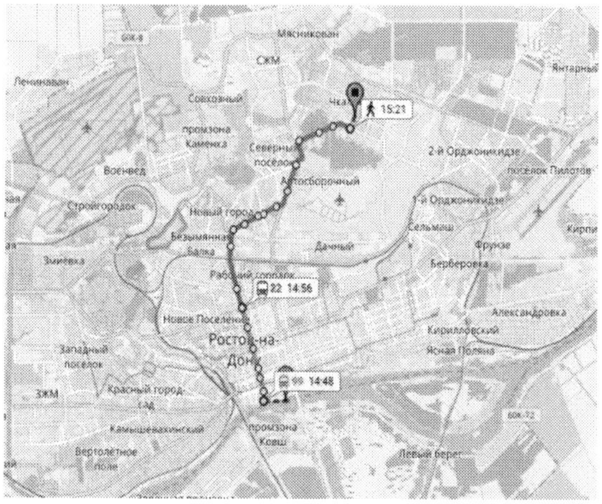

Fig. 1. The example of passengers travelling with a changing

An example of the rolling stock schedule for stops, used to estimate the number of vehicles of a particular route passing through one stop, is shown in Fig.2.

Fig. 3, 4 presents a graphical interpretation of the passengers travel time study going by public transport on a weekday.

The paper studies the directions from the city center to transportation districts in real time. These values are compared with the standard value, which is 40 minutes for the city of Rostov-on-Don.

* Corresponding author: author@e-mail.org

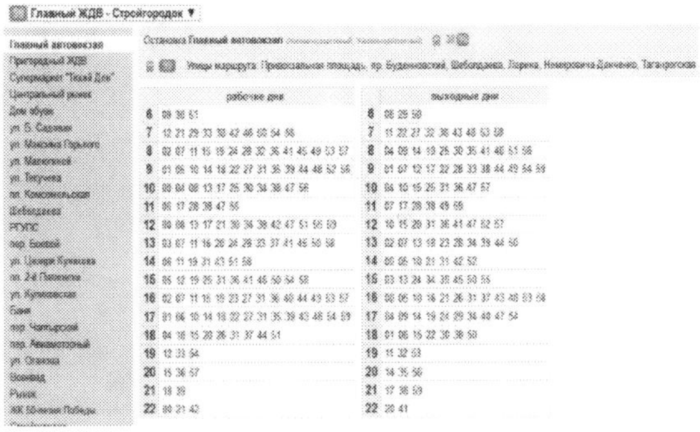

Fig. 2. Example of a timetable for the route № 99

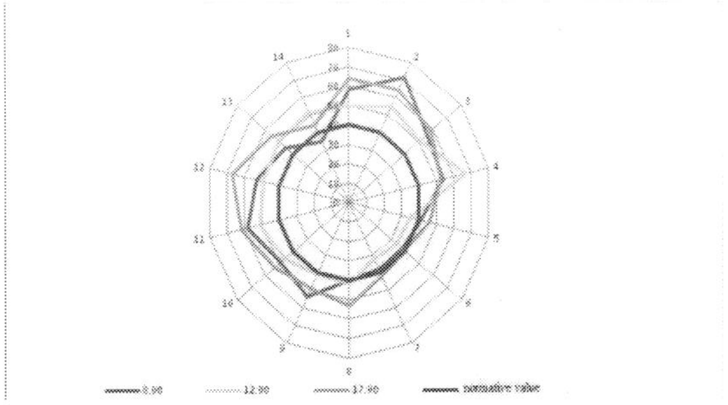

Fig. 3. Analysis of the passengers real time travel compared with the standard value on a weekday from the city center to transportation districts

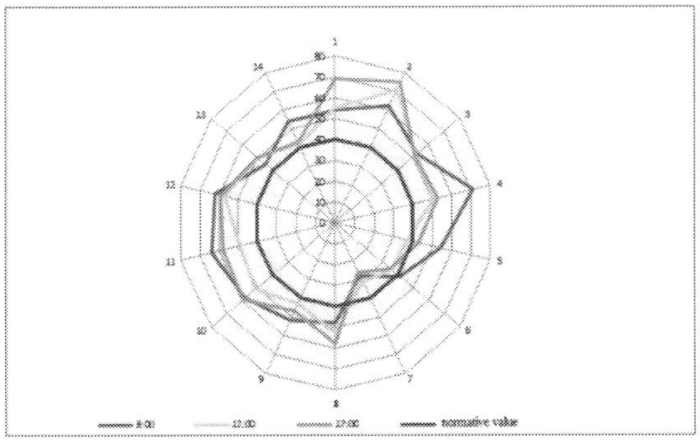

Fig. 4. Analysis of the passengers real time travel compared with the standard value on a weekday from the transportation districts to city center

Fig. 5, 6 shows the values of the passengers travel time by public transport on a day off from the city center to transportation districts in real time.

3 Conclusion

The undertaken study indicates that the travelling on the same routes at the weekend is faster because of less and faster traffic. The change in the urban transport speeds on different sections of the road network at different time periods has also been determined during our research.

On the basis of the study, recommendations for improving the routing network in Rostov-on-Don are proposed, which allow to reduce the average travel time and improve the quality of passenger service.

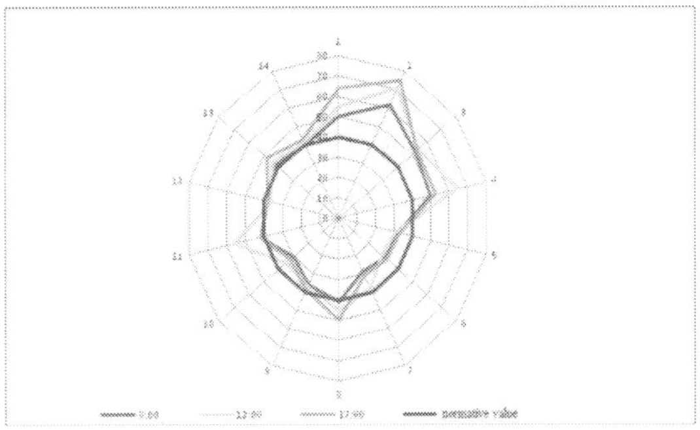

Fig. 5. Analysis of the real time travel of passengers compared to the normative value on a day off from the city center to transport areas

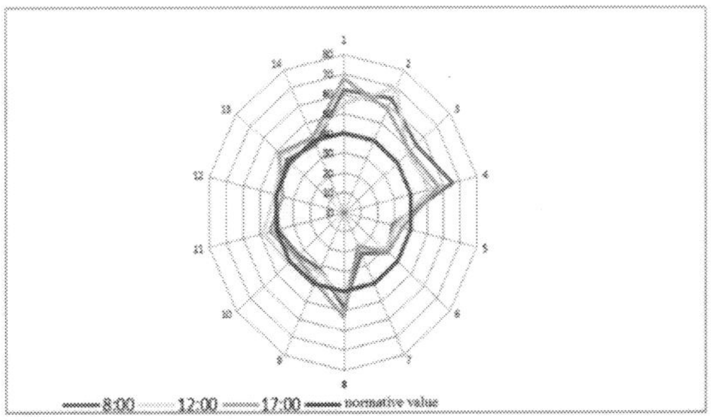

Fig. 6. Analysis of the real time travel of passengers compared to the normative value on a day off from the transport areas to city center

References

1. State Standard P 51004-96. Transport services. Passenger Transportation. Nomenclature of quality parameters. Moscow, 1997, 8 p. [in Russian]
2. E. Semchugova, V. Zyryanov, N. Negrov, A. Nikitina, *Transportation Research Procedia*, no. 20, 584 – 590 (2017)
3. E.Yu. Semchugova, *Sbornik nauchnykh trudov IX Rossiysko-Germanskogo simpoziuma po transportnoy politike i ekonomike*, 110-112, (2009)
4. V.V. Zyryanov, E.Yu. Semchugova, A.M. Skrynnik Don Engineer Bulletin, no.4, part 1 (2012) [in Russian], available at: http://www.ivdon.ru/magazine/archive/n4p1y2012/1083
5. E.Yu. Semchugova, N.S. Negrov, *Nauchnoe obozrenie* [Scientific Overview], no.9, part 3, 1022-1026 (2014)
6. E.Yu. Semchugova, N.S. Negrov, Don Engineer Bulletin, no.4 (2013) [in Russian], available at: http://www.ivdon.ru/magazine/archive/n4y2013/2192
7. V. Zyryanov, R. Sanamov, International Journal of Transport Economics, **XXXVI**, no. 1, 83-95 (2009)

8. N.O. Brailovsky, V.P. Granovsky, *Modeling of transport systems* (Transport, Moscow, 1978)
9. P.P. Volodkin, E.Yu. Semchugova, N.N. Martynenko Internet-journal "NAUKOVENIE", №4 (2012), available at: http://naukovedenie.ru
10. P.P. Volodkin, E.Yu. Semchugova, I.O. Zagorsky, A.N. Nikitina, Internet-journal "NAUKOVENIE", №3 (2012), available at: http://naukovedenie.ru
11. A. E. Gorev, *Fundamentals of the theory of transport systems: Textbook. allowance* (SPbGASU, St. Petersburg, 2010)
12. V.V. Zyryanov, E.Yu. Semchugova, *Quality of transport services: Textbook* - Growth. state. builds. Univ., Rostov n / a (2013)
13. V.V. Zyryanov, E.Yu. Semchugova, A.A. Litvin, Bulletin of Saratov State Technical University, №2 (71), issue 2, 349-353 (2013)
14. V.V. Zyryanov, E.Yu. Semchugova, A.A. Mironchuk *Safety, road, children: practice, experience, prospects and technologies: materials of the forum*, 121-124, (2015)
15. A.A. Litvin, E.Yu. Semchugova, *Science, education, society: problems and development prospects: a collection of scientific papers on the materials of the international scientific and practical conference,* (Publishing house "Business-Science-Society", Tambov), 188-190 (2013)
16. A.P. Lopatin, *Modeling of the transportation process on the GPT* (Transport, Moscow, 1985)
17. R. Mickus, Software development for Urban Public transport Planning, available at: http://www.merakas.lt/paslaugos/pikas
18. N.S. Negro, E.Yu. Semchugova, Scientific Review No. 11, Part 3, 945-948 (2014)
19. A.N. Nikitina, E.Yu. Semchugova, Electronic scientific journal "Engineering Herald of the Don", № 4, part 1, (2012), available at: http://www.ivdon.ru/magazine/archive/n4p1y2012/1114
20. E.Yu. Semchugova, *"Construction-2013": materials of the International Scientific and Practical Conference*, (Growth. state. builds. Univ., Rostov n/a), 11-13 (2013)
21. E.Yu. Semchugova, A.A. Mironchuk, A.V. Dombalyan, *Safety, road, children: practice, experience, prospects and technologies: materials of the forum*, 118-121 (2015)

Hardware in the loop simulation of objects internal structures inhomogeneities in multi-angle electrical impedance tomography

Grayr Aleksanyan [1,*], *Nikolay* Gorbatenko[1], *Artem* Kucher [1], and *Ivan* Shcherbakov [1]

[1]SRSPU (NPI), Department of Information and Measuring Systems and Technologies, 346428 Novocherkassk, Russia

Abstract. The present work is devoted to the development of an algorithm for automatic analysis of images as the results of biological object internal structures conductivity field distribution reconstruction for the automated generation of not only qualitative but also quantitative estimates of inhomogeneities in the tasks of multi-angle multi-frequency electrical impedance tomography. The paper describes the modeling of biological object internal structures inhomogeneities of a given size as a way of obtaining initial data for approbation of the algorithm and its further debugging and optimization. In the work an interactive analysis of the images obtained as a result of modeling is carried out, the results of its work are compared with the results of the developed algorithm of automatic analysis. Based on the analysis, conclusions are drawn about the applicability of the developed algorithm in the problems of multi-angle multi-frequency electrical impedance tomography.

1 Introduction

Electrical impedance tomography (EIT) is a non - invasive effective method to reconstruct and visualize the conductivity distribution in biological objects from the results of electrical measurements on its surface [1-13]. In the general case, the investigated object (IO) is connected to one or more sources of high frequency electrical current through electrodes placed on its surface, and measured the voltages at the electrodes (where i is the number pair of injecting electrodes, j is the number of measuring electrode) [14]. To convert measurement information from the surface of IO in the conductivity distribution in the internal structures of the object is in process of reconstruction, which can be divided into static (absolute), and dynamic (differential). [14,15]

Absolute (static) reconstruction is image reconstruction, which uses one dataset for one point in time and returns absolute value of the conductivity in the object. [1,2]

Differential (dynamic) reconstruction is reconstruction, which uses two sets of data at two points in time and returns change of conductivity in the object between the start and end time. [14,15]

When conducting multi-angle multi-frequency electrical impedance tomography (MMEIT), as in any other medical imaging technique, an important role is played by the analysis of the images. Automatic (machine) analysis of images, in contrast to the interactive, allows to obtain quantitative characteristics of the studied objects, and can be used in computer systems of diagnostics, and, what is the most important thing, in decision-making systems.

The purpose of this paper is to develop algorithms for analyzing the results of the reconstruction of conductivity of the internal structures in the task MAIT and debug to re-the results of hardware-software modelling of inhomogeneity of the internal structures of a biological object of a given size.

2 Materials and Methods

2.1 Experimental design

For modeling inhomogeneity and internal structures of biological object in the phantom [16] (container with polyethylene non-conductive walls with 16 electrodes located along the perimeter) positioned a cylinder of diameter d equal to 13, 18 and 27 mm non-conductive (polyethylene polymer) and conductive (stainless steel) material was placed. Cylinders that simulating inhomogeneity, were placed at fixed distances from the electrode 1 (20, 40, and 80 mm) according to a printed target, glued to the bottom of the phantom, as shown in figure 1(a). Figure 1(b-f) shows described above variety of inhomogeneities size and position in the phantom.

To simulate BO internal structure, 0,454 kg of chicken leg quarter minced in industrial grinder (to avoid metal inclusions that may affect on resulting impedance) with temperature of 4.2°C was used, at an ambient temperature of 24.8°C, atmospheric pressure of 754 mm Hg; and air humidity of 43%.

To conduct hardware and software simulation used the layout of the hardware-software complex electrical impedance imaging [17-20]. Potential differences $\Delta\varphi$ between all neighboring electrode pairs at a frequency of

* Corresponding author: graer@yandex.ru

injecting current f_I = 50 kHz and the amplitude of I_M = 5 mA were measured. These injected current parameters are chosen because of their proven safety for biological object [17], as the simulation of inhomogeneity in biological object is carried out.

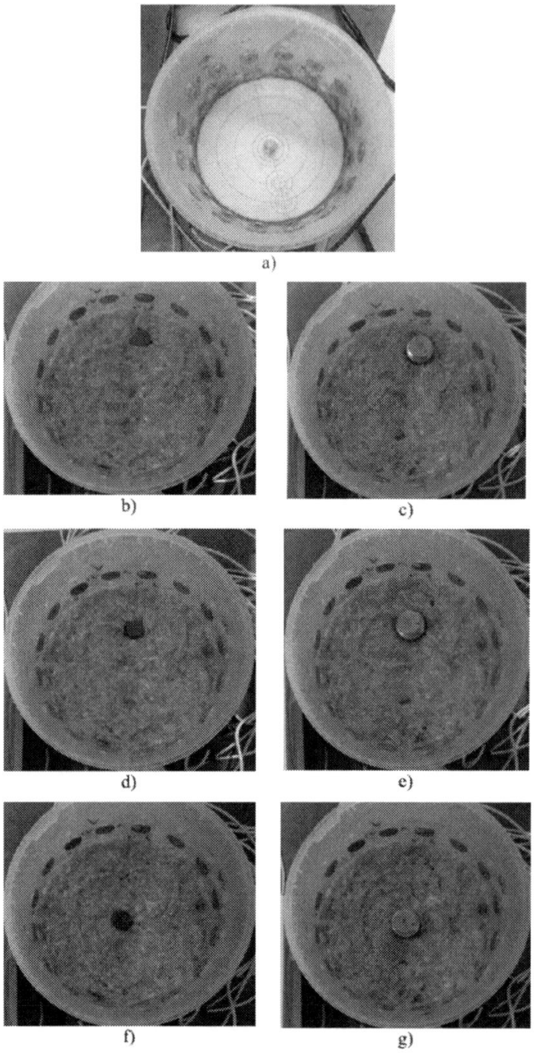

Fig. 1. a) printed target mounted in a bottom of a phantom for correct positioning; b) a non-conductive object with a diameter of 18mm at a distance of 20 mm from the electrode; c) a conductive object with a diameter of 27mm at a distance of 20 mm from the electrode; d) a non-conductive object with a diameter of 18mm at a distance of 40 mm from the electrode; e) a conductive object with a diameter of 27mm at a distance of 40 mm from the electrode; f) a non-conductive object with a diameter of 18mm at a distance of 80 mm from the electrode; g) a conductive object with a diameter of 27mm at a distance of 80 mm from the electrode.

Eleven repeated experiments carried out for each value of the diameter inhomogeneity and the distance between the electrode and center of inhomogeneity and type of inhomoheneity with total count of 198 measurements. As a result, the measurements taken are average values of the above parameters obtained during repeated experiments to reduce measurement inaccuracy. For each successful experiences with the layout of hardware and software, electrical impedance tomography reconstructed image of the matrix with the selected color map each numerical value of the matrix element (figure 2 (a-d)), and implemented interactive (manual visual computer-aided inspection) of the resulting image for later comparison with the results of the developed algorithm.

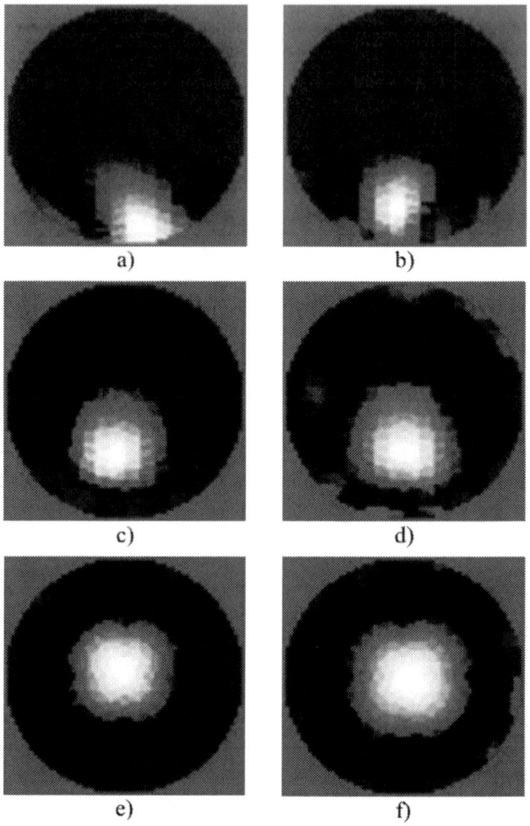

Fig. 2. a) reconstruction: a non-conductive object with a diameter of 18mm at a distance of 20 mm from the electrode; b) reconstruction: a conductive object with a diameter of 27mm at a distance of 20 mm from the electrode; c) reconstruction: a non-conductive object with a diameter of 18mm at a distance of 40 mm from the electrode; d) reconstruction: a conductive object with a diameter of 27mm at a distance of 40 mm from the electrode; e) reconstruction: a non-conductive object with a diameter of 18mm at a distance of 80 mm from the electrode; f) reconstruction: a conductive object with a diameter of 27mm at a distance of 80 mm from the electrode.

To carry out interactive analysis, a free and open-source raster graphics editor (GNU General Public License (GNU GPL v3) GIMP was used [21]. Any other available graphics editor with "measure tape" function can be used. The "measure tape" function allows to recognize the distances in pixels between the points of investigated image. Information about the size of the heterogeneity (showed white on fig.2) and the distance from its center to the electrode is obtained by clicking and holding the mouse button down, the number of pixels between the click point and the current position of the mouse pointer is determined [22]. The inaccuracy of this type of measurement is determined by the resolution of the image (as a result of the resolution of the reconstruction algorithm used in the EIT tasks) and the width and contrast of the region of the inhomogeneity boundary (showed blue on fig.2): the lower the contrast of this region, the worse the eye distinguishes the position of the boundary.

Since the resolution of the reconstructed image in pixels and the diameter of the phantom in millimeters are known, the size and position of the inhomogeneities can also be calculated from the pixels into metric units (millimeters). Measured data on the size of the inhomogeneity and the distance from its center to the electrode and their measurement errors are recorded for each repeated experiment and averaged.

The block diagram of the algorithm of automatic analysis of investigated object internal structures inhomogeneity presented in fig. 3.

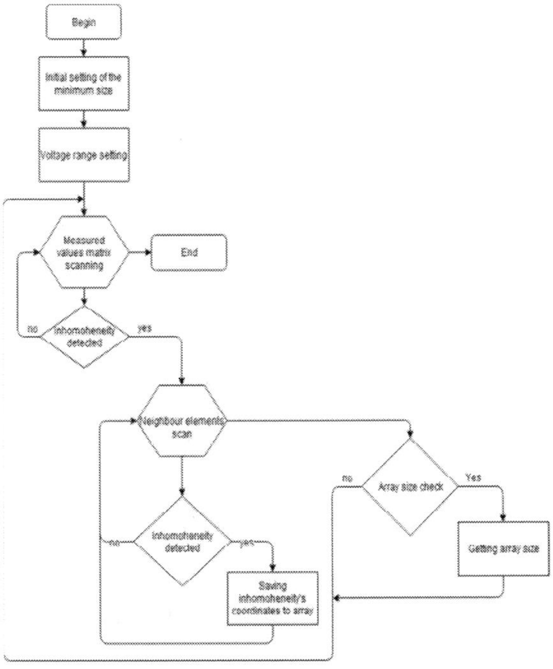

Fig. 3. Block diagram of the developed algorithm for automatic analyzing inhomogeneities of internal structures of the IO.

The principle of its work consists in: the initial setting of the minimum size threshold and voltage ranges (correlated with impedance in term of current source using[20]) of inhomogeneity.

Then the image is scanned as a matrix of voltage values (which are correlating with impedance values) in every pixel of reconstructed image to search for values that match the range of impedances. Upon detection of the first element of inhomogeneity, the values of neighboring elements in the matrix are also tested for inclusion in the range of impedances. If the neighboring elements are also parts of inhomogeneity, the array for storing the coordinates of the positions of the elements of inhomogeneity is created. After all the neighboring elements are tested the size of detected inhomogeneity is determined. If the size of the detected inhomogeneity exceeds a predetermined minimum size, the generated array is stored in memory and the algorithm continues scanning the matrix until find s a new inhomogeneity or finishes their work. The results of the algorithm which also calculated automatically from pixels of reconstructed image to millimeters are displayed in user-friendly form. The inaccuracy of this type of measurement is determined by the resolution of the image too (as a result of the resolution of the reconstruction algorithm used in the EIT tasks), but error caused by width and contrast of the region of the inhomogeneity boundary can be eliminated in this case by correct potential difference (voltage) range settings.

3 Results

Table 1 presents the results of the study.

Table 1. Results of the study.

Obj. type	Physical dimensions of inhomogeneity		The results of the interactive analysis		The results obtained by developed algorithm	
	d, mm	*l, mm*	*d, mm*	*l, mm*	*d, mm*	*l, mm*
Conductive	13±0,5	20±0,5	19±4	17±5	15±2,5	19±2,5
	13±0,5	40±0,5	20±4	25±7	11±2,5	32±2,5
	13±0,5	80±0,5	25±5	65±15	17±2,5	70±2,5
	18±0,5	20±0,5	28±5	12±10	22±2,5	20±2,5
	18±0,5	40±0,5	26±8	22±15	23±2,5	35±2,5
	18±0,5	80±0,5	28±12	76±5	16±2,5	77±2,5
	27±0,5	20±0,5	24±14	10±8	25±2,5	20±2,5
	27±0,5	40±0,5	33±5	34±4	29±2,5	33±2,5
	27±0,5	80±0,5	37±4	94±10	25±2,5	90±2,5
Non-conductive	13±0,5	20±0,5	19±4	17±5	15±2,5	19±2,5
	13±0,5	40±0,5	20±4	24±8	14±2,5	32±2,5
	13±0,5	80±0,5	19±5	56±15	15±2,5	68±2,5
	18±0,5	20±0,5	28±5	12±9	22±2,5	20±2,5
	18±0,5	40±0,5	25±8	22±12	23±2,5	35±2,5
	18±0,5	80±0,5	28±12	76±5	16±2,5	77±2,5
	27±0,5	20±0,5	23±13	8±6	24±2,5	19±2,5
	27±0,5	40±0,5	32±6	34±4	28±2,5	32±2,5
	27±0,5	80±0,5	39±5	89±8	23±2,5	86±2,5

Conducted analysis of the data presented in table 1 shows that the spread of the values of the inhomogeneity parameters (diameter and position) measured in an interactive (manual visual, computer-aided) way is greater in all cases than the spread of the values obtained as a result of physical measurements, and in most cases exceeds the spread of the values obtained with the developed algorithm, what can be explained by the impossibility of registering with the human eye individual shades of the pixel colors in the conditions of fuzzy boundaries of the object on the reconstructed image. Also worth noting is the lower accuracy of the results of the developed algorithm compared to the data obtained as a result of physical measurements of real objects, not least due to the relatively low accuracy of the results of the reconstruction algorithms (low resolution), that is input data for the image analysis algorithm.

4 Conclusion

Hardware and software modeling of inhomogeneities internal structures of objects using the layout of hardware and software electrical impedance tomography was conducted. The algorithm of automatic analysis of inhomogeneity of the internal structures of the object was developed.

A comparison of measured information consisting of the parameters of the inhomogeneity (the diameter and distance to electrode) and measurement errors was conducted on data:

- obtained as the result of the physical measurements of real objects (diameter of phantom, inhomogeneity cylinders samples);

- obtained as the result of the interactive (manual and visual, computer-aided, with graphics editor) analysis of the reconstructed image;

- obtained as the result of the developed algorithm of automatic analysis of inhomogeneity of the internal structures in the object.

The analysis of the comparison shows higher accuracy of the measurement of inhomogeneity with developed algorithm comparing to results of interactive image analysis, and higher efficiency of its use in problems of multi-angle multi-frequency electrical impedance tomography as a tool for the automatic analysis of the inner structures of IO. At the same time, also worth noting is the lower accuracy of the results of developed algorithm compared to the data obtained as a result of physical measurements of real objects, not least due to the relatively low accuracy of the results of the reconstruction algorithms that is input data for the image analysis algorithm.

Several conclusions can be conducted from above:

- developed algorithm has inaccuracy of inhomogeneity positioning due to inaccuracy of reconstruction algorithm;
- -developed algorithm has less factors affecting accuracy of measurement comparing to traditional interactive analysis (even computer-aided) which is widely used in clinical practice;
- developed algorithm of automatic analysis of investigated object internal structures can be used in tasks of multi-angle multi-frequency electrical impedance tomography;
- developed algorithm can be used in a decision-making systems with some accuracy limitations (which is the subject of further research).

Work is performed within the grant of President of Russian Federation for state support of young Russian scientists MK-196.2017.8 "Development of theoretical foundations and algorithms for multi-view systems are three-dimensional electrical impedance tomography for non-invasive medical imaging".

References

1. S.Martin, C.T.M. Choi, Sci. Rep., **7(1)**,7212 (2017)
2. J. Zhang, B.Yang, , H. Li, X.Dong, , M. Dai, Sci. Rep., **7(1)**,4608 (2017)
3. N. Eronia, T.Mauri, E. Maffezzini, etc., Annals of Intensive Care, **7(1)**, 76 (2017)
4. Y. Zhou, X. Li, Measurement: J Int. Meas. Confed., **110**, 27-42 (2017)
5. M.G. Crabb, Inverse Prob. in Sci. & Eng., **25(10)**, 1397-1422 (2017)
6. M.G. Crabb, Inverse Prob. in Sci. & Eng., **25(10)**, X (2017)
7. T.N. Tallman, J.A. Hernandez, NDT and E International, **91**, 156-163 (2017)
8. A. Motamed Dashliborun, M. Hamidipour, F. Larachi, AIChE J, **63(9)**, 4161-4176 (2017)
9. J.S. Martins, C.S. Moura, R.M.F. Vargas, Inverse Prob. in Sci. & Eng., 1-21 (2017)
10. S. Białka, M. Copik, K. Rybczyk, H. Misiołek, Anaesthesiology Intensive Therapy, **49(3)**, 222-226 (2017)
11. Y. Liu, Y. Deng, M. Zhang, P. Yu, Y. Li, Meas. Sci. & Tech., **28(9)**,095301 (2017)
12. G. González, V. Kolehmainen, A. Seppänen, Comp. & Math. with Appl., **74(3)**, 564-576 (2017)
13. Y. Yang, J. Jia, Rev. of Sci. Instrum., **88(8)**, 085110 (2017)
14. Y.S. Pecker, K.S. Brazovskii, V.Y. Usov, M.P. Plotnikov, O.S. Umansky, *Electrical impedance tomography* ("NTL" Publishing House, Tomsk, 2004)
15. D.S. Holder *Electrical Impedance Tomography. Methods, History and Applications* (Institute of Physics Publishing, Bristol and Philadelphia, 2005).
16. G.K. Aleksanyan, N.I. Gorbatenko, A.I. Kucher, Int. J Appl. Eng. Res., **10 (19)**, 40580-40584 (2015)
17. G.K. Aleksanyan, I.D. Shcherbakov, A.I. Kucher, V.V. Demyanov, J Eng. & Appl. Sci., **12 (8)**, 2129-2137 (2017)
18. G.K. Aleksanyan, N.I. Gorbatenko, A.I. Kucher, Int. J Appl. Eng. Res., **10 (19)** (2015)
19. G.K. Aleksanyan, A.I. Kucher, A.D. Tarasov, N.M. Cuong, C.N. Phong, Int. J Soft Comp., **10 (6)**, 462-467 (2015)
20. G.K. Aleksanyan, I.D. Shcherbakov, A.I. Kucher, J Eng. & Appl. Sci., **12 (3)**, 587-592 (2017)
21. A. Peck, Beginning GIMP: From Novice to Professional. (Apress Inc, 2006)
22. Carey Bunks. Grokking the Gimp. (New Riders Publishing, 2000)

Research of the conductivity of organic and inorganic media in multi-angle multi-frequency electrical impedance tomography

Grayr Aleksanyan [1*], *Ivan* Shcherbakov [1], *Artem* Kucher [1], and *Maria* Priyma [1]

[1]SRSPU (NPI), Department of Information and Measuring Systems and Technologies, 346428 Novocherkassk, Russia

Abstract. The work is devoted to the study of the impedance of organic and inorganic media containing inhomogeneities at different frequencies of the injecting current in order to determine the conditions for using the device for multi-angle multi-frequency electrical impedance tomography. The paper provides a method of obtaining data and analysis of the obtained measurement results. Based on the analysis, the dependences of the integral parameter of the change in the potential difference correlating with the total impedance on the frequency of the injected current are determined. To estimate the repeatability of the obtained measurement results, an analysis of the measurement data array was carried out by calculating the standard (root-mean-square) deviation for each obtained in measurement value. Based on the analysis of deviations, the dependences of the standard deviation of the obtained integral parameters on the frequency of the injected current are also determined.

1 Introduction

Electrical impedance tomography (EIT) is a non-invasive method for visualizing and evaluating the conduction field of internal structures of objects (both technical and biological) [1,2]. One of the new directions in the development of EIT is multi-angle multifrequency electric impedance tomography (MMEIT). In the course of the MMEIT, the frequency of the injected current applied to the biological object (BO) plays an important role. Due to the fact that the electrical properties of BO tissues depend on the frequency of the injecting current, the use of MMEIT in clinical practice is an important tool for obtaining diagnostic information. [3-17]

The aim of this work is to study the influence of the frequency of the injected current on the result of measurements, that is, on the impedance. Since devices for carrying out MMEIT operate with a limited range of load resistance, which in this case is the impedance of the BO, the results of this work can be applied to determine the conditions for using this type of device.

2 Materials and Methods

2.1 Experimental design

To estimate the impedance, around the perimeter of the container with polyethylene non-conductive walls containing the media under study (phantom [18]), 16 electrodes from the copper-nickel alloy were placed [18], current injection was performed clockwise. To model the inhomogeneity, cylinders of 18 mm in diameter from non-conductive (polyethylene polymer) and conductive (stainless steel) material were placed in the media under investigation. For correct positioning of inhomoheneity samples, constant relative to electrodes, printed target mounted in a bottom of a phantom.

During the measurements a model of the hardware-software complex of electrical impedance tomography [19] was used, in which the possibility of connecting a current source to any of the 16 electrodes and measuring the potential relative to a common point on any of the 16 electrodes is realized. In present study, the potential difference $\Delta\varphi$ was measured between all neighboring electrodes in pairs at frequencies f_i = 50 ... 250 kHz in 50 kHz increments at the constant amplitude of the injected current I_M = 5 mA. Since the device uses the current source developed in [20], and the impedance Z is calculated by the formula [4]:

$$Z = \frac{\Delta\varphi}{I_M}, \qquad (1)$$

where Z is the impedance value; $\Delta\varphi$ is the measured potential difference; I_M is the amplitude of the injected current, the potential difference in a predetermined frequency range correlates with the impedance of the sample under investigation.

Potential differences $\Delta\varphi$ for all selected potential-measuring points is summed and normalized to obtain the integral parameter Θ, which also correlates with the global impedance Z of the investigated object.

The measurements were carried out with two types of `media: inorganic (tap water, total salt content of 452 mg/l, temperature of 16.8°C) and organic (chicken leg quarter minced in industrial grinder to avoid metal inclusions that may affect on resulting impedance, with

* Corresponding author: graer@yandex.ru

temperature of 4.2°C), at an ambient temperature of 24.8°C, atmospheric pressure of 754 mm Hg; and air humidity of 43%. Photos of phantom with the placed inhomoheneity samples and electrodes connected to the measuring device are shown in Fig. 1 (a is an inorganic media with a conducting object, b is an inorganic media with a non-conducting object, c is an organic media with a conductive object, and d is an organic media with a non-conducting object). Eleven repeated experiments were performed for each organic and inorganic media, conductive and non-conductive sample and frequency f_i values with total count of experimental measurements of 220. As a result of the measurement, the average value of integral parameter Θ of repeated experiments was taken to measurement inaccuracy minimization.

To estimate the repeatability of the experimental results, the standard (root-mean-square) deviation σ of Θ, calculated by formula (2), is applied:

$$\sigma = \sqrt{\frac{1}{n}\sum_{1}^{n}(x_i - \bar{x})^2} \quad , \qquad (2)$$

where n is the number of measurements,

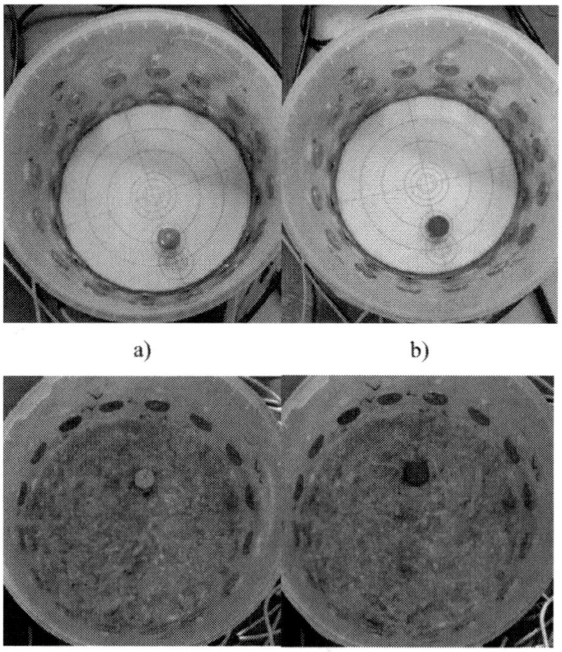

a) b) c) d)

Fig. 1. Phantom with connected electrodes with:
a) an inorganic media with a conducting object; b) an inorganic media with a non-conducting object; c) an organic media with a conducting object; d) an organic media with a non-conducting object.

xi is the value obtained in measurement,

$\bar{x}$ - the arithmetic mean, calculated by the formula:

$$\bar{x} = \frac{1}{n}\sum_{i=1}^{n} x_i \; . \qquad (3)$$

3 Results and discussion

Figure 2 shows the results of integral parameter Θ, obtained from measured on electrodes potential differences, that proportional to the change in global impedance of investigated object Z, on the frequency fi of the injected current.

Figure 2 shows the following dependencies:

- Fig. 2 (a) diagram of the dependence of Θ on fi for an inorganic media and a conductive sample;
- Fig. 2 (b) diagram of the dependence of Θ on fi for an inorganic media and a non-conducting sample;
- Fig. 2 (c) diagram of the dependence of Θ on fi for an organic media and a non-conductive sample;
- in Fig. 2 (d) diagram of the dependence of Θ on fi for an organic media and a non-conducting sample;

In order to estimate the repeatability of the research results, the results of the change in the standard (root-mean-square) deviation σ of the integral parameter Θ from the frequency fi of the injected current I_M are presented in Fig. 3. The following dependencies are shown in the figure:

- Fig. 3 (a) diagram of the dependence of σ on fi for an inorganic media and a conductive sample;
- Fig. 3 (b) diagram of the dependence of σ on fi for an inorganic media and a non-conducting sample;
- Fig. 3 (c) diagram of the dependence of σ on fi for an organic media and a conductive sample;
- in Fig. 3 (d) a diagram of the dependence of σ on fi for an organic media and a non-conducting sample;

Analysis of the data presented in Figures 2 shows that for the inorganic media, both with the conducting object and with the non-conducting as the inhomogeneity, with increasing fi the parameter Θ increases, which indicates an increase in global impedance Z.

At the same time, for an organic media, both with a conducting and non-conducting object as an inhomogeneity, the parameter Θ decreases with increasing fi, which indicates a drop in global impedance Z.

The analysis of the data obtained from the measurement of the standard deviation σ of the obtained integral parameter Θ shown in Figures 3 indicates an increase in the deviation of measured parameters with increasing frequency in all cases both of organic and inorganic media with both conductive and non-conductive samples, which indicates an increase in the spread of the measured parameters with increasing frequency, that measurement inaccuracy may be due to an insufficiently high sampling frequency of analog to digital converter, which is part of the layout of electrical impedance tomography. In any case this effect is a subject of further studies.

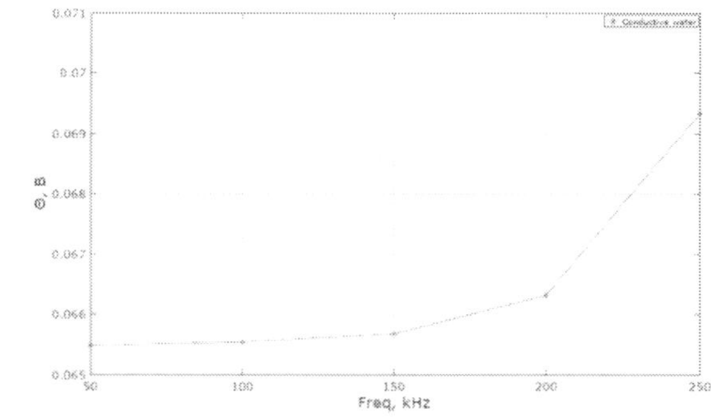

a)

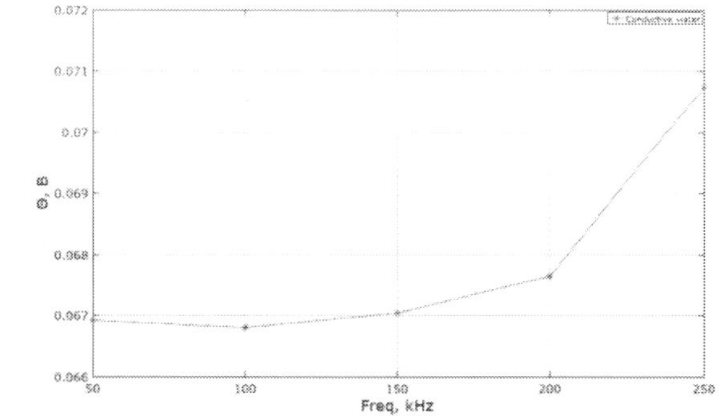

b)

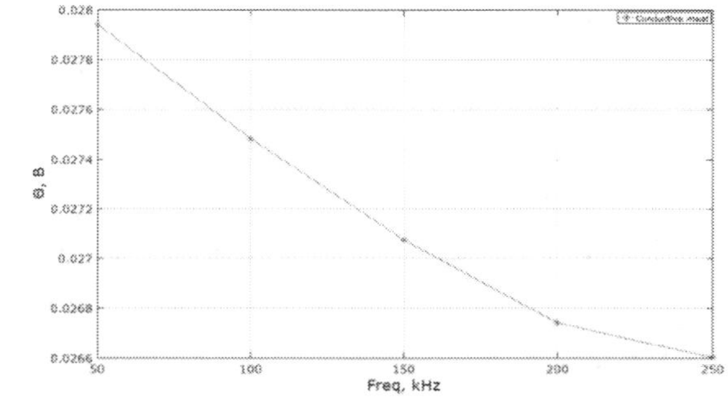

c)

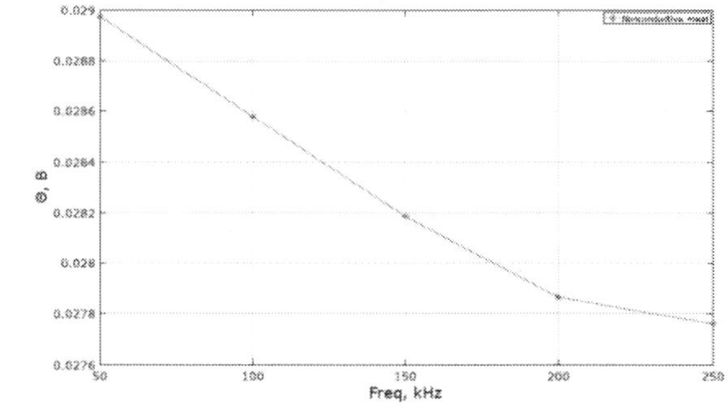

d)

Fig. 2. a) the dependence of Θ on fi for an inorganic media with a conducting object; b) the dependence of Θ on fi for an inorganic media with a non-conducting object; c) the dependence of Θ on fi for an organic media with a conducting object; d) the dependence of Θ on fi for an organic media with a non-conducting object.

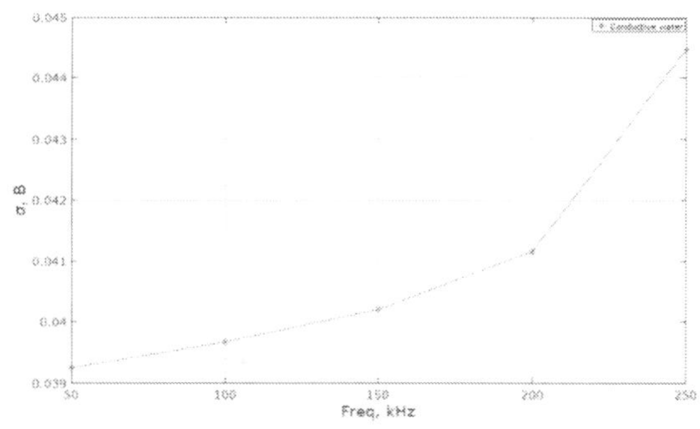

a)

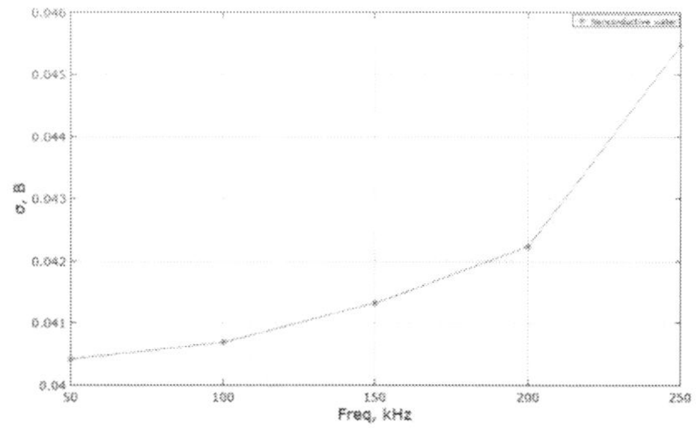

c)

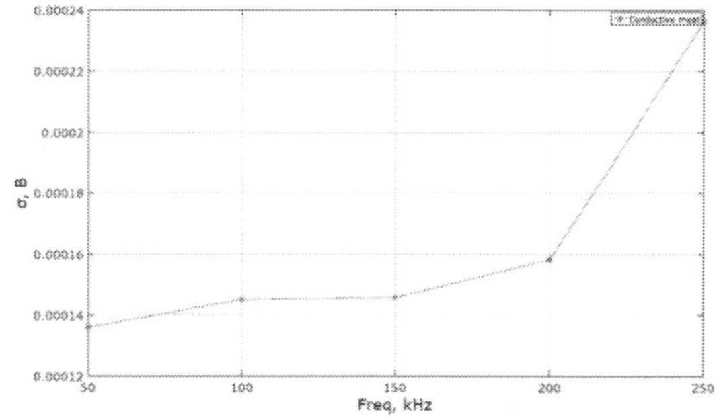

b)

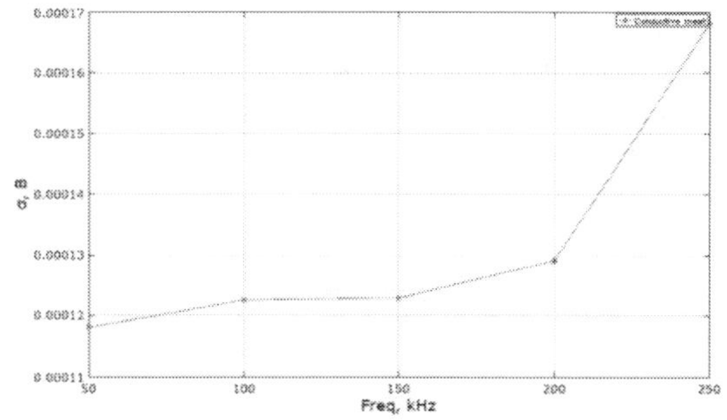

d)

Fig. 3. a) dependence of σ on fi for an inorganic media with a conducting object; b) the dependence of σ on fi for an inorganic media with a non-conducting object; c) the dependence of σ on fi for an organic media with a conducting object; d) the dependence of σ on fi for an organic media with a non-conducting object.

4 Conclusion

As a result of the study, the dependences of the integral parameter Θ (that correlating with impedance Z) on frequency fi was found for variety of conductive and non-conductive samples of predetermined size and position in organic and inorganic media. Using the mock-up of the hardware-software complex of electrical impedance tomography on a phantom in the range of frequencies fi = 50…250 kHz with a increment of 50 kHz and constant amplitude I_M = 5 mA of injected current, the values of mentioned parameter Θ were obtained.

On the basis of the measured data, the dependencies of the integral parameter Θ on the frequency fi at MMEIT are obtained.

An analysis of the dependencies obtained allows us to state that the impedance Z of an inorganic media with conducting and non-conducting inhomogeneities as the test sample of predetermined sizes increases slightly with increasing frequency fi of the injection current I_M.

At the same time, for an organic media with both conductive and non-conducting inhomogeneities as the test sample of the same sizes, the impedance Z decreases with increasing frequency fi of the injection current I_M.

The obtained dependencies show that the effectiveness of multi-angle multi-frequency electrical impedance tomography as a visualization tool for non-invasive investigation of internal structures of a biological object (which can be considered, of course, as a organic media) can be improved by varying the frequencies of the injected current.

At the same time, conducted research shows that varying the frequencies of the injected current when investigating inorganic media gives no positive effect to conductance: impedance Z slightly increases with increasing frequency of injection current. On other hand, such effect can give possibility to improve contrast of boundaries of reconstructed EIT-image [20] when inorganic conductive inhomoheneity is located in organic media (to clarify position and size of cavity containing liquid in a lung, or a bullet inside a body, for example).

Also conducted analysis showed a increase of standard deviation of integral parameter Θ that indicates a lower repeatability of the measurement results with increasing frequency, which is the subject of further research that will be conducted to find factors that affect this.

Work is performed within the grant of President of Russian Federation for state support of young Russian scientists MK-196.2017.8 "Development of theoretical foundations and algorithms for multi-view systems are three-dimensional electrical impedance tomography for non-invasive medical imaging".

References

1. Y.S. Pecker, K.S. Brazovskii, V.Y. Usov, M.P. Plotnikov and O.S. Umansky, *"Electrical impedance tomography"* ("NTL" Publishing House, Tomsk, 2004)
2. D.S. Holder, *Electrical Impedance Tomography. Methods, History and Applications* (Institute of Physics Publishing, Bristol and Philadelphia, 2005)
3. Y. Zhou, X. Li, Measurement: J Int. Meas. Confed., **110**, 27-42 (2017)
4. M.G. Crabb, Inverse Prob. in Sci. & Eng., **25(10)**, 1397-1422 (2017)
5. M.G. Crabb, Inverse Prob. in Sci. & Eng., **25(10)**, X (2017)
6. T.N. Tallman, J.A. Hernandez, NDT and E International, **91**, 156-163 (2017)
7. G.K. Aleksanyan, N.I. Gorbatenko, V.V. Grechikhin, T.N. Phong, T.D. Lam, ARPN J of Eng. & Appl. Sci., **11(9)**, 5871-5875 (2016)
8. G.K. Aleksanyan, N.I. Gorbatenko, A.I. Kucher, K.M. Shirokov, C.N. Phong, Biosciences Biotechnology Research Asia, **12**, 709-718 (2015)
9. G.K. Aleksanyan, N.I. Gorbatenko, A.I. Kucher, T. N. Phong, N. M. Cuong, Int. J of Appl. Eng. Res., **10(19)**, 40580-40584. (2015)
10. G.K. Aleksanyan, N.I. Gorbatenko, A.D. Tarasov, Biosciences Biotechnology Research Asia, **11**, 85-91 (2014).
11. G.K. Aleksanyan, A.I. Kucher, A.D. Tarasov, N.M. Cuong, C.N. Phong, Int. J of Soft Comp., **10 (6)**, 462-467 (2015)
12. K.S. Brazovskii, *Methods and tools to evaluate the functional state of the human brain based on the electrical measurements* (Dr. tehn. Sciences, Tomsk, 2015)
13. N.I. Gorbatenko, *Full-scale-model testing of products made of ferromagnetic materials* (Publishing House SKNTS VSH, Rostov n / D, 2001)
14. A.M. Lankin, M.V. Lankin, G. K. Aleksanyan, N.D. Narakidze, Int. J of Appl. Eng. Res., **10(3)**, 6489-6498 (2015)
15. D.V. Nikolaev, A.V. Smirnov, I.G. Bobrinskaya, S.G. Rudnev *Bioimpedance analysis of the composition of the human body* (Nauka, Moscow, 2009)
16. A.L. Zuev, V.Y. Mishlanov, A.I. Sudakov, N.V. Shakirov, A.V. Frolov, Russ. J of Biomechanics, **16-1(55)**, 110-120 (2012)
17. G.K. Aleksanyan, I.D. Shcherbakov, A.I. Kucher, V.V. Demyanov, J of Eng. & Appl. Sci., **12 (8)**, 2129-2137 (2017)
18. G.K. Aleksanyan, N.I. Gorbatenko, A.I. Kucher, Int. J Appl. Eng. Res., **10 (19)**, 40580-40584 (2015)
19. G.K. Aleksanyan, A.I. Kucher, A.D. Tarasov, N.M. Cuong, C.N. Phong, Int. J Soft Comp., **10 (6)**, 462-467 (2015)
20. G.K. Aleksanyan, I.D. Shcherbakov, A.I. Kucher, J Eng. & Appl. Sci., **12 (3)**, 587-592 (2017)

Modelling of Fuzzy Expert Information in the Problem of a Machine Technological Adjustment

Valery Dimitrov[1,*], *Lyudmila* Borisova[2], *Inna* Nurutdinova[3]

[1]Department of Quality Management, Don State Technical University, Rostov-on-Don, 346500, Russia
[2]Department of Economics and Management in Engineering, Don State Technical University, Rostov-on-Don, 346500, Russia
[3]Department of Applied Mathematics, Don State Technical University, Rostov-on-Don, 346500, Russia

Abstract. The paper considers the problem of stating fuzzy expert information for intelligent systems of decision support. The application of such systems is worthwhile when complex agricultural machines operate under the changing field conditions, characterized by fuzziness of information concerning external environment and parameters of a machine. One of the important stages of making expert systems for agricultural machines is modelling of fuzzy information regarding input and output parameters of the system and also their interrelations. The paper presents the problem solution of linguistic representation of the adjustable parameters of a combine harvester: speed of the combine movement, rotational speed of the threshing drum, and rotational speed of the separator fan. These parameters are most important for providing qualitative indices of technological process in combine harvesting, and they are characterized by efficiency of control and accessibility of updating. As a result of the analysis, linguistic variable, basic and extended term-sets have been determined. On the basis of the information, obtained from four experts, membership functions have been constructed with the use of standard functions of triangle and trapezoidal types. The indices of consistency of the expert models have been calculated, and the optimal models have been selected on the basis of their analysis.

1 Introduction

Intelligent information systems of decision support are widely used while performing control of complex technical systems. An urgent character of the development of such systems for providing control of harvesting agricultural crops is specified by complexity, inertia, diversity of external factors and parameters of the technical system, multiple connections and insufficient study of interrelations of the object domain attributes, difficulty of decision-making when eliminating abnormal situations connected with substandard quality of harvesting. Thereby, the problem of making machines of new, intelligent type is distinguished in the strategy of developing agricultural engineering in our country as one of the most important. Grain combine is the main machine used for harvesting many crops. Efficiency and high quality of harvesting directly depend on optimal solution of the control problem of combine harvesting technological process.

For harvesting machines, working in changing external conditions, the problem of setting optimal values of adjustable parameters is especially urgent. It is rather difficult under field conditions to assess certain values of external factors influencing the technological process, and find out their change in due time. Estimation of external factors has an approximate, frequently qualitative (subjective) character; this stipulates reference for the mathematical apparatus of the theory of fuzzy sets [1, 2]. In [3, 4] there is a review of the available approaches to the problem of a combine harvester technological adjustment; among the considered approaches there are regression models, experimental and statistical methods, automated facilities for control and management. The first two represent tedious mathematical constructions that don't always adequately reflect reality, their use in real time and complex conditions is difficult; the last one mainly solves the problems of a combine harvester movements control and doesn't provide decision support while carrying out technological adjustment of the combine harvester working tools. Complexity and low efficiency of the conventional mathematical models of a combine operation, and also fuzzy character of information concerning external factors and their interrelations with technological parameters stipulate the use of mathematical apparatus of the theory of fuzzy sets [1, 2] as the main one for describing the process of decision-making in complex multilevel hierarchic systems that grain combine harvesters refer to. The suggested approach of fuzzy modeling of decision-making processes, when performing technological adjustment and updating the parameters of the combine harvester

* Corresponding author: kaf-qm@donstu.ru

working tools, meets the requirements of the system analysis. It provides integrity of consideration of the complex hierarchy system (combine harvester) on the basis of accounting basic elements and processes in the system and relations between them. It also has sufficient degree of simplification when modelling, allowing to reflect adequately a real process and take into account the determining factors in this system.

Intelligent information systems of decision support on the basis of fuzzy control [5] are at present used in agricultural industry. But, as a rule, they are connected with the analysis of patterns, weather conditions, sorting or processing products, yield checking, they are applied for identification of weeds, definition of the combine harvester location in the field, its monitoring and movements etc. [6-13]. The problems of automatic setting some parameters of the combine operation on the basis of quality factors when using production rules are also considered [14, 15].

2 Problem definition

In fuzzy models for formalizing real systems a linguistic approach is applied, and the dependency between the sets of input and output variables is described on the qualitative level in the form of statements by means of production rules. The mechanism of fuzzy logic inference in fuzzy models is significantly based on expert knowledge. Authenticity and correspondence of expert information to real conditions essentially influence the accuracy of solution; this establishes high requirements to organization of the base of expert information.

Modelling of decision-making in fuzzy expert systems consists of three stages: fuzzification, composition and defuzzification. Generation of the expert information unit takes place at the stage of fuzzification. At this stage, an object domain is studied, significant factors are determined and the conditions of the problem solution are presented in a linguistic form. In accordance with the logic-linguistic approach, the models of input and output attributes are developed in the form of semantic spaces and the corresponding them membership functions (MFs).

In the problem of technological adjustment of the grain combine harvester parameters the input parameters include crop yield, humidity, content of impurities, haulm density and some others. In [16-17] we considered the problem of setting optimal parameters of the models of the mentioned linguistic variables (LVs) that characterize external factors.

The output LVs for the problem under consideration are adjusting parameters. For technological adjustment of the main units of the grain combine harvester (header, grain separator) 54 adjustable parameters are used.

The most important ones with regard to operational efficiency, accessibility and degree of impact on the quality of harvesting are speed of the combine movement, rotational speed of a threshing drum and rotational speed of a separator fan. The present paper considers establishment of fuzzy expert information for the indicated adjusting parameters.

3 Problem solution

The performed analysis of the object domain "Combine harvester technological adjustment" made it possible to set up basic and extended term-sets of the linguistic variables Speed of a harvester (SH) (for the crop yield of approximately 40 q/ha), Rotational speed of a threshing drum (RSTD) (for the crop yield of approximately 50 q/ha) and Rotational speed of a separator fan (RSSF) (for harvesting barley). To provide assessment, four experts were given 5-term and 7-term models. Typical functions of trapezoidal and triangle types have been applied for describing the terms [18]. To choose the optimal quality of terms we applied the technique [16, 17] which is based on the analysis of indices of general and pair consistency of models.

General consistency of models of expert estimation of an attribute is determined by additive k and multiplicative $\tilde{k}$ indices [18, 19].

$$k = \frac{1}{m}\sum_{l=1}^{m}\frac{\int_0^1 \min\limits_{\forall i=1,2,\dots,n}\mu_{il}(x)dx}{\int_0^1 \max\limits_{\forall i=1,2,\dots,n}\mu_{il}(x)dx};\ \tilde{k} = \sqrt[m]{\prod_{l=1}^{m}\frac{\int_0^1 \min\limits_{\forall i=1,2,\dots,n}\mu_{il}(x)dx}{\int_0^1 \max\limits_{\forall i=1,2,\dots,n}\mu_{il}(x)dx}} \quad (1)$$

where l=1,2,...,m - is the term number, i=1,2,...,k - is the expert number, $\mu_{il}(x)$ - is MF which was assigned by the i -th expert for the l-th term.

The index of difference d_{ij}^l between the models of two experts, i-th and j-th, within the framework of l-th term is determined as a linear distance (by Hemming) between fuzzy sets with MF $\mu_{il}(x)$ and $\mu_{jl}(x)$ [20].

$$d_{ij}^l = \int_0^1 \left|\mu_{il}(x) - \mu_{jl}(x)\right| dx \quad (2)$$

The values d_{ij}^l form a matrix of fuzziness indices D^l for l-th term.

The index of consistency between the models is determined by the value k_{ij}^l [18].

$$k_{ij}^l = \frac{\int_0^1 \min[\mu_{il}(x),\mu_{jl}(x)]dx}{\int_0^1 \max[\mu_{il}(x),\mu_{jl}(x)]dx} \quad (3)$$

Then we construct the pair consistency matrix K^l of the models of i-th and j-th experts within the framework of l-th term. It is evident that on the main diagonal of the matrix D^l there are nulls, and on that of the matrix K^l there are unities and symmetric matrices.

On the basis of matrix of fuzziness indices and matrix of pair consistency of the models for all terms, matrix of fuzziness indices D and matrix of consistency K of the models for all terms are determined. Their elements are determined by the formulae [18, 19].

$$d_{ij} = \frac{1}{m}\sum_{l=1}^{m} d_{ij}^{l}, \quad k_{ij} = \frac{1}{m}\sum_{l=1}^{m} k_{ij}^{l} \tag{4}$$

A subsystem of obtaining and updating of intelligent information systems knowledge has been developed for computations automation on estimating consistency of fuzzy expert information [21]. When introducing new and updating available linguistic variables a computation of additive and multiplicative indices of common consistency among the models is performed.

Results of calculations of additive k and multiplicative $\tilde{k}$ indices by the formulae (1) for all the models are given in table 1.

Table 1. Results of calculations of indices k and $\tilde{k}$.

LV	Model	k	$\tilde{k}$
SH	5-term	0,721	0,716
	7-term	0,892	0,890
RSND	5-term	0,697	0,692
	7-term	0,765	0,753
RSSF	5-term	0,734	0,727
	7-term	0,729	0,728

The analysis of consistency of expert models made it possible to choose optimal models of the linguistic variables: 7-term models for Speed of a harvester and Rotational Speed of a threshing drum and a 5-term model for Rotational speed of a separator fan. As a result we obtained the following linguistic description of the adjustable parameters under consideration.

The tuple of the output LV "Speed of the harvester-40" (i.e. for the crop yield of approximately 40 q/ha) has the following form:

< SPEED OF HARVESTER, km/h {Very low, Low, Lower than nominal, Nominal, Higher than nominal, High, Very high}, [2,5 – 5,5] > SH = {VLSH, LSH, LNSH, NRTH HNSH, HSH, VHSH, km/h}.

The tuple of the output LV "Rotational speed of the threshing drum for wheat-50" (i.e. for the crop yield of approximately 50 q/ha) has the following form:

<ROTATIONAL SPEED OF THRESHING DRUM, rev/min {Very low, Low, Lower than nominal, Nominal, Higher than nominal, High, Very high}, [620 – 940] >
RSTD = {VL, L, LN, N, HN, H, VH, rev/min}.

The tuple of the output LV "Rotational speed of the separator fan - barley" (i.e. for barley harvesting) has the following form:

<ROTATIONAL SPEED OF SEPARATOR FAN, rev/min {Low, Lower than nominal, Nominal, Higher than nominal, High}, [550 – 800] >
RSSF = { L, LN, N, HN, H, rev/min}.

For the selected optimal models the parameters of the generalized MFs have been calculated. When averaging, we used equal weighting factors, and the reason for this was rather high indices of pair consistency (3) for all experts. As an example we provide matrix K (4) of pair consistency for all terms for the optimal model of the LV Rotational speed of the separator fan.

$$K = \begin{pmatrix} 1 & 0,992 & 0,735 & 0,907 \\ 0,992 & 1 & 0,795 & 0,832 \\ 0,735 & 0,795 & 1 & 0,813 \\ 0,907 & 0,832 & 0,813 & 1 \end{pmatrix}$$

The graphs of the generalized MFs in question are presented in figures 1 - 3.

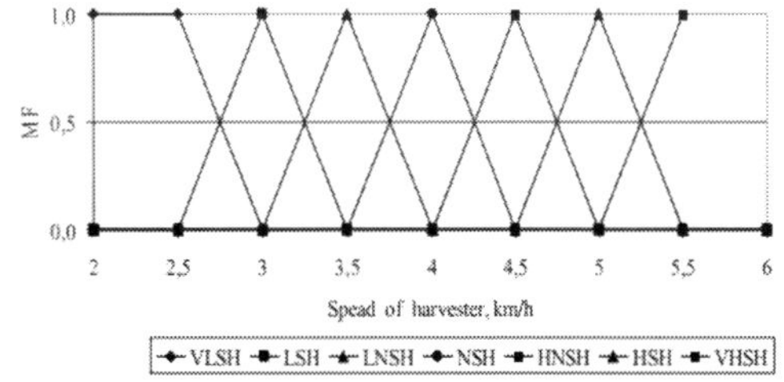

Fig. 1. Membership functions of LV terms Speed of the harvester-40.

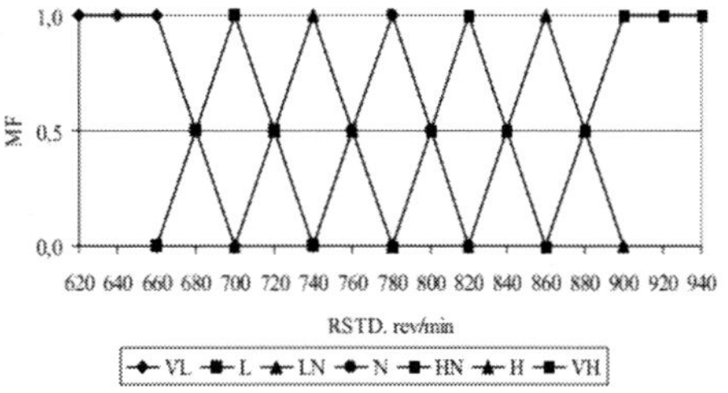

Fig. 2. Membership functions of LV terms Rotational speed of the threshing drum for wheat.

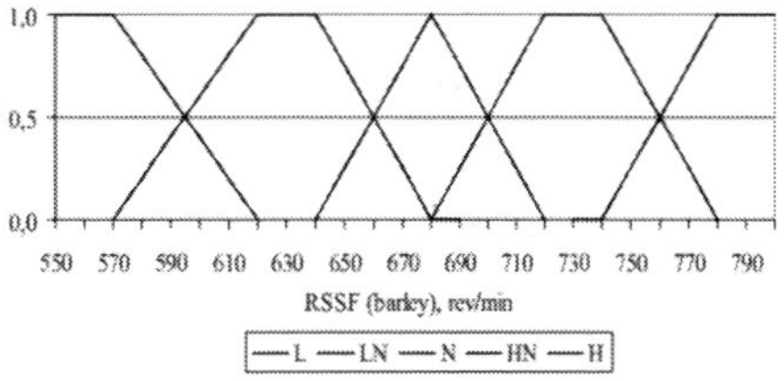

Fig. 3. Membership functions of LV terms Rotational speed of the separator fan.

The generalized membership functions are applied further in the system of fuzzy logical inference [22] for the determination of exact values of adjustable harvester parameters.

4 Conclusion

The analysis of the published papers on theoretical and applied aspects of the theory of fuzzy sets illustrates that the application of fuzzy models allows us to represent correctly the properties of the initial system and simplify the process of its analysis. The adopted degree of abstraction is the most reasonable as more sweeping generalization will be fruitless from the point of view of practice, and insufficient degree of abstraction will result in infeasibility of the problem solving – it will be too complicated.

The application of the mentioned approach to the problem of technological adjustment of the combine harvester in the field made it possible to formalize the object domain, obtain linguistic description of the adjustable parameters, select optimal term-sets, and to construct the generalized membership functions of the linguistic variables. Therefore, the unit of fuzzy expert information has been prepared for realizing fuzzy logical inference of the expert system, which provides setting-up optimal adjustable parameters of the combine harvester. The application of such expert systems in the field environment makes it possible to increase output of the combine harvester and reduce grain losses.

References

1. L.A. Zadeh, Information and Systems, **8**, 338-353 (1965)
2. L.A. Zadeh, *An Introduction to Fuzzy Logic Applications in Intelligent Systems*, 1-25 (Springer Science+Business Media, New York, 1992)
3. L.V. Borisova, I.N. Nurutdinova and V.P Dimitrov, *International Conference on Mathematical Models and Methods in Applied Science (MMAS'14)*, Saint Petersburg, Russia, 226-229 (2014)
4. L.V. Borisova, I.N. Nurutdinova and V.P. Dimitrov, International Journal of Fuzzy Systems and Advanced Applications, **1**, 88–93 (2014).
5. K. Asai, D. Vatada, S. Sugeno, *Applied fuzzy systems* (Mir, Moscow, 1993).
6. M. Omid, Expert Systems with Applications, **38**, 4339–4347 (2011)
7. R.F, Chevalier, G. Hoogenboom, R.W. McClendon, and J.O. Paz, Environmental Modelling & Software, **35**, 84–91 (2012).
8. M. Sujaritha, S. Annadura, J. Satheeshkuma, S.K. Sharan and L. Mahesh, Computers and Electronics in Agriculture, **134**, 160–171 (2017)
9. W. Goodridge, M. Bernard, R. Jordan and R. Rampersad, Computers and Electronics in Agriculture, **133**, 80–87 (2017)
10. J.T. Walls III, P. Caciagli, J.F. Tooker, J.M. Russo, E.G. Rajotte, and C. Rosa, Computers and Electronics in Agriculture, **127**, 775–786 (2016)
11. H. Zareiforoush, S. Minaei, M.R. Alizadeh, A. Banakar and B.H. Samani, Computers and Electronics in Agriculture, **124**, 14–22 (2016)
12. H. Navarro-Hellín, J. Martínez-del-Rincon, R. Domingo-Miguel, F. Soto-Valles and R. Torres-Sánchez, Computers and Electronics in Agriculture, **124**, 121–131 (2016)
13. M.A. Márquez-Vera, J.C. Ramos-Fernández, L.F. Cerecero-Natale, F. Lafont, J.-F Balmat and J.T. Esparza-Villanueva, Computers and Electronics in Agriculture, **124**, 168–174 (2016)
14. M. Omid, M. Lashgar, H. Mobli, R. Alimardani, S. Mohtasebi and R. Hesamifard, Expert Systems with Applications, **37**, 7080–7085 (2010)
15. G. Craessaert, J. de Baerdemaeker, B. Missotten and O. Saeys, Biosystems Engineering, **106**, 103–111 (2010)
16. L.V. Borisova, I.N. Nurutdinova and V.P. Dimitrov, Wseas Transaction on Systems, **14**, 278–285 (2015)
17. L.V. Borisova, I.N. Nurutdinova and V.P Dimitrov, *Proceedings of the 19th International Conference on Systems (part of CSCC'15)*, Zakynthos Island, Greece, 460-463 (2015)
18. A. Kofman, *Introduction in the theory of fuzzy sets*, (Radio and communication, Moscow, 1982)
19. V.P. Dimitrov, L.V. Borisova and I.N. Nurutdinova, Vestnik of DSTU, 1-2(62), 46–50 (2012)
20. A.N. Borisov, A.V. Alekseev, G.V. Merkurieva, N.N. Sliads and I.V. Glushkov, *Processing of fuzzy data in the decision-making support systems*, (Radio and communication, Moscow, 1989)
21. V.P. Dimitrov, L.V. Borisova, I.N. Nurutdinova, and E.V. Bogatyreva, Vestnik of Don State Technical University, **11**, 83–90 (2011)
22. L.V. Borisova, I.N. Nurutdinova and V.P. Dimitrov, Vestnik of DSTU, **15**, 2(61), 100-107 (2015)

Optimization of the automated colorimetric measurement system for pH of liquid

Oleg Katin[1,*], *Aleksander* Lukyanov[1], and *Ksenia* Goryanina[1]

[1]Don State Technical University, Rostov-on-Don, Russia

Abstract. This article considers the issue of automatic control of the acidity of the aquatic environment, which is relevant in such branches of agriculture as hydroponics and aeroponics. A method for measuring the pH of a liquid using a potentiometric method is considered. This method allows to obtain the most accurate pH values, but has some drawbacks. In particular, the article analyzes the advantages of this method over the use of a universal indicator paper and a color sensor for pH determination, describes the components, the conditions of their operation and storage, describes the basic principles of the measuring system operation. A method for converting a signal received at a pH electrode into a microcontroller suitable for processing is also contemplated. The main goal of the development and application of the measuring system described in the article is to achieve a high degree of autonomy and automation of aeroponic and hydroponic greenhouse complexes.

1 Introduction

Currently, all-possible methods of growing plants in greenhouse facilities are gaining popularity. Aeroponics, hydroponics and similar methods are developing most intensively [1-3]. Since the plant develops, obtaining the necessary substances from the liquid, rather than the soil, the issue of maintaining a balance of nutrients can be called one of the most important in these industries. From its solution directly depends on the quality of the harvest and productivity. Automation of monitoring the state of solution, reducing energy costs, reducing operating costs and increasing the degree of system autonomy.

This article is devoted to the analysis and refinement of the method of measuring acidity in the automatic pH control system described in the same name article [4].

Previously, it was proposed to use a universal indicator paper to measure the acidity of the liquid [5]. This method is effective in cases when the required accuracy of measurement does not exceed integral units.

During the operation and testing of the mentioned system, it was decided to refine the measuring part in terms of increasing the accuracy. The objectives of such refinement are: elimination of measurement errors, elimination of identified shortcomings, expansion of the target market segment.

2 Application of the pH electrode

As a measuring device, it is proposed to use a pH electrode that combines in itself a measuring electrode and a reference electrode (Fig. 1).

Fig. 1. Combined pH electrode

The short principle of its operation is to determine the potential of the measuring electrode in the liquid under investigation with respect to the potential of the reference electrode [6, 7]. It is not possible to measure directly the absolute value of the potential at the present time.

The voltage at the sensor outputs is small, you need to convert it for processing by the controller [8, 9]. There are many ways of this transformation. Specialized modules are used to solve this problem. The disadvantage of such modules is the cost.

Self-assembly of the signal conversion circuit will reduce costs. There are many options for building converters. They differ in the accessories, but in all cases the result is the possibility of processing the readings by the microcontroller.

We chose a circuit (Fig. 2 and 3), supporting the I2C interface, which greatly simplifies the integration of the sensor into the measuring system.

* Corresponding author: okatin96@yandex.ru

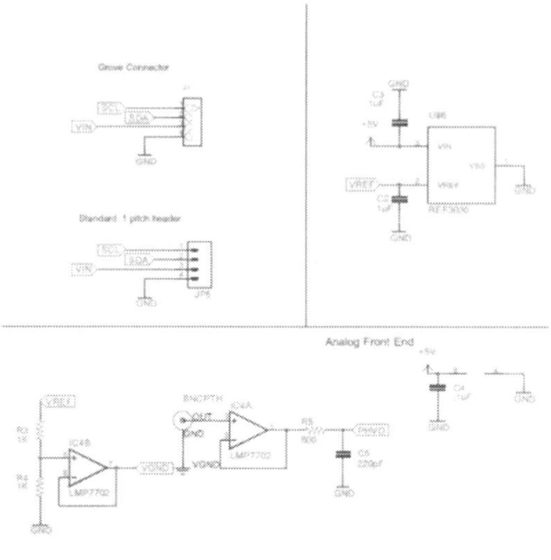

Fig. 2. Scheme of the signal converter (part 1)

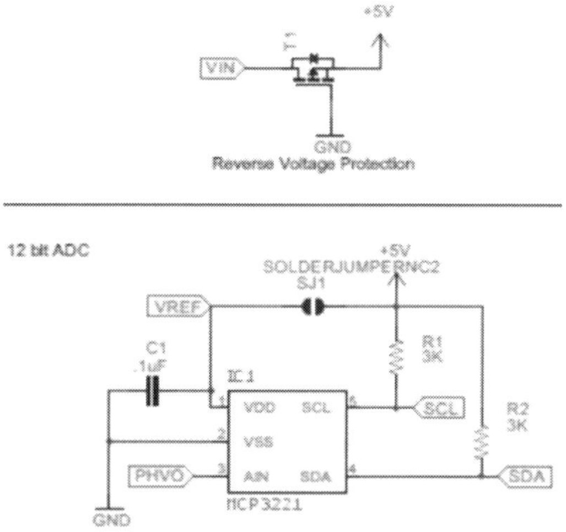

Fig. 3. Scheme of the signal converter (part 2)

3 I2C interface

Let's consider in more detail the I2C interface and its capabilities. Connection of peripheral devices is simplified, the number of necessary wires is reduced. With the exception of power lines, two connections are required.

The interface provides the ability to connect a large number of compatible devices at the same time, which in the future can be used to monitor pH values in several containers with a nutrient solution by a single controller [10]. The benefits of such an opportunity are most noticeable in the automation of a large greenhouse complex with a complex branched structure. In such conditions, the cost of connecting a plurality of sensors and their organization into a single automatic system becomes the most significant.

4 Arduino Uno

Arduino Uno controller (Fig. 4) is used in this system, which has sufficient capabilities for solving the task [11, 12]. It is also easy to learn, does not require special skills for working with microprocessor technology, it is often used for teaching students. The programming language supported by this controller allows the user libraries to be used, which greatly increases the functionality and adapts the controller to its needs. Distributed and various programs to simplify the interaction with the controller. Some of them are at the level accessible even for the pupils of the junior school.

Fig. 4. Arduino Uno

5 Analysis of method of measurement

Measurement is suggested to be performed by moving the electrode from the buffer solution to the test solution [13]. The use of a buffer solution is necessary to comply with the conditions of storage and operation of pH electrodes.

These conditions also include the calibration of the electrode in two, and to minimize the measurement error at the three reference pH values, in order to avoid distortion of the readings during long-term operation of the sensor [14-16]. For these purposes, solutions that support clear preset pH during its shelf life. Depending on the desired accuracy of the measurements, calibration can be made at two or three points. They usually correspond to pH values of 4, 7, 10 (Fig. 5).

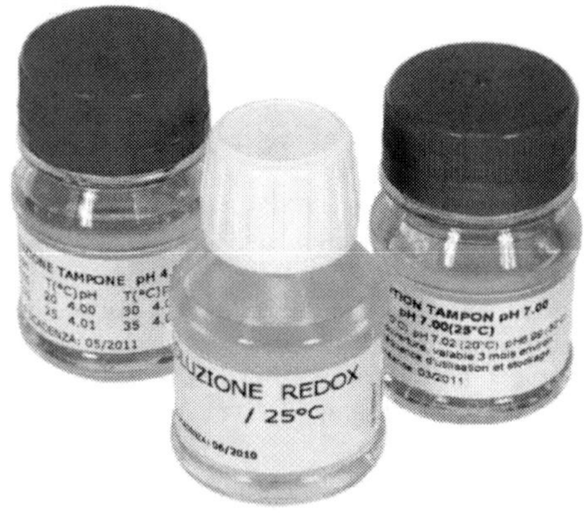

Fig. 5. Calibration solutions

Violation of the storage conditions will entail a loss in the accuracy of the measurement and will approximate the time of the output of the electric power [16]. However, even if all the requirements are met, the possibility of an electrode failure is not ruled out. To avoid negative consequences, it is necessary to conduct periodic testing of sensors.

The pH value of the meter depends on the temperature. If this factor is not taken into account, the result of the measurements will be far from reality. Any temperature-sensitive temperature sensor with acceptable accuracy can be used for thermal compensation.

Moving is carried out according to the following scheme:

1) the electrode moves from the buffer solution to the test solution, with the help of a specialized drive;

2) there are measurement and transmission of data to the controller;

3) the electrode is returned to the buffer solution, where it waits for the next measurement cycle;

4) the controller processes the received data.

The result of the measurement is the pH value, determined to within hundredths of a fraction. To use the system in aeroponics, hydroponics, fish farms, this is more than enough [17]. Let us consider aeroponics as examples.

To obtain a good yield, the pH of the nutrient solution needs to be maintained in the range of 5.5-6.5 units. At the same time, the balance of the substances necessary for the plant is ensured, and, consequently, the quality of the crop. Proceeding from this, we conclude that the accuracy of the proposed system is sufficient.

This aspect is especially important in the cultivation of fruit plants, which are more demanding of the nutrient medium than others. In this case, the fruit's balance depends on the balance of trace elements contained in the solution, which directly affects the competitiveness of such a product. It should be noted that some plants require a variable pH value during their life cycle. That is, it is necessary to regulate the amount of mineral additives depending on the growth phase.

The advantages of the potentiometric measurement method have been described above. Let us turn to the consideration of disadvantages.

First, the cost of pH electrode and consumables exceeds the cost of measurement using universal indicator paper. There is also the possibility that the measuring electrode will fail. In this case, its replacement is required, which will entail additional costs.

Secondly, it is necessary to observe the conditions of storage and operation. This means that it is required to monitor the state of buffer and calibration solutions. It is also important to test the electrode.

6 Conclusion

Application pH electrode is more promising than universal indicator paper [18-20]. The decision to use the pH electrode was taken after testing the layout of the system with a universal indicator paper (Fig. 6).

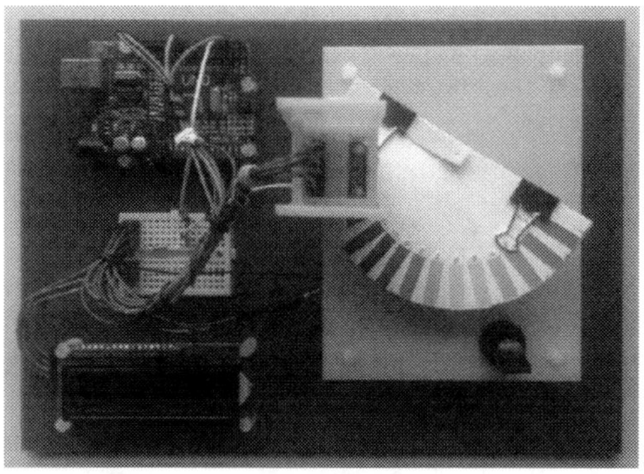

Fig. 6. Layout of measuring systems using universal indicator paper

When selecting components, their advantages and disadvantages were determined, the principles of work were studied, and the instruments were tested. The following is noted:

1) the pH measuring system using an electrode provides greater accuracy;

2) the possibility of using the I2C bus greatly simplifies the connection and operation of the system;

3) the use of thermal compensation and calibration are mandatory.

Based on the results of the study, a model of the measuring system is being developed.

The cost of maintaining the system has increased, but such growth is offset by new opportunities. And since its design used components with relatively low cost, the system is quite competitive in the market of automatic equipment for hydroponics and aeroponics. At this point, testing and development of the system continues. In fact, the scope of its application goes beyond aeroponics and hydroponics. However, they remain prioritized.

References

1. E. V. Kirillova, *Hydroponics* (Rosmen, Moscow, 2005) [in Russian]
2. Aeroponics, available at: http://strmnt.com/sad/l-diz/s-other/aeroponika.html (accessed 20.07.2017) [in Russian]
3. A guide to growing vegetables and herbs on hydroponics, available at: http://step2nature.ru/books/Hydroponics.pdf (accessed 20.07.2017) [in Russian]
4. O. I. Katin, International Student Scientific Bulletin, **1**, №5, 117-118 (2017) [in Russian]
5. O.I. Katin, A.D. Lukyanov, *Modern informatization problems in economics and safety: Proceedings of the XXII-th International Open Science Conference*, 16 – 21, (2017) [in Russian]
6. M. J. Goldcamp, Journal of Chemical Education, 87(11), 1262-1264 (2010)
7. D. J. Graham, Journal of Chemical Education, 90(3), 345-351 (2013)

8. J. F. Cox, *Fundamentals of linear electronics: integrated and discrete* (Cengage Learning, 2001)
9. B. A. Shenoi, Introduction to digital signal processing and filter design (Hoboken: John Wiley & Sons, 2006)
10. I2C Interface, available at: http://www.i2c-bus.org/i2c-interface (accessed 01.08.2017) [in Russian]
11. V. A. Petin, *Projects using the Arduino controller* (BHV-Petersburg, St. Petersburg, 2015) [in Russian]
12. Arduino, available at: https://www.arduino.cc/en/Tutorial/HomePage (accessed 25.07.2017)
13. O. I. Katin, *Aspects of the development of science, education and industrial modernization: materials of the All-Russian scientific-practical conference with international participation. Technical Sciences*, Rostov-on-Don, 295-297, (2017) [in Russian]
14. D. V. Gough, ACS Applied Materials & Interfaces, **1**, 4(11), 6247-6251 (2012)
15. G. Cepria, Journal of Chemical Education, 91(4), 524-530.
16. A. P. Nechiporenko, *Physico-chemical (instrumental) methods of analysis. Electrochemical methods. Potentiometry and conductometry: the teaching method.* Allowance. NIU ITMO IKH i BG, SPb, 2013) [in Russian]
17. O. I. Katin, *State and Perspectives of the Development of Agricultural Machinery: a Collection of Articles of the Xth International Scientific and Practical Conference*, Rostov-on-Don, 604-607 (2017) [in Russian]
18. G. Held *Introduction to light emitting diode technology and applications* (CRC Press, Worldwide, 2008)
19. G. Brooker *Introduction to sensors for ranging and imaging* (ScitTech Publishing, 2009)
20. R. Gonzalez, R. Woods, *Digital image processing* (Technosphere, Moscow, 2012) [in Russian]

MATEC Web of Conferences 132, 04011 (2017) DOI: 10.1051/matecconf/201713204011
DTS-2017

Model implementation of the simulation environment of voting algorithms, as a dynamic system for increasing the reliability of the control complex of autonomous unmanned objects

Igor Kovalev[1], *Vasiliy* Losev[1], *Mikhail* Saramud[1,*], and *Mariam* Petrosyan[1]

[1]Reshetnev Siberian State University of Science and Technology, Russian Federation

Abstract. In the article the question of application of redundant schemes for the development of fault-tolerant software, in particular multiversion programming, is considered with reference to the task increasing the reliability of the control complex of autonomous unmanned objects. Modifications of the basic voting algorithms by an agreed majority and its fuzzy version are proposed. The software implementation of the simulation environment that implements the described approach and the proposed modified algorithms is considered. The results of the simulation the dependence of the reliability indicators of the system on its input parameters are shown by authors.

1 Introduction

With the development of robotics in recent years, autonomous unmanned objects, including unmanned aerial vehicles (UAV), have become increasingly important. The impetus for this is the development of the technical base - the emergence of compact controllers with higher processing power, accurate sensors, cameras with high resolution, and the emergence of the market for finished products. For example, a circular laser scanner (lidar) is currently an affordable consumer product and number of manufacturers is already on the market. At the same time, electronic components are becoming more compact and energy-efficient every year - they consume less electrical power and produce less heat, thanks to the transition of semiconductor manufacturers to a more subtle process technology and the improvement of microarchitecture of chips. The second factor is the development of the algorithmic base, autonomous management has been an urgent task for several years already, and this direction is actively developing, new algorithms, methods and ready-to-use software solutions are emerging. One of the most important tasks in autonomous control is image recognition, since it is the output of the pattern recognition unit that is the main source of information about the world for the system, based on which a decision is made about the behavior of the entire object. There is also active development of unmanned vehicles, for which the correct recognition of road signs and markings is critical, which, in turn, are not always in favorable conditions for recognition (pollution, installation in different positions, at different angles to the roadway, partial overlapping). The faults in the recognition of images can lead to serious negative consequences, including the destruction of the controlled object; therefore, it is necessary to minimize the possibility of faults, thereby maximizing the reliability of the system. In recent years, many different algorithmic approaches developed, as well as hardware solutions for machine vision and pattern recognition. The presence of a variety of different image recognition algorithms, allows you to develop a redundant system that includes several versions of software modules with different algorithms. These modules, based on the same or different hardware capabilities (different cameras, lidars, range finders, pyrometers, etc.) make a decision in the classification of the image, and the execution environment makes the final decision based on the outputs of these versions. In this case, expediently in the similar system uses the multiversion programming scheme (MVP) with versions that perform the same functionality, but with considering different methods and a decision block that chooses the "correct" output. This block makes a decision on the outputs correctness based on the voting algorithms: the voting by an absolute majority, agreed by the majority, their fuzzy and weighted modifications, median voting.

2 Voting Algorithms

For improving of voting algorithms stability to the related faults of program modules [1], in comparison with unweighted analogs, are suggested the modifications of the basic voting algorithms agreed by a majority and a fuzzy vote by an agreed majority [2].

For each software version in a system creates the Boolean stack, of fixed length, in which "0" is added in the event, that the voting block decides that the version gives the wrong answer and "1" if it gives the right answer or in case of a fuzzy vote if the value of the version's output belongs to the winning class > 0, i.e. all

* Corresponding author: msaramud@gmail.com

 DOI: 10.1051/matecconf/201713204011

versions that have added weight to the class that won the vote will be marked as correct.

As the stack length is fixed, new data will replaces the most oldest, i.e. the stack operates according to the principle FIFO (if the stack comes first, the stack will leave first). This action lets to input the forgetting element to the depth of a stack. For example, if the depth of a stack is 100 to the results of a work version are older that for 100 voices and won't be counted.

The forgetting element is necessary to ensure the operative response of the system to changes in the behavior of versions when a certain version can significantly change its reliability in case of changing the input data stream [3], therefore is necessary to change rapidly its rating of version for correct distribution of the weights during voting [4]. The rating is determined by summing of all the elements of the stack. For example, if in stack of 100 elements length 97 values of "1" and 3 values of "0" are recorded, the weight of this version will be equal to 0.97.

We have made only one restriction with the software implementation; the version weight cannot be equal to one, what can happen in practice. Quite reliable enough versions give the correct answer 100, 1000, 10000, etc. times in a row. And without a limit they would get the whole stack of units (or TRUE), which would give them a rating of one, but these situations should not be allowed, since in case of an incorrect answer by this version, it would get a weight of one. Although the right answer of remaining N-1 versions on a weight only approaches one and as a result loses the vote. If a reliability rating of the version is equal to one, it isn't make sense, because if we have a reliable software module, the sense of the system will be lost [5]. Thus, we have made restrictions in calculating the rating of the version with each vote.

2.1 Software implementation of the simulation environment

When voted by an agreed majority, the voting block receives responses from each version (or in our program is simulation). If the output value doesn't match the previously obtained one, a new class is created, if the value matches the existing class [6-7], then this class weight will be recalculated as

$$P_{total} = P_{class} + (1 - P_{class}) * P_{version} \tag{1}$$

After all the versions gave an answer (response time limit in this model is absent [8]), the weights of the resulting classes are compared. Thus, the class with the greatest weight wins, but as it is seen, is not always a class, which the largest number of versions "voted". The weight of each version has a value, which allows considering the reliability of each separate version separately [9].

After determining the correct exit version of the version, those who voted for it get a weight "1" on the stack, and the versions voted "0" differently.

In case of fuzzy voting by an agreed majority, while voting one more pass occurs in the program when versions whose output value have not been matched the class value. This value is different for it no more than the tolerance "E" and, in this way, it belongs to the class is equally >0 that adds weight to the class:

$$P_{total} = P_{class} + (1 - P_{class}) * P_{version} * K_{implement} \tag{2}$$

where

$$K_{implement} = 1 - \left(\left| X_{class} - \frac{X_{version}}{E} \right| \right) \tag{3}$$

X - Class value and version, which gave non-matching answer, but a significance that enters the tolerance E [10].

In this studied system, the version simulations give three types of faults: a random fault, a simulating fault in the module, as related fault, and an inaccuracy to the answer is close to the correct one, remote from it is no more than the tolerance, but not equal to it. This type of a fault simulates "inaccuracy" is connected with rounding faults with a lack of bit capacity, inaccurate digitization of analog sensor outputs, etc. We take into account the situation when the algorithm version worked correctly, but it gave an inaccurate response due to rounding faults, digitization, etc. [11] The fault appears with the probability given for each version and each data stream. In the event of a fault, the following checks occur: if it's not first fault in current voting, in this case an related fault will be generated with predetermined probability, i.e. a value matching the value of the past fault returns.

Further, an "inaccuracy" is generated with a given the probability and the output is not equal to right has been returned, but remote from it by no more than a given deviation of E. If the previous probabilities do not work, then a random fault will returned, simulating a malfunction in the current version of the module.

From the results of the program, it is seen that in the absence of "inaccuracies", there is no difference in the voting methods; a random fault is always too "far" from the correct answer to change the weight of classes [12]. Thus, we can conclude, that if a system does not have possible places of occurrence of "inaccuracies" that is digitization of analog signals, lack of digit capacity for the mathematical operations, etc. then using of a more resource-intensive fuzzy voting algorithm will not give advantages [13].

However, if the possibility of such "inaccuracies" is present that the fuzzy algorithm will increase the reliability of a system. The only drawback will be the same evaluation for both the versions giving the ideally correct output, and the versions algorithmically correct, but having "inaccuracies".

The developed system stimulates the multiversions separate function that returns with in advance set probabilities either the correct answer or numerical result of one of three types of mistakes to "inaccuracy", "failure" and "related".

These probabilities are specified by the program for 3-successively varying input data streams, allowing to model different versions of the response to a change in the input data stream. After a certain iteration, the probability distribution changes, and it is possible to track the system's response to this [14], check the correctness of the operation, the reaction rate depending on the algorithm chosen, the stack length and other system parameters.

The system permits to change the version numbers N in the range from three to night, the probability P of each version for 3 different data streams, the depth of the weight stack of versions. It influences the speed of the system response to the change in version work, set the number of iterations for each thread, the probability of the related faults, inaccuracies, and the E tolerance for a fuzzy voting algorithm. As a result of the simulation, the system gives the number of faults allowed by each algorithm, for fuzzy voting, the "failures" and "inaccuracies" are calculated separately, the sum of the weights of the winning voting classes and the number of version related faults.

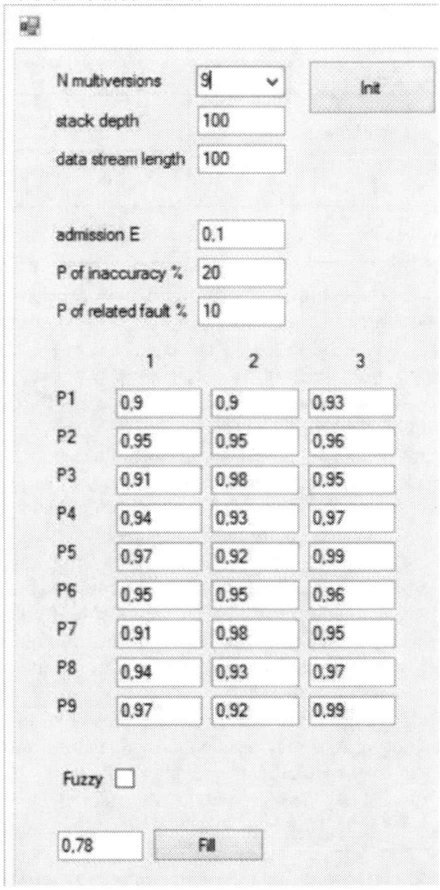

Fig. 1. The interface of the simulation environment: simulation parameters.

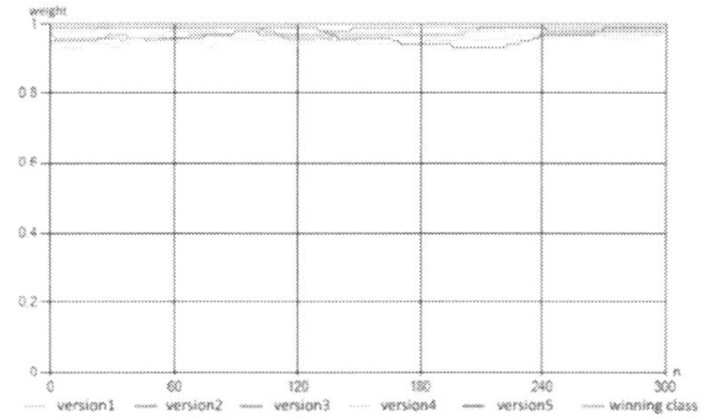

Fig. 2. Results of the simulation model implementation of the fuzzy voting algorithm by an agreed majority (versions are potentially reliable).

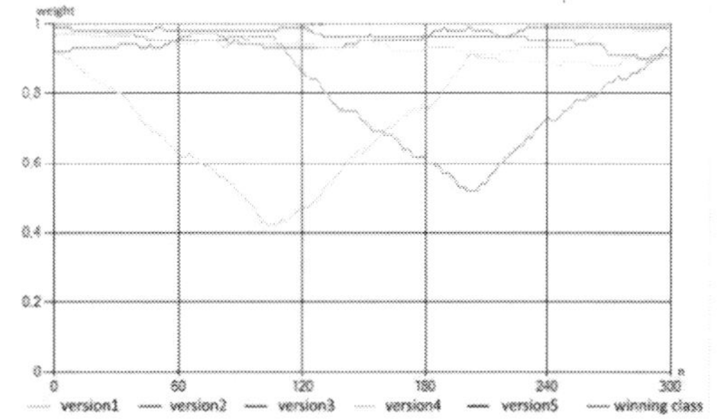

Fig. 3. Results of the simulation model implementation of the voting algorithm by an agreed majority (in the first version of the first data set a failure and in the second version of the second data set with 50% probability).

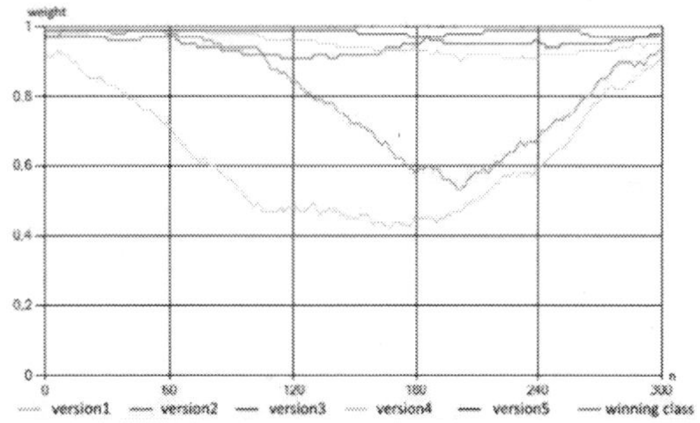

Fig. 4. Results of the simulation model of the implementation of the voting algorithm by an agreed majority (in the first version of the first data set a failure and in the second version of the second data and in the second version only second data with 50% probability).

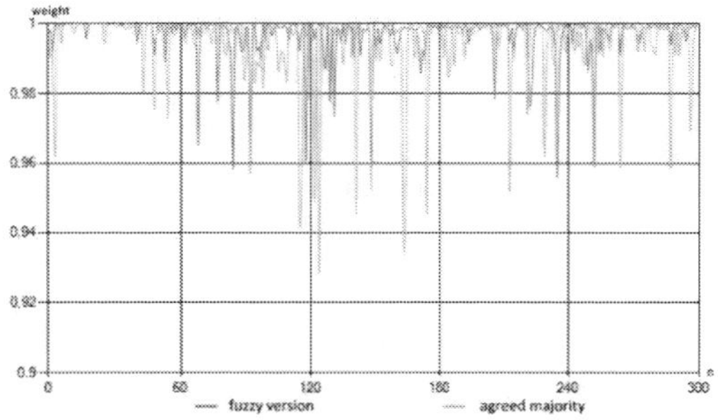

Fig. 5. Results of the simulation model: weights of the winning classes for all voting algorithms.

The system's reaction to the change in version behavior is visible on the graphs (Figures 2-4) when a reliable version starts to give faults or on the contrary, a low-weight version no longer makes a mistake. In an article shows the graphics for a stack of 100 deep and three consecutive data streams of 100 votes for each. Overall, it is evident that the system responds quickly to changes in the behavior of the versions and, through a number of votes equal to the depth of the version stack, receive estimated weights that are accurate in their probability of a correct answer.

Based on the simulation results, the program selects the optimal algorithm based on the total number of errors allowed and the sum of the weights of the classes. Output numerical results are also presented on the form: the total number of errors, the number of failures and inaccuracies for a fuzzy voting algorithm, the sum of the weights of the winning classes.

For the convenience of debugging on the form, the system parameters are displayed at the last failure: the iteration number, the values of the winning class, its weight. This can help in identifying inadequate operation of voting algorithms on a certain set of input parameters.

If the system can uniquely choose the best voting algorithm based on the simulation results, the user is given a pop-up window with the selected result, if the best algorithm can not be uniquely determined on the basis of the received data, the user is asked to make a decision independently based on the modeling data presented by the system on the form.

This simulation environment supports further refinement, it is possible to add new decision algorithms, for example - t/(n-1) algorithm, in order to compare it with classical algorithms and algorithms modified by us under the same simulation conditions.

2.2 Modeling results

Table 1. Results of the system performance for different quantities of multiversions (N 3-6).

N		3	4	5	6
Precise	Number of faults	49	22	7	0
	Total weights	269,43	286,53	296,05	299,21
Imprecise	Number of faults	88	45	13	2
	Total weights	279,45	289,04	297,70	299,70
Related faults		16	22	42	48

The system's reaction to the change in version about the last recorded fault at the output of the voting block to the iteration number, the output value, and the weight of this class. The system permits to build the weight graphics of each version and the winning classes along the axes of the iteration number and weight. In addition, we have an opportunity to change a scale for clarity, for example when examining the weights of the winning classes, the values do not fall below 0.9 and on a scale from zero to one represent a practically straight line flat in the field of one. For better perception in this mode, the scale changes to a range from 0.9 to one along the weight axis.

A form for convenience shows the information about the last recorded fault at the output of the voting block to the iteration number, the output value, and the weight of this class. The system permits to build the weight graphics of each version and the winning classes along the axes of the iteration number and weight. In addition, we have an opportunity to change a scale for clarity, for example when examining the weights of the winning classes, the values do not fall below 0.9 and on a scale from zero to one represent a practically straight line flat in the field of one. For better perception in this mode, the scale changes to a range from 0.9 to one along the weight axis.

Table 2. Results of the system performance for different quantities of multiversions (N 7-9).

N		7	8	9
Precise	Number of faults	3	0	0
	Total weights	298,67	299,21	299,21
Imprecise	Number of faults	4	2	2
	Total weights	299,46	299,70	299,70
Related faults		84	89	123

Tables 1-2 summarizes the simulation results for different numbers of multiversions from 3 to 9, with the probability of all versions in all streams equal to P = 0.7, the depth of the stack 100, the tolerance E = 0.1, the probability of inaccuracy P = 20%, the probability of related faults P = 10%.

The low reliability index of all versions as 0.7 is taken for clarity of simulation results, since with quite reliable versions (P> 0.95) even with the number of multiversions N = 3, both voting algorithms give no faults for 300 votes [15-16]. What we have seen data is a kind of a perfect example of the system reliability increases overall, despite each models extremely unreliable software modules [14,17,18] throw an fault of 30% cases, we observe that in a system with 9 versions all 300 responses are correct.

The fuzzy algorithm shows more faults, because sometimes voting is won a class with a value close to the right one (no further than in E from the correct one), but not equal to it [19]. This answer is considered by the system as a fault. However, the versions that added weight to the winning class still get to stack one, increasing their own weights. Except that a large number of faults, the sum of the weights of the fuzzy algorithm is higher.

3 Conclusions

The results obtained with the help of developing simulation environment of modeling demonstrate the effectiveness of the proposed voting algorithms. In addition, it proves the possibility of creating a reliable system of the unreliable software modules and it is very important in the issue of pattern recognition and classification. Since in current time, there are not fault-free algorithms in this field and the multiversion approach will significantly increase the accuracy of recognition and classification. This approach increases the reliability of the on-board software management system for autonomous unmanned objects.

The reported study was funded by Russian Foundation for Basic Research, Government of Krasnoyarsk Territory, Krasnoyarsk Region Science and Technology Support Fund to the research project № 16-47-242143.

References

1. D.E. Eckhardt, L.D. Lee, *IEEE Trans. Soft. Eng.*, **SE-11**, no. 12, 1511-1517, (1985)
2. J. L. Gersting, R. L. Nist, D. B. Roberts, R. L. Van Valkenburg, *Proceedings of the Twenty-Fourth Annual Hawaii International Conference on System Sciences*, 253 – 262, **2** (1991)
3. K. E. Grosspietsch, *Proceedings 16th International Parallel and Distributed Processing Symposium*, (2002)
4. L. Chen, A. Avizienis, *Proc. Int. Symp. Fault-Tolerant Computing FTCS-8*, 3-9, (1978)
5. J.-C. Laprie, J. Arlat, C. Beounes, K. Kanoun, *IEEE Computer*, **23**, no. 7,39-51 (1990), Reprinted in Fault-Tolerant Software Systems: Techniques and Applications, H.Pham (ed.), *IEEE Computer Society Press*, 5-17 (1992)
6. Z. Aghajani, M. A. Azgomi, *2009 International Conference on Innovations in Information Technology*, 304 – 308 (2009)
7. G. Latif-Shabgahi, S. Bennett, *Proceedings 25th EUROMICRO Conference. Informatics: Theory and Practice for the New Millennium*, **2**, 113 – 120, (1999)
8. I.V. Kovalev, P.V. Zelenkov, V.V. Losev, D.I. Kovalev, N.V. Ivleva, M.V. *IOP Conf. Series: Materials Science and Engineering*, 173 (2017) 012025.
9. J. Knight, B. Littlewood (eds.), *IEEE Softw.*, January (1994)
10. Wang Ping A *2007 8th International Conference on Electronic Measurement and Instruments*, 2-666 - 2-669, (2007)
11. S.S. Brilliant, J.C. Knight, N.G. Leveson, *IEEE Trans. Soft. Eng.*, **16**, no.2, 238-247 (1990)
12. J.C. Knight, and N.G. Leveson, *IEEE Transactions on Software Engineering*, **SE-12**, no. 1, 96-109, (1986)
13. R. Frullini, A. Lazzari, *Proceedings of the International Conference on Railway Safety Control and Automation Towards 21st Century*, London, U.K., 292-299 (1984)
14. M.R. Lyu (ed.), *Software Fault Tolerance* (John Wiley & Sons, Chichester, U.K., January 1995)
15. I. Lee, D. Tang, R.K. Iyer, and M.C. Hsueh, *IEEE Trans. Rel.*, **42**, no. 2, 238-249 (1993)
16. A. Athavale, "Performance Evaluation of Hybrid Voting Schemes," M.S. thesis, North Carolina State University, Department of Computer Science, (1989)
17. A. Avizienis, *IEEE Trans. Soft. Eng.*, **SE-11**, no. 12, 1491-1501 (1985)
18. A. Avizienis, and L. Chen, *Proceedings COMPSAC*, 77, 149-155 (1977)
19. D.F. McAllister, C.E. Sun, and M.A. Vouk, *IEEE Trans. Rel.*, **39**, no. 5, 524-534, (1990)

MATEC Web of Conferences 132, 04012 (2017) DOI: 10.1051/matecconf/201713204012
DTS-2017

Artificial Intelligence Method for Electric Drives Mode Operating and Technical Condition Determination

Tatyana Kruglova[1,*], *Alexey* Bulgakov[2], *Alexander* Vlasov[1], and *Ivan* Shmelev[1]

[1]South-Russian State Polytechnic University, 346428 Novocherkassk, Russia
[2]Southwest State University, 505040 Kursk, Russia

Abstract. This paper proposes artificial intelligence method to determinate the status of electromechanical equipment by analysing the changes in the readings o parameters of its operating mode. As experimental results have revealed the dependence of wavelet transformation coefficients on the characteristic scales of functional and faulty motors under different loads, based on which a neural classification network is developed to reveal the current state of the electromechanical equipment. Further studies have shown that any mother wavelet function can be used to implement the proposed method. The researches of the state of the drive under various loads confirm the correctness of the theoretical calculations and the adequacy of the model.

1 Introduction

Long-term operation of electrical equipment can arise various kinds of defects as a consequence of the wear and tear process. Operating the motors in poor technical condition of leads to financial losses related to unpredictable failure of equipment and a consequent violation of the technological process. One of the ways to solve the problem of sudden equipment failures is to monitor and evaluate its current status, identified by means of diagnostic methods, allowing to classify the current technical condition of the motor to one of the predetermined classes of diagnoses and to determine the locations of the defects that have arisen. The functional approach allows to identify the state of the drive during its operation. The selected diagnostic parameters should identify all the object faults that have occurred, should be simple to measure and analyse the outcomes. In contrast, these requirements are met by vibration, noise and current consumption. The traditional method of analysing such parameters is the Fourier transform [1], which has a number of significant drawbacks [2], not allowing its implementation for automatic diagnostics of electric drives operating under dynamic loads. As for main challenge stands the large amount of non-formalized qualitative information that cannot be used with conventional modelling methods. This issue can be addressed and solved using artificial intelligence methods.

2 Materials and methods

Long operation activity of the electrical motors at the large reversed loads can generate faults. One of the most simple and accessible methods of diagnosing the drive condition is the method of the spectral analysis of stator current signals allowing to carry out diagnostics of the electrical equipment and the connected mechanical devices. The received data are converted into the frequency domain using the Fourier transformation [1]. This transformation is widely used, however it has the following disadvantages:

1. The Fourier transform provides the frequency information, which is contained in the signal, however it cannot determine the time of occurrence of this frequency.
2. Limiting descriptiveness analysis of non-stationary signals: there is no possibility to analyze characteristics of the signal because the frequency-domain signal drop occurs over the entire frequency range of the spectrum. In addition the noise generated from "parasitic" high-frequency components render the readings very difficult to analyze.
3. Harmonic basis functions decomposition is not able to display signals with slope type rectangular pulses.
4. Simultaneously, producing temporal and frequency analysis is impossible.
5. Frequency analysis obtained using the Fourier transformation is complex and requires experienced specialist to understand it, which entails additional expenses.

These drawbacks do not make it possible to distinguish the faulty state of the electric drive from the change in its operating mode.To solve this problem, it is possible to apply wavelet analysis of signals, which considers the analysed time functions in terms of oscillations localized in time and frequency, providing a two-dimensional sweep of one-dimensional signals. In this case, the frequency and coordinate are considered as independent

* Corresponding author: tatyana.kruglova.02@mail.ru

variable, which makes it possible to analyse the signals in two spaces at once. The wavelet functions of the basis allow ~~us~~ to concentrate the attention on certain local features of the analysed processes, which cannot be detected using traditional Fourier and Laplace transforms.

The wavelet transformation of a signal is represented in the form of a generalized series or the Fourier integral over a system of basis functions [3-5]

$$\psi_{ab}(t) = 1/\sqrt{a} \cdot \psi(\frac{t-b}{a}), \qquad (1)$$

constructed from the mother (original) wavelet $\psi(t)$ possessing certain properties, due to the time shift operations b and the time scale change a. The factor $1/\sqrt{a}$ ensures that the norm of these functions is independent of the scaling number a. Small values, a correspond to small scales $\psi_{ab}(t)$, or high frequencies (ω~1/a), large parameters a – to large scale $\psi_{ab}(t)$, i.e. stretching the parent wavelet $\psi(t)$ и and compressing its spectrum. Thus, the wavelet scale, as a unit of the time-frequency representation of the signals, is inversely proportional to the frequency (Fig. 1).

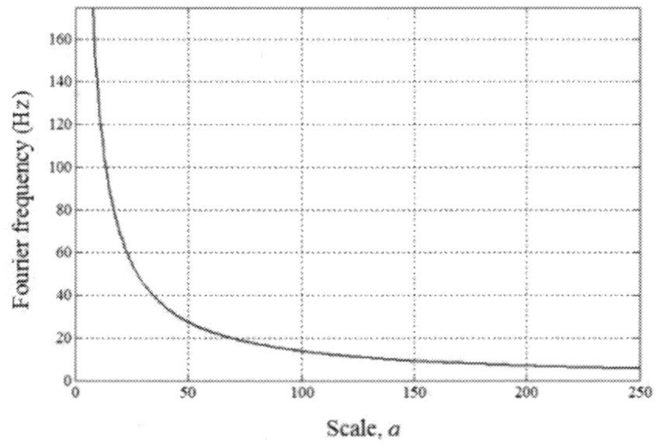

Fig. 1. Value for Fourier frequency with reference to the wavelet scale

The recalculation of the Fourier frequencies of the spectrum into the scale of the wavelet, according to the dependence in fig. 1, will allow signals analysis into wavelet space [6] and find new signal characteristics that cannot be detected using Fourier analysis.

2.1 Experimental researches

To analyse the state of the drive, the current signal, phase voltage and vibration speed of the new healthy motor, which is taken as the reference are required. Troubleshooting in the spectrum of the current signal is carried out at characteristic frequencies [6] by ~~i~~ comparing the current spectrum with the reference spectrum using artificial intelligence described later in the paper. For example, for a DC motor, the characteristic of Fourier transform frequencies are shown in Table 1.

Table 1. Characteristic frequencies of the current signal

DC motor faults	Current (voltage) signal frequency
Commutation defects	$2 \cdot k \cdot p \cdot f_r$
Rotor defects	$2 \cdot p \cdot f_r$, $k \cdot f_r \pm 2 \cdot p \cdot f_r$
Voltage ripple	$k \cdot f_s$
Stator defects	$k \cdot f_r$

where f_s – frequency of the network supplying the rectifier, (Hz); f_r – motor rotor speed, (Hz); $k = 1,2,3$ – number of current harmonic; p – the number of poles.

The decomposition of the time current, voltage or vibration signals into a Fourier series and the analysis of the amplitudes at these characteristic frequencies reveals the observed object's faults. However, the analysis process is very laborious and will not distinguish the faulty state of the engine from a change in its operating mode. Therefore, for further analysis, it is necessary to recalculate the characteristic frequencies of the Fourier analysis into the scale of the wavelet. Any mother function can be chosen considering that of for a wavelet is achieved. Table 2 shows the result of the conversion to the scale of the Morlet wavelet for brushless DC motors PITTMAN 5413 at 3 Hz (Table 2).

Table 2. Ratio of the characteristic frequencies of the Fourier transform and the scale of the wavelet

DC motor faults	Fourier spectrum frequencies	Morlet wavelet scale
Commutation defects	24	48
	48	24
	72	16
Rotor defects	24	48
	27	43
	30	39
	33	35
Voltage ripple	50	23
	100	12
	150	8
Stator defects	3	386
	6	193
	9	129

A comparative analysis of the wavelet coefficient functions at characteristic frequencies for different operating modes reveals the general rule to diagnostic DC and AC motors using the aforementioned method (fig. 2). From the provided readings, it is visible that the wavelet coefficients of a serviceable unloaded engine have insignificant fluctuations at motor start-up and then the process is practically linearized. When a load appears, the oscillatory process at the start of the motor is clearer, but then decreases. However, it should be noted that the process is repeated with certain periodicity Although the process is stable, there is no significant increase in the amplitude of the oscillations with time. The coefficients of the wavelet transform of the faulty motor are much lower than those of the serviceable one and have constant oscillations-increasing with the load.

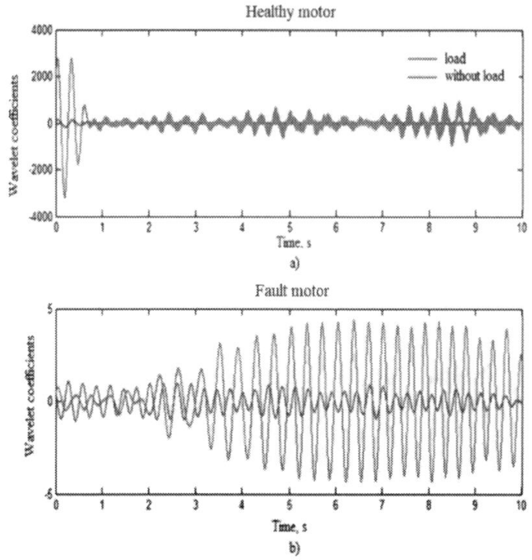

Fig. 2. Wavelet stator current signal: (a) healthy motor, (b) faulty motor

The carried-out analysis shows which wavelet values coefficients of current, voltage and vibration at the scales corresponding to malfunction has an identical characteristic appearance that can be used for electric drives mode operating and technical condition determination.

Fig. 3 illustrates the non-characteristic wavelet values. From this diagram it is visible, the signal has high density and small values wavelet coefficients. At the same time, the signal is regular and completely repeats with the given frequency. This type of signals is identical to all frequencies indicating failure in independence of technical condition of the motor

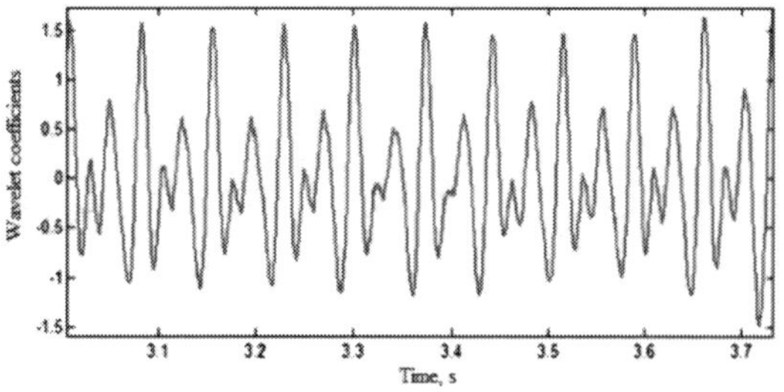

Fig. 3. The wavelet coefficients of the feeding tension of the serviceable and faulty engine at an non-characteristic scale

By analysing the results, we receive five characteristic signals for diagnosing. Signals of characteristic frequencies for the operational off-load and loaded engines (fig. 2), and also an unrepresentative signal (fig. 3), it is expedient to carry to a class "is serviceable" while faulty loaded and off-load to a class "is faulty. This information can be used for simulation of a neural network to classify the technical condition of the engine.

2.2 Identification of neural network for status classification

For automatic detection of technical condition of the The purpose of the introduced neural network is to carry out automatic detection of technical condition of the electric drive without involvement of the specialist expert [7-19]. As basic data values, wavelet coefficients at a scale are used (fig. 2), characteristic of malfunction, and an uncharacteristic signal (fig. 3). As an entrance is set the matrix containing five lines of characteristic signals. The output of network is the diagnosis class: "1" – an object is serviceable, "2" – an object is faulty.

Fig. 4 illustrates the neural network used to classify the status of the drive

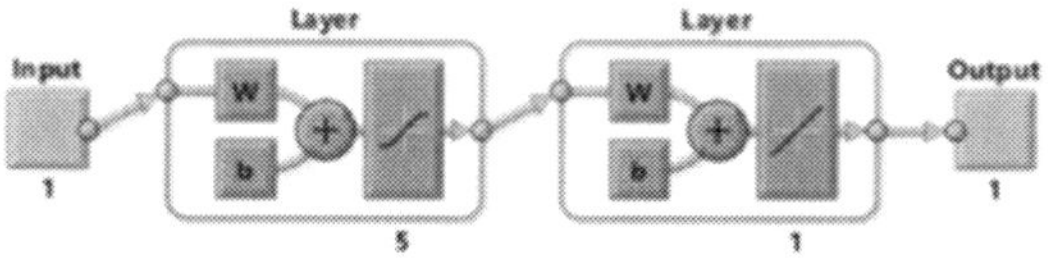

Fig. 4. Neural network for electric drive technical condition classification

The network consists of three layers: input, hidden and output. The hidden layer has five neurons with tangential function of activation, the output – one linear neuron.

For training of neural network is achieved using the Levenberg-Markvardt [11] algorithm intended for optimization of parameters of nonlinear regression models. The algorithm consists in consecutive approach of ~~the~~ set initial values of parameters to a required local optimum.

A training selection is a set of multiple couples of free variable of $x \in X^M$ (network inputs) and dependent variable of $y \in Y^M$ (a target vector). The functional dependence is the set representing a regression $y = f(\omega, x_n)$ continuously differentiable in the field of $W \times X$. The parameter ω is a vector of weight factors.

It is required to find such value of a vector ω, which would deliver a local minimum of function of an error.

$$E_D = \sum_{n=1}^{N} (y_n - f(\omega, x_n))^2 \qquad (2)$$

Before work of an algorithm is set the initial vector of weight coefficients ω.

On each step of iteration, vector $\omega + \Delta\omega$ replaces the initial one. For an increment assessment of $\Delta\omega$ is used the linear approach of function:

$$f(\omega + \Delta\omega, x) \approx f(\omega, x) + J\Delta\omega, \qquad (3)$$

where J – Jacobean of function $f(\omega, x_n)$ in ω. The matrix can be presented visually in the form:

$$J = \begin{bmatrix} \frac{\partial f(\omega, f_1)}{\partial \omega_1} & \ldots\ldots & \frac{\partial f(\omega, f_1)}{\partial \omega_R} \\ \ldots\ldots & \ldots\ldots & \ldots\ldots \\ \frac{\partial f(\omega, f_N)}{\partial \omega_1} & \ldots\ldots & \frac{\partial f(\omega, f_N)}{\partial \omega_R} \end{bmatrix} \quad (4)$$

where $\omega = [\omega_1 \ldots \omega_R]^T$ is the vector of parameters.

The increment $\Delta\omega$ in a point ω, the delivering minimum of E_D is equal to zero. Therefore for finding ~~of~~ the subsequent value of an incrementприращения $\Delta\omega$ we will equate to zero a vector of private derivative производных E_D on weight ω. For this purpose we will present (2) in the form

$$E_D = \left| y - f(\omega + \Delta\omega) \right|^2, \quad (5)$$

where $y = [y_1, \ldots y_N]^T$ and

$$f(\omega + \Delta\omega) = [f(\omega + \Delta\omega, x_1), \ldots f(\omega + \Delta\omega, x_N)]^T. \quad (6)$$

By transforming and differentiating equation (6):

$$\begin{aligned} \left| y - f(\omega + \Delta\omega) \right|^2 &= (y - f(\omega + \Delta\omega))^T y - f(\omega + \Delta\omega)) = \\ &= f^T(\omega + \Delta\omega) f(\omega) - 2y^T f(\omega + \Delta\omega)) + y^T y \end{aligned} \quad (7)$$

we can write the following:

$$\frac{\partial E_D}{\partial \omega} = (J^T J)\Delta\omega - J^T (y - f(\omega)) = 0 \quad (8)$$

Hence, to find value $\Delta\omega$ it is necessary to solve system of the linear equations

$$\Delta\omega = (J^T J)^{-1} J^T (y - f(\omega)). \quad (9)$$

As the number of conditionality of a matrix of $J^T J$ is a square of number of conditionality of a matrix of J, a matrix $J^T J$ can be degenerate.

$$\Delta\omega = (J^T I)^{-1} J^T (y - f(\omega)), \quad (10)$$

where I – single matrix.

This parameter is appointed on each iteration of an algorithm. If the value of an error E_D decreases quickly, λ reduces this algorithm to the Gauss-newton algorithm. The algorithm stops in case the increment $\Delta\omega$ in the subsequent iteration is less than present value or if the vector of weight coefficients delivers an error E_D, smaller the set size or if the number of cycles of training of neural network is exhausted.

The value of a vector ω on the last iteration is considered required.

The learning disabilities of the neural network are presented in fig. 5

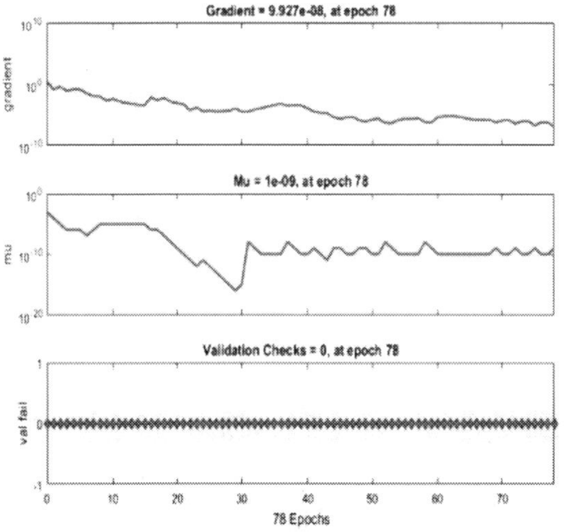

Fig. 5. Neural network training result for technical condition classification

The unloaded and loaded condition analysis of the drive (DC motors PITTMAN 5413, rotating speed of the motor of resistance of 1 Hz) showed that 386, 193 and 129 scales of wavelet expansions belong to the class "2" and all remaining scales to the class "1", which demonstrates the existence of failure of the stator.

Similar tests were executed on the serviceable and faulty motor (Frequency of 0,2 Hz, 0,4 Hz, 0,6 Hz 0.8 Hz.) A similar experiment was made for Daubechies wavelet of 10-th order, Haar, the Mexican hat and Gauss. Recalculation of frequencies of Fourier of conversion to scales of these wavelet is executed. Substitution wavelet coefficients on data of characteristic scales in the trained neural network allowed to identify failure of an object unmistakably. The results certify to correctness of theoretical calculations and adequacy to the developed neural network model of diagnosing.

The provided neural network algorithm ensure the distinguishing between functional and faulty engine irrespective of the load.

To determine the technical condition and operating mode of the motors, an advanced neural network of the following structure is introduced (fig. 6).

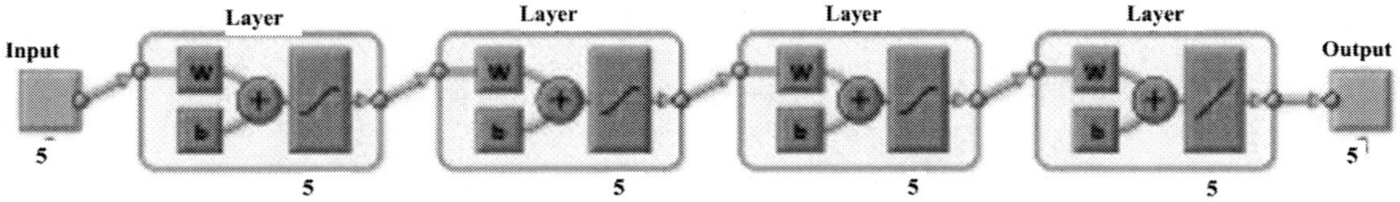

Fig. 6. Neural network for electric drive technical condition and mode operation classification

The neural network s consists of five inputs, four hidden layers of with five neurons in each and five outputs. The goal of the neural network training is to distribute the five input signals (fig. 2, 3) to four classes: «11» – functional unload, «12» – functional loaded, «21» – faulty unload, «22» – faulty loaded. For neural network training the Levenberg-Markvardt [20] algorithm was used. The learning results of the neural network are presented in table 3 and fig. 7 respectively.

Table 3. Results of neural network for technical condition and operating mode classification

Signals	Network Results
Functional unload, (fig. 2,a)	11
Functional loaded, (fig. 2,a)	12
Uncharacteristic signal (fig. 3)	11
Faulty unload, (fig. 2,b)	21
Faulty loaded, (fig. 2,b)	22

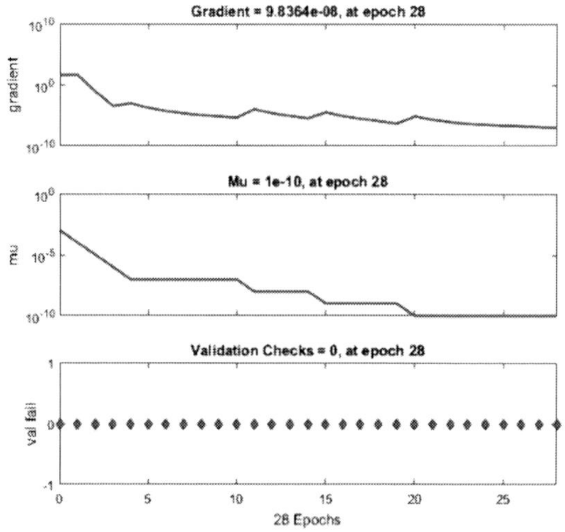

Fig. 7. Neural network training result for technical condition and operation mode determination

3 Conclusions

The artificial intelligence method of diagnosing of the electric drive allows with a high precision to find malfunctions of an observed electromechanical object. The realization of this method is based on wavelet transformation and neural networks. The validity of the theoretical calculations and adequacy of model are endorsed with simulation results based on specific electromechanical system.

References

1 Körner T.W. *Fourier Analysis*, CUP (1988)

2 Daubechies I. *The wavelet transform time-frequency localization and signal analysis*, IEEE TIT 961–1004 (1992)

3 Kruglova T.N. *Intelligent Diagnosis of the Electrical Equipment Technical Condition,* PE, **129**, 219-224 (2015)

4 Peter W. Tsea, Wen-xian Yangb, Tama H.Y., JSV **277,** 1005-1024 (2004)

5 Liu B., Ling S.F., MSSP, **13 (1)**, 145-162 (1999)

6 Nandi S., Hamid A. T., and. Li X., *IEEE Conf J & M.*, **20 (4),** 719-729 (2005)

7 Yoon-Seok Lee, Kyung-Tae Kim, JinHur., *IEEE TM*, **50(2),** 7022004-7022004 (2015)

8 Shakouhi M. Mohamadian M. Afjei E., IEEE TIE, **62 (3)**, 1400-1409 (2015)

9 Penga Z.K, Peter W. T., Chub F.L., MSSP, **19,** 974-988 (2005)

10 Awadallah M.A. Morcos M.M., *IEEE TECC,* **20(1),** 2460-247 (2005)

11 Shen Z., Chen X., Zhang X., He Z., Measurement, **45 (1)**, 30-40 (2012)

12 Kliman G.B. Stein J., EMPS, **20 (5)**, 463–474 (1992)

13 Jiang L., Liu Y., Li X., Tang S., Measurement, **44 (7)**, 1284-1292 (2011)

14 Betta G., Liguori C., Pietrosanto A., Measurement, **30 (2),** 129-138 (2001)

15 Dister C.J., Schiferl R., *Proceedings of the IEEE – IAS AMC,* 1, 312-318 (1994)

16 Toliyat H.A., Lipo T.A., *IEEE TEC,* **10**, 241-247 (1985)

17 Toliyat H.A., Lipo T.A., *IEEE TEC,* **10**, 241-247 (1985)

18 Siddique A, Yadava GS, Singh B., *EC IEEE Trans,* **20(1)**, 106-114 (2005)

19 Medoued A, Lebaroud A, Boukadoum A, Boukra T, Clerc G. (SDEMPED), *2011 IEEE Int. symp. IEEE,* 525-528 (2011)

20 Andersen T.J. Wilamowski B.M., WCNN, **1** 687-690 (1995)

* Corresponding author: tatyana.kruglova.02@mail.ru

Investigation of the influence of the microcapillary structure of natural skins on relative humidity in vacuum-sorption humidification

Ludmila Larina[1], *Dmitryi* Ruslyakov[1,*], *Olga* Tikhonova[1]

[1]Don State Technical University, 344000 Rostov-on-Don, Russian Federation

Abstract. The article presents the results of studies confirming the law of the gamma distribution of microcapillaries in the structure of materials having a stochastic structure, for example, natural skins. The relative humidity of the latter depends on the mass of moisture collected by the microcapillaries, with heat and mass transfer under vacuum conditions. Numerical values of relative humidity in this case may differ from those recommended by footwear manufacturing technologies and should be considered as an integral part of the phenomenon of high-intensity heat and mass transfer under vacuum conditions and be determined by the proposed models.

1 Introduction

Studies of the processes of heat and mass transfer through a porous medium in materials to which the natural skin [1] refers are based on the method of studying the diffusion of a liquid or a steam mixture flowing through a single capillary tube simulating a porous medium. Such an approach makes it possible to calculate the diffusion coefficient, the rate of mass transfer over a single capillary, but because of the complexity of the structure of the porous medium, the experimental data differ significantly from the calculated parameters. With high-intensity heat and mass transfer, for example, under vacuum conditions [2], the mechanism of mass transfer by molecular diffusion changes to effusive, in the form of vapor [3].

Its further condensation in the pores (microcapillaries) commensurate with the size of its molecules requires consideration of the stochastic structure of the porous medium and the laws of distribution of microcapillaries in it, which will allow calculating not only the diffusion coefficient, the mass transfer rate, but also the moisture gain in the materials under consideration, due to sorption phenomena, their relative humidity [4, 5]. So, for example, the relative humidity of the skin with intensified hygrothermal action [6, 7] can differ from the values of 26 ... 30%, which are recommended by the standard technologies of making footwear. Thus, according to the data given in the work of R.V. Lutsyk [8], for a half-skinned chrome-tanned type of tanning, the moisture of the microcapillaries is from 22 to 27%, and for a heat-resistant yufti, from 16 to 21%. In this connection, obviously, with intensified hygrothermal action [9, 10], it becomes necessary to consider the scatter of indicators of relative skin moisture, caused by microcapillary porosity Π under experimental studies [11], as an integral part of the phenomenon and to study its nature in more detail. The nature of the observed scatter should be analyzed by statistical methods for the subsequent use in recommendations based on the statistical conclusions about the trend and dispersion, the indicators of the relative humidity of the skin.

2 Experimental part

To study the law of distribution of relative moisture indices of leather samples obtained from the results of experimental studies, a sample set of values for each batch of samples was considered $n = 30 \cdot 2 = 60$, as is known [12], for sample size $n \geq 60$ statistical criteria give fairly accurate and unbiased recommendations at the level of significance $\alpha = 5\%$.

For statistical studies of the sample set, a program written using the Maple 9.5 package commands was used.

Based on the results of calculations of the sample set of relative humidity for the leather samples (HVAC), a statistical series was compiled. The number of digits of the statistical series is calculated by the Stürgers rule [6]:

$$k = [1 + 3{,}32 \cdot \lg n] + 1 \tag{1}$$

here *[k]* is the operation of computing the integer part of a number.

Thus, with the volume of the sample population $n = 60$, the number of intervals of the statistical series is $k = 6$.

* Corresponding author: ruslyakof@yandex.ru

The boundaries of the categories of the statistical series with the frequencies of sample values falling into the corresponding category are presented in table 1.

Table 1. Statistical series for indicators of relative skin moisture

Ranks	1	2	3	4	5	6
Borders ranks	23,1 25,567	25,567 28,033	28,033 30,5	30,5 32,967	32,967 35,433	35,433 37,902
Frequencies	2	6	9	16	17	10

Graphical representation of the statistical series (histogram) is shown in fig. 1.

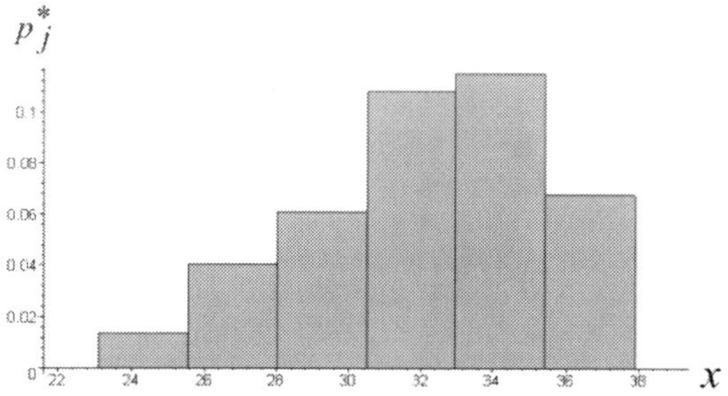

Fig. 1. Histogram of indices of relative humidity samples of leather

On the horizontal axis, in fig. 1, the boundaries of the discharges of the statistical series are plotted, on the vertical axis – the relative frequencies, divided by a step. Here: *mj* is the number of sample values (Table 1) that fall into the *j*-th category $j = 1,2,...6$; $x_{\max} - x_{\min}$=14,8 – sampling range; $h = \frac{x_{\max} - x_{\min}}{k} = 2{,}467$ – sampling step; $p_j^* = \frac{m_j}{n \cdot h}$ relative frequencies of sample values, divided by a step (values of the empirical density function).

By sample $\{x[k]\}_{k=1}^{n}$ observed values of a random variable, the basic numerical characteristics of the empirical distribution law are calculated:

1) the average of the sample $x_{\text{выб}} = \frac{1}{n}\sum_{k=1}^{n} X[k]$;

2) sample variance

$$D_{\text{выб}}(\text{ОВК}) = \frac{1}{n}\sum_{k=1}^{n}\left(X[k] - x_{\text{выб}}\right)^2 ;$$

3) selective standard deviation

$$\sigma_{\text{выб}}(\text{ОВК}) = \sqrt{D_{\text{выб}}(\text{ОУК})};$$

4) selective coefficient of asymmetry

$$As_{\text{выб}} = \frac{\frac{1}{n}\sum_{k=1}^{n}\left(X[k] - x_{\text{выб}}\right)^3}{\sigma_{\text{выб}}^3(X)};$$

5) selective kurtosis coefficient

$$\text{E}_{\text{выб}} = \frac{\frac{1}{n}\sum_{k=1}^{n}\left(X[k] - x_{\text{выб}}\right)^4}{\sigma_{\text{выб}}^4(X)} - 3 .$$

The selective average for the obtained series of observations is $x_{\text{выб}} = 32{,}0567$.

The sample mean is a consistent, unbiased and efficient estimate for the mathematical expectation of a hypothetical distribution law. The sample mean indicates the average value of the distribution of the skin's relative skin densities.

The value of the sample variance is $D_{\text{выб}} = 9{,}9418$.

The sample variance is a consistent estimate for the theoretical variance of the hypothetical distribution law.

The selective mean square deviation is $\sigma_{\text{выб}} = 3{,}1531$.

The value $\sigma_{\text{выб}}$ indicates the average variance of the observed values of the relative skin moisture values around the mathematical expectation.

The value of the selective asymmetry coefficient for a given sample $As_{\text{выб}} = -0{,}743$ is substantially less than zero and indicates the left-sided slowness of the histogram relative to the normal distribution law with the same mathematical expectation.

The value of the selective kurtosis coefficient was $\text{E}_{\text{выб}} = 0{,}1453$ and points to the sharpness of the empirical density function with respect to the density of the normal distribution law with the same parameters of mathematical expectation and variance.

To test statistical hypotheses about the distribution law of the random variable under study, according to the method of moments, the unknown numerical characteristics of the hypothetical distribution law are equated to their empirical estimates:

$$\begin{aligned} &\text{M}(\text{ОВК}) = x_{\text{выб}};\\ &D(\text{ОВК}) = D_{\text{выб}};\\ &As = As_{\text{выб}};\\ &\text{E} = \text{E}_{\text{выб}}. \end{aligned} \qquad (2)$$

3 Approach

It is known from the above theoretical studies [13,14] that the distribution of skin relative humidity parameters must obey the gamma distribution, and accordingly we will test the hypothesis of the gamma distribution of the sample population under study. A description of the main parameters of this distribution and the form of the corresponding density function is given in [15].

The practical value of the criterion χ^2 when checking statistical hypotheses on the distribution law is calculated by the formula:

$$\chi_{\text{пр}}^2 = \sum_{j=1}^{k_1} \frac{\left(m_j - np_j\right)^2}{np_j} \qquad (3)$$

where n is the sample size; m_j – the number of sampled values that fall into the j-th category, p_j – probability of falling into the j-th category of the random variable of the hypothetical distribution law; k_1 – the number of digits of the statistical series is recalculated by the Pearson criterion from the condition:

$$np_j > 5 \qquad (4)$$

Those bits for which condition (8) is not satisfied are combined with neighboring ones until the condition is satisfied.

The value of the criterion statistics χ^2 for the gamma distribution of the samples is:

$$\chi^2_{пр} = 3{,}316\,.$$

The theoretical value of the criterion $\chi^2_{\alpha,r}$ is calculated depending on the number of degrees of freedom r and the significance level of the criterion α. The level of significance of a statistical criterion is the probability of accepting an incorrect basic hypothesis, it is set by the experimenter and is equal to 0.05. The number of degrees of freedom according to [9] is equal to the number of digits of the statistical series after recalculation minus the number of constraints imposed on the frequencies:

$$r = k_1 - 1 - 2 \qquad (5)$$

For the gamma distribution $k_1 = 5$.

Links superimposed on frequencies:

1) the sum of the relative frequencies is unity,

2) two unknown parameters of the gamma distribution are equated to their empirical estimates.

Thus, the theoretical value of the criterion χ^2 at the significance level $\alpha = 0{,}05$ will be equal to

$$\chi^2_{\alpha;r} = \chi^2_{0{,}05;2} = 5{,}991.$$

Because the $\chi^2_{пр,вдоль} = 3{,}316 < \chi^2_{\alpha;r} = 5{,}991$, then there is no reason to reject the hypothesis of gamma distribution, and it is accepted as plausible.

The gamma distribution is given by the density function [7]

$$f(x) = \frac{x^{a-1} s^a}{\Gamma(a)} e^{-sx}, \qquad (6)$$

$\Gamma(a) = \int_0^\infty t^{a-1} e^{-t} dt$ – gamma function.

The parameters a and s satisfy the following system of equations:

$$\begin{cases} x_{выб} = \dfrac{a}{s}, \\ D_{выб} = \dfrac{a}{s^2}. \end{cases} \qquad (7)$$

Solving the system (7) with respect to a and s, we obtain:

$$a = \frac{x^2_{выб}}{D_{выб}} = 101{,}64; \quad s = \frac{x_{выб}}{D_{выб}} = 3{,}1707. \qquad (8)$$

The graph of the hypothetical density function in combination with the histogram is shown in fig. 2 and table 2.

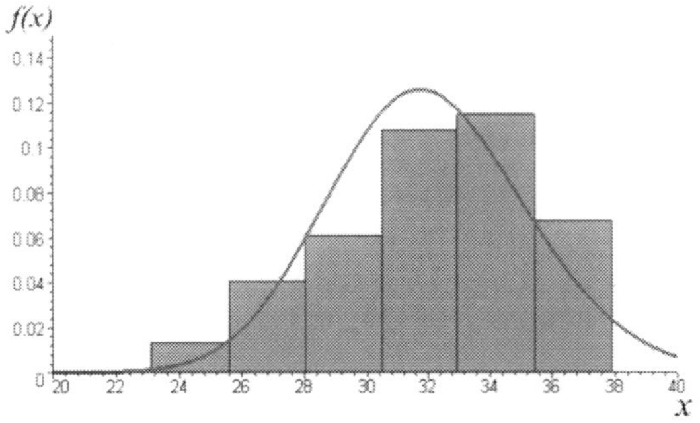

Fig. 2. Histogram of relative humidity indices for skin samples and a hypothetical function of the gamma distribution density

Table 2. Values of the gamma distribution density function for the samples

Borders ranks	Values function
23,1	$0{,}128 \cdot 10^{-2}$
25,567	0,0140
28,033	0,0598
30,5	0,1164
32,967	0,1171
35,433	0,0669
37,902	0.0234

From the function of the density of the gamma distribution, calculate:

$$P(ОВК < 31{,}9516) = \int_0^{31{,}9516} f(x)\,dx = 0{,}5\,.$$

Find the boundaries of the interval $\gamma = 0{,}95$

$$P(|ОВК - M(ОВК)| < 6{,}2155) =$$
$$= P(25{,}841 < ОВК < 38{,}272) = 0{,}95\,.$$

The index of the relative skin moisture for the samples, calculated from the statistical series, according to the rule of two sigma changes in the range (25,841; 38,272) with a confidence probability of 0,95.

4 Conclusions

The results obtained indicate a scatter in the relative humidity of the skin due to the different microcapillary porosity $П$ of samples cut from different parts of the skin and therefore the results of the intensified hygrothermal treatment should be evaluated, in our opinion, not by the numerical index of the relative humidity, but by the time necessary to fill a single microcapillary of the skin as a result of sorption phenomena and heat exchange between the skin and steam according to the equations given in [16]:

$$m_c = \pi \cdot r_0^2 L \cdot \rho_в \cdot \left(1 - e^{-\frac{2D\rho_п}{L^2 \cdot \rho_в} \cdot \tau} \right) \qquad (9)$$

where r_0 is the radius of the microcapillary at the initial instant of time; L – length of the capillary, taking into account its tortuosity; ρ_n is the vapor density corresponding to the process conditions (pressure and temperature); $\rho_в$ is the density of water; D is the diffusion coefficient, and τ is the processing time.

$$m_t = M_к \cdot \frac{C_к}{4{,}19} \cdot \left[e^{0{,}00176 \cdot (T - T_в)} - 1 \right] \cdot 100\% \qquad (10)$$

Having set by the specific heat of skin of S_k equal 1,61 kJ / (kg To) [4], we will receive numerical values of an additional weight of moisture in skin weight depending on temperature.

The numerical values of mass of the condensed moisture as a result of capillary and sorption processes and heat exchange received on the equations (9,10) and corresponding to them values of relative humidity of skin are given in table 3.

Table 3. Values of relative humidity of samples of skin depending on parameters and time of vacuum and sorption moistening

Key parameters vacuum and sorption moistening	Moistening time, s			
	120	240	360	480
Average value of pressure, Pa	$25 \cdot 10^3$	$30 \cdot 10^3$	$40 \cdot 10^3$	$50 \cdot 10^3$
Average value of temperature, °C	55	57	58	59
Average value of density of steam, kg/m³	$16 \cdot 10^{-2}$	$19 \cdot 10^{-2}$	$26 \cdot 10^{-2}$	$36 \cdot 10^{-2}$
Average speed of the thermal movement molecules, m/s	$6{,}23 \cdot 10^2$	$6{,}24 \cdot 10^2$	$6{,}24 \cdot 10^2$	$6{,}26 \cdot 10^2$
Average length of a free run molecules, m	$0{,}4 \cdot 10^{-6}$	$0{,}37 \cdot 10^{-6}$	$0{,}28 \cdot 10^{-6}$	$0{,}22 \cdot 10^{-6}$
Coefficient of diffusion,.m²/ s	$0{,}83 \cdot 10^{-4}$	$0{,}76 \cdot 10^{-4}$	$0{,}58 \cdot 10^{-4}$	$0{,}45 \cdot 10^{-4}$
Volume of moisture, m³	$3{,}51 \cdot 10^{-14}$	$5{,}25 \cdot 10^{-14}$	$5{,}84 \cdot 10^{-14}$	$6{,}09 \cdot 10^{-14}$
The calculated mass of moisture of microcapillaries as a result of the sorption phenomena (9), %	9,5	13,5	15,3	16,0
The mass of moisture calculated by a formula as a result of heat exchange (10), %	2,5	2,6	2,7	2,8
The total	12,0	16,1	18,0	18.8
calculated mass of moisture, %				
The calculated relative humidity of samples of skin, %	18,0+12,0= =30	18,0+16,1= =34.1	18,0+18,0= =36,0	18,0+18,8= =36,8

The time of maximum filling of the microcapillary with moisture calculated according to the formula (9,10) will be 480s [10] (the relative humidity of the leather samples is 36,8% and this value can be recommended with intensified hygrothermal treatment of natural skins in the corresponding equipment [17-20].

References

1. L.V. Larina, N.Yu. Leshina Technical and technological problems of service **3**, 42-44 (2014)
2. L.V. Larina, *Proceedings of IGTA conference*, (Progress-2012), Ivanovo, 163-164 (2012)
3. L.V. Larina, Research of the process and development of the installation for vacuum-sorption humidification of the upper parts of the shoe: dis.... cand. tech. Sciences, Moscow, (1991)
4. M.U. Kuramysov, Yu.V. Bondareva, E.I. Bitus Izv. universities. Technology of the textile industry, No. 3m, 110-112. (2015)
5. V.M. Janpayzova, Izv. universities. Technology of the textile industry, No. 6. 124-126 (2013)
6. L.V. Larina V.A. Pershin, Engineering bulletin of Don, No. 3. (2012) Access mode: http://www.ivdon.ru/ru/magazine/archive/n3y2012/902. - The mass media registration certificate Al No. FS77-27308 from 2/22/2007
7. L.V. Larina, V.V. Smirnov, *Design of mechanisms of cars of systems of service and light industry: a laboratory practical work for bachelors* (ISOiP (branch) of DSTU in Shakhty, Shakhty, 2014), ISBN 978-5-93834-911-7, Access mode: http://www.libdb.sssu.ru
8. R.V. Lutsyk, Development of methods of study, analysis of interrelation and forecasting of heat and mass exchange and physico-mechanical properties of textile and leather-shoe materials: dis.... Doct. tech. Sciences, Kiev (1987).
9. L.V. Larina *Vacuum and capillary gigrotermichesky processing of skin of L.V. Larin* (ИСОиП (branch) of DGTU, Shakhty, 2014)
10. L.V. Larina, N. Yu. Leshina, V.V. Smirnov, *Current problems of the equipment and technology: proceedings*, Shakhty, 158-162 (2014)
11. L. In Larina, V. A., Pershin Smirnov V. In, Dzheliyev Yu.O Patent No. 2447822 Russian Federation, MPK A43D 11/10
12. G. Korn *Handbook on Mathematics for Scientists and Engineers* (Nauka, Moscow, 1970)

13. Larina L.V. Izvestiya Vuzov. North-Caucasus. region. Technical science,Novocherkassk: YURSTU, 140-145 (2011)
14. L.V. Larina, Izvestiya Vuzov. North-Caucasus. region, technical sciences, Novocherkassk: YURSTU, No. 1, 129-132 (2012)
15. G. Kramer, *Mathematical methods of statistics*, (Mir, Moscow, 1975).
16. L.V. Larina, V.V. Karaschenko, World Applied Sciences Journal, **29**, N 12, (2014) Access mode: http:www.idosi.org/wasj/wasj/29%2812%292014.htm
17. LV. Larina, A.A. Zagoskin, A.A. Small, etc. Pat. 2312574 Russian Federation IPC6, A 43 D 95/10.
18. LV. Larina, V.V. Smirnov, B.I. Surmilov et al. Pat. 2414832 Russian Federation IPC6, A 43 D 11/10.
19. Larin L.V., Smirnov V. V., Tikhonov O. B., Smirnov E. I. Pat. No. 2537160 the Russian Federation, MPK A43D 11/00
20. Larina L.V. *ITNO-2017: Proc. V International scientific and practical conference DSTU*, 410-415 (2017)

Development of the control algorithm of processes of intensive hygrothermal impact on capillary and porous materials in the conditions of the vacuum

Ludmila Larina[1,*], *Dmitriy* Ruslyakov[1], *Olga* Tikhonova[1], and *Sergey* Morozov[1]

[1]Don State Technical University, 344002 Gagarin Square, 1, Rostov-on-Don, Russian Federation

Abstract. Objective of this research is creation of an algorithm of a control system of the modes of the intensive hygrothermal influence (IGI) in the conditions of a vacuum when performing the corresponding operations: moistening; the subsequent, if necessary, cyclic drying from within preparation of top of footwear; damp thermal treatment on universal installation with adjustable parameters of a working environment. For assessment of quality of the intensified hygrothermal impact on preparations of top of footwear the integrated criteria of efficiency of processes were used. Ensuring automatic control of parameters of processes of IGV on preparations of top of footwear in universal vacuum installation will allow to control quality of preparations upon transition from performance of one operation to another according to standard manufacturing techniques of footwear.

1 Introduction

The Intensive hygrothermal Influence (IGI) on capillary and porous materials to which also genuine leather belongs provides change to the best physicomechanical indicators of the processed materials [1]. It is known [2, 3] that for achievement of the indicators of physicomechanical properties of previously vacuumized skin demanded by manufacturing techniques [30] of footwear except the coordination of indicators of temperature and pressure of a working environment providing her required relative humidity it is necessary to regulate also time of processing. And, depending on the purpose of the carried-out hygrothermal influence – increases in molding capacity of skin or a form stability of finished products it is required to change to the best either the conditional module of elasticity of material, or size of residual deformation, and their best values after pumping out and the subsequent hygrothermal processing fall on various values of time of processing in the range from 2 up to 8 minutes in table 1 [5].

As appears from experimental data of dependence, full and residual lengthenings at 10 MPas from time of vacuum and sorption moistening of skin of chromic tanning have obviously expressed maximum corresponding to moistening within 4 minutes and to a moisture additional weight within 18-19%. At a traditional way of moistening the moisture additional weight in 15% Excess of the optimum humidity received at a vacuum and sorption way of moistening for 3-4% is optimum is explained, obviously, by features of penetration and effect of moisture on internal structure of skin at this way of moistening.

Table 1. Change of Physicomechanical Indicators of Samples of Skin of Chromic Tanning depending on duration of vacuum and sorption moistening

Indicators	Time of moistening, min.			
	2	4	6	8
Additional weight of moisture, %				
in longitudinal samples	15,0±1,0	18,8±1,2	19,0±1,2	11,6±0,7
in cross samples	16,4±1,0	19,7±1,8	16,2±1,0	11,6±0,7
average	15,7±1,0	19,3±1,5	17,6±1,1	11,6±0,7
Residual lengthening at 10 MPas, %				
longitudinal	14±1	18±2	16±1	14±1
cross	18±1	28±2	23±2	13±1
average	16±1	23±2	19,5±1,5	13,5±1
Full lengthening at 10 MPas, %				
longitudinal	22±2	29±2	25±2	25±2
cross	28±2	52±3	42±3	34±2
average	25±2	40,5±2,5	33,5±2,5	29,5±2
Loading at a gap, H				
longitudinal	290±16	240±9	230±17	215±14
cross	236±15	180±8	145±11	140±9
average	263±15	210±8	188±14	180,5±11,5
Strength at stretching, MPa				
longitudinal	33,3±2,1	30,0±1,2	28,3±2,0	28,9±1,7
cross	29,5±1,8	23,5±1,1	22,0±2,0	18,9±1,5
average	31,4±2,0	27±1,1	25,0±2,0	23,9±1,6
Lengthening at a gap, %				
longitudinal	66±4	66±5	63±4	59±4

*Corresponding author: Ludmila-larina2010@mail.ru

cross	75±5	75±5	72±4	55±3
average	71±5	71±5	68±4	57±4
Conditional module of elasticity, MPa				
longitudinal	40,0±2,5	27,7±1,7	40,4±2,5	33,8±2,1
cross	28,1±1,7	15,2±0,9	17,5±1,0	23,7±1,5
average	34±2,1	21,5±1,3	29±1,8	28,8±1,8

Therefore, depending on the processing purpose, it is required to maintain not only the required parameters of processes [5,6], but also to fix time of processing [7] taking into account a prevalence of these or those indicators in total defining processes and products of processing as systems with the set properties.

For assessment of quality of the intensified hygrothermal impact on preparations of top of footwear in works [8], the integrated criteria of efficiency of processes $\prod_{i=1}^{k} \pi_i^1$ received by method of similarity of functioning of technical systems [9] were used.

At the same time were compared calculated for a reference sample – an analog value of the generalized criterion of similarity – the material analog establishing functional connection between indicators of physicomechanical properties, and parameters of its processing and value of the generalized criterion of the similarity calculated $\prod_{i=1}^{k} \pi_i^1$ for other studied material excellent on the physicomechanical to properties from material analog, but with the same structure of parameters of processing, and in case of inequality of the calculated values of criteria of similarity, other parameters of processing of the studied material before equality performance were selected

$$\prod_{i=1}^{k} \pi_i = \prod_{i=1}^{k} \pi_i^1 \text{ [5].}$$

2 Experimental part

Research problem at this stage is ensuring automatic control with the IGV modes when performing the corresponding operations: moistening; the subsequent, if necessary, cyclic drying from within preparation of top of footwear; damp thermal treatment [10].

Object of researches is the installation realizing a serial-parallel way of hygrothermal processing of the preparations found from skin of chromic tanning in the form of circles with a diameter of 100 mm with the concentric circles with a diameter of 40 and 60 mm applied on them for the subsequent installation on punches. They were exposed to hygrothermal processing on the developed universal installation with adjustable parameters of a working environment [11]. (which design is given further), providing consecutive (with dredging of the samples which are on a nonperforated punch) and serial-parallel, with application of the punched punch [12], performance of operations of moistening, drying and damp thermal treatment in the conditions of a vacuum and without him.

Experimental installation [13] which schematic diagram is provided on the figure 1 has been developed for conducting pilot studies of the modes of hygrothermal processing (moistening, damp and thermal fixing, drying and damp thermal treatment).

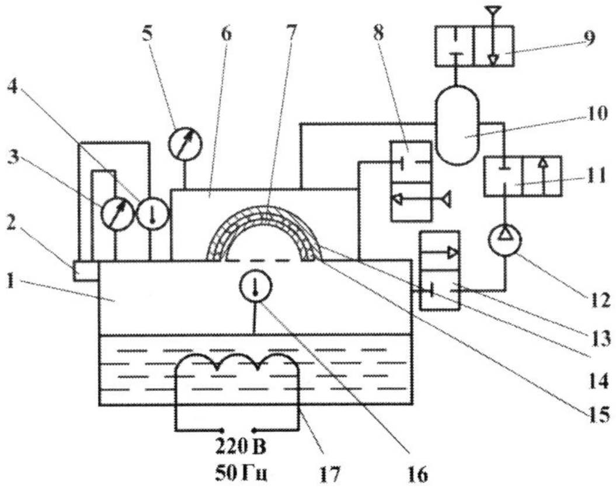

Fig. 1. The Schematic diagram of installation for gigrotermichesky impact on preparations of top of footwear: 1 – the lower vacuum chamber with water; 2 – hardware of the experimental installation; 3, 5 – pressure sensors; 4, 16 – temperature sensors; 6 – upper camera; 7 – the punched baffle and a punch; 8, 9 – the air supply valve; 10 – air accumulator; 11 – air pumping valve; 12 – vacuum pump; 13 – connective valve; 14 – airfast film; 15 – skin sample; 17 – TEN.

Are a part of installation upper 6 and the lower 1 vacuum chambers with a total amount of 0,09 m^3 which are partitioned punched by a baffle and a punch 7 which the sample of skin 15 is put on and then it is sealed from the upper camera by an airfast film 14.

Evacuation of the lower camera is carried out by the vacuum pump 12 (type 2NVR-5DM, with a productivity of 5 p/a, with outlet pressure from atmospheric to 0,02 MPas, electromotor of 0,55 kW) via the piston valve of control 13. The pumped-out air gathers in the accumulator of air 10 and moves in the upper camera for the additional fixing which is realized as required.

Heating of water in the lower camera realizes the tubular electric heater 16 (the 1 kW TEN). Tension given on TEN was regulated in the diagram of the hardware of the experimental installation 2 (a figure 1).

Steam to samples in case of a research of processes of hygrothermal influence arrived in case of evaporation of the warmed-up water which is in the lower part of the camera after preliminary evacuation last up to 0,02 MPas.

Ratios between temperature and pressure [14] were supported automatically: when moistening $T = 2028 \cdot P^{0,31}$ when drying $T > 2028 \cdot P^{0,31}$ at damp thermal treatment $T < 2028 \cdot P^{0,31}$.

3 Approach

The control system which algorithm is given in the figure 2 is developed for ensuring automatic control with the modes of intensive hygrothermal influence when performing the corresponding operations according to requirements of technological process of production of footwear of genuine leather.

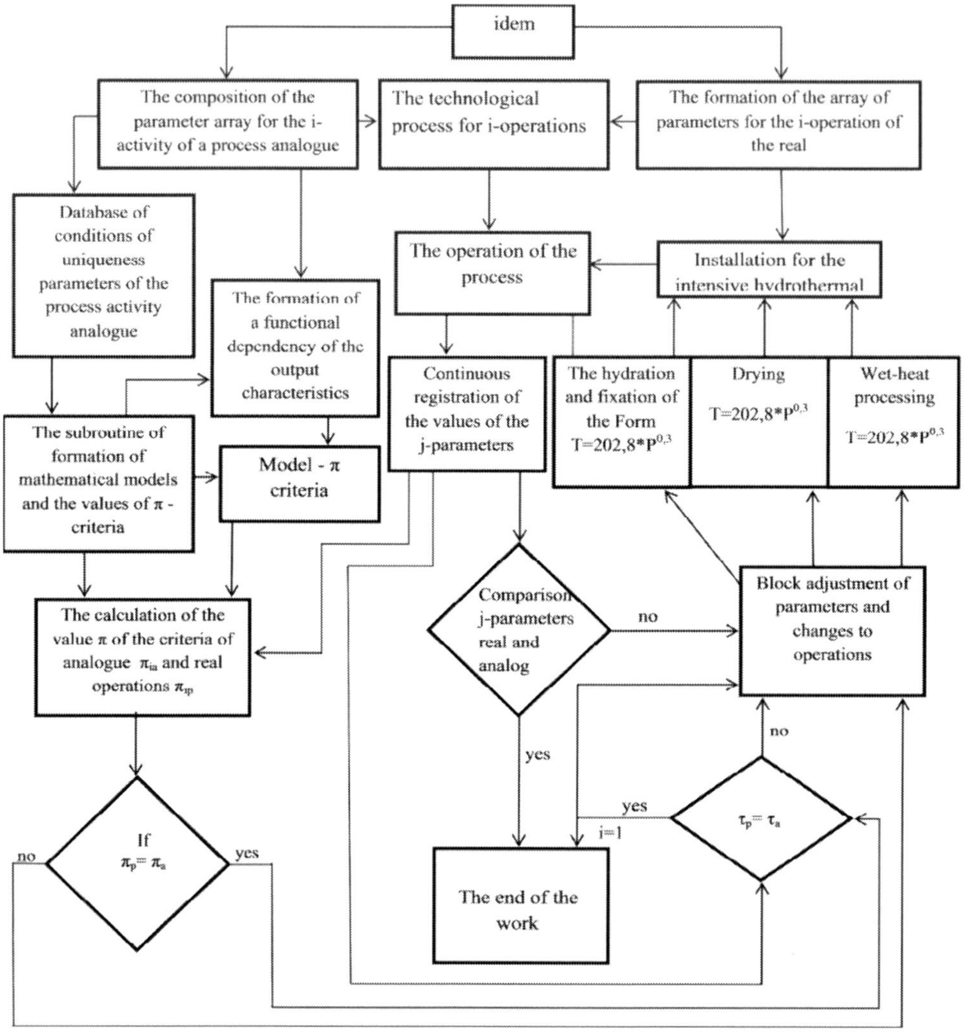

Fig. 2. Scheme of an algorithm of a control system technological processes of IGV

Depending on the chosen technological process of IGV two structures of the massif – for process of an analog and real process which include the parameters defining physicomechanical properties of material, parameters of a working environment and installation are entered into a control system. The number of parameters has to be equal in both massifs. The massif of an analog has to be made on the basis of unambiguity conditions which allow to define value of the parameters corresponding to a certain type of processes of IGV.

On the basis of the program for creation of mathematical models and values π criteria the functional dependences of output characteristics of processes of IGV from the parameters defining its status are created.

Based on these data the subprogramme calculates values π-criteria of an analog – π_{ia} and real process – π_{ip} if the received values match, then the modes corresponding to process analog are applied also to real process. In this case the signal arrives on the unit of comparing where after an output of parameters T and P of real process on the curve set by the equation $y = 202{,}8 \cdot x^{0,3}$ when moistening and corresponding to the maximum saturation by moisture of the steam-air environment which is in the accepted volume, the ratio of these parameters is supported to constants during moistening time.

When drying the ratio of temperature and pressure in the set limits during drying time according to inequality is supported $y > 202{,}8 \cdot x^{0,3}$.

At damp thermal treatment the ratio of temperature

and pressure in the set limits during time of damp thermal treatment according to inequality is supported $y < 202,8 \cdot x^{0,3}$.

In that case when values π – criteria aren't equal, the signal on the block of regulation of parameters which changes temperature of processing and pressure corresponding to her to the values providing equality π_{ia} and π_{ip} is given. Then process is carried out as it is described above.

4 Conclusions

1) When using of the offered algorithm automatic control of parameters of processes of IGV on preparations of top of footwear in universal vacuum installation upon transition [15] from performance of one operation to another according to standard manufacturing techniques of footwear will be provided [16-18].

2) On a management information cover of skilled installation for hygrothermal processing of skin the certificate of the Russian Federation on the state registration of the computer program No. 2010617180 [19, 20] is received.

References

1. L. V. Larina, N.Yu. Leshina Technical and technological problems of service, n 3, 42-44 (2014)
2., K.M. Zurabyan [etc.] News of higher education institutions. Technology of light industry. No 5, 75-80 (1990)
3. L. V. Larina, V. A.Pershin, Engineering bulletin of Don No. 3., (2012) Access mode: http://www.ivdon.ru/ru/magazine/archive/n3y2012/902.
4. L. V. Larina, Pershin V. A., Smirnov V.V., Dzheliyev Yu.O. News of higher educational institutions. North Caucasus region. Technical science, No., 2012, 129-132
5. L. V. Larina, V.I. Yurchenko, News of higher education institutions. Sowing, No. 6. 140-145, (2011)
6. L. V. Larina, News of higher education institutions. Sowing, No. 1, 129-132 (2012)
7. L.V. Larina, V. V. Karaschenko World Applied Sciences Journal, **29**, N. 12. (2014) Access mode: http://www.idosi.org/wasj/wasj/29%2812%292014.htm
8. L.V. Larina, V.A. Pershin [etc.] The Engineering Bulletin of Don, No. 3 (2012) Access mode: http://www.ivdon.ru/ magazine/latest/n3y2012/896/
9. V. A. Pershin, Methodology of similarity of functioning of technical systems (monograph) (YuRGUES, Shakhty) (2004)
10. L.V. Larina *Vacuum and capillary gigrotermichesky processing of skin* (ISoIP (branch) of DGTU Shakhty, 2014)
11. L. V. Larina, V.V. Smirnov, B.I. Surmilov, V.A. Soroka, S.A. Tryapichkin, Pat. 2414832 Russian Federation, MPK6 And 43 D 11/00.
12. Larina L.V., Smirnov V. V., Tikhonova O. B., Smirnova E. I. Pat. No. 2537160 Russian Federation, MPK A43D 11/00/
13. L.V. Larina, B.I. Surmilov, A.A. Zagoskin, A.A. Maly, 2312574 Dews. Federation of MPK6, A 43 D 95/10
14. S.L. Rivkin, A.A. Alexandrov, *Heatphysical properties of water and water vapor* (Moscow, Energy, 1980)
15. L.V. Larina, *Proceedings of IGTA conference*, (Progress-2012), Ivanovo, 163-164 (2012)
16. Larina L.V. Modern science: current problems of the theory and practice. It is gray.: Natural and technical science. No. 3-4, 53-55 (2014)
17. L. V. Larina, Leshina N.Yu., Smirnov V.V. *Current problems of the equipment and technology: proceedings*, Shakhty, 158-162 (2014)
18. L. V. Larina, *Household appliances, technology and equipment of the housing and public utilities enterprises, service and mechanical engineering*: proceedings, Shakhty: FGBOU VPO of "YuRGUES", 48-49 (2013)
19. L. V. Larina. Smirnov V.V., Cherunova I.V., Lebedeva of E.O. Merkulov A. V., Modern problems of science and education, No. 6 (2013) Access mode: http://online.rae.ru/1656.
20. L.V. Larina, I.V. Tarara, V.V. Smirnov, Certificate of the Russian Federation on the state registration of the computer program No. 2010617180

Model of Motion of the Combined Manipulator

Vladimir Poletaykin[1], *Pavel* Kolesnikov[2]

[1]Department of Technologies and machinery of environmental engineering of FSBEI HE "Reshetnev Siberian State University of Science and Technology", Krasnoyarsk, Russia.
[2]Department of Vehicles, tractors and forestry machines of FSBEI HE "Reshetnev Siberian State University of Science and Technology", Krasnoyarsk, Russia.

Abstract. The combined manipulators composed of telescopic booms and rotary supports with movable in longitudinal vertical plane columns which provide an increased outreach of operative parts and a bigger service space are widely used in different branches of industry (engineering, transport). The article considers development of mathematical model of kinematic links motion of the combined manipulator with a column movable in longitudinal vertical plane. The mode of load lifting in a synchronous movement of boom and column from the position of picking the load up to the transport position was analyzed.

1 Introduction

Combined hydraulic manipulators are widely used in production of weight-handling, loading and other machinery. Production and improvement of such machines should be based on studying the dynamics of structural elements. Herewith methods of mathematical modeling are considered to be the most effective. On this basis, researches with the use of mathematical models of combined manipulator links motion mode dynamics focused on validation of kinematics parameters and design of the processing equipment, should be considered as relevant.

2 Foundation of the calculation model

Fig. 1 shows the calculation model of the "processing equipment – load" system. The analyzed mode can be used when the manipulator operates as a processing equipment of forestry, road-building, load-lifting and other machines.

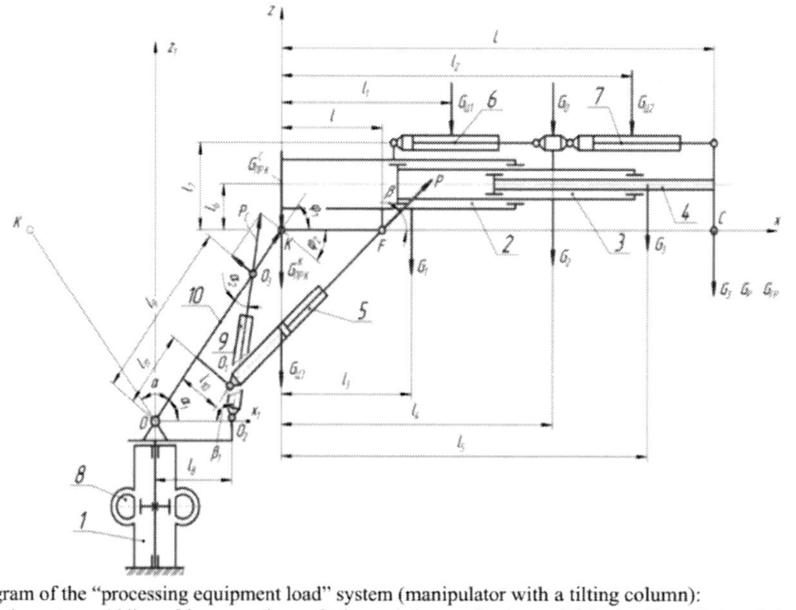

Fig. 1. Calculation diagram of the "processing equipment load" system (manipulator with a tilting column):
1 – rotary support; 2, 3, 4 – outer, middle and inner sections of telescopic boom; 5 – boom lift hydraulic cylinder; 6, 7 – hydraulic cylinders of sections' telescoping mechanism; 8 – mechanism of manipulator's rotation in horizontal plane; 9 – hydraulic cylinder of column rotation; 10 – column; O1K = C; OO2= ℓ8 = C1

* Corresponding author: atlm@sibsau.ru

When working element picks the load up, the telescopic boom brings it to the machine by retraction of sections, then the load is set to the transport position by actuation of column rotation hydraulic cylinder (CRHC) and boom lift (BL). Meanwhile the boom rotates around K-axis, and the column rotates around O-axis. Rotation angle of the boom is φ (relative motion), rotation angle of the column is α (transportation motion). Reference point of the angle φ is the lower limit position of the boom; reference point of the angle α is the rightmost position of the column.

Following symbols are used in Fig. 1.

G_1, G_2, G_3: gravitation forces of outer, middle and inner sections of telescopic boom;

$G_{Ц1}$,$G_{Ц2}$, G_0: gravitation forces of hydraulic cylinders while sections' telescoping and of sections' telescoping mechanism;

$G_З$, $G_{ГР}$, G_P: gravitation forces of gripping device, load and rotator;

$G_{Ц3}$, $G_{Ц4}$: gravitation forces of hydraulic cylinders while column rotation and boom lift;

$G_{ПР.C}$: total gravitation force of structural elements of the boom and the load brought to point *C*;

P, P_C: forces on rods of the column rotation and boom lift hydraulic cylinders;

L: boom dimensions at retracted sections;

ℓ_1, ℓ_2, ℓ_3, ℓ_4, ℓ_5: distances from *K*-axis of boom rotation to the centers of structural elements gravity

ℓ_6, ℓ_7, ℓ_8, ℓ_9, ℓ_{10}, ℓ_{11}: dimensions of manipulator elements.

Constructing the equations of motion for the "processing equipment – load" system

The boom rotates in ZKX-plane, the column rotates in Z1OX1-plane. The rotation angles α and φ clearly determine the positions of this elements of the system in the rotation planes. If dimensions of the boom *L* and the column *LK* are known, we can find position of any point by means of specified parameters. Based on this, the system can be considered as a system with two degrees of freedom (*K* = 2) with generalized coordinates α and φ.

To formulate equations of motion for this mechanical system the Lagrange equations of the 2-nd kind was used. According to the number of degrees of freedom of the system two Lagrange equations were formulated

$$\frac{d}{dt}\left(\frac{\partial T}{\partial \dot{\alpha}}\right) - \left(\frac{\partial T}{\partial \alpha}\right) = Q_\alpha,$$
$$\frac{d}{dt}\left(\frac{\partial T}{\partial \dot{\varphi}}\right) - \left(\frac{\partial T}{\partial \varphi}\right) = Q_\varphi \qquad (1)$$

where Q_α, Q_φ are generalized forces corresponding to generalized coordinates α and φ respectively.

The motional energy of the system is equal to the sum of motional energies of the column and the boom, i.e. the sum of the motional energies of relative and transportation motions

$$T = T1 + T2, \qquad (2)$$

where T1 is motional energy of the given column mass with the given masses of the structural elements mounted on it (column rotation and boom lift hydraulic cylinders and other elements of the hydraulic actuator); T2 is motional energy of the given mass of the column and the load.

When the telescopic boom rotates, its sections do not telescope, and the dimensions of the boom L do not change, therefore positions of the mass centers of the structural elements of the boom relatively to the K-axis (radius of mass inertia) remain constant. In this case in order to simplify the determination of the motional energy of the system masses of structural elements of the telescopic boom were brought to point C (point of suspension of the rotator to the boom). Masses of structural elements of the column were brought to K-axis (mounting axis of the boom). Calculation of the mass of structural elements of the boom mПР.C is based on the condition that motional energy of the reduced mass is equal to the sum of the motional energies of the masses it replaces. Therefore,

$$\frac{G_{ПР.C}\dot{\varphi}^2 L^2}{2g} = \frac{\dot{\varphi}^2}{2g}\cdot\left[\left(G_1\ell_3^2 + G_2\ell_4^2 + G_3\ell_5^2 + +G_{Ц1}\ell_2^2 + G_0\ell_4^2 + G_P + G_З + G_{ГР}\right)\right] \qquad (3)$$

On this basis, the boom mass brought to point C is

$$m_{ПР.C} = \frac{G_{ПР.C}}{g} = \frac{G_1}{g}\cdot\frac{\ell_3^2}{L^2} + \frac{G_2}{g}\cdot\frac{\ell_4^2}{L^2} + \frac{G_3}{g}\cdot\frac{\ell_5^2}{L^2} + \frac{G_{Ц1}}{g}\cdot\frac{\ell_1^2}{L^2} + + \frac{G_{Ц2}}{g}\cdot\frac{\ell_2^2}{L^2} + + \frac{G_0}{g}\cdot\frac{\ell_4^2}{L^2} + \frac{G_P + G_З + G_{ГР}}{g} \qquad (4)$$

or

$$m_{ПР.C} = m_1\frac{\ell_3^2}{L^2} + m_2\frac{\ell_4^2}{L^2} + m_3\frac{\ell_5^2}{L^2} + +m_{Ц1}\frac{\ell_1^2}{L^2} + m_{Ц2}\frac{\ell_2^2}{L^2} + m_0\frac{\ell_4^2}{L^2} + m_P + m_З + m_{ГР} \qquad (5)$$

where m_1 is mass of the outer section of the boom; m_2 is mass of the middle section of the boom; m_3 is mass of the inner section of the boom; $m_{Ц1}$, $m_{Ц2}$ are masses of hydraulic cylinders of the sections' telescoping mechanism; m_0 is mass of the sections' telescoping mechanism; m_P, $m_З$, $m_{ГР}$ are masses of gripping device, load and rotator.

Mass of the column and structural elements mounted on it brought to point *K* can be calculated with the following equation:

$$\frac{G_{ПР.K}\dot{\alpha}^2 L_K^2}{2g} = \frac{G_K\dot{\alpha}^2(0{,}5L_K)^2}{2g} + \frac{0{,}5G_{Ц3}\dot{\alpha}^2(0{,}5\ell)^2}{2g} + + \frac{0{,}5G_{Ц4}\dot{\alpha}^2(0{,}5\ell_8)^2}{2g} \qquad (6)$$

The equation (6) is based on the assumption that gravitation forces of cylinders GЦ3 and GЦ4 are evenly distributed between the boom and the column as well as between the column and the base of the rotary support; points of their application are 0,5ℓ and 0,5ℓ8 respectively from the mounting axis of the boom (point *K*).

According to the equation (6) mass of the column and structural elements mounted on it brought to point *K* is

$$m_{ПР.K} = \frac{G_K}{g}\cdot\frac{(0{,}5L_K)^2}{L_K^2} + \frac{0{,}5G_{Ц3}}{g}\cdot\frac{(0{,}5\ell)^2}{L_K^2} + \frac{0{,}5G_{Ц4}(0{,}5L_K)^2}{gL_K^2} = 0{,}25\frac{G_K}{g} + 0{,}125\frac{G_{Ц3}}{g}\cdot\frac{\ell^2}{L_K^2} + 0{,}125m_{Ц4}\cdot\frac{\ell_8^2}{L_K^2} \qquad (7)$$

or $m_{ПР.K} = 0{,}25\, m_K + 0{,}125m_{Ц3}\cdot\frac{\ell^2}{L_K^2} + +0{,}125m_{Ц4}\cdot\frac{\ell_8^2}{L_K^2}$ (8)

According to (2) the motional energy of the system is

$$T = T1 + T2 = \frac{I_0\,\dot{\alpha}^2}{2} + \frac{1}{2}(m_{ПР.C}V_{ac}^2 + I_C\dot{\varphi}^2), \qquad (9)$$

where I_0 is moment of inertia of the column relative to O-axis, I_C is central moment of inertia of the boom assembly; V_{ac} is an absolute motion velocity of e point C (point of boom mass brining).

The theorem on composition of velocities was applied, according to which absolute velocity of point C is equal to geometric sum of its transportation and relative velocities

$$V_{ac}^2 = V_{ec}^2 + V_{rc}^2 + 2V_{ec} \cdot V_{rc} \cos\gamma,$$

where V_{ec}, V_{rc} are velocities of transportation and relative motion of point C,

γ is an angle between vector directions of transportation and relative motion of point C.

Fig. 2 shows a diagram for determining velocity of an absolute motion of point C - V_{ac}, point of boom and the load mass bringing.

$$V_{ec} = OC \cdot \dot{\alpha}\ ;\ V_{rc} = L \cdot \dot{\varphi}.$$

According to OKC triangle

$$OC^2 = L_K^2 + L^2 - 2L_K \cdot L\cos(\varphi_H + \varphi).$$

According to the same triangle $L_K^2 = L^2 + OC^2 - 2L \cdot OC \cdot \cos C$.

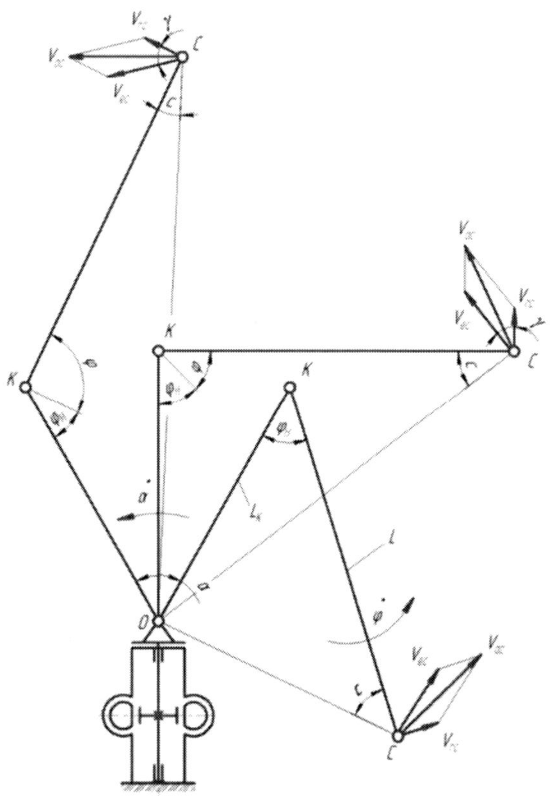

Fig. 2. Diagram for determining velocity of an absolute motion of point C in synchronous movement of the boom and the column: φн is an initial angle between the boom and the column; φ is rotation angle of the boom (relative motion); α is rotation angle of the column (transportation motion); $\dot{\alpha}$, $\dot{\varphi}$ are angular velocities of the column and the boom, respectively

Fig. 2 shows, that angles γ and C are equal being angles with mutually perpendicular sides. Thus,

$$\cos C = \cos\gamma = \frac{L^2 + OC^2 - L_K^2}{2L * OC}$$

Absolute motion velocity

$$V_{ac}^2 = \dot{\alpha}^2 \cdot OC^2 + \dot{\varphi}^2 \cdot L^2 + 2\dot{\alpha} \cdot \dot{\varphi} \cdot L \cdot OC \cdot \cos\gamma.$$

$\cos\gamma = \cos C$ is expressed by $\cos(\varphi_H + \varphi)$.

$$\cos\gamma = \frac{L^2 + OC^2 - L_K^2}{2L \cdot OC}$$

$$= \frac{L^2 + L_K^2 + L^2 - 2L_K \cdot L \cdot \cos(\varphi_H + \varphi) - L_K^2}{2L \cdot OC} = \left[\frac{L^2 - L_K \cdot L}{L \cdot OC}\right] \cdot \cos(\varphi_H + \varphi).$$

After substituting

$$V_{ac}^2 = \dot{\alpha}^2 OC^2 + \dot{\varphi}^2 \cdot L^2 + 2\dot{\alpha} \cdot \dot{\varphi} \cdot L \cdot OC \cdot \left[\frac{L^2 - L_K * L}{L * OC}\right] \cdot \cos(\varphi_H + \varphi) \quad (10)$$

In accordance with (10), equation of motional energy looks like the following

$$T = \frac{m_{\Pi P.K} \cdot L_K^2 \cdot \dot{\alpha}^2}{2} + \frac{m_{\Pi P.C} \cdot OC^2 \cdot \dot{\alpha}^2}{2} + \frac{m_{\Pi P.C} \cdot L^2 \cdot \dot{\varphi}^2}{2} + m_{\Pi P.C}$$

$$\cdot L \cdot OC \cdot \dot{\alpha} \cdot \dot{\varphi} \cdot \left[\frac{L^2 - L_K \cdot L}{L \cdot OC}\right] \cdot \cos(\varphi_H + \varphi) + \frac{I_C \cdot \dot{\varphi}^2}{2} \quad (11)$$

Equations of motion for the "processing equipment – load" system

By differentiating the equation of the motional energy (11) for the components of the Lagrange equations (1), the following system of equations of motion of the dynamic system of the manipulator is expressed:

$$m_{\Pi P.K} \cdot L_K^2 \cdot \ddot{\alpha} + m_{\Pi P.C} \cdot OC^2 \cdot \ddot{\alpha} + m_{\Pi P.C} \cdot OC \cdot L \cdot \ddot{\varphi} \left[\frac{L^2 - L_K \cdot L}{L \cdot OC}\right]$$

$$\cdot \cos(\varphi_H + \varphi) - m_{\Pi P.C} \cdot OC \cdot L \cdot \dot{\varphi}^2 \cdot \left[\frac{L^2 - L_K \cdot L}{L \cdot OC}\right]$$

$$\cdot \sin(\varphi_H + \varphi) = Q_\alpha$$

$$(m_{\Pi P.C} L^2 + I_C)\ddot{\varphi} +$$

$$+ m_{\Pi P.C} \cdot OC \cdot \ddot{\alpha} \cdot L \cdot \left[\frac{L^2 - L_K \cdot L}{L \cdot OC}\right] \cdot \cos(\varphi_H + \varphi) = Q_\varphi \quad (12)$$

Determination of generalized forces Qα and Qφ corresponding to the generalized system coordinates φ and α.

To determine generalized forces Qφ and Qα the principle of possible motions of the system in direction of the increase of generalized coordinates φ and α - ΔφиΔα was used. For calculation of the generalized force Qφ one considers Δα = 0, and for calculation of the generalized force Qα Δφ = 0. Gravitation forces of structural elements and load Gi, and forces on rods of the hydraulic cylinders PC and P are considered as active forces. Generalized force is considered as a factor while expressing sum of elementary works of active forces in the direction of possible movement:

$$\delta A_\varphi = Q_\varphi \cdot \Delta\varphi;\ \delta A_\alpha = Q_\alpha \cdot \Delta\alpha. \quad (13)$$

Let us determine the sum of elementary works of active forces in direction of the generalized coordinate φ. Herewith $\Delta\varphi \neq 0, \Delta\alpha = 0$.

To determine the generalized force Qφ equations (4), (5) of mass of the boom mΠP.C brought to point C in a horizontal position of the boom are used.

Then $\Sigma\delta A_{\varphi} = (P\ell \sin\beta - m_{\text{ПР.С}}gL)\Delta\varphi$

$$Q_{\varphi} = P\ell \sin\beta - m_{\text{ПР.С}}gL \quad (14)$$

Let us determine the sum of elementary works of active forces and gravitation forces of structural elements in direction of generalized coordinate α. Herewith $\Delta\varphi = 0$; $\Delta\alpha \neq 0$. In the direction of generalized coordinate α active force PC and gravitation forces of structural elements of the boom and the column perform an operation. In order to simplify the equation of generalized force Qα gravitation forces of structural elements of boom, load and column to the point K are reduced. In this connection horizontal position of the boom is considered. Bringing the forces to selected points is performed on condition of equality of moments of the reduced force to the sum of moments of reduced forces relative to any point on the plane (Varignon theorem).

In order to determine gravitation force brought to point K, equations of mass of the boom brought to the point C in a horizontal position of the boom (4), (5) mПР.C are use.

Let us formulate an equation of moments of gravitation force of the boom brought to point C ($G_{\text{ПР.С}}$) and gravitation force of the boom brought to point K ($G^{C}_{\text{ПР.К}}$) relative to O-axis.

$$m_{\text{ПР.С}}g(L + L_{K}\cos\alpha_1) = G^{C}_{\text{ПР.К}} \cdot L_{K}\cos\alpha_1 .$$

Therefore, $G^{C}_{\text{ПР.К}} = \frac{m_{\text{ПР.С}}g(L + L_{K}\cos\alpha_1)}{L_{K}\cos\alpha_1} = \frac{m_{\text{ПР.С}}gL + m_{\text{ПР.С}}gL_{K}\cos\alpha_1}{L_{K}\cos\alpha_1}$

$$= \frac{m_{\text{ПР.С}}gL}{L_{K}\cos\alpha_1} + m_{\text{ПР.С}}g \quad (15)$$

Now it is necessary to determine gravitation force of structural elements of the column briught to point K and gravitation force of hydraulic cylinders of GC, GЦ4, GЦ3actuators drive. In doing so equations (7), (8) for determination of reduced mass of the said structural elements to the point K are used:

$$G^{K}_{\text{ПР.К}} = m_{\text{ПР.К}}g \quad (16)$$

Then

$$\Sigma\delta A_{\alpha} = \left[P_C \cdot \sin\alpha_2 \cdot \ell_9 - \left(G^{K}_{\text{ПР.К}} + G^{C}_{\text{ПР.К}}\right)L_{K}\cos\alpha_1\right]\Delta\alpha$$

$$Q_{\alpha} = P_C \cdot \sin\alpha_2 \cdot \ell_9 - \left(G^{K}_{\text{ПР.К}} + G^{C}_{\text{ПР.К}}\right)L_{K}\cos\alpha_1 .$$

Since the transportation motion of the system is rotational and while simultaneous rotation of the column and the boom, the distance OC (Fig. 3) is constantly increasing, so there appears repeated rotating (Coriolis) acceleration ωC and the Coriolis force FC.

Coriolis acceleration is calculated by the equation

$$\omega_C = 2\dot{\alpha} \cdot V_{rc} = 2\dot{\alpha} \cdot \dot{\varphi} \cdot L$$

According to the moment of Coriolis force the generalized force Qα is calculated in the following way:

$$Q_{\alpha} = P_C \cdot \sin\alpha_2 \cdot \ell_9 - \left(G^{K}_{\text{ПР.К}} + G^{C}_{\text{ПР.К}}\right)L_{K} \cdot \cos\alpha_1 -$$

$$-2m_{\text{ПР.С}} \cdot \dot{\alpha} \cdot \dot{\varphi} \cdot L \cdot L_{K} \cdot \cos\gamma_1 . \quad (17)$$

According to the equations (14) and (17) motions of the considered dynamical system are determined as follows:

$$m_{\text{ПР.К}} \cdot L_K^2 \cdot \ddot{\alpha} + m_{\text{ПР.С.}} \cdot OC^2 \cdot \ddot{\alpha} +$$

$$+m_{\text{ПР.С}} \cdot OC \cdot L \cdot \ddot{\varphi}\left[\frac{L^2 - L_K \cdot L}{L \cdot OC}\right] \cdot \cos(\varphi_{\text{Н}} + \varphi) - m_{\text{ПР.С}} \cdot OC$$

$$\cdot L \cdot \dot{\varphi}^2 \cdot \left[\frac{L^2 - L_K \cdot L}{L \cdot OC}\right] \cdot \sin(\varphi_{\text{Н}} + \varphi) =$$

$$= P_C \cdot \sin\alpha_2 \cdot \ell_9 - \left(G^{K}_{\text{ПР.К}} + G^{C}_{\text{ПР.К}}\right)L_{K}\cos\alpha_1 -$$

$$-2m_{\text{ПР.С}} \cdot \dot{\alpha} \cdot \dot{\varphi} \cdot L \cdot L_K \cdot \cos\gamma_1 . (m_{\text{ПР.С}}L^2 + I_C)\ddot{\varphi} +$$

$$+m_{\text{ПР.С}} \cdot OC \cdot \ddot{\alpha} \cdot L \cdot \left[\frac{L^2 - L_K \cdot L}{L \cdot OC}\right] \cdot \cos(\varphi_{\text{Н}} + \varphi) =$$

$$= P\ell \sin\beta - m_{\text{ПР.С}}gL \quad (18)$$

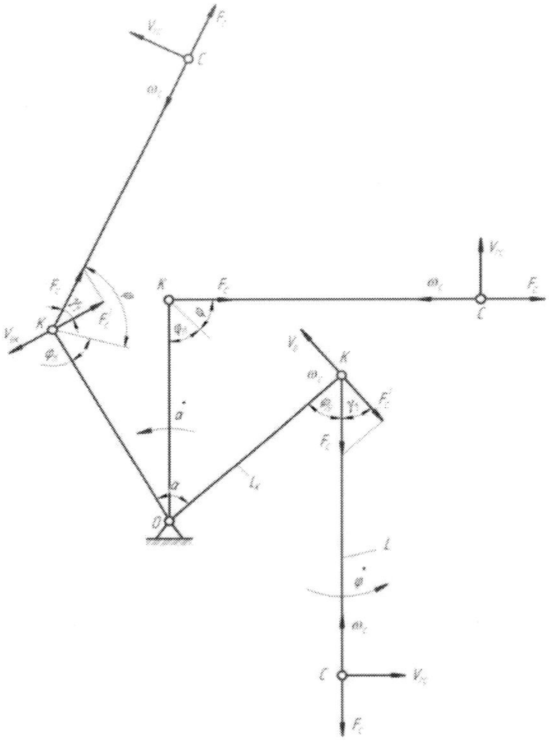

Fig. 3. Diagram for determination of Coriolis acceleration and Coriolis forces: ωC is Coriolis acceleration, FC is Coriolis force, Vℓc, Vrc are linear velocities of transportation and relative motion of point C

3 Conclusion

The result of the performed work is a system of inhomogeneous differential equations of secondary order, which are the basis of mathematical models of combined manipulator in the mode of load lifting with synchronous movement of the boom and the column. The obtained system of equations allows analyzing influence of parameters of kinematic scheme on the level of loads on processing equipment in the process of modeling ($\ddot{\varphi}$, $\ddot{\alpha}$, L, LK , $\dot{\varphi}, \varphi, \dot{\alpha}, \alpha$, OC, l9 , $\alpha_2, \beta, l, \gamma_1$), as well as analyzing weight value of load and processing equipment (mПР.C, mПР.К). Besides, by changing the rotation rate of shafts of hydraulic pumps motion rate of cylinder rods are changed and, consequently, an absolute velocity of point of reduction of mass of the boom and the load (point C), angular velocity and angular acceleration of the boom and the column are changed as well. The equations allow estimating influence of these factors on amount of load on structural elements of processing equipment.

References

1. V. Poletaykin, P. Kolesnikov, A. Savelyev, G. Moiseev. Mechanization of Construction, **77**, no. 12, 41-45 (2016).
2. V. Poletaykin, P. Kolesnikov, A. Savelyev, G. Moiseev. Construction Mechanization, **77**, no. 5, 36-40 (2016).
3. D. Chernik, P. Kolesnikov. Journal of international scientific researches, 3(28), 123-128 (2016).
4. V. Poletaykin, P. Kolesnikov, Systems. Methods. Technologies, no. 2 (22), 50-55 (2014).
5. P. Kolesnikov, G. Moiseev. New materials and technologies in mechanical engineering, no. 18, 151-153 (2013)
6. V. Poletajkin, P. Kolesnikov. Mathematical modeling of technological processes of rotary loggers with a variable departure of freight (International Scientific Research, 2015).
7. V. Poletajkin Applied mechanics of forest hoisting-and-transport cars. (Monograph) (2010)
8. P. Kolesnikov Simulation of operating modes of the logger with variable embarkation of a load. (Monograph) (2007)
9. V. Muratov, V. Ilyin, V. Poddubny. *Dynamic loadings of a hydraulic cylinder of the logger and adjustment of the safety equipment* (TsNIIME, 1980).
10. Z. Emtyl, Improvement of kinematics, dynamics and design of timber industry hydraulic manipulators (2001).
11. G. Moiseyev, S. Teleshev, P. Kolesnikov, Relevant directions of scientific research, **2**, no 5-4 (10-4), 200-204 (2014)
12. V. Poletajkin, P. Kolesnikov., Coniferous boreal zone, **XXVI**, no 2, 278-283 (2009)
13. V. Poletaikin, Forest magazine, №6, 34-37, (2002)
14. V. Poletaikin, Bulletin of the Krasnoyarsk state agricultural university, № 11, 185-192 (2012)
15. I. Kilin, M. Kanun. *Fundamentals of mechanics of robots. Dynamics of industrial robots* (STI, Krasnoyarsk, 1992)
16. A. Yablonsky. *Course of theoretical mechanics.*, P. 2.: Dynamics. (Moscow, 1966).
17. V. Poletaykin, I. Goncharov, Bulletin of the Krasnoyarsk state agricultural university, № 1, 46-52 (2014)
18. M. Bat, G. Dzhanelidze, A. Kelzon. *Theoretical mechanics in examples and tasks*, **2**.: Dynamics (Moscow, 1968).
19. V. Poletaykin, S. Guskov, Systems. Methods. Technologies, (2014).
20. D. Volkov. Dynamics and durability of excavators (Mechanical engineering, 1965).

Complex of parallel programs for modeling oil products transport in coastal systems

Alena Semenyakina[1] and *Sophia* Protsenko[2]

[1]Southern Federal University, Taganrog, Russia Russia
[2] Taganrog Institute of A.P. Chekhov (branch) of the Rostov State University of Economics, Taganrog, Russia

Abstract. This paper covers the model of oil products transport in view the evaporation of light, neutral and no-evaporating pseudofractions of oil slick, dissolution of oil slick and biodegradation. Approximation of diffusion-convection problem was performed on the basis of schemes of high-order of accuracy. Experimental software was designed for mathematical modelling of possible scenarios of development of ecosystems of shallow waters for oil spills on the example of the Azov-Black Sea basin, based on multiprocessor computer systems. Decomposition methods of grid domains have been used for computationally laborious diffusion-convection tasks in parallel implementation. Maximum acceleration was equal to 228.36 times on 512 computational nodes.

1 Introduction

The influence of oil pollution on the water is manifested in the next: the deterioration of the physical properties of the water (turbidity, change in color, taste, smell); the dissolution of toxic substances in water; the formation of oil surface film and sludge at the bottom of the reservoir, which decreased the oxygen in the water.

Characteristic smell and taste appear at a concentration of oil and oil products in the water 0.5 mg/dm^3, and naphthenic acids – 0.01 mg/dm^3. Significant changes in chemical parameters of water occurs when the content of oil and oil products more than 100-500 mg/dm^3. The oil film on the surface of water basin worse the gas exchange with the atmosphere and slow the rate of aeration and removal of carbon dioxide, formed by oxidation of the oil. When the oil film is equaled to 4.1 mm and the oil concentration in water is equaled to 17 mg/dm3, the amount of dissolved oxygen is reduced by 40% per 20-25 days.

Analysis of the numerical solution of the model problem of transport materials showed that time costs are reduced for the explicit scheme at the increasing of grid dimension. Modification of the explicit scheme is the introduction of the second-order derivative of the difference with the factor - regularizater - can significantly ease restrictions on the permissible value of the time step [1]. In addition, explicit regularized schemes showed real-time cost advantages (10-15 times and more), compared to the previously used conventional implicit and no-regularized explicit schemes [2].

A variant of the finite volumes method in the case of account of filling control domains was proposed in [3]. Algorithm of calculation in view the partial «fill» cells without flows associated with a stepwise boundary of representation on a rectangular grid. The proposed method was applied to solve three-dimensional hydrodynamic problems [4]. Flow fields, which are used to calculate the transport of oil products, have been calculated on the basis of this model.

Schemes of high-order of accuracy are used in solving the problem of oil products transport. It should be noted that in solution the problem of diffusion model was able to increase the accuracy in 66.7 times, and for convection-diffusion problems – in 48.7 times [5].

2 Mathematical model of hydrodynamics

The input data of the problem of transport of oil products are fields of the speed vector of the water flow, which required for mathematical modelling of water environment. The initial equations of hydrodynamics of shallow waters are [4]:

– motion equation (the Navier – Stokes equation)

$$u'_t + uu'_x + vu'_y + wu'_z = -p'_x / \rho + (\mu u'_x)'_x + (\mu u'_y)'_y + (\nu u'_z)'_z + 2\Omega(v\sin\theta - w\cos\theta),$$

$$v'_t + uv'_x + vv'_y + wv'_z = -p'_y / \rho + (\mu v'_x)'_x + (\mu v'_y)'_y + (\nu v'_z)'_z - 2\Omega u\sin\theta, \quad (1)$$

$$w'_t + uw'_x + vw'_y + ww'_z = -p'_z / \rho + (\mu w'_x)'_x + (\mu w'_y)'_y + (\nu w'_z)'_z + 2\Omega u\cos\theta + g(\rho_0 / \rho - 1);$$

– continuity equation in the case of variable density can be written as follows:

* Corresponding author: sukhinov@gmail.com

$$\rho_t' + (\rho u)_x' + (\rho v)_y' + (\rho w)_z' = 0, \quad (2)$$

where $\bar{u} = \{u, v, w\}$ are components of the speed vector; p is the overpressure on the hydrostatic pressure of the unperturbed liquid, ρ is a density; Ω is the angular velocity of the earth's rotation, θ is the angle between the vertical and angular velocity; μ, ν are horizontal and vertical components of the turbulent exchange coefficient.

The system of equations (1)-(2) is considered in the following boundary conditions:

– at the entrance (the mouth of the Don and Kuban rivers):

$$u(x,y,z,t) = u(t), v(x,y,z,t) = v(t),$$

$$p_n'(x,y,z,t) = 0, \bar{u}_n'(x,y,z,t) = 0,$$

– the lateral boundary (the bank and the bottom):

$$\rho_v \mu (u')_n (x,y,z,t) = -\tau_x(t), \ \rho_v \mu (v')_n (x,y,z,t) = -\tau_y(t),$$

$$\bar{u}_n(x,y,z,t) = 0, \ p_n'(x,y,z,t) = 0,$$

– the upper boundary:

$$\rho\mu(u')_n(x,y,z,t) = -\tau_x(t), \ \rho\mu(v')_n(x,y,z,t) = -\tau_y(t),$$

$$w(x,y,t) = -\omega - p_t' / \rho g, \ p_n'(x,y,t) = 0, \quad (3)$$

– output (Kerch Strait):

$$p_n'(x,y,z,t) = 0, \quad \bar{u}_n'(x,y,z,t) = 0,$$

where ω is the rate of evaporation of the liquid, τ_x, τ_y are tangential stress components (Van Dorn law), ρ_v is the density suspension.

The components of the tangential stress for the free surface:

$$\tau_x = \rho_a C_p(|\bar{w}|) w_x |\bar{w}|, \ \tau_y = \rho_a C_p(|\bar{w}|) w_y |\bar{w}|,$$

where $\bar{w}$ is the wind speed vector relatively of the water; ρ_a is the density of atmosphere; $C_p(x)$ is the dimensionless coefficient.

Components of the tangential stress for the bottom based on this notation can be written as follows:

$$\tau_x = \rho C_p(|\bar{u}|) u |\bar{u}|, \ \tau_y = \rho C_p(|\bar{u}|) v |\bar{u}|.$$

On the basis of the measured speed fluctuations the discussed below approximation is required to the developing the coefficient of vertical turbulent exchange, heterogeneous by the depth [6]:

$$\nu = C_s^2 \Delta^2 \frac{1}{2} \sqrt{\left(\frac{\partial u}{\partial z}\right)^2 + \left(\frac{\partial v}{\partial z}\right)^2}, \quad (4)$$

where Δ is the characteristic scale of the grid; C_s is the dimensionless empirical constant, whose value is generally determined based on the calculation of the decay of homogeneous isotropic turbulence.

The grid method is used to solve (1)-(3) problem [7]. The approximation of equations by the time variable is determined on the basis of splitting schemes by physical processes [8] in the form of an amendment method of pressure.

3 Mathematical model of oil products transport

The system of equations was used to describe the process of oil products transport, subject to the evaporation of light, neutral and no-evaporating pseudofractions of oil spot, dissolution of oil spot and biodegradation [9]:

$$(c_i)_t' + u(c_i)_x' + v(c_i)_y' = \left(\mu^*(c_i)_x'\right)_x' + \left(\mu^*(c_i)_y'\right)_y' -$$

$$-\left(\frac{K_E P_i}{RT} + K_D S_i\right) X_i M_i^m - \frac{1}{q}\frac{\mu_m c_i}{c_i + K_s} M, \quad (c_i)_n'\Big|_{(x,y)\in\gamma} = 0,$$

$$M_t' + uM_x' + vM_y' = \left(\mu M_x'\right)_x' + \left(\mu M_y'\right)_y' + \quad (5)$$

$$+\frac{\mu_m c_i}{c_i + K_s} M - \lambda M, \ M_n'\Big|_{(x,y)\in\gamma} = 0,$$

$$(\varphi_i)_t' + u(\varphi_i)_x' + v(\varphi_i)_y' + w(\varphi_i)_z' = \left(\mu(\varphi_i)_x'\right)_x' +$$

$$+\left(\mu(\varphi_i)_y'\right)_y' + \left(\mu(\varphi_i)_z'\right)_z',$$

$$(\varphi_i)_n'\Big|_{(x,y,z)\in\Gamma\setminus(z=0)} = 0, \ (\varphi_i)_z'\Big|_{z=0} = K_D S_i X_i M_i^m,$$

where c_i is the concentration of the i-th oil fractions; ρ_ω, ρ are the water and oil density respectively; $\mu^* = \mu + (\rho - \rho_\omega) g h^3 / \mu$ is the function describing the process of pollutant decomposition; μ is the diffusion coefficient; g is the acceleration; $h = \sum_1^i c_i$ is the width of oil slick; $i = \overline{1,k}$; $M_i^{(m)}$ is the molar mass of i-th component, [kg/mol]; $K_E = 2{,}5 \cdot 10^{-3} U^{0.78}$ is the mass transport coefficient for hydrocarbon, [m/s]; U is the wind speed, relatively to the water, [m/s]; X_i is the mole fraction of the component with the number i, equaled to $v_i / \Sigma v_i$; v_i is an amount of substance i-th components, [mol]; P_i is the vapor pressure of the i-th component, [Pa]; R = 8,314 J/mol; K is the universal gas constant; T is an ambient temperature over the surface of the spot, K, $K_D = kK_{D0}$ is the coefficient of mass transport of dissolution; K_{D0} is the initial value of the mass transport coefficient of dissolution; k is the coefficient, depending on the sea state; S_i is the dissolubility of i-th component in water, [kg/m^3]; A is the area of the oil spill, [m^2]; M is the concentration of microorganisms; μ_m is the maximum speed of the growth of microorganisms; K_s is the coefficient of saturation; λ is the rate of cell death; q is the proportion coefficient between the amount of bacteria and the absorbed substrate.

Changes of the oil initial solubility are described by the equation:

$$S = S_0 e^{-0{,}1t},$$

where S_0 is an initial oil solubility; t is the time, [day].

The coefficient of horizontal turbulent diffusion depends on the hydrodynamic and climatic conditions in which the process takes place. The coefficient of horizontal turbulent diffusion will be subject to the law of the «four thirds» by Richardson for difficult hydrodynamic and climatic conditions of the Azov-Black Sea area [10]:

$$\mu \approx \varepsilon^{1/3} L^{4/3}, \qquad (6)$$

where L is the characteristic size of diffusing spots; ε is the rate of dissipation of turbulent energy. It is equaled to the order of $1\text{-}10^{-1}$ cm^2/s^3 at the surface and decreased at the average with depth to values of the order of $10^{-3}\text{-}10^{-4}$ cm^2/c^3.

Boundary and initial conditions for the one-shot volley oil spill were determined for solving the above systems of equations:

$$c\big|_{t=0,(x,y)\in S_0} = c_0,\; c_{t=0,(x,y)\notin S_0} = 0,$$

where S_0 is an area, covered by the spot; c_0 is the oil concentration in the considered area.

4 Approximation problem of oil products transport

A two-dimensional convection-diffusion problem is considered for implementation the oil products transport model in the form:

$$c'_t + uc'_x + vc'_y = \left(\mu c'_x\right)'_x + \left(\mu c'_y\right)'_y + f \qquad (7)$$

with the boundary conditions: $c'_n(x,y,t) = \alpha_n c + \beta_n$,

where u, v are the velocity components, μ is the coefficient of turbulent exchange, f is a function, describing the intensity and distribution of sources.

We introduced a uniform rectangular grid:

$$w_h = \{t^n = n\tau, x_i = ih_x, y_j = jh_y; n = \overline{0..N_t}, i = \overline{0..N_x},$$
$$j = \overline{0..N_y};\; N_t\tau = T, N_xh_x = l_x, N_yh_y = l_y\}$$

where τ is the time step; h_x, h_y are spatial steps; N_x, N_y are the spatial boundary; N_t is the upper time boundary.

B.N. Chetverushkin have been proposed the idea of use of regularized schemes to increase the safety factor of explicit schemes [1]. A modified equation is used to construct the explicit regularized scheme for the equation (7):

$$c'_t + \frac{\tau^*}{2} c''_t + uc'_x + vc'_y = \left(\mu c'_x\right)'_x + \left(\mu c'_y\right)'_y + f, \qquad (8)$$

where $\tau^* \approx h/c^*$ is the regularization parameter; h is the grid step; c^* is the speed of sound in water. Adding the regularize-term (the second derivative of the time difference with the factor τ^*), the restriction on the time step variable is $\tau \le O\left(h^{3/2}\right)$ is sufficient for the stability of the explicit scheme. This condition is less rigid in comparison with the condition for explicit no-regularized scheme is $\tau \le O\left(h^2\right)$.

We conducted the discretization of convective and diffusive transport operators of the second order of approximation error in the case of partially filled cells as follows:

$$\left(q_0\right)_{i,j} uc'_x \simeq \left(q_1\right)_{i,j} u_{i+1/2,j} \frac{c_{i+1,j} - c_{i,j}}{2h_x} + \left(q_2\right)_{i,j} u_{i-1/2,j} \frac{c_{i,j} - c_{i-1,j}}{2h_x},$$

$$\left(q_0\right)_{i,j} \left(\mu c'_x\right)'_x \simeq \left(q_1\right)_{i,j} \mu_{i+1/2,j} \frac{c_{i+1,j} - c_{i,j}}{h_x^2} - \qquad (9)$$
$$-\left(q_2\right)_{i,j} \mu_{i-1/2,j} \frac{c_{i,j} - c_{i-1,j}}{h_x^2} - \left|\left(q_1\right)_{i,j} - \left(q_2\right)_{i,j}\right| \mu_{i,j} \frac{\alpha_x c_{i,j} + \beta_x}{h_x},$$

where q_i is the coefficient, describing the occupancy of computing domaines [11].

The approximation of convective transport operator uc' by the difference scheme, which has the fourth order of accuracy, is given by [12]:

$$\left(q_0\right)_i L(c) = -\left(q_1\right)_i \frac{u_{i+1/2}}{12h} \frac{\left(q_1\right)_{i+1}}{\left(q_0\right)_{i+1}} c_{i+2} - \left(-\left(q_1\right)_i \frac{u_{i+1/2}}{12h}\left(2 + \frac{\left(q_1\right)_i}{\left(q_0\right)_i}\right) + \right.$$
$$\left. + \left(q_2\right)_i \frac{u_{i-1/2}}{12h} \frac{\left(q_1\right)_i}{\left(q_0\right)_i} + \left(q_1\right)_i \left(-\frac{u_{i+1/2}}{2h} + k_i^{(1)} + k_i^{(2)}\right)\right) c_{i+1} +$$
$$+ \left(-\left(q_1\right)_i \frac{u_{i+1/2}}{12h}\left(2 + \frac{\left(q_2\right)_{i+1}}{\left(q_0\right)_{i+1}}\right) + \left(q_2\right)_i \frac{u_{i-1/2}}{12h}\left(2 + \frac{\left(q_1\right)_{i-1}}{\left(q_0\right)_{i-1}}\right) + \right.$$
$$+ \left(q_2\right)_i \frac{u_{i-1/2}}{2h} - \left(q_1\right)_i \frac{u_{i+1/2}}{2h} - \left(\left(q_2\right)_i - \left(q_1\right)_i\right) k_i^{(1)} +$$
$$\left. + \left(\left(q_2\right)_i + \left(q_1\right)_i\right) k_i^{(2)}\right) c_i - \left(-\left(q_1\right)_i \frac{u_{i+1/2}}{12h} \frac{\left(q_2\right)_i}{\left(q_0\right)_i} + \right.$$
$$\left. + \left(q_2\right)_i \frac{u_{i-1/2}}{12h}\left(2 + \frac{\left(q_2\right)_i}{\left(q_0\right)_i}\right) + \left(q_2\right)_i \left(\frac{u_{i-1/2}}{2h} + k_i^{(2)} - k_i^{(1)}\right)\right) c_{i-1} -$$
$$- \left(-\left(q_2\right)_i \frac{u_{i-1/2}}{12h} \frac{\left(q_2\right)_{i-1}}{\left(q_0\right)_{i-1}}\right) c_{i-2}, \qquad (10)$$

where $k_i^{(1)} = \left(\frac{\left(q_1\right)_i}{\left(q_0\right)_i}\left(u_{i+1} - u_{i,}\right) - \frac{\left(q_2\right)_i}{\left(q_0\right)_i}\left(u_i - u_{i-1}\right)\right) / (8h)$,

$$k_i^{(2)} = \frac{\left(q_1\right)_i}{\left(q_0\right)_i} \frac{u_{i+1} - u_i}{8h} + \frac{\left(q_2\right)_i}{\left(q_0\right)_i} \frac{u_i - u_{i-1}}{8h}.$$

The approximation of diffusive transport operator by the difference $\left(\mu c'\right)'$ schemes, which has the fourth order of accuracy, is given by:

$$\left(q_0\right)_i \left(L(c)\right) \simeq -\left(q_1\right)_i \frac{\mu_{i+1}}{12h^2} \frac{\left(q_1\right)_{i+1}}{\left(q_0\right)_{i+1}} c_{i+2} + \left(\left(q_1\right)_i \frac{\mu_{i+1/2}}{h^2} c_{i+1} + \right.$$
$$+ \left(q_1\right)_i \frac{\mu_{i+1}}{12h^2}\left(\frac{\left(q_1\right)_i}{\left(q_0\right)_i} + 2\right) + \left(q_2\right)_i \frac{\mu_{i-1}}{12h^2} \frac{\left(q_1\right)_i}{\left(q_0\right)_i} - \left(\left(q_1\right)_i \frac{\mu_{i+1/2}}{h^2} + \right.$$
$$+ \left(q_2\right)_i \frac{\mu_{i-1/2}}{h^2} + \left(q_1\right)_i \frac{\mu_{i+1}}{12h^2}\left(\frac{\left(q_2\right)_{i+1}}{\left(q_0\right)_{i+1}} + 2\right) + \qquad (11)$$
$$+ \left(q_2\right)_i \frac{\mu_{i-1}}{12h^2}\left(\frac{\left(q_1\right)_{i-1}}{\left(q_0\right)_{i-1}} + 2\right) + \left(q_2\right)_i \left(\frac{\mu''_i - \mu''_{i-1}}{12} - k_i\right) -$$
$$\left. - \left(q_1\right)_i \left(\frac{\mu''_{i+1} - \mu''_i}{12} + k_i\right)\right) c_i + \left(\left(q_2\right)_i \frac{\mu_{i-1/2}}{h^2} c_{i-1} + \right.$$

$$+(q_1)_i \frac{\mu_{i+1}}{12h^2}\frac{(q_2)_i}{(q_0)_i}+(q_2)_i\frac{\mu_{i-1}}{12h^2}\left(\frac{(q_2)_i}{(q_0)_i}+2\right)+$$

$$+(q_2)_i\left(\frac{\mu_i''-\mu_{i-1}''}{12}-k_i\right)\Bigg)c_{i-1}-(q_2)_i\frac{\mu_{i-1}}{12h^2}\frac{(q_2)_{i-1}}{(q_0)_{i-1}}c_{i-2},$$

where $\mu_i''=\left(\frac{(q_1)_i}{(q_0)_i}c_{i+1}-2c_i+\frac{(q_2)_i}{(q_0)_i}c_{i-1}\right)/h^2$,

$$k_i=\frac{(q_1)_i}{(q_0)_i}\frac{\mu_{i+1}-\mu_i}{4h^2}-\frac{(q_2)_i}{(q_0)_i}\frac{\mu_i-\mu_{i-1}}{4h^2}.$$

5 Parallel implementation

Decomposition methods of grid domains are used for computationally laborious convection-diffusion tasks in parallel implementation, which are considered parameters of architecture and multiprocessor systems. The maximum performance of multiprocessor computer system (MCS) is equaled to 18.8 teraflops. 128 single-type of 16-core Blade-servers HP ProLiant BL685c were used as computational nodes, each of which has the four quad-core AMD Opteron processor 8356 of 2.3GHz and memory, equaled to 32 GB.

The time costs, which are required for the one time layer on various grids, acceleration and effectiveness values for different numbers of MCS cores are given in the Table 1.

Table 1. Acceleration and Efficiency of the Number of Processors

		100x100	500x500	1000x 1000	5000x 5000
1	Time	0.000271	0.00846	0.03608	1.633
	Speed-up	1	1	1	1
	Efficiency	1	1	1	1
4	Time	0.000052	0.00341	0.00978	0.533
	Speed-up	5.212	2.481	3.689	3.064
	Efficiency	1.303	0.62	0.922	0.766
16	Time	0.000025	0.00054	0.00628	0.142
	Speed-up	**10.84**	15.667	5.745	11.5
	Efficiency	0.677	0.979	0.359	0.719
64	Time	0.000125	0.00016	0.00110	0.044
	Speed-up	2.168	**52.875**	32.8	37.114
	Efficiency	0.034	0.826	0.513	0.58
128	Time	-	0.00040	0.00048	0.017
	Speed-up	-	21.15	**75.167**	96.059
	Efficiency	-	0.165	0.587	0.75
512	Time	-	-	0.00129	0.0072
	Speed-up	-	-	27.883	**228.36**
	Efficiency	-	-	0.054	0.446

6 Software description

The experimental software was developed on the basis of MCS, which is intended for mathematical modelling of possible scenarios of ecosystems of shallow waters on the example of the Azov-Black Sea basin. The software «Azov3d» was designed for construction of operational flow forecasts turbulence of the water environment - the speed field of water environment on grids with high resolution. This complex is used to calculate the three-dimensional speed vector of water environment of the Azov Sea. The complex takes into account such physical parameters as the Coriolis force, the turbulent exchange, the complex geometry of the bottom surface and coastline, evaporation, river flows, wind-surges, wind currents and friction bottom, and provides the following functions: the calculation of the speed field without pressure; the calculation of hydrostatic pressure (used as an initial approximation for the hydrodynamic pressure); the calculation of hydrodynamic pressure; the calculation of three-dimensional velocity field of water flow.

The output parameters of this software are the next: steps by spatial coordinates, error of calculation of grid equations, the grid dimension, the time interval, the evaporation intensity, the initial distribution of the components of the speed vector of water environment and pressure.

We can add new calculation functions to the developed complex. Particularly, program modules were designed and integrated to this complex for calculation the oil products transport in view the evaporation of light, neutral and no-evaporating pseudofractions of oil spot, dissolution oil spot and biodegradation. The water flow fields, calculated on the basis of the mathematical model (1)-(4), are included to the input data for the oil products model (5).

Results of natural experiments for researching the destruction of crude oil in seawater are given in [10]. The experimental results are shown that only 3-15% of the original amount of crude oil subject to oxidation, biodegradation, photochemical reactions, while from 10 to 40% of substance is evaporated. The peak of its solubility is occurred on the 10th day of exposure (Fig. 1).

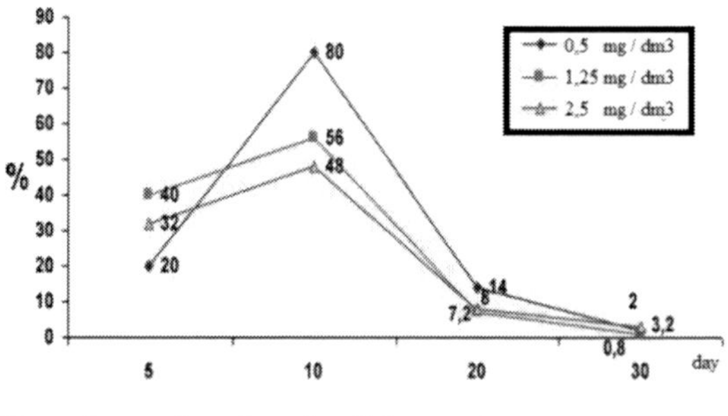

Fig. 1. Dynamics of crude oil destruction

According to [11], upon receiving a message about the oil and oil products spill, the containment time should not exceed 4 hours at spill in the area, and 6 hours – at spill on the ground since the discovery of the oil and oil products spill or from the providing the information about the spill. We considered the case where measures are not taken to localize oil spills. The results of natural experiments are shown that the calculated time interval must be equal to 20-30 days.

The wind speed in the range of 3-8 m/s is an ideal for the localization of oil pollution: in this case, the slicks appear as dark spots on the bright (rough) surface of the sea (see Fig. 2).

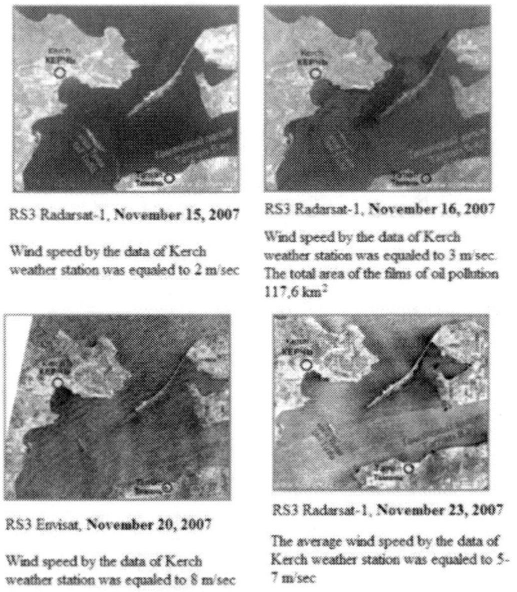

Fig. 2. The radar image of area during the catastrophic oil spill

The data about the wind speed and direction during the extreme storm, occurred in November 2007, is given in Fig. 3. The highest wind speed was fixed in November 11th, 2007 in Kerch Strait and amounted to 24 m/s according to the Gismeteo [12].

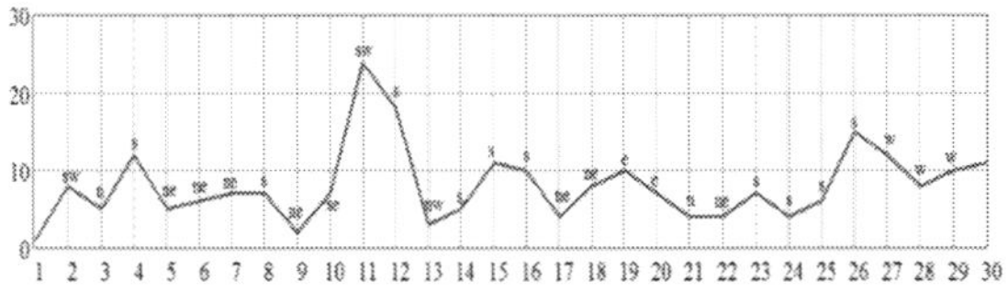

Fig. 3. The wind speed in the Kerch Strait in November, 2007

Results of numerical experiments of modelling the light oil transport in the Kerch Strait on November 16th, 2007 are given in Fig. 4. The calculation was performed on the basis of the developed software. The results was used to test the efficiency of this complex.

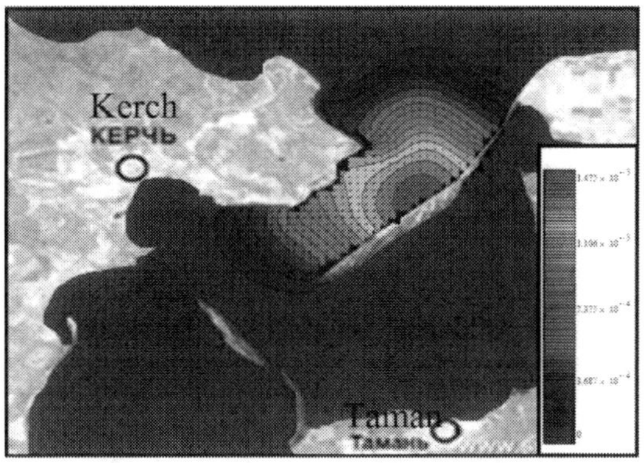

Fig. 4. The concentration field of light oil products

Further, we schedule to develop a model for calculation the transport of bottom materials [13, 14], as well as for calculation the transport of dissolved substances and oil vapor. The three-dimensional mathematical model of surface aerodynamics is required to develop a model of transportation of oil vapor. In the modelling of oil spills it is also important to consider the influence of the dissolved oil to the nature and course of hydrobiological processes in the water [15, 16]. Comparing the calculation results of the concentration of light oil products, which are given in Fig. 4, with the results of radar images, where a catastrophic oil spill (Fig. 2) was occurred, we can see their qualitative conformance. The forecast time was equaled to 4 days after the spill.

7 Conclusion

The three different models for describing the oil products transport in view of evaporation of light, neutral and no-evaporating pseudofractions of oil spot, dissolution of oil spot and biodegradation were developed in [9]. The developed model was presented in this paper that describes all of the above processes. The usage of regularized schemes [1] for increasing the safety factor of explicit schemes has been proposed by B.N. Chetverushkin. The approximation of convection-diffusion problem was performed on the basis of schemes of high-order of accuracy. The experimental software was developed on the basis of multiprocessor computer system for mathematical modeling of possible scenarios of ecosystems of shallow waters on the example of Azov-Black Sea basin in oil spills. The decomposition methods of grid domains are used for computationally laborious convection-diffusion problems in parallel implementation, which are considered parameters of architecture and multiprocessor systems. The maximum acceleration was equaled to 228.36 times on 512 computational nodes. The advantages of the developed software is also the use of hydrodynamic models, included the motion equation for the three coordinate directions.

This paper was partially supported by the grant No. 17-11-01286 of the Russian Science Foundation.

References

1. B.N. Chetverushkin, Mathematical Models and Computer Simulations, **5**, No. 3, 266–279 (2013)
2. A.I. Sukhinov, E.A. Protsenko, A.E. Chistyakov, S.A. Shreter, Computational Methods and Programming. **16**, No. 3., 328–338 (2015)
3. A. I. Sukhinov, A. E. Chistyakov, E. F. Timofeeva, A. V. Shishenya, Mathematical Models and Computer Simulations. **5**, No. 2., 122-129 (2013)
4. A.I. Sukhinov, A.E. Chistyakov, Computational Methods and Programming, **13**, No. 1, 290–297 (2012)
5. A.I. Sukhinov, A.E. Chistyakov, A.A. Semenyakina, A.V. Nikitina Computational Methods and Programming. **16**, No. 2., 256–267 (2015)
6. O. M. Belotserkovskii, *Turbulence: New Approaches* (Nauka, Moscow, 2003).
7. Samarskii A.A.The theory of difference schemes (Science, Moscow, 1989)
8. O.M. Belocerkovskij, V.A. Gushhin, V.V. Shhennikov, USSR Computational Mathematics and Mathematical Physics, **15**, No. 1, 190-200 (1975)
9. I.Y. Gluhenky, A.V. Lavrent'ev, G.G. Popov, Safety in Technosphere, No. 6., 3-6 (2011)
10. A.S. Abdusamadov, A.P. Panarin, A.K.Magomedov L.D. Kovalenko, B.R. Huseynov, A.M. Dohtukaeva, L.A. Dudurhanova, Geography and Geoecology. Southern Russia: ecology, razvitie., №1., 165-166 (2012).
11. Resolution of the Government of the Russian Federation on April 15, 2002 № 240 "On the procedure for organizing the activities for prevention and liquidation of oil spills in the Russian Federation "
12. https://www.gismeteo.ru/diary/5211/2007/11/
13. A.I. Sukhinov, A.E. Chistyakov, E.A. Protsenko, Mathematical Models and Computer Simulations, **6**, No. 4, 351-363 (2014)
14. A.I. Sukhinov, A.E. Chistyakov, E.A. Protsenko, Computational Methods and Programming., **15**, 610-620 (2014)
15. A.I. Sukhinov, A.V. Nikitina, A.E. Chistyakov, Mathematical Models, **24**, No. 9., 3–21 (2012)
16. A.V. Nikitina, A.I. Sukhinov, G.A. Ugolnitsky, A.B. Usov, A.E. Chistyakov, M.V. Puchkin, I.S. Semenov, Mathematical Models and Computer Simulations, 9 (1), 101-107 (2017)
17. A.I. Sukhinov, A.E. Chistyakov, I.I. Levin, I.S. Semenov, A.V. Nikitina, A.A. Semenyakina, *5th ICIEV*, 1128 – 1133 (2016)
18. A.I. Sukhinov, A.E. Chistyakov, A.V. Nikitina, A.A. Semenyakina, I. Korovin, G. Schaefer, *5th ICIEV*, 1134 – 1139 (2016).
19. A.I. Sukhinov, A.E. Chistyakov, A.A. Semenyakina, A.V. Nikitina, Computer researches and modeling. **8**, no. 1., 151 – 168 (2016)
20. A.I. Sukhinov, D.S. Khachunts, A.E. Chistyakov, Computational Mathematics and Mathematical Physics. **55**, No. 7, 1216-1231 (2015)
21. A.I. Sukhinov, A.E. Chistyakov, A.V. Shishenya, Mathematical Models and Computer Simulations, **6**, No. 3., 324-331 (2014)
22. N. Buzalo, P. Ermachenko, T. Bock, A. Bulgakov, A. Chistyakov, A. Sukhinov, E. Zhmenya, N. Zakharchenko, Proc. Eng., **85**, 84-93 (2014)
23. A.I. Sukhinov, A.E. Chistyakov, E.F. Timofeeva, A.V. Shishenya, Mathematical Models and Computer Simulations, **5**, No. 2., 122-129 (2013)
24. A.I. Sukhinov, A.E. Chistyakov, Mathematical Models and Computer Simulations, **4**, Iss. 4, 398-409 (2012).

DOI: 10.1051/matecconf/201713204017

Automated calculation of the absorber temperature field of an inflatable collector

Olga Smirnova[1], *Yuri* Avetisyan[1], *Nadezhda* Dmitrienko[1], *and Oxana* Balabaeva[1]

[1]Don state Technical University, 344000, Rostov-on-Don, Russian Federation

Abstract. This article deals with calculating temperature field of the inflatable collector made of polymer materials. The efficiency of the considered power system depends on the intensity of the absorber heating. In its turn, the process of heating coolant is affected by a great number of external environmental factors. The developed mathematical model and program product let the researchers consider various influence degree of each factor separately and in a complex.

1 Introduction

Using heat energy generated by alternative air power systems is in great demand in agriculture, tourism, drying operations of various production processes and house heating. The development of applied technologies in the chemical industry made it possible to develop and explore a new type of a collector- an inflatable one, made of polymer materials. Its effectiveness is largely determined by thermophysical properties of applied materials and accessories,constructive design, the quality of connecting details and the level of heat loss in energy transmission from the device to the heating object.

2 Problem formulation

To ensure maximum heat transfer from a useful area of the collector, a new design consisting of series-connected cylindrical segments is proposed in the article (Figure 1).

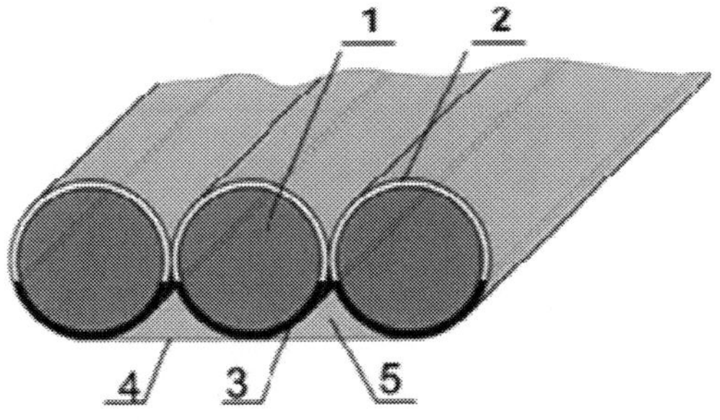

Fig. 1. An air inflatable collector in section
1 – a cylindrical segment of a collector; 2 – light-transmitting transparent layer of a segment; 3 – heat-absorptive layer of a segment (absorber); 4 – heat-insulation layer; 5 – coolant area.

Segments of the collector are made of two half-cylinders: the upper is light-transmitting and the lower is heat-absorbing, i.e. the absorber. Form stability of cylindrical segments is provided by filling the inner space with air. The interval between the segments is designed to accommodate the coolant.

Effective maintenance of the inflatable collector is provided by thermo-physical characteristics of the applied materials. Due to high rates of heat resistance and light transmission (0,85), the upper layer of the semi cylinder is supposed to be made of polyethylene terephthalate. The absorber is likely to be made of polymer material of black color to ensure high heat conductivity, heat capacity, and thermal conductivity. The authors propose to use polyethylene of high density for the production of the absorber [1-3].

The study of the collector is fulfilled with the help of natural and laboratory research and let the authors determine a lot of factors influencing the intensity of the absorber heating[4-8]. The main environment factors are known to be intensity of insolation, the direction of solar radiation effect, the power of heat flow, wind speed, air humidity, and others.

To determine the impact degree of each of them is hard due to large number of variables. To solve this problem is possible if to apeal to mathematical modeling. Developing the mathematical model, the authors used the method of finite elements. As a research object a single cylindrical segment was selected, and its estimated area is presented on a figure 2.

Due to using a high light transmission material for upper semi cylinder in a mathematical modeling a mathematical admission was based on the fact that the upper layer does not influence distribution of the temperature field of the absorber [8-12].

*Olga Smirnova: smir_nova@inbox.ru

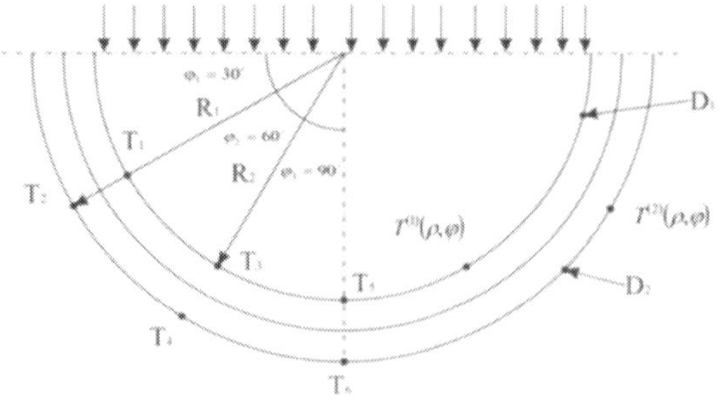

Fig. 2. An estimated area of the collector cylindrical segment

The basic data for mathematical simulation: air temperature Tat = 23°C; heat flux, $q = q_1 \sin\varphi$, $q_1 = 310$ W/м2; materials thermal conduction of the absorber: low layer – $\lambda_1 = 0{,}03$ W/м·K; upper layer – $\lambda_2 = 0{,}34$ W/м·K; heat-transfer coefficient $\alpha_T = 9{,}5$ W/м2·K; $R_1 = 0{,}05$ м; $R_2 = 0{,}0492$ м; $\varphi_1 = 30°$; $\varphi_2 = 60°$; $\varphi_3 = 90°$.

It is necessary to find distribution of temperature ranges in a set mode in the following points during mathematical simulation: $T_1(R_1,\varphi_1)$; $T_2(R_3,\varphi_1)$; $T_3(R_1,\varphi_2)$; $T_4(R_3,\varphi_2)$; $T_5(R_1,\varphi_3)$; $T_6(R_3,\varphi_3)$.

The mathematical model of the task has the following form:

$$\text{area } D_1 \quad \operatorname{div}\left(\lambda_1 \operatorname{grad} T^{(1)}\right) = 0; \tag{1}$$

$$\text{area } D_2 \quad \operatorname{div}\left(\lambda_2 \operatorname{grad} T^{(2)}\right) = 0. \tag{2}$$

Consider the following boundary conditions: $\rho = R_1$,

$$\lambda_2 \frac{\partial T^{(2)}}{\partial \rho} = -\alpha_1 \left[T^{(2)}(R_1,\varphi) - T_{\text{окр}}\right];$$

$$\rho = R_2,$$

$$T^{(1)}(R_2,\varphi,t) = T^{(2)}(R_2,\varphi,t); \; \lambda_1 \frac{\partial T^{(1)}}{\partial \rho} = \lambda_2 \frac{\partial T^{(2)}}{\partial \rho}.$$

On the basis of this mathematical model a software product to facilitate and visualize the main mathematical calculations was developed. The interface of the program is given on the figure 3.

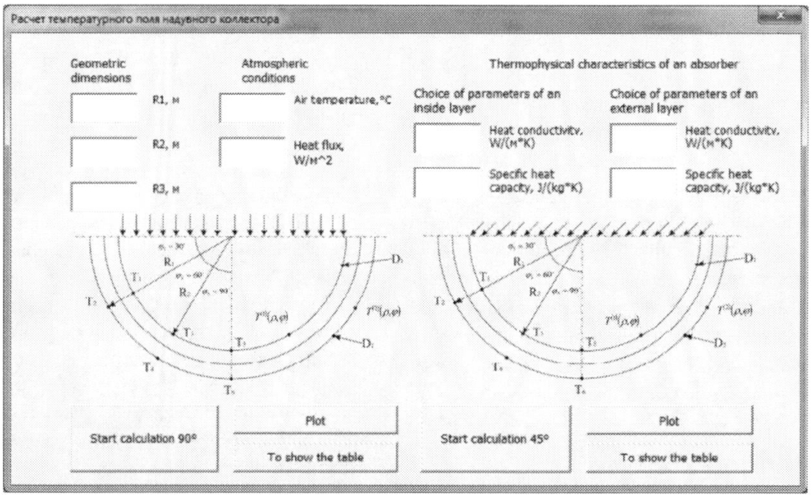

Fig. 3. Interface of the automated program calculation of the absorber temperature field of an inflatable collector

Varying the parameters of ambient temperature, changing thermal physical characteristics of applied materials [13, 14] and geometrical constructions of cylindrical segment of the collector, it is possible to calculate the distribution of the collector temperature field. The results of the calculations can be presented in the table or in the form of graphic dependences.

Let us consider the temperature distribution of the collector as it is supposed to be depend on the ambient air temperature and the heat flux maintenance intensity of the collector in different seasons. Constructive characteristics and properties of applied materials are left without changes.

Climate data [15], used in program calculation are given in the table 1.

Table 1. Climatic parameters of the collector maintenance in various seasons

Season	Air temperature, °C (an average maximum during the season)	Heat flux, *W*/sq.m (average monthly values during the season)
Spring	19,7	500
Summer	35,4	623
Autumn	18,09	353,3

By means of using the program, temperature values of the absorber heating material were received on semi cylindrical surface at each 10°.

The diagrams of the absorber temperature distribution in three seasons of its maintenance are given in the figure 4. In the considered case, the light source is located at the right angle to diameter of the collector segment.

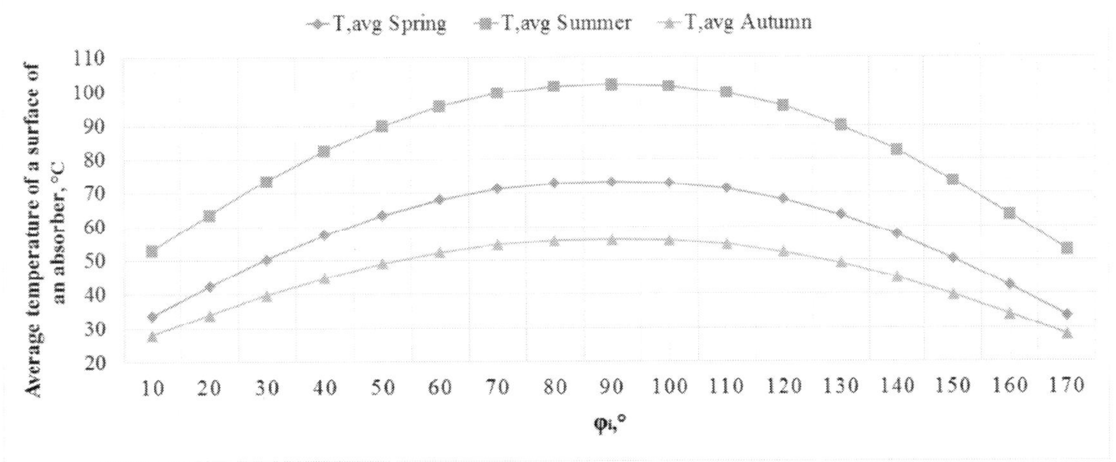

Fig. 4. The diagram of the absorber's average temperature distribution on a semi-cylindrical surface of a segment depending on error angle φ in case of direct sunlight impact

It is necessary to point out that for the calculation various maximum parameter values of season maintenance are selected. It is natural that the maximum values of heating are stated during summer period as they are 102 °C, more than 70 °C they reach in spring, and then they show significant decline in autumn to 20 °C in comparison with spring temperature values. These temperature values are typically observed at noon under the influence of direct sun light, so it is necessary to guide or fix the collector to follow the sun position. As the considered collector design is characterized by low cost, high heating rates can be achieved due to the geometric structure design when using cylindrical segments in which the absorber is laid out in the form of a half-cylinder. Let us consider the distribution of the absorber temperature field under indirect effect of solar radiation. To meet this aim corresponding graphic dependencies were built by means of the above said software product.

In case of indirect influence of radiation, the distribution of the absorber temperature field has some specific features: higher temperature values are typical for lit quarter of a segment while lower values are typical for shadow side. In accordance to the figure 5, the difference between shadow and lit absorber quarters laid out in the form of a semi cylinder with deviations from 15 to 23 °C. At the same time, even minimum temperature values of the absorber, they are still higher than ambient air temperature values, at least, to 20 °C.

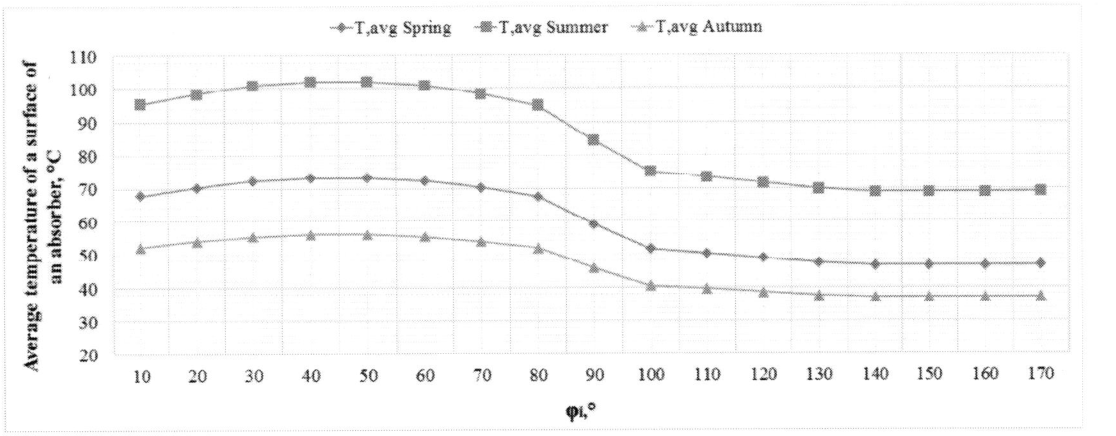

Fig. 5. the diagram of average temperature distribution of the absorber on a semi cylindrical surface of a segment depending on error angle φ in case of sunlight impact at an angle 45°

3 Analytical solution

Thus, results of simulation modeling allow the authors to come to the conclusion that the authors' construction and the design of the inflatable collector provide heating of the coolant.

The study of various environmental factors affecting the efficiency of the collector is to be continued, the constructive decision is optimized, confectioning of polymer materials is executed by means of the software product.

References

1. O.A. Smirnova, I.V. Vashchinskaya, M.A Goncharova, Yu.E. Avetisyan, Innovations in Science: Sat. Art. by mater. LVI Intern. scientific-practical. Conf. No. 4 (53). Part II. - Novosibirsk: SibAK, 52-58 (2016)
2. O.A. Smirnova, M.A. Goncharova, *Actual issues of technical sciences in modern conditions*: Collection of scientific papers on the results of the international scientific and practical conference. St. Petersburg, 11/01/2016 - Innovation Center for the Development of Education and Science (2016)
3. A.N. Nikiforov, *Optimization methods*: Textbook. allowance (YURSTU (NPI), Novocherkassk, 2007)
4. GOST 28205-89 «Basic environmental testing procedures.Part 2.Tests. Guidance for solar radiation testing».
5. GOST P 51910-2002 «Research and testing strategy by accelerated metods of environmental conditions influence for industrial produets longevity and storege adility».
6. GOST 26883-86 «Exposure factors. Terms and definitions».
7. GOST 28202-89 «Basic environmental testing. Part 2. Tests. Test Sa: Simulated solar radiation at ground level».
8. O.A. Smirnova, M.A Goncharova, Yu.E. Avetisyan. *Concept of complex assessment of operational indexes of an inflatable collector.* Innovative researches: problems of introduction of results and direction of development. Perm, (2017)
9. V.V. Kostenyuk, A.V. Doroshenko, M.A. Glauberman, Khodililna tehnika te tehnologii, №8, 34-40 (2010)
10. A.N. Gorin, A.V. Doroshenko *Solar energy (theory, development, practice)*, (Nord-Press Donetsk, 2008)
11. O.A. Smirnova, I.V. Vashchinskaya, Yu.E. Avetisyan, *New science: current state and ways of development: International scientific periodical on the results of the International Scientific and Practical Conference*, Part 2: RIC AMI, Sterlitamak (2015)
12. Yu.A. Bakhvalov, N.I. Gorbatenko, V.V. Grechikhin *Inverse problems of electrical engineering*: monograph (Izd-vo of the journal Izv. Universities. Electromechanics, Novocherkassk, 2014)
13. Encyclopedia of polymers 3 vol. V.A. MS Kabanov Akutin. Ed. Soviet Encyclopedia, 599-600 (1977)
14. O.A. Smirnova, I.V. Vashchinskaya, Yu.E. Avetisyan, International Scientific Journal, №8 76-78, ISSN 2410-6070 (2015)
15. NASA Surface meteorology and Solar Energy https://eosweb.larc.nasa.gov/cgi-bin/sse/grid.cgi?&num=221138&lat=47.709&submit=&hgt=100&veg=17&sitelev=&email=&p=grid_id&p=midday_dwn&p=swv_dwn&p=sol_noon&p=ret_psh0&p=mnavail1&p=surplus1&p=TSKIN_MN&p=toa_dwn&step=2&lon=40.214

 DOI: 10.1051/matecconf/201713204018

Ecological and hydrophysical research of impact the vertical turbulent exchange coefficient on the concentration of dissolved oxygen in the bottom layer of shallow water

Alexander Sukhinov[1,*], *Alla* Nikitina[2], *Yulia* Belova[1] and *Tatyana* Bednaya[1]

[1]Don State Technical University, Rostov, Russia
[2]Southern Federal University Taganrog, Russia

Abstract. The paper presents the results of hydrochemical observations obtained during the expedition on the scientific-research vessel "Deneb" in the Eastern part of the Azov Sea and the Taganrog Bay in July 2017. In the ecological and hydrophysical researches we have found abnormally low levels of dissolved oxygen in the bottom layer of the Azov Sea. Among the various approximations of the vertical turbulent exchange coefficient, the algebraic subgrid models are the most applicable. These models are based on the determination of turbulent flows as averaged multiplication of deviations over space or time (correlation) of the velocity components of the currents and a transferring physical quantity. The numerical experiment for determination of vertical turbulent exchange coefficient for some points of the Azov Sea was performed. Direct dependence between values of the vertical turbulent exchange coefficient and the concentration of dissolved oxygen in the bottom layer was obtained.

1 Introduction

Human impact on nature along with the abnormal climate changes lead to the violation of the natural functioning of the hydro-ecosystems, eutrophication, decreasing of biological diversity. Pollutants accumulated in bottom sediments. During the summer, sediments are heated, and the sludge microorganisms release substances that immediately cause the release of hydrogen sulfide. The decomposition of huge amounts of organic matter in a short time period leads to a significant reduction in the concentration of dissolved oxygen in the water, causing fish kills and, as a result, leads to water quality deterioration. Among all hydrochemical parameters influencing on the biological productivity of the Azov Sea, the dissolved oxygen has a special attention as the source of the respiration of aquatic organisms and the factor determining the speed and completeness mineralization of organic matter in the water. A special feature of the oxygen regime of the Azov Sea is in its instability, which is defined by permanent and seasonal factors: the absorption of atmospheric oxygen by the surface water layer, the uneven inflow of rivers and Black Sea waters, the development and decay of organisms, circulation processes determining the possibility of oxygen penetration in the bottom layers, and other processes. The concentration of dissolved oxygen in sea water depends on several factors, the main ones are: the gas exchange between water and atmosphere; the intensity of the photosynthesis process; the process by which bacterial and chemical oxidation metabolites of marine organisms, remains of dead organisms of plant and animal origin; the intensity of convective mixing and the flow of oxygen-depleted waters; water advection of masses, biochemical oxygen demand of bottom sediments [1].

In summer under conditions of density stratification occurs enhanced reduction of oxygen concentration in bottom horizons of the Azov Sea (in some areas up to 60 %). The leading factors of formation of the oxygen regime in surface waters are primarily production processes, in the bottom – the biochemical decomposition of bottom sediments. The oxygen in the process of invasion, penetrates into the underlying layers through turbulent exchange, mainly vertical turbulent diffusion in stratified waters [2-6].

The aim of this paper is verification the influence of hydrodynamic conditions on the change of dissolved oxygen concentration.

2 Expidition route, equipment

The scientific-research vessel "Deneb" (SRV "Deneb") of the Southern Scientific Center of the Russian Academy of Science RAS (SSC RAS) was used to performed scientific researches in the Azov Sea.

The researches were performed at 17 stations in the Eastern part of the Azov Sea (Fig.1).

* Corresponding author: sukhinov@gmail.com

 DOI: 10.1051/matecconf/201713204018

a)

b)

Fig. 1. Scientific researches in the Azov Sea (July 2017):
a) the expedition route; b) SRV "Deneb".

SRV "Deneb" equipment:

- SEACAT SBE19 hydrological CDT-probe;
- RCM 9LW currents registrar;
- SES-2000 light (Innomar Technologie GmbH) escaladieu parametric profilograph;
- 13.540 B gravity dirt tube with the possibility of establishing piston system (Piston/Gravity corer Model 13.540B);
- San++ automatic flow analyzer with the sampler SA1100, containing 2 x 50 sample positions;
- deep-sea sampling carousel complex;

The equipment of the vessel "Deneb" for hydrological and lithological researches:

- the Molchanov's and Niskin's samplers for sampling water;
- the Petersen's and Van-Veena's bottom samplers for sampling of bottom sediments;
- the bentonia dredge;
- the Epstein's and Jedi's plankton nets for sampling plankton;
- the caviar net for sampling ichthyoplankton;
- the set of nets and dredges for ichthyological researches;
- dirt direct flow pipe with the possibility of taking a column of precipitation 2-2.5 m.

The hydrochemical laboratory is located on the board of the vessel "Deneb" due to which we can performed all complex of fiend analyzes (oxygen, nutrients, pH, sulfide, plant pigments, production-destruction).

The equipment of the vessel "Deneb" for ichthyological research:

- bottom bentrawl;
- spacer pelagic trawl (28 m – in horizontal, 8 m – in vertical);
- the SBE19 plus hydro-probe;
- the SBE43 dissolved oxygen sensor;
- PC (personal computer) with software for hydrophysical probes;
- bathometers 3 l и 5 l, complete equipment for the measurement of oxygen concentration by Winkler method;
- (ADCP) profiler WHS 600.

3 Work content

The expeditions were carried out in the framework of the RSF project No. 17-1101286 "Predictive modeling of hydrophysics and biological kinetics processes in inland water and marine systems of the South of Russia on supercomputers", the purpose of which is to obtain systematic information of the influence of chemical and physical water properties on the ecological situation of the Azov Sea [7-14].

Oceanographic, hydrochemical, hydrobiological observations were carried out according to standard procedures and guidelines for conducting oceanographic researches with the modern expedition equipment. Complex researches of the Azov Sea and Taganrog Bay are included: meteorological, hydrological and hydrochemical observations; research of peculiarities of spatial distribution of plankton and benthos. Hydrometric and hydrological works are included: monitoring the temperature of the water, air, depth, transparency and color of the water, the excitement, the speed and direction of wind, visibility and cloudiness, velocity and oxygen concentration analysis of the species composition and concentration of planktonic populations; measurement of water temperature and selection of its sample, followed by determination of salinity using the samplers at each of the stations less than 7 m from two, and in depths greater than 7 m – from three horizons (surface, 5 m, near the bottom). Upon detection of temperature gradients, the measurements of temperature and water sampling for salinity were performed in additional horizons.

Quality observation program of marine waters are included the following activities:

- the determination of chemical compounds concentration: petroleum hydrocarbons, mg/dm^3 (mg/l); dissolved oxygen, mg/dm^3 (mg/l, %); pH; chlorinated hydrocarbons, including pesticides, mkg/dm^3 (mkg/l); heavy metals – mercury, lead, cadmium, copper, mkg/dm^3 (mkg/l); phenols, mkg/dm^3 (mkg/l); synthetic surface-active substances,mkg/dm^3 (mkg/l);
- the determination of indicators and nutrients, specific to the selected area: the nitrite nitrogen, mkg/dm^3 (mkg/l); the silicon, mkg/dm^3 (mkg/l); the water salinity, %; the water and air temperature,°C; the wind speed and direction, m/sec; water transparency (color points); sea rough (points);

- the performing visual observations of the sea water object surface.

4 Pulsation measurements of the velocity vector of water flow

The parameterization problem of turbulent exchange coefficient occurs in modeling the water flow fields. Turbulent exchange influence to the intensity of such processes as seabed erosion, sediment transport and transport of impurities [15-18]. The ability of water to cleanse itself depends on mechanisms of turbulent mixing.

Stochastic methods are often used for description turbulent flows. Various fluctuating quantities are considered as random functions. Turbulence has a complicated statistical structure due to the strong intermittency at the dissipative scales.

Mechanisms of vertical turbulent exchange are suppressed in numerical modeling on large vertical grids. This fact determines the necessary using small scale vertical resolutions.

Data of velocity pulsations in some points of shallow systems were obtained during the expedition with using ADCP (Acoustic Doppler Current Profiler) WHS600 Sentinel. The ADCP-profiler was fixed at free water surface. The measurements were carried out vertically from the free surface to the bottom (Fig. 2). The ADCP-profiler works on the Doppler's effect, transmitting the audio signal at a fixed frequency and receiving echoes of returned signal using the sound emitter through the entire depth of the water column. The emitter radial velocity (along-beam) is calculated from the Doppler's echo. Ripple velocity components (u', v', w') are calculated from the values of radial velocities.

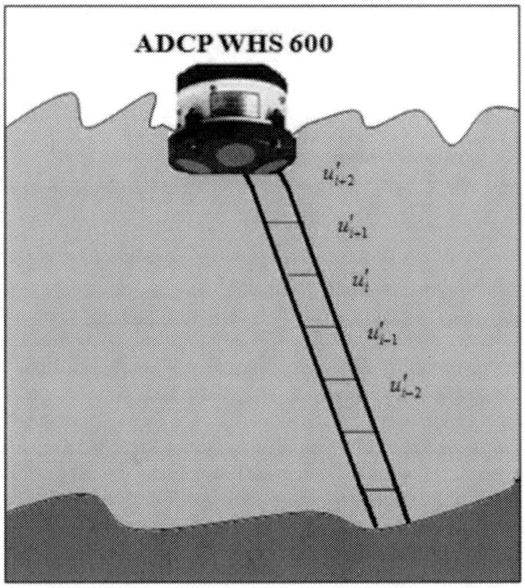

Fig. 2. ADCP-profiler position, fixed on the free water surface

The barometrically selection of samples was done, and the analysis of water composition was performed during the researches. The table, contained the measured values of dissolved oxygen concentration in the horizons of the researched water at the stations the expedition route, are shown below (Table 1).

Table 1. Dissolved oxygen concentration in the Azov Sea

Station No.	Station coordinates	Measured values of dissolved oxygen			
1(49)	46° 39, 569′ N	depth, m	0	4	11,5
	37° 20, 743′ E	O_2, mg/l	9,02	8,53	5,32
2(48')	46°29,725′ N	depth, m	0	6	12
	36°59,786′ E	O_2, mg/l	12,69	13,03	0,66
3(44')	46° 19, 693′ N	depth, m	0	6	12
	36° 59, 683′ E	O_2, mg/l	5,75	4,94	0,61
12(42)	45° 59, 906′ N	depth, m	0	5,5	11
	37° 20, 128′ E	O_2, mg/l	9,21	9,87	1,41

5 The vertical turbulent viscosity coefficient

The D. Smagorinski's model is also an algebraic model of the "mixing length", however, it is also known as subgrid model of the vertical turbulent exchange coefficient.

The principle is the following: in the case if the vertical resolution of the computational grid allows to reproduce all mechanisms of the simulation process to the scale of viscous dissipation of very small vortices, the turbulence is directly calculated from the Navier-Stokes equations and is naturally obtained in numerical simulations. Thus mechanisms of vertical turbulent exchange are suppressed in numerical modeling on large vertical grids. This fact determines the necessary using small scale vertical resolutions. The D. Smagorinski's idea consists in adding the turbulent viscosity to molecular viscosity. The turbulent viscosity, in turn, is determined by the mixing length, which corresponds to the size of small eddies.

Method for determine the coefficient of vertical turbulent viscosity are divided into two categories, depending on the observational data: the distribution data of average speed and data of pulsation measurements of the Reynolds stresses. The coefficient of vertical turbulent viscosity, defined by the first method, changes in interval of $0.1\text{-}10^3$ cm^2/s. The second method requires the use of sophisticated equipment, which can be used to measure the velocity fluctuations of water flow. The turbulent viscosity coefficient can be defined as the ratio of the Reynolds stresses to the gradient of average velocity (according to the A.S. Monin). It is known that the density stratification is virtually absent in shallow waters, and hydrodynamic processes in the vertical direction are essentially nonlinear. So we have to resort to non-standard third method of measurement and evaluation the coefficient of vertical turbulent viscosity, based on the calculation of the average velocity gradient of the water flow.

Among the various approximations of the coefficient of vertical turbulent exchange the algebraic subgrid models are the most applicable. These models are based on the determination of turbulent flows as averaged multiplication of deviations over space or time (correlation) of the velocity components of the currents and a transferring physical quantity.

Evaluation methods of turbulence characteristics on the movement or scattering markers (indicators, buoys, etc.) are subgrid methods. The large measured data are necessary to evaluate the characteristics of vertical turbulence for direct subgrid methods. This is directly related to the duration and high cost of expidition researches.

Due to numerical simulation of turbulence we can obtained the coefficient of vertical turbulent exchange, inhomogeneous in depth, on the basis of measured velocity pulsations of water flow.

In works [19-24] it was shown that the best results, closest to experimental field data, were obtained on the basis of the O.M. Belotserkovsky parametrization.

In the case if the vertical resolution of the computational grid allows to reproduce all mechanisms of the simulation process to the scale of viscous dissipation of very small vortices, the turbulence is directly calculated from the Navier-Stokes equations and is naturally obtained in numerical simulations. Thus, the dissipation rate can be expressed through the average strain rate of cell scale in numerical modeling on large vertical grids :

$$\bar{s} = 2\bar{s}_{ij} \cdot \bar{s}_{ij},$$

where $\bar{s}_{ij}$ is the averaged tensor of deformation rate: $\bar{s} \sim \varepsilon^{2/3}\Delta^{-4/3}$.

The expression for the turbulent viscosity has the form:

$$\nu = C_s^2 \Delta^2 \bar{s}^{\,1/2},$$

where C_s is the dimensionless empirical constant, whose value is determined based on the calculation process of the decay of homogeneous isotropic turbulence. Particularly, the selected constant value C_s must ensure the conformity of simulation results with experimental measurements. As a rule, the values of the C_s constant in most cases, vary from 0.1 (in the flow channel) to 0.2 (isotropic turbulence).

In the case if the grid size is selected sufficiently small, the average rate of strain cell scale $\bar{s}$ is equaled to the average deformation rate in the Prandtl's parameterization D. Thus, we obtained the Prandtl's analogue in the form:

$$\nu = C_s^2 \Delta^2 \sqrt{2 D_{ij} D_{ij}},$$

Neglecting horizontal gradients $\bar{u}$ и $\bar{v}$ and gradients of the vertical velocity component $\bar{w}$, we obtained the following:

$$\nu = C_s^2 \Delta^2 \frac{1}{2} \sqrt{\left(\frac{\partial \bar{u}}{\partial z}\right)^2 + \left(\frac{\partial \bar{v}}{\partial z}\right)^2}.$$

The D. Smagorinski's model of turbulent viscosity sometimes called as a theory of subgrid scale of the first order. It is easy to analyzed the algebraic models, discussed above, and this model.

6 Numerical calculation of vertical turbulent viscosity coefficient

The numerical experiments were performed on the basis of the above approaches for all points, $\nu \sim \varepsilon^{1/3} \cdot \Delta^{4/3}$, in which field measurements were conducted during the expedition researches. In the results the coefficients distribution of vertical turbulent exchange for the Azov Sea were obtained (Fig. 3).

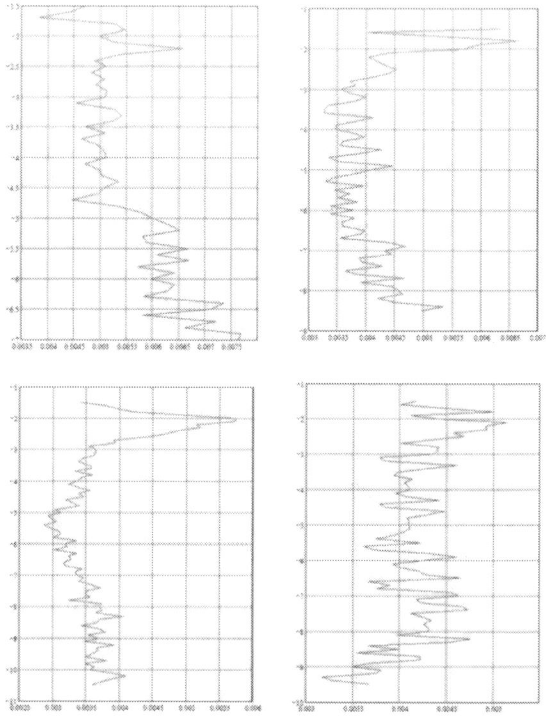

Fig. 3. The vertical turbulent exchange coefficient in points 1, 2, 3, 12 (in horizontal – values, m²/sec, in vertical – the water level from the free surface to the bottom) (according to the D. Smagorinski).

The impact of changes in the values of vertical turbulent exchange coefficient on the concentration of dissolved oxygen in the bottom layer of a shallow water was researched. The analysis of simulation results of hydophysics processes is shown that the highest concentration of dissolved oxygen in the bottom layer of the Azov Sea is corresponded to the highest value of vertical turbulent exchange coefficient (point 12), and the lowest value of vertical turbulent exchange coefficient is corresponded to the lowest value of dissolved oxygen concentration (point 3). The values of vertical turbulent exchange coefficient is close to zero at the depth of about 3 meters and below. This means that the turbulent exchange is reduced in the vertical in this region, and explains the phenomenon of anoxia in the bottom layer of the Central-Eastern part of the Azov Sea.

7 Conclusion

The concentrations of dissolved oxygen and data of velocity pulsations of water flow in some points of shallow systems were obtained during the expedition researches with using ADCP WHS600 Sentinel. Areas with low dissolved oxygen content were discovered in the bottom water layer.

The distributions of vertical turbulent exchange coefficient with data of velocity component pulsations, measured during the expedition, were obtained experimentally.

The conformity of calculation results based on the statistical velocity field data of water flow and the subgrid Smagorinski's models with field measurements was obtained.

The impact of changes in the values of vertical turbulent exchange coefficient on the concentration of dissolved oxygen in the bottom layer of the shallow water was researched.

This paper was partially supported by the grant No. 17-11-01286 of the Russian Science Foundation.

References

1. B.N. Chetverushkin, Mathematical Models and Computer Simulations, **5**, No. 3, 266–279 (2013)
2. A.I. Sukhinov, E.A. Protsenko, A.E. Chistyakov, S.A. Shreter, Computational Methods and Programming. **16**, No. 3., 328–338 (2015)
3. A. I. Sukhinov, A. E. Chistyakov, E. F. Timofeeva, A. V. Shishenya, Mathematical Models and Computer Simulations. **5**, No. 2., 122-129 (2013)
4. A.I. Sukhinov, A.E. Chistyakov, Computational Methods and Programming, **13**, No. 1, 290–297 (2012)
5. A.I. Sukhinov, A.E. Chistyakov, A.A. Semenyakina, A.V. Nikitina Computational Methods and Programming. **16**, No. 2., 256–267 (2015)
6. O. M. Belotserkovskii, *Turbulence: New Approaches* (Nauka, Moscow, 2003).
7. Samarskii A.A.The theory of difference schemes (Science, Moscow, 1989)
8. O.M. Belocerkovskij, V.A. Gushhin, V.V. Shhennikov, USSR Computational Mathematics and Mathematical Physics, **15**, No. 1, 190-200 (1975)
9. I.Y. Gluhenky, A.V. Lavrent'ev, G.G. Popov, Safety in Technosphere, No. 6., 3-6 (2011)
10. A.S. Abdusamadov, A.P. Panarin, A.K.Magomedov L.D. Kovalenko, B.R. Huseynov, A.M. Dohtukaeva, L.A. Dudurhanova, Geography and Geoecology. Southern Russia: ecology, razvitie., №1., 165-166 (2012).
11. Resolution of the Government of the Russian Federation on April 15, 2002 № 240 "On the procedure for organizing the activities for prevention and liquidation of oil spills in the Russian Federation "
12. https://www.gismeteo.ru/diary/5211/2007/11/
13. A.I. Sukhinov, A.E. Chistyakov, E.A. Protsenko, Mathematical Models and Computer Simulations, **6**, No. 4, 351-363 (2014)
14. A.I. Sukhinov, A.E. Chistyakov, E.A. Protsenko, Computational Methods and Programming., **15**, 610-620 (2014)
15. A.I. Sukhinov, A.V. Nikitina, A.E. Chistyakov, Mathematical Models, **24**, No. 9., 3–21 (2012)
16. A.V. Nikitina, A.I. Sukhinov, G.A. Ugolnitsky, A.B. Usov, A.E. Chistyakov, M.V. Puchkin, I.S. Semenov, Mathematical Models and Computer Simulations, 9 (1), 101-107 (2017)
17. A.I. Sukhinov, A.E. Chistyakov, I.I. Levin, I.S. Semenov, A.V. Nikitina, A.A. Semenyakina, *5th ICIEV*, 1128 – 1133 (2016)
18. A.I. Sukhinov, A.E. Chistyakov, A.V. Nikitina, A.A. Semenyakina, I. Korovin, G. Schaefer, *5th ICIEV*, 1134 – 1139 (2016).
19. A.I. Sukhinov, A.E. Chistyakov, A.A. Semenyakina, A.V. Nikitina, Computer researches and modeling. **8**, no. 1., 151 – 168 (2016)
20. A.I. Sukhinov, D.S. Khachunts, A.E. Chistyakov, Computational Mathematics and Mathematical Physics. **55**, No. 7, 1216-1231 (2015)
21. A.I. Sukhinov, A.E. Chistyakov, A.V. Shishenya, Mathematical Models and Computer Simulations, **6**, No. 3., 324-331 (2014)
22. N. Buzalo, P. Ermachenko, T. Bock, A. Bulgakov, A. Chistyakov, A. Sukhinov, E. Zhmenya, N. Zakharchenko, Proc. Eng., **85**, 84-93 (2014)
23. A.I. Sukhinov, A.E. Chistyakov, E.F. Timofeeva, A.V. Shishenya, Mathematical Models and Computer Simulations, **5**, No. 2., 122-129 (2013)
24. A.I. Sukhinov, A.E. Chistyakov, Mathematical Models and Computer Simulations, **4**, Iss. 4, 398-409 (2012).

Subject-independent modeling and representation data on the formation and distribution of innovative value

Aleksey Frolov[1] *and Irina* Maslova[2,*]

[1]Orel State University, Department of software engineering, 302020, Orel, Russia
[2]Orel State University, Department of economic security, 302020, Orel, Russia

Abstract. Creating an innovation environment is shown in the context of interaction of economic agents in the creation and consumption of innovative value-based infrastructure approach. The problem of the complexity of collecting heterogeneous data on the formation and distribution of innovative value in the conditions of the dynamic nature of the research object and the environment is formulated. An information model providing a subject-independent representation of data on innovation value flows is proposed and allows to automate the processes of data collection and analysis with the minimization of time costs. The article was prepared in the course of carrying out research work within the framework of the project part of the state task in the field of scientific activity in accordance with the assignment 26.2758.2017 / 4.6 for 2017-2019. on the topic "System for analyzing the formation and distribution of the value of innovative products based on the infrastructure concept".

1 Introduction

For modern economic development, financial instruments are needed, more flexible, maneuverable and stimulating the innovation process than those used in the traditional economy.

According to official statistics, in the Russian Federation, the cost of technological innovation in 2015 was 1,200,363.8 million rubles [22].

The analysis of the dynamics of organizations engaged in innovative activity in the Russian Federation shows that in 2015 only 9.3% (0.6% less than in 2014) of the organizations surveyed engaged in innovative activity. At the same time, the share of organizations that carried out technological innovation was 8.3% (against 8.8% in 2014) [22].

Traditionally in the Russian Federation, as in many countries of the world, the state was the most significant source of funding for science and innovation. However, recently in western countries there has been a decrease in the share of public financing due to the increase in the share of the private corporate sector against the background of the strengthening of indirect stimulation of innovation processes (tax breaks, concessional lending, support of innovative business).

For private investors, the possibility of obtaining transparent and reliable information on the formation of value added of innovative products and its market actualization is of particular importance.

2 Literature review

Among the classics of economic theory laying the methodological foundation of the theory of value and the mechanism of pricing, it is necessary to name O. Bem-Bawerka, J. Gelbright, J. Debreu, JM Keynes, V. Leontiev, K. Marx, A. Marshall, B. Petty, A. Pigou, D. Ricardo, J. Robinson, A. Smith, I. Schumpeter, P. Samuelson, L. Thurow, M. Friedman, I. Fischer, J. Hicks, and others.

Domestic economists also made a significant contribution to the development of the theory of value, value, the mechanism of pricing and valuation, the most outstanding among them: L.I. Abalkin, V.M. Ageev, V.A. Bazarov, K.K. Walthuh, A.S. Gusarov, V.P. Dyachenko, L.V. Kontorovich, I.D. Kondratiev, I.K. Larionov, B.C. Nemchinov, A.V. Novichkov, N.S. Perekalina, K.N. Plotnikov, V.T. Smirnov, S.G. Strumilin, N.P. Fedorenko, P.N. Shulyak, Yu.V. Yakovets and others.

In the field of international financial reporting standards, the following works are noteworthy: D. Alexander, A. Britton, H. Grüning, E. Jorissen, M. Cohen, Karsberg B. Mikerin G. et al.

The problems of controlling the structural dynamics of complex systems are reflected in the works of BV Sokolov, RM Yusupova. and other researchers of the St. Petersburg Institute of Informatics and Automation of the Russian Academy of Sciences.

The following researchers consider modeling and creating adaptive and adaptive software based on relevant information models: Cangussu J. W., Miller S. D., Cai K., Mathur A. P., C. Dellarocas, M. Klein, H. Shrobe, P. Oreizyetal, J. Palsberg, C. Xiao, and K. Lieberherr, Y. Diao, J.L. Hellerstein, S. Parekh, R. Griffith, G.E. Kaiser и др.

* Corresponding author: tera_27@mail.ru

3 Hypotheses and methods of research

In the process of research, collection and processing of information, classical empirical and theoretical methods of research were used: study of literature, observation, survey, monitoring, generalization of experience, dialectical method, hypothesis construction, proof, analysis, synthesis, comparison, abstraction, concretization, generalization, induction , deduction, analogy, modeling, system approach.

4 Results of the study

Innovative activity of the enterprise is a system of measures, united in a logical chain. Scientific research, experimental design and technical development, innovative, financial, commercial and industrial activities are subject to the same goal – the creation and implementation of innovations. Innovative activity is characterized by frontality, high level of uncertainty and risk, complexity of forecasting results, requires taking into account the time factor at different stages of the innovation process.

The process of creating innovations is associated with the costs of labor and material resources and as a result, like any productive process, leads to the formation of socially useful value embodied in innovative products.

Now in Russia at the state level (administrative and public infrastructure) prerequisites for a civilized process of creating innovations are created. The existing state regulation and support of the intellectual market should nevertheless be improved at all levels of the innovative macro-climate, affecting the internal innovation environment of the enterprise.

In the innovation process, the time factor plays a big role, since a significant amount of time can pass between the innovation and the release of an innovative product on its basis. Thus, the degree of readiness of the result of the scientific and innovative process to commercialization is important. Market actualization of value – bringing the value in line with the state of the economic environment through commercial transactions, expressed in the purchase by the buyer of natural media of value and the formation of the seller of the monetary equivalent of value [11] (figure 1).

In accordance with the Methodology for Determining the Levels of Technology Preparedness within the Framework of the Projects of the Federal Targeted Program "Research and Development in Priority Areas for the Development of the Scientific and Technological Complex of Russia for 2014-2020", the following ranges of funding are allocated at the expense of the state target program (figure 2) [18].

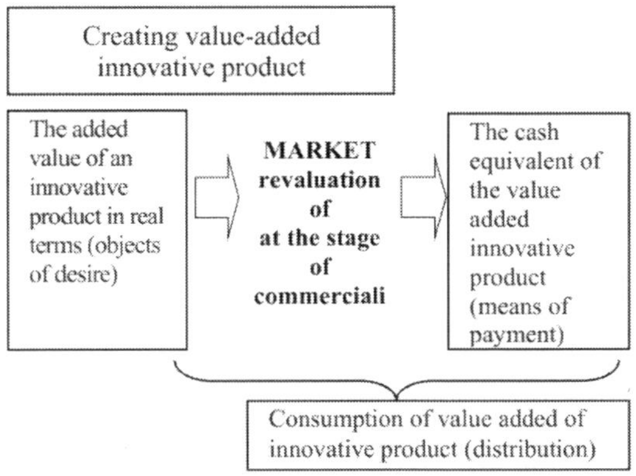

Fig. 1. Market's actualization of the added value of an innovative product

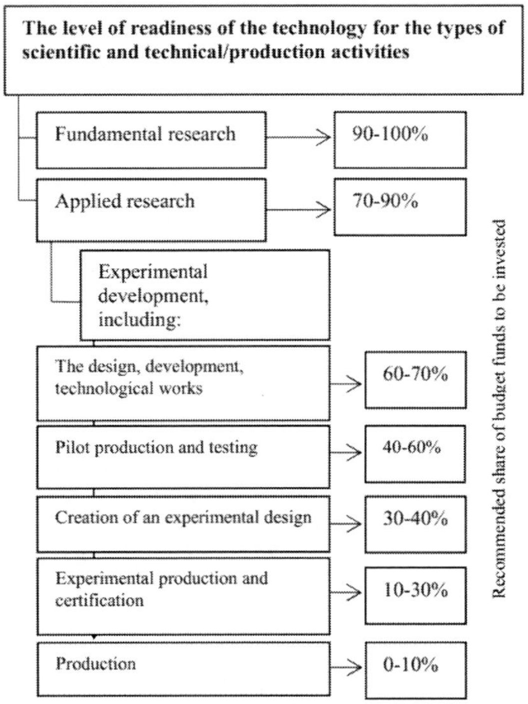

Fig. 2. Potential share of budgetary funds in the cost of innovation, depending on the level of availability of technology

The current practice in Russia of reflecting the cost of innovative products in the accounting and analytical registers of enterprises is characterized by the lack of integrity of the results of the compilation of accounting data from the point of view of carrier uniformity and by type of value, which makes it difficult to reliably estimate innovative value and innovative value added (created and actualized on the market).

Innovation activity has a significant impact on all aspects of financial and economic activity, requires the investment of own and borrowed funds in significant amounts and therefore, transparent information on the flow of innovative value should be reported. However,

DOI: 10.1051/matecconf/201713204019

in the accounting reporting there are no such concepts as innovations and objects of innovative activity taking into account its stages of a full life cycle.

IFRS regulates the accounting and reporting of revenues, expenses, assets and liabilities, providing a factual information base for making managerial decisions at all levels. The relationship between the national multi-level accounting system and IFRS is shown in the figure 3.

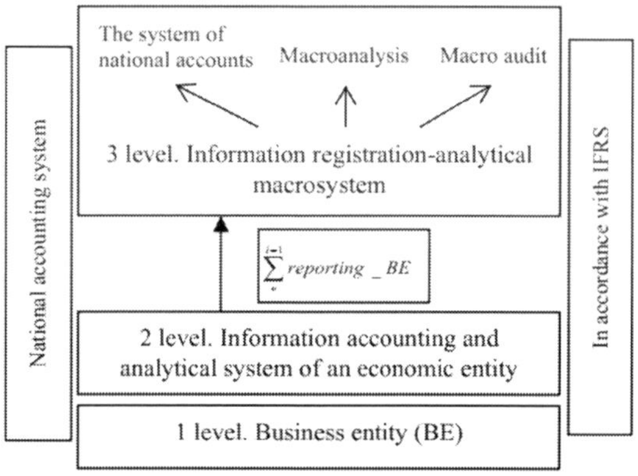

Fig. 3. Interrelation of the accounting and analytical system of an economic entity with an accounting and analytical macro system in the national accounting system in accordance with the transition to IFRS.

The complexity of accounting and analytical processes requires the use of modern tools for collecting, storing and analyzing data. And when considering each specific implementation of the accounting and analytical process, the structure and composition of the system of indicators varies. The reasons for this can be the dynamism of both the accounting object itself, which is a complex system, and the environment. In the works [5,6,7,17], "formalization and research of a new class of applied problems, namely, problems of managing the structural dynamics" of technical and organizational-technical systems (OTS) was carried out. According to their works, structural dynamics can manifest themselves in the form of the following processes:

- changes in the ways, purposes of functioning of the OTS, their content, sequence of implementation;
- movement in the space of individual elements of the OTS;
- redistribution and decentralization of functions, tasks, control algorithms, information flows between management levels;
- management of reserves;
- Reconfiguration of OTS structures in case of its degradation.

Data collection under such conditions is usually called monitoring. The type of monitoring, involving the prompt collection of information on the state of the control object from various sources (manual data entry by the operator, import from the databases of the MRP/ERP subsystems) will be called administrative. The implementation of new specialized or modification of existing software for data collection and storage in such cases is inefficient due to their one-time use or use a small number of times. At the same time, the emerging time costs do not allow the operative solution of the assigned task.

Thus, it becomes urgent to use the system of administrative monitoring, which provides an adaptive organization of data collection and data storage processes, that is, a process-oriented system built on the basis of the software accounting and analytical activity that is adaptable to the task. Studies in the field of software adaptation are reflected the works [8,13,14,20, 21,23,24,]. They address the choice of architecture for adaptive software, the development of control algorithms for adaptive software; about quality criteria for the evaluation of adaptive software.

This report proposes an approach to structural adaptation, based on a two-level abstraction of the information model of the accounting object. The proposed data storage model provides an invariant representation of the accounting object and flexible customization when requirements for the data collection and storage procedure change [19].

Data on the types of accounting objects arrive at the input of the system at the stage of adaptation of the monitoring system to the requirements of accounting and analytical processes (see Figure 4).

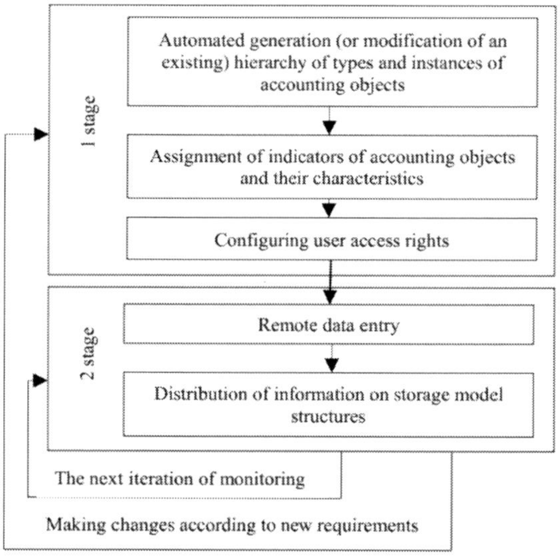

Fig. 4. General methodology of organization of processes collection and storage of data

Since the monitoring system supports not only linear connectivity of the types of accounting objects, but also the network type, the input types are structured. That is, each newly entered type of accounting object is assigned possible types of parents and possible types of descendants. The set of types forms the input set Vt.

A set of parent type relationships is expressed by a subset of the Cartesian product

$Vt \times Vt$: $Et = \{(t1,t2) \mid (t1 \subset Vt) \,\&\, (t2 \subset Vt)\}$. (1)

Accounting objects (AO) are compared to the types of accounting objects in many respects because of the need to conduct different monitoring of them, tracing changes or current states for certain sets of indicators. For this, certain sets are mapped to sets of indicators. It is for these sets of indicators that all monitoring objects of this type will be monitored. We denote the set of all indices entering the input of the system P. Then the set of all possible sets of exponents:

$$2P = \{XP \mid XP \subset P\} \quad (2)$$

Then $Rtp \subset 2P \times Vt$, represents the indicators inherent in each type of accounting object. This relation can be represented as a mapping from the Boolean 2P to the set Vt: $Rtp = \{(XP, t) \mid XP \subset P \;\&\; t \in Vt\}$.

Data on instances of accounting objects are entered at the stage of filling in the model of links of instances of accounting objects.

Data processing of this type includes the generation of hierarchical and network links between the added and existing instances, the control of the correspondence between the format of the model object to be added to the accounting object, the control of the authorization of changes in the structure of the links of the objects of the accounting objects.

Eto – the object typing relation. The element of this relation is an instance of the op-amp with its associated type (t, o). Such typed instances are linked in pairs by the "child-parent" relationship, thus creating ordered pairs of elements of the set Eto. The model of object links:

$$Eo = \{(Eto1, Eto2) \mid Et1=(t1,t2) \;\&\; Eto1=(o1,t1) \;\&\; Eto2=(o2,t2). \quad (3)$$

That is, typed AO Eto1 is a descendant of a typed copy Eto2 if there is a pair «type- descendant → type-parent » Et1=(t1,t2) such that in a pair Eto1 to a copy o1 mapped type t1, and in a pair Eto2 to a copy o2 mapped type t2. The relation Eo can be represented by an array of arcs of the graph.

Elementary unit of the dialog input interface is a mathematical object that is associated with some instance of the accounting object and an indicator whose value will be entered into the system. The ratio of typed accounting objects to the sets of indicators for which monitoring is conducted: $Eto \circ Rtp$:

$$Rpo = Eto \circ Rtp = \{(o1, XP) \mid o1 \in Vo \;\&\; XP \subset P \;\; \exists\, t \in Vt \;\; o1Etot \;\&\; tRtpXP\}. \quad (4)$$

Each element of this relation represents a set of interface elements for registering the values of the indicators of a certain accounting object. Thus, the element of the relation is the ordered pair of an instance of the accounting object o1 and its associated set of indicators XP, $o1 \in Vo$ & $XP \subset P$, provided there is a certain type of accounting object from a set of types such that the ratio Eto can map an instance of o1 to type t and the ratio Rtp can be obtained from type t associated with it XP.

This ratio allows us to describe the set of generated input interfaces for registering the values of the metering object indicators, which is the result of the adaptation of the monitoring system to the monitoring task.

Each implementation of the introductory interface for an accounting object instance for the input of indicator values is represented by the function i (o, Xp) from the arguments: an instance of the accounting object and a set of indicators by which the values are registered. In the case of using indicators with a limited general population of possible values, the model expands. Each indicator of this type is associated with a set of possible values from a set of values available for input to users of the system, which is expressed by the ratio:

$$Rps = \{(p1,Xs) \mid p1 \subset P \;\&\; Xs \subset 2Xs\}. \quad (5)$$

Then the composition of relations $Rpo = Eto \circ Rtp$ will change to $Rvpo = Eto \circ Rtps$, where Rtps – the ratio of the accounting object to the set of indicators with a limited general population of values, and will be compiled with the participation of the Rps ratio defined above instead of P in one of its components:

$$Rtps = \{(Rps1, t1) \mid Rps1 \subset Rps \;\&\; t1 \in Vt\} \quad (6)$$

$$Rvpo = Eto \circ Rtps = \{(o1, Rps1) \mid o1 \in Vo \;\&\; Rps1 \subset Rps \;\; \exists\, t \in Vt \;\; o1Etot \;\&\; tRtpRps1\} \quad (7)$$

Thus, the element of the relation is the ordered pair of an instance of the accounting object o1 and its associated set of indicators, each of which is associated with a set of possible values of Rps1, provided there is a certain type of accounting object from a set of types such that the ratio Eto can map an instance of o1 to type t and the relation Rtp it is possible to obtain from the type t the indicators associated with it with associated sets of possible values.

When working in the indicator values monitoring mode, the registered value of the indicator v Vs (Vs – the set of indicator values) is compared to the interface element (i). The interface element model is an element of the relationship between typed accounting objects and sets of indicators. Let us express this with the attitude:

$$Rspo = \{(i, v) \mid (i \subset Rpo \;\|\; i \subset Rvpo) \;\&\; v \in Vs\}. \quad (8)$$

The information is entered into the monitoring system in two modes:

- mode of preparation (adaptation) of the system to a specific accounting and analytical process;
- a mode of collection of values of indicators of objects of the account.

At present, practical implementation of adaptable data storage structures and data collection interfaces has been carried out. The prototype of the system is tested on particular tasks.

5 Conclusions

The report presents elements of the concept of constructing an accounting and analytical system of innovation value by the stages of its formation and distribution of value added. Methodological model of accounting, analysis and control of innovative value is presented on the example of own sources of financing of the innovation process.

Practical implementation of operational accounting of innovative value is faced with technological limitations, due to the complexity, including the dynamic nature of the research object and environmental factors.

The proposed subject-independent model for presenting data on the formation and distribution of innovation value, based on two-level abstraction, provides the ability to quickly adapt the data collection and storage system to changing conditions.

The implementation of the subsystem for the collection and storage of data in the form of a subsystem in the accounting and analytical system will make it possible to fully and effectively implement the proposed approach to the accounting and analysis of the flow of the value of an innovative product.

References

1. A.S. Alisenov, *Tax stimulation of innovations in the Russian economy*: monograph. (Prospekt, Moscow, 2015)
2. A. Pelin, K. Kala, EUROPEAN ECONOMIC REVIEW, **91**, 89-117 (2017)
3. A. Amrein-Beardsley, M. Pivovarova, T. J. Geiger, PHI DELTA KAPPAN, **98**, Iss. 2, 35-40 (2016)
4. B.G. Maslov, Management Accounting, No. 11, 23-29 (2009)
5. B.V. Sokolov, E.G. Tsivirko, R.M. Yusupov, Proceedings of SPIIRAS., Iss. 11, 11-49 (2009)
6. B.V. Sokolov, R.M. Yusupov, Problems of Control and Computer Science, No. 5, 24-41 (2002)
7. B.V. Sokolov, R.M. Yusupov, *Sat. sci. tr. Intern. sci. School "Modeling and analysis of safety and risk in complex systems* (IADB-2008)", 146-155 (2008)
8. C. Dellarocas, M. Klein, and H. Shrobe. In Proc. *The Third International Software Architecture Workshop*, 29-32 (1998)
9. Caraballo, Jose G .; Jiang, Xiao Value-Added, JOURNAL OF ECONOMIC ISSUES, **50**, Iss. 1, 288-296 (2016)
10. E.G. Dedkova, Management accounting, No. 10., 70-76. (2011)
11. I.A. Maslova, Managerial Accounting, No. 1, 17-25, (2016)
12. I.A. Maslova, Finance and credit, №4 (2006)
13. J. Palsberg, C. Xiao, and K. Lieberherr, ACM Transactions on Programming Languages and Systems, 17 (2), 264-292 (1995)
14. J. W. Cangussu, K. Cooper, and C. Li, *In Proc. of the 19th Annual ACM Symposium on Applied Computing* (March 2004).
15. L.V. Popova I.A. Maslova, B.G. Maslov, Actual problems of economy, №6, 390-402 (2015)
16. Lind, C. Holmstrom, Kang, Olivia H., Management international review, **57**, Iss.4, 571-602 (2017)
17. M.Yu. Okhtilev, B.V. Sokolov, R.M. Yusupov, *Intellectual technologies for monitoring and managing the structural dynamics of complex technical objects* (Nauka, Moscow, 2006)
18. Methodology for determining the levels of technology readiness in the framework of the projects of the federal special-purpose program "Research and development in priority areas of development of the scientific and technological complex of Russia for 2014-2020" [Electronic resource] / Access mode: http://fcpir.ru/upload/ medialibrary / 955 / gt_57_14vn_metodika-ugt-_002_.pdf
19. N.A. Kravtsova, Automation of the processes of data collection and storage during administrative monitoring: Dis .... kand. ... cand. those. sciences. Eagle., 164 with. (2012)
20. P. Lin, A. MacArthur, and J. Leaney, *In Proc. The 2005 Australian Software Engineering Conference. IEEE Computer Society* (2005).
21. P. Oreizyetal, IEEE Intell. Syst., 54-62 (1999)
22. Science and innovation, access mode: http://www.gks.ru/wps/wcm/connect/rosstat_main/rosstat/ru/statistics/science_and_innovations/science/#
23. J. W. Cangussu, S.D. Miller, K. Cai, and A. P. Mathur, Software Cybernetics, Encyclopedia of Computer Science and Engineering, John Wiley & Sons, **4**, 2575 - 2583 (2009).
24. Y. Diao, J. L. Hellerstein, S. Parekh, R. Griffith, G. E. Kaiser, and D. Phung, IEEE J. Sel. Areas Commun., 23 (12): 2213-2222 (2005).

The calculation problem of thermodynamic processes in a steam turbine

Alexander Chistyakov[1,*], *Alla* Nikitina[2] and *Elena* Protsenko[3]

[1]Don State Technical University, Rostov, Russia
[2]Southern Federal University, Taganrog, Russia
[3]Taganrog Institute of A.P. Chekhov (branch) of the Rostov State University of Economics, Taganrog, Russia

Abstract. This paper covers the calculation of aerodynamic processes in a steam turbine using the modern information technologies and computational methods, that are contributed to the increasing the accuracy of the calculations. The practical significance is the development and application of the model of aerodynamic processes in a steam turbine, and determination of limits and prospects with using of proposed model. The aerodynamic processes in a turbine are characterized by the air and heat flow non-uniformity, significantly affect the reliability and efficiency of the turbine. The calculation was performed taking into account the complex geometry of the turbine, and can be applied to any turbine of similar design with minor changes.

1 Introduction

At present, the question about the optimization problem of installation and exploitation of steam turbines is the actual. With the development of modern technology and industry needs to the exploitation of turbines, requirements are the more stringent associated with the reliability and efficiency of their operation. A large number of existing turbines is practically close to the elaboration its resource. Therefore, the introduction of more modern units is required. The fundamentals of the theory of heat transfer and analysis results of transfer processes are required to assess the reliability and efficiency of the facility. Therefore, these data should be taken into account in designing of steam turbines. Thermal systems modelling include the problems of optimal control of thermal modes, due to which we can choose the best from different implementations. The optimization of thermal modes is reduced to the solution of heat conduction problem. Mathematical modelling of thermal processes in technogenic systems is relevant at the present. Due to it, we can check the correctness of engineering ideas and correct errors at the stage of designing by the simple, low-cost means. The mathematical model is represented by the scheme «model – algorithm – program», and must contain the structure, characteristic features of the process and should be described by the equation system or functional relations. After the designing, it is necessary to determine the real values of temperature at significant points of the steam facility and analyze the compliance with required values.

2 Problem statement

The main equations of air dynamics are:
– Navier-Stokes motion equations

$$\frac{\partial v_x}{\partial t}+v_x\frac{\partial v_x}{\partial x}+v_y\frac{\partial v_x}{\partial y}+v_z\frac{\partial v_x}{\partial z}= -\frac{1}{\rho}\frac{\partial P}{\partial x}+\mu\left(\frac{\partial^2 v_x}{\partial x^2}+\frac{\partial^2 v_x}{\partial y^2}+\frac{\partial^2 v_x}{\partial z^2}\right)+f_x,$$

$$\frac{\partial v_y}{\partial t}+v_x\frac{\partial v_y}{\partial x}+v_y\frac{\partial v_y}{\partial y}+v_z\frac{\partial v_y}{\partial z}= =-\frac{1}{\rho}\frac{\partial P}{\partial y}+\mu\left(\frac{\partial^2 v_y}{\partial x^2}+\frac{\partial^2 v_y}{\partial y^2}+\frac{\partial^2 v_y}{\partial z^2}\right)+f_y,$$

$$\frac{\partial v_z}{\partial t}+v_x\frac{\partial v_z}{\partial x}+v_y\frac{\partial v_z}{\partial y}+v_z\frac{\partial v_z}{\partial z}= =-\frac{1}{\rho}\frac{\partial P}{\partial z}+\mu\left(\frac{\partial^2 v_z}{\partial x^2}+\frac{\partial^2 v_z}{\partial y^2}+\frac{\partial^2 v_z}{\partial z^2}\right)+f_z, \quad (1)$$

– continuity equation

$$\rho_t'+(\rho u)_x'+(\rho v)_y'+(\rho w)_z'=0. \quad (2)$$

– state equation

$$P=\varepsilon\frac{\rho}{M}RT, \quad (3)$$

where $\mathbf{V}=\{v_x,v_y,v_z\}$ are velocity vector components; p is a pressure; ρ is a density; M is the molar mass; R is the universal gas constant; μ is the turbulent exchange coefficient; T is the temperature, ε is the coefficient describing the deviation of pressure superheated steam from the ideal gas.

* Corresponding author: cheese_05@mail.ru

We assume that the air environment is initially in quiescence. Therefore, the initial conditions have the form:

$$u = 0,\ w = 0,\ P = P_a, \quad (3)$$

where P_a is the initial pressure.

The system of equations (1) and (2) are considered with the following boundary conditions:

– on an impervious boundary

$$\rho_w \eta \frac{\partial v_x}{\partial n} = \tau_{x,b}(t),\ \rho_w \eta \frac{\partial v_y}{\partial n} = \tau_{y,b}(t),\ \rho_w \eta \frac{\partial v_z}{\partial n} = \tau_{z,b}(t),$$

$$\mathbf{V}_n = 0,\ P'_n = 0;$$

– on the lateral permeable borders

$$\frac{\partial v_x}{\partial n} = 0,\ \frac{\partial v_y}{\partial n} = 0,\ \frac{\partial v_z}{\partial n} = 0,\ \frac{\partial P}{\partial n} = 0;$$

– on the source

$$v_x = U,\ v_y = V, v_z = W,\ \frac{\partial P}{\partial n} = 0;$$

where U, V, W are components of velocity vector on the source; τ_x, τ_z are components of tangential shear stress.

The system of Navier-Stokes' equations in the cylindrical coordinate system has the form:

$$\frac{\partial v_r}{\partial t} + v_r \frac{\partial v_r}{\partial r} + \frac{v_\theta}{r}\frac{\partial v_r}{\partial \theta} - \frac{v_\theta^2}{r} + v_z \frac{\partial v_r}{\partial z} = -\frac{1}{\rho}\frac{\partial P}{\partial r} +$$
$$+\mu\left(\frac{\partial}{\partial r}\left(\frac{1}{r}\frac{\partial(r v_r)}{\partial r}\right) + \frac{1}{r^2}\frac{\partial^2 v_r}{\partial \theta^2} - \frac{2}{r^2}\frac{\partial v_\theta}{\partial \theta} + \frac{\partial^2 v_r}{\partial z^2}\right) + f_r,$$

$$\frac{\partial v_\theta}{\partial t} + v_r \frac{\partial v_\theta}{\partial r} + \frac{v_\theta}{r}\frac{\partial v_\theta}{\partial \theta} + \frac{v_\theta v_r}{r} + v_z \frac{\partial v_\theta}{\partial z} = -\frac{1}{r\rho}\frac{\partial P}{\partial \theta} +$$
$$+\mu\left(\frac{\partial}{\partial r}\left(\frac{1}{r}\frac{\partial(r v_\theta)}{\partial r}\right) + \frac{1}{r^2}\frac{\partial^2 v_\theta}{\partial \theta^2} + \frac{2}{r^2}\frac{\partial v_r}{\partial \theta} + \frac{\partial^2 v_\theta}{\partial z^2}\right) + f_\theta,$$

$$\frac{\partial v_z}{\partial t} + v_r \frac{\partial v_z}{\partial r} + \frac{v_\theta}{r}\frac{\partial v_z}{\partial \theta} + v_z \frac{\partial v_z}{\partial z} = -\frac{1}{\rho}\frac{\partial P}{\partial z} +$$
$$+\mu\left(\frac{1}{r}\frac{\partial}{\partial r}\left(r\frac{\partial v_z}{\partial r}\right) + \frac{1}{r^2}\frac{\partial^2 v_z}{\partial \theta^2} + \frac{\partial^2 v_z}{\partial z^2}\right) + f_z. \quad (4)$$

The system of equations (4) in the case of axial symmetry has the form:

$$\frac{\partial v_r}{\partial t} + v_r \frac{\partial v_r}{\partial r} - \frac{v_\theta^2}{r} + v_z \frac{\partial v_r}{\partial z} = -\frac{1}{\rho}\frac{\partial P}{\partial r} +$$
$$+\mu\left(\frac{\partial}{\partial r}\left(\frac{1}{r}\frac{\partial(r v_r)}{\partial r}\right) + \frac{\partial^2 v_r}{\partial z^2}\right) + f_r,$$

$$\frac{\partial v_\theta}{\partial t} + v_r \frac{\partial v_\theta}{\partial r} + \frac{v_\theta v_r}{r} + v_z \frac{\partial v_\theta}{\partial z} =$$
$$= \mu\left(\frac{\partial}{\partial r}\left(\frac{1}{r}\frac{\partial(r v_\theta)}{\partial r}\right) + \frac{\partial^2 v_\theta}{\partial z^2}\right) + f_\theta,$$

$$\frac{\partial v_z}{\partial t} + v_r \frac{\partial v_z}{\partial r} + v_z \frac{\partial v_z}{\partial z} = -\frac{1}{\rho}\frac{\partial P}{\partial z} +$$
$$+\mu\left(\frac{1}{r}\frac{\partial}{\partial r}\left(r\frac{\partial v_z}{\partial r}\right) + \frac{\partial^2 v_z}{\partial z^2}\right) + f_z. \quad (5)$$

The continuity equation (2) in the cylindrical coordinate system has the form:

$$\frac{\partial \rho}{\partial t} + \frac{1}{r}\frac{\partial(\rho r v_r)}{\partial r} + \frac{1}{r}\frac{\partial(\rho v_\theta)}{\partial \theta} + \frac{\partial(\rho v_z)}{\partial z} = 0,$$

The equation (5) in the case of axial symmetry has the form:

$$\frac{\partial \rho}{\partial t} + \frac{1}{r}\frac{\partial(\rho r v_r)}{\partial r} + \frac{\partial(\rho v_z)}{\partial z} = 0. \quad (6)$$

3 The state equation

The main equation for the superheated steam is the equation for the specific Gibbs energy, consisting of two parts – relating to ideallocation as γ^0 and describing real component γ^r:

$$g(\mathrm{p},\mathrm{T}) / RT = \gamma(\pi,\tau) = \gamma^0(\pi,\tau) + \gamma^r(\pi,\tau) \quad (7)$$

$$\gamma^0 = \ln \pi + \sum_{i=1}^{9} n_i^0 \tau^{J_i^0} \quad (8)$$

$$\gamma^r = \sum_{i=1}^{43} n_i \pi^{I_i} (\tau - 0{,}5)^{J_i} \quad (9)$$

where $\pi = p / p^*$ and $\tau = T^* / T$, $p^* = 1MPa$, $T^* = 540\,\mathrm{K}$.

From the relation for the specific volume

$$vP / RT = \pi \gamma_\pi,\ \gamma_\pi = \left(\frac{\partial \gamma}{\partial \pi}\right)_\tau$$

follows the next:

$$PM = \left(1 + \sum_{i=1}^{43} n_i I_i (p / p^*)^{I_i} (T^*/T - 0{,}5)^{J_i}\right)\rho RT.$$

4 Turbulence model

We used the model of Abramovich-Sekundow, which takes into account very important factors such as the presence of hard walls, the prehistory of the flow, convective and diffusion turbulent pulsation transfer:

$$\frac{\partial v_{турб}}{\partial t} + \sum_{i=1}^{3} v_i \frac{\partial v_{турб}}{\partial x_i} = \sum_{i=1}^{3} \frac{\partial}{\partial x_i}\left(\left(v_{мол} + k v_{турб}\right)\frac{\partial v_{турб}}{\partial x_i}\right) +$$
$$+ v_{турб} f\left(\frac{v_{турб}}{8 v_{мол}}\right) D - \gamma S,\ S = \frac{v_{турб}\left(v_{мол} + \beta v_{турб}\right)}{L_{\min}^2},$$
$$D = \sqrt{\sum_{i=1}^{3}\sum_{j} \frac{\partial v_i}{\partial x_j}\left(\frac{\partial v_i}{\partial x_j} + \frac{\partial v_j}{\partial x_i}\right)}, \quad (10)$$
$$f(z) = 0{,}2\frac{z^2 + 1{,}47z + 0{,}2}{z^2 - 1{,}47z + 1},$$

where $k = 2$, $\gamma = 50$, $\beta = 0{,}06$, $L_{\min}$ is the minimum distance to a solid wall; $v_{мол}$ is the molecular viscosity, $v_{турб}$ is the turbulent viscosity.

Turbulence model (10) is considered with the boundary condition $\left(v_{турб}\right)'_n (t,x,y,z)\big|_{(x,y,z)\in\Gamma} = 0$.

Complete model (10) with initial condition: $\left(v_{турб}\right)(t,x,y,z) = v_0$.

The next equation is used to calculate values of dynamic viscosity $v_{мол}$:

$$v_{мол} = \bar{v}_0(\tau)\cdot\bar{v}_1(\tau,\delta)\cdot v^*, \quad (11)$$

where v^* = 55,071·10^{-6} Pa·s; $\tau = T/T^*$; T is the temperature, [K]; T^* = 647,226 K; $\delta = \rho/\rho^*$; ρ is a density, [kg/m^3]; ρ^*=317,763 kg/m^3; $\bar{v}_0 = v_0/v^*$; v_0 is the dynamic viscosity of water vapor in the limit of zero density

$$\bar{v}_0(\tau) = \tau^{0,5}\left(\sum_{i=0}^{3} H_i\tau^{-i}\right)^{-1}, \quad (12)$$

where $H_0 = 1$, $H_1 = 0,978197$, $H_2 = 0,579829$, $H_3 = -0,202354$.

$$\bar{v}_1(\tau,\delta) = \exp\left(\delta\sum_{i=0}^{5}\sum_{j=0}^{6} H_{i,j}(1/\tau-1)^i(\delta-1)^j\right). \quad (13)$$

5 The heat conductivity problem

Thermal processes in turbine $\bar{G}$ are described by the heat conduction equation:

$$c\rho\frac{\partial T}{\partial t} + v_x\frac{\partial T}{\partial x} + v_y\frac{\partial T}{\partial y} + v_z\frac{\partial T}{\partial z} = \frac{\partial}{\partial x}\left(\lambda\frac{\partial T}{\partial x}\right) + \frac{\partial}{\partial y}\left(\lambda\frac{\partial T}{\partial y}\right) + \frac{\partial}{\partial z}\left(\lambda\frac{\partial T}{\partial z}\right) + q_v(P), \quad (14)$$

which can be written in the case of axial symmetry as: $\rho crT'_t + rv_rT'_r + rv_zT'_z = r(\lambda T'_z)'_z + (\lambda rT'_r)'_r + rq_v(P)$. (15)

In the system (14), (15): T is the temperature, $[K]$; λ is the coefficient of conductivity of water; ρ is the density of the metal; c is the coefficient of the metal conductivity; r is the polar radius; $q_v(P)$ is the source function. We will consider equation (15) with boundary conditions of third order:

$$T'_{\mathbf{n}}(x,r,t) = \alpha_{\mathbf{n}}T + \beta_{\mathbf{n}}, \quad (16)$$

where $\mathbf{n}$ is the normal vector to the surface $\bar{G}$.

We will consider equation (15) with the initial condition:

$$T(x,r,0) = T_0(x,r), \ (x,r)\in\bar{G}. \quad (17)$$

The next equation is used to calculate the coefficient of conductivity of the water and steam in the international practice:

$$\lambda = \lambda_0(\tau) + \lambda_1(\delta) + \lambda_2(\tau,\delta) \quad (18)$$

where λ is the conductivity, W/(m·K); $\tau = T/T^*$; T is the absolute temperature, K; T^* = 647,256 K; $\delta = \rho/\rho^*$; ρ is a density, kg/m^3; ρ^* = 317,7 kg/m^3. The conductivity of water steam in the state of ideal gas $\lambda_0(\tau)$ is determined by the equation:

$$\lambda_0(\tau) = \tau^{0,5}\sum_{\kappa=0}^{3} a_\kappa\tau^\kappa, \quad a_0 = 0{,}01028; \quad a_1 = 0{,}02996;$$

$$a_2 = 0{,}01561; \quad a_3 = -0{,}004224.$$

The function $\lambda_1(\delta)$ is defined:

$$\lambda_1(\delta) = b_0 + b_1\delta + b_2\exp\{B_1(\delta+B_2)^2\},$$

where b_0 = –0,39707; b_1 = 0,4; b_2 = 1,06; B_1 = –0,171; B_2 = 2,392, and the function $\lambda_2(\tau,\delta)$ is defined:

$$\lambda_2(\tau,\delta) = \left(\frac{d_1}{10} + d_2\right)\delta^{9/5}\exp\left[C_1(1-\delta^{14/5})\right] + d_3S\delta^Q\exp\left[\left(\frac{Q}{1+Q}\right)(1-\delta^{1+Q})\right] + d_4\exp\left(C_2\tau^{3/2} + \frac{C_3}{\delta^5}\right),$$

where Q and S are functions of the magnitude $\Delta\tau = |\tau-1| + C_4$:

$$Q = 2 + C_5/\Delta\tau^{0,6}; \quad S = \begin{cases} 1/\Delta\tau & for\ \tau\ge 1; \\ C_6/\Delta\tau^{0,6} & for\ \tau < 1. \end{cases}$$

The coefficients d_i and C_i have the next values:

d_1 = 0,0701309; d_2 = 0,0118520; d_3 = 0,00169937; d_4 = –1,0200; C_1 = 0,642857;

C_2 = –4,11717; C_3 = –6,17937; C_4 = 0,00308976; C_5 = 0,0822994; C_6 = 10,0932.

The equation (14) is applicable with the following values of the temperature and pressure: $p \le 100$ MPa for $0 \le T \le 500$ ºC; $p \le 70$ MPa for $500 < T \le 650$ ºC; $p \le 40$ MPa for $650 < T \le 800$ ºC.

The error values in the liquid field at the temperatures 25-200 ºC and pressures up to 5 MPa is equaled to 1,5 % at higher temperatures up to 300ºC – 2 %. For the water steam at temperatures up to 550 ºC under pressure 0,1 MPa error is equal to 1,5 %, under pressures up to 40 MPa – 3 %. The equation (18), in contrast to the theoretical conclusions, determines not infinite, but finite conductivity value at the critical point, which does not allow to estimate the error values near its critical point.

6 Splitting schemes into physical processes

The initial aerodynamic model are divided into three subtasks according to the of the pressure amendment method [1–4]. The first subtask is representative by the diffusion-convection problem on the basis of which the velocity field are calculated at the intermediate time step:

$$\frac{\tilde{v}_r - v_r}{\tau} + v_r\frac{\partial v_r}{\partial r} - \frac{v_\theta^2}{r} + v_z\frac{\partial v_r}{\partial z} = \mu\left(\frac{\partial}{\partial r}\left(\frac{1}{r}\frac{\partial(rv_r)}{\partial r}\right) + \frac{\partial^2 v_r}{\partial z^2}\right) + f_r,$$

$$\frac{\hat{v}_\theta - v_\theta}{\tau} + v_r\frac{\partial v_\theta}{\partial r} + \frac{v_\theta v_r}{r} + v_z\frac{\partial v_\theta}{\partial z} = \mu\left(\frac{\partial}{\partial r}\left(\frac{1}{r}\frac{\partial(rv_\theta)}{\partial r}\right) + \frac{\partial^2 v_\theta}{\partial z^2}\right) + f_\theta,$$

$$\frac{\tilde{v}_z - v_z}{\tau} + v_r\frac{\partial v_z}{\partial r} + v_z\frac{\partial v_z}{\partial z} = \mu\left(\frac{1}{r}\frac{\partial}{\partial r}\left(r\frac{\partial v_z}{\partial r}\right) + \frac{\partial^2 v_z}{\partial z^2}\right) + f_z.$$

Calculation of pressure distribution (the second subtask):

$$\frac{\partial \rho}{\partial t}+\frac{1}{r}\frac{\partial(\rho r\tilde{v}_r)}{\partial r}+\frac{\partial(\rho\tilde{v}_z)}{\partial z}=\frac{\tau}{r}\frac{\partial}{\partial r}\left(r\frac{\partial P}{\partial r}\right)+\tau\frac{\partial^2 P}{\partial z^2}$$

or with consideration the state equation

$$r\frac{\partial(PM/\varepsilon RT)}{\partial t}+\frac{\partial(\rho r\tilde{v}_r)}{\partial r}+r\frac{\partial(\rho\tilde{v}_z)}{\partial z}=$$

$$=\tau\frac{\partial}{\partial r}\left(r\frac{\partial P}{\partial r}\right)+\tau r\frac{\partial^2 P}{\partial z^2}.$$

The simplified hydrostatic model of water motion is used as the initial approximation for this problem. It is significantly decreased the calculation time.

The third subtask is determined the distribution of the steam velocity at the upper time layer by the explicit formulas:

$$\frac{\hat{v}_r-\tilde{v}_r}{\tau}=-\frac{1}{\rho}\frac{\partial P}{\partial r},\ \frac{\hat{v}_z-\tilde{v}_z}{\tau}=-\frac{1}{\rho}\frac{\partial P}{\partial z}.$$

where τ is a time step; v is a value of velocity field on the previous time layer; $\tilde{v}$ is a value of velocity field on the intermediate time layer; $\hat{v}$ is on the current time layer.

7 Software description

The input model parameters are the next: rotation frequency of the turbine, the range of output speeds at the turbine blades, number of blades, the number of nozzle channels, the width of the nozzle channel to the impeller, the vapor pressure of the working wheels. The calculations were performed in three functional areas 1 – 3. The location of areas which were determined by the velocity field is given in Fig. 1.

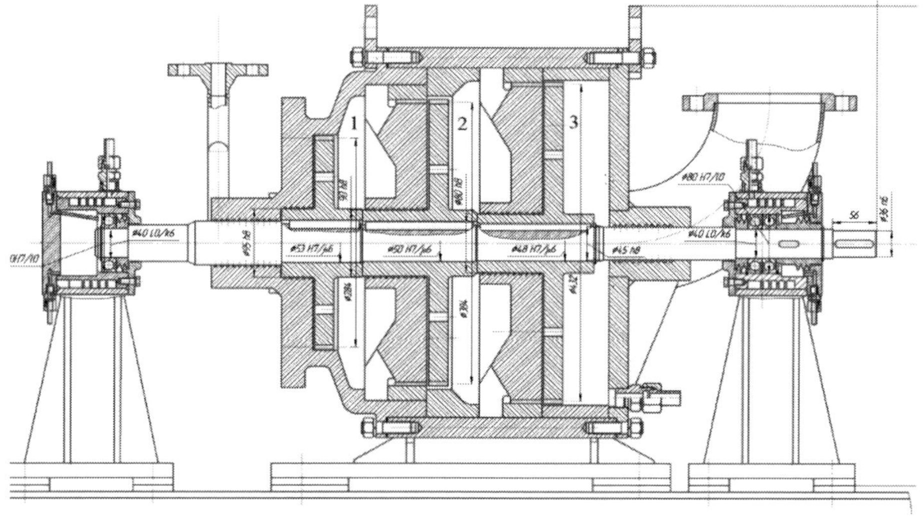

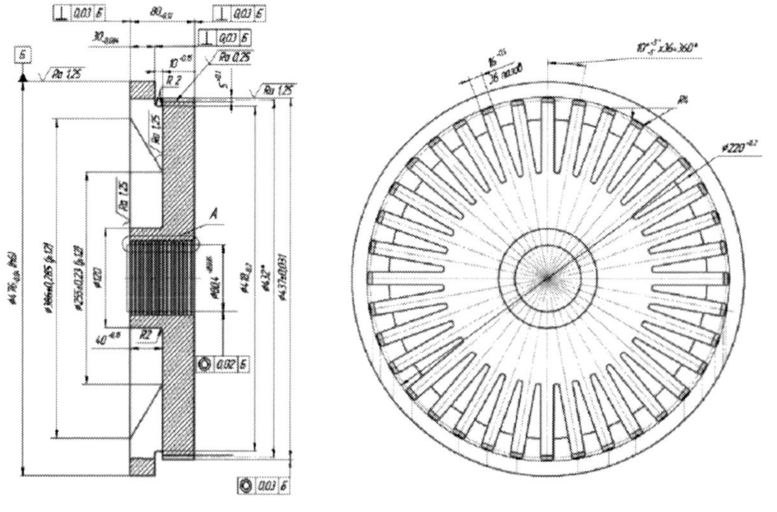

Fig. 1. The geometry of turbine housing and the nozzle grid.

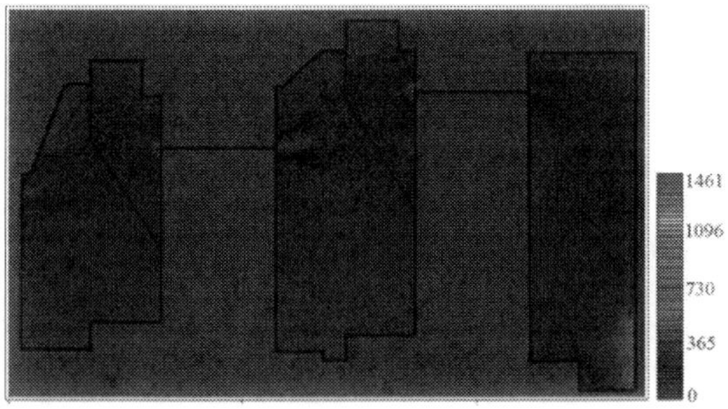

Fig. 2. The velocity of steam movement in working chamber.

The flow pattern inside a pair of working chambers (shown in a longitudinal and radial component of the velocity vector) is given in Fig. 2.

Color reflects the speed of movement of the steam. The units of measurement are meters per second. The developed software allows to calculate the movement of air environment within areas with cylindrical symmetry and allows to identify patterns of velocity fields and pressure within axial steam turbines. The permeability of the medium is used in the mathematical model to describe the nozzle geometry of the lattice. The developed software can calculate the steam movement in multiple stages at the same time.

Distribution fields of the main physical values calculated for inside the first working chamber (1 is the longitudinal and radial component of the velocity vector, m/s; 2 is the rotation speed, rev/min; 3 is the pressure, atm.; 4 is the turbulent viscosity, Pa·s; 5 is the temperature, °C; 6 is the density, kg/m^3) are given in Fig. 3.

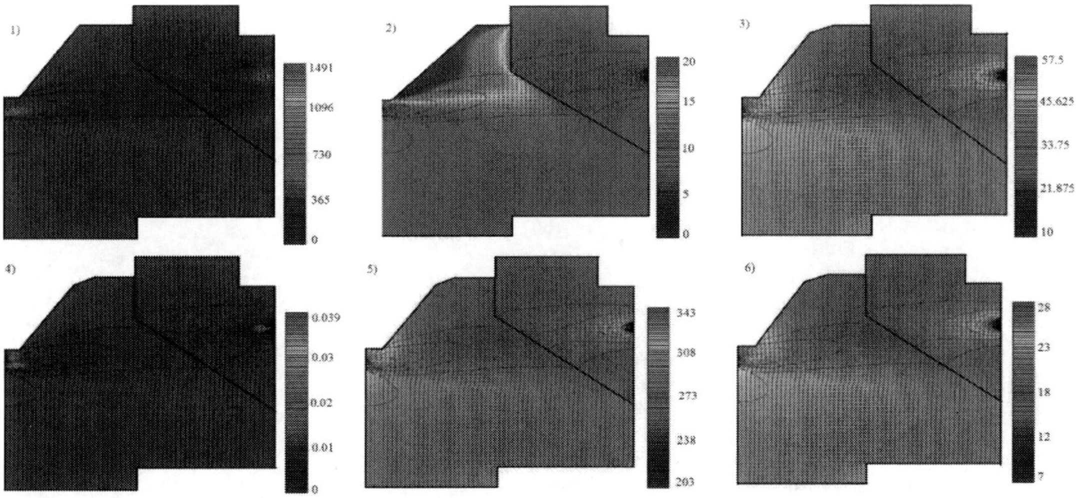

Fig. 3. Distribution fields of the main calculated physical quantities inside for the first working chamber

8 Conclusion

The mathematical model of aerodynamic processes in the steam turbine was developed. The main equations are the next: the Navier-Stokes equations, continuity equations and state equations. The initial system of equations was determined in cylindrical coordinate system taking into account the axial symmetry of steam flow field in the turbine. The splitting schemes into physical processes (the pressure amendment method) was used for approximation of problem at the temporary variable. According to this method, the solution problem is calculated in three steps. The solution of diffusion-convection problem with high Peclet numbers is required in calculation of velocity field components without pressure. New difference schemes were obtained on the basis of the modification of the "cabaret" scheme and used for this problem solution. Owning to these difference schemes, the accuracy of diffusion-convection problem solution is increased in 2,5 – 3 times, compared to the traditional "cabaret" schemes, which are effective for solution such problems. The software implementation was performed on the basis of the developed algorithms. Results of numerical calculations of aerodynamics processes in the steam turbine were obtained.

This work was financially supported by the FTPRD «Research and Development for the Development of the Scientific and Technological Complex of Russia for 2014-2020», Project No. 14.577.21.0260.

References

1. A.A. Samarsky, E.S. Nikolaev, *Methods of solving grid equations* (M.: Science, 1978)
2. A.I. Sukhinov, A.E. Chistyakov, J. Mathematical modeling, **24**, no. 1, 3-20 (2012)
3. A.N. Konovalov, J. Differential equations, **40**, no. 7, (2004)

* Corresponding author: cheese_05@mail.ru

4. A.N. Konovalov. Siberian mathematical journal, **43**, no. 3, 552 (2002)
5. A.I. Sukhinov, A.E. Chistyakov, A.V. Shishenya J. Mathematical modeling, **25**, no. 11, 53-64 (2013)
6. A.I. Sukhinov, A.E. Chistyakov, N.A. Fomenko, J. Izvestiya SFedU engineering sciences, no. 4, 87-98 (2013)
7. A.A. Samarskiy, *Theory of difference schemes*. (Nauka, Moscow, 1989)
8. A.A. Samarskiy, A.V. Gulin, *Numerical methods* (Nauka, Moscow, 1989)
9. K.A. Beklemishev, I.B. Petrov, A.V. Favorsky, J. Mathematical modeling, **25**, no. 11, 3-16 (2013)
10. I.B. Petrov, A.V. Favorsky, A.V. Sannikov, I.E. Kvasov, J. Mathematical modeling, **25**, no. 2, 42-52 (2013)
11. A.I. Sukhinov, A.E. Chistyakov, A.A. Semenyakina A.V. Nikitina, J. Computational methods and programming: new computing technology, **16**, no. 2, 256-267 (2015)
12. A.V. Nikitina, A.A. Semenyakina, A.E.Chistyakov, E.A. Protsenko, I.V. Yakovenko, J. Fundamental research, no. 12-3, 500-504 (2015)
13. A.E. Chistyakov, D.S. Hachunts, A.V. Nikitina, E.A. Protsenko, I.Yu. Kuznetsova, J. Modern problems of science and education, no. 1-1, 1786 (2015)
14. A.I. Sukhinov, A.V. Nikitina, A.A. Semenyakina, E.A. Protsenko, J. Engineering journal of Don, **38**, no. 4 (38), 52 (2015)
15. A.V. Nikitina, Y.A. Abramenko, A.E. Chistyakov, J. Informatics, computer science and engineering education, no. 3 (23), 49-55 (2015)
16. A.E. Chistyakov, A.V. Nikitina, G.A. Ougolnitsky V.M. Puchkin, I.S. Semenov, A.I. Sukhinov, A.B. Usov, J. Game Theory and Applications, **17**, 37-48 (2015)
17. A.E. Chistyakov, A.V.Nikitina, V.V. Sumbaev, J. Herald of computer and information technologies, no. 8 (146), 3-7 (2016)
18. A.V. Nikitina, A.A. Semenyakina, A.E. Chistyakov, J. Herald of computer and information technologies, no. 7 (145), 3-8 (2016)
19. A.I. Sukhinov, A.E. Chistyakov, A.A. Semenyakina A.V. Nikitina, J. Computer researches and modeling, **8**, no. 1, 151 – 168 (2016)
20. A.I. Sukhinov, A.V. Nikitina, A.A. Semenyakina, A.E. Chistyakov, *CEUR Workshop Proceedings*, 10th Annual International Scientific Conference on Parallel Computing Technologies, PCT 2016; **1576**, 308-319, (2016)
21. A.V. Nikitina, A.I. Sukhinov, G.A. Ugolnitsky, A.B. Usov, A.E. Chistyakov, M.V. Puchkin, I.S. Mathematical Models and Computer Simulations. – 2017. 9 (1), 101-107.
22. A.I. Sukhinov, A.E. Chistyakov, I.I. Levin, I.S. Semenov, A.V. Nikitina, A.A. Semenyakina *5th International Conference on Informatics, Electronics and Vision (ICIEV)*, 1128 – 1133 (2016)
23. A.I. Sukhinov, A.E. Chistyakov, A.V. Nikitina A.A. Semenyakina, I. Korovin, G. Schaefer *Conference on Informatics, Electronics and Vision (ICIEV)*. 1134 – 1139, (2016)
24. A.I. Sukhinov, A.E. Chistyakov, A.A. Semenyakina, A.V. Nikitina Computer researches and modeling. **8**, no. 1., 151 – 168 (2016)
25. A.I. Sukhinov, D.S. Khachunts, A.E. Chistyakov, Computational Mathematics and Mathematical Physics, **55**, No. 7., 1216-1231 (2015)
26. A.I. Sukhinov, A.E. Chistyakov, A.V. Shishenya Mathematical Models and Computer Simulations. **6**, No. 3.,324-331 (2014)
27. N. Buzalo, P. Ermachenko, T. Bock, A. Bulgakov A.Chistyakov, A. Sukhinov, E. Zhmenya, N. Zakharchenko Procedia Engineering, **85**, No. C., 84-93 (2014)
28. A.I. Sukhinov, A.E. Chistyakov, E.F. Timofeeva, A.V. Shishenya, Mathematical Models and Computer Simulations, **5**, No. 2, 122-129, 2013
29. A.I. Sukhinov, A.E. Chistyakov, Mathematical Models and Computer Simulations, **4**, Iss. 4, 398-409, (2012)

Regulator synthesis for the self-sensing control system of the proportional electromagnet dc based on reduced-order models

Danil Shaykhutdinov[1,*], *Nikolay* Gorbatenko[1], *Yuri* Manackov[1], and *Konstantin* Shirokov[1]

[1]SRSPU (NPI), Department of Information and Measuring Systems and Technologies, 346428 Novocherkassk, Russia

Abstract. Modern DC proportional solenoid control systems use current values in magnetizing coils, or specialized sensors for position determination. These methods do not provide the possibility of accurate control and diagnostics, in case of aiming to the miniaturization of finished devices. In this article, it is proposed to use methods of self-sensory identification of the moving element position based on the method of the full-scale-model experiment. The functioning of the method is based on the electromagnet model obtained by the reduced-order model approach. These models have an advantage in the calculation speed in comparison with finite element models and have an advantage in accuracy in comparison with analytical models. Ansys Electronics is used to obtain the model. The electromagnetic control system is proposed. Its model is implemented in the system Matlab Simulink. Synthesis of PID-regulator parameters using Matlab is performed. The results of a study of a control system for a given displacement with the aid of the obtained control system are presented.

1 Introduction

An important problem of modern science is the provision of high reliability of industrial and household technical systems based on electromagnetic mechanisms, for example [1, 2]. This is one of the main conditions for preventing system accidents and catastrophes. Existing methods and tools of monitoring, control and diagnosis these systems do not use achievements of modern information technologies based on modelling and methods of data mining. Integration of diagnostic functions into the device control system can significantly improve accuracy and reliability. The solution of this problem requires the development of the foundations for the functioning of systems for the diagnosis of electromagnetic mechanisms. Solving the problem of creating theoretical foundations and modern intelligent diagnostic mechanisms will help lay the foundation for the further development of intelligent systems in the field of smart factoring [3, 4], so relevant at the present time [5, 6].

To solve this problem, as the first step, it is proposed to develop new control systems with diagnostics functions based on reduced-order mathematical models for assessing the efficiency of device operation.

Purpose of work is to increase the accuracy and efficiency of the operation of electrical systems, improve the reliability of the devices operating in facilities with increased responsibility and those industries where electrical products operate under difficult operating conditions.

Electromagnets, as a transformers, are the most common representatives of electromagnetic structures. Proportional electromagnets are the most complex system. This type of electromagnetic drives requires systems that can control the parameters of the movement of the electromagnets of the elements. In the development of new products, it is important to adjust the regulators and make a comprehensive testing of electromagnets [7, 8]. In addition, high reliability requirements are designated for this class of devices when it is used as part of increasing the responsibility of systems. The complexity and continuous improvement of magnetic systems currently require a new high-precision approach to controlling the position of moving elements [9]. Automated systems controlling the production of intelligent electric drives are necessary to implement an effective approach. One of the main elements of production systems are the stands for diagnostics and adjusting the parameters of the regulator. When developing the experimental production of new proportional DC electromagnets, this stand can be presented in the form of an automated system for research and complex tests. These operations are designed to improve the performance of processes for optimizing the electromagnets of structures, including using new materials [10] or materials known in limiting regimes [11].

In case of a high complexity of the control object (for example, a current source having a limit of power or configuration of the electromagnetic system of an

* Corresponding author: d.v.shaykhutdinov@gmail.com

electromagnet has a high geometric complexity) or in the absence of complete a priori information about the object, it becomes necessary to apply intelligent control technologies [12, 13]. Currently, a very wide range of publications on intelligent technologies, which confirms the relevance and popularity of these control systems.

2 Materials and Methods

2.1 Self-sensing control method

The proposed approach is based on the following assumptions. The position of the electromagnet movable element x defines a degree of magnetic circuit closure d. The degree of of magnetic circuit closure d determines the level of magnetic flux F at a preset value of magnetopropellent Iw. The level of a magnetic flux has direct interrelation with the flux linkage of the winding of magnetic conductor ψ. Thus, there is physical interrelation between the weber-ampere characteristic of electromagnet $\psi(Iw)$ and the position of its movable element, x: $x \rightarrow d \rightarrow F(Iw) \rightarrow \psi(Iw)$. Based on this assumptions, the automatic control system (Fig. 1) is proposed.

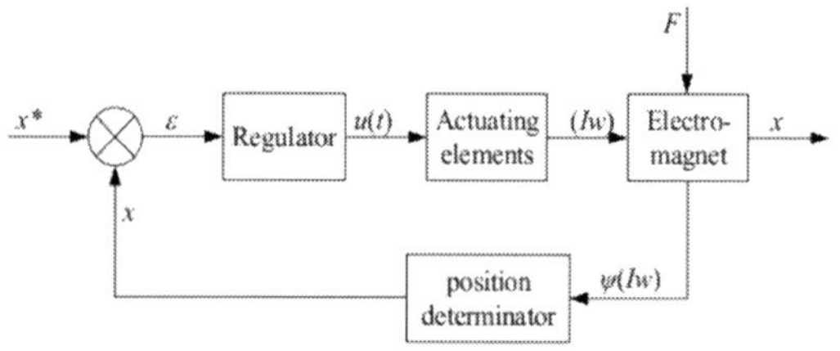

Fig. 1. Automatic control system

The system (Fig. 1) consists of: Regulator – the element that forms the control actions, Actuating elements – the aggregate of the current source and the magnetizing coil of the electromagnet, Elecromagnet – the proportional electromagnet of direct current, position determinator – the element for determining the position of the movable element of the electromagnet. For the formation of the latter, the application of the method of full-scale-model tests is proposed.

2.2 Full-scale-model tests method

Full-scale-model tests method refers to a group of iterative methods for solving inverse problems. Its peculiarity is that the results of the experiment are used both as input data for modeling the field of the system, and as a criterion for estimating the accuracy of the numerical solution of this problem. In general, the iterative process is implemented as a change in the configurable parameter of the model by summing the value of the given parameter obtained in the previous step and some correction value. At that, to determine the value of the parameter to be adjusted at the zero iteration, the value of the results of the physical experiment is used. The correction value is found as a function of the difference between the result of the physical experiment for measuring the initial parameter and the result of the model experiment for determining the same parameter when the value of the parameter being tuned into the model is substituted into the model. The iterative process is completed, if the necessary accuracy of the numerical solution of the identification problem is reached, defined as the difference between the result of the physical experiment in measuring the initial parameter and the result of the model experiment for determining the same parameter (Fig. 2).

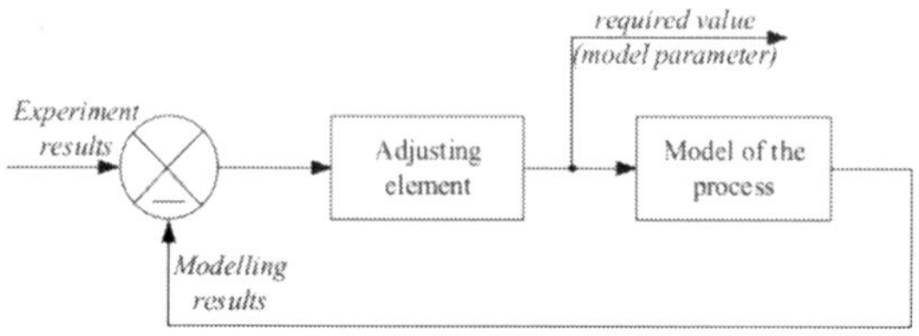

Fig. 2. Realization of measurement based on full-scale-model tests method

We proposed the method for obtaining the measurement information about the state of the technical objects being studied, which differs from the known ones in that the operation for calculating the results of measurements is replaced by the operation of solving the inverse problem of identifying the model of the system under investigation with the subsequent determination of the sought value from the simulation results [14]. In the case under consideration, the initial data (experimental results) and simulation results will be the weber-ampere characteristic values. As the output data – the position of the moving element of the electromagnet. To calculate these parameters, in the first case it is necessary to have a mathematical description of the control object.

2.3 Electromagnetic field mathematical modelling method

The methods of modeling magnetic fields on the basis of the equations of magnetic circuits, analytical methods for calculating the field and the equations of the electromagnetic field are considered. For carrying out of experimental researches the method of mathematical modeling of an electromagnetic field of objects on the basis of system of Maxwell's equations is chosen. In developing the mathematical model, the following assumptions were made: as a result of the investigation of magnetic DC systems or industrial frequency current, it is assumed that there are no free charges in the system; the field is assumed to be quasistationary, which implies the absence of a delay in the electromagnetic wave in the dielectric; the losses due to hysteresis when the ferromagnetic metals are heated are not taken into account because of their insignificance in comparison with the eddy current losses [15]. The magnetic field equation has the form of

$$\operatorname{rot}\vec{H}=\vec{\delta};\ \operatorname{div}\vec{B}=0, \qquad (1)$$

wherein $\vec{B}$, $\vec{H}$ – vectors of the magnetic flux density and magnetic field strength, respectively, $\vec{\delta}$ – the known current density in the excitation winding V_{obm}.

For the system (1) $\vec{B}=\mu_0\vec{H}$ – in the air V_0 and in the excitation winding V_{obm}; $\vec{B}=\mu_{Fe}\vec{H}$ – in ferromagnets V_{Fe}; $B_n^+=B_n^-$; $H_\tau^+=H_\tau^-$.

On the basis of the equation $\operatorname{div}\vec{B}=0$, we introduce the magnetic vector potential $\vec{A}$ by relations $\operatorname{rot}\vec{A}=\vec{B}$; $\operatorname{div}\vec{A}=0$. Using $\vec{A}$, we transform the system (1). As a result, we obtain

$$\operatorname{div}\left(\frac{1}{\mu_0}\operatorname{grad}\vec{A}\right)=\begin{cases}\vec{\delta}-\text{ in } V_{obm};\\ 0-\text{ in } V_0;\end{cases} \qquad (2.1)$$

$$\operatorname{div}\left(\frac{1}{\mu}\operatorname{grad}\vec{A}\right)=0-\text{ in } V_{Fe}. \qquad (2.2)$$

At the interface we have

$$A^+=A^-;\ \frac{1}{\mu^+}\frac{\partial A^+}{\partial n}=\frac{1}{\mu^-}\frac{\partial A^-}{\partial n}. \qquad (3)$$

The solution of system (2) with condition (3) can be conveniently found by using the method of finite differences, the method of boundary elements, the variational version of the finite element method (FEM) [6]. The main criterion in the analysis, due to the complexity of the objects under study, is the possibility of solving nonlinear problems for bodies of complex shape. A finite element method was chosen that provides ample opportunities for describing the complex geometry of the objects under study, taking into account existing nonlinearities in the analysis of field problems, constructing calculation algorithms and automating the process of processing simulation results in modern software complexes. Minimization of the functional corresponding to problem (2), (3) leads to a system of algebraic equations.

$$\sum_k\frac{1}{\mu^k}\sum_j\beta_{ij}^kA_i=\frac{1}{3}\sum_k\mu_0\delta^kS^k,\ k=1,2,\dots,N, \qquad (4)$$

where $\beta_{ij}^k=\iiint_{V^k}grad\psi_i^k\,grad\psi_j^k dV$, ψ_i^k and ψ_j^k – the basic functions of the k-th finite element; S^k – area of the face of the k-th finite element.

But solving the system (4) requires experimentally large amounts of time and computing resources. Thus, this approach can't be directly applied in problems of high-speed systems control, for example proportional electromagnets DC. To solve this problem the using of reduced-order models approach is proposed.

2.4. Reduced-order models

The problem of reducing the number of variables of the finite element model to a number of main components can be solved using the reduced-order model (ROM) approach [16-22]. This approach allows us to take into account only the most important features of the modeling object with the help of several state variables and several ordinary differential equations. As a result of applying this approach, the original matrix of the system (4) is transformed into a simplified one. A simplified matrix can be more effectively used to solve real-time control problems. The obtained results can be integrated into the MatLab Simulink system, for example, to automate the synthesis of the control system regulator. In general view, a decrease in the order of the FEM model can be described as the transformation of some initial system:

$$\begin{cases}\dot{z}(t)=Az(t)+Bg(t);\\ y(t)=Cz(t),\end{cases}$$

wherein z – the state vector, f – the input vector, y – the output vector, to the system of equations of:

$$\begin{cases}\dot{z}_{rom}(t)=A_{rom}z_{rom}(t)+B_{rom}g(t);\\ y(t)=C_{rom}z_{rom}(t),\end{cases} \qquad (5)$$

It should be noted, that the dimension of the vectors of input actions and output values, for system (5) is preserved. To obtain ROM from the full FEM models, various methods are used, which are described in detail in [22]. In particular, the ANSYS Electronics system, which performs finite element modeling of objects [15], allows to obtain ROM using the built-in algorithm based on modal decomposition.

By applying the approach of reduced-order models and solving the result system, we find B_i^0 $(i=1,2,\dots,N)$, where N – a number of measuring transducers. Then, we will following the next calculations sequence:

1. Calculate the square of the norm of the difference $J_i=\left\|B_i^0-B_i^*\right\|^2$.
2. Check the condition $J_i\le\varepsilon$, where ε – accuracy of determining the minimum of the functional J_i.
3. If the condition of point 4 is satisfied, the solution is obtained and it equals $\vec{f}^0$. Otherwise, using the gradient method of minimizing the functional J, we find the next approximation $\vec{f}^1$. Here the ratio $\vec{f}^{k+1}=\vec{f}^k-[\alpha]\operatorname{grad}J^k$ is used, wherein $[\alpha]$ – a diagonal matrix of steps.
4. Return to point 3, using the new $\vec{f}$ values.

3 Results and discussion

3.1. Experiment description

To realize a system for the study of proportional direct current magnets, the National Instruments platform was chosen. The industrial computer is based on the NI PXIe-1078 chassis under the control of the NI PXI-8133 controller, with the addition of the NI PXIe-6341 multifunctional device, NI PXIe-4071 multimeter, NI

PXI-4130 power supply. To control the operation of the electromagnet, a dedicated NI sbRIO-9636 board is used, operating on the basis of FPGA and a real-time controller. To determine the reference values of the position of the moving element, a distance sensor of type LS5-15 / 10-232-2-V- (12-24) -A is used. The object of research is an electromagnet of the KTS type P25A00-24.

To implement the software, the NI LabVIEW programming environment is used. To implement the model-based tests, Ansys Electronics software package was used, which uses the finite element method to calculate the magnetic fields. In Fig. 3 shows the appearance of the electromagnet under investigation and its model.

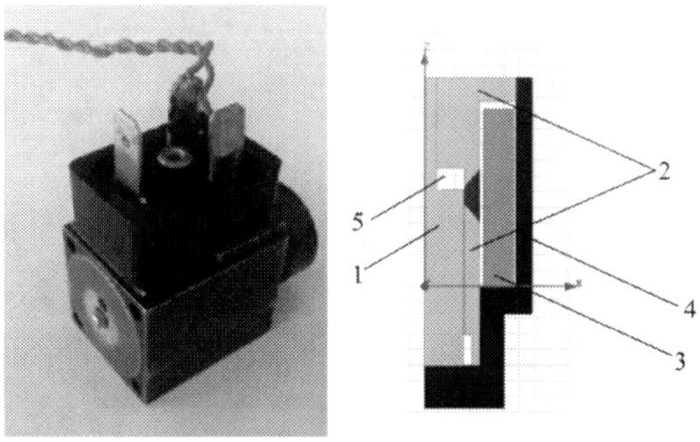

a b

Fig. 3. Proportional electromagnet DC: a – real look, b – two-dimensional axisymmetric model: 1 – anchor, 2 – magnetic core, 3 – magnetizing coil, 4 – housing, 5 – air gap

The experiment was carried out as follows. After connecting the electromagnet's electrical contacts to the system for research, the NI PXIe-4071 measures the electrical parameters of its windings. At the second stage, a pulse signal is sent to the working winding of the electromagnet (using the sbRIO-9636 FPGA, by means of a controlled current source) sufficient to make the complete armature stroke, senseless measurement of its weber-ampere characteristic (without setting additional measuring winding – and voltage on the working winding by means of the NI PXIe-6341 ADC with subsequent calculation of the weber-ampere characteristic in NI PXI-8133) and fixation of the anchor position versus time (using LS5 and NI PXI-8133). At the next stage, the magnetic parameters of the electromagnet model are tuned according to the results of the method of the full-scale-model experiment.

3.2. Experiment results

As a result of the physical experiment, the dependence of the electromagnet movable element displacement x at the time t was obtained (Fig. 4). This dependence is obtained under the condition that there is no purposeful regulation of the current values in the magnetizing coils of the electromagnet.

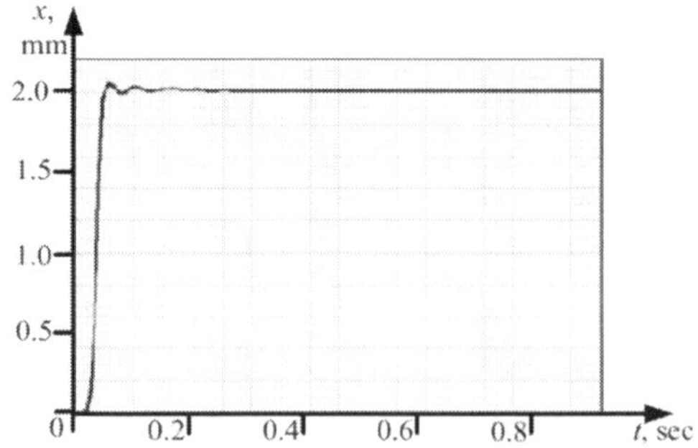

Fig. 4. The variation of the movable element position. The electromagnet operating cycle

As a result of the physical experiment, a weber-ampere characteristic of the electromagnet operating cycle was also obtained. The Weber-ampere characteristic of the electromagnet operating cycle was subjected to a study using the full-scale-model test method described above. An example of the result of applying the method of full-scale-model tests to obtain a model adequate to the results of measuring the weber-ampere characteristic of the electromagnet operating cycle is shown in Fig. 5.

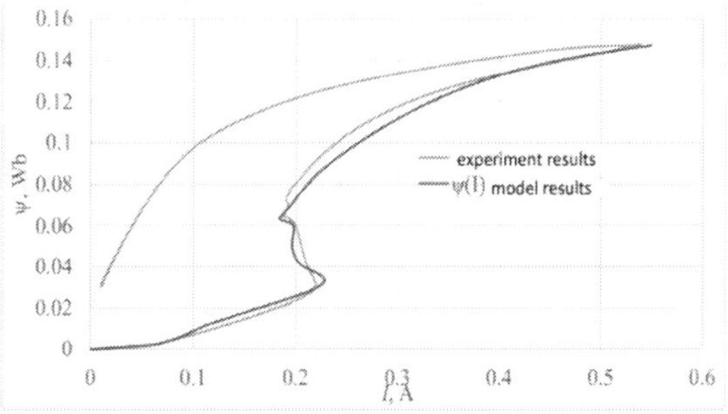

Fig. 5. The model adjustment using the method of full-scale-model tests. Results example

As it seen in Fig. 4 the final phase of the motion has an oscillatory character. This point is dangerous, as it reduces the duration of equipment operation. Where is a requirement to reduce the oscillation.

On the basis of the obtained mathematical model of an electromagnet, a regulator of a control system of a proportional DC electromagnet is synthesized. To configure its parameters, it is suggested to use the Matlab package. The controller is selected as the type - PID controller. To solve this problem, the following actions must be performed. The variables *Kp*, *Ki*, *Kd* are initialized in the Matlab command window according to the Ziegler-Nichols method. Next, the parameters of the PID Controller block are set by entering the *Kp* variable in the Proportional parameter field, in the Integral – *Ki* field, and in the Derivative – *Kd* field. The restrictions imposed on the output of the system (the Transfer Fcn block) are specified. A corridor is installed, within which the input signal of the NCD Outport must be located. Next, automatic adjustment of the control parameters is performed by optimization. At the end of the optimization process, the resulting values of the custom variables are used in Matlab Simulink.

The result of modeling the moving element displacement is shown in Fig. 6.

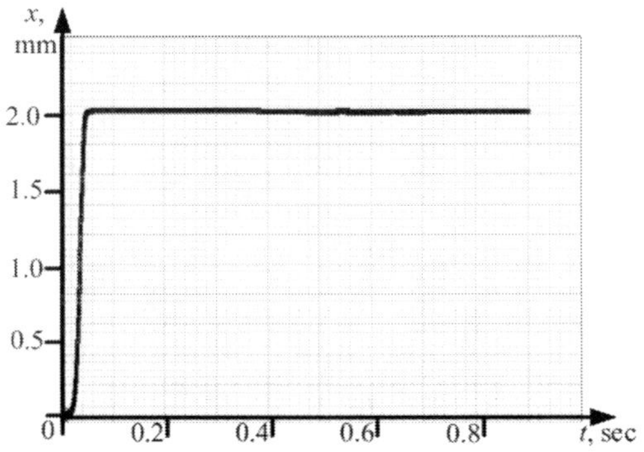

Fig. 6. Modeling the moving element displacement. Results example

3 Conclusion

The problem of self-sensing control of proportional DC electromagnets is considered. Self-sensor principle allows to exclude control basing on the device sensors, without additional sensors. To determine the position of the moving element, the weber-ampere characteristic of the electromagnet operating cycle is used. The relationship between the position of the moving element and the weber-ampere characteristic is established using the method of full-scale-model tests. To implement this method, it is proposed to use the finite element method. The corresponding system of equations undergoes a transformation using the approach of reduced-order models. This approach allows us to reduce the time spent on solving the system of equations and increase the efficiency of controlling the electromagnet.

This article is based on results of the project No. SP 4108.2015.1, implemented under the program "Scholarship provided by the President of the Russian Federation for young scientists and graduate students engaged in advanced research and development in priority areas of modernization of the Russian economy" using equipment of shared facility "Diagnosis and energy-efficient electrical equipment" (NPI).

References

1. V. I. Dubrov, D. V. Shaykhutdinov; K. M. Shirokov, S.V. Akhmedov, N. I. Gorbatenko, *2015 International Siberian Conference on Control and Communications, SIBCON 2015*, 7147317 (2015).
2. V. I. Dubrov, D. V. Shaykhutdinov, K. M. Shirokov, Pro. Eng., **129**, 184-190 (2015).
3. D. Shaykhutdinov, R. Leukhin, I. Stensenko, *2016 2nd international conference on industrial engineering, applications and manufacturing (ICIEAM)*, 7910945 (2016).
4. D.V. Shaykhutdinov, N.I. Gorbatenko, S.V. Akhmedov, R.I. Leukhin, *2015 International Siberian Conference on Control and Communications, SIBCON 2015*, 7147329 (2015).
5. V. Dubrov, R. Oganyan, D. Shaykhutdinov, *2016 2nd international conference on industrial engineering, applications and manufacturing (ICIEAM)*, 7910999 (2016).
6. R.G. Oganyan, N.D. Narakidze, D.V. Shaykhutdinov, N.I. Gorbatenko, S.O. Maksuta, *International conference on mechanical engineering, automation and control systems 2016*, 012083 (2017).
7. R.I. Leukhin, D.V. Shaykhutdinov, V.I. Leukhin, Pro. Eng., **150**, 347-353 (2016).
8. D.V. Shaykhutdinov, R.I. Leukhin, K.M. Shirokov, Pro. Eng., **129**, 749-753 (2015).
9. D.V. Shaykhutdinov, N.I. Gorbatenko, N.D. Narakidze, A.S. Vlasov, I. A.Stetsenko, *International conference on mechanical engineering, automation and control systems 2016*, 012091 (2017).
10. N.I. Gorbatenko, V.V. Grechikhin, D.V. Shaikhutdinov, Metal Sci & Heat Treatment, **56 (11-12)**, 609-613 (2015).
11. D.V. Shaikhutdinov, N.I. Gorbatenko, K.M. Shirokov, Metal Sci & Heat Treatment, **56 (11-12)**, 618-620 (2015).
12. R.I. Leukhin, D.V. Shaykhutdinov, K.M. Shirokov, N.D. Narakidze, A.S. Vlasov, *International conference on mechanical engineering, automation and control systems 2016*, 012008 (2017).
13. D. Shaykhutdinov, N. Gorbatenko, R. Leukhin, R J Appl. Sci., **10 (12)**, 845-847 (2015).
14. N.D. Narakidze, D.V. Shaykhutdinov, K.M. Shirokov, N.I. Gorbatenko, S.G. Yanvarev, *International conference on mechanical engineering, automation and control systems 2016*, 012112 (2017).
15. A. Bulgakov, D. Shaykhutdinov, N. Gorbatenko, S. Akhmedov, Pro. Eng., **123**, 94-100 (2015).
16. X. Xie, D. Wells, Z. Wang, T. Iliescu, J Comp. & Appl. Math., **328**, 12-29 (2018).
17. P. Bhopale, P. Bajaria, N. Singh, F. Kazi, Adv. Intell. Sys. & Comp., **629**, 60-71 (2018).
18. P.S. Reddy, K.Y. Rani, S.C. Patwardhan, Comp. & Chem. Eng., **106**, 40-56 (2017).
19. Z. Deng, H. Deng, L. Yang, Y. Cai, X. Zhao, Energy, **138**, 509-519 (2017).
20. F. Massa, I. Turpin, T. Tison, Mech. Sys. & Sign. Proc., **96**, 291-302 (2017).
21. Y. Zhang, A. Combescure, A. Gravouil, Int J Num. Methods Eng., **111(2)**, 176-200 (2017).
22. A. Buscarino, L. Fortuna, M. Frasca, M.G. Xibilia, Int. J Control, **90(7)**, 1437-1445 (2017).

Investigation of electrostatic fields by modeling methods

Alla Merkulova[1,*], *Irina* Cherunova[1], *Irina* Danilenko[1]

[1]Don State Technical University, Rostov, Russia

Annotation. The environment is filled with electrostatic fields of various origins. The negative influence of the electrostatic field requires the provision of electrostatic safety standards. So modeling of electrostatic fields is a topical research task. The electrization of objects is complex, depending on the many parameters process. The article considers the existing methods of studying electrostatic fields. The authors have justified the criterion for assessing the safety of an electrostatic field. The article proposes the developed advanced mathematical model describing the electrization of a complex, multi-layered object. Variable values in the proposed mathematical model are the main characteristics of the materials forming the layers of the modeled object, and the parameters of the condition of the environment. The constructed mathematical model of electrization of a complex object has undergone a successful experimental verification in field trials. Analysis of the experimental data confirmed the expected condition of the system, which is a complex object.

1 Introduction

The investigation of electrostatic fields surrounding human in everyday life and at work is a topical scientific task. Active use of the electric current energy leads to the formation of electrical charges on the external elements of household appliances and production structures. Another reason for static electricity charges on objects used by a person is the use of new materials. Modern materials are materials that are complex in composition and structure, which, as a general rule, have high electrical resistance. Surface electric charge appears on household items. Surface electric charge appears even on clothing. Electrization of materials in this case occurs during the operation of products.

Electric charge is a negative production factor. The limiting values of the parameters determining the magnitude of the electric charge are limited by normative documents [1].

Spark gas discharge is one of the processes of the accumulated electric charge from the surface of the material. Spark gas discharge can cause primary injuries associated with the passage of electric current through the human body. They include: a burn with a spark of human skin, the ignition of hazardous substances in the environment. A huge temperature (up to 20000 K) is concentrated in the spark discharge channel. This powerful heat flow causes the instantaneous ignition of the materials of the clothing package, severe burns and is the main damaging factor when exposed to an electric discharge directly on a person. Spark gas discharge can cause secondary injuries. They include: bruises and injuries in the fall. In this case, even weak discharge currents can cause a breakdown in the nervous system and worsen a person's overall condition, if their action doesn't stop for a long time.

As is known, the control system of a living organism is built on neuro-electrical impulses. Therefore, the effect of an external electrostatic field on a living organism is very complex and can cause functional disorders of the nervous and cardiovascular systems.

However, the main danger of static electrization is associated with discharges of static electricity. Static electricity discharges can ignite flammable gases, vapors or dust in the air. Therefore, in many enterprises, the conditions of a person's professional activity are associated with an electrostatic danger [2]. For example, at enterprises of the oil and gas complex, the environment contains flammable liquids, their vapors and gases. The requirements to ensure for a person safe living and working conditions [3, 4] determine the direction of scientific researches. Therefore, the investigation of the conditions for the appearance and change of the characteristics of electrostatic fields is a topical scientific task.

2 Similarity method for the study of electrostatic fields

The similarity method is often used in the study of electrostatic fields. This method is a physical simulation of the phenomenon. When studying real electrostatic fields, equivalent fields are created. For example, the electrostatic field between charged conductive bodies replaces the study of the current field between the same

* Corresponding author: avmerkulova@mail.ru

system of conductors. The electrical potentials of the conductors are maintained constant.

The similarity method asserts that if the sizes of the electrodes creating the electrostatic field and all the distances between these electrodes vary in the same proportion, the structure of the electrostatic field will remain the same.

Electrostatic fields in electrostatic lenses, in electronic tubes are very complex and unique electrostatic fields. Such electrostatic fields do not describe analytically. When studying similar electrostatic fields, the similarity method is used. The similarity method is rationally used in describing the electrostatic fields of stable electrified systems.

Such a study is experimental. In this case, by means of electrical measuring instruments, a large number of differences in the space in which the electrostatic field is distributed are measured. Then the researchers build the surfaces of an equal potential and the line of tension of the electrostatic field under study. Visualization of the results of the research is usually carried out with the help of special programs for computer simulation of electric fields.

However, the similarity method allows studying only the given electrostatic field created by a real system of charged bodies and at a particular time. In this case, it is impossible to make estimates of the dynamics of the change in the parameters of the electrostatic field. The similarity method does not allow to use the received scientific information on similar electrified systems, the parameters of which differ slightly.

3 Modeling of electrification processes

In everyday life and production human uses materials, most of which are dielectrics. They are electrified by contact with each other and with conductors.

On the surface of the dielectric, an electric charge is accumulated. It is practically impossible to reduce the electric charge accumulated on the dielectric surface. As a result of these processes, an electrostatic field is created in the surrounding electrified materials.

The processes of the appearance of electric charge and the influence of electrostatic fields on production and on the human body were considered in the works of a number of researchers [5-7].

Researcher Loeb [7] notes the difficulty in interpreting and comparing the experimental results obtained in the study of electrization. He explains the emerging problem by the variety of processes of interaction between contacting bodies and material environments into which materials are placed. Friction, compression, stretching, shearing, etc., increase the contact area of two bodies. Therefore, the rate of exchange processes changes and the value of the electrostatic charge formed on the contacting surfaces increases. All researchers note the great role of mechanical influences in electrization processes.

Investigations of the contact electrization of materials have shown that the contact of clean surfaces in the process of electrization is played by electrons. If materials are contaminated or active friction occurs, in the process of electrization the main role is played by ions or charged particles of matter. The resultant surface charge density is determined by the type of charge carriers.

The model of the electrization process at the contact of two dielectrics with different energies of the Fermi level is provided by the electron transition and considered in [8]:

$$\sigma = \pm\frac{1{,}77\cdot 10^{-9}(\Delta\Phi - \Delta\Psi)}{e(\upsilon_1/\varepsilon_2 + \upsilon_2/\varepsilon_1)},$$

where σ – is the surface electric charge, C/m^2; $\Delta\Phi$ – is the difference of energy at the Fermi level, eV; $\Delta\Psi$ – is the difference in the electron affinity energy, eV; e – is the value of the elementary electric charge, C; υ_1 and υ_2 – the depth of injection of charge carriers with respect to the interface of dielectrics (for the first and second material, respectively), m; ε_2 and ε_1 – are the dielectric permittivities of the contacting materials (respectively, the first and second substances).

The model of the electrification process, based on the transition of ions from one surface to another, was proposed in [9]:

$$\sigma = \frac{\varepsilon_0\varepsilon}{h}\left[(W_1 - W_2) - \frac{kT}{e}\ln\frac{n_1}{n_2} \pm hE\right] \quad (1)$$

where ε_0 – is the electric constant (equal to $8{,}85\cdot10^{-12}$ F/m); ε – is the minimum of two (ε_2 and ε_1) dielectric permittivity; h – distance between surfaces (unevenness height), m; W_1 (and W_2) – is the binding energy of charge carriers to a solid, eV; k – is the Boltzmann constant (equal to $1{,}38\cdot10^{-23}$ J/K); T – is the absolute temperature, K; n_1 (and n_2) – is the concentration of ions or particles, m^{-3}; E – external electrostatic field strength, V/m.

Static electrization consists of two processes: the accumulation of static electric charges and the draining of electric charges over time. If the processes of generation of electric charges and their decay are continuous, then one can go to infinitesimals and integrate:

$$\sigma(t) = \int\frac{d\sigma_{рост}}{dt}dt - \int\frac{d\sigma_{спад}}{dt}dt$$

Experiments show that the process of electric charge flow can be considered continuous. The process of the flow of electric charge is determined by the electrical conductivity of the material of the electrified object and the emission of a charge from its surface into the environment. Generation of electric charges is usually a step-by-step process. Generation of electric charges is determined by friction, compression of contacting surfaces during electrization. When modeling the standard processes of object electrization, it is possible to generate and drain the electric charge continuously.

4 Construction of a model for the electrization of a multilayered package of materials

From the electrization processes, electrization by friction is more common. Electrization by friction is accompanied by friction, squeezing and deformation. Analysis of the developed models of electrization showed that they are studying, as a rule, contact electrification. We have considered a more complicated process of electrization. Our model studies the electrization of a multilayered object taking into account friction, deformation and emission of electric charge into the environment [10].

The friction process leads to an increase in temperature at the points of contact. The increase in temperature leads to an increase in the rate of metabolic processes, including electrization. When the contacting objects are squeezed, deformation of their surfaces arises. The deformation of the surfaces leads to an increase in the contact area of the surfaces. The growth of the area of contact increases the degree of electrification of objects. Emission of electric charge is determined by the electrical conductivity of the environment. The emission of electric charge reduces the value of the electric charge accumulated on the surfaces of the contacting objects.

The development of common numerical algorithms that make it possible to calculate three-dimensional electrostatic fields with the necessary accuracy is an topical research task. The problem of constructing models of electrostatic fields arises for inhomogeneous and nonlinear material environment and complex forms of interfaces for material environment.

We considered triboelectrization of a multilayer object. In the proposed model of electrization, we assessed the contribution to the electrification process of a complex object of electrostatic charges accumulated by separate surfaces of structural layers.

The electrostatic field is a potential force field. At each point in space, the electrostatic field is determined by the values of the tension vector $\vec{E}$ and the potential φ of the electrostatic field.

Restrictions on occupational safety [11] require that, $W \le kW_{min}$, where W_{min} – minimum energy of ignition of gas mixtures, k – safety factor (k=0,4).

The limiting values of the parameters of the electrostatic field in the air are subject to Paschen's law [12]. The energy of the electrostatic field is given by:

$$W = \frac{\varepsilon\varepsilon_0 E^2 Sd}{2},$$

where E – is the intensity of the electrostatic field, V/m.

It's obvious that $E \sim \sqrt{W}$. This is the rationale for choosing the parameter for assessing the electrostatic safety of environmental conditions. The value of the electric field strength can be used as a parameter for assessing the electrostatic safety of environmental conditions.

We have obtained an estimate of the critical value of the electric field strength: $E_{кр} \sim \sqrt{0.4}E_{пр}$, here $E_{пр}=3\cdot 10^6$ V/m is the limiting value of the electrostatic field strength for the air medium under normal conditions [5]. Thus, the critical value of the electric field strength is equal to $E_{кр}=1{,}90\cdot 10^6$ V/m.

We obtained an analytical solution for calculating the characteristics of the electrostatic field formed by the electric charges distributed in space. We simplified the geometric form of the electrified bodies under study. In the mathematical model of the electrostatic field, we switched from real objects possessing non-linear characteristics to an ideal one, i.e. linear. The external environment in which the electrostatic charges that create the electrostatic field are distributed is inhomogeneous. We used the method of secondary sources to move to a homogeneous medium. Tozoni [13]. has developed the basis of the secondary sources method. The method of secondary sources introduces additional bound charges, the values of which are determined by the characteristics of the contacting surfaces and the parameters of the media separating them. We moved to a homogeneous medium, adding additional secondary sources distributed along the media interface.

In our mathematical model of electrification, the surface electric charge of the i-th layer of a multilayer object is determined by the equation:

$$\sigma_i(t) = \sigma_{фрик\_i}\exp(-t/\varepsilon_0\varepsilon\rho_V) - \sigma_{эм} \qquad (2)$$

here $\sigma_{фрик\_i}$ – is the value of the generated surface electrostatic charge (determined by the refined (1)

$$\sigma_{фрик} = \frac{a\varepsilon_0\varepsilon}{h}\left[(W_1 - W_2) - \frac{kbT}{e}\ln\frac{n_1}{n_2} \pm hE\right] \qquad (3)$$

here the coefficients a and b take into account friction and compression, respectively; $\sigma_{эм}$ – surface electric charge flowing as a result of emission into the environment.

Now the value of the resultant surface electric charge on the surface of the i-th layer of the multilayer object is determined by the equation:

$$\sigma_i = \sigma_{св\_i} + \sigma_i(t) \qquad (4)$$

here $\sigma_{св\_i}$ – the surface electric charge created by injected padding latent electric charges.

Electrostatic intensity, created by a multilayer electrified object, in any point of space is defined by the equation:

$$E_n = \frac{1}{\varepsilon_0\varepsilon R}\sum_{i=1}^{N}\sigma_i r_i \qquad (5)$$

The intensity of the resultant electrostatic field is:

$$\vec{E} = \vec{E}_n + \vec{E}_{внеш} \qquad (6)$$

here $\vec{E}_{внеш}$ – is the strength of the external electric field. The safety requirements limit the modulus of the resulting electric field E. The strength of the electric field should not exceed the critical value $E_{кр}$.

The proposed improved mathematical model of the electrization process is described by a system of equations (2-6). The main characteristics of the materials of the layers of the simulated complex object and the parameters of the state of the environment are variable values in this system of equations [14, 15].

This system of equations simulates the electrization of a multilayer object. Each layer of the object is materials with different characteristics. Researchers and engineers can evaluate the modulus of the electrostatic field strength near the outer surface of a complex object. The proposed mathematical model of the electrization process allows us to predict the provision of electrostatic safety in the vicinity of the object under study.

5 Conclusion

The constructed mathematical model of the electrization of a complex object has undergone a successful experimental verification in tests [16]. We conducted experiments on the electrization of a multilayered package of heat-protective clothing [15]. We estimated the magnitude of the electrostatic field strength modulus near the surface of the heat-shielding suit. We subjected the triboelectrization in real conditions of operation to a heat-protective suit.

Analysis of the experimental data confirmed the expected condition of the heat-shielding suit. Preliminary mathematical modeling of the process of electrifying a heat-shield suit should be carried out when constructing a costume design. Preliminary mathematical modeling of the process of electrization a heat-protective suit will help to select materials for it. Preliminary mathematical modeling of the process of clothing electrization will provide a stable protection of a person from static electricity.

The results of optimization of the design of a special anti-electrostatic suit for protection from low temperatures based on the mathematical model of electrization of a complex object, developed by us, are confirmed by patents [17-20].

References

1. GOST 12.0.003-2015 Occupational safety standards system. Dangerous and harmful working factors, Date of issue 2017-03-01, (Standartinform, Moscow, 2016)
2. V.A Veryovkin, V.I. Gorshkov, V.A. Bondar, B.G. Popov, *Static electricity in polymers: reports of a seminar «A static electricity in polymers»*, 42-45, (Chemistry, Leningrad, 1968).
3. GOST 12.4.045-84 Occupational safety standards system. Electrostatic fields. Tolerance levels and methods of control at working places Date of issue 30.06.85. – (Standartinform, Moscow, 1985)
4. GOST 12.4.124-83. Occupational safety standards system. Means of the protection against static electricity . General technical requirements Date of issue 01.01.1984. (Standards Publishing House, Moscow, 1983)
5. P.L. Gefter, *The electrostatic phenomena in processes of processing of chemical fibers* (Legprombytizdat, Moscow, 1989).
6. G.A. Lushcheykin, *Research techniques of electric properties of polymers* (Chemistry, Moscow, 1988)
7. L. Loeb *Static electrization* Gosenergoizdat (Moscow-Leningrad, 1963)
8. D.R. Davis, Inst. Phys. Conf. Ser., N 4, 29-36. (1963)
9. P. S. Henry, H. J. Text. Inst., N 1., **48**, 1-7 (1957)
10. Ch.M. Badmayev, etc. The Modern problems of science and education, №6., available at: http://www.science-education.ru/106-7972 (2012)
11. GOST 12.1.044-89 Occupational safety standards system. Fire and explosion hazard of substances and materials. Nomenclature of indices and methods of their determination Date of issue 01.01.1991. (Standards Publishing House, Moscow, 1989).
12. L. D. Landau, E.M. Lifshits, *Theoretical physics in 10 volumes. Volume 8. Electrodynamics of continuous mediums: manual*, Fizmatlit, Moscow, 2005)
13. O.V. Tozoni, *Mathematical methods for calculation of electric and magnetic fields*: (Naukova dumka, Kiev, 1964).
14. O.V., Kononova, A.V. Merkulova, I.V. Cherunova The Modern high technologies, №8., **1**, 31-32 (2013)
15. A.V. Merkulova, Research and development of express anti-electrostatic clothes for protection against the under temperatures: abstract of Candidate of Technical Sciences, Shakhty (2007)
16. A.V. Merkulova, I.V. Cherunova, I.N. Danilenko, Measurement Techniques, **52**, №4, 402-404 (2009)
17. I.V. Cherunova, A.V. Merkulova, I.Ju. Brink, V.V. Gorchakov, Patent № 2355266 Russian Federation, C2, MPK A 41D 13/008 Date of publication 20.05.2009 Bull. №14. (2006.01).
18. E.A. Shchenikova, I.V. Cherunova, A.V. Merkulova Patent № 2426483 Russian Federation, C1, MPK A 41D 13/008 Date of publication 20.08.2011 Bull. №23. (2006.01).
19. E.A. Shchenikova, I.V. Cherunova, A.V. Merkulova N.V. Kornev, The useful model № 128969 Russian Federation, U1, MPK A41D 13/008 Date of publication 20.06.2013 Bull. № 17. (2006.01).
20. E.A. Shchenikova, I.V. Cherunova, A.V. Merkulova N.V. Kornev, The useful model № 128968 Russian Federation, U1, MPK A41D 13/008 Date of publication 20.06.2013 Bull. № 17. (2006.01).

DOI: 10.1051/matecconf/201713204023

The hydrodynamic dust retention modeling process with a foam layer for open linear dynamic technological systems at construction industry enterprises

Vadim Bespalov[1,*], *Oksana* Gurova[1], *Irina* Tsarevskaya[1] and *Ludmila* Alekseenko[1]

[1]Don State Technical University, Rostov, Russia

Abstract. The physical essence of the reducing air pollution hydrodynamic method by means of the foamy method as the most effective dust retention technology for open long sources at construction industry enterprises is presented in the article. The mathematical description of this technology has been performed, taking into account the physical and chemical properties of the dust and foam bubble, as well as the parameters of the return of dust particles to technological raw material due to various physical mechanisms for their capture. The obtained parametric dependence of dust retention efficiency provides the possibility of its predicted calculation as a resultant parameter for the considering technology, taking into account the interaction of dust with a foam layer.

1 Introduction

At present, the operation of construction industry enterprises, for which belt conveyors are one of the main types of linear dynamic systems of process equipment, is accompanied by intensive formation and release of particles [1,2].

Analysis of modern studies results [3] allows us to assume that the most effective technology for the retention of inorganic dust PM2,5 and PM10 with a $SiO_2 > 70\%$ for belt conveyors at construction industry enterprises is the hydrodynamic method based on the foam layer application. However, the dust retention management technology which also applies to very dynamic systems is possible only on the scientific justification basis and modeling of both its individual parameters and efficiency as its resulting parameter which will ensure maximum dust concentration in the air of working zones and the surface layer of the atmosphere.

In this case it is necessary to take into account that the nature of the interaction process in three-phase system "dust-liquid-air" is influenced by hydrodynamic, aerodynamic, chemical and physical factors.

A number of methods for implementing the hydrodynamic method, characterized by the lowest liquid consumption, include a foam method for controlling dust, which achieves the highest value of dust retention efficiency (dust shielding). The efficiency of the dust retention process can be controlled studying and making purposeful changes in the properties of the foam [3,4].

Foam, which is a coarsely dispersed colloidal system, is characterized by the following main parameters: density, dispersity, thermal conductivity, electrical conductivity, optical properties, and stability and multiplicity, which are the main parameters [5].

The foam density is characterized by the density of the foam solution and the air enclosed in the bubbles of the foam and depends on the ratio of the liquid and gaseous phases and can vary within half the value of the density of the liquid phase to value close to 0.

The dispersion of the foam used for dust suppression is estimated by the average diameter of the bubbles, the values of which vary over a wide range – from 0.01 to 30 mm.

The thermal conductivity of the foam is characterized by a corresponding coefficient and is determined by the formula:

$$\lambda = 0{,}67 \cdot (\alpha \cdot \lambda_s) + (1-\alpha) \cdot \lambda_a,\ W/(m \cdot K), \quad (1)$$

where α is the degree of foam filling by the liquid phase; λ_s, λ_a – respectively, the coefficients of thermal conductivity of the foam solution and air, $W/(m \cdot K)$.

The electrical conductivity of the foam is given by:

$$\beta_f = 2 \cdot \beta_l / 3 \cdot K_f,\ C / m^3, \quad (2)$$

where β_f, β_l – respectively, the specific electrical conductivity of the foam and the intra-film liquid, C/m^3.

The optical properties of the foam are determined respectively by the coefficients of reflection, refraction, and absorption of the light flux by the sides (films) of the foam bubbles and depend on the thickness of the films themselves, as well as the foam layer.

* Corresponding author: izos-rgsu@mail.ru

DOI: 10.1051/matecconf/201713204023

In order to characterize the state of foam in practice, one of the main parameters uses the foam resistance τ, defined by the formula:

$$\tau_n = Q_r \cdot K_f / V_f, \text{c}, \quad (3)$$

where Q_r is the flow rate of the foamer solution, m^3 / s; V_f is the specified volume of foam, m^3.

Persistence is characterized by the time of individual bubbles existence or a certain volume of foam and is determined mainly by the nature of the foaming agent, its concentration. Depending on the nature of the foaming agent, the stability of the foam can reach a maximum value at some other rates of concentration of the foaming agent, after which it decreases almost to zero, which is typical for low-molecular compounds.

The effects of Marangoni and Gibbs are among the factors that ensure the stability of the foam. The Marangoni effect improves resistance, manifesting itself the ability of thin films to react to local thickness changes. The Gibbs effect develops itself when under the influence of mechanical or thermal perturbations, causing the film to stretch, local surface tension differences arise and cause the flow of surfactant molecules in the adsorption layer toward a lower concentration-toward the stretched part of the film [5,6].

Another basic parameter characterizing the state of the foam is its multiplicity K_n, which is in a certain dependence on the resistance and is the ratio of the volume V_n of the foam to the volume V_p of the liquid that went into its formation:

$$K_f = V_f / V_s = 0.78 \cdot D_2 \cdot \pi \cdot g \cdot \upsilon_s / Q_r, \quad (4)$$

where D_{fg} is the diameter of the barrel of the foam generator, m; υ_c – speed of passage of foam through an output mouth of the foam generator, m/s.

In addition to the listed basic parameters for the characteristics of the foaming process, two additional criteria are additionally used:

- the first criterion of foaming (Dyakonov's criterion) D_1, characterizing the dependence of foam stability on the degree of dynamic external conditions:

$$D_1 = \tau_d / \tau_{st}, \quad (5)$$

where τ_d, τ_{st} – dynamic and static resistance of foam, s.

- the second criterion of foaming D_2, characterizing the physicochemical properties of foaming agents solutions and allowing to describe the foaming process in both dynamic and static conditions:

$$D_2 = K_f \cdot \tau_{st}, \quad (6)$$

Thus, the efficiency of the foam method of air dedusting can be controlled by changing the parameters discussed above with the appropriate organization of foam technology and the supply of foam to the dust suppression zone to provide a shielding effect.

2 Materials and Methods

Mathematical modeling of the dust-holding process with a foam layer was performed by us taking into account the peculiarities of source dust shielding with a foam layer, the properties of foam bubbles, screened dust particles and the dynamic conditions of the physical mechanisms display of their interaction.

3 Results

As a result, taking into account the probabilistic realization of the successive interdependent physical mechanisms of dust particles interaction [7,8] with a foam layer, the dust suppression efficiency is described by the following dependence:

$$E_{ef\text{-}dr} = 1-(1-E_{ef(1)\text{-}dr})\cdot(1-E_{ef(2)\text{-}dr})\cdot(1-E_{ef(3)\text{-}dr}), \quad (7)$$

where $E_{ef(1)\text{-}dr}$, $E_{ef(2)\text{-}dr}$, $E_{ef(3)\text{-}dr}$ – respectively efficiencies caused by the action of inertia forces; forces of elastic interaction and electric forces.

The efficiency $E_{ef(1)\text{-}dr}$ depends on inertial sedimentation coefficient Kin and is determined by the formula:

$$E_{ef(1)\text{-}dr} = 1-K_{in}, \quad (8)$$

The coefficient K_{in} is determined [9]:

- under the condition that the velocity of the bubbles is less than the velocity at which the bubbles are fractionized, that is, υb < υfr, according to the formula:

$$K_{in} = \text{Stk} / (\text{Stk} + A_1), \quad (9)$$

- in case $\upsilon_b \geq \upsilon_{fr}$, then the process of bubbles fragmentation is of a mass character and $E_{ef(1)\text{-}dr} = 0$.

The value of Stk conf depends on the foam volume shape, located in the form of the layer above the source of dust emission [10]. In addition, the configuration of the dust source itself also affects the Stk value.

The Stokes number Stk is determined by the formula:

$$\text{Stk} = d_p \cdot \upsilon \cdot \rho_h / 18 \cdot \mu_e \cdot l_1, \quad (10)$$

where d is the particle diameter, m; υ_{en} – speed of unperturbed air stream, running on an obstacle in the form of a foam layer, m/sec; $\rho_ч$ is the dust material density, kg/m^3; μ_v is the coefficient of dynamic air viscosity equal to $1.82 \cdot 10^{-5}$ Pa · s; l_1 is the characteristic size of the streamlined obstacle in the form of a foam layer, m, which for a foam layer (at $l_n = 0$) becomes l.

4 Discussion

In the evaluationof the coefficient A_1 various discrepancies are observed in different authors' works [11, 12], that can be explained by the dependence of this coefficient on the particle sizes, the motion velocity, the bubbles velocity, and a number of other factors.

Regarding the foamy method for the implementation of the dust retention process, it is possible to extract that dust particles with a size dcr can not reach the foam layer, since their motion is so slowed that they lose their inertia and under the influence of the velocity components perpendicular to the direction of the main dust particle flow get demolished in parallel with the inner foam layer.

Taking into account the experimental data and the method presented in [13], the dependence of the coefficient A_1 on the ratio of the actual dimensions dp of dust particles and their critical size dcr was obtained for the foamed method:

$$A_1=2\cdot 10^{n1}, \quad (11)$$

where n_1 is a coefficient that takes into account the ratio of the actual dimensions of dust particles dp and their critical size d_{cr}.

The value of Stk_{cr}. depends on the shape of the foam volume located in the form of a layer above the source of dust emission. In addition, the configuration of the dust source itself also affects the Stk value. Assuming inertial deposition of particles on the foam layer similar to deposition on a plate, we assume for the dust retention process:

- $Stk_{cr.}$= 0,28 on the source of dust emission of a rectangular section (which is typical for the conveyor belt surface);
- Stk_{cr} = 0,1 on the source of dust emission of square or round section (which is a characteristic of the feed port or the delivery orifice of the overload nodes).

The process of shielding dust particles with a foam layer is characterized by the efficiency $E_{ef(2)\text{-}n}$ which value depends mainly on the properties of the foam and the speed of the meeting of the dust particles with the foam bubbles. Taking into account the results of the experimental and theoretical studies presented in [14, 15] with respect to the foam layer the quantity $E_{ef(2)\text{-}dr}$ is determined by the formula:

$$E_{ef(2)\text{-}dr}= 1–\exp\,[–0.693 \cdot (\upsilon^s_{cr}/ \upsilon)2)\, 2], \quad (12)$$

where υ^s_{cr} is the critical velocity of a dust particle at which the foam bubble can collapse under a shielding (elastic) interaction, m/s.

The value of υscr can be determined on the basis of [16] taking into account the elastic properties of the bubble film:

$$\upsilon^s_{cr}= [2\pi\cdot\sigma\cdot dp/m_p\cdot[\delta/\delta_{cr}\cdot(1+d_p/d_n\cdot(d_p+\delta_{cr})) -1], \quad (13)$$

where δ, δ_{cr} – respectively, the average and critical thickness of the foam bubbles films, m; m_p is the mass of a dust particle, kg.

It is known that the thickness of the foam bubbles films varies with time due to evaporation, the outflow of the solution from the film, the entrainment of the solution by the air stream, etc., and depends on the film deflection wavelength, surface tension and other factors, the bubble does not collapse before and during the interference fringes emergence and is destroyed after their disappearance, that is, with the film thickness up to $7{,}6\cdot 10^{-7}$m, and also up to $4\cdot 10^{-7}$m [17].

The mass of a dust particle mp, the value of which depends on the density of its material and dimensions, can be determined by the formula:

$$m=\pi\cdot d_p^3\rho_p/6. \quad (14)$$

Capture of dust particles by foam under the action of electric forces [17, 18] is characterized by the efficiency $E_{ef(3)\text{-}dr}$.

Analysis of the experimental data [19, 20] made it possible to obtain the dependence of the efficiency of dust particles capture under the action of electric forces:

$$E_{ef(3)\text{-}dr} =1{,}7\cdot 10^{-3}\cdot q^{0{,}65}, \quad (15)$$

where q_0 is the specific electrical charge of foam, C / m.

It should be mentioned that with artificial foam electrification the specific charge q_0 depends on the method of foam recharging and the potential value.

With allowance for equations (8)-(15) the dependence (7) of the efficiency of dust retention by a foam method using a foam layer takes the form:

$$E_{ef(3)\text{-}dr.}= –d_p\cdot\upsilon\cdot\rho_p/(d_p\cdot\upsilon\cdot\rho_p+360\cdot\mu\cdot l)\cdot\exp[–7{,}48\cdot\sigma/\upsilon^2\cdot\rho_p\cdot$$
$$\cdot(l/d_p+2{,}1\cdot d_p/d_f\cdot(d_p+4\cdot 10^{-7}))]\cdot(1{,}7\cdot 10^{-3}\cdot q^{0{,}65}). \quad (16)$$

Conclusions

Thus, the performed mathematical modeling of the complex dynamic technical system "belt conveyor technology of dust retaining" provides the possibility of predictive calculation of the efficiency of dust holding with a foam layer, taking into account the peculiarities of its interaction with dust aerosol.

References

1. Bespalov V.I., Gurova O.S., Yudina N.V. *Improvement of methods and means for dust removal of the air environment at concrete mixing departments of factories of reinforced concrete products and structures* (Rostov State Civil Engineering University, Rostov-on-Don, 2015) [in Russian]
2. Bespalov V.I., Gurova O.S., Samarskaya N.S., Paramonova O.N. *Industrial ecology* (Rostov State Civil Engineering University, Rostov-on-Don, 2009) [in Russian]
3. Bespalov V.I., Gurova O.S., J. Scientific review, no.6, 193-196 (2012) [in Russian]
4. Dyakonov G.K. *Questions of the theory of similarity in the field of physical and chemical processes* (USSR Academy of Sciences, Moscow, 1956) [in Russian]
5. V.I. Bespalov, N.A. Strahova, *Dedusting in the construction: intercollegiate collection,* (RISI, Rostov-on-Don, 1990)
6. Bespalov V.I., Gurova O.S., Samarskaya N.S., Ingenerniy vestnik Dona, no. 4 (2013) [in Russian]
7. Ann T.W. Yu, Yuzhe Wu, Bibo Zheng, Xiaoling Zhang, Liyin Shen, Habitat International, no.44, 77-185 (2014)
8. Tikhomirov V. K. *The foam. Theory and practice of its obtaining and destruction*, (Chemistry, Moscow, 1983) [in Russian]
9. Abramzon A. A. *Surface-active substances. Properties and applications* (Chemistry, Leningrad, 1981) [in Russian]

10. Raist, P. *Aerosols. Introduction to theory,* (Mir, Moscow, 1987)
11. Bespalov V.I., Gurova O.S., Samarskaya N.S., Lysova E.P., Mishchenko A.N., Biosciences biotechnology research Asia, no.3, 1615-1620 (2014)
12. Bespalov V.I., Gurova O.S., Ingenerniy vestnik Dona, no. 3 (2013)
13. Daniela Vallero. *Fundamentals of Air Pollution fourth edition,* (Civil and Environmental Engineering Department Pratt School of Engineering Duke University, Durham, 2007)
14. V. I. Bespalov, O. S. Gurova, N. S. Samarskaya and others, J. Biosciences Biotechnology Research Asia, no.2, 1587-1596 (2015)
15. Romaniuk T.F., *Technology of building materials production,* (Federal Agency for Education, Tomsk State University. Architectural and Construction Universit, Tomsk, 2006)
16. Bespalov V.I., Gurova O.S., Meshcheryakov S.V., *Processes and apparatus for reducing air pollution* (Rostov State Civil Engineering University, Rostov-on-Don, 2014)
17. V.I. Bespalov, O. S. Gurova, I*nt. Scient. Pract. Conf.*, 35 (2015).
18. V. I. Bespalov, O. S. Gurova, E. P. Lysova, J. Biosciences Biotechnology Research Asia, no.2, 1459-1470 (2015)
19. Gurova O. S., Bespalov V.I., Paramonova O.N., J. Naukovedenie, no. 6, (2015)
20. Gurova O. S., Bespalov V.I., Paramonova O.N. J. Naukovedenie, no. 6 (2015)

Modeling of dynamic conditions of operation of wind turbines (WT) and choice of environmentally efficient and energy economic technology of wind power energy transformation

Vadim Bespalov[1,*], *Oksana* Gurova[1], *Oksana Akay*[1] and *Mikhail* Izotov[1]

[1]Don State Technical University, 344000 Rostov-on-Don, Russia

Abstract. Wind energy usage is one of the promising areas among renewable energy sources. The article considers a methodical approach to the modeling and selection of environmentally efficient and energy-efficient wind turbines at the design stage, taking into account the characteristics of the natural-territorial complex and the specific anthropogenic load on the territory of the WT location.

1 Introduction

The practical use of renewable energy sources (RES) is growing in many countries of the world [1]. The leading place among them is occupied by wind power plants (WPP) capable of generating sufficiently large amount of electricity for consumers of different scale varying from single or group installations up to networked ones (wind power plant).

The main advantages of wind turbines are the inexhaustibility of wind energy, the relative environmental safety of the energy received and the possibility of placement at any location. The shortcomings that limit the wide use of windmills include a narrow range of wind speed characteristics use from 2 to 15 m/s, variable work depending on the wind direction change, and a long payback term of the installation.

Further progress in the development of wind turbines, in our view, should be based on optimizing the management of wind turbines through modeling a dynamic and rather complex technology for converting wind energy into electrical energy [2,3].

In the current conditions for the wind turbines development, it becomes necessary to develop criteria for assessing and selecting the type of installation and optimizing the technological parameters of power generation [4,5]. The study of the most frequently used methods of assessment and selection of optimal wind turbines parameters taking into account the necessary volumes of electricity received and the facility location made it possible to identify their shortcomings:

- a need to pre-assign a specific type of equipment;
- lack of sufficient energy characteristics of equipment based on the complex impact on the environment state and the comfort of living of the population in the area of WT location.

At present, in connection with the existing variety of wind turbines, it becomes necessary to manage the process of selecting the type of installation, based on the principles of energy efficiency and environmental efficiency. By the term «energy efficiency» we mean the share of energy (useful energy) used by the WT regarding to the entire wind potential of the territory. By the term «environmental efficiency» we mean the degree of reduction of negative factors (noise, vibration, etc.) that are provided by the WT to the environment during its operation in a particular area [6].

The proposed technique allows to simulate the conversion technology of wind energy into electric energy in various natural and climatic conditions, to evaluate various options for implementing this technology and to select the optimal option with a subsequent calculation of the WT technological parameters.

2 Materials and Methods

The main goal of our research is to provide the ability to manage a very complex dynamic technology for converting wind energy into electricity based on the technology simulation, calculating and selecting the optimal technological parameters of WT based on the energy efficiency and environmental efficiency criteria. This methodological approach allows to choose the most environmentally efficient and energy-efficient technology. To achieve this goal, we have accomplished the following tasks:

- system analysis of natural and climatic factors and landscape characteristics of the planned territory for the WT location has been done;
- the parameters describing the main characteristics of the WT have been singled out and formalized;

* Corresponding author: izos-rgsu@mail.ru

- a mathematical description of the environmental and energy efficiency as resulting evaluation criteria for the technology of WT conversion to electricity has been performed;
- the technology, based on the calculated resulting criteria has been chosen;
- a methodology for assessing and selecting the most environmentally and energy-efficient technology for converting wind energy into electrical energy with subsequent calculation of design parameters of wind turbines for certain territorial conditions has been developed.

3 Results

The performed analysis of the technical and economic characteristics of the WT currently in operation on the territory of the Russian Federation showed their compliance with modern indicators of foreign power plants.

However, it should be noted that any modern WT are subjected to constant technical improvement, and therefore, when choosing a particular type of such plants, their quality should be considered at the time of the intended use. The cost of such wind turbines does not exceed the cost of foreign analogs [7,8].

As a result of research of WT different types, we can state that, despite their wide application, they still have two main drawbacks: the wind utilization factor does not exceed 0.4; their use in the immediate vicinity of the consumer is hampered by the increased environmental hazard of these installations (in particular, the level of sound volume that exceeds the standard values).

Further improvement of WT (the capacity unit increase, the wind energy use coefficient increase to 0.5 and more, and improvement of their ecological characteristics) is possible on the basis of the use of slow-moving orthogonal WT with a vertical axis [9-14].

Thus, at present there is a sufficient number of WT on the energy market that can be used in both autonomous and centralized power supply systems.

We examined the main parameters that determine, in the final analysis, the design perfection, ecological efficiency and energy efficiency of operation of the WT.

Thus, the kinetic energy of the air (wind) flow per time unit can be determined by the formula:

$$K=m/V^2,\ \text{Вт}, \quad (1)$$

where m is the mass air flow rate, kg / s; V – wind speed, m / s.

In this case, the mass flow rate of air flow through the plane of the wind wheel is determined by the formula:

$$m=\rho\cdot V\cdot S,\ \text{kg/s}, \quad (2)$$

where ρ – air density, kg/m^3; S – area brushed by the wind wheel, m^2.

The area brushed by the wind wheel can be determined for the axial wind turbines with the formula:

$$S=\pi\cdot D^2/4,\ \text{m}^2, \quad (3)$$

and for orthogonal wind turbines according to the formula:

$$S=DH,\ \text{m}^2, \quad (4)$$

where D is the diameter of the axial or orthogonal wind wheel, m^2; H – height of the orthogonal wind wheel, m.

Wind power:

$$P=\rho V^3 S/2,\ \text{W}. \quad (5)$$

Thus, the energy supplying the wind wheel is proportional to the cube of the incoming air flow speed (wind speed).

The main characteristics of the wind wheel are:
- power factor, determined by the formula:

$$C_p=P_{ww}/P, \quad (6)$$

- coefficient of the peripheral forces moment defined by the formula:

$$C_m=2M/q\cdot F\cdot D \quad (7)$$

where P_{ww} – the stream power used by a wind wheel during its rotary motion, W; M is the moment created by the peripheral force, N · m; q – speed head, defined by the formula:

$$q=\rho\cdot V^2/2,\ \text{Н/м}^2. \quad (8)$$

The coefficient Cp shows how much of the air flow power can profitably be used by the wind wheel, and is also called the coefficient of wind energy use ξ. The value of ξ first of all depends on the type of the WT, the shape of its wings and the quality of their production [15].

For high-speed wind turbines with wings of streamlined aerodynamic profile ξ is in the range from 0.42 to 0.46. This means that the wind wheels of such engines can be profitably used, i. e. transform into mechanical work 42 - 46% of the energy that the wind flow passing through the wind wheel has. For slow-moving wind turbines, the values of ξ can be in the range of 0.27-0.33.

The dimensionless quantities ξ and Cm are functions of a dimensionless parameter – suction specific speed equal to the ratio:

$$Z=U/V, \quad (9)$$

where U is the peripheral speed on the outer side of the wind wheel, determined by the formula:

$$U=\omega\cdot D/2=\omega\cdot R,\ \text{m/s}. \quad (10)$$

The speed coefficient is determined by the formula:

$$Z=\pi\cdot n\cdot D/60\cdot V=2\pi\cdot n\cdot R/V. \quad (11)$$

The parameters Cp, ξ, Cm and Z are the main operational parameters determining the design perfection and the efficiency of the operation of the wind turbine.

Thus, the completed simulation of wind energy conversion technology into electric technology is the basis of our proposed methodology, which includes the following main stages.

At the first stage grouping of the natural climatic characteristics and area landscape features of the structure arrangement is carried out regarding the type of landscape, including surface roughness, calculation of the wind potential of the territory in the speed range from 2-13 m / s taking into account annual and month average wind speed and number of prevailing wind direction hours. The aggregate of these parameters determines, in the final analysis, the total wind power (W1) per area unit perpendicular to the direction of the wind flow in the analyzed area.

In the second stage, according to the power demanded by the consumer (W2), various plant variants are selected (by types of wind turbines). Technical characteristics and environmental parameters are determined for each type of WT.

Technical characteristics of the installation are: the diameter of the wind wheel, the height of the mast, the type of the rotation axis, the number of blades and materials they are made of, the lead of a screw. Ecological parameters wind turbines impact, characterizing environmental safety in the area of the facility location: mechanical noise, vibration, infrasound, maintenance intervals, wear resistance of the structure and its elements.

At the third stage the characteristics of the types of wind turbines preliminarily selected in the second stage are compared in order to select the units for the minimum values of wind energy loss (ΔWi), capable of providing the power required by the customer:

$$\Delta W_i = W_1 - W_2 \cdot \xi, \tag{12}$$

ΔW_i means losses of wind specific energy, W.

At the fourth (final) stage, one of installations having passed the third stage of selection, is selected and that is optimal for ecological and energy efficiency. At this stage, as the resultant criteria for WT we propose ecological and energy efficiency [16-20].

At the same time it is suggested that we choose the optimal choice for the environmental efficiency of wind turbines on the basis of calculating this parameter using the formula:

$$E_{ef}^{WT} = (A_1 - A_2)/A_1 \cdot 100\%, \tag{13}$$

where A_1 is the level of impact of a particular environmental factor on the state of the territory without wind turbine; A_2 – the level of impact of a separate environmental factor on the state of the territory, taking into account the impact of wind turbines.

In case several variants of the wind turbine will have the same values of environmental efficiency, we suggest the further selection of the optimum wind turbine to be based on the energy efficiency of each option under consideration based on the calculation of the energy-consuming parameter of the wind turbine as a criterion for estimating the energy efficiency:

$$E_e^{WT} = W_2/W_1 \cdot 100, \%, \tag{14}$$

where W_2 is used (converted) by WT wind power specific wind energy unit («useful energy»), W; W_1 is the total available specific wind energy in the territory under consideration («spent energy»), W.

4 Discussion

As a result of our studies, we proposed a technique for choosing the technology for converting wind energy into electric power and selecting the design characteristics of wind turbines based on modeling various options for such technologies, as well as selecting optimal parameters of wind turbines based on the principles of ensuring maximum values of ecological and energy efficiency. At the same time, starting from the third stage, in case only one wind turbine remains as a result of the selection the realization of the subsequent stages is not required and the choice of the optimal wind turbine can be completed at the appropriate stage of the procedure.

5 Conclusions

Thus, the methodology proposed in the given article makes it possible to select an ecologically and energy efficient wind turbine at the design stage, taking into account the characteristics of the natural-territorial complex and the specific anthropogenic load on the area of the WT location.

References

1. Solomin E.V., Sirotkin E.A., Seradskaya O.V., International Scientific Journal "Alternative Energy and Ecology", no. 12, 33-34 (2017)
2. Kozhukhov Y.V., Lebedev A.A., Danilishin A.M., Davletgareev E.V., J. CAD/CAM/CAE Observer, no. 7 (107), 81-87 (2016)
3. Popel O.S., Frid S.E., Shpilrain E.E., Scheglov V.N., Izosimov D.B., Tumanov V.L. J. Heat power engineering,no. 3, 42-50 (2006)
4. Popel O.S., Frid S.E., Shpilrain E.E., Izosimov D.B., Tumanov V.L. Prospects of power engineering, no.10, 77-90 (2006)
5. Popel O.S., Evald E., Shpilrain E.E., *International Conference "Renewable Energy*, 9-13 (2006)
6. Redko I.Y. Multifunctional energy technology complex (METC) is the basis for power supply to remote regions of the Russian Federation: http://pt21.ru/docs/pdf/46.pdf
7. Wind Energy - the facts, (EWEA, London, 2009)
8. Kvitko A.V., Khitskova A.O. Multidisciplinary network electronic scientific journal of Kuban State Agrarian University, no. 97 (2014)
9. Starkov A.S., Grigoriev A.V., Idrisova A.R. Journal of the Magistrate's Bulletin, no.1, 63-64 (2016)
10. Shefter Y.I., *Use of wind energy* (Energoatomizdat, Moscow, 1983)
11. Twaidel J., *Renewable energy sources* (Energoatomizdat, Moscow, 1990)

12. P.P. Bezrukih, Y.D. Arbuzov, G.A. Borisov, *Resources and efficiency of renewable energy sources use* (Nauka, SPb, 2002)
13. A. Sankovsky.*Perspectives for the development of renewable energy in Russia. The results of the project TACIS EuropeAid* (ATMOGRAPH, Moscow, 2009)
14. Nikolaev V.G. Ecology and Life, no. 9 (94), (2009)
15. Shishkin N.D. *Small energy-economic complexes with renewable energy sources* (Gothic, Moscow, 2000)
16. Marchenko O.V., J. Izvestia RAS. Power Engineering, no. 2, 91-103 (2000)
17. Eremin L.M., J. Energetika, no. 3, 22-24 (2003)
18. Grigorash O.V. *In Proceedings of KubSAU*, no. 6, 192-196 (2011)
19. Grigorash O.V. J. Industrial Power Engineering, no 4, 37-40 (2001)
20. Lyashkov V.I., Kuzmin S.N. *Non-traditional and renewable energy sources* (Tambov State. Tech. University, Tambov, 2003).

MATEC Web of Conferences 132, 04025 (2017)
DTS-2017
DOI: 10.1051/matecconf/201713204025

Bluetooth API Implementation into Android

Sergey Konev[1], *Anastasia* Stus[1], *Elena* Kasyanenko[1] and *Vasiliy* Dolgov[1]

[1]Don State Technical University, Chair "Production Process Automation", 344000, Gagarin square, Rostov-on-Don, Russia

Abstract. Bluetooth is a popular method of communication between devices. Many smartphones today have the capability to communicate using Bluetooth. Android developers sometimes need to use Bluetooth in their projects. Android OS provides a powerful API for Bluetooth that allows to simplify scanning the environment for devices, pairing and connecting, data transfer between devices and more. However, utilizing the Bluetooth API can be difficult for first-time users. The objective of this article is to demonstrate the key points of implementing Bluetooth API in the Android application.

1 Introduction

Bluetooth API is located in the android.bluetooth package [1]. It contains more than ten classes. The package includes specific classes for medical Bluetooth devices, audio headsets and data transfer. This article will only deal with the basic classes implementing Bluetooth basic functionality; without them, it is impossible to create any Bluetooth-supporting application [2]:

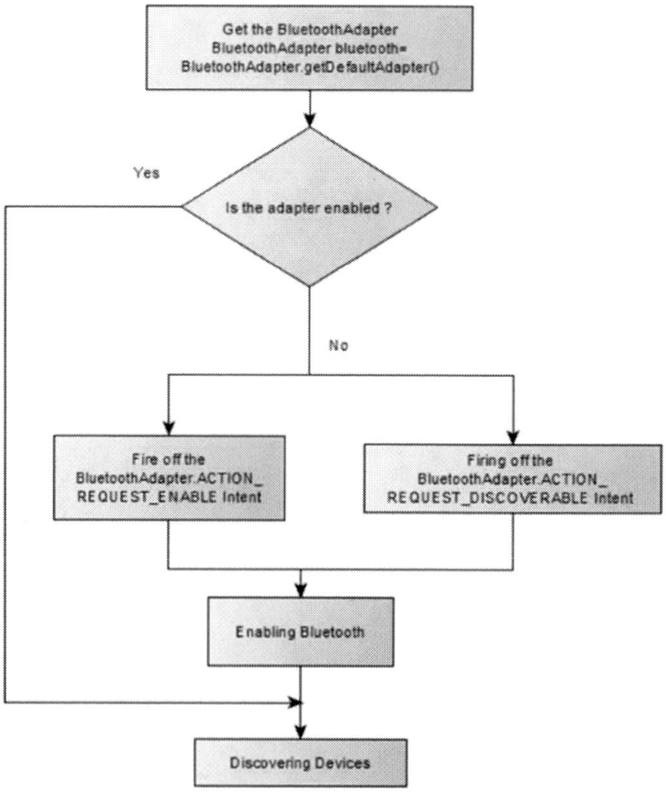

Fig. 1. Enabling Bluetooth flowchart

* Corresponding author: lex1998@rambler.ru

- The Bluetooth Adapter class represents the Bluetooth radio hardware on the

local device. The Bluetooth Adapter is the entry-point for all Bluetooth interaction.

- The Bluetooth Device class represents a remote Bluetooth device.

- The Bluetooth Server Socket class is used to open a socket to listen for incoming connections and provides a Bluetooth Socket object when a connection is made. Represents an open server socket that listens for incoming requests (similar to a TCP Server Socket).

- The Bluetooth Socket class is used by the client to establish a connection to a remote device.

Let's start with the description of the process of the application development for Android. We will study the main stages of Bluetooth implementation in Android application in more details

2 Enabling Bluetooth

At the very beginning of the work, before the connection, it is necessary to check whether your device is equipped with a Bluetooth unit. If the answer is positive, the user may be offered to enable the unit or make the device visible for other users, and that will enable it automatically. The test of the unit and its performance, as well as enabling is carried out with the use of methods of Bluetooth Adapter class (Fig. 1).

3 Discovering devices

The search of devices for connection is also performed with the use of Bluetooth Adapter class. You can also start scanning or request the list of paired, i.e. already known devices.

The search of new available Bluetooth devices is performed during scanning. If a device with Bluetooth enabled is within the reach, it will send some identifying information in response to the request: name, class, MAC-address. Based on this information you can arrange connection and data transfer. To start scanning, you need to call the start Discovery method. Scanning is performed in a separate asynchronous flow. This method requires the BLUETOOTH_ADMIN permission. To retrieve information on the devices found, the application has to register a broadcasting receiver for the following Intents:

ACTION_DISCOVERY_STARTED: Occurs when the discovery process initiates;

ACTION_FOUND: Occurs each time a remote Bluetooth device is found;

ACTION_DISCOVERY_FINISHED: Occurs when the discovery process completes.

The search requires many resources, so as soon the suitable device has been found, it is necessary to stop the scanning process. Besides, the search process may also narrow gating width of the radio channel significantly, that's why it's better to refrain from search of new devices when the connection is established.

Before you start the environment scanning, it's reasonable to show the list of already known and paired devices to the user. It is necessary to make difference between paired and connected devices. Paired devices have a reference key, which they can use for authentication, and they are able to create an encrypted connection with each other. Connected devices share one radio channel and are able to transfer data to each other. You can use the Bluetooth Adapter to query for available Bluetooth devices to connect to. The get Bonded Devices method returns a set of Bluetooth Device objects that represent the list of paired devices.

4 Establishing connections between devices

Connection through Bluetooth is arranged according to the "client-server" principle and is similar to the connection through TCP-sockets [3]. To connect two devices, server and client parts of the code have to be written. One of the devices shall open a listening server socket for waiting the request for incoming connection, and the second shall initiate the connection, i.e. to give the client-socket back using MAC-address of the server. Server and client are considered connected when both of them have an active Bluetooth Socket object on one and the same radio channel [4]. After this data exchange can be initiated.

To set up a server socket and accept a connection, complete the following sequence of steps:

1. Get a Bluetooth Server Socket by calling listen Using Rfcomm With Service Record. The string is an identifiable name of your service, which the system automatically writes to a new Service Discovery Protocol (SDP) database entry on the device. The name is arbitrary and can simply be your application name. The UUID is also included in the SDP entry and forms the basis for the connection agreement with the client device.

2. Start listening for connection requests by calling method accept. This is a blocking call. It returns when either a connection has been accepted or an exception has occurred. A connection is accepted only when a remote device has sent a connection request containing a UUID that matches the one registered with this listening server socket. When successful, accept returns a connected Bluetooth Socket.

3. Unless you want to accept additional connections, call method close. This method call releases the server socket and all its resources, but doesn't close the connected Bluetooth Socket that's been returned by accept.

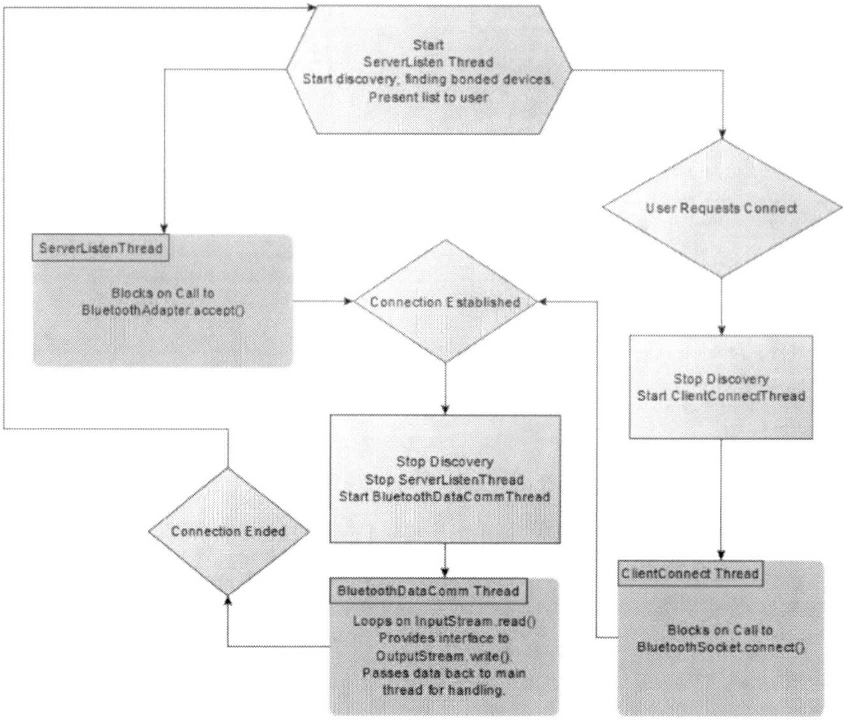

Fig. 2. Diagram showing behaviour flow for a Bluetooth application on Android

In order to initiate a connection with a remote device that is accepting connections on an open server socket, client needs to obtain a Bluetooth Device object that represents the remote device. Consider now the steps of connecting to the server from the client's point of view:

1. Using the Bluetooth Device, get a Bluetooth Socket by calling create Rfcomm Socket To Service Record (UUID). The UUID passed here must match the UUID used by the server device when it called listen Using Rfcomm With Service Record (String, UUID) to open its Bluetooth Server Socket.

2. Initiate the connection by calling connect. Note that this method is a blocking call. After a client calls this method, the system performs an SDP lookup to find the remote device with the matching UUID. If the lookup is successful and the remote device accepts the connection, it shares the RFCOMM channel to use during the connection, and the connect method returns. The method connect is a blocking call, you should always perform this connection procedure in a thread that is separate from the main activity (UI) thread.

After a successful connection, each of the connected devices has a Bluetooth Socket object with which it is easy to implement reception and transmission of data:

1. Using the get Input Stream and get Output Stream methods, get the Input Stream and Output Stream objects that control the transmission through the socket.

2. Read and write data to the streams using read (byte) and write (byte).

Fig. 2 shows a reasonable layout for a Bluetooth implementation

4 Conclusion

Bluetooth provides users with a quick and easy way of exchanging data between a wide range of different devices. Bluetooth technology is a robust, easy-to-use wireless solution for mobile applications. Utilizing Bluetooth in Android applications can be daunting for those unfamiliar with the process. This article can help Android developers gain experience in developing an Android application with Bluetooth.

References

1. Android's Bluetooth API developer guide. Retrieved September 19, (2017).
2. Conder S., Darcey L. *Android Wireless Application Development. Second edition*. (Addison-Wesley, 2011).
3. Gratton D.A. *The Handbook of Personal Area Networking Technologies and Protocols*. (Cambridge University Press, Cambridge, 2013).
4. Bluetooth Core Specification v.5.0. Retrieved September 19 (2017).

The development of a mobile monitoring system for agricultural object

Sergey Konev[1], *Miroslav* Vernezi[1], *Daniil* Donskoy[1], and *Alexander* Lukyanov[1]

[1]Don State Technical University, Chair "Production Process Automation", 344000, Gagarin square, Rostov-on-Don, Russia

Abstract. The article studies the questions of system development for mobile control of data received through wireless networks. For practical implementation, the task has been chosen of control, visualization and saving measurable values received through Bluetooth connection in devices controlled by Android OS. The goal of using such a system may be remote control of temperature, pressure, and other measurable values.

1 Introduction

The task of remote control of temperature, pressure, and humidity is relevant for many industrial and laboratory applications, for example, in the field of thermal processing, in drying equipment, in the food industry, in tasks of maintenance of a micro-climate of premises for growing plants, etc. Conventional systems of remote monitoring are based on the use of wire systems and PC. Such systems have significant disadvantages: they are bulky, and inconvenient to service and reconfigure. Implementation of wireless technologies in this field enables one to refuse to be bound to a certain stationary work place, and to increase mobility thanks to the use of mobile and wearable devices in the control system as a human-computer interface tool, providing additional convenience for operating personnel and increasing its efficiency [1].

2 System architecture

To implement a designed system, it is necessary to select development tools. Based on the task and functions of the system being developed, the following tools can be identified, which are necessary for microclimate control system development:

1. Sensor for temperature measurement
2. Controller for the collection and processing of measured data
3. Device for data transfer to the device for monitoring of readings
4. Client device for temperature monitoring, storage, and visualization of data

2.1 Temperature sensor

DHT22 digital temperature and sensor manufactured by Adafruit Industries is used as a measuring device. The sensor consists of a capacitive humidity sensor, a thermistor and contains an ADC for converting analogue values of humidity and temperature. Determination of humidity - from 0 to 100% with 2% accuracy. Determination of temperature - from -40 to 120 °C with an accuracy of ±0.5°C. Sampling frequency - not exceeding 0.5 Hz (no more than 2 times per second) [2].

Fig. 1. Connection diagram of the Arduino controller with DHT22 sensor and HC06 module.

* Corresponding author: lex1998@rambler.ru

2.2 Controller for data collection and processing

As the device for the primary collection of sensor readings, the most popular controller of the Arduino family is used - Arduino UNO, which continues the very first controller product line. The Arduino Uno controller can be used to create systems for receiving and processing signals from various digital and analogue sensors, and controlling various actuators.

The Arduino Uno controller can be used to create devices and systems for receiving and processing signals from various digital and analogue sensors that can be connected to it, and controlling various actuators [3]

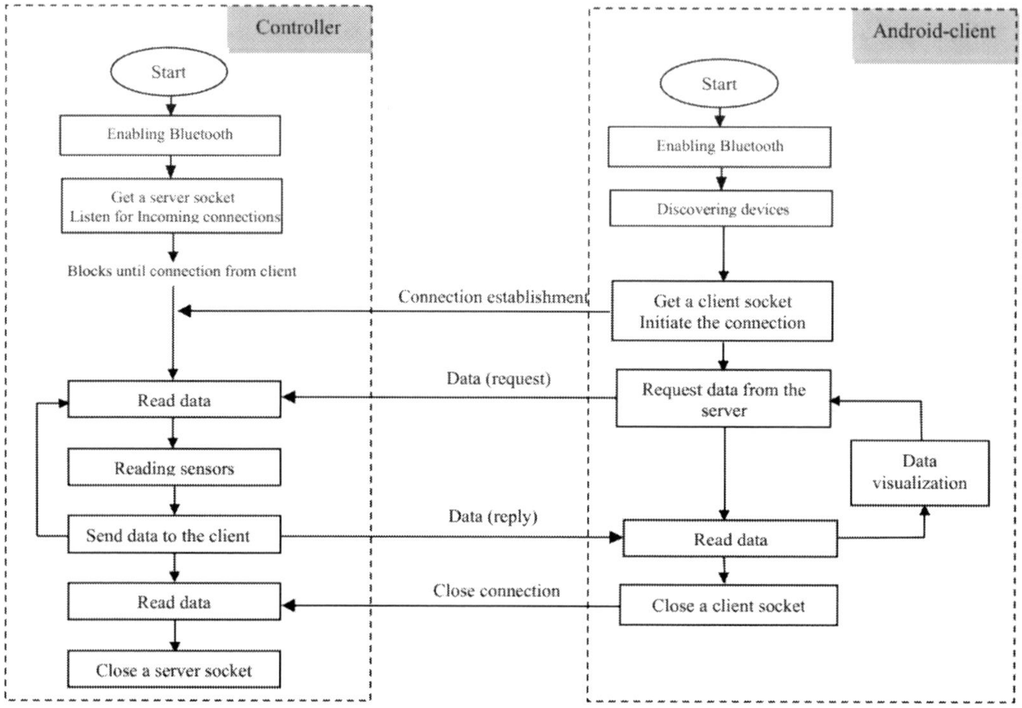

Fig. 2. System operation algorithm.

2.3 Wireless communication protocol and module

Bluetooth protocol is used as the technology for wireless communication between devices. This protocol uses a low power radio interface and was developed as a replacement for existing serial cable interfaces in consumer electronics and industrial applications[4].

Bluetooth technology is a flexible and easy-to-use wireless solution. Operating in the same 2.4 GHz free ISM band (industrial, scientific and medical bands) as other wireless protocols, Bluetooth has optimal characteristics in terms of requirements for stability, reliability, and interaction with wireless LANs, specified by the industry [5].

Data transfer through Bluetooth between the controller and Android-based device is conducted with the help of a connectable external Bluetooth module HC-06 manufactured by Olimex. It is an open-frame monoboard module intended for surface mounting. The board provides support for the Bluetooth 2.0 EDR protocol stack at the hardware level. Built-in flash memory capacity for storing firmware and setup settings is 1 MB. Maximum signal range is 30 m [6]. Figure 1 shows the connection diagram of the Arduino controller with temperature sensor and Bluetooth module.

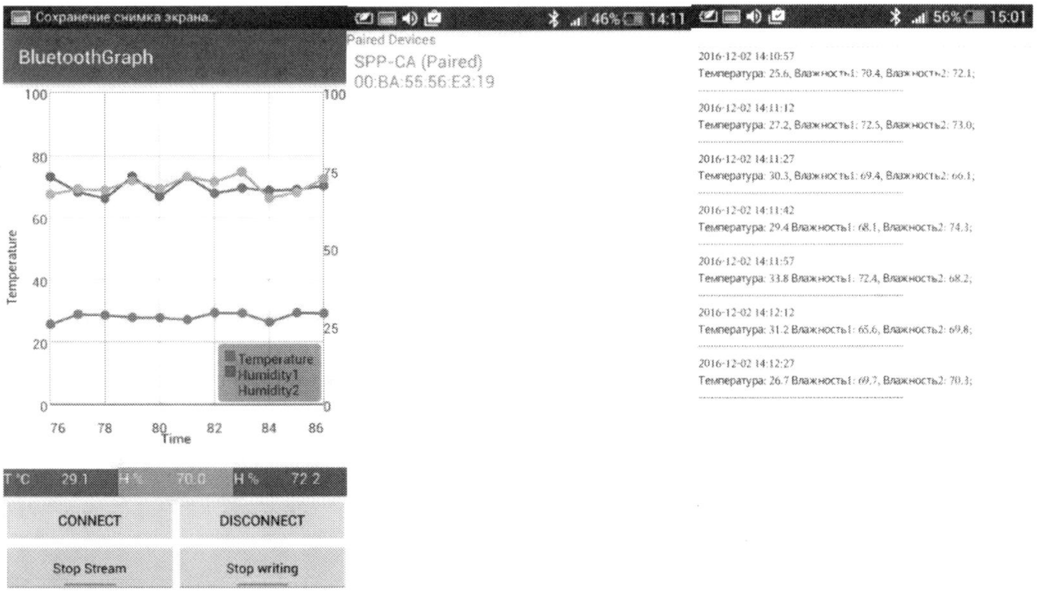

a) b) c)

Fig. 3. Mobile application interface: a) visualization of data received through Bluetooth, buttons for control of the connection and recording of data in a file; b) the list of coupled devices ready for connection and data transfer; c) data received through Bluetooth and saved in a text file.

2.4 Mobile device

An Android-based smartphone is used as the mobile device for remote control of measurable parameters. This is a multi-level environment based on the Linux kernel with rich functionality. Android possesses a wide range of options for connectivity, covering Wi-Fi, Bluetooth and protocols of data transfer over a cellular network (GPRS, EDGE, 3G, 4G, etc.) [7].

3 Operation algorithm

The memory of the Bluetooth module includes a special program, which provides for the performance of all procedures for detection of Bluetooth devices and establishing a connection. This program starts immediately after the controller is turned on, and the module switches to connection standby mode. The Bluetooth connection is arranged according to the "client-server" principle [8]. The Android application is a client that initiates a connection, requests data, and terminates a connection, and the application on the controller is the server that receives the client's commands and provides the data. Variation of values of measurable parameters is displayed on the smartphone in real time in the form of graphs.

3.1. Practical implementation

Practical implementation of the system consists of two programs: a program for the Arduino controller for data reading and processing and transfer of data received from temperature and pressure sensors, and a mobile application for processing, visualization in real time and storing of information received.

The mobile application consists of two parts, of two classes. The first class implements a Bluetooth connection to the remote device, sending and receiving data. The second one implements the user interface and provides for building graphs of measurable values on the main screen of the application and their storage. Accumulated data can be saved in a file on the SD card of the mobile device in a text data format with the .txt extension. Figure 3 represents the application interface and received data recorded in a file.

For visualization of data received through Bluetooth, a GraphView external library is used. This is a library with open source code for building various graphs and diagrams in Android applications [9]. A GraphView library enables visualization and updating of data, which change in real time, and the building of several graphs of various types simultaneously on one screen.

4 Conclusion

This article provides an example of creation of a mobile system for real time monitoring of environmental parameters, based on the use of an Android-based smartphone as a device for remote monitoring, a controller for collection and processing of data received from measuring devices, and Bluetooth wireless protocol for data transfer to the portable device. Practical implementation of a client-server application, operating on an Android-based mobile device, for building real time graphs based on data transferred by an Arduino controller via a Bluetooth connection is presented.

References

1. R. Nilsson, Industrial wireless: Bluetooth can be robust, easy to use (2013) Retrieved: September 22, 2017
2. DHT22 Temperature and humidity sensor. Retrieved September 21, 2017
3. Microcontroller Arduino UNO Retrieved September 22, 2017
4. D.A. Gratton, *The Handbook of Personal Area Networking Technologies and Protocols.* (Cambridge University Press, Cambridge, 2013)
5. Bluetooth Core Specification v.5.0. Retrieved September 19, 2017
6. Bluetooth-Serial module HC-06. Retrieved September 19, 2017
7. S. Conder, L. Darcey, *Android Wireless Application Development. Second edition.* Addison-Wesley, (2011)
8. Android's Bluetooth API developer guide. Retrieved September 19, 2017
9. GraphView – graph-plotting library for Android. Retrieved September 19, 2017

DOI: 10.1051/matecconf/201713204027

Structural-parametric optimization of the experimental data decomposition into approximated fragments

Rudolf Neydorf[1,*], *Orkhan* Iarakhmedov[1]

[1]Don State Technical University, Rostov-on-Don, Russia.

Abstract. In the process of modeling non-linear objects are often used piecewise approximation, spline methods or other methods based on their experimental data fragmentation. To effectively use fragmentation, it is necessary to develop and examine all the described stages of the approach. In this paper, a new approach to the selection of the structure of the decomposition of matrix data has been developed and proposed. It is based on the method of the ant colony, specially adapted for solving similar problems. Along with the characteristic properties of ant algorithms, special elements of interaction between agents of different colonies were introduced into this modification, for the efficient operation of the algorithm. The approach is considered as a tool for implementing the initial stage of any methods using data fragmentation. In the process of work, a software tool was designed and implemented in C #. The main data structures used in the software are described. Adjustment of the tuning parameters and the graphic output interface of the work process are implemented in the software tool. The graphical output was implemented by built-in tools. Test cases in detail describe the stages of data fragmentation under different initial parameters.

1 Introduction

In the process of modeling various objects and phenomena, many difficulties arise. They often occur in connection with a significant non-linearity of the test object [1-2]. In this case, approximative approaches are used, oriented to fragmentation of an array of experimental data (ED), for constructing a mathematical model (MM), because the fragments can be described by exactly considerably less complex MM. The most popular in solving these problems is the use of piecewise approximation [3], spline approximation [4-5], and methods of radial basis functions [6]. A comparatively recently developed of "Cut-Glue" approximation method [7-9], which has proved to be a very effective and promising method of mathematical description of nonlinear dependencies, is oriented towards fragmentation too. An obvious advantage of the "Cut-Glue" method over the others is the description of data by a single function, which is also analytical.

The initial stage of any of the listed methods is the splitting of the ED into fragments with the least pronounced nonlinearity. This dramatically increases the possibility of an accurate description of each ED. Therefore, the task of efficient fragmentation of data is extremely important, because is an important preparatory stage to any method focused on fragmentary approximation.

In connection with the fact that combinatorial methods for the formation and comparison of variants of ED fragmentation clearly refer to NP-completeness, in this paper, one of the possible approaches based on the biological phenomenon of rivalry of ant colonies was proposed and investigated [12-13], since preliminary studies have shown promising results.

1.1 Formulation of the problem

It is necessary to develop an search suboptimization algorithm of dividing the area of ED into fragments with a phased use of the scenario of seizure of territories by developing ant colonies.

2 Algorithmic model

As a domain of disputed territories, an ED matrix is considered. To divide it into fragments, the rival ant colonies method (RACM) is used, which simulates the behavior of ant colonies in the division of their habitats [10-16], and adapted for the task. It is based on the mechanisms of interaction of ants with each other, with the environment, as well as colonies as societies among themselves.

In the process of partitioning, we can distinguish the following stages: initialization, agent migration, pheromone laying and checking the break condition, which are also in the classical ant algorithm.

All investigated space is traditionally divided by a grid, but not conditional, but rigidly tied to segments of step-by-step variation of factors when obtaining ED [14-16]. The number of colonies is varied by the "Cut-Glue" approximation algorithm, because it affects the number of fragments obtained after the partition. At their casual arrangement there are a number of restrictions. The colony can not be located on the boundaries of ED. This condition is established for a more even distribution of

* Corresponding author: ran_pro@mail.ru

territory between colonies. Also, the standard normal distribution is used in a random arrangement. Each colony generates a certain initial number of ants in the range that the operator specifies. This is done to ensure that the colonies differ in numbers, and therefore in capture capabilities.

There are several features of territory development by ants, as well as their interactions inside and outside the colony. Since the splitting as a result must consist of a set of rectangular-shaped fragments, a rule for preserving the rectangular shape of the territory under study is provided.

Moving virtual ants is carried out on a probabilistic basis [14-16]. At each iteration, each agent considers 8 possible directions to move as shown in Figure 1. Since the ants can expand the territory of the colony either by mastering unoccupied territories, or by capturing strangers, the corresponding stages are envisaged.

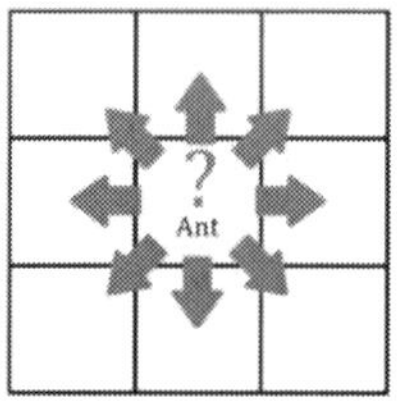

Fig. 1. Available options for moving

Expanding one of the four boundaries of the colony, the ants test it for the presence of someone else's pheromone. If the new border area is completely free, it is endowed with the pheromone of the colony, and the extension continues in a similar way. In the event that an someone else's pheromone is found on the territory under development, the ants return to their assigned territory and are looking for new ways of expansion.

The pheromone is used to highlight areas of the colonies. Its updating occurs at each moving of ants. Finished fragments after the splitting pass through the optimization and gluing stages.

In conditions when each colony has 4 boundaries that form a rectangle or a square, there can be a free space between the colonies, which is not occupied by any of the colonies (wasteland). To solve such situations, the algorithm includes the mechanism of interspecific competition of colonies bordering on the wasteland. The essence of this mechanism is that the disputed territory begins to retreat to that colony that occupies a smaller area. The introduced priority is due to the fact that when it is expanded the changes in the fragmentation structure will be minimal. Too small fragments during the partitioning are often inefficient, both from the position of the quality of the approximation, and from the position of the number of fragments. If the stage of "reclaiming the wasteland" leads to the formation of a new wasteland, the procedure for assessing neighboring fragments is repeated and a new applicant is appointed to capture it.

The break condition at the stage of partitioning is the satisfaction of the integrity condition and rectangularity of all the fragments obtained, as well as the absence of wasteland in the fragmented space.

Since fragments should have adjacent boundaries, it is necessary to work out a rule in which these faces will be determined. In this case, each fragment will include one more vector of values to the right and the bottom of the fragment, unless from these sides it is not limited by the value of the matrix.

Therefore, the result of a single execution of RACM is the variant of data fragmentation. Multiple launch of RACM allows you to generate a sample of options to search for the optimal or suboptimal by the total accuracy of the approximation data under study. These variants go through further stages of approximation, "gluing" and estimating the error of describing the entire dependence in accordance with the "Cut-Glue"approximation paradigm.

2 Conseptual-mathematical model

In connection with the task, it is necessary to build a RACM MM, radically different from the classical route model [10-11]. This is due to the fact that ant colony algoithm was first applied to the fragmentation problem solution. Nevertheless, some principles for the movement of agents remain unchanged; are the basis of the very principle of the "sequential-kinematic" solution of the problem.

Consider ED as a set of D numbers that are the values of the responses (output variables) of the described object and are associated with the points of the r-dimensional factor space (FP) of the input (independent) variables. In this paper, we consider a two-dimensional version of a phase transition.

For any fragmentation methods and algorithms (FA), the following two basic conditions must always be fulfilled, ensuring the correctness of the procedure.

1) The combination of all the obtained experimental data fragments (EDF) forms the entire investigated set of ED:

$$D_1 \cup D_2 \cup \cdots \cup D_n = D \quad (1)$$

where $D_i,\ i = \overline{1,n}$ – fragments of a partition; n – number of fragments; D - set of ED.

2) The intersection of all the obtained EDF forms a subset of ED, which is the union of the boundaries of all fragments:

$$D_1 \cap D_2 \cap \cdots \cap D_n = G_1 \cup G_2 \cup \cdots \cup G_n \quad (2)$$

where $G_1, i = \overline{1,n}$ – the edges of fragments, which in the general are the partition.

The first condition is obvious. The second condition excludes the overlapping of fragments in which the boundary data of some fragments coincide with the internal data of other fragments. When the EDF merges ("gluing together") with the fulfillment of these conditions, the distortions in the transition between them are the least significant [7-9].

As already noted, the early agents master the space of experimental data presented in the form of a matrix, while moving, they are guided by a certain behavior and direction, which has both random and meaningful components.

Meaningful components are determined by the overall goal of FA and depend on many factors: the state of the boundaries of their colony, the state of the nearest border areas and the territory's fullness with a pheromone and etc.

Random components of displacement are used to expand the range of search for various solutions of the problem, greater uniformity of development of the territory in all directions, and so on.

The behavior of agents depends on many factors: the fulfillment of the rectangle condition, the availability of adjacent faces for expansion, the occupancy of adjacent faces in the expansion of the territory, the state of the integrity of boundaries, and others.

Under different statuses of these conditions, it is possible to distinguish certain features of the colony's behavior.

For example, with a "peaceful" expansion of borders, agents randomly choose the direction and check the whole new face for the possibility of moving and filling it with their pheromone. If the fragments do not contain a foreign pheromone and there are no other restrictions, the territory passes under the possession of the colony and is marked with a certain pheromone (color), as shown in Figure 2 (a) and 2 (b).

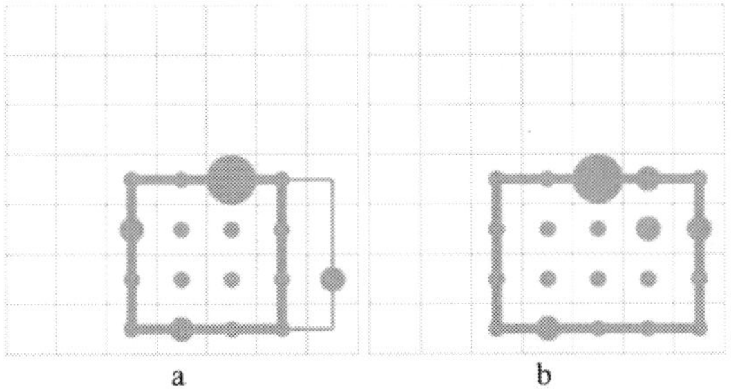

a b

Fig. 2. Process of capturing an unoccupied territory

In a conflict ("aggressive") expansion, agents intentionally shift the boundaries of their colony and mark the entire territory with their pheromone, including those fragments that belonged to another colony. An example of this process is shown in Figure 3 (a) and 3 (b).

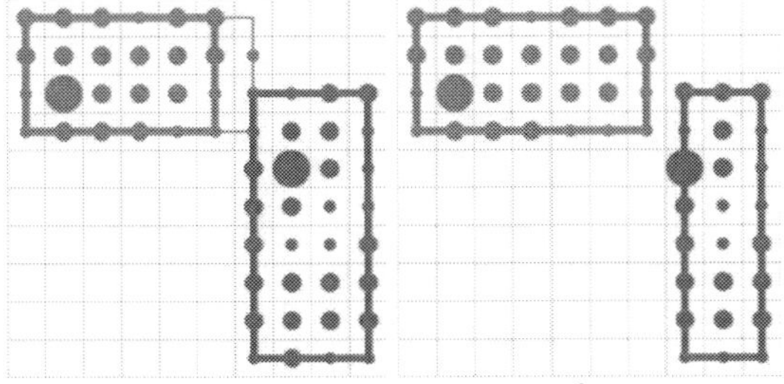

a b

Fig. 3. Process of capturing someone else's territory

Such behavior is caused by the appearance of all kinds of collisions between the basic principles that are mandatory for fragmentation (for example, conditions (1) and (2), the requirement of rectangular fragments, etc.), and the real configurations, for example, the appearance of the previously described "wastelands". This mechanism is illustrated in detail in the test case later in this work.

In the process of struggle between the colonies, their structure changes because of "wastelands," and the basic rule of rectangularity is violated. In these conditions, agents need to restore their maximum possible territory by retaining the rectangle rule. To do this, all agents return to the colony, and re-expand the boundaries of the territory already marked with their pheromones, as far as possible (see Figure 4 (a) and 4 (b)). At the same time, the pheromone on the side that is partially or completely captured, evaporates, thus a new "wasteland" and the opportunity to change the structure of fragmentation appear.

There are also other collisions, the analysis of which is beyond the scope of this article.

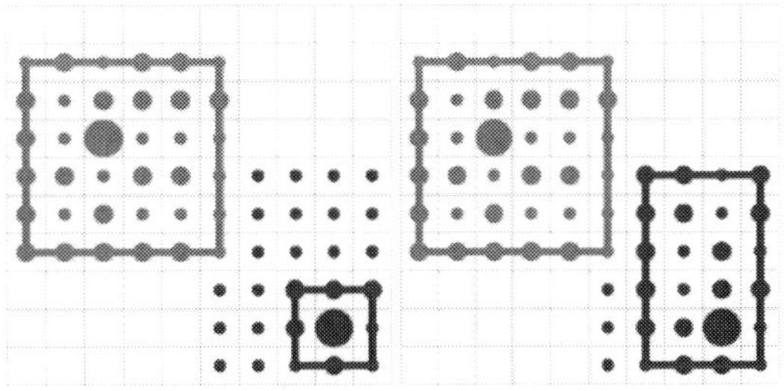

a b

Fig. 4. The process of restoring borders

Based on the developed algorithm, its mathematical and conceptual models (1-2), the "Experimental Data Fragmentation by Rival Ant Colonies Method" (EDF_RACM) software tool (ST) was designed and developed.

3 Demo software tool "EDF_RACM"

3.1 Data structures

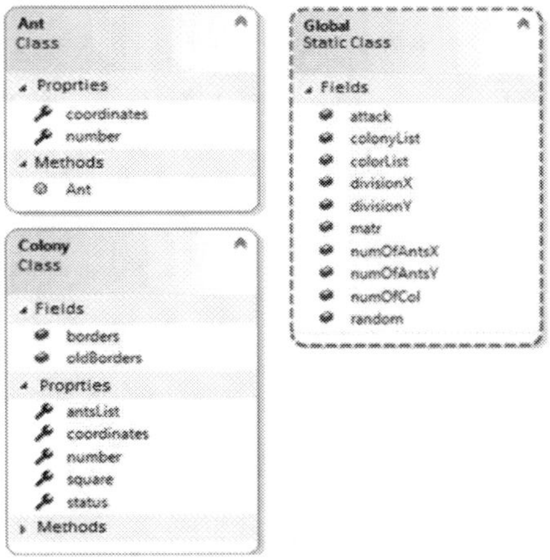

Fig. 5. UML-diagram of EDF_RACM

Figure 5 shows the UML-diagram of the ST classes.

Figure 5 shows that this ST consists of 3 main classes:

- class Ant – class describing an ant as an object;
- class Colony – class describing the colony as an object;
- class Global – class used to store global values and methods.

3.2 GUI

Figure 6 shows the main window of the software tool.

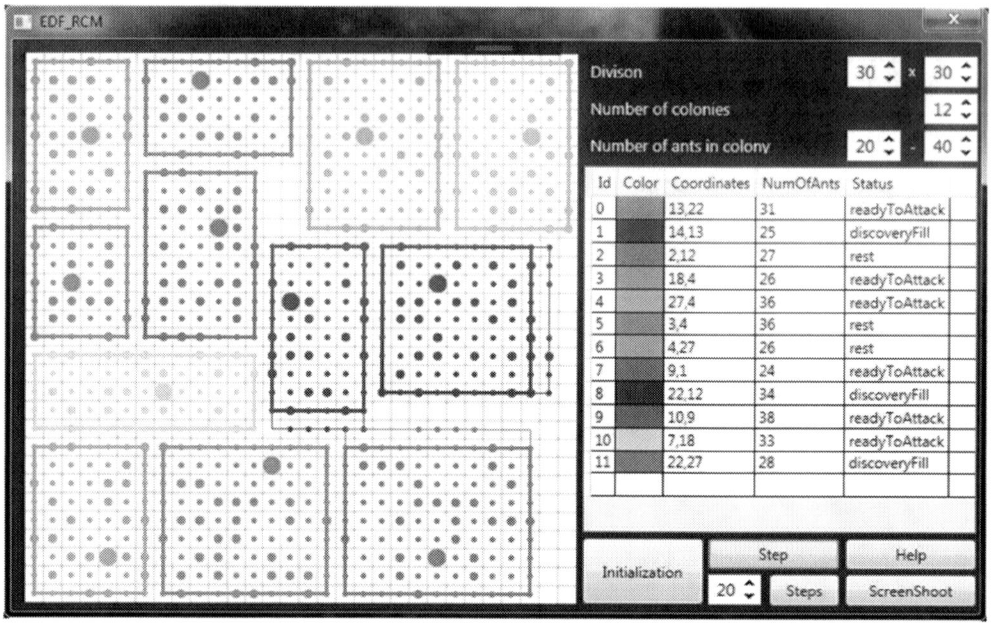

Fig. 6. GUI

The main window of the ST consists of a graphical output window to the right and a panel for setting parameters and outputting information to the left. Thanks to the graphical interface, you can visually follow all the steps of fragmenting experimental data, and using the settings panel you can select a variety of source tasks. Colonies are marked with different colors, and their boundaries are indicated by bold lines.

In the settings panel, the operator can specify the partition, the number of colonies and the range of generated ants. There is also an information area in which the current status of the colonies is reflected. At the bottom of the tuning panel there are process control buttons.

3.3 Experiments

As a demo, the space of 30x30 dimension was considered. It was generated 8 colonies in a random order. Figure 7 (a) shows the initialization stage, which shows the location of the 8 colonies. Figures 7 (b) and 7 (c) illustrate the intermediate stage and the resulting result in the form of a partition.

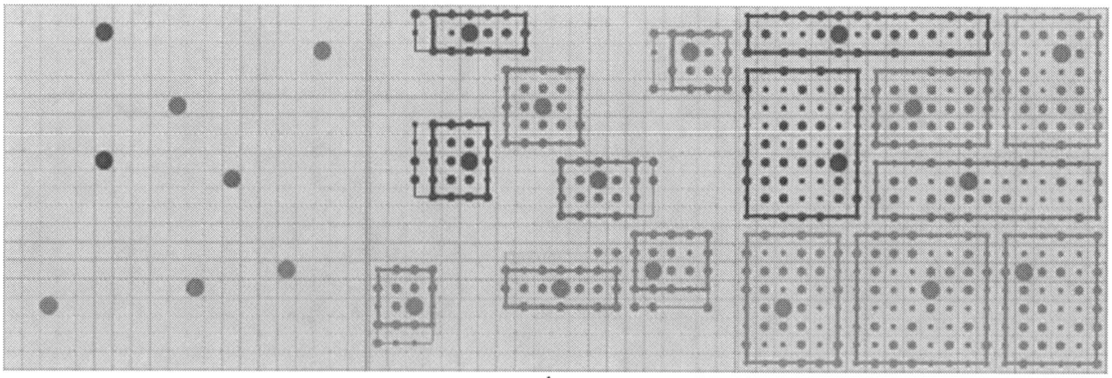

a b c

Fig. 7. Stages of the program: a – the initialization stage; b – intermediate stage; c – result

Experiments with matrices of various dimensions were carried out. The results of partitioning the matrices of dimension 10x10, 30x30 and 40x40 are shown in Figure 8 (a), 8 (b) and 8 (c).

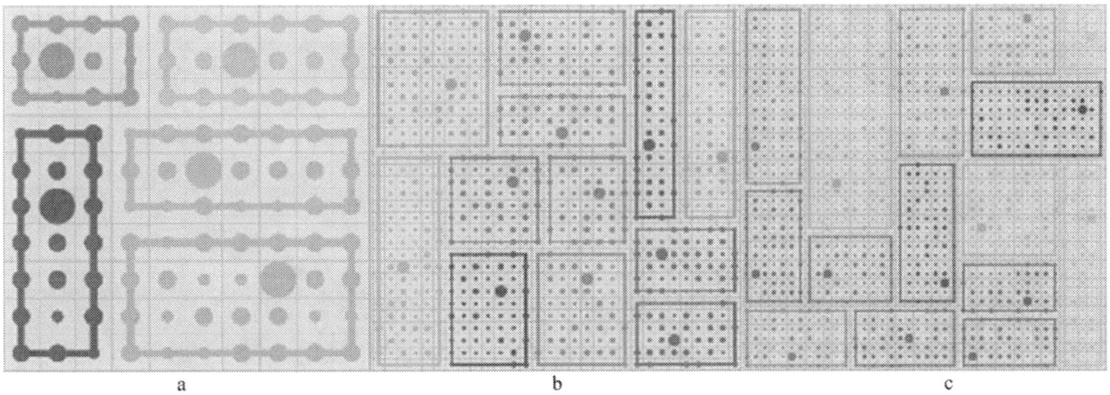

Fig. 8. Examples of partitions: a – partition 10x10; b – partition 30x30; c – partition 40x40.

3.4 Additional experiments

To demonstrate the mechanism of getting rid of "wastelands", an additional experiment was conducted. 6 colonies were generated in a random order. Figure 9 (a) shows the stage in which it is seen that all free space has been investigated, with the exception of one controversial fragment, and a colony with a minimal territory begins to capture this fragment, advancing its right border. This change entailed a change in the boundaries of another colony and the liberation of a new territory for development (see Figure 9 (b)). The following stages are successively shown in Figure 9 (c), 9 (d), 9 (e) and 9 (f).

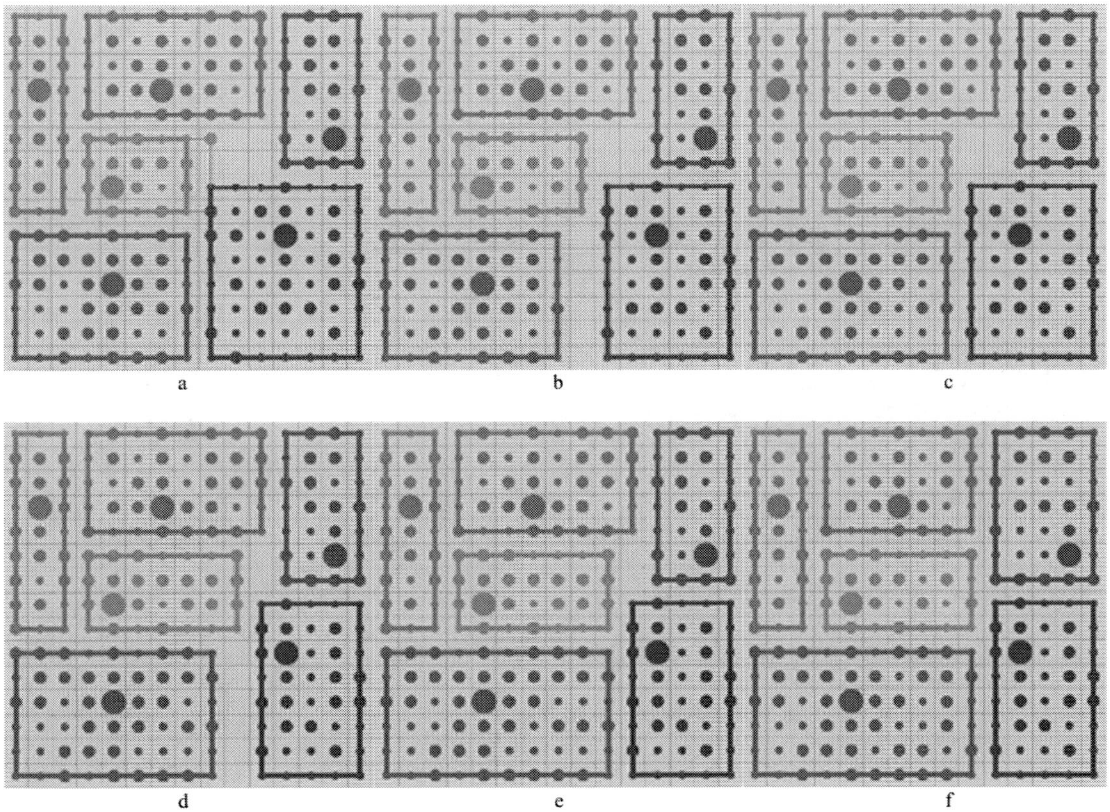

Fig. 9. Stages of the program in the occurrence of "wasteland"

* Corresponding author: ran_pro@mail.ru

3.5 Computational resources and performance

This ST was developed in the Microsoft Visual Studio 2017 environment using the C # programming language. The solution of the tasks was carried out on the PC with the AMD Phenom II P960 process and 6Gb of RAM.

4 Conclusion

The developed modification of the ant algorithm in the RACM variant showed acceptable qualitative and quantitative results in solving problems of varying complexity. The check was carried out on tasks with a partition from 10 to 50. In all tests, the algorithm did not allow errors during fragmentation. The results of it tests showed the expediency of continuing research in the chosen direction. It can be concluded that the developed ST meets all the necessary conditions and allows you to find the required partitioning.

It should be noted that the algorithm requires the involvement of some additional mechanisms to improve performance [14-16].

References

1. G. Boeing, Visual Analysis of Nonlinear Dynamical Systems: Chaos, Fractals, Self-Similarity and the Limits of Prediction. Systems. **4** (4): 37. (2016).
2. A.V. Panteleev *Nonlinear control systems: description, analysis and synthesis* (University book, Moscow, 2008)
3. P.-J. Laurent, *Approximation and optimization* (Hermann, Paris, 1972).
4. C. De Boor, *A practical guide to splines* (Springer, 1978)
5. Z. Cieselski, J. Musielak. Spline bases in function spaces. Approximation theory, Reidel, 49 (1975).
6. M.D. Buhmann. *Radial Basis Functions: Theory and Implementations*, (Cambridge University Press, 2003).
7. R. Neydorf, *Proceedings of the ASME 2014 International Mechanical Engineering Congress and Exposition*. (Montreal, Quebec, Canada, 2014).
8. R. Neydorf, SAE Int. J. Aerosp. **8**(1) 47-54 (2015)
9. R. Neydorf, A. Neydorf, SAE Technical Paper, 2016-01-2035 (2016).
10. M. Dorigo, L.M. Gambardella, IEEE Trans. Evol.Comput., **1**(1) 53 (1997)
11. M. Dorigo, G. Di Caro, L. M, Artificial Life, 5(2) 45 (1999)
12. K.P. Fatemeh, F. Fardad, S.N. Reza, Journal of Academic and Applied Studies. **3**(2), 15 (2013).
13. M.D. Toksari, Applied Mathematics and Computation. 176, 6 (2006).
14. R. Neydorf, I. Chernogorov, V. Polyakh, O. Yarakhmedov, J. Goncharova, D. Vucinic, *Proceedings of The X International Conference on Advanced Engineering Computing and Applications in Sciences ADVCOMP2016*, 7 (2016).
15. R.A. Neydorf, O.T. Yarakhmedov, V.V. Polyakh, I.V. Chernogorov, D. Vucinic *ACEX2016-10th International Conference on Advanced Computational Engineering and Experimenting, ACEX280*, (2016).
16. C.Y. Pang, X. Li, *5th IEEE Conf. on industrial Electronics and Applications*, 4 (2009).

* Corresponding author: ran_pro@mail.ru

Dynamic document object model formation technique for corporate website protection against automatic coping of information

Vasily Galushka[1,*], *Daniil* Marshakov[1], and *Vladimir* Fathi[1]

[1]Don State Technical University, Computer systems and information security department, 344000 Gagarina sq. 1, Rostov-on-Don, Russian Federation

Abstract. The article describes solution path of the problem of information automatic copying from web-sites in the Internet, which is implemented using parsing techniques based on regular expressions or function libraries. To protecting against this type of information security threat, it is proposed to dynamically generate and periodically change the object model of the HTML document when generated and sent to the browser. These changes should affect the values of the identifying tag attributes and the structure of the object model tree. As attribute values it is offered to use character sets of limited length obtained as a result of random numbers hashing; change of the structure of the object model should be done by adding of additional tags at the corresponding levels of the hierarchy of the tree representing it. The simultaneous application of these methods excludes the possibility of algorithm compile for the necessary information extraction in the overall structure of the web page.

1 Introduction

Currency, that is the correspondence to the current moment of time, is an inalienable property of any information, however, in some cases it acquires special significance. This is the case of information available on the websites on the Internet and has value for short time interval, for example, news or stock-exchange quotations short-term forecasts. In addition to standard security threats, such websites are subject to another kind of negative impact - automatic copying of information, carried out with the hardware and software help without human intervention, with minimal resources, often in real time. Such process is called parsing [1-5] and is realized on the basis of the regular expressions [5-6], or with specialized program functions libraries use [7].

Notwithstanding that the information resources above as an example are objects of intellectual property and should be protected by copyright law, unfortunately, the practice shows [8] the complexity of legal methods applying at Internet security providing, as well as epy problems of author rights attribution and confirming, which leads to the need to develop technical and software security tools.

2 Parsing

The methods of information automatic copying on web-sites are based on the syntactic and semantic analysis of the HTML- page source code. This is due to the fact that access to any other information can not be obtained from outside, at least when proper security tools used by the site, web server or hosting administrator. For technique of protection against automatic copying development it is necessary to establish the reasons and vulnerabilities allowing to make this breach of website's security, and for this purpose to analyze of HTML language features and to reveal those whichever allow to realize copying of the information contained on web pages with software and without human involvement.

It is obvious that it's impossible to do without the human involvement in this process, so the phrase "without the human involvement" means only the process of direct data transfer from the site-victim to the site of the attacker at the user request. In this case, human intervention at the stage of parsing system setting up is necessary. Human intervention functions includes manual analysis of the structure of the HTML code, definition of blocks with useful information and their separation from the rest of the document, as well as cleaning of the content from the tags that specify the markup, because on the website on which information is copied, the markup can, and, most likely, will be completely different. Then, based on the analysis, parsing rules will be compiled and written in the programming language, most often using any library that facilitates the work with the pages source code.

It is important that these actions, although they are rather difficult and require high qualification of the person performing them, but are executed only once — at a stage of website creation. At the same time, there is

* Corresponding author: galushkavv@yandex.ru

no method to hide the HTML code from such an attacker, since in compliance with the standards it must be transmitted as the clear text to the browser. It follows from the HTML language assignment — the markup of a document is its main function, that is, document dividing into parts or blocks, each of which must be logically separated from the rest and have all the necessary information about the final representation of its contents, as well as the content itself. The HTML language does not provide any built-in protection. It is assumed that if the information on the sites is intended for open access, then it makes no sense to complicate both the language itself, and the means of its interpretation by additional information security features.

Regardless of the implementation method, parsing uses the object model (DOM) of the HTML document to search for the information of interest to it and to separate the content from the design defined using HTML tags and their attributes [9-11]. Therefore, protection from this type of information security threats should be based on changing objects included in the DOM, in order to make difficult, and ideally impossible, the algorithmic analysis of the object model by software.

The object model is a universal and generalized document structure description and has the form of tree whose the tags are nodes of a tree, and the branches are the connections between them [12]. The document object model is built by the web browser on the analysis of the HTML page source code and serves as an internal intermediate representation formed prior to direct draw of the page

The simplest and, the most frequently parsing quite effective way is the search for the tag by its attributes: "id" and, more rarely, "class". These attributes allow the most exact identification of the tag containing the required information and are present at any HTML-design template, and their values are associated with the corresponding identifiers in the CSS file for use of the styles described in it. To eliminate the possibility of program creation capable to execute search of the tag by these attributes, it is necessary to assign random values to them at each page load.

3 Object model automatic generation with dynamic attributes use

In accordance with the CSS standard [13], the first characters of identifiers can not be the numbers or special characters such as commas, colons and the like, and they are usually written as text values. However, the random number generator in PHP, as well as in any other programming language, is focused on the numbers formation only. The MD5 hashing use to text formation from numbers is proposed. Although this algorithm already some time is not considered reliable and strong [14-15], its application for this problem is justified, since the hashing algorithm functions, in this case, reduce only to the transformation of values randomly provided by the random number generator [16]. Since the algorithm MD5 generates an alphabetic and numerical sequence of 32 characters in length, 2 problems arise: to shorten the sequence length and to exclude the digits appearance in the beginning of sequence.

To solve these problems, the names formation will be made in accordance with the following rule:

$name := prefix + substring_{0,5}(md5(rand(0, PHP\_INT\_MAX)))$, where

- *prefix* — text constant, defined for the site, in whole;
- $substring_{k,n}$ — is the function of part of a string selection starting with the *k*-th position with the length of *n* symbols;
- *md5* — is the function of the hash sum computing;
- *rand(l,m)* — is the function generating a random number in the range from *l* to *m*;
- *PHP_INT_MAX* — the predefined constant, meaning the maximum integer value supported by this PHP version.

The prefix presence allows to prevent the use of the value starting with the digit resulting from the hashing as the identifier, and the assignment of a 5-character substring (each of which can be one of 26 letters of the English alphabet or one of 10 digits) will provide $(26 + 10)^5 = 60\ 466\ 176$ different versions of identifiers, which is enough for most sites, and, this amount can be increased, if necessary.

The software implementation of this technique can be divided into several stages.

1. Separation of the CSS file into static and dynamic parts

2. Random values generation for properties of tags that meet the requirements for CSS-identifiers.

3. The received values writing to a file with the dynamic part of the CSS.

4. The received values assignment to tags with information to be protected.

In this form the algorithm is applicable at relatively small load on the site, since, with a large number of access to the site, CSS file overwriting will be performed too often and can lead to the fact that the page with one set of identifiers of tags will be sent to the user, at this time another request will come, new random values not corresponding to those sent will be written to the file. In this regard, the presented algorithm should be applied separately for each user, the names of CSS-files should be stored in the database, and the files themselves should be dynamically generated on request.

4 Dynamic object model automatic formation based on a change in document tree structure

The above described method does not provide protection against tag search by its position respectively to other tags. So for DOM tree, shown in Figure 1 as an example, the conditional path to the contents of the first paragraph would be written without any attributes use in the form of the following parsing rule:

HTML → BODY → P_1, where

P_1 — is the first <P> tag from the set of all tag data.

In general, the rule can be written as:

Root → ... → Parent → Child → ... → $Element_i$

This record realizes the passage through the tree in depth and is the fastest way to access to the tree finite elements (leaves), which, as applied to the structure of the HTML document, usually contain all the useful information [17-18].

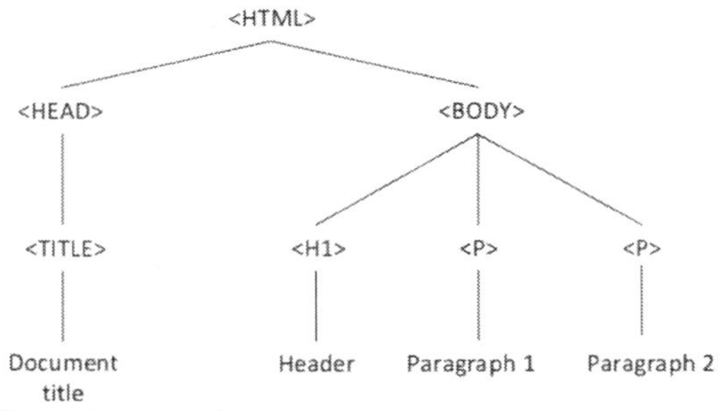

Fig. 1. Document object model tree.

The possibility of such calls to the content for its automatic copying can be prevented by structure of the DOM changing by adding of some random number of additional <P> tags before the information in which it is necessary to be protected. Thus, in the object model tree, a new "extra" element will be added to the same hierarchy level as the protected one. We will designate such approach as "increase in width of tree" (Figure 2).

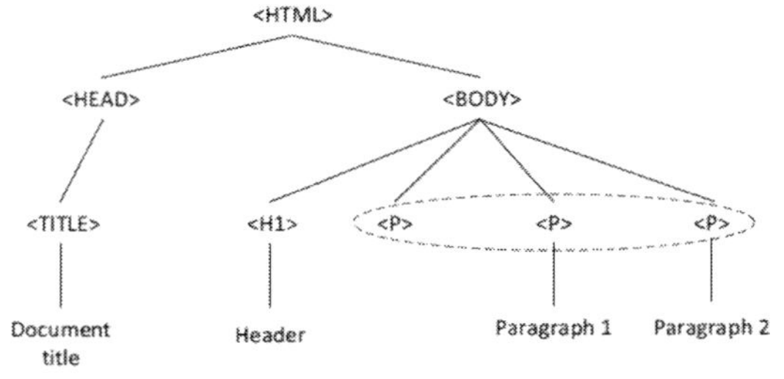

Fig. 2. Increase in width of tree.

Another approach — increase in depth of tree, involves the new nesting levels adding to the HTML document [19-20] block structure i.e., new hierarchy levels to the DOM tree adding (Figure 3).

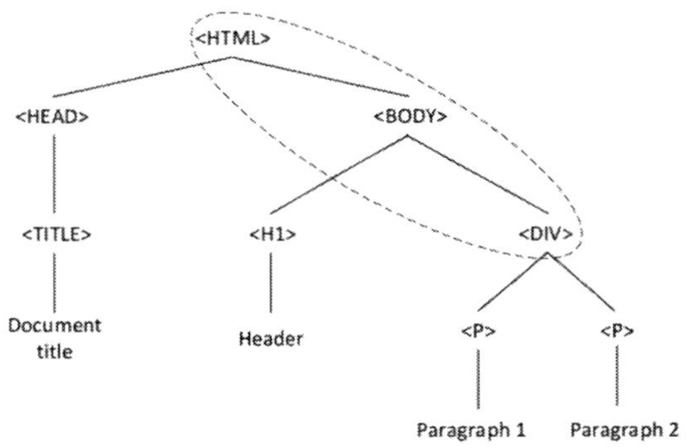

Fig. 3. Increase in depth of a tree.

It is important to correctly determine the branch of the tree to which the elements will be added, as well as their number and set of properties, and the properties of the added elements should not allow them to be distinguished from the real ones in the process of parsing. Let's consider the listed conditions in order.

When determining the tree branch, it should be guided by the following considerations:

– it should be on the same tree level as the node that contains protected information

– branch can be one level above the node with protected information, provided that it includes one of the ancestor node

The next parameter is the number of added elements, which should be a random value. This will allow to complicate parsing having excluded a possibility of path instructions on the object model tree to required element The best option is to add from 1 to 3 additional elements what on the one hand ensures that the tree will be changed at least by 1 element, and not too complicate the structure of the document and will not lead to its distortion by the browser, on the other hand.

In the process of implementation of these methods we must not forget that changes in the structure of the document must not affect its display in the browser and the user's perception, which imposes certain restrictions on the set of admissible changes. So, for example, already considered the <P> tag adding may cause too much padding for the following paragraph, and adding of the <DIV> tag can lead to a complete change in the page appearance. To avoid such situations add tags CSS-property "visible" with a value of "false" or property "display" with value "none" should be set.

5 Conclusion

The combination of the described methods of the document dynamic object model development one of which is based on use of accidental values of identification attributes of tags, and the second on change of the by increase in its width and depth, deprives of the malefactor of an opportunity to formulate and write an algorithm of website pages parsing for separation in them significant units of up-to-date information that in turn makes impossible its automatic copying.

References

1. J. Tidal. The Code4Lib Journal, no.18 (2012)
2. P. Patil, P. Chawan, P. Chauhan. International Journal of Advanced Research in Computer Engineering & Technology, no.1 (2012)
3. H. Kim, K. Doh, D. Schmidt. *Lecture Notes in Computer Science: 20th International Symposium*, 7935 (2013)
4. Z. Zhao, M. Bebenita, D. Herman, J. Sun, X. Shen. ACM Transactions on Architecture and Code Optimization, no.4 (2013)

5. Y. Han, S. Oh. Journal of the Korean Society for information Management, no.2 (2010)
6. S. Levithan, J. Goyvaerts *Regular Expressions Cookbook, 2nd Edition* (O'Reilly Media, Sebastopol, 2012)
7. Z. Chun-lin, S. Fang-hao. Computer Programming Skills & Maintenance, no.12 (2014)
8. G. Kunanbaeva. Bulletin of the Kazakh-American free university, no.4 (2009)
9. C. Lindley *DOM Enlightenment: Exploring JavaScript and the Modern DOM*, (O'Reilly Media, Sebastopol, 2013)
10. R. Győrödi, C. Győrödi, G. Pecherle, G. Cornea. WSEAS Transactions on Computers, no. 9 (2010)
11. S. Jensen, M. Madsen, A. Møller. *In proc. 19th ACM SIGSOFT symposium* (2011)
12. Document Object Model (DOM) Technical Reports, available at: https://www.w3.org/DOM/DOMTR
13. Cascading Style Sheets Level 2 Revision 1 (CSS 2.1) Specification, available at: https://www.w3.org/TR/2011/REC-CSS2-20110607/.
14. S. Kumar, E. Gupta. International Journal of Computer Science and Information Technologies, no. 5 (2014)
15. X. Zheng, J. Jin. *9th International Conference on Fuzzy Systems and Knowledge Discovery (FSKD) (2012)*
16. J. Trampert, A. Fichtner, J. Ritsema. Geophysical Journal International, no. 2, (2013).
17. L. Sang-Un. Journal of the Korea Society of Computer and Information, no. 7 (2014)
18. Z. Tao. Journal of Chinese Computer Systems, no.1 (2014)
19. PHP Documentation, available at: http://php.net/manual/en/function.rand.php
20. HTML 4.01 Specification, available at: https://www.w3.org/TR/1999/REC-html401-19991224/

Fusion of Deep Features and Weighted VLAD Vectors based on Multiple Features for Image Retrieval

Yanhong Wang[1,2], *Yigang* Cen[1,2], *Liequan* Liang[3,*], *Linna* Zhang[4], *Viacheslav* Voronin[5], *Vladimir* Mladenovic[6]

[1]School of Computer and Information Technology, Beijing Jiaotong University, 100081, Beijing, China
[2]Key Laboratory of Advanced Information Science and Network Technology of Beijing, 100081, Beijing, China
[3]Institute of Electronic Commerce, Guangdong University of Finance & Economics, 510320, Guangzhou, China
[4]College of Mechanical Engineering, Guizhou University, 550025, Guiyang, China
[5]Department of Radio-electronic systems, Don State Technical University, 346500, Rostov-on-Don, Russia
[6]Faculty of Technical Sciences University of Kragujevac, 32000, Cacak, Serbia

Abstract. In traditional vector of locally aggregated descriptors (VLAD) method, the final VLAD vector is reshaped by summing up the residuals between each descriptor and its corresponding visual word. The norm of the residuals varies significantly, and it can make "visual burst". This is caused by a fact that the contribution of each descriptor to VLAD vector is not the same. To address this problem, we add a different weight to each residual such that the contribution of each descriptor to the VLAD vector becomes even to a certain degree. Also, traditional VLAD method only uses the local gradient features of images. Thus it has a low discrimination. In this paper, local color features are extracted and used to the VLAD method. Moreover, we fuse deep features and the multiple VLAD vectors based on local gradient and color information. Also, in order to reduce running time and improve retrieval accuracy, PCA and whitening operations are used for VLAD vectors. Our proposed method is evaluated on three benchmark datasets, i.e., Holidays, Ukbench and Oxford5k. Experimental results show that our proposed method achieves good performance.

1 Introduction

In this paper we consider the task of large-scale image retrieval. In the past few years, Bag-of-Visual-Words (BOW) [1] [2] method has achieved great effect in image retrieval area. Generally, in order to ensure retrieval recall, a relatively large vocabulary will be required. Thus, it will lead to a low efficiency of retrieval time and high memory consumption.

Recently, Jégou et al [3] proposed vector of locally aggregated descriptors (VLAD) model, which aggregates descriptors based on a locality criterion in feature space. In fact, VLAD is a kind of representation of Fisher vector without probability. Its implementation is very similar to the BOW model. Also, VLAD is very cheap in consumptions of time and memory. In traditional VLAD method, the final VLAD vector is reshaped by summing up the residuals between each descriptor and its corresponding visual word. The norm of the residuals varies significantly, thus it can make "visual burst" [4]. To address this problem, we add a weight to each residual such that the contribution of each descriptor to the VLAD vector becomes even to a certain degree.

Originally, the SIFT [5] descriptors are adopted in VLAD method, and has shown good performance. As we all known, the SURF descriptor [6] is faster than the SIFT descriptor. Moreover, the performance of SURF and SIFT is comparative in most cases. [7] verified that the SURF descriptor was not only more efficient but also leading to higher accuracy than SIFT and rootSIFT descriptors. However, both the SIFT and SURF descriptors represent only local gradient information, which miss important color information. In order to solve this problem, many works combine the gradient and color information. For examples, in [7] CSURF feature was proposed, which are SURF-based color information; In [8], the author fused the CLOG [9] features and the SURF features at the stage of similarity measurement; In [10] the author proposed "color-SURF" descriptors which combined SURF with the approximate color local kernel histograms. In this paper, Color names (CN) [11] and SURF features are used in VLAD method.

In recent years, deep features are popular for image processing, such as image classification [12], object detection [13] and speech recognition [14] etc. In this paper, we adopt the pre-trained networks to obtain the deep features of images. Also, in order to improve retrieval accuracy, multiple VLAD vectors and image representation based on deep features are fused.

* Corresponding author: lianglq@gdufe.edu.cn

2 Methodology

2.1 Framework of our proposed method

The framework of our proposed method is shown in Fig.1. Vocabulary based on local features (SURF or CN) are trained from an independent training dataset. For an image, SURF and CN features are extracted and quantized on corresponding vocabulary respectively. Here, we improve the traditional VLAD method by adding a weight for each residual, called as "weighted VLAD". The CN and SURF features are adopted to weighted VLAD method, respectively. Then, weighted VLAD vectors based on the two features are obtained. Moreover, the deep features are extracted from the image, and image representation based on deep features is computed. Then the vectors are fused into a vector to represent the image. Finally, similarity scores of the query image and dataset images are measured, and the retrieval results are returned.

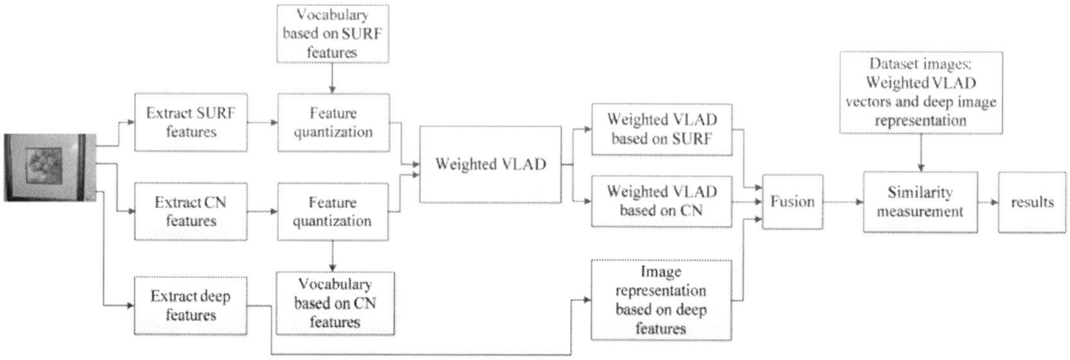

Fig. 1. The framework of our proposed retrieval method

2.2 Weighted VLAD

For an image, the generation process of VLAD vector is as follows: (1) Detect interest regions of the image and extract local descriptors, denoted as $x = (x_1, x_2, \ldots, x_n) \in \mathbb{R}^{d \times n}$. (2) The local descriptors are quantized on the vocabulary $C = (c_1, c_2, \ldots, c_K)$ with K visual words $(L = K)$, denoted as $NN(x_i) = \arg\min_k \|x_i - c_k\|$. (3) The residuals of a visual word and the descriptors that are quantized to this word are computed, formulated as $V_k = \sum_{i:NN(i)=k} (x_i - c_k)$. Then, the residuals are summed. A vector of length $L = K * d$ is obtained, which is called as the VLAD vector. (4) Power-law normalization is adopted for the vector obtained at the step (3). It contains two steps: firstly, there is the square root with symbol, formula as $V_i = |V_i|^{\alpha} \times sign(V_i), 0 \le \alpha < 1$; secondly, the vector is normalized by the $L_2 - norm$, denoted as $V = V / \|V\|_2$.

According to above step (3), it may cause "visual burst" phenomenon because contribution of each descriptor for the VLAD vector is not same, i.e., the closer from center of the cluster, the greater the contribution is, and vice versa. In the similarity measurement stage, this residual will be reflected in the contribution since the Euclidean distance is used. To address the problem, we add a different weight for each residual. Here, the weight is set to be the normalized distance of the descriptor and its nearest visual word, denoted as Eq. (1) and Eq. (2), i.e., the smaller (greater) is the distance between the descriptor and its nearest visual word, the smaller (greater) is the weight. The validity is verified in the experimental section (Section 3). It should be noted that a same weight is added to the residuals corresponding to each visual word in [4]. But in our algorithm, different weights are added to the residuals respectively. This makes our algorithm become more flexible and adaptive for image retrieval.

$$V_k = \sum_{i:NN(i)=k} \omega_i (x_i - c_k) \tag{1}$$

$$\omega_i = \frac{d(x_i, c_k)}{\sum_{i:NN(i)=k} \|x_i - c_k\|} \tag{2}$$

where ω_i is a weight coefficient. Also, $d(x_i, c_k)$ represents the distance of the descriptor x_i and the visual word c_k.

For a given image, SURF descriptors are extracted, denoted as $x^{SURF} = (x_1^{SURF}, x_2^{SURF}, \ldots, x_{n_1}^{SURF})$. For the SURF-based vocabulary $C^{SURF} = (c_1^{SRUF}, c_2^{SURF}, \cdots, c_{K_1}^{SURF})$, according to Section 2.2, a SURF-based VLAD vector can be obtained, denoted as V^{SURF}. Its length is $L^{SURF} = K_1 * d_1$. In addition, each CN descriptor is extracted from an image patch of size $p \times p$, denoted as x_i^{CN}. Specifically, for each pixel of a patch, an 11-D CN descriptor is extracted. Then, the average of all descriptors in the patch is regarded as the color descriptor of the patch. Thus, a set of CN descriptors for the image are obtained, denoted as $x^{CN} = (x_1^{CN}, x_2^{CN}, \cdots, x_{n_2}^{CN})$. For the CN-based vocabulary $C^{CN} = \left(c_1^{CN}, c_2^{CN}, \cdots, c_{K_2}^{CN}\right)$, similarly, the CN-based VLAD vector is computed, denoted as V^{CN}. Also, the length is $L^{CN} = K_2 * d_2$. In order to improve retrieval accuracy, deep features are extracted by using pre-trained deep convolutional neural networks (CNN) models, and the length is L_3.

2.3 Feature fusion and similarity measurement

For an image, the three image representations are fused to a vector, denoted as:

$$V = [\lambda_1 * V^{SURF}, \lambda_2 * V^{CN}, \lambda_3 * V^{CNN}] \tag{3}$$

where λ_1, λ_2, λ_3 are the weight parameters, and $\lambda_1 + \lambda_2 + \lambda_3 = 1$. The best values fitting λ_1, λ_2, λ_3 are selected by loop iteration.

Euclidean distance is used to compute similarities between the query image and the dataset images. To reduce running time, we adopt PCA and whitening method which suppressed the co-occurrence problem with the dimensionality reduction [15].

Our proposed algorithm is summarized as follows:

Algorithm 1 Fusion of deep features and weighted VLAD vectors based on multiple features

1) Off-line

- Train vocabularies C^{SURF}, C^{CN} on the training dataset.
- Extract the dense SURF and CN descriptors from each image of the dataset.
- Compute weighted VLAD vectors V_I^{SURF}, V_I^{CN} and extract deep feature V_I^{CNN} for each image I in the dataset.
- Fuse V_I^{SURF}, V_I^{CN} and V_I^{CNN} by Eq. (3), denoted as V_I.
- Reduce dimensionality of V_I, and the result is denoted as V_{IR}.

2) On-line

- Extract the dense SURF and CN descriptors from a query Q.
- Compute weighted VLAD vectors V_Q^{SURF}, V_Q^{CN} and extract deep feature V_Q^{CNN} for the query image.
- Fuse V_Q^{SURF}, V_Q^{CN} and V_Q^{CNN} by Eq. (3), denoted as V_Q.
- Reduce dimensionality of V_Q, and the result is denoted as V_{QR}.
- Compute similarity score.
- Return images of the dataset with the high similarity scores.

3 Experimental results

In this section we verified our proposed method on three benchmark datasets, i.e., Holidays [16], Ukbench [17] and oxford5k [18]. In addition, Paris60k [19] is used to train vocabularies for Oxford5k. Vocabularies are trained from Mirflickr25k [20] for other datasets. All experiments are implemented on a computer with 8GB memory and 3.3GHz CPU (Intel(R) Core(TM) i5-4590).

3.1 Selection of parameters

In our experiments, the dense SURF descriptors are extracted for each image. Moreover, each CN descriptor is obtained in an image patch of size 4×4. Also, the CN-based and SURF-based vocabularies of size 64 are used. Moreover, CNN features of images are obtained by the VGG-f model [21]. Here, a CNN-based representation is obtained from the second fully-connected layer of convolutional networks for each image, it is a 4096-D vector.

α is a power of the absolute value of VLAD vectors matrix. However, we find that the best value of α is between 0.1 and 0.6. The accuracies with different α by using weighted VLAD based on different features on three datasets are shown in Fig. 2.

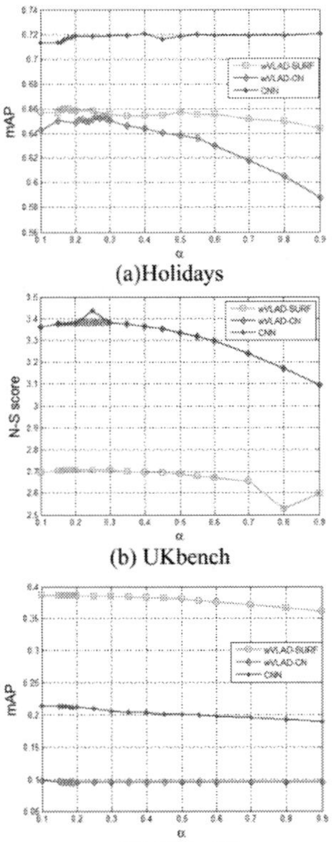

(a)Holidays

(b) UKbench

(C) Oxford5k

Fig. 2. The comparison with different values of α by using CNN-based vectors and VLAD vectors based on SURF features and CN features. In (a) and (c), the mAPs are shown on the Holidays and Oxford5k dataset. In (b), the N-S scores on the Ukbench dataset are shown.

Here, weighted VLAD vectors based on SURF and CN features are denoted as wVLAD-SURF, wVLAD-CN, respectively. On Holidays, we select α=0.19 for wVLAD-SURF, α=0.25 for wVLAD-CN, and α=0.2 for CNN-based vectors. On UKbench datasets, α is set to be 0.19, 0.25, 0.25 for wVLAD-SURF, wVLAD-CN and CNN-based vectors, respectively. Moreover, α is set to be 0.1, 0.15, 0.1 on Oxford5k , respectively.

In Eq. (3), λ_1, λ_2, λ_3 are the weight parameters of SURF-based and CN-based VLAD vectors and CNN-based vectors for feature fusion. On Holidays they are respectively set to be 0.3, 0.3, 0.4. Also, they are set to be 0.2, 0.35, 0.45 and 0.6, 0.05, 0.35 on the UKbench and Oxford5k.

3.2 Effectiveness of weighted VLAD

In this subsection, we verified the effectiveness of proposed weighted VLAD model. In Fig.3, two examples on UKbench dataset are shown. It can be seen that the results of wVLAD-SURF are better than traditional VLAD-SURF. In addition, we compare our weighted VLAD with [4] (VLAD-LCR-RN) in Table 1. On Holidays, it can be seen that the results are the same, but the length of our vector is only about a half of VLAD-LCR-RN. On Oxford5k when vectors are reduced to 128D, wVLAD-SURF achieves a better result.

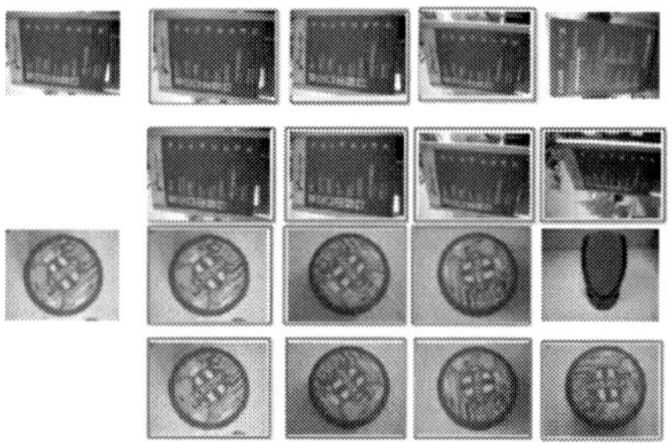

Fig. 3. Two examples by using the VLAD vectors based on SURF features on the UKbench dataset. The left images are queries, and the first row represents the results obtained by the traditional VLAD based on SURF descriptors, where those with green boxes are the correct results. Also, the second row is the results obtained by using our proposed weighted VLAD based on the SURF.

Table 1. Comparison between VLAD-LCR-RN and weighted VLAD-SURF.

Methods	VLAD-LCR-RN[4]	wVLAD-SURF
Holidays (mAP/d)	0.658/8192	0.6581/4096
Oxford5k (mAP/d)	0.517/8192	0.3877/4096
Oxford5k (mAP/d)	0.322/128	0.4050/128

3.3 Fusion of multiple features

In experiments, wVLAD-SURF and wVLAD-CN are fused, denoted as wVLAD-SURF+wVLAD-CN. The multiple weighted VLAD vectors and deep features are fused into a vector, denoted as wVLAD-SURF+wVLAD-CN+CNN. In Table 2, the retrieval accuracies on different datasets are listed, where $L = 128$ denotes the length of VLAD vectors obtained by PCA and whitening operations. When $L = 128$, on Holidays, it can be seen that the mAP obtained by wVLAD-SURF + wVLAD-CN increased by nearly 13% compared to wVLAD-SURF and wVLAD-CN respectively. Also, the mAP achieves 0.8306 for wVLAD-SURF+wVLAD-CN+CNN. Also, On UKbench and Oxford5k, the N-S score and the mAPs reach respectively 3.6916 and 0.4322 for wVLAD-SURF+wVLAD-CN+CNN. Thus, feature fusion make retrieval accuracies obviously improved.

Table 2. Comparison of results obtained by using different features and by reduction on Holidays, Oxford5k and UKbench datasets.

Methods	L	Holidays (mAP)	Ukbench (N-S score)	Oxford 5k (mAP)
wVLAD-SURF	4096	0.6581	2.7074	0.3877
	128	0.6902	2.6268	0.4050
wVLAD-CN	704	0.6515	3.3825	0.0962
	128	0.6698	3.3766	0.0964
CNN[21]	4096	0.7179	3.4375	0.2146
	128	0.7402	3.4199	0.2146
wVLAD-SURF+ wVLAD-CN	4800	0.7702	3.5572	0.3907
	128	0.7954	3.4774	0.4066
wVLAD-SURF+ wVLAD-CN+CNN	8896	0.8220	3.7421	0.4311
	128	0.8306	3.6916	0.4322

Table 3. Comparison between various methods and our proposed method.

Methods	L	Holidays (mAP)	Oxford 5k (mAP)	Ukbench (N-S score)
VLAD-CSURF [6]	128	0.738	0.293	3.50
Triangulation embedding [22]	128	0.617	--	3.40
wVLAD-SURF+wVLAD-CN	128	0.7954	0.4066	3.4774
Deep fully conneted+VLAD [23]	512	0.783	--	--
OUR	128	0.8306	0.4322	3.6916
OUR	256	0.8435	0.4455	3.7215

We compare our method with various methods in Table 3. Specially, in reference [22], the authors constituted "hand-crafted" features like SIFT which is called triangular embedding. It can be seen that the mAP of wVLAD-SURF+wVLAD-CN is higher than the mAPs of other methods on Holidays. We fuse VLAD vectors and deep features. In reference [23], they proposed the multi-scale orderless pooling (MOP-CNN) scheme which combined the deep features and VLAD. The results compared with [23] are listed in Table 3. It can be seen that our method achieves the better results on three datasets.

4 Conclusion

Since the contribution of each descriptor to the VLAD vector is not the same in traditional VLAD method, it will result in visual burst phenomenon. To address the problem, we added a different weight for every residual to balance the contribution of each descriptor for the VLAD vector. The SURF features describe local gradient information of an image, while the CN features represent local color information. Thus, to improve the image retrieval accuracy, we proposed a simple and effective method that fused our proposed weighted VLAD vectors based on local texture features and local color features. Moreover, in order to improve the accuracy further, deep features are extracted and fused with the multiple weighted VLAD vectors. In order to reduce running time, the PCA and whitening operations were adopted in this paper. Finally, our experiments obtain the better results when compared with other methods.

This work was supported by National Natural Science Foundation of China (61572067); International (Regional) Project Cooperation and Exchanges of National Nature Science Foundation of China (61611530710); Beijing Municipal Natural Science Foundation (4162050); The Natural Science Foundation of Guangdong Province (2016A030313708) and the Fundamental Research Funds for the Central Universities (2017JBZ108).

References

1. J. Sivic, A. Zisserman, *Proceedings Ninth IEEE International Conference on Computer Vision*, **2**,1470-1477 (2003)
2. J. Philbin, M. Isard, *IEEE Conference on Computer Vision and Pattern Recognition, CVPR*, 1-8 (2007)
3. H. Jégou, M. Douze, C. Schmid, *IEEE In Proc. CVPR*, 3304-3311(2010)
4. J. Delhumeau, P. H. Gosselin, *ACM International Conference on Multimedia*,653-656 (2013)
5. D. Lowe, International Journal of Computer Vision, IJCV, no.2, 91-110 (2004)
6. H. Bay, A. Ess, T. Tuytelaars, L.V. Gool, Computer vision and image understanding, no.3, 346-359 (2008)
7. E. Spyromitros-Xioufis, S. Papadopoulos, I.Y. Kompatsiaris, IEEE Transactions on Multimedia, no.6, 1713-1728(2014)
8. Q. S. Chen, Y. Y. Ding, H. Li, J. Wang, X. Deng, *IEEE International Conference on System, Man, and Cybernetics*, 2391-2396(2014)
9. D. A. R. Vigo, F. S. Khan, J. V. D. Weijer, *20th International Conference on Pattern Recognition*, 1549-1553(2010)
10. P. Fan, A. Men, M. Chen, *IEEE International Conference on Network Infrastructure and Digital Content*, 726-730(2009)
11. J. V. D. Weijer, C. Schmid, J. Verbeek, IEEE Transactions on Image Processing a Publication of the IEEE Signal Processing Society, no.**7**, 1512-23(2009)
12. A. Krizhevsky, I. Sutskever, G. E. Hinton, Advances in Neural Information Processing Systems, 25, 1097-1105 (2012)
13. M. D. Zeiler, R. Fergus, *In Proc. ECCV*, 8689, 818-833(2014)
14. A.S. Razavian, H. Azizpour, J. Sullivan, S. Carlsson, *In Proc. CVPR Workshops*, 806-813 (2014)
15. H. Jégou, O. Chum, *In Proc . ECCV*, 774-787(2012)
16. H. Jégou, M. Douze, C. Schmid, *In Proc. ECCV*, 304-317 (2008)
17. D. Nister, H. Stewenius, *In Proc. CVPR'06,* 2161-2168 (2006)
18. J. Philbin, O. Chum, M. Isard, *In Proc. CVPR* (2007)
19. J. Philbin, O. Chum, M. Isard, J. Sivic, A. Zisserman, *In Proc. CVPR*(2008)
20. M.J. Huiskes, M.S. Lew. *Proceeding MIR '08 Proceedings of the 1st ACM international conference on Multimedia information retrieval*, 39-43 (2008)
21. K. Chatfield, K. Simonyan, A. Vedaldi, A. Zisserman, *In proc. BMVC*(2014)
22. H. Jégou and A. Zisserman, *In Proc. CVPR*, 3310–3317 (2014)
23. Y. Gong, L. Wang, R. Guo, and S, Lazebnik. *In Proc. ECCV*, 392–407(2014).

Planning the trajectory of manipulator gripper with a parallel structure and flexible links

Vadim Dubovskov[1,*], *Yuriy* Valyukevich[1]

[1]Don State Technical University, 344000 Rostov-on-Don, Russia

Abstract. Traditionally, the tasks of the objects transportation in working space are solved using manipulation robots with a sequential kinematic structure. However, the increasing requirements for the mechanical rigidity, load-carrying capacity, mass-size and speed parameters caused the need to develop a new classes of manipulation robots with a parallel connecting circuit. The closed kinematic scheme of such manipulators provides a more higher rigidity of the entire structure. The reduced mass of moving parts reduces the load on electric drives and thus provides an increase in manipulator gripper dynamics and the accuracy of positioning. Finally, the energy efficiency of the process increases. The problems of planning the trajectory of manipulator gripper with a parallel structure and flexible links are considered within the paper.

1 Parallel structure manipulator

As the name implies, a cable-driven parallel manipulator, or cable robot, is a type of parallel manipulator in which the platform is driven by a set of cables in place of traditional rigid links. These cables are sometimes referred to as tendons, or wires. A winch, consisting of a tensioning motor and spool, is used to adjust the length or tension in each cable. Coordinated retraction and extension of the cables allows for the position and orientation of the platform to be controlled. Cable-driven systems offer many advantages over traditional parallel manipulators. Replacement of the heavy prismatic actuators with relatively light cables facilitates performance at higher accelerations and allows for potentially vast workspaces. Mounting of the winches to the fixed base further reduces the system's moving inertia. Compared to the moving platform, the cables have relatively low mass and inertial properties and thus, their effects are oftentimes considered negligible – an assumption which greatly simplifies the modeling and analysis. The simple design of cable-driven robots also gives rise to a system that is reconfigurable, less expensive to construct and maintain, and easily transported, disassembled, and re-assembled.

The limitations in parallel robot mechanism paved a path for the cable robots. The research was directed towards developing a mechanism that would keep the characteristics of parallel robot intact and reduce the drawbacks, especially small workspaces. The abilities of parallel robots to move with high speed and accelerations, stiff structure, high force carrying capacity were important to be kept intact.

Workspace computation is an important aspect of robot design, as it gives an idea of reachability of the robot.

The kind of workspace required depends largely on the application of the system. As discussed in [6] of the paper, there are many ways to analyse the workspace of a robot. The force feasible workspace approach is adopted for this study as it is based primarily on the force values. A feasible workspace is considered as a set of positions and orientations of a robot in which the EEs are controllable, tensions in cables are positive, forces values lie within the pre-set bounds and the EE is far from singularity.

Despite their numerous advantages, there are several challenges associated with cable-driven robots. The fact that cables can pull but not push gives rise to a unilateral constraint wherein the cables must always be maintained in tension. As a result, the modeling and analysis methods that have been developed for conventional rigid-link manipulators cannot be directly applied to cable-driven systems. Furthermore, this unilateral property gives rise to the need for actuation redundancy – that is, more cables than degrees of freedom (DoF) – in order to completely restrain and control the platform. In fact, if the number of DoF in the system is denoted by m, and the number of cables is denoted by n, it has been shown [9-10] that a fully-restrained system requires at least $n=m+1$ cables. This concept can be illustrated through a simple example [5]. Consider a rigid body in the horizontal plane that is subjected to linear motion, i.e. a system with one degree of freedom. We wish to control the motion of the rigid body using cables. If we place a cable on one side of the mass – say, to the left-hand side – it is obvious that the mass can easily be pulled to the left. However, in such a configuration, the mass cannot be moved to the right, as

* Corresponding author: dybovskov@mail.ru

DOI: 10.1051/matecconf/201713205003

the cable cannot exert a compressive force. Any attempt to do so will result in the cable becoming slack. By adding one more cable on the opposite side of the rigid body, motion along both directions can now be controlled. This idea can be extended to systems with greater degrees of freedom.

2 Schematic of the manipulator

Schematic of the manipulator with flexible links is presented in fig. 1.

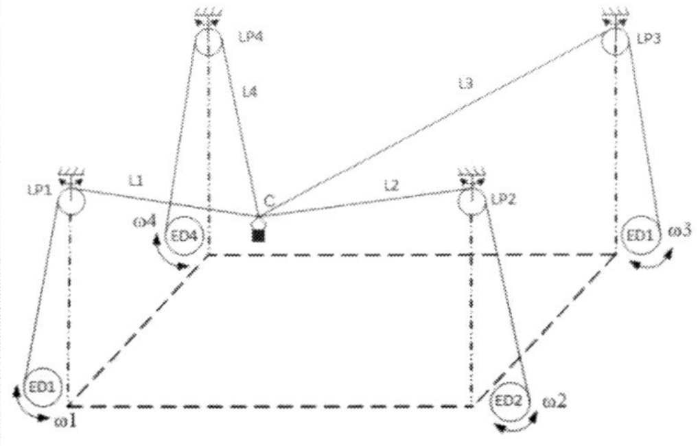

Fig. 1. Schematic of the manipulator with flexible links

The following notations are accepted for the fig. 1: ED1-ED4 – electric drives of the manipulator links № 1-4 respectively; LP1-LP4 – link pulleys № 1-4 respectively; L1-L4 – links (cables) № 1-4 respectively; C – the manipulator gripper; ω1- ω4 – angular velocities of the electric drives № 1-4 respectively; φ1-φ4 – angles of rotation of the link pulleys № 1-4 respectively.

The solution of direct and inverse kinematics of position and speed for the structure presented in fig. 1 is available in [1,2]. Based on it the two methods of trajectory planning are proposed. The first one have a significant methodological error in accuracy. The second one requires a high speed of the hardware and software complex that implements the manipulator control system. The task of developing the control actions on electric drives of links can be formulated as follows:

- the trajectory segment is given as a straight line segment (fig. 2);
- the absolute coordinates of gripper position and the generalized coordinates of all links at the beginning and end of the segment are known;
- the current values of generalized coordinates are could be measured at any point of motion trajectory;
- the speed of motion and the sampling frequency of control actions on speed of the electric drives links are given.

The method of planning the trajectory is based on the calculation of instantaneous velocity in the direction of each of generalized coordinates as a function of velocity vector oriented along the displacement trajectory in absolute coordinates at regular intervals. Fig. 2 shows an example of the relative position of vectors with a given velocity magnitude along the trajectory and the velocity vector of one of the generalized coordinates.

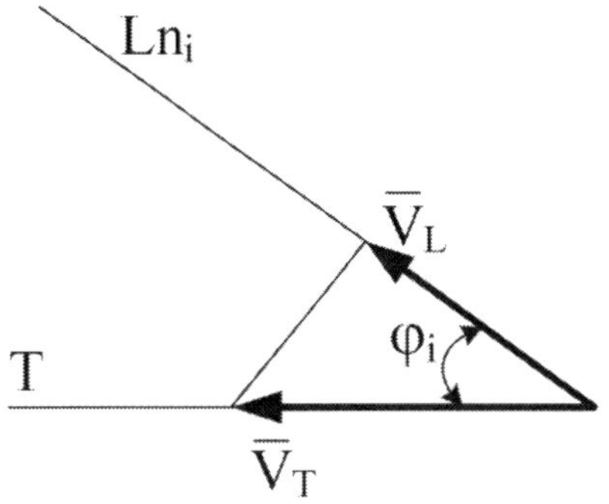

Fig. 2. Velocity vector diagram of the gripper motion along given trajectory

The following notations are accepted for the fig. 2: $\overline{V}_T$ – velocity vector of gripper motion along the trajectory; $\overline{V}_L$ – velocity vector of the manipulator link motion; φ_i – angle between the vectors $\overline{V}_T$ and $\overline{V}_L$.

The module of instantaneous velocity for each manipulator link is determined from the relation:

$$V_{Lni} = V_T \cos f_{ni} \tag{1}$$

The index n corresponds to the link number and the index i – to the number of current iteration.

The angle φ_i can be found by the cosine theorem from the triangle $\Delta ONiM$ (fig. 3), which is formed by a segment of trajectory NiM, initial (segment ONi) and final (OM) positions of the manipulator link n (fig. 1).

$$\cos\phi_{li} = \frac{Ln_{Ni}^2 + L_{MNi}^2 - Ln_M^2}{2Ln_{Ni} \cdot L_{NMi}} \tag{2}$$

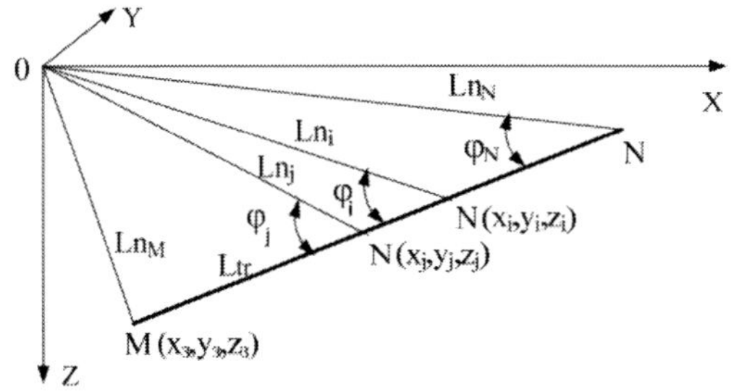

Fig. 3 – An example of the gripper motion trajectory in absolute coordinates x,y,z

The final values of generalized (point M of the trajectory segment) and Cartesian coordinates are known from the condition of problem. The current value of generalized coordinate (segment ON of the triangle) is known from the position sensor data. The current values of Cartesian coordinates of point Ni can be determined by solving the direct problem of manipulator's kinematics, providing that the values of remaining generalized ones are known from condition of the problem. The current length of segment *NMi* is common for all links and determined by the Cartesian coordinates of current and final points of trajectory in the manipulator working space.

The motion from one reference point to the other within a trajectory is performed with a fixed generalized velocity along each coordinate. Estimated number of the trajectory segment division, can be determined from the relation:

$$K = \frac{L_{tr}}{V_T \cdot \tau} \tag{3}$$

Where the *Ltr* – initial length of trajectory segment; τ – the time per iteration.

The denominator in Equation 3 corresponds to an elementary displacement Δl during the time of one iteration τ. The step per one iteration for each Cartesian coordinate is determined from the relation:

$$\Delta x = \frac{X_N - X_M}{K};$$

$$\Delta y = \frac{Y_N - Y_M}{K};$$

$$\Delta z = \frac{Z_N - X_M}{K} \tag{4}$$

In equations 4, the index *N* corresponds to the initial point coordinates and the index *M* – to the final point coordinates within the trajectory segment.

The error of proposed planning method can be divided into two components: a deviation error from desired trajectory and an absolute error due to difference between the instantaneous velocity vectors along the trajectory and the generalized coordinate, as shown in fig. 4.

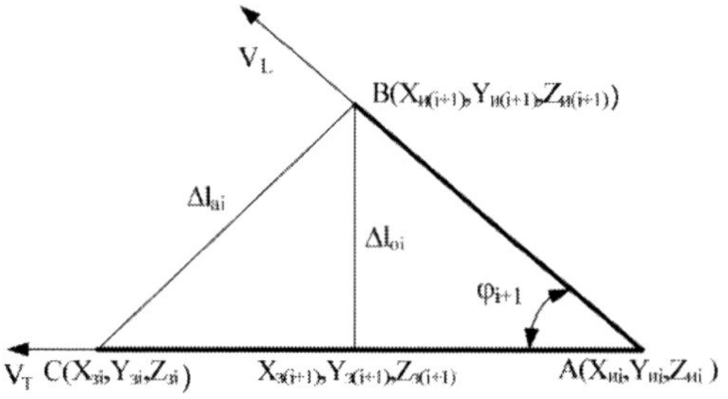

Fig. 4 – Methodological errors of the planning method

Estimating the accuracy of motion along desired trajectory the absolute error is not significant since it's corrected at each step due to true values of initial point coordinates used for each iteration - $X_{u(i+1)}$, $Y_{u(i+1)}$, $Z_{u(i+1)}$. The error magnitude is determined by misalignment of the true and desired velocity values along the trajectory. This is a specific property of proposed method which is not the scope of this paper.

The value of relative error Δloi, providing that the ΔABC vertices coordinates (fig. 4) are known can be determined quite simply by the methods of analytical geometry. It should be noted that the absolute coordinates $X_{u(i+1)}$, $Y_{u(i+1)}$, $Z_{u(i+1)}$ are determined based on the three generalized coordinates. Thus the current initial point coordinates within the segment are corrected relating to the ideal case presented in the fig. 2 at every step of motion along the trajectory. It is obvious from the equation 3 that the magnitude of relative error is depends on ratio between the motion velocity along trajectory V_T and the sampling frequency $1/\tau$ of control actions on the electric drive of manipulator link.

3 Planning method tested

The presented planning method was tested on the real model of manipulator performed in accordance with the kinematic scheme shown in fig. 1. The working space of manipulator's layout is represented by a parallelepiped 2100x1600x2500mm. The control system is a hierarchical two-level one.

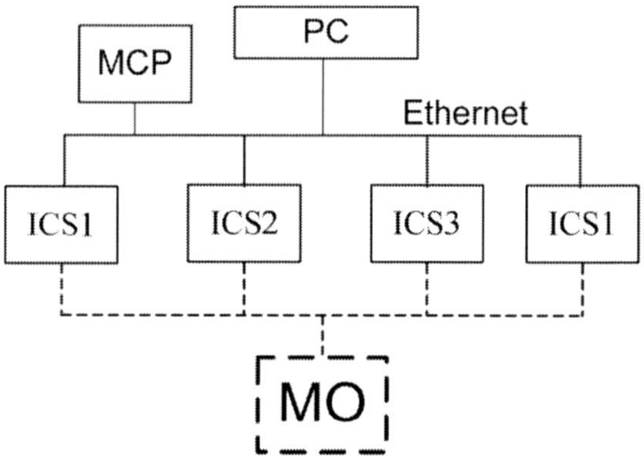

Fig. 5 – General functional scheme of manipulator control

The following notations are accepted for the figure 5: PC – personal computer; ICS1-ICS4 – information-control systems of the manipulator links 1-4 respectively; MO – moving object; MCP – manual control panel.

It is obvious from the fig. 5 that the control system is a hierarchical two-level one. The first level is implemented on a personal computer in software way and solves the following range of control tasks:

–the initial initialization of the control system;

–collection and analysis of the state parameters of the control systems 3in1-3in4;

–performing the control actions on the electric drives ACS of the manipulator links;

–displaying state of the ACS parameters during the moving process.

The second level includes the ACS for each of the links and for the information support of control process of the upper hierarchy level. Communication between all devices is based on a wired link in the Ethernet standard, protocol UDP.

The upper control level problem have been solved using a program shell of information control, implemented in C ++ language under the operating system Windows 7.

Fig. 6 presents a screenshot of the manipulator control program user interface in the mode of moving by a given trajectory. The current gripper position in frontal and horizontal projection of working space is continuously displayed in the corresponding graphical windows of screen interface. The numerical values of parameters mentioned above are displayed in the corresponding tabular windows of screen interface. Each desired position of gripper is determined by position of the sliders. The desired trajectory is also displayed in graphical windows in the frontal and horizontal projections of working space. In the mode of motion, the true trajectory is superimposed on the desired one and the current values of Cartesian and generalized coordinates as well as the relative error are displayed in the corresponding cells of the tables.

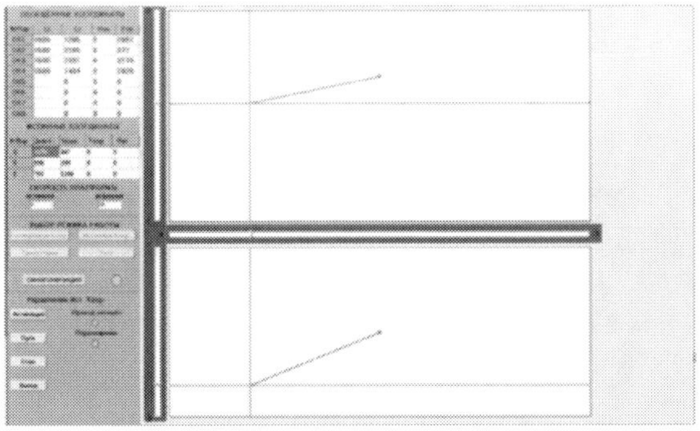

Fig. 6. Screenshot of the manipulator control interface window.

The trial operation showed a full efficiency of proposed method for planning the trajectory of moving manipulator gripper. Maximum deviation from a desired trajectory for the V_T = 200 mm/s and τ = 20 ms was $\Delta lomax$ = 5 mm. For the V_T = 50 mm/s and τ = 20 ms the value of relative error was within plus or minus 4 discretes of measured generalized coordinates (one discrete is equal to 0.154 mm).

4 Conclusions

In this paper, the methods of planning the trajectory of moving manipulator gripper was considered. The method based on the calculation of instantaneous velocity value for each of generalized coordinates and on the formation of control actions on ACS of the links electric drives. The magnitude of error and the factors affecting the method are determined. The results of full-scale modeling on the manipulator real model base are presented.

References

1. Yu.A. Valyukevich, A. V. Alepko, Technical science, no. 6 (2011) [in Russian]
2. Manikandan Bakthavatchalam, François Chaumette, and Eric Marchand, *IEEE International Conference on Robotics and Automation*, 5521– 5526 (2013)
3. Marco Carricato and Jean-Pierre Merlet, *IEEE International Conference on Robotics and Automation*, 3011–3017, (2011)
4. Jordan B., Batalin M., and Kaiser W., *IEEE International Conference on Robotics an Automation*, 144–150 (2017)
5. Lumia, Ronald, *In proc. International Symposium on Robotics and Manufacturing, Maui, HI*, (1994)
6. Riechel A., Bosscher P., Lipkin H., and Ebert-Uphoff I., *In Proc. 10th International Topical Meeting on Robotics and Remote Systems for Hazardous Environments*, (2004)
7. Shanmugasundram A.P. and Moon F.C., *ASME International Mechanical Engineering Congress and Exposition*, **57**, 55-65 (1995)
8. Shiang W., Cannon D., and Gorman J., *In Proc. 1999 IEEE International Conference on Robotics and Automation* (1999)
9. Takemura F., Enomoto M., Tanaka T., Denou K., Kobayashi Y., and Tadokoro S., *In Proc. 2005 IEEE/ASME International Conference on Advanced Intelligent Mechatronics*, 658–663 (2005)
10. Takemura F., Maeda K., and Tadokoro S., *In Proc. IEEE/RSJ International Conference on Intelligent Robots and Systems*, 3504–3509 (2007)
11. Williams R. L., Gallina P., and Vadia J., Journal of Robotic Systems, no.3 (2003)
12. Williams R. L. and Gallina P., In Proc. *2001 ASME Design Technical Conference* (2001)

Algorithm for controlling parameters of linear filter based on fuzzy expert system

Alexandr Eliseev[1,*], *Alexandr* Prygunov[1]

[1]Don State Technical University, 344000 Gagarin square, 1, Rostov-on-Don, Russia

Abstract. The problem of linear filtration of parameters of the vector of an maneuvering object state is considered. The problems to detect the maneuver and control the filter coefficients are solved. A fuzzy expert system is used to adapt the filter. The efficiency of the filter is shown in the test case.

1 Introduction

In the organization of traffic of civil aircrafts the control authorities widely use trajectory information coming from the radio surveillance equipment. The surveillance equipment intended for monitoring the position and movement parameters of aircraft include the multipurpose radar systems, automatic dependent surveillance equipment, and multilateration systems. One of the most important parameters characterizing the quality of a radar is the error in estimating the parameters of the aircraft trajectory, especially under the conditions of maneuvering. This error depends on many factors, including the methods and algorithms used for the secondary processing of the radar measurements. In the secondary processing system the dynamic filtering algorithms with maneuver detection are widely used [1-25]. These algorithms consist of three main modules: maneuver detection module; module for adjusting filter parameters; and filter module [1, 2]. The existing variety of dynamic filtering algorithms is due to the features of the implementation of these modules. For example paper [6] proposes an algorithm for adjusting the Kalman filter by changing the value of the elements of the matrix of the intensities of the forming noises. However, the feature of this algorithm is the iterative adjustment, which leads to the dynamic filtering error in the transition period. The purpose of the work is to reduce the dynamic error of the $\alpha-\beta$-filtering of the trajectory parameters of the maneuvering aircraft. The problem to be solved is the development of the algorithm to adjust $\alpha-\beta$-filter parameters

2 Mathematical formulation of the problem

Let the parameters of the trajectory of the maneuvering aircraft on the time interval $[t_0, T]$ be described by the difference equation [1, 2]:

$$\mathbf{X}(t_{j+1}) = \begin{cases} \mathbf{\Phi}(j+1,j)\mathbf{X}(j) + \mathbf{\Gamma}(j+1,j)\mathbf{N}_x(j), \\ t_j < t_{BM}; \\ \mathbf{\Phi}(j+1,j)\mathbf{X}(j) + \\ +\mathbf{\Gamma}(j+1,j)\big(\mathbf{A}_x(j) + \mathbf{N}_x(j)\big), \\ t_{BM} \le t_j < t_{EM}; \\ \mathbf{\Phi}(j+1,j)\mathbf{X}(j) + \mathbf{\Gamma}(j+1,j)\mathbf{N}_x(j), \\ t_j \ge t_{EM}, \end{cases} \quad (1)$$

and the observation equation has the form

$$\mathbf{Z}(j) = \mathbf{H}(j)\mathbf{X}(j) + \mathbf{N}_z(j), j = 0, 1, 2, \quad (2)$$

where $\mathbf{X}(j)$ is the vector of the process parameters; $\mathbf{Z}(j)$ is the measurement vector; $\mathbf{\Phi}(j+1,j)$, $\mathbf{\Gamma}(j+1,j)$, $\mathbf{H}(j)$ are known functional matrices; t_{BM}, t_{EM} are the start and end time of the maneuver respectively; $\mathbf{A}_x(j)$ is the vector of the intensity of the process (maneuver) changes with unknown elements; $\mathbf{N}_x(j)$, $\mathbf{N}_z(j)$ are random noises of the object (1).

It is required: to obtain according to the results of current observations $\mathbf{Z}(j)$ the optimal (in terms of the mean-square) estimate $\hat{\mathbf{X}}(j)$ of the filtration of the state vector (1) under conditions of a priori uncertainty with

* Corresponding author: eliseev_av65@mail.ru

respect to the values of the elements of the matrix $\mathbf{A}_x(j)$.

3 Solution to the problem

To improve the speed of the estimation $\hat{\mathbf{X}}(j)$ of the parameters vector and to reduce the requirements for the amount of memory and performance of the digital computer complex of the subsystem for secondary processing of trajectory measurements, we will consider the $\alpha-\beta$-filter as tracking filter [1]:

$$\hat{\mathbf{X}}(j+1)=\hat{\mathbf{X}}(j+1\mid j)+ \\ +\mathbf{K}(j+1)\left[\mathbf{Z}(j+1)-\mathbf{H}(j+1)\hat{\mathbf{X}}(j+1\mid j)\right], \quad (3)$$

where $\hat{\mathbf{X}}(j+1\mid j)=\mathbf{\Phi}(j+1,j)\hat{\mathbf{X}}(j)$,

$$\mathbf{K}(j)=diag\left[\mathbf{K}_{Xi}(j), i=\overline{1,q}\right],$$

$\mathbf{K}_{Xi}(j)=\left[\alpha_i(j), \beta_i(j)/\Delta T\right]^T$, $\Delta T=t_j-t_{j-1}$ is the time interval between measurements $\mathbf{Z}(j)$, $j=1,2,3,\ldots$, $\alpha_i(j)$, $\beta_i(j)$ are elements of the matrix of the amplification coefficients $\mathbf{K}_{Xi}(j)$, q is the dimensionality of the state vector $\hat{\mathbf{X}}(j)$.

In the general case, when the state vector of a non-objectifying object is filtered the coefficients $\alpha_i(j)$, $\beta_i(j)$ vary according to the following law [1]:

$$\alpha_i(j)=\frac{2(2j-1)}{j(j+1)},\ \beta_i(j)=\alpha_i^2(j)\Big/\left(2-\alpha_i(j)\right).$$

Thus, in order to adapt the $\alpha-\beta$-filter, it is necessary to solve consistently two main problems: to detect the beginning or ending of the maneuver and to control the values of the amplification coefficient elements $\mathbf{K}(j)$.

Let us consider successively the possible solutions to these problems.

The problem to detect a maneuver. To solve this problem, it is necessary to enter a maneuver detector (MD) [1, 2] in the adaptive $\alpha-\beta$-filter, which should form a maneuver sign $SM(t_j)=SM(j)\in\{0,1\}$ that assumes a value of "1" in case of detecting the beginning of maneuver (continuation of maneuver), or "0" in case of the maneuver end (lack of maneuver).

Let us introduce, by analogy with [6], two additional characteristics of residual: the modulus of the arithmetic mean of the residual $\left|\overline{\varepsilon(j)}\right|$, and the arithmetic mean of the modulus of the residual value $\overline{\left|\varepsilon(j)\right|}$ (for simplicity, without loss of generality, we assume that the residual is scalar)

$$\left|\overline{\varepsilon(j)}\right|=\frac{\left|\sum_{i=0}^{n-1}\varepsilon(j-i)\right|}{n},$$

$$\overline{\left|\varepsilon(j)\right|}=\frac{\sum_{i=0}^{n-1}\left|\varepsilon(j-i)\right|}{n} \quad (4)$$

where n is the number of consecutive subsequent measurements used to detect maneuvering, $\varepsilon(j)=\mathbf{Z}(j)-\mathbf{H}(j)\hat{\mathbf{X}}(j\mid j-1)$.

The criterion for detecting a maneuver, taking (4) into account, takes the form

$$K_{OM}\equiv\begin{cases}(SM=1),\ if\ \left|\overline{\varepsilon(j)}\right|=\overline{\left|\varepsilon(j)\right|},\\ (SM=0),\ if\ \left|\overline{\varepsilon(j)}\right|<\overline{\left|\varepsilon(j)\right|}.\end{cases} \quad (5)$$

The problem to control the values of the amplification coefficient elements $\mathbf{K}(j)$. To adapt the $\alpha-\beta$-filter to change of the model of the movement of an object we use the operatively advising expert system (OAES), which is based on the application of fuzzy logic inference [26-28]. We will assume that the process of the filter functioning can be represented in the form of a tuple of some problem situations (PrS). Any PrS can be described by a situational vector $\mathbf{sv}=\left[sv_k, k=\overline{1,K}\right]^T$, each coordinate of which sv_k is a linguistic variable with a given set of terms $\left\{SV_k^l, l=\overline{1,m_k}\right\}$. We think that for some concrete realizations of the situational vector $\mathbf{sv}^*$ there are precedents of the successful solution of the current PrS, characterized by some precedent vector $\mathbf{pv}=\left\{pv_m, m=\overline{1,m_{PV}}\right\}$, each coordinate pv_m of which is a linguistic variable with a given set of terms $\left\{PV_m^p, p=\overline{1,n_m}\right\}$. Suppose that some PrS, which occurs during the dynamic filtering of measurements, can be described by a situational vector $\mathbf{sv}=\left[sv_k, k=\overline{1,2}\right]^T$ with the following elements: sv_2 – "the value of the relative error module $\left|\delta\varepsilon(j)\right|$", where $\left|\delta\varepsilon(j)\right|$ is determined by the formula

$$\left|\delta\varepsilon(j)\right|=\frac{\sum_{i=0}^{n-1}\left|\varepsilon(j-i)\right|}{n\sigma^*},$$

where σ^* is some allowable mean-square error $\varepsilon(j)$

Let these variables be described by the following term sets: $\mathbf{SV}_1=\{\text{"PE"}, \text{"IDE"}\}=\left\{SV_1^l, l=\overline{1,2}\right\}$;

$$\mathbf{SV}_2 = \begin{Bmatrix} "VSE","SE","AE", \\ "BE","VBE" \end{Bmatrix} = \{SV_2^l, l = \overline{1,5}\},$$

where "PE", "IDE" are the terms "permissible error" and "invalid dynamic error", respectively, "VSE" is "very small error", "SE" is "small error", "AE" is "average error", "BE" is "large error", "VBE" is "very large error".

We assume that the linguistic variable sv_1 is given on a unified universal set $E = \{0, 1\}$, the linguistic variable sv_2 on a universe $E = [0, |\delta\varepsilon_{max}|]$, and their terms are described by membership functions $\mu_{SV_k} \in \{\mu_{SV_k^l}, l = \overline{1, m_k}\}$.

Let the terms of a linguistic variable sv_1 be described by membership functions of the singleton type:

$$SV_1^1 : \mu_{SV_1^1} = \begin{cases} 1, if\ x = \overline{x}_1^1 = 0 - "PE"; \\ 0, если\ x \neq \overline{x}_1^1; \end{cases}$$

$$SV_1^2 : \mu_{SV_1^2} = \begin{cases} 1, if\ x = \overline{x}_1^2 = 1 - "IDE; \\ 0, if\ x \neq \overline{x}_1^2. \end{cases}$$

The terms of the linguistic variable sv_2 are described by the following membership functions [3]:
the terms SV_2^1, SV_2^5:

$$\mu_{SV_2^l}(|\delta\varepsilon|) = 1 - \frac{1}{1 + \exp(-a_{2l}(|\delta\varepsilon| - c_{2l}))}, l \in \{1, 5\}, (6)$$

the terms SV_2^2, SV_2^3, SV_2^4:

$$\mu_{SV_2^l}(|\delta\varepsilon|) = \exp\left(\frac{-(|\delta\varepsilon| - c_{2l})^2}{a_{2l}}\right), l \in \{2, 3, 4\}, \quad (7)$$

where a_{2l}, c_{2l}, $l = \overline{1,5}$ are the membership functions parameters. Taking into account that the controlled parameter of the $\alpha-\beta$-filter, as it was said before, is the matrix of amplification coefficients $\mathbf{K}(j)$, and in the particular scalar case it is the coefficients $\alpha(j)$, $\beta(j)$, we introduce a precedent vector $\mathbf{PV} = \{pv_m, m = \overline{1, m_{PV}}\} = \{pv_1\}$ consisting of one variable pv_1 – "value of the coefficient $\alpha(j)$". In addition we assume that the type of the precedent vector is determined by the value of the element of the situation vector sv_1 – "filtering error type".

Thus when $sv_1 =$ "PE", the value $\alpha(j)$ is determined based on the following precedent

$$pv_1 : \alpha(j) = \begin{cases} 0.743 - 0.043j, при\ j \leq 8, \\ 0.4\ при\ j > 8. \end{cases} \quad (8)$$

In the case when sv_1 takes the linguistic meaning "invalid dynamic error", the precedent vector contains a linguistic variable pv_1 specified on the universe $E = [0, 1]$, and its terms are described by membership functions $\mu_{PV_1} \in \{\mu_{PV_1^p}, p = \overline{1, n_m}\}$.

Let the linguistic variable pv_1 be described by the following term sets:

$$\mathbf{PV}_1 = \begin{Bmatrix} "VSV\ \alpha",\ "SV\ \alpha", "AV\alpha", \\ "BV\ \alpha", "VBV\alpha" \end{Bmatrix} = \{PV_1^p, p = \overline{1,5}\},$$

where "VSV α" is "very small value of the parameter $\alpha(j)$", "SV α" is "small value of the parameter $\alpha(j)$ ", "AV α" is "average value of the parameter $\alpha(j)$", "BV α" is "large value of the parameter $\alpha(j)$", "VBV α" is "very large value of the parameter $\alpha(j)$".

We define the terms of the set $\mathbf{PV}_1$ by membership functions similarly to (6), (7).

The system of rules describing the mechanism for solving the current PrS arising during the operation of the $\alpha-\beta$-filter takes the form:

$$r = 1:\ \text{if}\ \left(sv_1 = "PE"\left(\mu_{SV_1^1}\right)\right),\ \text{then}\ \left(pv_1 = \begin{cases} 0.743 - 0.043j\ \forall\ j \leq 8, \\ 0.4\ \forall\ j > 8. \end{cases}\right),$$

$$r = 2:\ \text{if}\ \left(sv_1 = "IDE"\left(\mu_{SV_1^2}\right)\right)\text{and}\left(sv_2 = "VSE"\left(\mu_{SV_2^1}\right)\right),\ \text{then}\ \left(pv_1 = "VSV\alpha"\left(\mu_{PV_1^1}\right)\right),$$

$$r = 3:\ \text{if}\ \left(sv_1 = "IDE"\left(\mu_{SV_1^2}\right)\right)\text{and}\left(sv_2 = "SE"\left(\mu_{SV_2^2}\right)\right),\ \text{then}\ \left(pv_1 = "SV\alpha"\left(\mu_{PV_1^2}\right)\right),$$

$$r = 4:\ \text{if}\ \left(sv_1 = "IDE"\left(\mu_{SV_1^2}\right)\right)\text{and}\left(sv_2 = "AE"\left(\mu_{SV_2^3}\right)\right),\ \text{then}\ \left(pv_1 = "AV\alpha"\left(\mu_{PV_1^3}\right)\right),$$

$$r = 5:\ \text{if}\ \left(sv_1 = "IDE"\left(\mu_{SV_1^2}\right)\right)\text{and}\left(sv_2 = "BE"\left(\mu_{SV_2^4}\right)\right),\ \text{then}\ \left(pv_1 = "BV\alpha"\left(\mu_{PV_1^4}\right)\right),$$

$$r = 6:\ \text{if}\ \left(sv_1 = "IDE"\left(\mu_{SV_1^2}\right)\right)\text{and}\left(sv_2 = "VBE"\left(\mu_{SV_2^5}\right)\right),\ \text{then}\ \left(pv_1 = "VBV\alpha"\left(\mu_{PV_1^5}\right)\right).$$

Using the "minimax" Mamdani-Zade rule [26-28], the expression for calculating the final membership function $\mu_{pv_{1\Sigma}}$ of the output linguistic variable pv_1 can be written as:

$$\mu_{pv_{1\Sigma}}\left(sv_1^*, sv_2^*\right) = \max_{r} \min_{pv_1}\left(\min_{k}\left(\mu^*_{SV_k(r)}\right), \mu_{PV_1(r)}\right),$$

where, $\mu_{SV_k(r)} \in \left\{\mu_{SV_k^l}, l = \overline{1, m_k}\right\}$,

$\mu_{PV_1(r)} \in \left\{\mu_{PV_1^p}, p = \overline{1, n_m}\right\}$ are the membership functions, respectively, of the linguistic variables sv_k and pv_1, which are part of the production rule with the number r, sv_1^*, sv_2^* are clear values of the input variables, $\mu^*_{SV_k(r)} = \mu_{SV_k(r)}\left(sv_k^*\right)$. To find the clear value pv_1^* of the output variable, we use the "center of gravity" method [26].

To assess the effectiveness of the developed algorithm, the mathematical simulation was carried out. We use the following model of aircraft movement:

$$\mathbf{X}(j+1) = \begin{cases} \mathbf{\Phi X}(j) + \mathbf{\Gamma} n_x(j), (0 \le j < 100) \vee \\ \vee (200 \le j < 400) \vee (500 \le j \le 600); \\ \mathbf{\Phi X}(j) + \mathbf{\Gamma}(1.5g + n_x(j)), 100 \le j < 200; \\ \mathbf{\Phi X}(j) + \mathbf{\Gamma}(-4g + n_x(j)), 400 \le j < 500, \end{cases}$$

$$\mathbf{\Phi} = \begin{bmatrix} 1 & \tau \\ 0 & 1 \end{bmatrix}, \ \mathbf{\Gamma} = \begin{bmatrix} \frac{\tau^2}{2} \\ \tau \end{bmatrix},$$

$$\tau = t_{j+1} - t_j = 1\text{c}, \ \ j = \overline{0, 600}.$$

As a result of the simulation it was established that the application of the developed adaptation algorithm allows to reduce the dynamic filtering error by 35% in comparison with the classical $\alpha-\beta$-filter.

References

1. A. Farina, F. Studer, Radio and communications, (1993)
2. H.W. Sorenson, *Kalman filtering: theory and application* (IEEE Press, New York 1985)
3. P. Zarchan, *Fundamentals of Kalman filtering - A practical approach* (American Institute of Aeronautics and Astronautics, USA 2005)
4. B.P. Gibbs, *Advanced Kalman filtering, least-squares and modeling: a practical handbook,* (Hoboken: John Wiley & Son, Canada 2011)
5. I.R Petersen., A.V. Savkin, *Robust Kalman filtering for signals and systems with large uncertainties.* (Birkhauser, Boston 1999)
6. M. Rodkin, In Proc. *VI All-Russian Conference Radiolocation and radio communication*, 125–128, (2012) [in Russian]
7. Y.G. Bulychev., A.V. Eliseev, Avtomatika i Vychislitel'naya Tekhnika, no.2, 26-35, (1998) [in Russian]
8. A.V. Eliseev, Aviakosmicheskoe priborostroenie, 30-38, (2006) [in Russian]
9. A.V. Yeliseyev, Radioelectronics and Communications Systems, 13-19 (2005) [in Russian]
10. K.V. Ramachandra, *Kalman filtering techniques for radar tracking,* (Marcel Dekker, New York 2000)
11. V. Gomez, A. Maravall, *Approximate Kalman filtering.* (World Scientific, Singapore 1993)
12. M. Verhaegen, P. Van Dooren, IEEE Transactions on Automatic Contro*l,* no.10, 907-917, (1986)
13. Q. Xia, M. Rao, Y. Ying, X. Shen, Automatica, no.8, 1333-1338 (1994)
14. Y. Geng, J. Wang, Journal of Beijing University of Aeronautics & Astronautics, no.5, 434-437, (2004)
15. M. Oussalah, J. Schutter, *In Proc. 25th Int. Conf. on Noise and Vibration Engineering,* 1225-1232 (2000)
16. R. A. Singer, IEEE Transactions on Aerospace and Electronic Systems, no.4, 473-483 (1970)
17. T. Simeonova, T. Semerdjiev, Inform. & Security. No.9, 154-165 (2002)
18. R. A. Best., J. P. Norton, IEEE Transactions on Aerospace and Electronic Systems, no.3, 1030-1037 (1997)
19. T.C. Wang, P. K. Varshney, IEEE Transactions on Aerospace and Electronic Systems, no.3,910-924 (1993)
20. Y.T. Chan, F. Couture, *IEEE Proc. on Radar and Signal Processing,* **1**, 21-28 (1993)
21. Y. Bar-Shalom, K. Birmiwal, IEEE Trans. Aerosp. Electron. Syst., no.5, 621-629 (1982)
22. H.B. Sui, X.Y. Fang, and Y. Wu, Radar Science and Technology, no.3, 202-205 (2008)
23. F.M. Sun, X.Q. Wu, K. Qi, Journal of Data Acquisition & Processing, no.2, 191-195 (2008)
24. X.R, Li,. V.P. Jilkov, IEEE Trans. Aerosp. Electron. Syst., no.1, 96-119 (2010)
25. X.R. Li,. V.P. Jilkov, In Proc. Signal and Data Processing of Small Targets, 6-8 (2010).
26. V. Gostev, Designing fuzzy logic controls for automatic control systems (BHV-Petersburg, SPb 2011)
27. A.V. Eliseev, Aviatsionnaya Tekhnika, no.3, 20-25, (2005)
28. A.V. Eliseev, Avtomatizatsiya i Sovremennye Tekhnologii, no.11, 3-12, (2005)

Algorithm for detecting violations of traffic rules based on computer vision approaches

Samir Ibadov[1], *Ragim* Ibadov[2,*], *Boris* Kalmukov [1], *Vladimir* Krutov[1]

[1]Institute of Services Sector and Entrepreneurship (branch), Don State Technical University, 344010, Rostov-on-Don, Russia
[2]Institute Radioengineering Systems and Control, Southern Federal University, 344006, Rostov-on-Don, Russia

Abstract. We propose a new algorithm for automatic detect violations of traffic rules for improving the people safety on the unregulated pedestrian crossing. The algorithm uses multi-step proceedings. They are zebra detection, cars detection, and pedestrian detection. For car detection, we use faster R-CNN deep learning tool. The algorithm shows promising results in the detection violations of traffic rules.

1 Introduction

According to statistics, in the Russian Federation in 2015, there were 58,221 road accidents (RA) in which pedestrians suffered. Of this number of accidents, pedestrians hit 56886 times. As a result, 52,306 people were wounded, and 6,991 people were killed. Pedestrians who crossed the road on crosswalks have been hit by drivers of vehicles of 19779 times. At crosswalks, for one the year 1233 people have died, and 19576 people are wounded. How many of the wounded became disabled is not specified. The number of accidents of running-down pedestrians at crosswalks in 2015 on comparisons with 2014 has increased by 2%.

Analyzing the statistics of road traffic violations in the Russian Federation in 2015, it can be concluded that the creation of automatic detection systems for traffic violations on unregulated crosswalks is necessary to improve safety on the roads of the country. Therefore, the development of an algorithm for fixing traffic violations on unregulated crosswalks is urgent and will be in demand.

The algorithm works in two stages. At the first stage, it is necessary to detect three classes of objects in the video sequence. First class crosswalk (zebra), the second class moving vehicles and the third class the pedestrians going on the crosswalk. After all the object classes are found on each frame of the video sequence, the second stage begins. It is necessary to impose video sequence frames with the allocated objects at each other. If on the same frame of the video sequence the pedestrian and the vehicle are is at the crosswalk, the vehicle means violates Traffic regulations, having done not pass the pedestrian.

2 Zebra Detection

At the first stage, we need to detect a zebra; the video sequence is for this purpose loaded. Further, there is a splitting the video sequence into frame and allocation of an object and background using tags by the user [1]. Stages of the offered approach are presented in Figure 1.

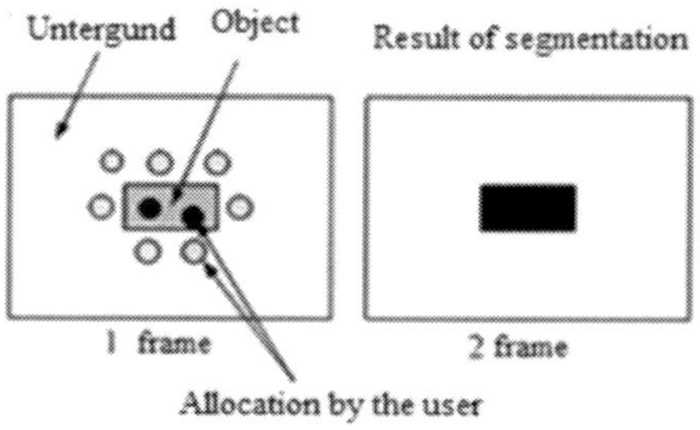

Fig. 1. Stages of segmentation.

To do this, first select the object frame is marked with a white background, the background is marked in black color the figure 2a, in the figure 2b an alpha the channel of the image received using a segmentation method on the received tags at the first stage is present. In the figure, 2c show the selected object on the original frame of the video sequence.

a) b) c)

Fig. 2. Stages of segmentation of a zebra:
a) marking of an object and background, b) alpha channel of the image, c) the allocated object.

Next, it is necessary to detect pedestrians on all frames of the video sequence.

* Corresponding author: ragim_ibadov@mail.ru

3 Car Detection Using Faster R-CNN Deep Learning

Faster R-CNN [2] is an extension of the R-CNN [3] and Fast R-CNN [4] object detection techniques. All three of these techniques use convolutional neural networks (CNN). The difference between them is how they select regions to process and how those regions are classified. R-CNN and Fast R-CNN use a region proposal algorithm as a pre-processing step before running the CNN. The proposed algorithms are typical techniques such as Edge Boxes [5] or Selective Search [6], which are independent of the CNN. In the case of Fast R-CNN, the use of these techniques becomes the processing bottleneck compared to running the CNN. Faster R-CNN addresses this issue by implementing the region proposal mechanism using the CNN and thereby making region proposal a part of the CNN training and prediction steps.

Load Dataset. This example uses a small vehicle data set that contains 295 images. Each image contains 1 to 2 labeled instances of a vehicle. A small dataset is useful for exploring the Faster R-CNN training procedure, but in practice, more labeled images are needed to train a robust detector.

Create a Convolutional Neural Network (CNN). A CNN is the basis of the Faster R-CNN object detector. Create the CNN layer by layer using Neural Network Toolbox™ functionality. Begin with the image input layer function, which defines the type and size of the input layer. For classification tasks, the input size is typically the size of the training images. For detection tasks, the CNN needs to analyze smaller sections of the image, so the input size must be similar in size to the smallest object in the data set. In this data set all the objects are larger than [16 16], so select an input size of [32 32]. This input size is a balance between processing time and the amount of spatial detail the CNN needs to resolve.

Next, define the middle layers of the network. The middle layers are made up of repeated blocks of convolutional, ReLU (rectified linear units), and pooling layers. These layers form the core building blocks of convolutional neural networks. Combine the input, middle, and final layers.

Configure Training Options. Trains the detector in four steps. The first two steps train the region proposal and detection networks used in Faster R-CNN. The final two steps combine the networks from the first two steps such that a single network is created for detection [2]. Each training step can have different convergence rates, so it is beneficial to specify independent training options for each step.

Train Faster R-CNN. Now that the CNN and training options are defined train the detector using trainFasterRCNNObjectDetector.

During training, image patches are extracted from the training data. The 'PositiveOverlapRange' and 'NegativeOverlapRange' name-value pairs control which image patches are used for training. Positive training samples are those that overlap with the ground truth boxes by 0.6 to 1.0, as measured by the bounding box intersection over union metric. Negative training samples are those that overlap by 0 to 0.3. The best values for these parameters should be chosen by testing the trained detector on a validation set. To choose the best values for these name-value pairs, test the trained detector on a validation set.

Fig. 3. Vehicle detection result.

After the network is trained, it shows reliable results.

4 Pedestrian Detection

We detect pedestrians using Motion-Based Multiple Object Tracking of a method.

Detecting of moving objects and tracking of their movement is an important element of computer sight [7].

Detection of moving objects uses the algorithm of subtraction of a background based on a Gaussian mixture of the model. Morphological operations are applied to a resultant mask of the foreground the figure 4a to the elimination of noise. Finally, the analysis reveals a group of related pixels that are likely to correspond to a moving object, the figure 4b.

The association of detection besides to an object is based only on the movement. The movement of each traced object is estimated using Kallman's filter. The filter is used for forecasting of tracking of the provision of objects in each shot.

a) b)

Fig. 4. Detecting of the pedestrian:
a) the foreground, b) the detected pedestrian.

The detect objects function returns average points and limits to a framework the found objects. It also returns a binary mask which has the same size, as an entrance shot. Pixels with value 1 correspond to the foreground, and pixels with value 0 corresponds to a background.

Function carries out segmentation of the movement, using the foreground detector. Then it carries out morphological operations over the received binary mask to remove noisy pixels and to fill the remained holes.

Kalman filter is used to predict the midpoint of each of the following structural frame and updates the bounding rectangle respectively.

5 Block diagram of the proposed algorithm

The block diagram of the proposed algorithm shows on figure 5. Algorithm for fixing a violation of the Traffic Rules does not allow pedestrians to pass at a pedestrian crossing by a vehicle in a video sequence.

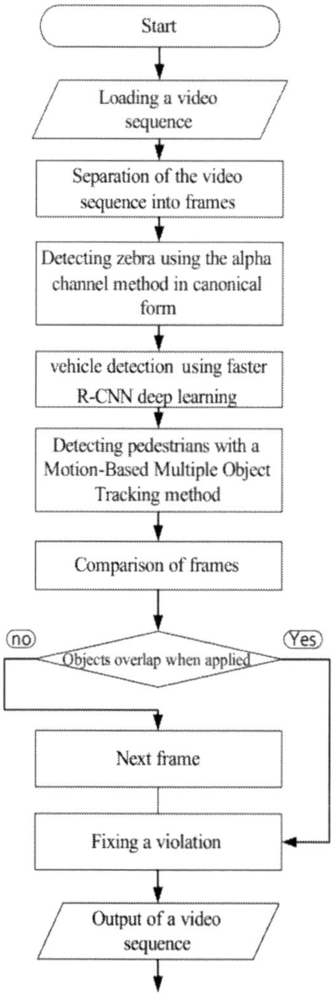

Fig. 5. Block diagram of the proposed algorithm.

The input is a video sequence. Next, the video is a separation of the video sequence into frames. After that, we need to detect three classes of objects. First class crosswalk (zebra), the second class moving vehicles and the third class the pedestrians going on the crosswalk. Detecting zebra occurs using the alpha channel method in canonical form. The vehicle is detected using Vehicle detection using faster R-CNN deep learning. Then, pedestrians are detected using the Motion-Based Multiple Object Tracking method.

The second stage of an estimated method of definition not of the admission by the vehicle of the pedestrian on a zebra consists in the comparison of the frame with the allocated objects. To reveal violation, not of the admission of the pedestrian on a zebra, it is necessary to impose video sequence frame with the allocated objects at each other. If on the same frame of the video sequence the pedestrian and the vehicle cross a zebra, the vehicle means violated traffic regulations, namely not the admission of the pedestrian on a zebra.

6 Comparison

For example, when comparing the 130th frame of the video sequence with the allocated vehicle the figure 5a the 1st frame of the video sequence with the allocated zebra the figure 6b we see that the vehicle crossed a zebra the figure 6c.

a) b) c)

Fig. 6. Zebra crossing by the vehicle
a) the allocated vehicle on the 130th frame of the video sequence, b) the allocated zebra on the 1st frame of the video sequence, c) zebra crossing by the vehicle on the 130th frame of the video sequence.

Further, we check for crossing the 130th frame of the video sequence with the allocated pedestrian the figure 6a, with the 1st frame of the video sequence with the allocated zebra the figure 7b we see that the pedestrian crossed a zebra the figure 7c.

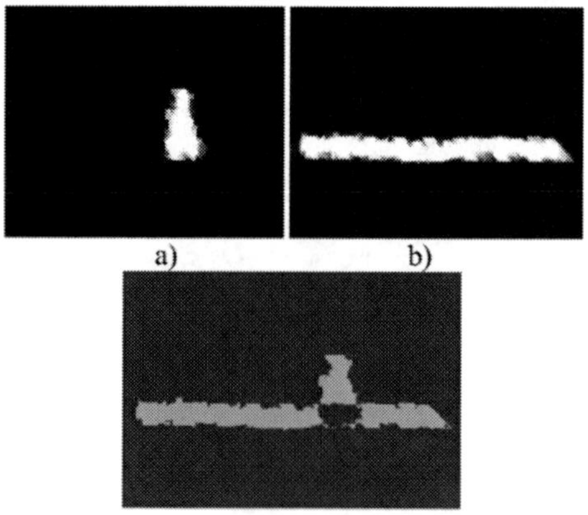

a) b)

c)

Fig. 7. Crossing by the zebra pedestrian:

DOI: 10.1051/matecconf/201713205005

a) the allocated pedestrian on the 130th frame of the video sequence, b) the allocated zebra on the 1st frame of the video sequence, c) crossing of a zebra by the pedestrian on the 130th frame of the video sequence.

Therefore, the vehicle violated traffic regulations, without having missed the pedestrian on the 130th frame of the video sequence.

7 Conclusion

We proposed the algorithm for automatic detect violations of traffic rules for improving the people safety on the unregulated pedestrian crossing. The algorithm uses multi-step proceedings. They are zebra detection, cars detection, and pedestrian detection. For car detection, we use faster R-CNN deep learning tool. The effectiveness of the new approach is shown in the examples demonstrating the effectiveness of the algorithm on test frames.

The reported study was supported by the Russian Foundation for Basic research (RFBR), research projects №15-07-99685 and №16-37-00386.

References

1. Voronin V.V., Gapon N.V., Sizyakin R.A., Ibadov R.R., Ibadov S.R., Semenishchev *E.A. In Proc. XI International scientific-technical forum / DSTU* (2014). [in Russian]
2. Ren, Shaoqing, et al. *Advances in Neural Information Processing Systems* (2015)
3. Girshick, Ross, et al. *In Proc. IEEE Conference on Computer Vision and Pattern Recognition* (2014)
4. Girshick, Ross. *In Proc. IEEE International Conference on Computer Vision* (2015)
5. Zitnick, C. Lawrence, and Piotr Dollar. *In Proc. ECCV*, 391-405 (2014).
6. Uijlings, Jasper RR, et al. *In Proc. IJCV* (2013): 154-171.
7. Motion-Based Multiple Object Tracking, available at:http://www.mathworks.com/help/vision/examples/motion-based-multiple-object-tracking.html/

Monitoring of the technical requirements compliance in the Solid Household Waste landfills functioning through the method of visual detection

Andrey Richter [1,*], *Maretta* Kazaryan [2], and *Michael* Shahramanyan [1]

[1]Researching Institute "AEROCOSMOS", 105064, Moscow, Russia
[2]Financial University at Government of Russian Federation Vladikavkaz branch, 362001 Vladikavkaz, Russia

Abstract. The paper describes the method of visual detection in the task of assessing compliance with the requirements for the design, operation and reclamation of landfills for solid domestic waste. An example of verification of requirements for landfills is given in accordance with the instruction on design, operation and reclamation. It is shown, in particular, the construction of a sanitary protection zone in the Google Earth program for assessing the permissibility of natural and anthropogenic objects it includes. The series of basic concepts associated with visual detection of solid waste landfills are given, and namely, a number of soil elements that can be detected by the method of visual detection are described.

1 Introduction

The solution of the problem of littering belongs to more than one generation of humanity, and its relevance will only grow with the time. In the study of the problem under consideration, a certain mathematical apparatus is used [6-10]. But you can start solving this long-term emergency situation right now by setting a point of support both inside yourself (i.e., changing the inner world, consciousness) and outside (ie changing the external world, i.e., the environment around you). Internal change is purely individual, and the external one can be given a logical explanation. The later, in particular, includes the monitoring of Waste Disposal Sites (WDSs) with the imparting of some power impulse (in a physical, economic, social aspect).

The advantage of classical algorithms of space monitoring is the automation of image processing, i.e. a large-scale survey of the area of observation using space images. But the reverse side of this "coin" is their main disadvantage - the limited possibilities for examining the area. Most of the information is not extracted from the image. In this regard, a method of visual detection of WDSs is suggested, in which a wider range of problems can be solved, but on a more narrow scale. For example, the state of the environment in a solid waste landfill from the point of view of visual detection can be assessed in various ways, in particular: 1) areas occupied by vegetation and their variation over time; 2) soil degradation: bogging, salinization, etc. [5]; 3) changes in the object composition of the territory; 4) abnormal zones in the vicinity of the WDS; 5) compliance with the Rules of Design, Operation and Reclamation (RDOR) of the WDS; 6) quality of vegetation.

Requirements for solid waste landfills, estimated by the method of visual detection.

Observance of many PDOR rules (see [4]) can be verified by visualization of cosmic images. Comparison of these data with field observations, of course, will give a more ambitious and qualitative result.

Table 1. Design Requirements.

Requirement	Control of compliance with the use of detection technologies	
	Visual	Automatic
Availability and layout of the main facilities of the landfill	+	+
Relation of the terrain and the polygon filling scheme	+	+
Drainage ditch, tree planting design	+	+
Designing of the foundation ditches of polygons	+	-
Designing of the base shields	+	-
Monitoring of the sanitary protection zone	+	+
Distance to the nearest town-planning objects	+	+

* Corresponding author: urfin17@yandex.ru

Table 2. Operating Requirements

Requirement	Monitoring the fulfillment of requirements using detection technologies	
	Visual	Automatic
Organization of storage (area, volume, order, rate of filling)	+	+
Availability of household, construction, industrial waste; volumes	+	+
Wetting of waste in fire hazardous periods	+	-
The order of work (development, backfilling, consolidation)	+	-
The presence and composition of intermediate screens (surfaces and slopes)	+	-
The presence of cluttering neighborhoods	+	+
Backfilling with vegetable soil, gardening		+
Quality of waste insulation from the environment	+	-
Availability of hazardous waste	+	-
Stage of degradation, humification, assessment of the composition of the landfill	+	-
Coating with final protective screens (surfaces, slopes)	+	+

Table 3. Requirements for reclamation

Requirement	Control of compliance with the use of detection technologies	
	Visual	Automatic
Availability of storage of waste	+	+
Backfilling with vegetable soil, gardening	+	+
Planting trees and shrubs on the slopes	+	+
Coverage of technological and biological remediation screens	+	-
Quality of technical reclamation	+	-
Quality of biological reclamation	+	-
Quality of work (agricultural, forestry and construction) for the further use of land	+	-

Let's consider some PDOR rules, which can be checked by the method of visual detection. Tables 1-3 show a number of requirements for the design, operation, and remediation, respectively, of landfills for solid domestic wastes (SDW) and Industrial Wastes (IW) using space monitoring technologies [1]. The sign "+" marks the developed technologies, the sign "-" - the technologies that require refinement.

One of the advantages of visual detection before automatic is the ability to test a larger set of PDOR rules (see tables 1-3). But at the same time there are restrictions on the number and quality of available satellite imagery. To verify many requirements, high resolution images and their high frequency (1 time per month) are required, which is possible in rare cases. Many images are of medium resolution, in many cases neighboring chronological images are separated from each other by several years according to the date of shooting.

2 Example of verification of compliance with the requirements of solid waste landfills

In accordance with [4], among the rules of Sanitary-Protective Zone (SPZ) are the following:

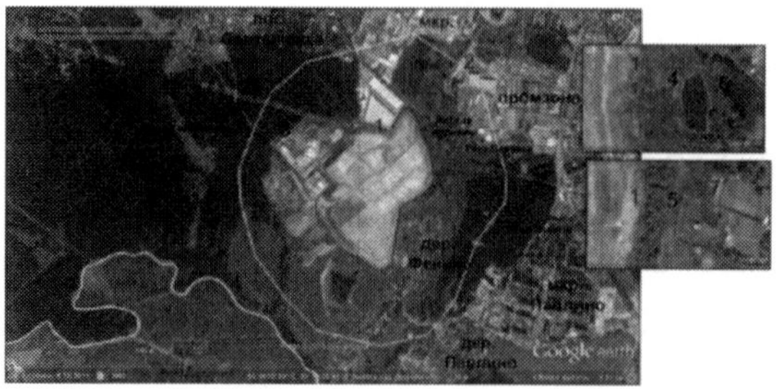

Fig.1 Identification of violations of the rules of the SPZ of the landfill of Kuchino

1. "Landfills are located outside of cities and other settlements. The size of the SPZ from residential buildings to the landfill boundaries is 500 m. Moreover, the size of the sanitary protection zone (SPZ) is specified in the calculation of gaseous emissions. The boundaries of the zone are established by isolating 1 of the maximum permissible concentrations, if it leaves the limits of the regulatory zone. The size of the zone less than 500 m not allowed "(paragraph 1.2). Source: [4];

2. " It is prohibited to place residential buildings , industrial wells and wells for drinking purposes in the SPZ of the landfill. In the absence of green plantations or earth embankments in the SPZ, cavaliers of the ground necessary for insulation at its closure are arranged along the perimeter of the landfill. The SPZ regime is determined by the current rules and regulations "(paragraph 1.29). Source: [4].

Thus, the nearest residential house and the source of drinking water must be located at a distance of at least 500 m from the boundaries of the storage site. Then conditionally, SPZ can be constructed parallel to the boundary of the storage site at a distance of 500 m from it (Figure 1). Then some violations / compliance with the SPZ rules will be identified, which can be counted individually or by the area of violations / observance of the SPZ rules.

In Figure 1: 1 - the boundary of the storage area [2], 2 - the boundary of the SPZ, 3 - the radius of 500 m of the SPZ. It can be seen from figure 1 that several water sources are lying inside the SPZ (the Pekhorka River, the Lesnoy Pond, etc.), as well as parts of several settlements. These include: the village. Feninoand the village. Saltykovka (insignificant part); a small part of the industrial zone located in the city Zheleznodorozhny; almost the entire area of the village. Fenino. In addition, the boundary of the SPZ of the polygon is close to the Moscow border and is approaching it with time. The minimum distances to objects of different classes are also violated, in particular, for water sources and residential buildings: 4 - distance to the nearest water source (40 m), 5 - distance to the nearest residential building (20 m), 6 - nearest water body, 7 - the nearest building.

An important factor in assessing compliance with requirements is the designation of the concepts governed in the document [4] on images. Studies have shown that these concepts can be divided into groups: Elements of the soil; Structural elements of the landfill; Operations at PDOR of Solid Domestic Waste landfills.

The main operations at any stage of the life of the landfill (design, operation or reclamation) are excavation [3]. This work includes the development, movement, laying and compaction of soil; the formation of embankments and depressions. Elements of the soil include:

- mound - an artificial form of relief in the form of an elevation of land and / or waste;
- excavation - an artificial form of relief in the form of a lowering of the earth;
- half-pouring (half-pit) - an artificial form of relief in the form of elevations from the earth, alternating with artificial depressions of the earth;
- shaft - high earthen mound, usually with a moat in front, erected for the purpose of localizing waste;
- career - a notch intended for filling with waste and being a constituent of a solid waste landfill;
- pit - a deep pit, a foundation dug for mining purposes;
- dump pit - a mound of excavated material removed from the excavation, laid behind the slope of the excavation, which is not used to construct the mound in another place;
- cascade - the structure of the depressions formed by the queue of operation;
- base (bottom) - part of the surface of the notch with a slope of less than 0.5%.

The base of the test site is the base (base) of a ditch or ravine dug for it; slope - part of the surface of the excavation and embankment with a slope of more than 0.5%; ravine is a natural form of relief in the form of relatively deep and steeply inclined unclosed troughs formed by temporary streams; trench - an open excavation in the soil of a trapezoidal section with a depth of 3-6 m, a width of 6-12 m and a length that is dependent on the temperature and amount of imported waste; ditch - recess length, much greater than the width used to drain the liquid; dam - a mound of length, much greater than the width, used as a temporary road. Separates the cascade notches.

- The base of the test site is the base (base) of a ditch or ravine dug for it;
- slope is a part of the surface of the excavation and embankment with a slope of more than 0.5%;
- ravine is a natural form of relief in the form of relatively deep and steeply inclined unclosed troughs formed by temporary streams;
- trench is an open excavation in the soil of a trapezoidal section with a depth of 3-6 m, a width of 6-12 m and a length that is dependent on the temperature and amount of imported waste;
- ditch is the recess length, much greater than the width used to drain the liquid;
- dam is a mound of length, much greater than the width, used as a temporary road. It separates the cascade notches.

The structural elements of the landfill include:

- the queue - part of the storage area, designed for operation by an economic entity for 3-5 years;
- tier - a layer of waste 2-2.5 m thick after the compaction operation;

• a temporary road is a road used by machines for storing solid waste.

The working card is the SDW storage area allocated for the given day. Width is 5 m (bulk map), 12 m (trench), length is 30-150 m; fence - is an artificial structure that describes the boundaries of a solid domestic waste landfill, an economic zone and / or a landfill site. Among others are:

- a slope,
- serpentine,
- screening,
- protective screens and other concepts.

Screening is the protection from the negative impact on the environment of certain dangerous or harmful factors, in particular, littering [3]. Screening is provided by protective screens, which, in turn, can be natural or artificial.

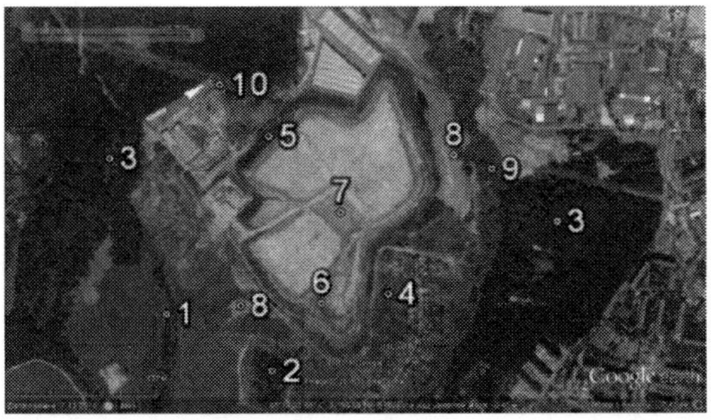

Fig.2 Shielding forms in the vicinity of the landfill of Kuchino

Natural screens are formed without the participation of man under the influence of the forces of nature. For example, a forest is a natural screen of solid waste landfills, which can grow on its own, if it is not cut down. The forest screen (see Figure 2) takes the form of a forest belt (1), a forest plant (2) or a forest in general (3), plantations (4), grass borders and trees of the storage site, vegetation cover of the slopes of the landfill (5). Covering old areas of storing with grass and / or forest vegetation (6) can also be attributed to natural screens.

Different screens have different composition, purpose and carry different protective functions, and not always favorable for environment. For example, the river 9, which runs along the landfill, screens the spread of the filtrate through the soil, but, in its turn, transports the filtration water. Forest plantations of 4 suburban areas delay the spread of the filtrate, but at the same time they experience significant anthropogenic load. Snow refers to natural time screens operating during the cold season. This screen has rehabilitating functions and restores annually, as far as possible, the soil structure [5].

To reduce the negative impact of littering on the environment, natural screens are not enough. Therefore, they resort to artificial ones. Figure 2 shows: 7 - covering of waste with a layer of soil, 8 - ground dump pit, also for delaying soil degradation, 10 - artificial fences for a solid waste landfill (fence).

The rules, as you know, can not only be violated, but also be respected. For example, backfilling with a dump pit of soil as an anti-filter screen (see 8 in Figure 2) is compliance with the rules for the operation of solid waste landfills. In the case of landfills in Russia, this is not common.

3 Conclusions

Despite the fact that visual detection of solid waste landfills (in the Google Earth program) is carried out manually, its aspects can be systematized into steps and procedures that carry a clear sequence of actions from one point (input data) to another (output, required results). With visual detection of litter objects, you can determine: tasks, methods and algorithms defined by certain flowcharts. Such an approach will provide a solution for a more wider range of tasks for monitoring solid waste landfills than for automated monitoring methods, together with greater operational efficiency and less time-consuming processing of data.

References

1. M.A. Shahramanian, A.A. Richter, *Methods and technologies of space monitoring of waste disposal facilities in the interests of ensuring the ecological safety of the territory: Teaching-methodical manual* (Center of the Russian State University of Oil and Gas, named after I.M. Gubkin, Moscow 2013)
2. A.A. Richter, M.A. Shahramanyan, M.L. Kazaryan, A.B., Fundamental research., no.2, 2866-2870 (2015)
3. A.A.Richter, M.A. Shahramanyan, M.L. Kazaryan, A.B., Fundamental research, no.2, 3095-3099 (2015)
4. K.D. Pamfilov, *Instructions for the design, operation and reclamation of landfills for solid domestic waste* (Ministry of Construction of the Russian Federation. Academy of Public Utilities, Moscow 1998) [in Russian]
5. V.G.Bondur, A.B. Murynin, A.A. Richter, M.A. Shahramanyan, Izvestiya SFU Technical science no.6, 131-134 (2012)
6. A.A. Richter, M.A. Shahramanyan, M.L. Kazaryan: Bulletin of the Vladikavkaz Scientific Center, no.3, 61-67 (2015)
7. A.V. Abrosimov, D.B. Nikolsky, L.V. Sheshukova, Ecology of urbanized territories, no. 1, 38-43 (2014)
8. S.S.Timofeeva, S.S.Timofeev., L.V. Sheshukova: Bulletin of the Irkutsk State Technical University, no. 2 84-88 (2013)
9. D.A. Lipilin, Fundamental research, 621-625 (2013)
10. A.V. Pogorelov, D.A. Lipilin. Proceedings of the Dagestan State Pedagogical University. Natural and exact sciences, no. 1, 114-121 (2014)
11. D.A. Lipilin. *In Proc. Fifth International Scientific and Practical Conference.* 186-189, (2013)

12. O.V. Brovkina, D.Yu. Skoropisov, Modern problems of remote sensing of the Earth from space, no.1, 153-155 (2012)

13. O.V. Brovkina: Earth exploration from space, no.6, 22-28 (2007)

14. I.V. Galitskaya, Geoecology, engineering geology, hydrogeology, geocryology no. 2, 144-147 (2005)

15. I.V.Galitskaya, Geoecology, engineering geology, hydrogeology, geocryology, no. 3, 225-237 (2007)

16. V.S. Putilina, I.V.Galitskaya, T.I. Yuganova: Ecology. A series of analytical reviews of world literature, no. 76, 1-100 (2005)

17. K.E. Levtin, Bulletin of the Siberian State Aerospace University. no. 3, 46-51 (2012)

18. V.I. Protsenko, N.L.Kazansky, P.G. Serafimovich, Computational Optics, no.4, 582-591 (2015)

19. V.M. Pavleichik, *The collection: Steppes of Northern Eurasia Materials of the VII International Symposium*, 603-608 (2015)

20. K.E. Levtin, Reshetnev Readings, no. 13, 514-515 (2009)

The synthesis of adaptive multi-mode regulators based on combined control of the combined maximum principle.

Andrey Kostoglotov[1], *Sergey* Lazarenko[2], *Vitaliy* Losev [2], and *Igor* Pugachev[2, *]

[1]Rostov State Transport University, 344038, Rostov-on-Don, Russia
[2]Don State Technical University, 344002, Rostov-on-Don, Russia

Abstract The problem of synthesis of the optimal controls using nonlinear dynamic systems in general form is not yet analytically solved. Fundamental results of L.S. Pontryagin leads to programs of optimal control, and the solution to the synthesis problem can be obtained using complex constructions only in special cases. The implementation of such solutions implies the need to take into account the sliding modes, the regimes of frequent switching, the negative influence of interference, and so on. This was the reason for the development of methods for synthesizing systems of variable structure, which allow to increase the accuracy of control, eliminate undesirable modes, and counteract the influence of external and parametric disturbances. The results of the research showed that using the combined maximum principle leads to obtaining a general structure of quasi-optimal control for different classes of constraints and allows to provide parametric and structural adaptation to of the state and the parameters due to the possibility of complex implementation of control principles with respect to deviation and disturbance.

1 Introduction

The trend of development of modern controlled mechanical systems is due to the fact that these objects become essentially the multimode multipurpose systems of wide use. Therefore, the natural situation is when the control objective, the dynamic parameters of the control object and the external environment are all continuously changing. Linear control laws, as a rule, ensure the achievement of the goal of the solution for one mode of motion [1]. Meanwhile, the real controlled objects are often characterized by several modes of motion [2, 3]. The theory of systems with variable structure leads to different controls obtained, each of that is constructed for the corresponding subsystem. As a result, we need to solve particular problems related to the choice of switching times, ensuring stability, reducing power consumption and increasing the performance [4, 5]. Overcoming these drawbacks can be carried out by changing the switching surfaces, for example, using fuzzy control methods or artificial neural networks [5].

The basis for the procedures to obtain the equations of motion in sliding mode is the physical approach [3]. Its use for solving the problem of synthesis of regulators for the systems which dynamics satisfies the Lagrange equation of the second kind leads to a combined maximum principle [5–8]. The controls built on this base are characterized by the property to adapt to the change in the dynamic parameters of motion [7, 6] and by nonlinear structure of controls. The variety of solutions of the inverse problems is related to the different behavior of dynamic measures of motion on the switching hypersurface. Combination of the controls of the combined maximum principle makes it possible to achieve the stated synthesis goal for a set of motion modes that satisfy the Hamilton-Ostrogradskii principle [8].

The purpose of the work is to perform the synthesis of adaptive multi-mode regulators based on combined control of the combined maximum principle.

2 Formulation of the problem

The principle of combined control includes the implementation of control processes for deviation and disturbance (see Figure 1).

* Corresponding author: bakut_8536@mail.ru

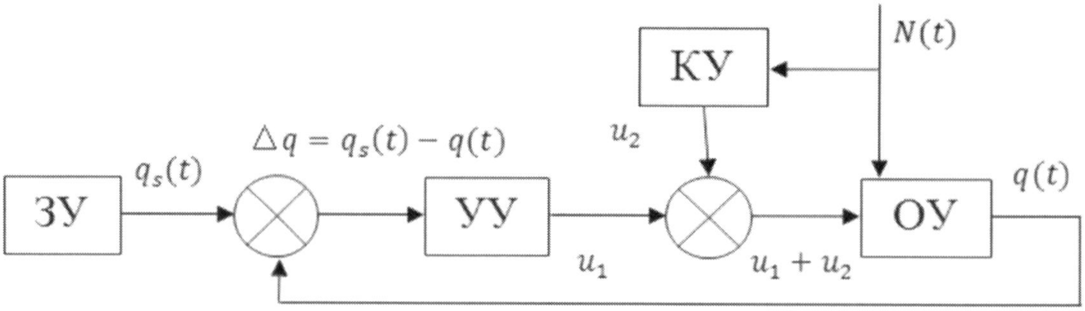

Fig. 1. The combined control

Where *ЗУ* is the driver unit; *УУ* is control unit; *КУ* is compensating unit; *ОУ* is object of control; $q_s(t)$ is a given trajectory; $q(t)$ is the trajectory of the *ОУ* motion; $N(t)$ is external uncontrolled influence; $u(t)$ is the control action (control): $u_1(t)$ is control for deviation; $u_2(t)$ is control for disturbance ; $\Delta q(t)$ is deviation.

The structure of the combined management is the sum of the control actions u_s defined on the sets $M_1 M_2 ... M_n$. It consists of the elements of the merged sets

$$\bigcup_{s=1}^{n} M_s = M_1 \bigcup M_2 \bigcup M_n \quad (1)$$

In this case, the controls are the set of allowed impacts, following one after the other

$$\bigcup_{s=1}^{n} u_s = \{u_1 \in [0, t_l) \wedge u_2 \in [t_1, t_2) \wedge ...\}. \quad (2)$$

In addition the combined control can be defined on the intersection of sets

$$\bigcap_{s=l}^{n} M_s = M_1 \bigcap M_2 \bigcap ... \bigcap M_n. \quad (3)$$

In this case, the controls are the set of allowed impacts, performed simultaneously and taken with certain weights, i.e. complementing each other. Such combined controls are complementary:

$$\bigcap_{s=1}^{n} u_s = \{u_k \in M_1 \vee u_k \in M_2 \vee ...\}. \quad (4)$$

3 The control of the pendulum oscillations in the presence of noise and disturbances

Let us consider a pendulum, to which the control and the external action are applied (see Figure 2) [7, 8]. The notation in Fig. 2 is as follows: $I = ml^2$ is moment of inertia; g is the gravity acceleration; m is the mass; l is the pendulum length; u is the control; P is the external influence.

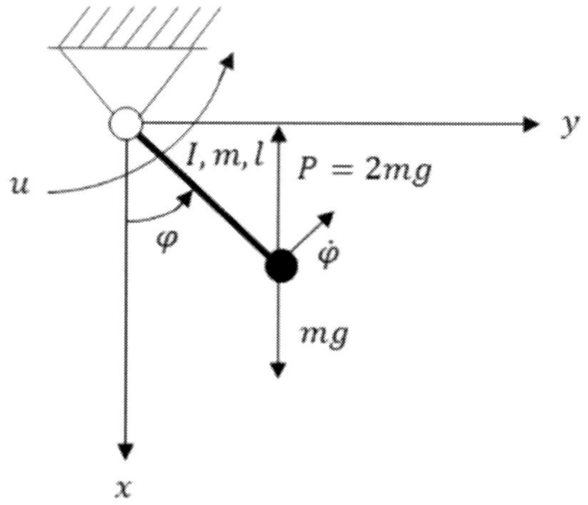

Fig. 2. The motion controls

where I is moment of inertia of the body; m is the mass; l is the length; u is the control; P – is the external influence.

The state of the pendulum is determined by the generalized coordinates $q = \varphi$, $\dot{q} = \dot{\varphi}$

The kinetic energy T and the momentum $p_\varphi = \frac{\partial T}{\partial \dot{\varphi}}$ of the moving pendulum are determined by formulas

$$T = \frac{1}{2} I \dot{\varphi}^2, p_\varphi = I\dot{\varphi}. \quad (10)$$

The potential energy is given by expression

$$\Pi = mgl(1 - \cos\varphi). \quad (11)$$

The work of generalized forces (control and disturbance) is

$$A = \int_0^q (u - mglsin\phi + Pl\sin\phi)dq. \quad (12)$$

The target functional is as follows

$$J = \frac{1}{2}\int_0^{t_k} (\phi - y(t))^2 dt = \int_0^{t_k} Fdt, \quad (13)$$

where $y(t)=\phi+\xi(t)$ is the observation, $\xi(t)$ is stochastic process with known characteristics.

The Lagrange equation of the second kind for $P = 0$ has the form

$$\ddot{\phi} = -csin\phi + u, \quad c = \frac{g}{l}. \tag{14}$$

4 The control laws for classes of measurable and piecewise-continuous controls

To obtain the conditions for a minimum of the target functional, we use the combined maximum principle, according to which we can write the extended Hamiltonian of the system as

$$H = \lambda\left[T + \int_{q_0}^{q} (-csin\phi + u)dq\right] + F. \tag{15}$$

The generalized power is determined by function of the form

$$\Phi = [\lambda(-csin\phi + u) + V]\dot{\phi} \to max, \tag{16}$$

where λ is the Lagrange multiplier,

$$V = \frac{\partial F}{\partial \phi} = (\phi - y(t)).$$

The control from (16) can be defined as follows:

$$u = \lambda^{-1}[\mu p_\phi - (V - \lambda csin\phi)], \tag{17}$$

where μ is synthesizing function.

The following variants to build the combined control are possible.

5 The control is chosen from the class of bounded measurable functions

The equation that defines the control switching line can be written as

$$u_0 = \lambda(-csin\phi) + (\phi - y(t)) = 0. \tag{18}$$

Along this line the Hamiltonian is constant

$$H_0 = \lambda\left(T + \int_0^{\phi} -csin\phi d\phi\right) + F = const.$$

The use of the operation of the Poisson bracket allows us to write:

$$\frac{\partial H_0}{\partial p_\phi}\frac{\partial H_0}{\partial \phi} - \frac{\partial H_0}{\partial \phi}\frac{\partial H_0}{\partial p_\phi} = 0,$$

where $\frac{\partial H_0}{\partial p_\phi} = \lambda\dot{\phi}$; $\frac{\partial H_0}{\partial \phi} = (\phi - y(t) + \lambda csin\phi)$;

$$\frac{\partial u_0}{\partial p_\phi} = \mu; \quad \frac{\partial u_0}{\partial \phi} = (1 - \lambda ccos\phi),$$

which yields

$$\mu = \frac{-|\dot{\phi}||1 - \lambda ccos\phi|}{\lambda^{-1}|\phi - y(t) - \lambda csin\phi|}.$$

The law of the combined control with respect to deviation and disturbance has the form

$$u = \lambda^{-1}\left[-\frac{|\dot{\phi}|\dot{\phi}|1 - \lambda ccos\phi|}{\lambda^{-1}|(\phi - y(t)) + \lambda csin\phi|} - ((\phi - y(t)) - \lambda csin\phi)\right]. \tag{19}$$

The motion equation has the form

$$\ddot{\phi} = \lambda^{-1}\left[-\frac{|\dot{\phi}|\dot{\phi}|1 - \lambda ccos\phi|}{\lambda^{-1}|(\phi - y(t)) - \lambda csin\phi|} - (\phi - y(t))\right]. \tag{20}$$

6 The control is chosen from the class of bounded functions

Compensation for the perturbation can be performed using the forces superposition principle. Then we can put $u = u_1 + u_2$, and here $u_1 = c\sin\phi$. Hence

$$\Phi = [\lambda u_2 + (\phi - y(t))]\dot{\phi} = max.$$

The conditions $u_2 = \lambda^{(-1)}[\mu p_\phi - (\phi - y(t))] = 0$, $u_1 + csin\phi = 0$ determine the switching line on which the Hamiltonian

$$H_0 = \lambda T + F = const.$$

Using the Poisson brackets

$$\frac{\partial H_0}{\partial p_\phi} = \lambda\dot{\phi}; \quad \frac{\partial H_0}{\partial \phi} = (\phi - y(t));$$

$$\frac{\partial u_0}{\partial p_\phi} = \mu; \quad \frac{\partial u_0}{\partial \phi} = 1$$

provides the synthesizing function

$$\mu = -\frac{|\dot{\phi}|}{\lambda^{-1}|(\phi - y(t))|}.$$

The law of the combined control with respect to deviation and disturbance has the form

$$u = u_2 + u_1 = \lambda^{-1}\left[-\frac{|\dot{\phi}||\dot{\phi}|}{\lambda^{-1}|(\phi - y(t))|} - (\phi - y(t))\right]. \quad (21)$$

The motion equation has the form

$$\ddot{\phi} = \lambda^{-1}\left[-\frac{|\dot{\phi}||\dot{\phi}|}{\lambda^{-1}|\phi - y(t)|} - (\phi - y(t))\right]. \quad (22)$$

7 The control is chosen from the class of discontinuous piecewise-continuous and relay controls

The piecewise constant equation

$$H = \lambda\left(T + \int_0^\phi -csin\phi d\phi + \int_0^\phi u d\phi\right) + F. \quad (23)$$

The generalized power function

$$\Phi = [\lambda(u - csin\phi) + (\phi - y(t))]\dot{\phi} = max. \quad (24)$$

To achieve the maximum we need to satisfy the following conditions:

$$|u| \geq c|sin\phi| = const, \quad u = |u| sign\dot{\phi}. \quad (25)$$

The substitution of (25) into (23) gives

$$H = \lambda[T - c(1 - cos\phi) + |u| sign\dot{\phi}] + \frac{(\phi - y(t))^2}{2},$$

which yields

$$u = |u| sign\left[-\frac{[\dot{\phi}^2 + 2c(cos\phi - 1) + F]sign\dot{\phi}}{2|u|} - (\phi - y(t))\right]. \quad (26)$$

The motion equation has the form

$$\ddot{\phi} = u. \quad (27)$$

The piecewise-continuous control.

On the basis of the principle of the independence of the action of forces in the Hamiltonian, the generalized force can be determined as the sum of the forces

$$Q = u - csin\phi.$$

Then

$$H = \lambda\left[T + \int_0^q Q dq\right] + F, \quad (28)$$

and the generalized power function

$$\Phi = (\lambda Q + (\phi - y(t)))\dot{\phi} = max \quad (29)$$

From the maximum condition it follows that

$$|Q| \geq |\phi - y(t)|, \quad Q = k|\phi - y(t)| \text{sign}\, \dot{\phi}, \quad (30)$$

where $k > 1$.

Satisfying the maximum principle (2.19), the piecewise continuous control will have the form

$$Q = k|\phi - y(t)| sign\left[-\frac{\dot{\phi}^2 sign\dot{\phi}}{2} - \frac{(\phi - y(t))^2 sign(\phi - y(t))}{2}\right]. \quad (31)$$

The motion equation has the form

$$\ddot{\phi} = Q, \quad (32)$$

the control

$$u = Q + csin\phi. \quad (33)$$

First we solve equation (32), then we solve equation (33).

The obtained basic solutions of optimal control can be extended on the basis of logical expressions

$$u = \bigcup_{s=1}^{n} \varphi_s(a) u_s,$$

where $\phi_s(a)$ is logical function;

$a(\phi, \dot{\phi}, t)$ is function of the generalized coordinates and the time.

The combined control allows to eliminate the shortcomings of various devices by means of simple expression

$$u = \sum_{s=1}^{n} a_s u_s \in \overline{G_u},$$

where a_s are weighting coefficients

$$\sum_{s=1}^{n} a_s = 1.$$

8 The study of the effectiveness of the combined controls

In the studies, we use the pendulum computational model (see Fig. 2).

The pendulum was subjected to loading by the reduced moment of gravity

$$M_0 = -c\cos\varphi, \quad c = 1.$$

and the control moment $u(t)$

In the process of controlling the pendulum, the measurements of the position of the pendulum are accompanied by noise with the characteristics $CKO = 0{,}5$, $M\,MO = 0{,}1$. Observation time 8; sampling period $\tau = 0{,}01$. All values are dimensionless.

The results of the studies are shown in Fig. 3–4, where the phase portraits and the transient processes for laws (19), (21) and (26) are shown. Fig. 5 shows the nature of the combined control for the laws (19), (21) and (26), respectively.

The results of the studies confirm the effectiveness of the synthesized combined controls for obtaining the required quality of the process to control a nonlinear

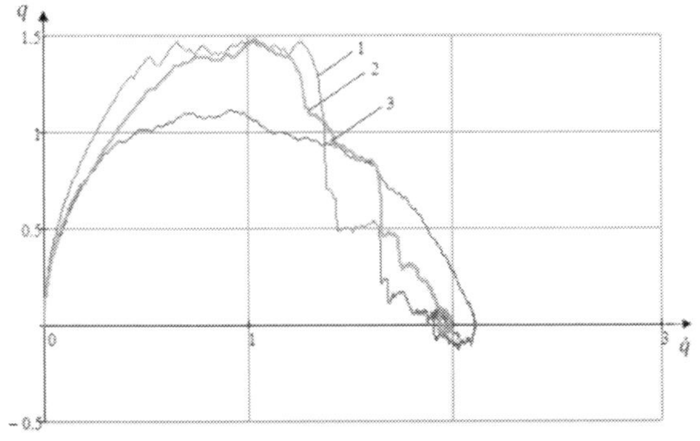

Fig. 3. The phase portrait: *1* – control (19); *2* – control (21); *3* – control (26)

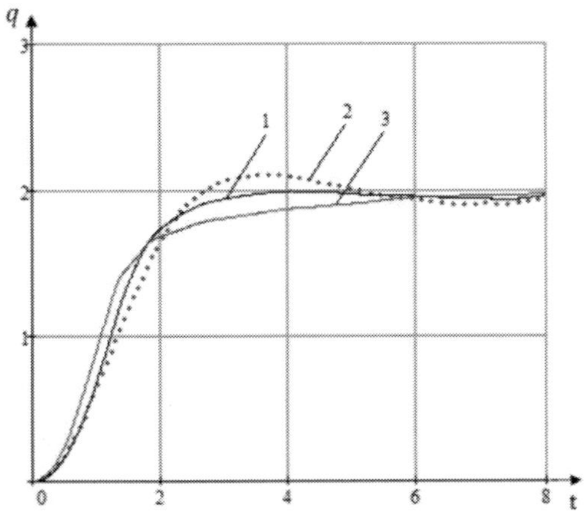

Fig. 4. The transient process: *1* – control (19); *2* – control (21); *3* – control (26)

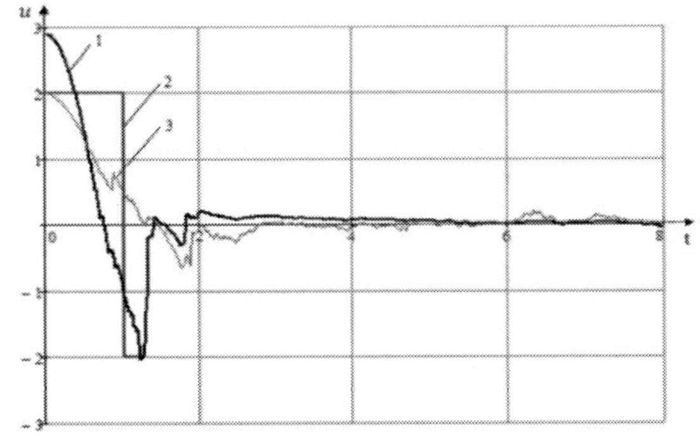

Fig. 5. The combined controls: *1* – control (19); *2* – control (21); *3* – control (26)

9 Conclusions

system — monotonic, aperiodic or oscillatory process with requirements for accuracy, overshoot, regulation time, etc.

On the basis of the theorem of the combined maximum principle, the basic laws to control the dynamic objects under the influence of external uncontrollable disturbances are obtained. The use of optimal laws to build the combined controls makes it possible: to exclude the chattering regimes generated by discontinuous (relay) controls in problems with the nonsmooth Hamiltonian; to ensure the stability of the object in an unbalanced state.

The paper has been accomplished with the support of Russian Federal Property Fund grants No. 15-08-03798.

References

1. R.A. Singer and K.W. Behnke, IEEE Transactions on Aerospace and Electronic Systems, no.1, 100 – 110 (1971)
2. X.R. Li and V. Jilkov, IEEE Transactions on Aerospace and Electronic Systems, no.3, 1333-1364 (2003)
3. A.A. Kostoglotov, S.V. Lazarenko. Journal of Communications Technology and Electronics, no.2, 121 – 125 (2017)
4. A.A. Kostoglotov, S.V. Lazarenko, Z.V. Lyaschenko. *In Proc. 2017 20th IEEE International Conference on Soft Computing and Measurements*, 568 – 571 (2017)
5. S.V. Lazarenko, I.V. Pugachev, A.A. Kostoglotov, I.V. Deryabkin, and D.S. Andrashitov. *In Proc. ICMSC 2017*, 292 – 297 (2017)
6. A.A. Kostoglotov, I.V. Deryabkin, D.S. Andrashitov, S.V. Lazarenko, and A.A. Kuznetcov. *In Proc. ICMSC 2017*, 367–372 (2017) [in Russian]
7. A.A. Kostoglotov, S.V. Lazarenko, D.S. Andrashitov, and I.V. Deryabkin. [Online resourse] *MATEC Web of Conferences*. – 2016.
8. A.A. Kostoglotov, I.V. Deryabkin, S.V. Lazarenko, I.V. Pugachev, and A.A. Kuznetsov. *In Proc. EWDTS* 310 – 314 (2016)

Model of multicomponent micro-Doppler signal in environment MatLab

Alexander Kucheryavenko[1,*], *Valentine* Fedosov[1]

[1]Institute Radioengineering Systems and Control, Southern Federal University, 344006, Rostov-on-Don, Russia

Abstract. The article illustrates the problem of measuring the speed glider component targets in the presence of a turboprop effect of the reflected signal in a pulse-Doppler radar, proposed a model turboprop signal component and an algorithm for its suppression

1 Turbine component

Modern radar systems make high demands on the processing of the received signals. This is especially true echo signals from aerodynamic objects.Therefore, the task of shaping the radiolocation portraits of aerodynamic objects are becoming increasingly important against the backdrop of expanding opportunities and increasing the performance of systems for digital processing.Modern methods of radio vision and formed on their basis radar portraits of aerodynamic objects can be used for [1]: recognition of classes or types of aerodynamic objects, including the air corridors of commercial airlines; the meteorologic formations, flocks of birds; providing dispatchers with information about the true average angle of the setting on landing of the aircraft in a strong crosswind and the availability of damages in hydroaerodynamics equipment.

When analyzing the echo signals in radar systems, the dynamics of the reflecting surface of the aerodynamic object, arising due to scattering from the rotating elements of the design, is evident in the spectrum of the reflected signal [2].

The phenomenon of turbo-propeller effect is expressed by the presence in the spectrum of the echo signal along with the body component additional spectral components, called turbine components.

The presence of the turbine component in the echo signal caused by the moving parts of the aerodynamic object [2]. The components of the spectrum caused by turboprop effect can mask the signal from a stationary part of the object and degrade the image on the radar. Therefore, the separation the parts of the spectrum caused by the stationary part of the object from the spectrum caused by moving, rotating or vibrating parts and is of interest for the research.

The fixed part of the aerodynamic object, on the radar picture is presented by glider leaves and rotating and vibrating parts of the object cause the turbine components.

Reflecting from a moving aerodynamic object, radar signals acquires additional modulation this is a manifestation of the turboprop effect. This is due to the presence in the composition of objects rotating elements (the turbine blades and compressor blades of the propellers).Secondary Doppler portrait as a spectrum of the reflected signal and an integral part of the two-dimensional radar image of the object may be distorted by the presence of additional harmonic components not related to the geometric characteristics of the airframe aerodynamic object.

The effects of turbine effects on the spectrum reflected from the aerodynamic object signal is expressed in the appearance of additional harmonics. The spectrum is always linear.

When processing signals from aerodynamic objects in the background noise one of the key problems is the allocation of the individual components of the useful signal [2].

In pulsed radar stations for obtaining a spectral Doppler portraits uses the values of the amplitudes and phases of radio pulses reflected from the aerodynamic object.This is equivalent to sampling the continuous reflected signal with a certain pulse repetition frequency.

2 Algorithm for suppressing the turboprop effect

It seems appropriate to develop the algorithm for suppression of turbine effect of the radar signal in pulse-Doppler radar system. This algorithm in pulse Doppler radar system can be based on the difference of the Doppler increment frequency for a body component of the spectrum of the echo signal and the modulation frequency due to the turbine effect, and the analysis of changes in the time of full amplitude and instantaneous phase of the signal propriety of the turbine effect [3].To develop algorithm of suppression of turbine effect of the echo signal from the aerodynamic object in a pulsed-Doppler radar system, following parameters should be specified: glider aerodynamic speed of the object; the wavelength of the probing signal radar station; the distance to the aerodynamic object; an effective dissipation surface of the

*Corresponding author: zetsu85@gmail.com

object; Doppler frequency of the turbine component; the level of the side components of the output signal relative to the carrier [3].

The algorithm should work with a priori unknown speed of rotation of the rotor or turbine rotor.

The comparison of the Doppler frequency increment for a body component of the spectrum and the modulation frequency due to the turbine effect, shows their essential difference, therefore, the time variation of the amplitude and the instantaneous phase of the full turbine effect is significantly slower than change these characteristics for a body component of the spectrum of the envelope of the echo signal.

Similarly, the change of the envelope there is a component unchanged in the process of formation of reflections from a body component of aerodynamic object and component, with the law changes over time, close to the law of change of the component of the total instantaneous phase, due to turbine effect.

Thus the full term of the instantaneous phase for a body component varies in time according to a linear law, and the summand corresponding to a turbine effect to a law close to harmonic, resulting in the appearance of the lateral components of the spectrum of the echo signal.

2 Development of models

As test signals it's necessary to develop a model of response signal from the aerodynamic object and the model of interference, which is a turboprop effect [3].

As a test signal it is recommended to use a bundle of linearly frequency modulated pulses.

Linear frequency modulated waveforms are used in radar as the method of formation and processing of the probe pulse. Application of linear frequency modulated signal allows to increase the measurement accuracy of radar.

The instantaneous value of the transmitted signal represented by the formula:

$$s(t) = \sum_{k=1}^{N} A\cos\left(\omega t + \frac{\alpha t^2}{2}\right), t \in [t_и], \quad (1)$$

where A is the pulse amplitude, ω – frequency filling pulse, α is the speed of linear frequency changes inside pulse t_и – pulse duration.

The instantaneous value of the received signal represented by the formula:

$$s(t) = \sum_{k=1}^{N} A[1 + M\cos(\Omega_T t)s_k(t)], \quad (2)$$

$$s_k(t) = \sum_{k=1}^{N} A\cos\left(\omega t + \frac{\alpha t^2}{2} + m_T\cos(\Omega_T t)\right), \quad (3)$$

where $t \in [t_и]$, M – the modulation ratio, Ω_T – the frequency of the turbine component, m_T – the modulation index of the turbine component.

Amplitude modulation occurs due to changes in effective surface scattering of the rotating blades of the turbine, and angular modulation due to the speed of rotation of these blades [3].

The optimum detector for coherent packets of radio pulses on a background of white noise is a serially connected: consistent with the packet filter, a detector and a threshold device. Complex frequency response (CFC) harmonized filter

$$k_c(j\omega) = \dot{S}(j\omega)e^{-i\omega T} = \dot{S}_0(j\omega)\dot{S}_n(j\omega)e^{-i\omega T} \quad (4)$$

is the complex conjugate function of the spectra of the bundles S(jω).

Here T – is the delay time of the signal in the filter; $\dot{S}_0(j\omega)$ – single pulse spectrum; $\dot{S}_n(j\omega)$ – spectrum of the bundle of δ-pulses.

The second equality in formula (4) makes it possible to divide the processing of the reflected signal during the period and between periods.

The task of signal detection against the background interference is correlated with the energy spectrum $G_n(\omega)$can be lead to the problem of signal detection against the background of white noise, if you first carry out a "whitewash" interference, since the interference is fed to the input of the detector to mix with the white noise of spectral density N0.

The "whitewash" filter must have a complex frequency response

$$K_{об}(j\omega) = \frac{K_0}{\sqrt{G_n(\omega)+N_0}} e^{j\varphi(\omega)}, \quad (5)$$

where K0 – constant; φ(ω) – arbitrary phase-frequency characteristic, which we believe zero.

Then the power spectral density of the mixture of the interference noise at the output of the filter

$$G_{вых}(\omega) = [G_n(\omega) + N_0]|K_{об}(j\omega)|^2 = K_0^2. \quad (6)$$

Signal with spectrumS(jω), passing through the "whitewash" filter acquires as a result of filtering facilities$K_{об}(j\omega)S(j\omega)$, therefore, the matched filter must have a complex frequency response

$$K_{сф}(j\omega) = \dot{K}_{об}(j\omega)\dot{S}(j\omega)e^{-i\omega T} = \frac{K_0\dot{S}(j\omega)e^{-i\omega T}}{\sqrt{G_n(\omega)+N_0}}. \quad (7)$$

The complex frequency response of series-connected whitewash and coherent filters

$$K_{опт}(j\omega) = \frac{1}{G_П(\omega)+N_0}\dot{S}(j\omega)e^{-i\omega T} \quad (8)$$

is the complex frequency response of an optimal filter of the detector quasideterminants signal on the background of correlated noise.

The expression corresponds to the series connected filter rejected passive interference with the complex frequency response

$$K_{опт}(j\omega) = \frac{1}{G_П(\omega)+N_0} \quad (9)$$

normal and agreed only with the signal filter

$$K_c(j\omega) = \dot{S}(j\omega)e^{-i\omega T} \quad (10)$$

Thus, the structural scheme of the optimal detector of known packs of samples consists of decorrelation noise, the unit weight of the accumulation and threshold device.

The actual sequence of radio pulses can be converted into a sequence of discrete samples by applying a matched filter followed by the temporal discretization [4].

If the initial phase of the burst pulse is unknown, the weighting processing is carried out in two quadrature channels, followed by the Association of quadratures.

Under unknown Doppler frequency of the aerodynamic object blocks matched filtering and the weight savings become multi-channel frequency. The number of frequency channels is determined by the resolution in frequency and a frequency spectrum.

The most common kind of disturbing effects in the communication channel is additive white Gaussian noise [4].

Based on these findings, we propose the following algorithm to reduce the influence of the turbine effect, comprising the following operations:

1. determination of the dependence of the modulus of the complex envelope and its argument in time within the counts of the bundle of the reflected signal after applying matched filtering in range;
2. the smoothing argument of the complex envelope along the axis of times each element of a range, it is necessary to use polynomial smoothing above 20 order;
3. the use of counts of the argument for smooth functions when restoring the complex envelope after the removal of the turbine effect on the argument in time. Modifying an argument caused by a turbine effect will be eliminated;
4. similarly, the smoothing module envelope in time and use the results of smoothing for the recovery module discrete values of the smoothed functions in time for each resolution element in range;
5. perform fast Fourier transformation on each element of a range.

This algorithm requires validation in modeling the problem of suppression of turboprop effect for a simple model of the echo signal.

Modeling complex envelope of the echo from a single object by the method of temporal smoothing provides full suppression of the turbine effect of a single source located at a constant distance.

3 Simulation program

To test the algorithm of suppression turboprop effect was applied to the test signals.

As the test signal the bundle of linear frequency modulated pulses have been taken.

The timing diagram of the emitted pulse packet is shown in figure 1.

This signal represents an even number of pulses with rectangular envelope and the frequency filling, which is a linearly frequency modulated radio signal.

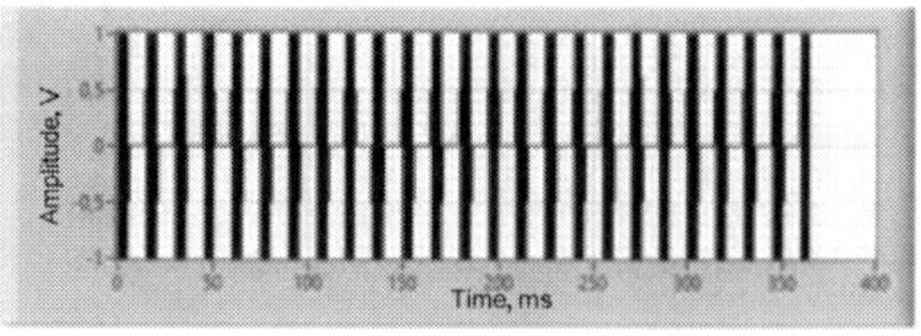

Fig. 1. Time diagram of the emitted burst of pulses.

The timing diagram of the envelope of the burst pulse is shown in figure 2. Obviously, the rectangular shape of the signals.

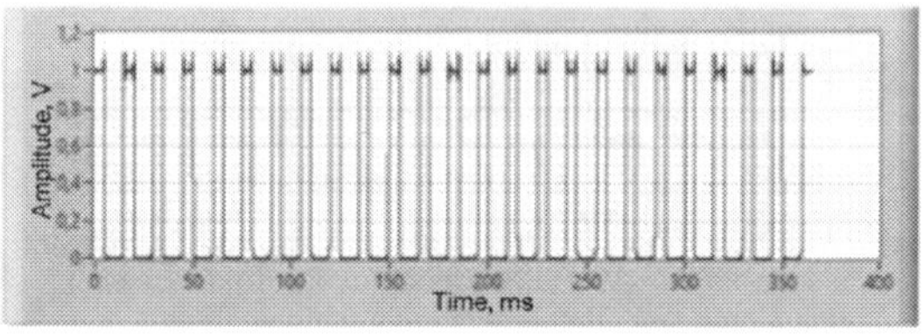

Fig. 2. Timing diagram of the envelope of the burst.

The spectrum of packets of pulses is presented in figure 3.

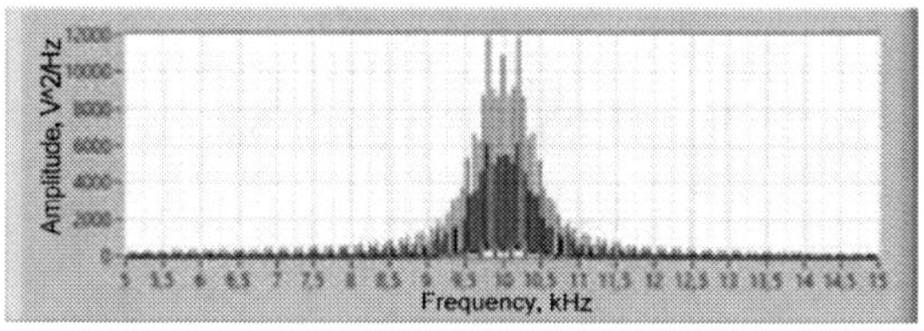

Fig. 3. The spectrum of the emitted burst of pulses.

This spectrum represents a typical spectrum of the linear frequency modulated signal.The timing diagram of the emitted pulse packet after applying matched filtering is presented in figure 4.

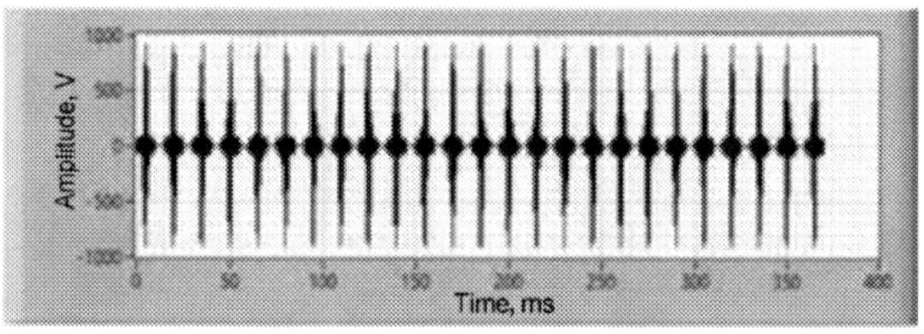

Fig. 4. Timing diagram of the emitted burst after consistent filtering.

The timing diagram of the envelope of the burst pulse after matched filtering is shown in figure 5.

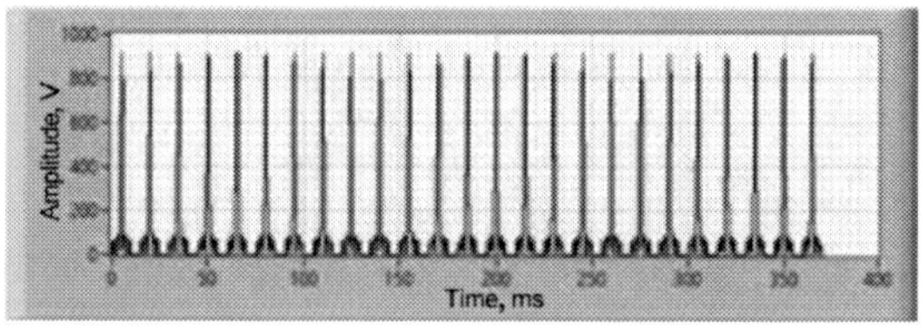

Fig. 5. Timing diagram of the envelope of the burst.

The spectrum of packets of pulses after matched filtering is presented in figure 6. This spectrum represents the spectrum of the linear frequency modulated signal.

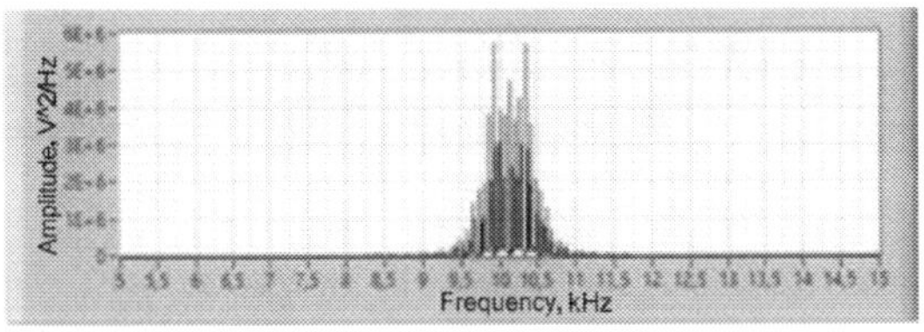

Fig. 6. The spectrum of the envelope of the burst.

The timing diagram accept packets of pulses is shown in figure 7.

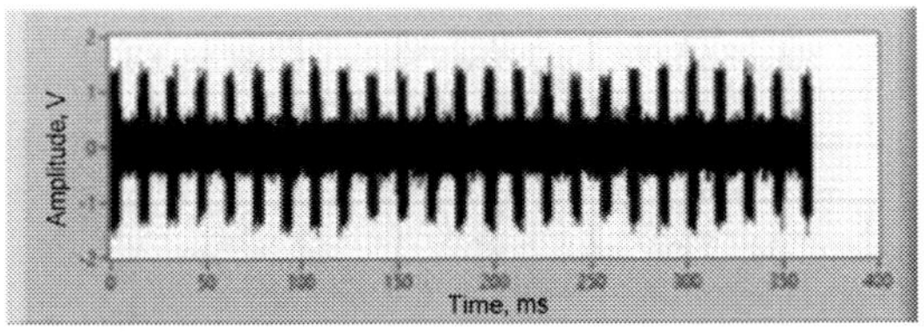

Fig. 7. Timing diagram of the receiving packet of pulses

In figure 7 one can notice the parasitic amplitude modulation caused by the rotating parts of the object. Figure 8 shows the spectrum of the received pulse packet.

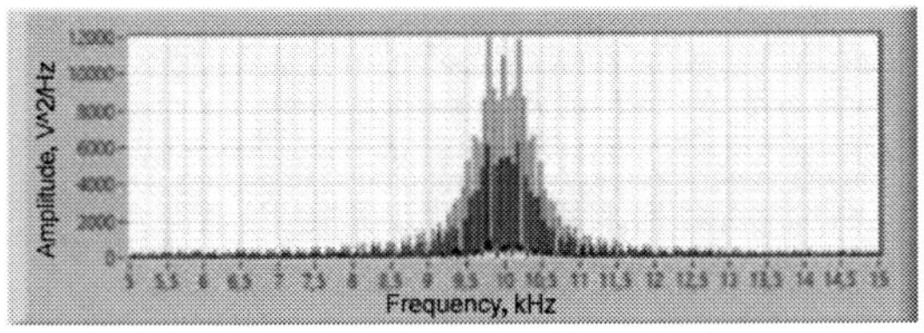

Fig. 8. The spectrum of the received burst of pulses

Further to the accepted signal is applied the procedure of matched filtering figure 9 and highlight the envelope of figure 10.

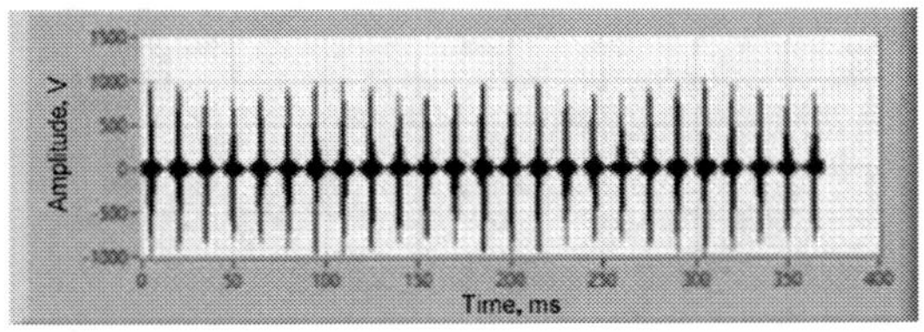

Fig. 9. Timing diagram of the received burst after consistent filtering.

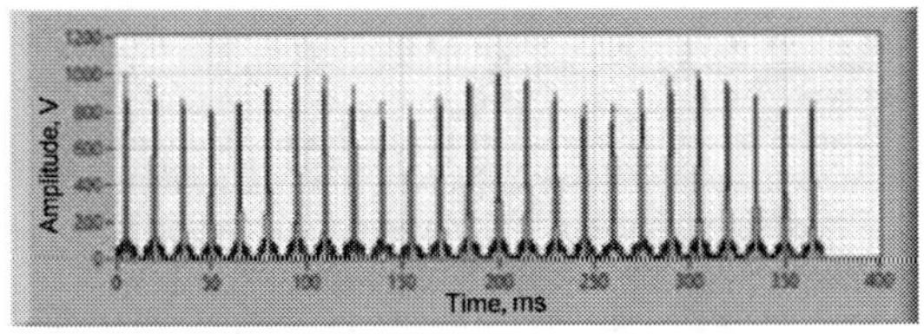

Fig. 10. Timing diagram of the envelope of the burst.

Then use the smoothing and reconstruction of a received signal. Due to the smoothing in the spectrum of the received signal remains one spectral component corresponding to the response signal from the aerodynamic object. The recovered spectrum of the signal shown in figure 11.

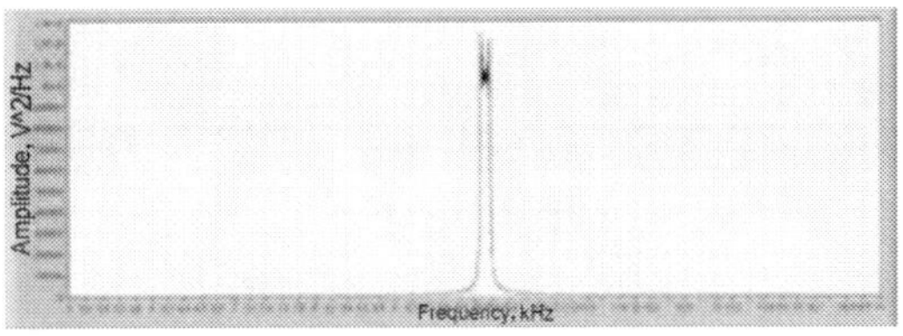

Fig. 11. Recovered signal spectrum.

When comparing the spectrum of the recovered signal (figure 11) and the spectrum of the received signal (figure 8), the reconstructed signal is smoothed by a parasitic modulation, which is a consequence of the presence of the turboprop component.

The study in the received signal contains only the useful component of the radio signal from aerodynamic facility, which is a body component. As a result of applying the proposed techniques were able to eliminate the noise components of the signal and interfering components, caused by the presence of turbo-propeller effect of the rotating parts of the aerodynamic object.

On the basis of the obtained results shows the effectiveness of smoothing algorithm on the proposed turboprop component models of the radiated radar signal.

References

1. E.O. Evdokimova, V.P. Fedosov, *In Proc. International Scientific Conference "Emission and scattering of EMV - IREMV-***2-13***"*. - (2013)
2. E.O. Evdokimova, Izvestiya SFU. Technical science, no. 5 ,(2013)
3. A.V. Kucheryavenko, *In Proc. VII International Scientific and Practical Conference,* no.2 (2015)
4. A.V. Kucheryavenko, V.P. Fedosov, *In Proc. XII All-Russian Scientific Conference of Students, Post-Graduates and Young Scientists "Information Technologies, Systems Analysis and Management (ITSAiU-2014)"* 192 p. (2014)

MATEC Web of Conferences 132, 05009 (2017) DOI: 10.1051/matecconf/201713205009
DTS-2017

The combined maximum principle in the problem of synthesis of an adaptive dynamic filter under conditions of disturbances in the measurement process

Andrey Kostoglotov[1], *Aleksandr* Kuzin[1] , *Sergey* Lazarenko[2] and *Igor* Pugachev[2,*]

[1] Rostov State Transport University, 344038, Rostov-on-Don, Russia
[2]Don State Technical University, 344002, Rostov-on-Don, Russia

Abstract. The article considers the problem of estimating the parameters of motion under conditions of disturbances in the measurement process, which are caused by missing data, misses in measurements, etc. in the operation of radar systems. The structure of the included filter for estimating the parameters of motion is determined by the mathematical model of the maneuvering target. At present time the kinematic models are widely used, but they do not fully correspond to the observed dynamics. This may lead to divergence of the estimation process and failure of the computational procedure. New dynamic filters of the combined maximum principle with the dynamic model of motion possess higher accuracy and stability and smaller amount of computational costs in comparison with common filters. The materials and methods We propose new mathematical model that determines the structure of the dynamic filter of the combined maximum principle. A numerical simulation of the operation of the proposed filter, as well as comparison of its efficiency with common filters by a number of criteria are performed. The results. The efficiency of functioning of radio engineering systems on the basis of the solution to the problem of synthesis of a dynamic filter using method of the combined maximum principle with construction of adaptive dynamic model of motion in vector-matrix form is increased. This makes it possible to eliminate the divergence of computational estimation procedures caused by the inadequacy of the structure of the mathematical model of motion. It is shown that the developed dynamic filter differs from the known filters in terms of the structure of the transition matrix. Discussion and conclusions The results of the mathematical modeling showed that in comparison with the Kalman filter and the "alpha-beta" filter the new dynamic filter under conditions of disturbances in the measuring process increase the number of performance indicators.

1 Introduction

The structure of the filterы for estimating the parameters of motion is determined by the mathematical model of the maneuvering target. Among them kinematic models, describing the simplest variants of motion with a constant speed or acceleration, are widely used [1]. More complex types of trajectories require to use the shaping filter [2]. Estimation of the motion parameters of the observed target implies the need to solve the multidimensional Stratonovich equation, which is quite difficult to perform in real time, so in practice the method of the Gaussian approximation of the probability density is used. The quasi-optimal filter (the extended Kalman filter) is very widely used. In this case, the additive combination of the observed coordinate and the white Gaussian noise with known intensity appears as the observation model. However, the real operating conditions are characterized by disturbances in the measurement process because of data skipping, rough measurement errors, and so on [3]. The kinematic model does not fully correspond to the observed dynamics, for example, in case of maneuvering the target. Then the misses are hidden and not excluded from observations [4]. All these factors lead to the divergence of the estimation process and hence to the failure of the computational procedure [1–4].

Thus, the operating of the filters under conditions of disturbances in measuring processes requires to adapt the filters. This leads to more complicated estimation procedures and to increase in the computational costs, which limits the use of such solutions for estimating multidimensional processes in real time [1, 4].

The work is aimed to solve the problem of decreasing of the effectiveness of radar systems under conditions of disturbances in the measurement process due to the divergence or failure of the estimation computational procedures because of the low adequacy of the mathematical model of motion which the filter is based on.

The filters of the combined maximum principle [5–12] with a dynamic motion model, which is obtained on the basis of the solution to the problem of the control synthesis [12], are distinguished by high accuracy and a

* Corresponding author: bakut_8536@mail.ru

small amount of computational costs in comparison with the common filters.

The aim of the research is to increase the efficiency of radar systems under conditions of disturbances in the measurement process based on the solution to the problem of synthesis of an adaptive dynamic filter using the combined maximum principle.

It is shown in [12] that the use of the hypothesis about the stationarity of the Hamiltonian on the switching surface considering the transversality conditions leads to a dynamic motion model. Its application provides the synthesis of the new dynamic filter. Mathematical modeling of the measurement of the target bearing measurement under perturbation conditions confirms its high efficiency in comparison with the Kalman filter and its clones (α-β filter) [1, 2, 12]. It is found that the application of the new approach provides an increase in the tracking factor, a decrease in the coefficient of false paths and the average lifetime of the false paths with high accuracy and reduction of the computational costs.

2 Formulation of the problem

In the observation space we define the target functional [5, 12]:

$$J=\frac{1}{2}\int_0^{t_1}[\mathbf{y}(t)-\mathbf{q}(t)]^T\mathbf{N}^{-1}[\mathbf{y}(t)-\mathbf{q}(t)]dt=\int_0^{t_1}F(\mathbf{y},\mathbf{q},t)dt\,, \quad (1)$$

where $\mathbf{N}^{-1}$ is the weight matrix characterizing the intensity of interference in the observation channel

$$\mathbf{y}(t)=\mathbf{q}(t)+\xi(t), \quad (2)$$

here $\mathbf{q}\in R^n$ is the vector of generalized coordinates, $\xi(t)\in R^n$ is the vector of random effects on the observation channel of known intensity, $t\in[0,t_1]\subset R$, n is the number of degrees of freedom of the dynamical system.

The observed dynamics satisfies the Hamilton-Ostrogradskii principle [11]. Let us write the extended action functional in the form [5–10]

$$S=\int_0^{t_1}[\lambda(T+A)+F]dt\,, \quad (3)$$

where $T=T(\mathbf{q},\dot{\mathbf{q}},t)=\frac{1}{2}\dot{\mathbf{q}}^T\dot{\mathbf{q}}$ is the kinetic energy, $A=\int_{q(0)}^{q(t_1)}\sum_{s=1}^{n}Q_s dq_s$ is the work of generalized forces $Q_s=Q_s(\mathbf{q},\dot{\mathbf{q}},t)$ at the true trajectory,

$$\mathbf{Q}\in\overline{G}; \quad (4)$$

here $\overline{G}$ is the set of generalized forces admitting the observed motion, $\lambda=\lambda(\xi,\mathbf{q},\dot{\mathbf{q}})$ is the Lagrange multiplier depending on the vector of random effects $\xi(t)$ and on the trajectory $(\mathbf{q},\dot{\mathbf{q}})\in R^{2n}$.

It is required to find the vector of generalized forces $\mathbf{Q}(\mathbf{q},\dot{\mathbf{q}},\xi,\lambda)$ as a function of the trajectory $(\mathbf{q},\dot{\mathbf{q}})\in R^{2n}$ and random effects $\xi\in R^n$.

The solution to this problem is based on the methodology of the combined maximum principle [5–12]. To refine the structure of the solution obtained, it is necessary to analyze the behavior of the invariants of motion on the characteristic trajectories in the phase space.

3 The synthesis of the dynamic filter

The condition of the minimum of the target functional can be represented in the form of the condition of the maximum of the generalized power [5–12]

$$\Phi(\mathbf{q},\dot{\mathbf{q}},\mathbf{Q},\lambda,\xi)=max\sum_{s=1}^{n}[\lambda Q_s+V_s]\dot{q}_s\,, \quad (5)$$

where $\mathbf{V}=gradF$;

under the transversality condition

$$H\big|_0^{t_1}=\lambda(A+T)+F\big|_0^{t_1}. \quad (6)$$

Expressions (5) and (6) serve as the basis to buils the quasideterministic motion model [8], which, taking into account the chosen form of kinetic energy, can be reduced to a quasilinear form with respect to the highest derivative [8, 11]

$$\ddot{q}_s=\left[-\sqrt{\lambda^{-1}}\dot{q}_s+\lambda^{-1}N_{ss}^{-1}[y_s-q_s]\right], s=\overline{1,n}. \quad (7)$$

The vector-matrix representation (7) leads to the first-order differential equation

$$\begin{aligned}&\dot{\tilde{\mathbf{q}}}=\tilde{\mathbf{\Phi}}\tilde{\mathbf{q}}+\mathbf{N}^{-1}\tilde{\mathbf{H}}^*[\mathbf{y}-\tilde{\mathbf{H}}\tilde{\mathbf{q}}]\\&\tilde{\mathbf{q}}(0)=\tilde{\mathbf{q}}^0,\end{aligned} \quad (8)$$

where

$$\tilde{\mathbf{q}}=\begin{bmatrix}q_1\\ \dot{q}_1\\ \vdots\\ q_n\\ \dot{q}_n\end{bmatrix},\ \tilde{\mathbf{H}}=[1\ \ 0\ \ \cdots\ \ 1\ \ 0],\ \tilde{\mathbf{H}}^*=\begin{bmatrix}0\\ \lambda^{-1}\\ \vdots\\ 0\\ \lambda^{-1}\end{bmatrix},$$

$$\tilde{\mathbf{\Phi}}=\begin{bmatrix}0 & 1 & \cdots & 0 & 0\\ 0 & -\sqrt{\lambda^{-1}} & \cdots & 0 & 0\\ \vdots & \vdots & \ddots & \vdots & \vdots\\ 0 & 0 & \cdots & 0 & 1\\ 0 & 0 & \cdots & 0 & -\sqrt{\lambda^{-1}}\end{bmatrix}.$$

4 The results

Let us consider the process of the bearing measurement in the discrete time

$$y(k)=\mathbf{H}\mathbf{x}(k)+\xi(k), \tag{9}$$

where $\mathbf{x}(k)$ is the extended state vector with structure determined by the mathematical model of motion, $\mathbf{H}$ is the matrix of the projection of the space of states onto the space of observations, $\xi(k)$ is the white Gaussian noise with known spectral density N, k is the current time.

The second order kinematic model has the form [1, 16]

$$\mathbf{x}(k+1)=\mathbf{\Phi}(\Delta T)\mathbf{x}(k)+\mathbf{G}v(k), \tag{10}$$

where $\mathbf{\Phi}(\Delta T)=\begin{vmatrix}1 & \Delta T\\ 0 & 1\end{vmatrix}$ is the transition matrix, $\mathbf{x}=\begin{bmatrix}x_1\\ x_2\end{bmatrix}$, $x_1(k)=q(k\Delta T), x_2(k)=\dot{x}_1(k)$, are the bearing and the speed of its change respectively, $\mathbf{G}$ is vector of the perturbations intensity, v the exciting function, ΔT is the data sampling period.

The Kalman filter equations (KF) can be represented by the expression [1, 2]

$$\begin{aligned}
&\hat{\mathbf{x}}(k+1|k)=\mathbf{\Phi}(\Delta T)\hat{\mathbf{x}}(k|k),\\
&\hat{\mathbf{x}}(k|k)=\hat{\mathbf{x}}(k|k-1)+\mathbf{K}(k|k-1)[y(k)-\mathbf{H}\hat{\mathbf{x}}(k|k-1)],\\
&\mathbf{K}(k|k-1)=\mathbf{P}(k|k-1)\mathbf{H}^T\left[\mathbf{H}\mathbf{P}(k|k-1)\mathbf{H}^T+N^2\right]^{-1},\\
&\mathbf{P}(k|k-1)=\mathbf{\Phi}(\Delta T)\mathbf{P}(k-1|k-1)\mathbf{\Phi}^T(\Delta T)+\mathbf{V}(k),\\
&\mathbf{P}(k|k)=\mathbf{P}(k|k-1)-\mathbf{P}(k|k-1)\mathbf{H}^T\left[\mathbf{H}\mathbf{P}(k|k-1)\mathbf{H}^T+N^2\right]^{-1}\mathbf{H}\mathbf{P}(k|k-1),
\end{aligned} \tag{11}$$

where $\hat{\mathbf{x}}(k+1|k)$ is the estimation $\mathbf{x}(k+1)$ based on all previous measurements, including the last k-th value (one step prediction), $\hat{\mathbf{x}}(k|k)$ is the estimation $\mathbf{x}(k)$ based on all previous measurements (filtering), $\mathbf{K}(k|k-1)$ is the feedback coefficient, $\mathbf{P}(k|k)$, $\mathbf{P}(k|k-1)$ are the matrices of estimation and one step prediction respectively,

$$\mathbf{V}=\begin{vmatrix}\frac{1}{3}\Delta T^3 & \frac{1}{2}\Delta T^2\\ \frac{1}{2}\Delta T^2 & \Delta T\end{vmatrix}$$

is covariant matrix of the perturbations vector [1, 2, 16]; $\mathbf{H}=\begin{vmatrix}1 & 0\end{vmatrix}$.

For the mathematical model (10) the expressions for the amplification coefficient and state covariance allow us to go to the α-β filter, where the Kalman filter amplification coefficients are determined by the coefficients α and β [1, 2, 11]:

$$\begin{aligned}
&\hat{\mathbf{x}}(k+1|k)=\mathbf{\Phi}(\Delta T)\hat{\mathbf{x}}(k|k),\\
&\hat{\mathbf{x}}(k|k)=\hat{\mathbf{x}}(k|k-1)+\begin{bmatrix}\alpha\\ \dfrac{\beta}{\Delta T}\end{bmatrix}[y(k)-\mathbf{H}\hat{\mathbf{x}}(k|k-1)]
\end{aligned} \tag{12}$$

The use of (8) leads to the new dynamic filter for estimating the parameters of motion in discrete time

$$\begin{aligned}
&\hat{\mathbf{x}}(k+1)=\mathbf{\Phi}(\Delta T)\hat{\mathbf{x}}(k)+\mathbf{H}^*[y(k)-\mathbf{H}\hat{\mathbf{x}}(k)],\\
&\mathbf{H}^*=\begin{bmatrix}0\\ \Delta T\lambda^{-1}N^{-1}\end{bmatrix},\ \mathbf{\Phi}(\Delta T)=\begin{bmatrix}1 & \Delta T\\ 0 & 1-\Delta T\sqrt{\lambda^{-1}}\end{bmatrix},
\end{aligned} \tag{13}$$

where sign ∧ denotes estimation.

The mathematical modeling results are shown in Figures 1–3. The performance indicators of the filters (11), (12), (13) are summarized in the Table 1.

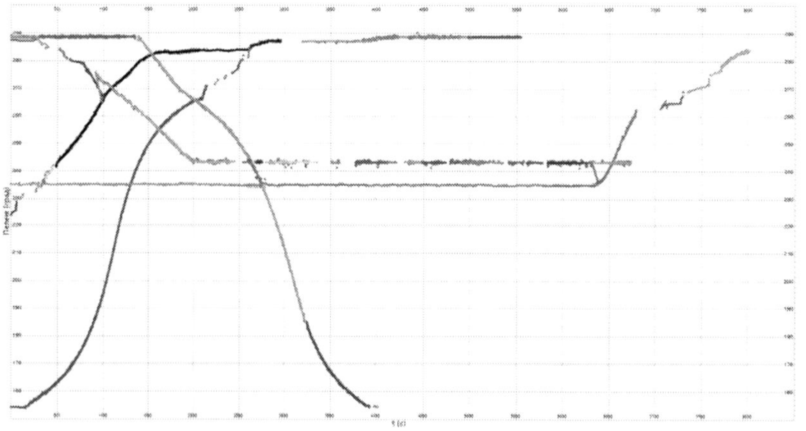

Fig. 1. Bearing estimates by the Kalman filter.

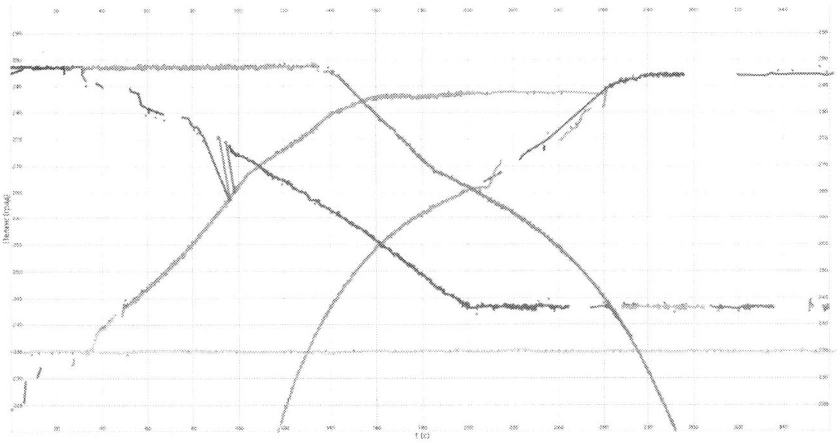

Fig. 2. Bearing estimates by the α-β filter.

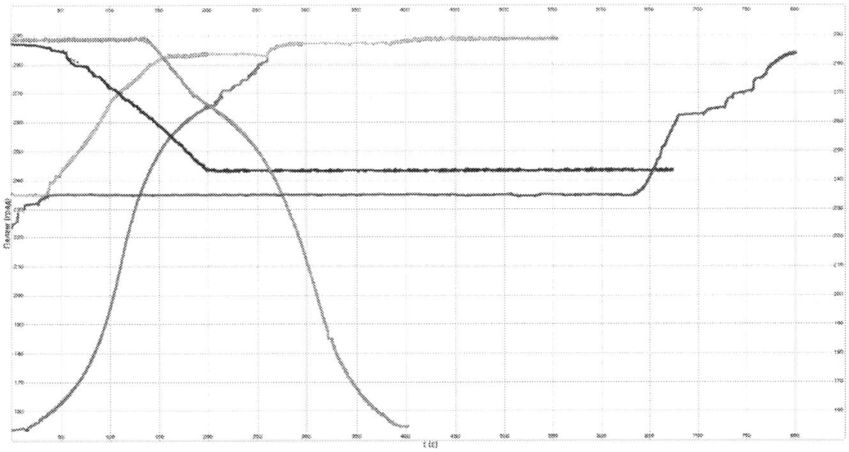

Fig. 3. Bearing estimates by the dynamic filter.

Table 1. Filters Characteristics

The performance indicators / Filters	The averaged estimation error, deg	The averaged tracking coefficient	The averaged coefficient of the false paths	The averaged lifetime of the false paths	The averaged lifetime of the paths discontinuity
The CMP filter	1.05	0.95	0.044	3	8.0
KF	1.34	0.86	0.130	15	8.4
α-β filter	1.17	0.77	0.225	8	7.2

5 Discussion and conclusions

The use of a dynamic motion model as the basis for the new filter provides an increase in the efficiency of its operating under conditions of disturbances in the measuring process with respect to a number of parameters in comparison with the Kalman filter and the "alpha-beta" filter [1, 2, 12].

The paper has been accomplished with the support of Russian Federal Property Fund grants No. 15-08-03798.

References

1. Y. Bar-Shalom, Li X. Rong, T. Kirubarajan. *Estimation with Applications to Tracking and Navigation* (John Wiley & Sons, New York, 2001)
2. R.A. Singer and K.W. Behnke, IEEE Transactions on Aerospace and Electronic Systems, no.1, 100 – 110 (1971)
3. A.A. Kostoglotov, S.V. Lazarenko, D.S. Andrashitov, and I.V. Deryabkin. Synthesis of Phase-locked Loop System Structure with Adaptation Based on Combined-maximum Principle, available at: http://www.matec-conferences.org/articles/matecconf/abs/2016/40/matecconf_icmmr2016_15002/matecconf_icmmr2016_15002.html (accessed 10.09.17).
4. A.A. Kostoglotov, I.V. Deryabkin, S.V. Lazarenko, I.V. Pugachev, and A.A. Kuznetsov. *In Proc. EWDTS*, 310 – 314 (2016)
5. A.A. Kostoglotov, S.V. Lazarenko. Journal of Communications Technology and Electronics, 121 – 125 (2017)
6. S.V. Lazarenko, I.V. Pugachev, A.A. Kostoglotov, I.V. Deryabkin, and D.S. Andrashitov. *In Proc. ICMSC 2017*, 292 – 297 (2017)
7. A.A. Kostoglotov, O.A. Kostoglotova, I.V. Deryabkin, I.E. Kirillov, and S.V. Lazarenko. *In Proc. ICMSC 2017*, 216 – 220 (2017)
8. A.A. Kostoglotov, I.V. Deryabkin, D.S. Andrashitov, S.V. Lazarenko, and A.A. Kuznetcov. *In Proc. ICMSC 2017*, 367 – 372 (2017)
9. A.A. Kostoglotov, S.V. Lazarenko, and Z.V. Lyaschenko. *In Proc. 2017 20th IEEE International Conference on Soft Computing and Measurements, SCM 2017*, 568 – 571 (2017)
10. C.C. Schooler, IEEE Transactions on Aerospace and Electronic Systems, no 6,1300 – 1306 (1975)
11. A.I.Lur'e. *Analytical mechanics* (GIFML, Moscow: 1961)
12. X. R. Li, V. P. Jilkov. IEEE Transactions on Aerospace and Electronic Systems, no.4, 1333 – 1364 (2003)

Research and comparative analysis of methods linear interpolation trajectory of the working tool of a two-tier mechanism

Ivan Naumov[1,*], *Anton* Paschenko.[1] *and Aleksandr* Koltcov.[1]

[1]Don state technical university (DSTU), Department of Radioelectronic and electrotechnical systems and complexes, 344002, Rostov-on-Don, Russia

Abstract. In this paper the comparative analysis of the interpolation methods (Bresenham's method and the least distance method) of the linear interpolation trajectory of the working tool of a two-tier mechanism is performed and its error is examined. The kinematic diagram of the two-tier mechanism is studied; the direct and inverse kinematics are determined. To perform the comparative analysis, the method of the mechanism's interpolation grid construction is studied.

1 Introduction

In most cases the flattened materials (metal, fabric, leather, stone and etc.) profiling operation by using the firmware reduces to segmentation (division) of the figure's compound contour to elementary figures, which can most be the rectilinear segment, considering the interpolation algorithms implementation simplicity. When dividing a compound contour to rectilinear segments the processing accuracy of a raw workpiece is determined by the number and the accuracy of the interpolation method applied.

The problem of an optimal compound contour division to the rectilinear segments is solved by most electric accounting machines (systems) and the interpolation problem rests to a CNC devices, which solution accuracy depends on its kinematic aspects, operating mechanism's accuracy and the interpolation methods.

There are many different methods of the linear interpolation, such as digital differential analyzer (DDA) and Bresenham's method [1], which are the most known methods at the present time.

The choice of the particular method to solve the linear interpolation problem depends on the kinematic characteristics of the equipment.

Due to this, here, we perform the comparative analysis of the rectilinear segment interpolation methods to study the interpolation error for every method.

It is known, that the DDA is more complicated in a way of hardware implementation, since it involves the recurrent division operations, so as the comparative methods in this work the Bresenham's method is regarded, as well as the method for the two-tier mechanism, suggested before [2].

2 Two-tier mechanism

The two-tier mechanism, which kinematic diagram is illustrated in fig. 1, is represented as the disk with radius R , which can turn around the center, located at coordinate origin of the basic coordinate system XOY through an angle , where the rotation angle $0 \leq \alpha < 2\pi$.

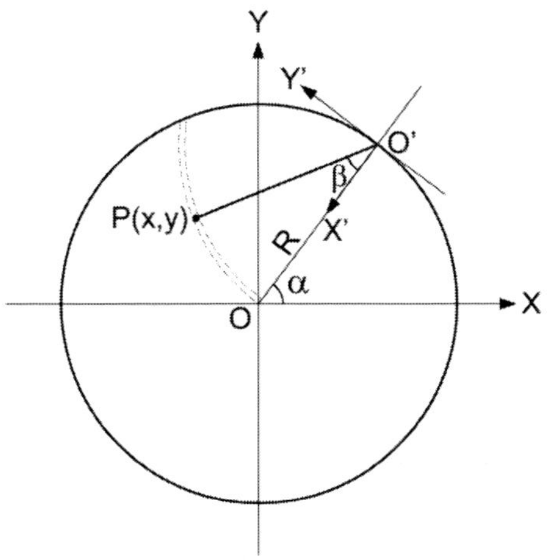

Fig.1. The kinematic diagram of the two-tier mechanism

At the edge of the disk the rod is fixed and its length equals to disk's radius R. The rod can turn through angle β on itself, which axis is located in the additional coordinate system $X'O'Y'$, where the rod's rotation angle $0 \leq \beta \leq \frac{\pi}{3}$. The working tool is fixed on the edge of the rod at the point $P(x, y)$ and goes through the disk, that's why disk is provided with a slot of the certain width. The slot strikes an arc of the circle with the radius R and the

* Corresponding author: naumov_ivan85@mail.ru

center in the origin of additional coordinate system $X'O'Y'$.

3 The solution of the inverse kinematic problem

The solution of an inverse kinematic problem [2] can be described as following:

$$\begin{cases} \alpha(t) = \arcsin\left(\frac{x(t)\cdot cos\beta(t) - x(t) - sin\beta(t)\cdot y(t)}{2\cdot R(cos\beta(t)+1)}\right) \\ \beta(t) = \arccos\left(\frac{-x^2(t)-y^2(t)+2\cdot R^2}{2\cdot R^2}\right) \end{cases} \quad (1)$$

The solution of the direct kinematic problem can be written in accordance with [2] as following:

$$\begin{cases} x(t) = R\cdot\sin(\beta(t)+\alpha(t)) - R\cdot\sin\alpha(t) \\ y(t) = R\cdot\cos(\beta(t)+\alpha(t)) - R\cdot\cos\alpha(t) \end{cases} \quad (2)$$

When solved the direct kinematics for the two-tier mechanism considering the angle resolution of the disk rotation $\Delta\alpha$ and the rod rotation angle resolution $\Delta\beta$, all connection joints of the interpolation grid coordinates can be defined. For this purpose, in MATLAB development environment the virtual prototype was actualized. The prototype allows calculating coordinates of all connection joints in the interpolation grid of the two-tier mechanism. The input parameters of the prototype are: disk radius R and the discrete pitch size of angles $\Delta\alpha$ and $\Delta\beta$. In accordance with the prototype's input parameters its algorithm creates a rectangular matrix of the joint coordinates M expressed in units $n\times m$, which has the following form:

$$M(i,j) = \begin{matrix} (\alpha_1;\beta_1) & (\alpha_2;\beta_1) & \dots & (\alpha_j;\beta_1) & (\alpha_{j+1};\beta_1) \\ (\alpha_1;\beta_2) & (\alpha_2;\beta_2) & \dots & (\alpha_j;\beta_2) & (\alpha_{j+1};\beta_2) \\ \dots & \dots & \dots & \dots & \dots \\ (\alpha_1;\beta_i) & (\alpha_2;\beta_i) & \dots & (\alpha_j;\beta_i) & (\alpha_{j+1};\beta_i) \\ (\alpha_1;\beta_{i+1}) & (\alpha_2;\beta_{i+1}) & \dots & (\alpha_j;\beta_{i+1}) & (\alpha_{j+1};\beta_{i+1}) \end{matrix} \quad (3)$$

where $\beta_{i+1} - \beta_i = \Delta\beta$ и $\alpha_{i+1} - \alpha_i = \Delta\alpha$.

In fig.2,a and fig. 2,b the interpolation grid of the two-tier mechanism with the radius $R = 500$ mm, $\Delta\alpha = 10°$ and $\Delta\beta = 5°$ is shown [3-10].

As seen in the fig.2,a and 2,b the specific arc-shaped curve is formed by the rod movement about the center in the point O' (fig. 1) in angle β. The set of arc-shaped curves is formed resulting from the rotation of the disk with the radius R about the center at point O in angle α. The set of concentric circles with the center at the point O, forms as the result of the angles α and β resolution. The set of cross-points of the concentric circles and the arc-shaped curves are the joint connections of the interpolation grid of the two-tier mechanism.

Let's locate a certain rectilinear segment in the two-tier mechanism's operating area, with endpoints at $M1(x1;y1)$ and $M2(x2;y2)$. If solving the inverse kinematics at the point $M1(x1;y1)$, we will have the values of the required parts rotation angles of the given mechanism which contain in the matrix (3). Such matrix element is denoted as $(\alpha_j;\beta_i)$, where $0 \le i < n$ и $0 \le j < m$. It is obvious, that $(\alpha_j;\beta_i)$ in the matrix (3) is surrounded by nearby elements:

$$\hat{M} = \begin{pmatrix} \dots & (\alpha_j;\beta_{i-1}) & \dots \\ (\alpha_{j-1};\beta_i) & (\alpha_j;\beta_i) & (\alpha_{j+1};\beta_i) \\ \dots & (\alpha_j;\beta_{i+1}) & \dots \end{pmatrix} \quad (4)$$

The summary of the method [2] is in the search for the nearby element from the matrix $\hat{M}$ upon criterion of the minimal distance from such element's coordinate to the endpoint of the rectilinear segment $\overline{M1M2}$. After such element is found, the search of its nearby elements upon the given criterion will be actualized again. Algorithm's iterative process will repeat until it will reach the end of the rectilinear segment $\overline{M1M2}$.

4 The result of computer simulation using the Brazenham algorithm

In fig.2 the result of the computer modelling in MATLAB is shown and it reflects the interpolation process of the rectilinear segment (line in fig. 2,a), indexed by two points, point $M1[-350;-100]$ and point $M2[160;280]$. Interpolation is fulfilled according to the the least distance method using the two-tier mechanism with disk radius $R = 500$ and $\Delta\alpha = 10°$, $\Delta\beta = 5°$.

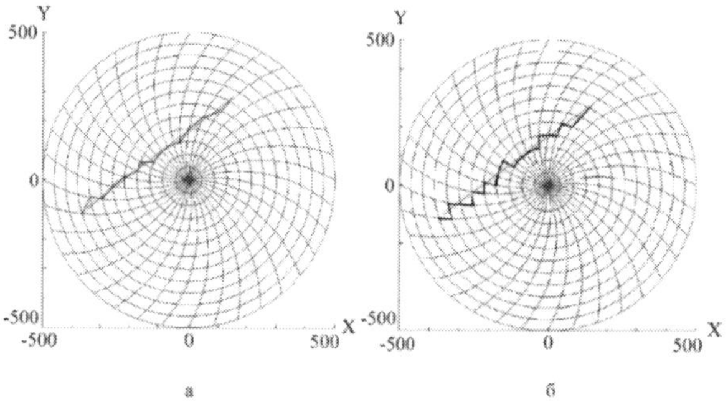

Fig.2. Results of the rectilinear segment interpolation: a - by the least distance method, b – Bresenham's algorithm

During the modelling it was discovered that root-mean-square deviation of the interpolation error by least distance method doesn't exceed the value of 9.914.

For the comparative analysis the rectilinear segment interpolation procedure by known Bresenham's algorithm was also actualized. The result of the computer modelling is presented in fig.2,b. This result was obtained with the same input parameters of the prototype [11-36].

5 Conclusion

Modelling showed that the root-mean-square deviation by the Bresenham's algorithm equals 15.799, which is significantly higher than the deviation with the least distance method, and this proves that the least distance method is expedient for the solving the interpolation problem in unorthogonal mechanisms.

References

1. Voltmander P.V. Mashinnaya grafika. *Osnovnye algoritmy. [Computer graphics. Basic algorithms]* (NGU, Novosibirsk 1997)
2. Valyukevich Y.A., Naumov I.I. Nauchno-tehnicheskie vedomosti Sankt-Peterburgskogo gosudarstvennogo politehnicheskogo universiteta, no.95, 177-181 (2010) [in Russian]
3. V.E. Gough, S.G. Whitehall, *In: Proc. 9th International Technical Congress, FISITA*, 177 (1962)
4. D. Stewart, *In Proc. Inst Mech. Eng., Pan I* 180 (15) 371-386 (1966)
5. M. Valenti, Machine tools get smarter, ASME Mech. Eng. 70-75 (1995)
6. F. Pierrot, P. Dauchez, A. Fournier, J. Robotic Syst. No.8 829-840 (1991)
7. .F. Behi. IEEE J. Robotics Autom, no.4, 561-565 (1988)
8. R.I. Alizade, N.R. Tagiyev, J. Duffy, Mechanism and Machine Theory, no. 1, 115-124 (1994)
9. J.W. Kimetal., *In: Proc. 1997 ASME IMECE Symposium on Machine Tools*, (1997)
10. G.R. Dunlop, T.P. Jones, Mechanism and Machine Theory, no.8, 903-920 (1997)
11. J. Nielsen, B. Roth, Int. J. Robotics Res., no. 12, 1147— 1160 (1999)
12. P. Nanua, K.J. Waldron, V. Murthy, IEEE Trans. Robotics Autom, no. 6, 438-444 (1990)
13. M.L. Musty, Mechanism and Machine Theory, no. 4 365-380
14. P. Dietmaier. Advances in Robot Kinematics: Analysis and Controll, 7-16 (1998)
15. Jack Bresenham. IBM Systems Journal, no.1, 25-30, (1965)
16. 14. Jack Bresenham. Communications of the ACM, no. 2, 100-106 (1977)
17. Jerry R. Van Aken, IEEE Computer Graphics & Applications.24-35 (1984)
18. . Michael Abrash. Dobbs Journal, no. 194, 171-176 (1992)
19. J.E. Bresenham, IBM Syst. J., no.4, 25–30 (1965)
20. J.E. Bresenham, Comput. J., no.25 116 –120 (1982)
21. J.E. Bresenham, Fundamental Algorithms for Computer Graphics, 59–104 (1985)
22. D. Field, ACM Trans. Graph., 1–11 (1985)
23. H. Freeman, IRE Trans. EC-102, 260 –268 (1961)
24. H. Freeman, Picture Processing and Psychopictories, 241–266 (1970)
25. K.Y. Fung, T.M. Nicholl, A.K. Dewdney, *Comput. Graph. Forum 3*, 267–277 (1992)
26. G.W. Gill, IEEE Comput. Graph., 66 –72 (1994)
27. P. Graham, S. Iyengar,*Iin: Proc. 1993 ACM Computer Science Conf.* (1993)
28. P. Graham, S. Iyengar, IEEE Comput. Graph. Appl., 49–53 (1994)
29. G.B. Regiori, *Digital computer transformations for irregular line drawings* (Department of Electrical Engineering and Computer Science, New York 1972)
30. J. Rokne, B. Wyvill, X. Wu, ACM Trans. Graph, no. 9 (4) 376 –388 (1990)
31. J. Rokne, C. Yao, ACM Trans. Graph., no.11(2) 183–192 (1992)
32. A. Rosenfeld, IEEE Trans. Comput. 1264 –1269 (1974)

Synthesis of an interpolation algorithm straight-line segment at planning the trajectory of a degenerate two-tier mechanism

Ivan Naumov [1,*], *Oleg* Tolstunov.[1] *and Larisa* Perekhrestenko[1]

[1]Don state technical university (DSTU), Department of Radioelectronic and electrotechnical systems and complexes, 344002, Rostov-on-Don, Russia

Abstract. In the work [1] the kinematic diagram of the two-tier mechanism-based device is described, its advantages under the mechanic technological processing of the different materials range and the scope of use are shown.

1 Introduction

The mechanism consists of the circle and the circular segment of the same radius which are the first and the second parts of the device (fig. 1). The circle rotates about axis placed at the origin of the basic coordinate system (point O in fig. 1). The circular segment is mounted to rotate at the origin of the additional coordinate system (point O' in fig. 1), placed at the edge of the circle.

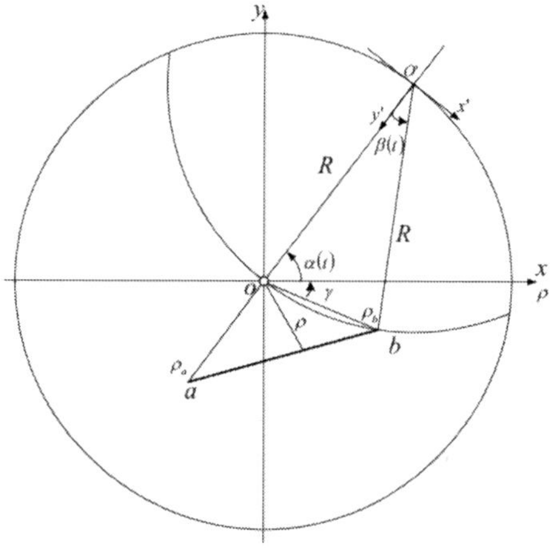

Fig. 1. The Kinematic diagram of the device.

In figure 1 the following notations are taken:

$\alpha(t)$ -the angle of circle's rotation about the basic coordinate system;

$\beta(t)$ - the angle of the rod's rotation about the additional coordinate system;

a,b – the endpoints of the segment ab;

R – the parts' radius;

ρ_a, ρ_b - the endpoints of the segment ab within the polar coordinate system respectively;

γ - the angle of the tool's current position within the polar coordinate system.

2 Formulation of the problem

Thus the general formulation of the rectilinear segment interpolation problem can be stated as follows: in accordance with the increments of the linear coordinates Δx and Δy (x `= 0,±1;y = x = 0,±1) known at every step, the increments of the mechanism parts' angular coordinates $\Delta\alpha$ and $\Delta\beta$.

The values of the unit angular increments $\Delta\alpha$ and $\Delta\beta$ are the constructible constant of the mechanism. The linear unit increments Δx and Δy (linear interpolation step) are chosen based on the desired precision of the linear interpolation and values of the $\Delta\alpha$ and $\Delta\beta$ according to regulations stated in [2].

In fig. 1 the mechanism's condition is shown, when the initial point of the rectilinear segment and the current position of the tool (radius vector end $\vec{\rho}_b$) coincide.

The current position of the angular and linear coordinates is known and also known the coordinates of the second part onset x1, y1 in the coordinate system which is related to the first part axis of rotation.

The operating system for the one computation cycle of the interpolation step can be described as following:

- the position measurement of the linear coordinates for the next segment point through the linear interpolation;
- the direction measurement and measurement of the distance movement in angular coordinate $\alpha(t)$;
- the measurement of the increment value and direction in coordinate $\beta(t)$;
- the implementation of the movements in real-time mode;
- the next interpolation step.

* Corresponding author: naumov_ivan85@mail.ru

3 Realization of linear interpolation based on the Brezenham algorithm

The linear interpolation procedure can be easily actualized as a matter of the Bresenham's line algorithm, used when constructing the lines by means of the bit-map graphics (for the PC monitors)[7].

To measure the increment in angular coordinate $\alpha(t)$ let's move to the polar coordinate system, the center of which is coincident.

To measure the increment in the angular coordinate $\alpha(t)$ let's pass on to the polar coordinate system, which center is coincident with the center of the basic Cartesian coordinate system xy and the polar axis is in the X-direction in Cartesian system; define the coordinates of the recurrent point position in this system, terms of the relations:

$$\begin{cases} \gamma_{i+1} = arctg\left(\dfrac{y_{i+1}}{x_{i+1}}\right) \\ \rho_{i+1} = \sqrt{x_{i+1}^2 + y_{i+1}^2} \end{cases} \quad (1)$$

Therefore, the total increment in the angular coordinate $\alpha(t)$ will equals:

$$\Delta\gamma_{i+1} = \gamma_i - \gamma_{i+1} \quad (2)$$

Here it should be pointed out that all calculations related to the definition of the angular coordinates $\alpha(t)$ and $\beta(t)$ are being performed using floating-point arithmetic when the control system functions with the fixed-point arithmetic, thus, the angular movement at every step of the interpolation can be defined from the relation [8-12]:

$$\Delta\alpha_{i+1} = n_\alpha \cdot \Delta\alpha \quad (3)$$

Where n_α - the number of steps of the unit movement in the angular coordinate $\alpha(t)$ at the current interpolation step, which defines as the quotient from:

$$\frac{\Delta\gamma_{i+1}}{\Delta\alpha} \quad (4)$$

Suggested calculation technique allows avoiding round-off errors at every step of the interpolation, since γ_i and γ_{i+1} are presented as the floating-point numbers so they aren't put through adjustment during the interpolation process [13-21].

4 Geometrical interpretation

In fig. 2 the geometrical interpretation of the increment calculation of angular coordinate $\beta(t)$ is shown.

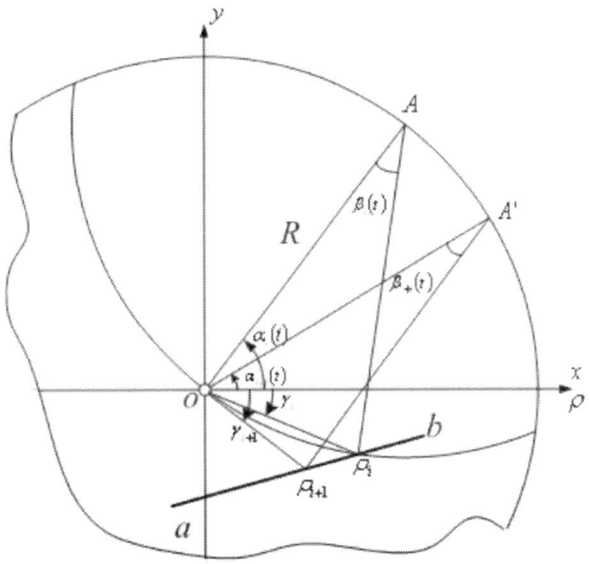

Fig.2. Geometrical representation of the angular coordinates increment.

Here is the position of the mechanism's parts at i-th and i+1th steps of the interpolation. The positions of the second part's onset at i-th and i+1th steps of the interpolation are denoted in figure as points A and A'. The initial point positions of the tool's second part for such cases are denoted as B and B' respectively. Hence, based upon the geometric parameters of the triangle OA'B', which three sides and angle are known in accordance with the previous parameter calculations stage for the next interpolation step, the absolute value of the angular coordinate $\beta(t)$ in X'OY' coordinate system (fig. 1) at i+1 interpolation step can be defined [3].

Having the known sine theorem applied for the triangle OA'B' this can be written in the following form:

$$\frac{OB'}{\sin(\beta_{i+1})} = \frac{A'B'}{\sin(\alpha_{i+1} + \gamma_{i+1})} \quad (5)$$

Where:

$$\begin{aligned} A'B' &= R^2, \\ OB' &= \rho_{i+1} \end{aligned} . \quad (6)$$

Value of the angular coordinate respectively:

$$\beta_{i+1} = \arcsin\left(\frac{\rho_{i+1} \cdot \sin(\alpha_{i+1} + \gamma_{i+1})}{R}\right) \quad (7)$$

The number of the elementary angular movements in coordinate $\beta(t)$ can be evaluated as the integer part of the formula similar to the definition of the elementary angular increments number in coordinate $\alpha(t)$:

$$m_{\beta} = \left\lfloor \frac{\beta_{ш} + \beta_{ш+1}}{\Delta\beta} \right\rfloor \quad (8)$$

The main advantage of the method suggested here is the constant static accuracy of the interpolation [5-6].

The finishing error of the suggested method doesn't depend on the number of multiline segments, which make the processing part's outline (profile) and doesn't exceed the value of the $\Delta\alpha$ and $\Delta\beta$ interpolation step in both coordinates in the entire working surface of the mechanism.

The suggested interpolation algorithm was actualized in C++ in Borland development environment.

The program testing showed full algorithm's service capability.

Interpolation method, suggested here, was tested on the prototype of the device.

Microprocessor control unit STM32F107 with Cortex M3 core, which provides the floating-point arithmetic, was used as the computing machine.

Experience has shown that the computation time of the one interpolation step equals 20ms, which is significantly less than the interpolation step time, calculated on the basis of the required movement's technologic speed of the tool – 2m per minute and the linear interpolation step $\Delta x = \Delta y = 0{,}05$ mm.

As a result, the suggested here interpolation method exceeds in computation expenses the suggested in the work [2] and can be recommended for applying in the technological devices control system based on the degenerate two-tier mechanism.

References

1. Valyukevich Y.A., Naumov I.I., Alepko A.V., e.a. Patent no 2399478, Russian Federation, 2010
2. Valyukevich Y.A., Naumov I.I. Nauchno-tehnicheskie vedomosti Sankt-Peterburgskogo gosudarstvennogo politehnicheskogo universiteta, no.95, 177-181 (2010) [in Russian]
3. Ahlfors, S.P., Ilmoniemi, R.J., Ha¨ma¨la¨inen, M.S., Electroencephalography and Clinical Neurophysiology, no.82, 225 – 236 (1992)
4. Ahlfors, S.P., Simpson, G.V., Dale, A.M., Belliveau, J.W., Liu, A.K., Korvenoja, A., Virtanen, J., Huotilainen, M., Tootell, R.B., Aronen, H.J., Ilmoniemi, R.J., Electroencephalography and Clinical Neurophysiology, no.82, 2545 – 2555 (1992)
5. Babiloni, F., Babiloni, C., Carducci, F., Romani, G.L., Rossini, P.M., Angelone, L.M., Cincotti, F., NeuroImage, no.19, 1 – 15 (2003)
6. Backus, G.E., Mathematical Problems in the Geophysical Sciences, 1 – 105 (1971)
7. P. Bao, J. Rokne, Comput. Graph., no.13 (4) 461–469 (1989)
8. J.E. Bresenham, IBM Syst. J. no.4, 25–30 (1965).
9. J.E. Bresenham, Comput. J. no. 25 116 –120 (1982)
10. J.E. Bresenham, Fundamental Algorithms for Computer Graphics, NATO Computer and Systems Series, 59–104 (1985)
11. D. Field, ACM Trans. Graph 1–11, (1985)
12. H. Freeman, IRE Trans. EC-102, 260 –268 (1961)
13. H. Freeman, Picture Processing and Psychopictories, 241–266 (1970)
14. K.Y. Fung, T.M. Nicholl, A.K. Dewdney, *Comput. Graph. Forum 3*, 267–277 (1992)
15. G.W. Gill, IEEE Comput. Graph. Appl. 66 –72, (1994)
16. P. Graham, S. Iyengar, *In: Proc. 1993 ACM Computer Science Conf.*, (1993)
17. P. Graham, S. Iyengar, IEEE Comput. Graph. Appl. 49–53 (1994)
18. G.B. Regiori, *Digital computer transformations for irregular line drawings* (Department of Electrical Engineering and Computer Science, New York., 1972)
19. J. Rokne, B. Wyvill, X. Wu, ACM Trans. Graph., no. 9 (4) 376 –388 (1990)
20. J. Rokne, C. Yao, ACM Trans. Graph., no.11(2) 183–192 (1992)
21. A. Rosenfeld, IEEE Trans. Comput, 1264 –1269 . (1974)

Multi-channel monitoring approach of E1 streams

Anton Nikitin[1], *Sergey* Makov [2], *Sergey* Klimenko [3]

[1]Don state Technical University, 344000, Rostov-on-Don, Russian Federation

Abstract. The object of development is a device for multi-channel monitoring of the E1 stream state. The subject of the development is the monitoring of the status of the channels and analysis of traffic in each time slot of the E1 stream. The purpose of this paper is to develop an approach that allows multi-channel monitoring of the E1 stream state.

1 Introduction

Transmission of telemetry data is an important task for a lot of companies. In most cases the data gathered by that equipment is extremely important, as proper systems' functioning and sometimes people's lives depend on its integrity. Thus, for the transmission of these data streams, it is advisable to use channel switches to which the E1 stream switches belong. The line data rate is 2048 kbps which is split into 32 timeslots, each being allocated 8 bits in turn, that makes the speed of each time slot 64 kbps. As a result, the total data speed rate in the E1 stream is 2048 Kbit / s. Out of 32 multiplexed channels two channels are used for frame synchronization and common channel signaling, where the service information necessary for the operation of telephone networks is transmitted. The remaining 30 channels are used to transmit digital information. Thus the application of E1 stream technology allows to organize 30 simultaneous independent communication channels in one physical stream. The relevance of the topic lies in the fact that solving the problem of monitoring the status of channels in the E1 stream and analyzing traffic in each time slot will allow to quickly detect violations in the equipment that generates incoming traffic E1 time slot analysis.

The modern digital primary network is built on the basis of three main technologies:

- Plesiochronous hierarchy (PDH);
- Synchronous hierarchy (SDH);
- Asynchronous transfer mode (ATM).

The PDH and SDH hierarchies interact through the multiplexing and demultiplexing of PDH streams into SDH systems.

The main difference between the SDH system and the PDH system is the transition to a new multiplexing principle. The PDH system uses the principle of plesiochronous multiplexing. According to which the procedure for equalizing the clock frequencies of incoming signals by the method of stuffing is performed for multiplexing, for example, four E1 (2048 kbit / s) streams into one E2 stream (8448 kbit / s). As a result, when demultiplexing, it is necessary to perform a step-by-step process of restoring the original channels.

The E1 channel is the primary channel of the plesiochronous hierarchy. It is the main channel used in secondary telephony networks and data transmission. In comparison with the other channels of the PDH hierarchy, this channel has several features, namely the supercycle structure and the signaling channel. The remaining channels of the plesiochronous hierarchy have only a cyclic structure.

The structure of the E1 transmission systems includes three levels of the OSI model: physical, channel and network. The physical layer describes the electrical interface of the E1 stream, as well as the E1 signal parameters. The link layer describes the procedures for multiplexing and demultiplexing channels of a lower hierarchy level (bcc 64 kbit / s and PM channels) into the E1 stream, the cyclic and supercyclous structure of the E1 stream in the primary network, and monitoring error parameters at the network level. Let us consider in more detail the structure of each of the three levels of the E1 system. [1].

Basic scientific developments in monitoring errors in the E1 Data stream are presented by the authors: A. Vainshtein, YJ. Stein, F. Inumaru, N. Sato, K. Wakabayashi and etc. [2-5]

2 Proposed approach

As the review of existing methods and devices shows, there are several ways of monitoring the E1 flow. There are solutions that allow monitoring the status of the channels in the E1 stream, but the equipment analyzing the traffic in each time slot and allowing to display the contained information in these channel intervals is absent. For this reason, there is a need to create a device that will analyze the traffic in each channel interval, which will allow us to quickly detect violations in the equipment of the forming traffic for the analyzed E1 channel.

[1]Nikitin Anton: Chaosvarp14@rambler.ru

Comparison of the methods used in cheap and expensive commutators for E1 stream analysis and the proposed method is shown in Table 1.

Table 1. Approaches overview.

Approach	E1 SYNC & MSYNC status	SNMP supported	The ability to track the status of time slots
Approach is used in cheap devices	yes	no	no
Approach is used in expensive devices	yes	yes	no
Proposed approach	yes	yes	yes

The device inputs to the transceiver block receive traffic from the E1 channels. Further, the obtained data is fed to the computing device for further analysis of the information contained in the received E1 channel. Thus, the error statistics in the E1 channels and the contents of their channel intervals are transmitted via the data bus to the SNMP server (Simple Network Management Protocol) [6].

The data received from the transceiver E1 is fed to the computing device further analysis is carried out of the information contained in the received E1 channel and a specific time interval. A survey of various materials has made it possible to distinguish three different ways of implementing a computing device:

- discrete digital elements of small and medium degree of integration;
- microcontroller or processor;
- digital microcircuit of a large degree of integration with a variable logic circuit.

Discrete digital elements of small and medium degree of integration include shift registers, coders, decoders, encoders, decoders, multiplexers, counters, etc. All these elements are implemented in the form of separate chips, and if a small project may require a small number of chips, then for a large project the number of necessary chips becomes large, which also requires a lot of external connections. As a result, the complexity of the design, the dimensions of the printed circuit board increases, which is an essential drawback of this method.

Microcontrollers and processors allow processing of digital signals and implement complex algorithms, but the lack of microcontrollers is the sequential reception and processing of data. Since the task is to monitor eight E1 streams, data reception and processing should occur in parallel.

A digital IC of a large degree of integration with a variable logic circuit in this case has the advantage that the desired structure of the device is obtained by programming the connections of switching matrices, by controlling the bit sequence with logical keys. The control bit sequence is written to the internal static RAM during configuration, which gives the name to this class of programmable logic integrated circuits as FPGA (Field Programmable Gate Array). The peculiarity of this solution is that every time the power is turned on, the structure of the device is restored by recording in the internal static RAM. The advantage of using FPGA in this case is the ability to create adaptive systems whose structure dynamically changes with time, which allows the implementation of various digital devices on a single chip.

Summarizing the advantages and disadvantages of the considered ways of implementing the computing device, the optimal solution is to use the programmable logic integrated circuit of the FPGA type to develop an adaptive structure on a single chip [7]. As a communication bus the SPI [8] was chosen, as it has higher throughput than its competitors and extremely simple hardware interfacing.

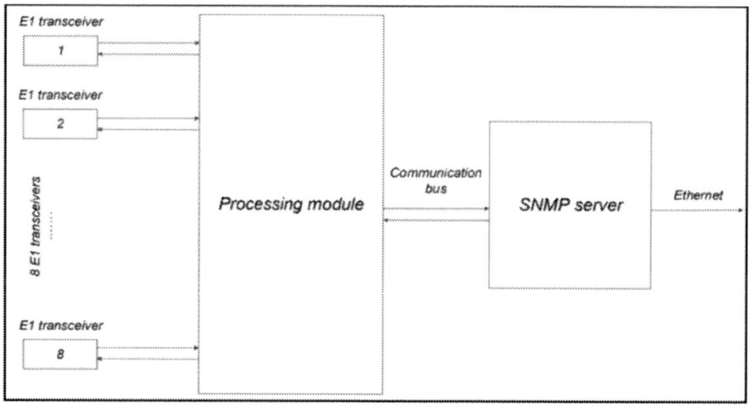

Fig. 1. Block diagram of the approach.

Based on SNMP server description, it has to have following capabilities:

-getting data from the communication bus;

-Ethernet transceiver, to send data to IP network;

-sufficient amount of RAM to store time slots errors and its content;

-fast processing unit for data processing.

Figure 2 shows a block of a single E1 port. The inputs of the unit receive the signal of the first channel E1, and clock frequency. Through the inversion, a signal indicating the signal loss is connected directly to the LED terminals. Thus, the green LED on the standard E1 connector will not light unless the E1_0_LOS signal is equal to one, which is equivalent to signal loss.

At the output of the E1 port block, there is the same clock frequency and E1 channel data, a bit signaling the remote synchronization error, a bit signaling the remote multiframe error, the sync bit, and the multi-frame clock bit. The other outputs are two outputs made in the form of a bus. One of these outputs is a 32-bit bus, and contains error statistics for 32-channel intervals. The second output is a 256-bit bus and transmits the contents of the 32-channel intervals of the first E1 channel.

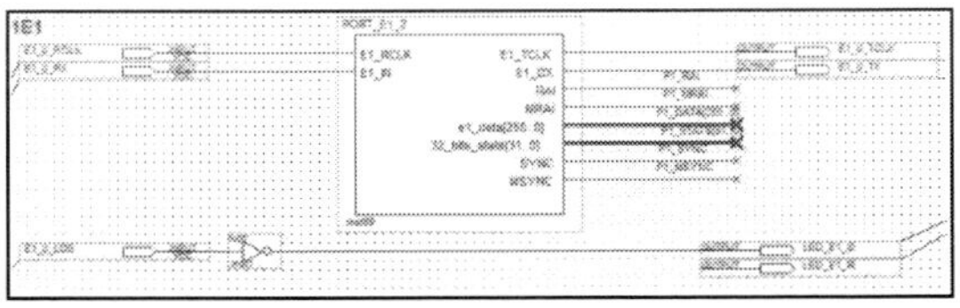

Fig. 2. Block of the first port E1.

Figure 3 shows the block of information extraction necessary for monitoring from the incoming E1 data stream. In this case, the outputs E1_TCLK and E1_DX are connected directly to the inputs E1_RCLK and E1_IN, and in other words simply duplicate the data and the clock frequency of the E1 stream.

Figure 4 shows the E1 flow processing unit that extracts the address from the E1 data stream as a 5-bit bus, and the data in the form of an 8-bit bus. The output of TS_WR_EN is responsible for the ability to record the data of the channel interval. Also, this block allocates a signaling bit from the stream data the multi-frame synchronization. Sync_found_si2000 sync block outputs the sync signal to the output and the multiframe clock.

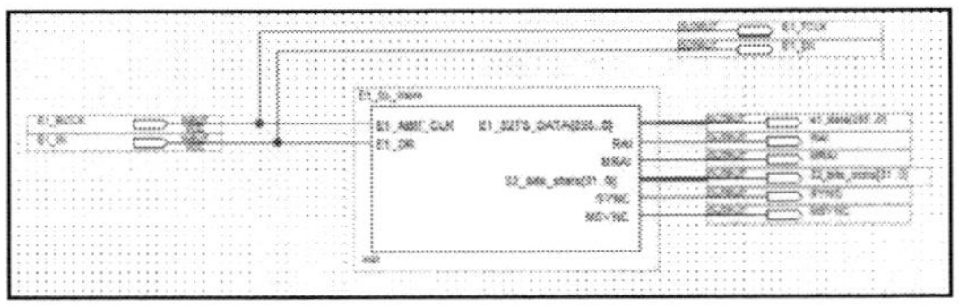

Fig. 3. The block of allocation of the information necessary for monitoring.

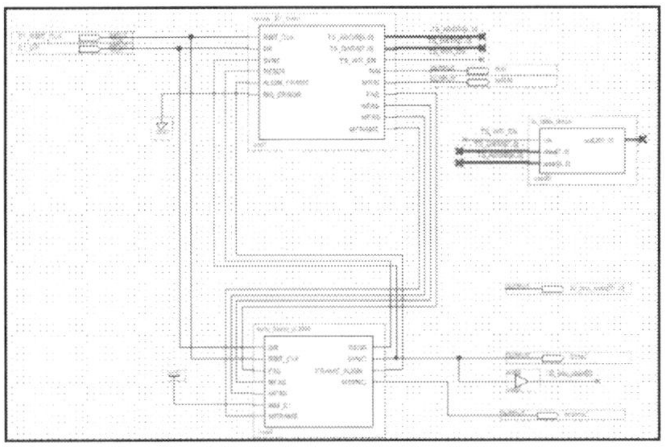

Fig.4. Flow processing block E1.

Figure 5 shows the time slot analysis block.

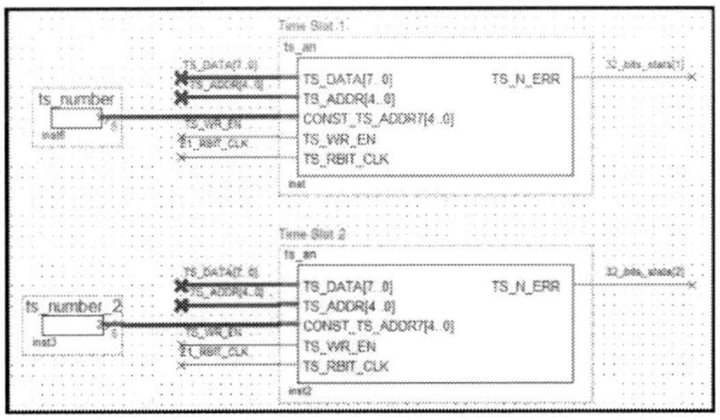

Fig. 5. Analysis blocks of the channel interval.

The input is provided with the data of the time slot, its address, the clock frequency, the recording resolution, and the constant in the form of a certain number, for further determination. The data of which particular time slot is transmitted at the time of the analysis. The output is either 0 or 1 indicating an error. The result is written to a 32-bit bus, into a bit number consistent with the number of the analyzed time slots.

3 Timeslot data analyzer

Based on the article "Design and implementation of E1 stream analyzer for CCSS7 protocol"[9] was developed by Timeslot data analyzer. And

Data and the clock frequency of the E1 stream are fed to the receiver's input. The receiver E1 allocates from the data stream, the clock frequency, and the data of the channel interval. Further, the data of the time slot is checked for the presence of "emergency sequences" that arise when the arrest is repeated for a long time in the E1 information stream, for example, a long repetition of the FFh or 00h bits. Thus, the E1 stream data, the error statistics of their channel intervals, the remote synchronization error and the multiframe, the sync signal and the multi-frame clock are transmitted to the SPI data bus. Figure 6 shows Block diagram of the time slot analyzer

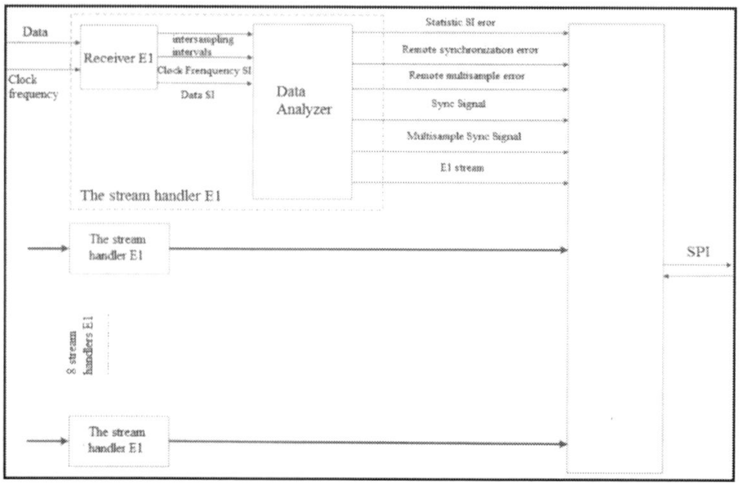

Fig.6. Block diagram of the time slot analyzer.

To test the functionality of the time slot [10] analysis unit, the built-in Quartus 2 modeling system was used[11]. The counter threshold is set as 5 data repetitions in one channel interval. Figure 7 shows that at the first repetition, the counter (count_repeats) increased by one, and became equal to 1.

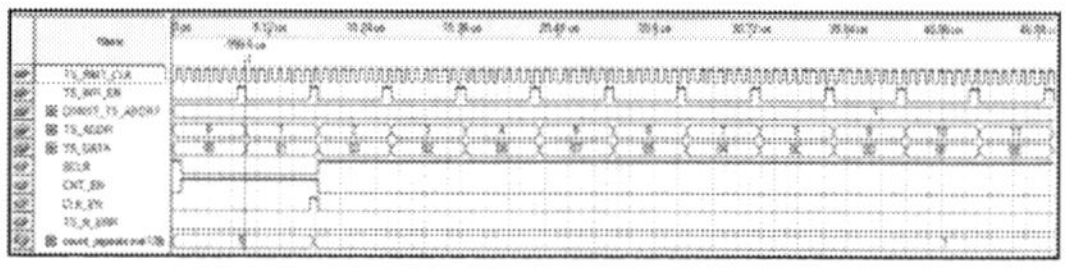

Fig. 7. Time diagram of the operation of the time slot analyser.

In Figure 8 seen that at the next meeting of the repeating information in the first channel interval, the counter increased, and reached a threshold value of 5, which made the error signal (TS_N_ERROR) equal to one.

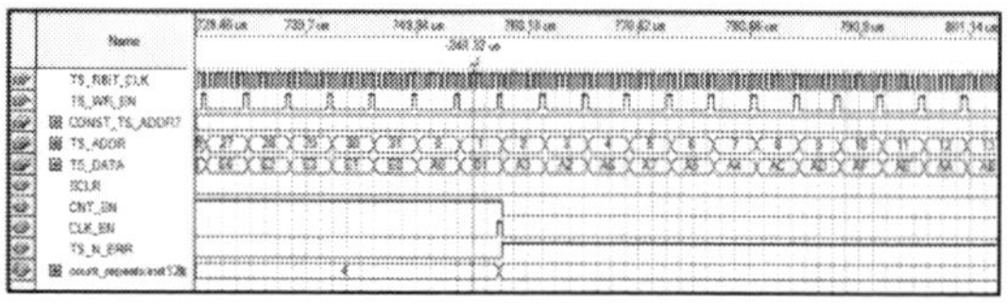

Fig.8. Time diagram of the operation of the time slot analyser.

The modeling process provide the correct operation of the time slot analyser.

4 Conclusion

As a result of this paper an overview of methods for detecting errors during the transmission of data through the E1 channel. The analysis of existing approaches used in devices that perform the monitoring functions of E1 stream and their shortcomings are revealed. A method was developed to monitor the state of the E1 stream based on the content in the time slots of the data stream.

References

1. E. Wright, D. Reynders, *Practical Telecommunications and Wireless Communications: For Business and industry* (IDC technologies, Oxford 2004)
2. ITU-T G.711. Pulse code modulation (PCM) of voice frequencies (STD.ITU-T RECMN G.711-ENGL 1988)
3. ITU-T Recommendation G.703. Physical/Electrical Characteristics of Hierarchical Digital Interfaces.(1972, the last correction in 1991).
4. ITU-T Recommendation G.704. Synchronous Frame Structures Used at Primary and Secondary Hierarhical Levels (1984., the last correction in 1990).
5. ITU-T Recommendation G.755. Digital multiplex equipment operating at 139 264 kbit/s and multiplexing three tributaries at 44 736 kbit/s.(1988, the last correction in 1993).
6. Simple Network Management Protocol (SNMP) Applications. Copyright (C) The Internet Society (2002)
7. Hussein Baher, *Signal Processing and Integrated Circuits.* (John Wiley & Sons, New York 2012)
8. Texas Instruments: SPI Interface User Guide. SPRUGP2A (2012)
9. A.Y.K. Malik, M.M. Ahmed, A. Azhar, *In Proc. INMIC* (2012)
10. O.Babatunde, S. Mbarouk, International Journal of Scientific Research Engineering & Technology (IJSRET), (2014)
11. Quartus: Prime Standard Edition Handbook Volume 1: Design and Synthesis. Altera Corparation.

DOI: 10.1051/matecconf/201713205013

A mathematical model for estimating and forecasting the state of a digital substation based on the wavelet transform method

Roman Oganyan [1*], *Danil* Shaykhutdinov[1], *Nuri* Narakidze[1], and *Dmitriy* Shurygin[1]

[1]SRSPU(NPI), Department of Information and measuring systems and technologies, 346428 Novocherkassk, Russia

Abstract. The article the possibility of applying the wavelet transform method in combination with a neural-fuzzy approach to solving problems of forecasting the state of digital substations is considered. The optimum level of the wavelet expansion of the time series corresponding to the change in the phase voltage for the day on the basis of the Hurst index is determined. The influence of the sample size and the type of the mother wavelet on the Hurst index is researched. It was revealed, that for wavelet decomposition, the use of the Daubechy wavelet as the mother wavelet is effective, which provides a smoother filtering of noise, compared to the Haar wavelet. Analysis of the original series does not allow to evaluate the optimal level of wavelet expansion if the noise level of the time series under consideration is low (less than 10%), since the Hurst index remains unchanged. However, using the logarithm of changing the time series allows for small fluctuations to be taken into account, which allows to determine the optimal level of the wavelet expansion for their smoothing.

1 Introduction

The technology "Digital substation" is a new level of automation of electrical substations. Based on the fact that this technology is relatively new, there is a need to develop new algorithms for diagnosing complex technical systems. The demand is confirmed by numerous researches and developments in this field [1-9].

As the basis of new algorithm for complex technical systems diagnosing (as example – for digital substation), that earlier proposed in [2], the method of full-scale-model test is used [10]. To diagnostics a computer model based on National Instruments technologies is developed [11]. The model allows to select the monitored parameter (current strength, phase/line voltage, frequency, etc.) and the monitored value (the numerical value of the monitored parameter, exceeding which is dangerous) during program operation. The model also allows to estimate the change of the monitored parameters within the time. Thus, it becomes possible to more accurate diagnose digital substation current state on the basis of an analysis of the parameters values.

However, the diagnosis of the substation current state does not always allow assessing the evolving defects in the equipment. Therefore, it becomes difficult to prevent the emergency operation of the equipment. To solve this problem, it is proposed to use methods of intellectual analysis to predict the substation state. In view of the fact that the data on the digital substation is permanently stored, there is access to a large amount of information provided from the sensors of power equipment. Thus, it becomes possible to use intellectual analysis (modern methods of signal processing, wavelet analysis, methods of artificial intelligence, machine learning) to diagnose equipment. The use of intellectual analysis allows to assess the current state of equipment, as well as to predict the pre-toxic states on the basis of an analysis of the values of the parameters characterizing the state of the equipment, as well as the relationships of the system parameters [3].

2 The wavelet transform method

To create a predictive model, the time series of parameters describing the state of the substation according to the *C2C power quality monitoring* project are taken. Data about a digital substation can be considered as random time series. The analysis of such time series reflects the behavior of the system in the past and makes it possible to predict its future behavior. The primary task in forecasting complex non-stationary time series is to decompose them as components. The structure of the time series is a set of components, such as the systematic component, the oscillations around the systematic component and the random component.

One of the most effective methods of intelligent data processing are neural networks. However, this method has drawbacks [12, 13]. The disadvantages of this method are partially eliminated by combining neural networks and fuzzy logic – a neuro-fuzzy approach [14]. According to numerous studies [15-22], due to the use of a neuro-fuzzy network, the forecast error is minimized. At the same time, a selection of the architecture of the neural-fuzzy network, taking into account the dynamics of the change in the rapidly changing and slowly

* Corresponding author: roman.work18@gmail.com

changing components of the time series, is a laborious process. In addition, accurate accounting for changes in the rapidly changing time series component leads to significant prediction errors. The solution of the above-mentioned problems, as well as an additional increase in the accuracy of the forecast, can be ensured by applying a wavelet transform. Forecasting is carried out according to the scheme shown in Fig. 1.

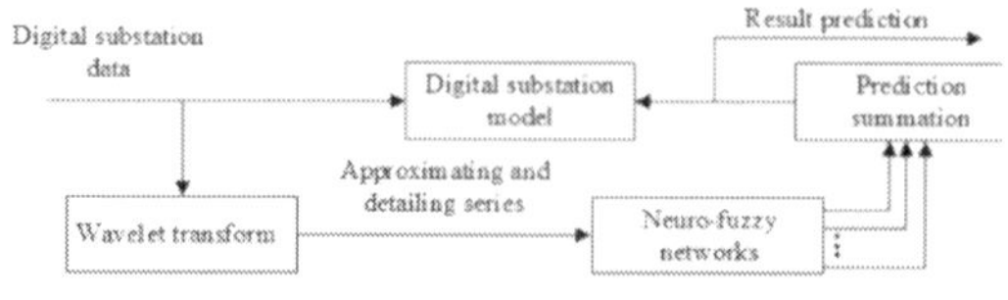

Fig. 1. Prediction diagram.

To solve the problem of predicting discrete time series, an algorithm of excessive discrete wavelet transform is chosen [23,24]. The choice is due to the fact, that other species have disadvantages that do not allow to solve the current task:

• continuous wavelet transformation - a large redundancy, leading to a large time-consuming analysis,

• fast wavelet transformation - presence of decimation operation, i.e. when the time series is shifted by at least one sample, the results of the whole decomposition will change, which will lead to the impossibility of forecasting).

3 The optimal level of detail based on the Hurst index

The prediction process (Fig. 1) includes the decomposition of a time series with a deterministic depth of detail, which is determined empirically from the analysis of the original time series, or on the basis of an analysis of the Hurst exponent [25]. This indicator characterizes the noisiness of the analyzed time series. The best decomposition depth is understood as the obtaining of uncomplicated quasiperiodic sequences at the output of the wavelet filtration with no sharp fluctuations and emissions at the approximating decomposition level (the Hurst index value is different but close to 0.5) [26].

For example, consider the expansion of the time series, which is a change in the phase voltage per day (1440 values, fixing the voltage every minute). The appearance of the change in voltage is shown in Fig. 2.

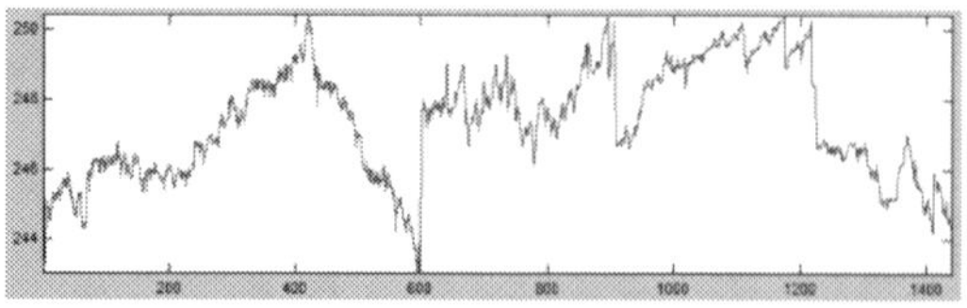

Fig. 2. The one day voltage level variations

To determine the optimal level of detail, an algorithm for calculating the Hurst index based on R / S analysis was implemented [27,28]. There are two ways of analyzing the time series:

• analysis of the original time series,

• analysis of variation in the original time series.

By variation of the original series N we mean either the difference between two neighboring terms of the series [29]:

$$n = N_{i+1} - N_i\ ,\ i = 1,2,3,\ldots,\ N\text{-}1 \qquad (1)$$

or the difference between the logarithms of two neighboring terms of the series [28,29]:

$$n = ln\ (N_{i+1} / N_i),\ i = 1,2,3,\ldots,\ N\text{-}1 \qquad (2)$$

According to the E. Neumann research "Calculation of the Hurst index for the purpose of revealing the trend (persistence) of financial markets and macroeconomic indicators," it is recommended to analyze the source series, rather than the original series variation. To calculate the Hurst exponent, used the formula [30] is used:

$$H = log\ (R / S) / log\ (\pi * N / 2), \qquad (3)$$

where H – the Hearst index; S – the standard deviation of a observations number; R – the magnitude of the accumulated deviation; N – the observation periods number.

The results of calculating the Hurst index for the initial series are presented in Table 1.

Table 1. The values of the Hurst index for the original time series.

Partition number	n value (the number of days for the period)	The number (t) of segments	The H value
1	72	20	0,5786
2	120	12	0,6633
3	180	8	0,7553
4	288	5	0,7514
5	480	3	0,7981
6	720	2	0,7876
	1440	1	0,7944

The value of the Hurst index, according to the recommendations [27], should be taken with $t \geq 5$. According [31], at small sample sizes, the Hearst index value is undergoes significant changes, and only starting from a certain sample size, the value of the Hearst index oscillates about a specific value. With respect to the time series under consideration, we take H = 0.7514, which indicates the presence of long-term memory in the analyzed series.

To test the possibility of further use of the obtained Hurst index, we select a test developed by Sheinkman and Le Baron [27]. The test consists in mixing the initial data with the subsequent recalculation of the Hurst index. If the Hurst index takes a value close to 0.5, we can conclude that the original series had a long-term memory, and as a result of mixing the original data, their structure was destroyed. In the case under consideration, after randomization of the initial data, the Hurst index became 0.4741. This result indicates the presence of long-term memory in the original series and, accordingly, indirectly confirms the correctness of the definition of the Hurst index.

To obtain the best decomposition depth, the Hurst index is calculated from the first level of detail, followed by an increase based on the Haar wavelet [32, 33]. The results are shown in Table 2.

Table 2. The Hurst exponent value for the Haar wavelet.

Type of Motherboard Wavelet	Level of detalisation	The H value
Haar wavelet	1	0,7524
	2	0,7533
	3	0,7548
	5	0,7656

It can be seen, that the value of the Hurst index undergoes minor variation with an increase in the level of detail. The next step to get the Hurst index as a function of the level of detail was tested, replacing the mother wavelet with the Daubechies wavelet with a number of zero moments equal to 4 [32,33], which provides a more smooth filtering of noise. The results of the wavelet decomposition are presented in Table 3.

Table 3. The Hurst exponent value for the Daubechy wavelet.

Type of Motherboard Wavelet	Level of detalisation	The H value
Daubechy wavelet	1	0,7521
	2	0,7533
	3	0,7553
	5	0,7623

It can be seen, that the value of the Hurst index also does not undergo significant changes. To study the dependence of the Hurst index on the level of detail, the sample size was increased to 4,096 values (Fig. 3).

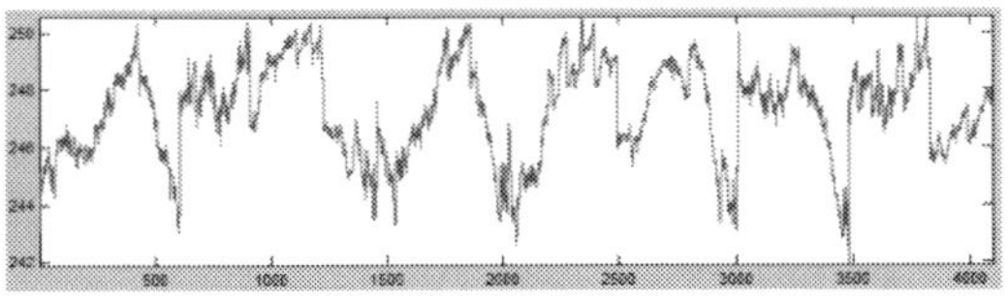

Fig. 3. The voltage level variations for 4096 datapoints

For the original time series, the value of the Hurst index is 0.7855. For the given sample size, the same procedures are repeated as for the previous time series. The results are shown in Table 4.

Table 4. The Hurst index value for the Haar and Daubechies wavelets with a sample size of 4096 values.

Type of Motherboard Wavelet	Level of detalisation	The H value
Haar wavelet	1	0,7859
	2	0,7863
	3	0,7873
	5	0,7914
Daubechy wavelet	1	0,7857
	2	0,7862
	3	0,7870
	5	0,7902

Thus, from the data, it is clear that to form a relationship between the level of detail and the Hurst index, it is necessary to analyze the logarithms of the change in the time series under consideration. This is due to the fact that due to the use of the properties of the logarithm, it will be possible to take into account the detailed changes in the time series under consideration. The results of calculating the Hurst index are presented in Table 5.

Table 5. The Hurst index value in the analysis of changes in the original time series.

Partition number	The n values (the number of days of the period)	The number (t) of segments	The H value
1	72	20	0,3747
2	120	12	0,4104
3	180	8	0,4576
4	288	5	0,4578
5	480	3	0,4448
6	720	2	0,4846
	1440	1	0,4381

According to the E. Neumann research, the value of the obtained Hurst index falls under the classification of a random series, or noisy. We carry out the wavelet decomposition using the mother wavelet Haar and Dobesi. The results are shown in Table 6.

Table 6. The Hurst exponent value for the Haar wavelet and Daubechies for different levels of detail.

Type of Motherboard Wavelet	Level of detalisation	The H value
Haar wavelet	1	0,4886
	2	0,4939
	3	0,4970
	5	0,4788
Daubechy wavelet	1	0,5161
	2	0,5639
	3	0,6068
	5	0,6955

According to the obtained data, it can be concluded that for the time series under consideration, the use of the Daubechy wavelet is more effective. This is due to the fact that using the Haar wavelet does not allow to smooth out the current noise level, but because of the specifics of this wavelet, it only eliminates the information content of the original signal (Fig. 4).

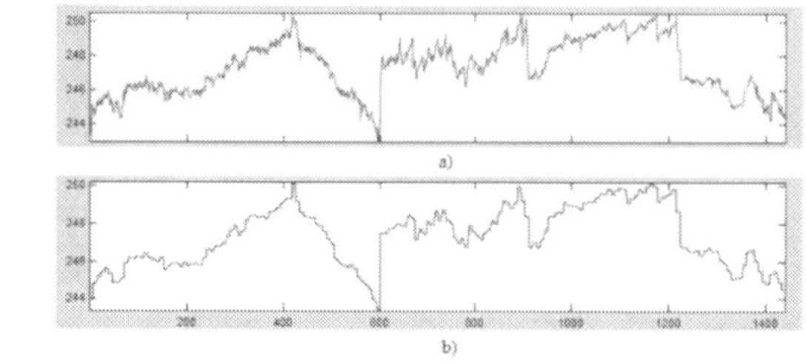

Fig. 4. Wavelet decomposition with a level of detail equal to three for the Haar mother wavelet: a) the original signal; b) an approximating signal

Based on the Daubechy wavelet, the results are similar to the "rough" estimate of the Hurst index without taking into account the logarithms of the change in the original series. It should be noted that with the increase in the level of detail, the Hurst index increases, respectively, indirectly confirms that the Hurst index is a measure of the complexity of the analyzed series.

Thus, the best depth of decomposition obtained as a result of approbation of different levels of detail for 4,096 values is 5. As the level of decomposition increases, a rough change in the configuration of the signal occurs, at a lower level of decomposition the noise component will not be sufficiently eliminated. As a result of the decomposition, we obtain 6 components-the approximating (a_5) and five detailed (d_5, d_4, d_3, d_2, d_1), which together give the original signal (Fig. 5):

$$S \cong a_5 + d_5 + d_4 + d_3 + d_2 + d_1, \qquad (4)$$

where S – source signal, a_5 –signal equal to approximation of 5; d_5, d_4, d_3, d_2, d_1 – the signal of detail levels 5, 4, 3, 2, and 1, respectively.

As can be seen from Fig. 5, the original signal is noisy. The use of a neural-fuzzy approach without wavelet transformation entails the appearance of an error, but taking an approximating time series, without taking into account the rapidly changing components, there will also be a significant prediction error [34].

Training is carried out for each level of decomposition separately, accordingly, in the case under consideration, 6 neural-fuzzy networks are formed. After the expansion of the time series, a prediction is carried out, which can be represented in the form of:

$$Y^*(t, l) = \varphi(t, l, Y(t), Y(t-1), \ldots, Y(2), Y(1)), \qquad (5)$$

where interval $[t,t+1]$ – the lead time interval, l – length of lead time interval, $Y^*(t, l)$ – pretiction, φ – prediction function, depends on the number of observations t, lehgth l and values of $Y(1), Y(2), \ldots Y(t)$. The received predictions $Y^*(t, l) \ldots Yn^*(t, l)$, are summarized, to abtain the result prediction value [35].

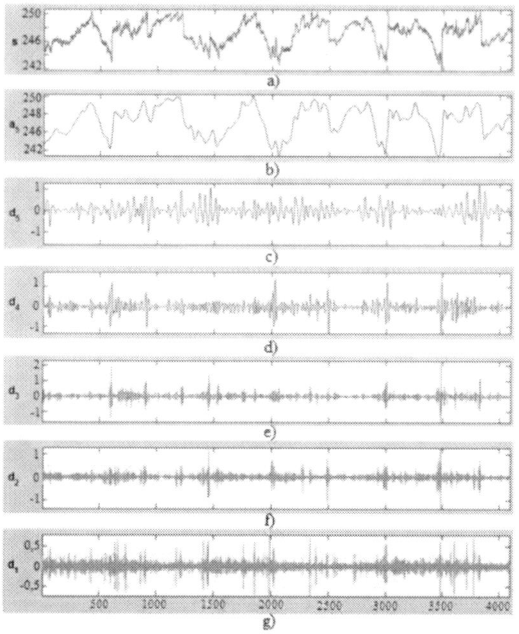

Fig. 5. The decomposition of a signal with a level of detail equal to three: a) the original signal; b) an approximating signal; c), d), e), f), g) the details of 5, 4, 3, 2 and 1 levels.

As a result of the research, it was revealed:

- For wavelet decomposition, the use of the Daubechy wavelet as the mother wavelet is effective, which provides a more smooth filtering of noise, in comparison with the Haar wavelet.
- The increase in the sample size does not have a significant effect on the Hearst index.
- Analysis of the original series does not allow us to estimate the optimal level of the wavelet decomposition if the noise level of the time series under consideration is low (less than 10%), since the Hurst index remains unchanged. However, using the logarithm of changing the time series allows for small fluctuations to be taken into account, which makes it possible to determine the optimal level of the wavelet expansion for smoothing them.

Thus, in order to solve the problems of predicting the state of a digital substation, the application of wavelet transform makes it possible to obtain a signal without noise, which is effective for further application in neural-fuzzy networks. This is due to:

• Eliminating the need to compile a neural-fuzzy network architecture,

• Lack of accurate accounting of the detailed components of the original time series, which leads to an increase in the accuracy of the forecast,

• Forecasting of individual rows after the decomposition of the original signal, which also leads to an increase in forecast accuracy.

This article is based on results of the project No. SP 4108.2015.1, implemented under the program "Scholarship provided by the President of the Russian Federation for young scientists and graduate students engaged in advanced research and development in priority areas of modernization of the Russian economy" using equipment of shared facility "Diagnosis and energy-efficient electrical equipment" (NPI).

References

1. V.I. Dubrov, R.G. Oganyan, D.V. Shaikhutdinov, E.V. Kirievsky, N.I. Gorbatenko, N.D. Narakidze, Modern high technology. **9**, 36-40 (2016)
2. V.I. Dubrov, R.G. Oganyan, D.V. Shaikhutdinov, E.V. Kirievsky, TN. Kruglova, N.D. Narakidze, Fundamental research. 9, 16-20 (2016)
3. L.A. Dar'yan, A.G. Mordkovich, G.M. Tsfasman, Electro. Electrical engineering, electric power industry, electrotechnical industry. **5**, 22-26 (2010)
4. Z. Ling, Y. Chaoran, H. Yanxiang, Mechatronic Science, Electric Engineering and Computer (MEC), 1374–1377 (2011)
5. J. Xin, Z. Liao, F. Wen, *Power System Technology (POWERCON), Inter. Conf.* 1–5 (2010)
6. P. Peng, Z. Liao, F. Wen, J. Huang, SCI 465, 153–165 (2013)
7. D. La Cascia, V. Di Dio, A. Lipari, Power Electronics and Motion Control Conf. and Expos (PEMC), 1300–1305 (2014)
8. R. G. Oganyan, N. D. Narakidze, D. V. Shaykhutdinov, N. I. Gorbatenko, S. O. Maksuta, IOP Conf. Ser.: Mater. Sci. Eng. **177**, 012083 (2017)
9. V.I. Dubrov, R.G. Oganyan, N.D. Narakidze, G.K. Aleksanyan, Journal of Engineering and Applied Sciences, **12**, 276-282 (2017)
10. N.I. Gorbatenko, *Full-scale model tests of products made of ferromagnetic materials* (Publishing house SKNC VSh, Rostov-on-Don:, 2001)
11. V.I. Dubrov, R.G. Oganyan, N.D. Narakidze, G.K. Alexanian, *Collection of Proceedings of the XV International Conference NIDays-2016*, 300-302 (2016).
12. S.Yu. Galkina, I.P. Kurochkina, Modern trends in the development of science and technology. **12**, 28-30 (2016)
13. L.A. Amaeva, Bulletin of the Kazan Technological University, **1**, 320-322 (2015)
14. S. Osovsky, *Neural networks for information processing* (Finance and Statistics, Moscow 2004)
15. E.S. Anushina, Intellectual systems, **1 (19)**, 84-90 (2009)
16. V.I. Alekseev, The Bulletin of the Ugra State University, **3 (30)**, 3–10 (2013)
17. V.Z. Manusov, E.V. Biryukov. Proceedings of Tomsk Polytechnic University. Engineering of geo-resources, **6**, 153-157 (2006)
18. A.A. Dudkin. System Analysis and Applied Informatics. **4**, 11-15 (2015)
19. O.P. Soldatova, I.A. Lyozin, I.V. Lyozina, A.V. Kupriyanov, D.V. Kirsch. Computer Optics, **5**, 787-794 (2015)
20. A.D. Danilov, Sh.M. Omar, A.L. Rutskov. Actual directions of scientific research of the XXI century: theory and practice, **6**, 70-77 (2016)
21. V.L. Burkovsky, V.N. Krysanov, A.L. Rutskov, O.Sh.M. Shukur. Bulletin of Voronezh State Technical University. **5**, 41-46 (2015)
22. O.Yu. Kuznetsova, A.A. Solomakh, V.I. Gorbachenko. Modeling and analysis of data. **1**, 86-90 (2014)
23. A.V. Pereberin. Computational methods and programming: new computational technologies. **3**, 15-40 (2001)
24. I. Dobesi, *Ten lectures on wavelets* (Izhevsk: SRC "Regular and chaotic dynamics", 2001)
25. H.E. Hurst, R.P. Black, Y.M. Simaika, *Long-term storage: An experimental study* (Constable, Leningrad 1965)
26. E.S. Filatova, D.M. Filatov. International Conference on Soft Computing and Measurement. **1**, 114 -117 (2014)
27. E. Peters. *Fractal analysis of financial markets: application of the theory of chaos in investment and economy* (Internet-trading, Moscow 2004)
28. E. Peters. *Chaos and order in the capital markets. A new analytical view of cycles, prices and market volatility* (Mir, Moscow 2000)
29. M.S. Chepeleva, S.A. Tkalich, S.A. Chepelev, Polytechnical network electronic scientific journal of the Kuban State Agrarian University. **74**, 326-337 (2011)
30. F.F. Pashchenko, E.Yu. Medvedev, Bulletin of the International Academy of System Studies. Informatics, ecology, economics. **1**, 74-80 (2013)
31. Yu. A. Kalush, V.M. Loginov. Siberian Journal of Industrial Mathematics. **5**, 4(12), 29-37 (2002)
32. O.V. Nagornov, V.G. Nikitaev, V.M. Prostokishin, S.A. Tyuflin, A.N. Pronichev, T.I. Bukharova, K.S. Chistov, R.Z. Kashafutdinov, V.A. Khorkin. *Wavelet analysis in examples: Textbook.* (NNIU MEPhI, Moscow: 2010)
33. A.N. Yakovlev, *Introduction to wavelet transforms: Proc. Allowance* (NSTU Publishing House, Novosibirsk 2003)
34. E.S. Filatova, D.M. Filatov, A.D. Stotskaya. International Journal of Applied and Fundamental Research. **10-1**, 46-50 (2015)
35. Filatova, D.M. Filatov, A.D. Stotskaya. International Journal of Applied and Fundamental Research. **11**, 26-29 (2014)

MATEC Web of Conferences 132, 05014 (2017) DOI: 10.1051/matecconf/201713205014
DTS-2017

Defect detection on videos using neural network

Roman Sizyakin[1,*], *Nikolay* Gapon[1], *Igor* Shraifel[1], *Svetlana* Tokareva[1], *Dmitriy* Bezuglov[1]

[1]DSTU, 344000, Rostov region, Rostov-on-Don, Gagarina 1, Russia

Abstract. In this paper, we consider a method for defects detection in a video sequence, which consists of three main steps; frame compensation, preprocessing by a detector, which is base on the ranking of pixel values, and the classification of all pixels having anomalous values using convolutional neural networks. The effectiveness of the proposed method shown in comparison with the known techniques on several frames of the video sequence with damaged in natural conditions. The analysis of the obtained results indicates the high efficiency of the proposed method. The additional use of machine learning as post-processing significantly reduce the likelihood of false alarm.

1 Introduction

With the rapid development of digital systems for reception, transmission and storage are a secondary analogue system of the video. Unfortunately, the analogue video sequence is such that over time the material recorded on them, are partially distorted. This is due to substandard equipment, improper storage of documents, as well as their physical damage. All this leads to the film abnormal values in the form of a single or group of pixels which value of brightness will significantly stand out from other pixels. The observation time of the abnormal data values, usually not more than a few frames in a video sequence, most often on one. Interference significantly degrades the visual quality of the video. Therefore, it is necessary to solve the task of automated detection of distorted values. In most cases, the problem of the digital restoration carried out by the artists for each frame manually, which significantly limits the speed of processing of archival recordings.

Dust removal based on applying some spatial or temporal processing algorithm frame. So, the spatial methods assume that the defect has a small area on the borders of the pronounced gradient values, that is, different from the rest of the image. The methods use grey-scale morphology (dilation, erosion, opening and closing) [1,2].

Some detectors are based on the analysis of some number of pixels, taken from the current, previous and subsequent frames. By this method, it is possible to carry SDIa, ROD and SROD detector. The peak detection index (SDI) is the most basic heuristic for detecting temporal discontinuities in the video sequence. This detector compares the brightness value of each pixel in the current frame $Y_{i,j}^{(k)}$ with the brightness values of the previous and subsequent frames of the video sequence. Then compute the minimum of the quadratic difference [3]. A method of ranking pixels (ROD) heuristic detector based on ordinal statistics [4]. A defect is detected as one of the ranked sequences exceeds a predetermined threshold T_l. The threshold T_l set by the user and determines the sensitivity of the detector. The difference is that the ROD from the SROD detector is that of the potential detector is not the median value and the difference between the local maximum and the local minimum [5]. The SROD detector compares the pixel value obtained as the result of motion compensation, the current pixel value. Damage is detected if the value of the current pixel is beyond the threshold.

Most of the defect detection methods using fixed threshold values to produce a mask with the presumed localisation of abnormal values. Incorrect assignment of a threshold may lead to an increase in the probability of false alarm at his fall, and to the insufficient detection while increasing its value.

2 Proposed method

The proposed method consists of three steps: motion compensated frames, the preprocessing of the detector, which is based on the ranking of pixel values, and classify all pixels with abnormal values using convolutional neural networks.

Video sequence assumes the following simplified mathematical model:

$$Y_{i,j}^{(k)} = (1-d_{i,j}^{(k)}) \cdot S_{i,j}^{(k)} + d_{i,j}^{(k)} \cdot c_{i,j}^{(k)} \qquad (1)$$

where $Y_{i,j}$ - the observed frame with strings $i=\overline{1,I}$ and columns $j=\overline{1,J}$, $k=\overline{1,K}$ - frame number, K - the number

* Corresponding author: roma_sz@mail.ru

of frames in a video sequence, $S_{i,j}$ - original (undamaged) image, $d_{i,j} \in \{0,1\}$ binary mask defects, which shows what site the image was corrupted, $c_{i,j}$ the mask, which contains values of brightness defects.

The scheme of the proposed algorithm is present in figure 1.

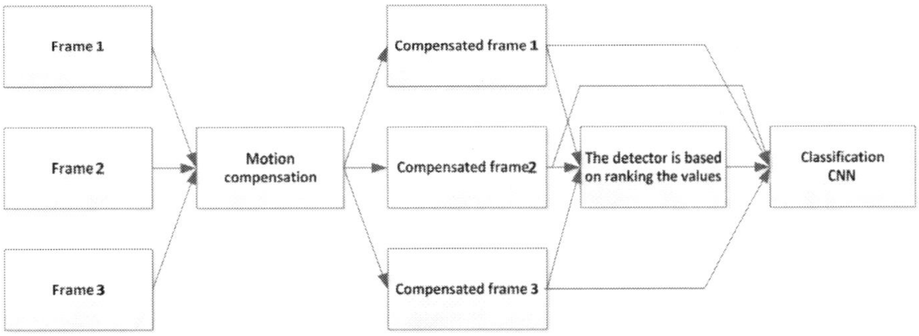

Fig. 1. The block diagram of defect detection on the video sequence.

Evaluation of motion compensation allows determining the direction of movement of the small blocks on the frame of the video sequence by finding them on adjacent frames. In this work as a metric to determine the similarity between the current and the desired block is used calculation evaluation of the sum of absolute differences (SAD):

$$SAD = \sum\sum\left|B_{current} - B_{required}\right| \rightarrow \min \qquad (2)$$

where $B_{current}$ - unit for which you want to estimate the motion vector, $B_{required}$ - a unit with a prospective new position.

This operation is schematically shown in figure 2.

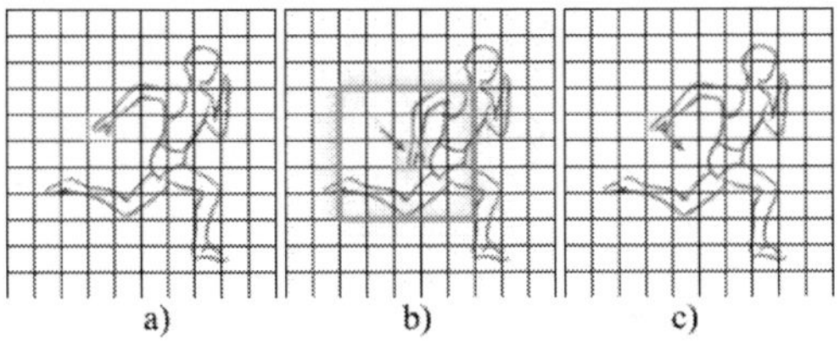

Fig. 2. Motion compensation a) current frame, b) - search for the similar block on the adjacent frame) - compensated frame.

Preprocessing is an important step when damage is detected. It allows to reduce the number of false alarms the detectors and to reduce the computational complexity of the algorithm since performance is an important indicator of the problem of damage detection for the dynamic signal.

As used pre-processing method proposed in [5]. The operation of this detector assumes that the brightness values in damaged areas of the frame that differs significantly from the brightness values in the same areas in adjacent frames. To determine the abnormal values, we use the ranking of pixel values in ascending order. This operation is schematically shown in Figure 3.

In the first stage of the method for each pixel of the current frame $Y^{(k)}$, six are selected spatially close pixels p_r, on adjacent frames $Y^{(k-1)}$ and $Y^{(k+1)}$. Then the selected pixels $p_r, r \in [1,6]$ ranked, where they are assigned sequence numbers d, moreover $d_1 \leq d_2 \leq ... \leq d_6$. Then the decision rule can be written as the following expression:

$$SROD = \begin{cases} d_1 - Y_{i,j}^{(k)}(R_{i,j}) & if\ d_1 - Y_{i,j}^{(k)}(R_{i,j}) > 0 \\ Y_{i,j}^{(k)}(R_{i,j}) - d_6 & if\ Y_{i,j}^{(k)}(R_{i,j}) - d_6 > 0 \\ 0 & else \end{cases} \qquad (3)$$

The condition under which the pixel value is assumed to be abnormal is written as:

$$L_{SROD}(R_{i,j}) = \begin{cases} 1 & if\quad SROD > T \\ 0 & else \end{cases} \qquad where\ T \geq 0 \quad (4)$$

where T - prior specified threshold value.

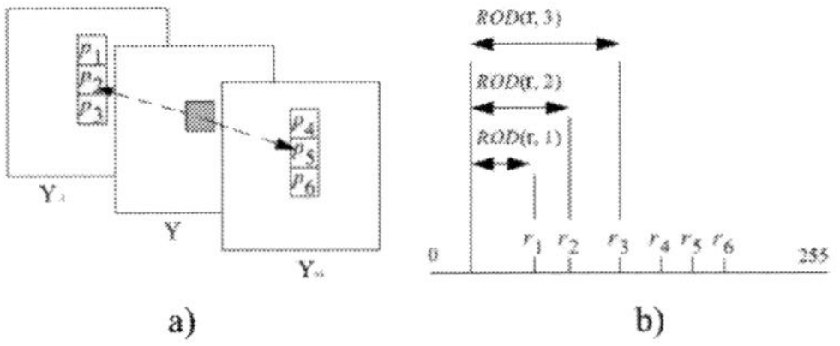

Fig. 3. A method of ranking the pixel values a) selection of pixel values spatially close to the alleged damage, b) – ranking of the selected pixel values.

The incorrect use of compensated frames what happens when the foreground object is moving or incorrectly chosen threshold value can lead to excessive false alarms. To reduce the number of false positives caused by these circumstances, and to reduce the dependence on the choice of the threshold value method is used for post-processing, based on machine learning.

The task of the classifier is to determine the class to which belongs each pixel of the marked unit on the mask obtained in the previous step.

The choice of convolutional neural networks as a post-processing based on two conditions: the first is attributed to the high-performance classification of two-dimensional signals, which is confirmed by modern studies [6], the second condition is that high speed, due to the lack of necessity for the computation of the special handles, which can take a considerable period.

To build a convolutional neural network this work uses a Google library the Flow Tensor [7]. This network has the convenient form of specifying variables and create the overall architecture has an interactive Visualizer logs, high speed of learning and classification using only CPU resources. Also, this convolutional neural network allows the use of an improved version of the method of error back-propagation "Adam optimiser" which allows you to achieve faster global extreme of a function.

At the initial stage of training, the neural network kernel in the convolutional layers is set to random values.

The architecture of the convolutional neural network is presented in figure 4.

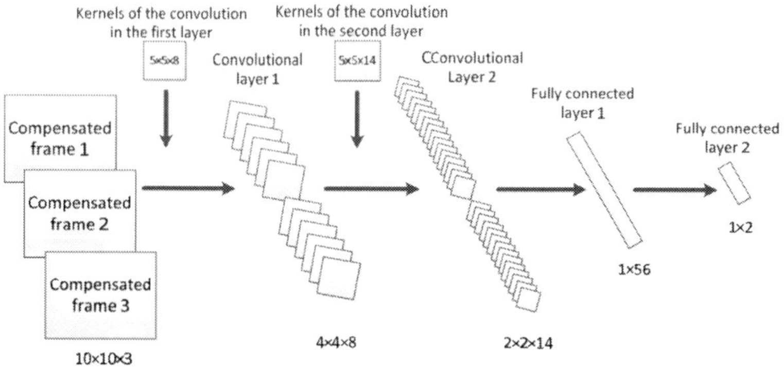

Fig. 4. The architecture of the convolutional neural network

Then for each input image are computed maps features by convolution of the image and a certain kernel. In a simplified form the function of a convolution can be represented as:

$$x_j^l = f\left(\sum_i x_i^{l-1} * k_j^l + b_j^l\right) \quad (5)$$

where x_j^l - the map features j (for a layer l), k_j^l - the convolution kernel size 10×10 pixels, $f()$ - the activation function of the hidden layer, b_j^l - the shear factor, x_i^{l-1} - the map of features of the previous layer.

In the output layer as activation function the linear rectification [8]:

$$f(x) = \max(0, x) \quad (6)$$

where x - the input card features.

As the activation function in layer two fully connected used "softmax" function, which allows determining the probability that a classified pixel between two classes. The activation function is written as:

$$y_j = \frac{e^{z_j}}{\sum_{i=1}^n e^{z_i}} \quad (7)$$

where $n = 2$ and z - the output value for the current patch and the corresponding function cross entropy:

$$C = -\frac{1}{m}\left[\sum_{l=1}^{m}\sum_{j=1}^{n} t_j \log\frac{e^{z_j}}{\sum_{i=1}^n e^{z_i}}\right] \quad (8)$$

where m the total number of patches, t_j - the indicator function.

Cross-entropy is used as an evaluation of loss of function in the method of backpropagation of error. We use a modification of the gradient descent method proposed in [9].

The essence of the modification is the addition of regularization parameter:

$$\Delta W_t = \eta(g_t + \rho \cdot W_{t-1}) + \mu\Delta W_{t-1}, \quad (9)$$

$$g_t = \frac{D_t}{1-\beta}\cdot\sqrt{\frac{1-\alpha}{S_t}}, \quad (10)$$

$$D_t = \beta\cdot D_{t-1} + (1-\beta)\cdot\nabla E_t;\quad D_0 = 0, \quad (11)$$

$$S_t = \alpha\cdot S_{t-1} + (1-\alpha)\cdot\nabla E_t^2;\quad S_0 = 0, \quad (12)$$

where ρ - regularization coefficient, α and β fixed values, equal 0.999 and 0.9.

Using this option helps to avoid overfitting by setting a "penalty" for excessive growth of the weights.

3 Experimental results

To validate the proposed method for automatic detection of lesions were used, the video sequence having damages obtained naturally. We use several frames taken from video sequence "Mistake engineer Cochin". Figure 5 shows the first test frame.

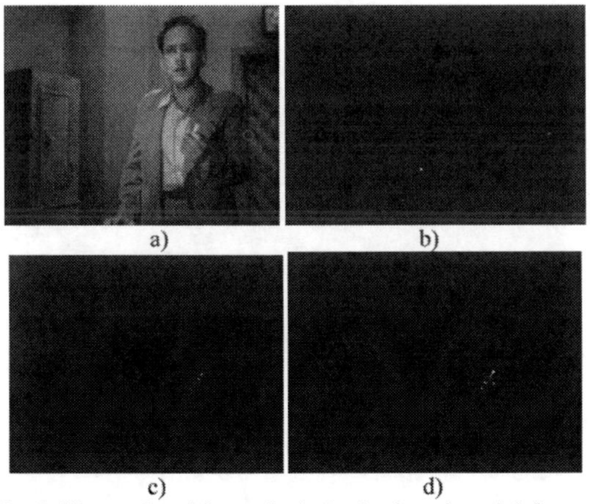

Fig. 5. The results of the methods for the detection of defects a) original frame, b) developed a method, c) - ROD detector d) SDI detector.

Results for the first test frame show that the proposed method allowed to detect more damage, in the absence of false alarms due to the possibility of using lower thresholds than in the known methods. The second test frame is shown in figure 6.

From these results, it can be concluded that known methods of damage detection are not always possible to detect interference with a low contrast compared to the background.

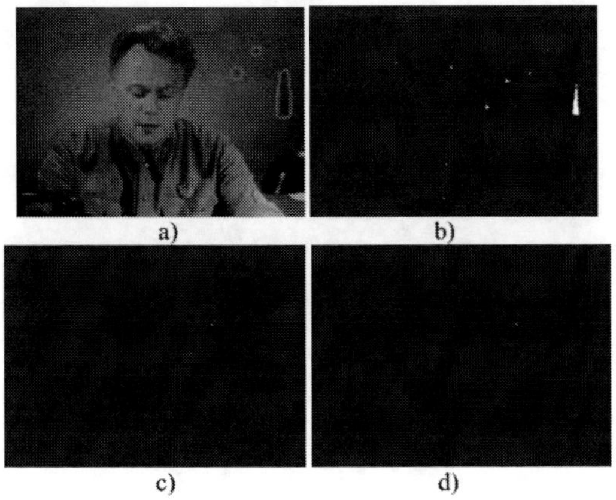

Fig. 6. The results of the methods for the detection of defects a) original frame, b) developed a method, c) ROD detector d) SDI detector.

The third test frame is shown in figure 7.

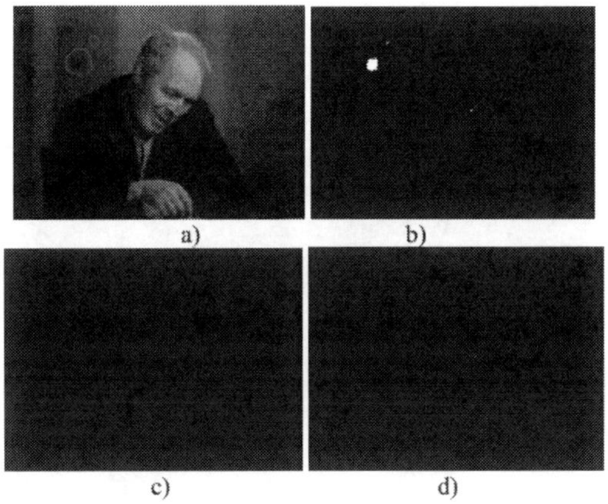

Fig. 7. The results of the methods for the detection of defects a) original frame, b) proposed method, c) ROD detector d) SDI detector.

Analysis of the obtained results indicates the high efficiency of the proposed method. The additional use of machine learning as a post-processing significantly reduces the error of the second kind. The error of the first kind, in most cases, the proposed method is lower than the known techniques of detecting the damage. The main part of the false alarms associated with the difficulty of motion compensation moving objects. This is because the object with fast motion is smeared, and find similar blocks in adjacent frames becomes almost impossible.

4 Conclusion

By conducted research we can draw the following conclusions:

- The method of defect detection on a video sequence showed high efficiency in the problem of damage detection with low contrast and complex shape.
- Use for additional classification on the post-processing made it possible to lower the threshold, which in turn led to a reduction in the probability of error of the second kind.
- Additional using classification based on machine learning reduced false alarm caused mistakes of the methods of motion compensation.

The reported study was supported by the Russian Foundation for Basic research (RFBR), research projects №16-37-00386, №16-37-00391 and №17-57-53192.

Reference

1. Tenze L., Ramponi G., and Carrato S., ICIP, Canada, pp. 660–663, 2000.
2. Joyeux L., Boukir S., Besserer B., and Buisson O., IVC journal, Elsevier, Vol 19 (8), pp. 503–516, May 2001.
3. Kokaram A.C., Springer Verlag, 1998.

4. Nadenau M. J. and Mitra S. K., Time–Varying Image Processing and Moving Object Recognition, Elsevier, pp. 27–35, 1997.
5. Roosmalen van P.M.B., Biemond J., and Lagendijk R. L., Signal Processing for Multimedia, 1999
6. Z. Lei, Y. Fan, D. Yimin, and J. Z. Ying, ICIP, 2016 IEEE International Conference, 2016.
7. Y. Wu, M. Schuster, Z. Chen, M. Wolfgang, M. Krikun, and etc., 2016, arXiv:1609.08144. [Online]. Available: https://arxiv.org/abs/1609.08144
8. A. Krizhevsky, I. Sutskever, and G. Hinton, Advances in Neural Information Processing Systems, p. 10971105, 2012.
9. P. D. Kingma and J. Ba, 2014, arXiv:1412.6980. [Online]. Available: https://arxiv.org/abs/1412.6980

MATEC Web of Conferences 132, 05015 (2017) DOI: 10.1051/matecconf/201713205015
DTS-2017

Image inpainting using a neural network

Nikolay Gapon[1*], *Mikola* Ponamorenko[2], *Marina* Pismenskova[1], *Olga* Tokareva[1]

[1] Don State Technical University 344000, Rostov-on-Don, Russia
[2] National Aerospace University, 61000 Kharkov, Ukraine

Abstract. The paper describes a new method of two-dimensional signals reconstruction by restoring static images. A new method of spatial reconstruction of static images based on a geometric model using a neural network is proposed, it is based on the search for similar blocks and copying them into the region of distorted or missing pixel values.

1 Introduction

Estimation of true pixel values of images to some extent is necessary for most digital imaging tasks. This problem is especially important for the automatic processing of images obtained in photosensitive matrices in the digital photo- and video cameras and computer vision systems. Currently, there are various methods of digital processing aimed at solving the problems of recovering partially lost image areas, but for their effective use, considerable amounts of prior information about the useful image are needed.

In connection this, the reconstruction of images is an important area of application of modern digital computer technology to obtain a reliable estimate for visual and especially automatic analysis.

2 Method of reconstruction

At the first step, for each pixel of the boundary $\delta S_{i,j}$ using the inversion method the shape of the domain for similarity search is adaptively determined by combining two adjacent homogeneous sub regions in the direction of the maximum gradient [3,4,9].

The second step calculates the priority value $P(\delta S)$ for each pixel value of the boundary, which consists of two factors (Figure 1):

$$P(\delta S) = C(\delta S) \cdot D(\delta S), \quad (1)$$

$$C(\delta S) = \frac{\sum_{l \in \Psi_{\delta S}} C(l)}{|\Psi_{\delta S}|}, D(\delta S) = \frac{|\nabla I_{\delta S}^{\perp} \cdot n_{\delta S}|}{\alpha}, \quad (2)$$

where: $\delta S_{i,j}$ - current pixel at the border of available pixels; $C(\delta S)$ - trust factor; $D(\delta S)$ - gradient coefficient; $\Psi_{\delta S}$ - pixel adaptive pixel block $\delta S_{i,j}$; $|\Psi_{\delta S}|$ - number of pixels of the adaptive block, $\nabla I_{\delta S}^{\perp}$ - vector orthogonal to the gradient at the point $\delta S_{i,j}$; $n_{\delta S}$ - vector orthogonal to the boundary δS at the point $\delta S_{i,j}$; α - normalized multiplier, which for eight bit images is equal to 255 [1,4].

First, it is assumed that the value of the confidence coefficient C for pixels from the area S is equal to 1, and for the domain $_{\eta}$ equal to 0

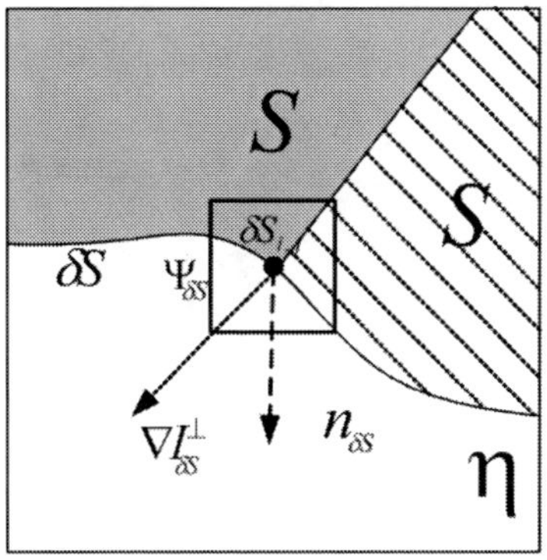

Fig. 1. Construction of orthogonal vectors.

Calculation of the priority using the expression (1) allows you to give more weight to the pixels that are on the differences in brightness (boundaries), thus restoring them in the first place. Accounting for the confidence factor $C(\delta S)$ allows you to assign less weight to the

* Corresponding author: nikolay-rt@mail.ru

DOI: 10.1051/matecconf/201713205015

restored pixels as the distance from the available pixels from the area increases S [2,5,7].

The pixel $p \in (i, j)$ is determined with the maximum priority value $\max(P(\delta S_{i,j}))$ [8,9,10] on the border δS, and the adaptive region Ψ_p is selected as belonging to this pixel (Figure 2). Use, which allows you to correctly consider the shape of the recovery area and do not capture unnecessary borders, which can lead to incorrect image reconstruction.

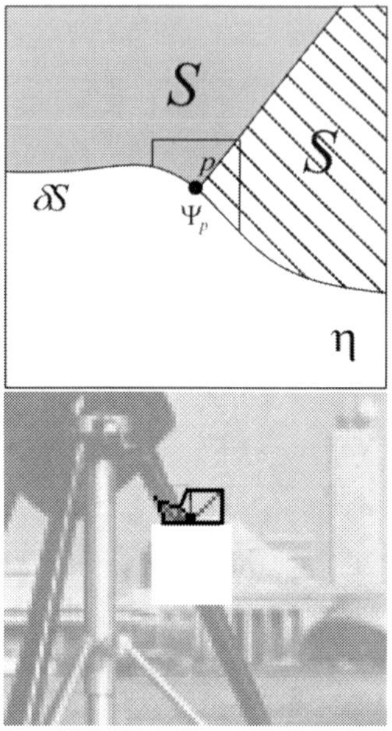

Fig. 2. Adaptive selection of the shape of the region to search for similarity.

At the third step, we determine blocks $\psi_{q^{(h)}}$, $h = \overline{1,R}$ in the area of available pixels S for which the Euclidean metric is minimal (Figure 3):

$$\sqrt{\sum(\Psi_p - \Psi_q)^2} \longrightarrow \min. \quad (3)$$

The number of similarity blocks R is determined using the confidence interval [6,11]:

$$\sqrt{\sum(\Psi_p - \Psi_q)^2}_{\min} + \alpha \cdot \sqrt{\sum(\Psi_p - \Psi_q)^2}_{\min} > \sqrt{\sum(\Psi_p - \Psi_{q^{(h)}})^2} \quad (4)$$

where: $h = \overline{1,R}$; α - significance level.

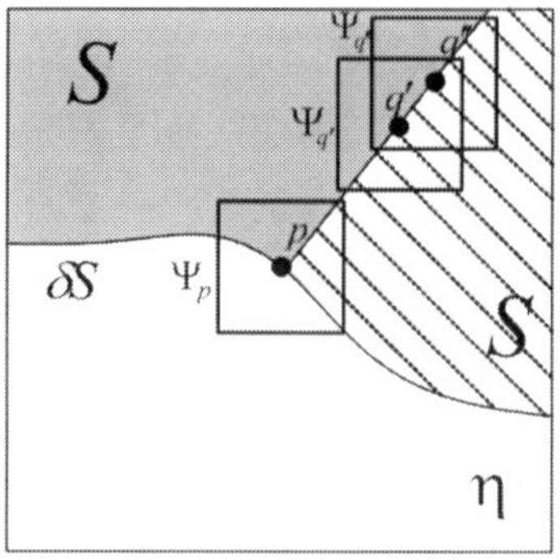

Fig. 3. Search for similar blocks.

The pixel values in the region η adjacent to the pixel with the highest priority p are reconstructed by averaging the corresponding pixels from chosen areas $\psi_{q^{(h)}}$ in the area of available pixels S using a neural network, in particular, a multi-layer perceptron.

The coefficient of trust C for restored pixels is assigned to the current value $C(p)$. After that, the procedure of priority correction and search of similar areas with subsequent replacement is repeated.

At the training stage of the neural network, pre-prepared data was fed to the input: a block with random coordinates 9x9 in size was allocated to the image, then the central pixel was removed, and five most similar blocks were found on the whole image comparing them by MSE.

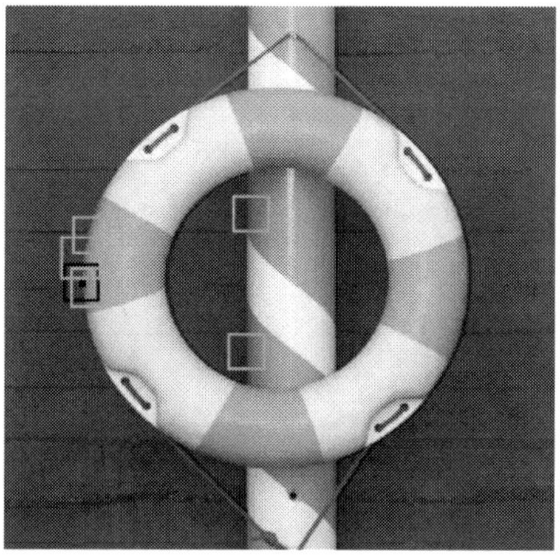

Fig. 4. Original 9x9 block with remote pixel and five similar blocks.

Then the procedure was repeated on 35 images, and 100,000 blocks were obtained, they were used as a training sequence for this network.

This network contains three layers: the first layer contains 20 neurons, the second layer also contains 20 neurons, the third layer contains ten neurons. The network created ten inputs, 5 of them were fed only the central pixels in the blocks found, and the other five inputs fed the MSE of these blocks.

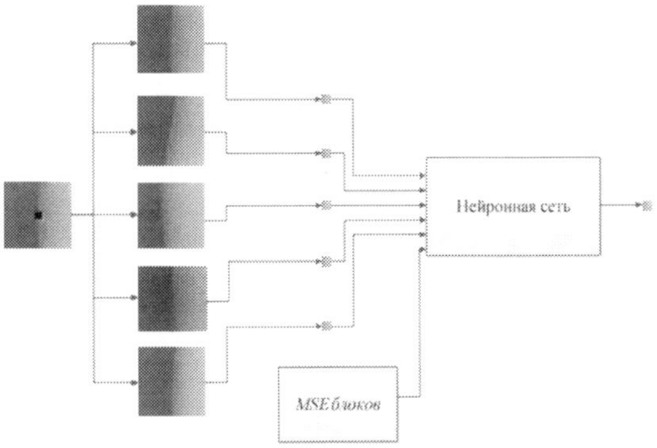

Fig. 5. Structure of the neural network training stage.

As an output, there were the central pixels of the original blocks. The TRAINLM training function is employed as network training function that modifies the values of weights and biases by the Levenberg-Marquardt optimization method. The activation function is TANSIG, a sigmoid function in the form of a hyperbolic tangent [5].

3 Study of the effectiveness of the proposed method

To test the effectiveness of the proposed reconstruction method 3 test images are selected. Figures 6-8 show the results of the proposed method.

a) 

b) 

c) 

Fig. 6. The result of the proposed method:a) the original image, b) the image with the mask of the distorted pixels, c) the image restored by the proposed method.

a) 

b) 

c) 

Fig. 7. The result of the proposed method:a) the original image, b) the image with the mask of the distorted pixels, c) the image restored by the proposed method.

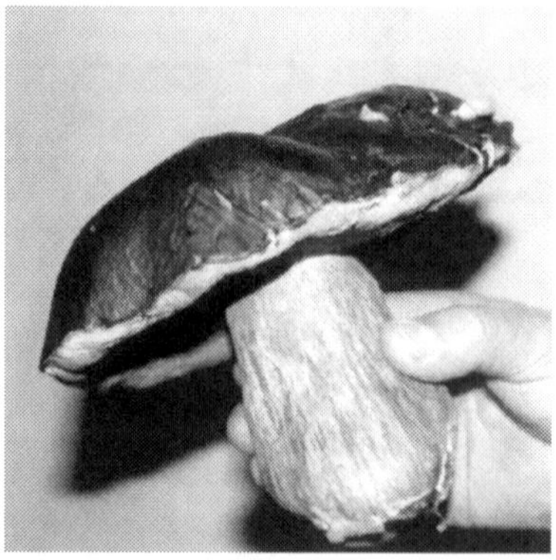
a)

b)

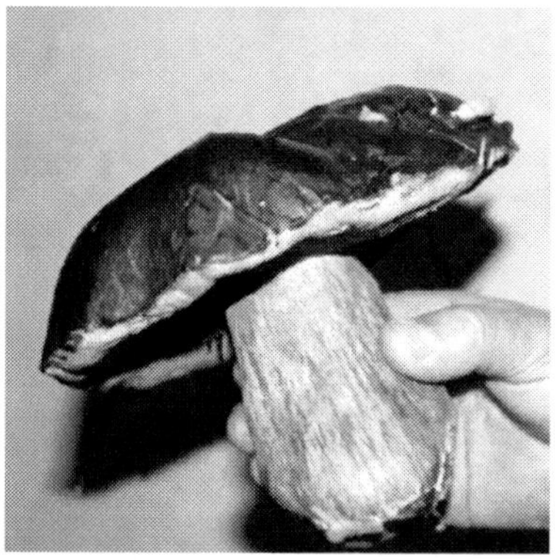
c)

Fig. 8. The result of the proposed method: a) the original image, b) the image with the mask of the distorted pixels, c) the image restored by the proposed method.

Analysis of the results shows that the proposed method allows you to effectively restore lost blocks, allows you to "correctly" restore the boundaries of the image. It should also be noted that the proposed method does not lead to blurring of texture and structure when restoring large areas with lost pixels

4 Conclusion

In conclusion, we can draw the following conclusions. The proposed method will provide a smaller recovery error than traditional methods by an average of 15-20%. The effectiveness of the new approach is shown in the examples demonstrating the effectiveness of the algorithm when restoring test images when objects are deleted.

The reported study was supported by the Russian Foundation for Basic research (RFBR), research projects №15-01-09092, №17-57-53192 and №16-37-00391.

References

1. A. Criminisi, P. P´erez, K. Toyama, IEEE transactions on image processing, no. 9 (2004)
2. Voronin V.V., Marchuk V.I., Gapon N.V., Sizyakin R.A., Sherstobitov A.I., Egiazarian K.O., *In Proc. SPIE-IS and Electronic Imaging - Image Processing: Algorithms and Systems XII*" (2014)
3. Gapon N.V., Sizyakin R.A., Pismenskova M.M., *In the collection: Modern Trends of Development of Science and Education materials of the international (correspondence) scientific and practical conference,* 113-117 (2016)
4. Pimenskovskaya M.M., Gapon N.V., Sizyakin R.A., *In the collection: Modern Trends of Development of Science and Education materials of the international (correspondence) scientific and practical conference,* 161-167 (2016)
5. Voronin, V.V., Marchuk, V.I., Petrosov, S.P., Svirin, I., Agaian, S., Egiazarian, K., *Proceedings of SPIE - The International Society for Optical Engineering*, (2015)
6. Frantc, V.A., Voronin, V.V., Marchuk, V.I., Sherstobitov, A.I., Agaian, S., Egiazarian, K. *Proceedings of SPIE - The International Society for Optical Engineering*, (2014)
7. Voronin, V.V. Marchuk, V.I., Sherstobitov, A.I., Semenishchev, E.A., Frantc, V.A., Pattern Recognition and Image Analysis, no.3, 553-562 (2015)
8. Voronin V., Shraifel I., Marchuk, V., Tokareva S., Sherstobitov A. *In Proc. ICSP* (2014)
9. Voronin, V.V., Marchuk, V.I., Gapon, N. V., Zhuravlev, A. V., Maslennikov, S., Stradanchenko, S. *Proceedings of SPIE - The International Society for Optical Engineering*, (2015)
10. Voronin VV, Marchuk VI, Franz VA, Gapon N.V. J. Successes of modern radio electronics, no.6, 31-34 (2012) [In Russian]
11. Voronin VV, Marchuk VI, Gapon N.V., Sizyakin R.A. J. Successes of modern radio electronics, no. 9, 16-20 (2011) [In Russian]

Classification of a two-dimensional pose using a human skeleton

Marina Pismenskova[1,*], *Oxana* Balabaeva[1], *Viacheslav* Voronin[1], and *Valentin* Fedosov[2]

[1]DSTU, Department of Radio-electronic and electrotechnical systems and complexes, 346500 Shakhty, Russia
[2]SFEDU, Department of Theoretical Foundations of Radio Engineering, 347922 Taganrog, Russia

Abstract. This article proposes an approach for the human pose recognition based on to preliminary prepared high-level data of the human skeleton. The coordinates of the feature points on the human skeleton is constructed by using a neural network. In the presented method, the coordinates of the feature points are a normalized relative of the human body height and the centre of gravity. We use these data for training the neural network.

1 Introduction

The task of human actions recognition is a classic task of computer vision and has a high relevance in such industries as intelligent video surveillance systems, analysis of monitoring cameras data, robotic vision systems, medical services, video analysis of sports games, etc. In recent decades, significant progress has been made in this area through the use of neural networks. The main current trend in the development of neural networks is the architecture of a convolutional neural network, proposed by Jan LeCun [1]. Methods based on convolutional neural networks have made a significant contribution to solving the problems of analyzing the video data of intelligent surveillance systems [2], but there are difficulties in their implementation in practice. Either recognition system requires an individual set of training, corresponding to a specific task. Preparation large sets of video data are required huge human resources, and training a neural network is an expensive process, both in terms of computation and time. Training sets consist of thousands of video sequences and has a significant amount of unnecessary information, which introduces error into the learning process. While images contain discriminative action poses, it improves of efficient to classify the pose [3].

2 Related works

Recently, many works have been prepared and methods have been implemented for recognizing actions and classification of the posture without using optical markers or any other motion sensors attached to parts of the human body. In [4], the authors use a continuous sequence of depth maps of the video sequence. The proposed online system extracts space-time information about human movement, and also information about the skeleton based on the detection of joints of the human body. For training, the authors use the hidden Markov model (HMM). In [5], methods for recognizing human actions based on a data stream of deep information are presented, which improves the performance of intelligent analysis systems Zanfir et al. [6] developed a descriptor of a moving pose, which takes into account both the information on the pose and the differential values (speed and acceleration) of the joints of the human body. Then the proposed descriptor is used in conjunction with the modified k-nearest neighbors (KNN) classifier. X. Yang and Y. Tian [7] proposed a function based on differences in the position of the three-dimensional joints and their own joints (Eigenjoints), which include information such as static posture, motion and displacement. They use the naive Bayesian classifier (NBNN) to classify. Xia et al. [8] used a compact representation of poses called HOJ3D, which characterizes human poses as histograms of the locations of three-dimensional joints in a modified coordinate system. Then they taught a discrete hidden Markov model to classify successive poses. In [9], the authors focused on understanding complex human activities. The activity consists of a set of actions with simple semantic information - elementary (atomic) actions (for example, a triple jump includes the following atomic actions: takeoff, jump and landing in the sand). As an interpretation of such a video sequence, which contains complex activities, a hierarchical description of the data containing the following information is assumed: the type of activity that takes place in the video clip, what atomic actions it includes and when these actions are performed. This method, in addition to obtaining specific recognition results on databases containing complex activities, also informatively describes the actions taking place on the video, captures the semantic concept and the temporal structure of elementary actions. In [10], proposed a modified global video descriptor for classification of realistic videos. This method detection of interest points, the extraction of local video descriptors and the quantization of descriptors into a codebook; it represents each video sequence as a single feature vector. The global descriptor is computed by applying a bank of 3-D spatiotemporal filters on the frequency spectrum of a

* Corresponding author: mpismenskova@mail.ru

video sequence, hence it integrates the information about the motion and scene structure.

3 The proposed approach

The purpose of the presented work is to classify a two-dimensional human pose from one image (frame). In this article, a method for recognizing a person's pose according to previously prepared high-level data of a human skeleton is proposed, which makes it possible to reduce the error of excessive information.

3.1 Representation of key point

As a method of key points representing of a human body, the recurrent neural network V. Belagiannis and A. Zisserman [11] is used. The method detects 16 key points on the skeleton of the human body, and namely: right ankle, right knee, right hip, left hip, left knee, left ankle, torso, neck, chin, top of the head, right wrist, right elbow, right shoulder, left wrist, left elbow, left shoulder.

A training set of images are frames obtained on the basis of a video data set "UCF101-Action Recognition Data Set" [13]. Three types of actions were used: Pull-ups, Squatting and push-ups from the wall (Wall Push Ups), in Figure 2 examples of frames of different actions are presented. Each video is divided into frames. As a result, the total training set was 8061 images.

3.2 Preparing high-level data

For each image, the coordinates of the singular points i_n were obtained, where n is the joint number (Figure 1). To prepare the data for the input of the neural network, the coordinates were normalized relative to the body length and relative to the center of gravity.

The length of the body (6) is calculated as the sum of the lengths of individual parts: the head (1), the torso (2), the leg from the hip to the knee and from the knee to the ankle (3,4), the maximal length of the right or left leg (5):

$$Length_{head} = \sqrt{(j_{10}(x) - j_9(x))^2 + (j_{10}(y) - j_9(y))^2} \quad (1)$$

$$Length_{torso} = \sqrt{(j_8(x) - j_7(x))^2 + (j_8(y) - j_7(y))^2} \quad (2)$$

$$Length_{leg\_right} = \sqrt{(j_3(x) - j_2(x))^2 + (j_3(y) - j_2(y))^2} + \sqrt{(j_2(x) - j_1(x))^2 + (j_2(y) - j_1(y))^2} \quad (3)$$

$$Length_{leg_{left}} = \sqrt{(j_4(x) - j_5(x))^2 + (j_4(y) - j_5(y))^2} + \sqrt{(j_5(x) - j_6(x))^2 + (j_5(y) - j_6(y))^2} \quad (4)$$

Fig. 1. An example of building a skeleton for an image from the dataset «Leeds Sports Pose Dataset» [12]

$$Length_{leg} = \max(Length_{leg\_right}, Length_{leg\_left}) \quad (5)$$

$$Length_{body} = Length_{head} + Length_{torso} + Length_{leg} \quad (6)$$

where $j_n(x)$ – x - the x coordinate of the n-joint, $j_n(y)$ – the y coordinate of the n – joint, $Length_{head}$ - is the head length, $Length_{torso}$ – is the length of the torso, $Length_{leg\_right}$ – is the length right leg, $Length_{leg\_left}$ – is the length left leg, $Length_{leg}$ – maximum leg length, $Length_{body}$ – total body length.

The center of gravity is calculated as:

$$Centr_x = \sum_n \frac{j_n(x)}{16} \quad (7)$$

$$Centr_y = \sum_n \frac{j_n(y)}{16} \quad (8)$$

At the next stage, the coordinates are normalized relative to the length of the body and the center of gravity:

$$i_n(x) = \frac{j_n(x) - Centr_x}{Length_{body}} \quad (9)$$

$$i_n(y) = \frac{j_n(y) - Centr_y}{Length_{body}} \quad (10)$$

where $i_n(x)$ – x – normalized coordinate n – joint, $i_n(y)$ – y – normalized coordinate n – joint.

Information on the distance between certain joints, which can characterize the features of the presented posture, namely the distance from the wrist to the shoulder

(11,12) and from the ankle to the hip (13,14), is also fed to the input.

$$Dist_{shoulder-wrist-r} = \sqrt{(j_{13}(x)-j_{11}(x))^2+(j_{13}(y)-j_{11}(y))^2} \quad (11)$$

neurons in the output layer (linear activation function). Figure 3 shows the described architecture of the neural network.

The validation dataset contains 1201 images corresponding to the types of actions. The presented implementation of the neural network and training data allow obtaining the correct recognition result in 85.5% of

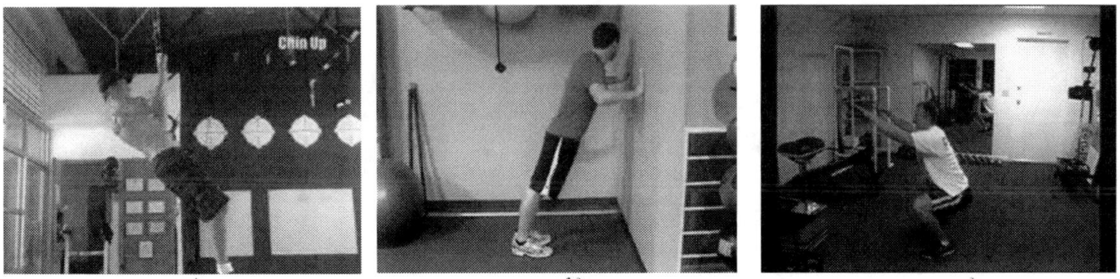

a) b) c)

Fig. 2. Examples of actions from the training set: a - pull-ups; b - squatting; c - push-ups from the wall

$$Dist_{shoulder-wrist-l} = \sqrt{(j_{14}(x)-j_{16}(x))^2+(j_{14}(y)-j_{16}(y))^2} \quad (12)$$

$$Dist_{hip-ankle-r} = \sqrt{(j_{3}(x)-j_{1}(x))^2+(j_{3}(y)-j_{1}(y))^2} \quad (13)$$

$$Dist_{hip-ankle-l} = \sqrt{(j_{4}(x)-j_{6}(x))^2+(j_{4}(y)-j_{6}(y))^2} \quad (14)$$

3.3 Configuration and learning the neural network

As the final stage of the recognition of actions, a neural network is used. In the input of the network is fed the prepared high-level data. Using descriptors, rather than the whole image, as input, allows you to reduce the excess of information and reduce learning time.

The data presented above are fed to the input of a neural network, which has 36 inputs. Because three types of actions were used to test the network, the output layer is represented by three neurons. During the experiments, various variants of neural network configuration were considered. The best result was achieved by using a perceptron consisting of 36 neurons on the input layer (logsig activation function), 64, 32 and 16 neurons on three hidden layers (logsig activation function) and three

cases.

Table 1 shows the percentage of correct recognition of individual actions

Table 1. correct recognition of individual actions.

Action	**Correctly (%)**
Pull Ups	97
Squatting	86
Wall Push Ups	78
All set	85

Errors in the recognition and construction of the skeleton take place in cases where an object is partially hidden (occluded), flipped over, there are several foreground objects.

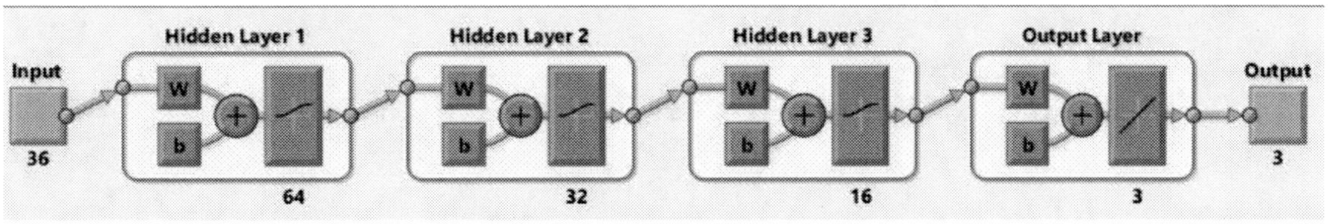

Fig. 3. The architecture of the neural network used

4 Conclusion and future work

We proposed an approach for the human pose recognition based using a neural network. In the future, there is planned to apply the presented approach to the video sequence in real time, to make the comparisons with state-of-the-art methods on public databases. There is supposed to conduct research into the effect of the visual data pre-processing step on the quality of action recognition, because various defects can take place during the creation and transfer of video information: incorrect focusing, lens shake, low resolution of images due to physical limits of cameras, scratches and dust on devices and etc. [14, 15].

The reported study was supported by the Russian Foundation for Basic research (RFBR), research projects №16-37-00386, №16-37-00391 and № 17-57-53192.

References

1. Y. LeCun. et al., The handbook of brain theory and neural networks (MIT Press, Cambridge 1995)
2. M. Pismenskova, N. Gapon, P. Sizykin, *Materials of the international scientific-practical conference. The Scientific and Publishing Center "The World of Science."*, 161-167 (2016)
3. S. Ma et al. *In Proc. CVPR*, 334-345 (2017)
4. A. Jalal et al. *In Proc. CVPR*, 295-308 (2017)
5. M. Pismenskova, V. Voronin, N. Gapon, L. Levina, Information Technology. Radioelectronics. Telecommunications, no. 6, 164-171 (2016)
6. M. Zanfir, M. Leordeanu, C. Sminchisescu. *Proceedings of the IEEE International Conference on Computer Vision*, 2752-2759 (2013)
7. X. Yang, Y. Tian. *In Proc. CVPRW*, 14-19 (2012)
8. Xia L., Chen C. C., Aggarwal J. K. *In Proc. CVPRW*, 20-27 (2012)
9. C. Liu, X. Wu., Y. Jia. International Journal of Computer Vision, 1-16 (2016)
10. T. Silkina, V. Voronin, V. Marchuk, M. Pismenskova, *Innovations, ecology and resource-saving technologies materials of the XI International Scientific and Technical Forum*, 1311-1315 (2014)
11. V. Belagiannis, A. Zisserman. *In Proc. 12th IEEE International Conference on Automatic Face & Gesture Recognition*, 468-475 (2017)
12. S. Johnson, M. Everingham, *In Proc. BMVC* (2010)
13. Khurram Soomro, Amir Roshan Zamir and Mubarak Shah, *In Proc. CRCV-TR* (2012)
14. S. Kundu, M. R. Raksha. 2016.
15. M. Ponomarenko, M. Pismenskova, K. Egiazarian, *In Proc. 25th European Signal Processing Conference (EUSIPCO)* (2017)

Method to reduce the effect of miagrafic and sensory noise with isolating the isoline on ECG signal

Evgeny Semenishchev[1,*], *Dmitry* Chernyshov[1], *Ilya* Svirin[2]

[1]Don state Technical University, 344000, Rostov-on-Don, Russian Federation
[2] CJSC Nordavind, 117545, Moscow, Russian Federation

Abstract: This paper deals with an approach to the analysis of ECG data, which allows to remove the noise component, to preserve the peaks characterizing the work of the heart, and to identificate floating isoline.

1 Introduction

Processes automation is one of the priorities of the modern world. Constant modernization of equipment requires large investments, as well as retraining of personnel, installation works and maintenance of new systems. In this connection, the most promising direction is the automation of data processing, which allows to reduce the number of personnel, and to improve the overall efficiency of the control system. One of the leading areas of data analysis is the development of the digital signals processing theory, based on the analysis of initial data.

The use of digital signal processing methods has found wide application [1]:

- in automation and control systems
- in modern antenna systems
- in the study of biomechanical parameters, biometric data collection systems located directly on the object under study
- in modern systems of computer vision and automatic processing of two-dimensional signals
- in economics and sociology in the study of trends;
- in information-measuring systems;
- in computer technology to increase accuracy of analog-digital transformation

Data, received with ECG, in addition to the useful information on the propagation of electrical impulses in the cardiac muscle, contains the noise component, which has a different source. Analog-digital convertion also adds random distortions to signal [2].

The main task in the signal processing is the separation of the useful component and attenuation of the noise component. In practice, a the mean-square error minimization criterion or the criterion of mean-absolute deviation is used to determine the processing quality. Each of this methods has advantages and limitations of use depending on the task and a priori information on the components of the input signal. In this connection, digital signal processing based on the objective function of the combined criteria becomes relevant objective. Of particular interest is the use of multicriterial methods for processing digital signals, that are represented by a single implementation with a limited amount of a priori information about the useful component function and statistical noise characteristics [3].

Figure 1 contains an algorithm for obtaining estimates using multi-criteria signal smoothing methods, based on minimizing the objective functions:

$$\varphi(\bar{s}_1, \bar{s}_2, ..., \bar{s}_n) = \alpha \sum_{k=1}^{n} (\bar{s}_k - Y_k)^2 + \sum_{k=1}^{n-2} (\bar{s}_k - 2 \cdot \bar{s}_{k+1} - \bar{s}_{k+2})^2$$

when α – adjusting factor.

When implementing the considered smoothing methods using machine modeling, in the form of programs, regions of values of the coefficients were obtained $0{,}01 \le \alpha \le 4.44$. ECG data processing requires real time analysis. It is done by finding estimates of the objective function in window k, and applying sliding window l for all values of the input signal. Here k is size of the window and l is step of sliding window.

The process of obtaining estimates in a sliding window is performed by parallel processing of the initial values of the multicriteria objective function in the window k, with various processing parameters α. The transition between estimates obtained with different parameters α is accomplished by the condition:

$$\bar{s}_k = \begin{cases} \bar{s}_k(\alpha_1) & (\bar{s}_k(\alpha_1) - \bar{s}_k(\alpha_2))^2 \le p \\ \bar{s}_k(\alpha_2) & (\bar{s}_k(\alpha_1) - \bar{s}_k(\alpha_2))^2 > p \end{cases}$$

Here $\bar{s}_k(\alpha_1)$, $\bar{s}_k(\alpha_2)$ are input realization estimates, received with parameters

* Corresponding author: sea.sea@mail.ru

$\alpha_1(\sigma_{ou})$ $\alpha_2(\sigma_{ou})$, p is a threshold value, determined experimentally with the RMS deviation of the additive noise component $\sigma_n < 0.2$ and equals $p = 0{,}15$ [3].

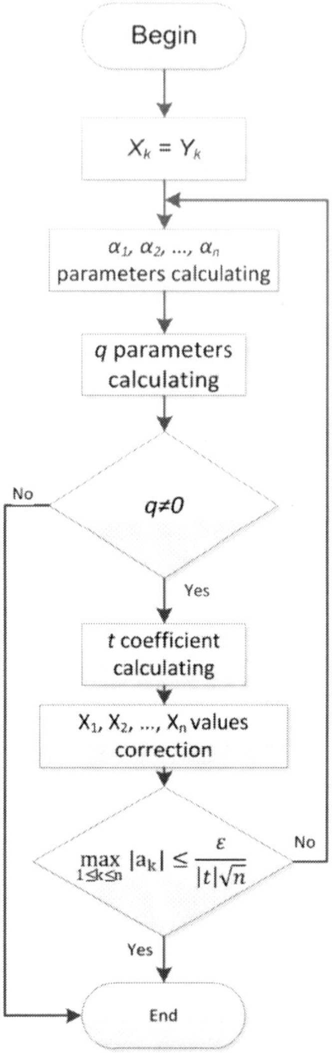

Fig. 1. Algorithm for obtaining estimates using multi-criteria signal smoothing methods.

The need to solve the isoline detection problem is due to the requirement to adjust the monitoring system to locate the report points recorded by the ECG device within the minimum error range of the analog-to-digital converter (ADC). To solve the problem of finding the midline, the algorithm presented in Figure 2 was developed.

Algorithm is shown on figure 2 and implemented as follows:

At the first step, the noise component, possibly with nonuniform distribution law is supressed. It is performed in order to exclude its influence on the middle line searching method.

In the second step, the R peak is suppressed using the R-R processing window.

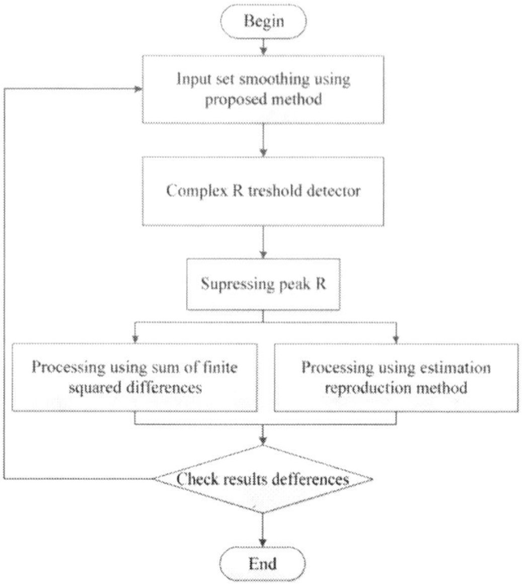

Fig. 2. Isoline searching algorithm.

At the final stage, parameters $\sigma_{ou}(\alpha)$ are fixed, and the entire set is processed. The choice of the filtering method parameter is done using the results shown in Figure 3.

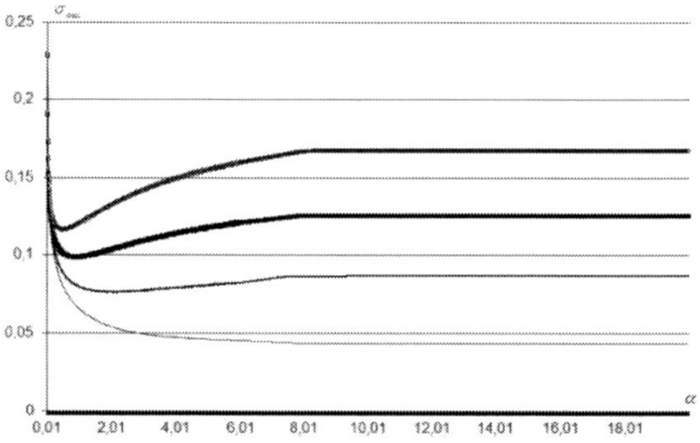

Fig. 3. A graph of the choice of the smoothing parameter for isoline selection method $\sigma_{ou}(\alpha)$.

Figure 3. - Example of the ECG isoline detection (1 - normalized ECG model, 2 - normalized ECG model with pathology, 3 - isoline).

The processing example shown in figure 3 is obtained with fixed parameters of the middle line search algorithm. The parameter of the processing window k is equal to the value of the R-R interval, the window step is fixed and is equal to half the processing window $l = k/2$, the filtering parameter $\alpha = 8$, i.e. the priority of the second summand of the objective function is eight times larger than the first.

2 Method of searching pathologies for ECG analysis

For detect pathologies by analyzing the electrocardiogram obtained by an automated mobile system and stored on a mobile device. We will develop a multi-criteria approach to the preliminary detection. Diagnosis by the means of the developed mathematical models of both a useful (ideal) ECG and possible pathologies.

As the first criteria, we will use the square measurement of the discrepancy between the developed mathematical model $s_{mat}(t_k) = s_{mat\,k}$ and the established cardiogram Y_k:

$$\left(s_{mat\,k} - Y_k\right)^2$$

As the second criteria, we will use a function representing the normalized correlation coefficient:

$$\frac{\Sigma(g_k - \bar{g}_k)(Y_k - \bar{Y}_k)}{\sqrt{\left(\Sigma(g_k - \bar{g}_k)^2\right)}\sqrt{\left(\Sigma(Y_k - \bar{Y}_k)^2\right)}}$$

where: g_k - model o.

Like a third criteria we will use resulting index, frequency area.

$$\sum_{i=0}^{n-1} Y_i e^{-\frac{2\pi j}{n} fft\_k*i}$$

Resulting multi-criteria function will have the form:

$$\psi(\bar{s}_1, \bar{s}_2, \ldots \bar{s}_n) = \delta\left(s_{mat\,k} - Y_k\right)^2 + \\ + \gamma \frac{\Sigma\left(g_k - \Sigma\frac{g_n}{n}\right)\left(Y_k - \Sigma\frac{Y_n}{n}\right)}{\sqrt{\left(\Sigma\left(g_k - \Sigma\frac{g_n}{n}\right)^2\right)}\sqrt{\left(\Sigma\left(Y_k - \Sigma\frac{Y_n}{n}\right)^2\right)}} + \tau\left(\sum_{i=0}^{n-1} Y_i e^{-\frac{2\pi j}{n} fft\_k*i}\right)$$

here: δ, γ, τ - parameters that establish weight for each of the criteria, n - set size, fft_k - frequency index, j - imaginary unit.

Each parameters of the criteria are established during the preparation phase using the experience of specialists in ECG analysis.

Figure 4 shows the pathology search algorithm for ECG analysis.

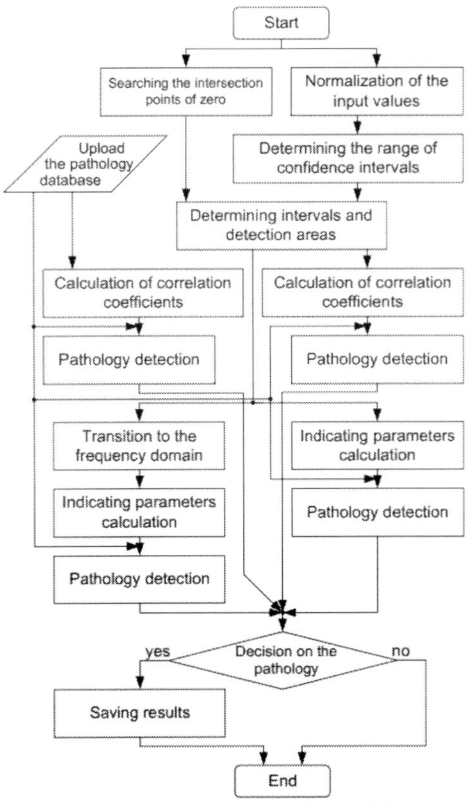

Fig. 4. Pathology search algorithm for ECG analysis.

The algorithm presented in Figure 4 is implemented as follows:

In the first step, the areas of intersection in established points of the cardiogram report with the isoline in the zones confined to confidence intervals are searched. At the same time, the input values are normalized in the range of R-R intervals.

In the next step, we make the determination of 5 peaks: P, Q, R, S, T and segments ST and R-R, the QRS complex. The determination is made according to the standard ranges [4] shown in Figure 5

In the third step, we use the ranges defined in the previous step to select intervals. The values for each selected range are normalized.

At the fifth step, the correlation coefficients are calculated for each pathology and for each examination area. Processing is performed in the sliding window. At the same time, we perform the transition to the frequency domain and the calculation of the main central moments for the normalized indicating parameters. The calculations of the central points are made using the analysis of the differential coefficient.

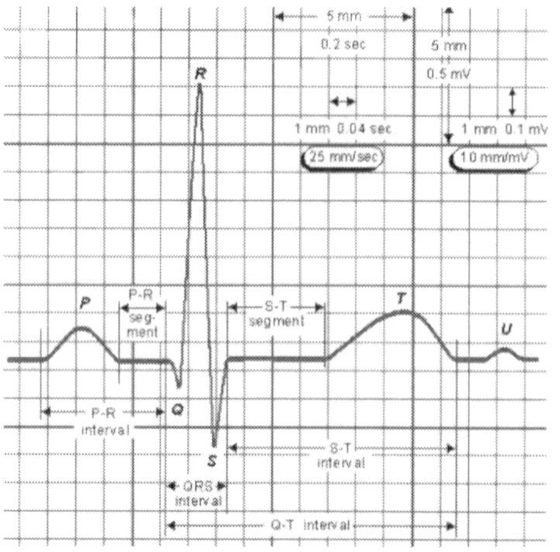

Fig. 5. The main complex of ECG.

On the sixth step the obtained frequency coefficients are compared with a mathematical model of each of the pathologies and the excess of a threshold value is detected. The calculation of the established threshold is also made by analyzing the divergence of the central moments and correlation coefficients. Pathology is considered detected if more than two indicators coincide and in case of repeated confirmation notification decision is made. In the case of coincidence of all criteria for ECG analysis and their coincidence with the investigated pathology, the patient is notified [5].

3 The implementation of the pathology search methods for the analysis of electrocardiograms

The first step is a searching for intersection of the fixed points of the cardiogram report with a baseline in that zones, which are limited by confidence intervals. We'll use a value of two quantization levels as confidence limit [6]. The device uses 10 bits analog-to-digital convertor or ADC and the range of standard QRS complex amplitudes for different areas of data sources should not exceed 22 mm and 25 mm for adults. This limit specifies the monitoring interval, which is limited in the range of 5mV. It means that the device indicates the intersection of the baseline when fixed values exceed the confidence intervals which are limited by two quantization levels and it's equal to ±0,02 mV. We will impose restrictions in the form of derivative of the change in direction for the obtained values. It will be used the criterion of exceeding the measurement sequence over the baseline for more than 3 reports as a simple pike detector. The restrictions on the excess of three reports are empirical, they are based on the minimum possible duration of restriction of the QRS complex. As a rule, this QRS complex lasts 0,08 seconds. For children under the age of 5 it lasts 0,09, and for children over 10 years – 0,10. When the QRS complex lasts more than 0,11 for adults, it is said of pathology and suggests ventricular hypertrophy or ventricular blockade. We get the minimum duration of QRS complex, which doesn't exceed 500 reports at a sampling frequency of the device limited to 5 kHz. This article introduced a restriction of the range of fixed changes, which is equal to 100 reports. It will be made a decision about the intersection of the baseline during these reports.

We will search for intervals of consecution of the pikes R-R according to the following procedure: we will use the complex approach in which the maximum value will be determined as the main condition for detecting the R wave, as secondary and confirming conditions we will also use the relative time and sequence of intersection intervals of the baselines corresponding to P and Q and the following ST pikes, as well as their relative intervals of consecution. We can see the ECG duration model at the pic.6. Along with the R-R pikes, it is also possible to detect the remaining pikes and their ranges according to the data, which was obtained in the previous stages.

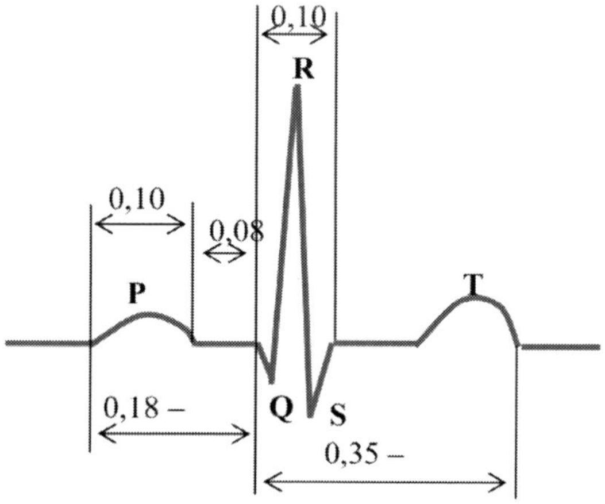

Fig. 6. The duration model ECG within normal limits.

The input values are normalized in the range of P-P intervals in the next step. The normalization process consists of fixing the maximum values at a given interval and dividing all values by a given value. Normalization of the values allows not to consider the scaling factor for the subsequent analysis and detection of pathologies. At the third stage, we produce the allocation of pike and segment intervals. We produce normalization of the values in the area of each obtained intervals. We define the group of central moments, such as mathematical expectation:

$$v_k = M(Y_k) = \sum Y_k / k$$

dispersion

$$D(Y_k) = M\left[(Y_k - M(Y_k))^2\right]$$

skewness ratio

$$\gamma_a(Y_k) = \frac{M\left[(Y_k - M(Y_k))^3\right]}{\left(\sqrt{D(Y_k)}\right)^3}$$

index of kurtosis

$$\gamma_a(Y_k) = \frac{M\left[(Y_k - M(Y_k))^4\right]}{\left(\sqrt{D(Y_k)}\right)^4} - 3$$

The calculation is made for each segment defined at this stage and the pathology specified in the mathematical model. After that, a comparison is made for a match in the parameters in the range defined by the rule 3σ. This calculation helps to define solution of the first criterion of the function.

We should set a research window $n = \frac{1}{3}k_d$ for realization of the second criterion of the function. The search for correlation coefficients and their comparison is performed in a sliding window in the range of the interval P-P.

$$cor_{pat} = \frac{\sum\left(g_k - \sum\frac{g_n}{n}\right)\left(Y_k - \sum\frac{Y_n}{n}\right)}{\sqrt{\left(\sum\left(g_k - \sum\frac{g_n}{n}\right)^2\right)}\sqrt{\left(\sum\left(Y_k - \sum\frac{Y_n}{n}\right)^2\right)}}$$

when k - elements of observations in the range of the processing window, cor_{pat} - correlation coefficient for each pathology.

When the threshold value is exceeded, a decision is made to detect the pathology, if it is confirmed for more than 30% of the searches in the processing window.

We'll make the transition to the frequency area using discrete Fourier transform for realization of the third criterion of the function.

$$X_{fft\_k} = \sum_{i=0}^{n-1} Y_i e^{-\frac{2\pi j}{n} fft\_k*i}$$

Comparison in the frequency area will be made according to the shape of the envelope of the frequency coefficients and the central moments calculated for them.

The envelope will be constructed using the method of least squares, taking into account the criterion $\sum(Y_n - f(x))^2 \to \min$. The function $f(x)$ is polynomial with the size of a polynomial not exceeding the 4th degree. The degree of the polynomial is bounded by the fourth order because of the computational complexity of the calculations of these indicators. In the case of coincidence of polynomial functions, or their discrepancy with an inaccuracy that does not exceed the threshold value, it is decided to confirm the diagnosis. Confirmation of the diagnosis in the frequency area is made in case of coincidence of the group of central moments with the central moments of the mathematical model of the supposed pathologies.

The diagnosis is made after the analysis of the results of each the components of the objective function.

4 Conclusion

As a result of the conducted researches the approach to the analysis of electrocardiograms is received. The proposed approach makes it possible to eliminate the noise component while retaining the signal pikes. According to the analysis of the received data, an algorithm for finding the diagnosis is proposed.

References

1. Chung, S. H., and R. A. Kennedy, Journal of neuroscience methods, 71-86 (1991)
2. Veeneman, D., and S. BeMent, IEEE transactions on acoustics, speech, and signal processing, no.2, 369-377 (1985)
3. Semenishchev, Evgeny, et al. *In Proc. EWDTS,* 444-449 (2016)
4. Stadler, Robert, et al. Patent No. 6115628 A, United States, A61B5/0452
5. Sameni, Reza, et al. Computers in Cardiology (2005)
6. Vorobyov, Sergiy, and Andrzej Cichocki, Biological Cybernetics. No. 4, 293-303 (2002)

DOI: 10.1051/matecconf/201713205018

Algorithm combination of deblurring and denoising on video frames using the method search of local features on image

Evgeny Semenishchev[1, *]

[1]Don state Technical University, 344000, Rostov-on-Don, Russian Federation

Abstract: In this paper, we propose an approach that allows us to perform an operation to reduce error in the form of noise and lubrication. To improve the processing speed and the possibility of parallelization of the process, we use the approach is based on the search for local features on the image.

1 Introduction

The primary data received by the cameras have a noise component. The cause of noise can be, sensitive elements of the matrix, cross-talk, electromagnetic interference, the ADC bit depth, the data transmission channel, primary codecs, etc. In the case of fixation of the frame by standard cameras having the possibility of changing the focusing, a blurring effect is observed on the secondary objects. The appearance of blur can also be related to the movement of the subject in the frame, the slow fixing speed, the appearance of disturbing factors and the shooting conditions. Elimination of these types of distortions makes it possible to improve the accuracy of subsequent processing and decision making.

In paper proposed many methods [1-5] to automatically reduce the noise component in two-dimensional signal. Some methods are fully automatic, but use a lot of many computing resources. These methods work well enough when it is possible to obtain data on noise. If this is not possible, then the operator's help is required to adjust to each type of scene. Otherwise, their effectiveness is sharply reduced. The result obtained depends on the variance of the noise component and the type of two-dimensional signal (image). Methods based on the first order square difference allow to reduce the effect of noise, but the differences blur the boundaries of the picture. Methods based on L2 norm, allow to keep the border between the differences of image brightness, but little suppress noise near the differences [6, 7].

The effectiveness of decision making depends on the result obtained at the first stages of data processing. For high-quality data processing, it is necessary to perform the following basic steps: elimination of the noise component, elimination of blurring, reconstruction and correction of images.

In this paper, we propose an approach that allows us to perform an operation to reduce error in the form of noise and lubrication. To improve the processing speed and the possibility of parallelization of the process, we use the approach is based on the search for local features on the image.

2 Mathematical model for image of blur and noise

A sequence of frames is a multidimensional array of data. The object that has fallen into the frame fixation can be non-stationary.

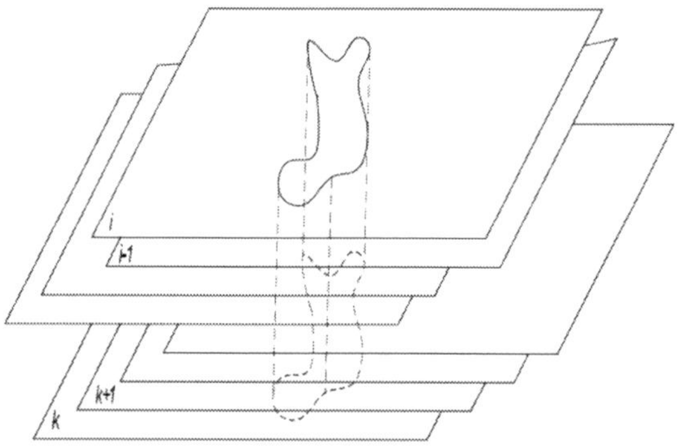

Fig. 1. An example of representing frames in an image.

Simplified mathematical model is presented in [1]:

$$Y_{m,n} = \xi \cdot S_{m,n} + \eta, \quad i = \overline{1,n}, \quad j = \overline{1,m}$$

where $Y_{m,n}$ - two-dimensional signal (image), $S_{m,n}$ - the original (undistorted) image, η - additive noise component, ξ - multiplicative blur component.

Often, the image can be divided by geometric features on several areas - texture, structure and contour. In solving the problem of denoising, main objective of is to reduce the additive noise in the areas of textures and

* Corresponding author: sea.sea@mail.ru

structures while maintaining the boundaries of the transitions between them. Often as an component additive noise, is considered a Gaussian noise with zero mean $\eta(0,\sigma_{noise})$ and variance σ_{noise}. The blur coefficient ξ has a Gaussian kernel. The difficult task is the reduction of the noise component in the absence of hard information about the signal (image) and noise. These restrictions prevent the use of classical methods in these conditions or require additional configuration by the operator.

2 Algorithm for eliminating blur and filtering

The step of reducing the effect of interference is realized by the algorithm shown in Figure 2

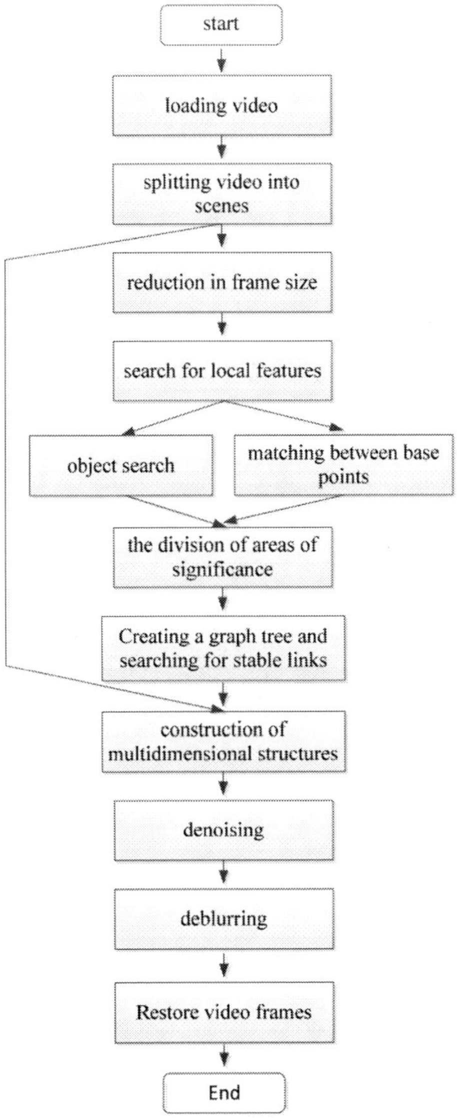

Fig. 2. Algorithm for eliminating blur and filtering.

Algorithm is shown on figure 2 and implemented as follows:

- At the first step the procedure for recording video data is performed;
- The next stage the video stream is divided into scenes. The separation is performed at time intervals and is specified by the operator;
- At the third stage, the operation of reducing the size of the video frames. This procedure involves combining the pixel blocks with the color priority reassignment;
- The next step is to look for local features. This stage includes the search for base points, the search for boundaries and angles on objects located in frames. This stage is discussed in Section 3;
- At the fifth stage, we perform in parallel the operations of searching for objects that are detailed or meaningful for humans and of the searching for correspondences between base points. The operations of searching for objects that are detailed or meaningful for humans is discussed in Section 4. For searching for correspondences between base points we use SURF detector;
- At the next stage, a multidimensional array of cut objects from each frame is assembled. The resulting multidimensional array carries data about structure and the change in the object;
- The next stage is related to the previous one. At the current step, there are stable links in the changes in the positions of the objects. Using this stage, you can determine the significance of objects in relation to the frame. The construction of trees is standard, the most stable branches are considered basic;
- At this stage, the results are transferred to the size of the video. A multidimensional structure is the constructed for a group of full frames;
- The next two steps perform denoising and debluring elimination procedures (Chapter 5);
- At the final stage of the implementation, the procedure for transferring the results to the final video and frame-by-frame restoring the video data.

3 Search for local features on images

An analysis of the literature revealed that the main methods of detecting for borders and corners in images used in practice are detectors: Roberts, Sobel, Prewitt, Laplace, Kirsch, Laplacian of Gaussian (LOG), Difference of Gaussians (DOG), Canny and Harris.

For blurred images, the best of the quantitative results within the SNR, PSNR, and MSE assessments shows Roberts' method. However the result has a large number of errors, almost complete loss of boundaries in low-contrast areas is observed. Similar results show Harris detector and LOG-detector. The task of detecting local features was handled by the detector Canny. This detector has not high quantitative indicators but visually allows you to find all the boundaries. Thes same detector has the best quantitative results on the noisy image and he is detecting edges in test image with presence low-contrast area. Results similar to the Canny detector, have detectors the Sobel and Prewitt.

On the smooth images, Canny detector shows the best quantitative results. Only this filter allows accurately detect the low-contrast boundaries. The worst quantitative indicators has Harris detector he is forming many false pixels and double border. Low-contrast area on the smooth images defines no one detector except Canny detector [8].

In the work, we applied the Canny method, which showed high efficiency in both smoothed and noisy images.

4 Definition of significance areas

From a large group of approaches performing data analysis, it is possible to single out approaches based on a mathematically light device, and methods requiring large computational costs, for example neural networks.

The first group includes the methods processing in window and simple methods for analyzing gradients. The following approaches can be distinguished from this group: "White/black edges", "number of transitions", "Density", LPI-ICI, etc.

The conducted studies [8] showed that the application of computationally simple algorithms has a number of limitations:

- the lowest efficiency, in comparison with other methods of selecting detailed objects, shows the method "Number of transitions". This method, finding detailed areas on the image, allocates a sufficiently large number of erroneous areas and requires an empirical selection of the coefficient of detail.
- The effectiveness of the method "Density" is high. This method more precisely defines the detailed objects and has a smaller error. But this method has a large sliding window size. Reducing the size leads to an increase in the erroneously found visually non-detailed areas. An important advantage of the "Density" method is that it does not require an empirical selection of the threshold, but determines it automatically for each image.
- the "White / Black Ratio" method are similar to those of the "Density" method. But this method has a smaller size of the sliding window, which minimizes the error in determining false regions and more locally determines the detailed areas. The main drawback of this method is the empirical choice of the coefficient of detail for each image.
- the LPI-ICI method makes it possible to effectively detect local stationary regions. Has low speed. Badly works on noisy images and images with high detail.

Methods based on the search for human attention cards, allow localizing areas that have a higher priority for the user. Most often these approaches are based on the modelling of the human visual system and machine learning. These approaches have a high computational complexity.

In this paper, as a block, offering simple computational algorithms and neural networks for localization of area.

5 Noise reduction and blur reduction operation

To perform the operation of reducing the effect of interference, it is proposed in the paper to combine the regions with the object into a three-dimensional array. Layered data generation allows you to use the resulting detail masks from one frame to the next. The use of layered structures also makes it possible to refine the boundaries of objects and accelerate the process of primary data processing.

To find the correspondences between layers, we applied the SURF method [9]. For each pair, cross check layers were assigned. Based on the resulting data set, a link tree is built. Stable branches are a basis.

The operation of eliminating noise and blurring is performed using the methods proposed in the works [10, 11]. The filtration method is based on simultaneous minimization L2 norm $\sum_i \sum_j \left(\bar{s}_{i,j} - s_{i,j}\right)^2$. Here is used measurement results $s_{i,j}$ of the assessments $\bar{s}_{i,j}$ and the first order square difference sequence of estimates row $\sum_i \sum_j \left(\bar{s}_{i,j} - \bar{s}_{i,j-1}\right)^2$ and columns $\sum_i \sum_j \left(\bar{s}_{i,j} - \bar{s}_{i-1,j}\right)^2$. The objective function can be written as follows:

$$\varphi\left(\bar{s}_{0,0}, \bar{s}_{0,1}, \ldots, \bar{s}_{0,n}, \bar{s}_{0,1}, \bar{s}_{1,1}, \bar{s}_{1,n}, \ldots, \bar{s}_{m,0}, \bar{s}_{m,1}, \ldots, \bar{s}_{m,n}\right) =$$
$$= \sum_{i=0}^{m} \sum_{j=0}^{n} \left(\bar{s}_{i,j} - s_{i,j}\right)^2 + \lambda \sum_{i=1}^{m} \sum_{j=0}^{n} \left(\bar{s}_{i,j} - \bar{s}_{i-1,j}\right)^2 +$$
$$+ \mu \sum_{i=0}^{m} \sum_{j=1}^{n} \left(\bar{s}_{i,j} - s_{i,j-1}\right)^2$$

where λ, μ are fixed positive constants.

The method of restoring the boundary. This method works as follows: A mask of blurred boundaries is drawn; Pixel-by-pixel displacement of the inner and outer borders to centre; The united boundaries are taken for the final border. The operation is performed step by step. As the recovery algorithm, the approach to suppressing the Gaussian blur. Stopping criterion is the fulfillment of the requirements for the minimum width of the pixels within the boundaries (n) for the maximum number of objects (k).

An example of restoring a frame of video data is shown in Figure 3.

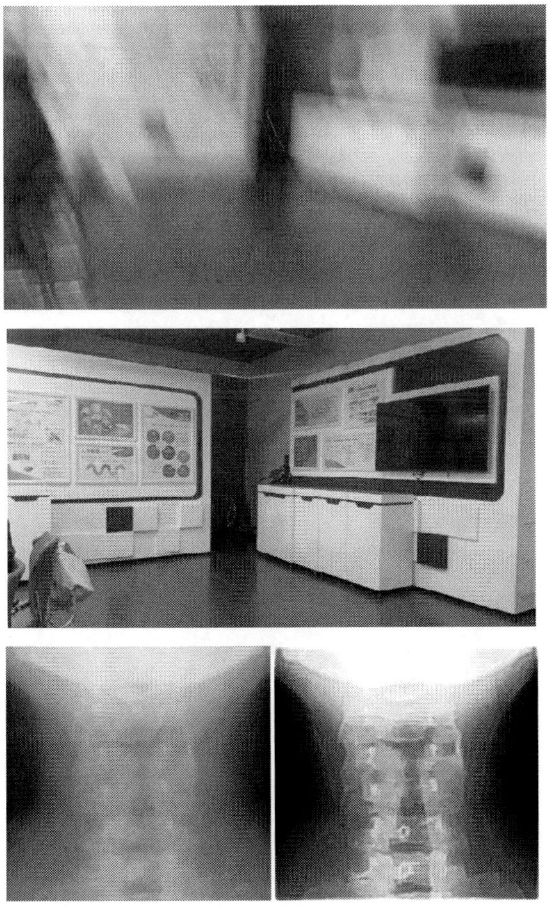

Fig. 3. Example of restoring a frame of video data.

5 Conclusion

As a result of the conducted researches, the approach to the analysis of frame the video. The proposed approach makes it possible to eliminate the noise component while retaining the edges. The algorithm is based on the following steps: the search and selection of domains with boundaries in areas of overlap; Finding reference points and their correlation; The transformation to reduce the action of visible distortion.

We hereby acknowledge the support of the Ministry of Science and Education of the Russian Federation, in accordance to the decree of the Government of April 9, 2010, №218, project number 03.G25.31.0234."

References

1. Gonzalez R.C. and Woods R.E., *Digital image processing* (Prentice Hall. Upper Saddle River, NJ, 2002)
2. Buades, A., Coll, B., Morel J. M., Society for Industrial and Applied Mathematics, no.52 (1), 113-147 (2010)
3. Fiard, J.M., Manteuffel, T.A., Mccormick, S.F., Society for Industrial and Applied Mathematics, no. 19 (6), 1958-1979 (1998)
4. Dabov, K., Foi, A., Katkovnik, V., and Egiazarian, K., *In Proc. Workshop on Signal Processing with Adaptive Sparse Structured Representations (SPARS'09)* ,(2009)
5. Kuan Darwin T., Sawchuk A.A., Strand T.C., Chavel P., *In Proc. Pattern Analysis and Machine Intelligence* **7** (2), 65-177 (2009)
6. Vorobyov, Sergiy, and Andrzej Cichocki. Biological Cybernetics, no.4, 293-303 (2002)
7. Wang Zhou, Zhang, D., *In. Proc. Analog and Digital Signal Processing IEEE*, **46 (1),** 78-80 (2002)
8. Russo, F., Ramponi, G., *Signal Processing Letters IEEE* **3 (1)**, 168-170 (2002)
9. Semenishchev, E. A., et al. *SPIE Sensing Technology+ Applications. International Society for Optics and Photonics,* (2015)
10. Rong, Wen, et al. *In Proc. 24th International Conference IVCNZ'09*, (2009)
11. Semenishchev, Evgeny, et al. *SPIE Commercial+ Scientific Sensing and Imaging. International Society for Optics and Photonics,* (2016)
12. Semenishchev S. E. et al. Electronic Imaging, no. 6, 90-96 (2017)

AUTHOR INDEX

Abramishvili Roman....53
Akay Oksana....262
Aleksanyan Grayr....183, 187
Alekseenko Ludmila....258
Atoyan Tatiana....8
Avetisyan Yuri....229
Babichev Anatoly....11
Bakhracheva Yulia....176
Balabaeva Oxana....229, 343
Batishchev Denis....45
Bednaya Tatyana....233
Belikov Sergey....87
Belov Aleksandr....117
Belova Yulia....233
Bespalov Vadim....258, 262
Bezuglov Dmitriy....334
Biryukov Yurii....135
Bolshenko Irina....45
Borisova Lyudmila....191
Bulgakov Alexey....204
Cen Yigang.....283
Chernova Anna....180
Chernyshov Dmitry....347
Cherunova Irina....254
Chistyakov Alexander....164, 243
Churyomov Alexandr....104
Chuveyko Mikhail....36
Danilenko Irina....254
Davydova Irina....8
Denisova Olga....53
Dimitrov Valery....191
Dmitrienko Nadezhda....229
Dolgov Vasiliy....266
Donskoy Daniil....269
Dub Vladimir....104
Dubovskov Vadim....288
Efimov Andrey....171
Egorov Ivan....79
Egorov Nikolay....79
Eliseev Alexandr....292
Fathi Vladimir....279
Fedosov Valentin....343
Fedosov Valentine....310
Fomin Valerii....27
Frolov Aleksey....238
Frolova Natalia....5
Galushka Vasily....279
Gapon Nikolay....334, 339
Gorbatenko Nikolay....183, 249
Goryanina Ksenia....49, 195
Grachev Vladimir....130
Grechikhin Valery....153
Grib Stella....87
Grushko Irina....100
Gurova Oksana....258, 262
Iarakhmedov Orkhan....273
Ibadov Ragim....296
Ibadov Samir....296
Ignatenko Vitaliy....68
Igumnov Leonid....95, 117
Illarionov Anatoliy....108
Illarionov Anatoly....87
Ivanov Vladimir....8, 11
Izotov Mikhail....262
Kalach Gennady....23
Kalmukov Boris....296
Karabanalov Maksim....108
Kasyanenko Elena....266
Katin Oleg....49, 195
Kazachek Nina....23
Kazaryan Maretta....300
Khanykova Catherine....112
Khashev Denis....32
Kireev Sergey....171
Kiryuschenko Evgeny V.....20
Klimenko Sergey....325
Klimenko Yuri....45
Kolesnikov Pavel....218
Koltcov Aleksandr....319
Konev Sergey....266, 269
Korchagina Marina....171
Korelin Andrey....108
Kostoglotov Andrey....305, 314
Kovalev Igor....199
Kozlov Alexander M.....20
Kozlov Andrey A.....20
Kruglova Tatyana....204
Krutov Vladimir....296
Kucher Artem....183, 187
Kucheryavenko Alexander....310
Kudryakov Oleg V.....91
Kuimov Denis....73, 145
Kuzin Aleksandr....314
Lankin Anton....153
Larina Ludmila....209, 214
Lazarenko Sergey....305, 314
Lebedev Valeriy....8
Liang Liequan....283
Loginov Yury....112
Losev Vasiliy....199
Losev Vitaliy....305
Lukinova Nataliya....83
Lukyanov Aleksander....195
Lukyanov Alexander....269
Lyakhnitskaya Olga....1
Lysenko Aleksey....32
Makov Sergey....325
Manackov Yuri....249
Marshakov Daniil....279
Maslova Irina....238

AUTHOR INDEX

Matrosov Audrey 83
Merkulova Alla 254
Merkushkin Alexey 141
Michail Lankin 153
Minkin Maksim 73
Minkin Maxim 145
Mladenovic Vladimir. 283
Mogilin Yaroslav 71
Moiseev Denis 15
Morozov Sergey 214
Mozgovoj Andrej 73
Narakidze Nuri 329
Naumov Ivan 319, 322
Nechaev Igor 130
Negrov Nikolai 180
Neydorf Rudolf 273
Nguyen Chiem X. 41
Nikitin Anton 325
Nikitina Alla 233, 243
Nizhnik Dariya 83
Nosachev Sergey 36
Novokreschenov Stepan 108
Nurutdinova Inna 191
Obukhov Pavel 71
Oganyan Roman 329
Paschenko Anton 319
Perekhrestenko Larisa 322
Petrosyan Mariam 199
Petrov Andrey 117
Petrov Sergey P. 63
Phan Hai N. 41
Pilipenko Aleksandr 58, 63
Pilipenko Anastasia 58
Pilipenko Olga 58, 63
Pismenskova Marina 339, 343
Pogorelov Nick 11
Poletaykin Vladimir 218
Ponamorenko Mikola 339
Popkov Anton 104
Priyma Maria 187
Protsenko Elena 243
Protsenko Sofiya 158
Protsenko Sophia 223
Prygunov Alexandr 292
Pugachev Igor 305, 314
Richter Andrey 300
Rozen Andrei A. 130
Rozen Andrei E. 130
Ruslyakov Dmitriy 214
Ruslyakov Dmitryi 209
Ryabov Aleksandr 95
Rybak Aleksandr 1
Ryndya Sergey 135
Ryzhkin Anatoliy 15
Saramud Mikhail 199
Semchugova Elena 180
Semenishchev Evgeny 347, 352
Semenyakina Alena 223
Serebryanaya Irina 83
Sevostianov Igor 149
Shahramanyan Michael 300
Shapovalova Antonina A. 91
Shaykhutdinov Danil 249, 329
Shcherbakov Ivan 183, 187
Shirokov Konstantin 249
Shishulin Denis 95
Shmelev Ivan 204
Shraifel Igor 334
Shurygin Dmitriy 329
Sidoryakina Valentina 164
Sizyakin Roman 334
Smirnova Olga 229
Smolyanskiy Alexander 135, 141
Soloviev Igor 104
Stepanov Stepan 112
Stus Anastasia 266
Sukhinov Alexander 164, 233
Sukhinova Tatyana 158
Svirin Ilya 347
Terekhina Yuliya 83
Tikhonova Olga 209, 214
Tokareva Olga 339
Tokareva Svetlana 334
Tolstunov Oleg 322
Tsarevskaya Irina 258
Tsybriy Irina 68
Uryanskaya Ekaterina 32
Valyukevich Yuriy 288
Varavka Valery N. 91
Vasiliev Andrey 121
Vasilyeva Svetlana 141
Vernezi Miroslav 49, 269
Vernigorov Yuriy 5
Vlasov Alexander 204
Volkov Ivan 95
Volkov Sergei 121
Volokitin Gennadiy 15
Volosnyh Georgi 23
Vorobtsov Igor 117
Voronin Viacheslav 283, 343
Wang Yanhong. 283
Yurovskikh Artem 87
Zabiyaka Igor Yu. 91
Zelentsov Vladimir 125
Zhang Linna. 283
Zheltova Anna 135, 141